International Federation of Automatic Control

INFORMATION CONTROL PROBLEMS
IN MANUFACTURING TECHNOLOGY
1989

IFAC Symposia Series, 1990. Number 13

IFAC SYMPOSIA SERIES

Editor-in-Chief
JANOS GERTLER, Department of Electrical Engineering,
George Mason University, Fairfax, Virginia 22030, USA

Other IFAC Publications

AUTOMATICA
the journal of IFAC, the International Federation of Automatic Control
Editor-in-Chief: G. S. Axelby, 211 Coronet Drive, North Linthicum,
Maryland 21090, USA

IFAC WORKSHOP SERIES
Editor-in-Chief: Pieter Eykhoff, University of Technology, NL-5600 MB Eindhoven,
The Netherlands

Full list of IFAC Publications appears at the end of this volume

NOTICE TO READERS

INFORMATION CONTROL PROBLEMS IN MANUFACTURING TECHNOLOGY 1989

Selected papers from the 6th IFAC/IFIP/IFORS/IMACS Symposium
Madrid, Spain, 26–29 September 1989

Edited by

E. A. PUENTE

Departamento de Automática, Ingeniería Electrónica e
Informática Industrial (DISAM)
Universidad Politécnica de Madrid

and

L. NEMES

CSIRO Division of Manufacturing Technology,
Victoria, Australia

Published for the

INTERNATIONAL FEDERATION OF AUTOMATIC CONTROL

by

PERGAMON PRESS
Member of Maxwell Macmillan Pergamon Publishing Corporation
OXFORD · NEW YORK · BEIJING · FRANKFURT
SÃO PAULO · SYDNEY · TOKYO · TORONTO

U.K.	Pergamon Press plc, Headington Hill Hall, Oxford OX3 0BW, England
U.S.A.	Pergamon Press, Inc., Maxwell House, Fairview Park, Elmsford, New York 10523, U.S.A.
PEOPLE'S REPUBLIC OF CHINA	Pergamon Press, Room 4037, Qianmen Hotel, Beijing, People's Republic of China
FEDERAL REPUBLIC OF GERMANY	Pergamon Press GmbH, Hammerweg 6, D-6242 Kronberg, Federal Republic of Germany
BRAZIL	Pergamon Editora Ltda, Rua Eça de Queiros, 346, CEP 04011, Paraiso, São Paulo, Brazil
AUSTRALIA	Pergamon Press Australia Pty Ltd., P.O. Box 544, Potts Point, N.S.W. 2011, Australia
JAPAN	Pergamon Press, 5th Floor, Matsuoka Central Building, 1-7-1 Nishishinjuku, Shinjuku-ku, Tokyo 160, Japan
CANADA	Pergamon Press Canada Ltd., Suite No. 271, 253 College Street, Toronto, Ontario, Canada M5T 1R5

First edition 1990

Library of Congress Cataloging in Publication Data

Information control problems in manufacturing technology 1989: selected papers from the 6th IFAC/IFIP/IMACS/IFORS symposium, Madrid, Spain, 26–29 September 1989/edited by E. A. Puente and L. Nemes.—1st ed.
    p. cm.—(IFAC symposia series: 1990, no. 13)
"Papers presented in the 6th IFAC/IFIP/IMACS/IFORS Symposium on Information Control Problems in Manufacturing Technology, Madrid, Spain on September 26–29 1989"—Pref.
1. Computer integrated manufacturing systems—Congresses.   2. Automation—Congresses.
I. Puente, E. A.   II. Nemes, L.   III. IFAC/IFIP/IMACS/IFORS Symposium on Information Control Problems in Manufacturing Technology (6th: 1989: Madrid, Spain)   IV. Series.
TS155.6.I54   1990   670'.285—dc20   90–7005

British Library Cataloguing in Publication Data

Information control problems in manufacturing technology 1989.
1. Production. Automatic control
I. Puente, E. A.   II. Nemes, L.   III. International Federation of Automatic Control
IV. Series
670.427

ISBN: 9780080370231

Transferred to Digital Print 2009

Printed and bound in Great Britain by
CPI Antony Rowe, Chippenham and Eastbourne

6TH IFAC SYMPOSIUM ON INFORMATION CONTROL PROBLEMS IN MANUFACTURING TECHNOLOGY 1989

Sponsored by
IFAC (International Federation of Automatic Control) Technical Committee
  on Manufacturing MANTECH

Co-sponsors
IFAC Technical Committees on:
  —Computers,
  —Education,
  —Components & Instruments;
IFIP (International Federation for Information Processes);
IMACS (International Association for Mathematics & Computers in
  Simulation);
IFORS (International Federation of Operational Research Societies).

Organized by
División de Ingeniería de Sistemas y Automática (DISAM)
Escuela Técnica Superior de Ingenieros Industriales Universidad Politécnica
  de Madrid

Supported by
Rectorado de la UPM
Dir. Gral. Investigación Científica y Técnica del MINER
El Corte Inglés S.A.

International Programme Committee

L. Nemes, Hungary (Chairman)
J. Aracil, Spain (Vice-chairman)
L. Basañez, Spain
J. Browne, Ireland
J. A. Buzacott, Canada
Chen Znen-Yu, PRC
G. Doumeingts, France
R. W. Gellie, Australia
M. C. Good, Australia
A. Halme, Finland
Hyung Suck Cho, Kampuchea
K. Iwata, Japan
Jiang Xinsong, PRC
T. Kanade, USA
I. Kato, Japan
L. Keviczky, Hungary
P. Kopacek, Austria

N. A. Kuznetsov, USSR
T. Lang, Czechoslovakia
M. Mansour, Switzerland
V. R. Milacic, Yugoslavia
G. Ollson, Sweden
P. K. Patwardhan, India
L. F. Pau, Denmark
E. A. Puente, Spain
K. N. Reid, USA
L. N. Reijers, The Netherlands
U. Rembold, FRG
M. G. Rodd, UK
M. Somalvico, Italy
G. Spur, FRG
D. R. Towill, UK
H. van Brussel, Belgium

National Organizing Committee

E. A. Puente (Chairman)
R. Aracil (Vice-chairman)
F. Aldana
C. Balaguer
A. Barrientos

E. Bautista
E. Camacho
P. Campoy
S. Dormido
J. Tornedo

PREFACE

This volume contains the papers presented in the 6th IFAC/IFIP/IMACS/IFORS Symposium on Information Control Problems in Manufacturing Technology which helds in Madrid (Spain) on September 26-29, 1989.

This aim of the Symposium is to present, discuss, and summarize research on new theories, as well as advanced applications of automatic systems in Computer Integrated Manufacturing (CIM). This Symposium is the main MANTECH-IFAC event and follows the 4th and 5th symposia held in Maryland (USA) in 1982 and Suzdal (USSR) in 1986.

A total of 110 papers were presented given in 48 sessions, covering the following topics:

- General aspects of CIM
- System management and planning
- Simulation of manufacturing processes
- Control problems
- Manufacturing networks
- AI and Expert systems in manufacturing
- Information system for manufacturing
- Sensor-based robots in manufacturing
- Advanced applications and case studies

In addition to these papers, three plenary sessions presented by invited speakers, covers the theory, methods, applications and future trends in manufacturing technology. Two round tables "Difficulties of introducing the automation in the industry" and "Education in Manufacturing and robotics", took also place during this symposium.

I would like to thanks the members of the International Program Committee for the effort in the selection of papers and the members of the National Organizing Committee for their support in the organization of this event.

I hope that the publication of these papers, which come from experts from 29 countries, will make a good contribution to the development field.

E.A.Puente
Editor

CONTENTS

SENSOR-BASED ROBOTS IN MANUFACTURING I

EXPERT SYSTEMS IN CIM I

INFORMATION SYSTEMS FOR MANUFACTURING I

ADVANCED APPLICATIONS AND CASE STUDIES I

MANUFACTURING MODELLING I

ROBOTS IN MANUFACTURING I

AUTOMATIC PRODUCTION PLANNING I

MOBILE ROBOTS I

MANUFACTURING MODELLING II

STATE OF THE ART AND TRENDS IN CIM

ROBOTS IN MANUFACTURING II

ADVANCED APPLICATIONS AND CASE STUDIES II

PRODUCTION SCHEDULING II

DESIGN METHODOLOGIES I

ROBOT ENVIRONMENT MODELLING

SENSOR-BASED ROBOTS IN MANUFACTURING II

AI TOOLS FOR MANUFACTURING II

INFORMATION SYSTEMS FOR MANUFACTURING II

AUTOMATIC PRODUCTION PLANNING II

PRODUCTION SCHEDULING III

INFORMATION SYSTEM FOR MANUFACTURING III

CONTROL STRATEGIES II

SIMULATION OF MANUFACTURING PROCESSES III

EXPERT SYSTEMS IN CIM II

ADVANCED APPLICATIONS AND CASE STUDIES IV

SENSOR-BASED ROBOTS IN MANUFACTURING III

CAD/CAM SYSTEMS IN CIM

MANUFACTURING NETWORKS

CONTROL STRATEGIES III

MOBILE ROBOTS III

DESIGN METHODOLOGIES III

AUTOMATION PRODUCTION PLANNING III

EDUCATION IN MANUFACTURING AND ROBOTICS

Organizer: P.Kopacek

Panelists: L.Nemes, J.Paiuk, J.Scrimgeour

Introductory statement by P.Kopacek

As frequently pointed out education in CIM and robotics is
a very difficult and complex subject. On one hand education in this field requires a lot of a priori knowledge in
different fields (e.g. mechanics, electronics, control, computer sciences...) on the other hand those who have to be
taught come from different educational levels. Consequently scientists and industrial engineers of various fields have
to cooperate for efficient teaching in CIM and robotics.

A first stage providing some general background
knowledge enables non specialists to understand the language of specialists and to get a feeling for what CIM components are able to do and what they are not able to do.
Therefore some basic knowledge about factory automation
must also be provided. One of the aims of CIM and
robotics education at this (low, but general) level will be to
enable the engineer to make a profound decision about the
incorporation of robots and manipulators in a special automation concept.

The next-higher level in CIM education is for a specialist
in application. In addition to the knowledge mentioned
above for the lower level he will be educated in more
detail e.g. in the structure of programming languages, in
CAD/CAM, in artificial intelligence, in pattern recognition,
safety problems and last not least in social effects and
economical considerations.

The highest level includes additional more theoretical
background about the topics listed above. The thus educated factory automation specialist has a deep insight in
this field as well as in application problems. One very important fact during the education is the connection between theory and practice. Theory and practice should
have equal rights in each educational program. This holds
also true for all other kinds of education (including continuing education). Laboratory experiments, demonstra-
tions, audiovisual facilities and visits to different factories
are absolutely necessary for this goal.

Statement by L.Nemes

Education from the research and development point of
view

Research and development activities on Computer Integrated Manufacturing need highly skilled engineers with
extensive knowledge on information technology, good understanding of manufacturing processes and with production engineering practice. It is required from these people
to work in a multidisciplinary team to solve complex tasks
in system design, machine and software research and
development etc.

In the early years the Mechanical Engineering Automation
Division, Hungarian Academy of Sciences recruited either
mechanical engineering or computer science graduates. Although they were the best students in their respective
fields, they just had general knowledge on complementary
subject areas.

Almost all these students have had some practice in research activities because they have acquired it during their
postgraduate courses. Because of degree examinations require individual achivements the students usually work
alone on well defined tasks thus they have had little experience how to work, interact in teams.

To be able to hire properly trained engineers upon their
graduation a few people from our staff accepted part time
teaching jobs at various departments of technical universities. They selected the brightest students, to whom we offered part time jobs in the Division. They have come to
work whenever they had time because they were highly
motivated to contribute to "real research" to have access to
better equipment what they have been offered at the
universities.

During their "incubational" periods they had collected
knowledge and expertise on necessary fields due to their
personalized training program. In most cases they have also
participated in factory implementation of R and D results
as well.

In every year we re-employed the best students and we invited new ones to replace the drop-outs. Those who managed to maintain good performance for years have had the privilege to be offered a research position in the Division.

Although this activity has been highly successful but the burden on the senior research staff has been significant. It was proposed therefore for the Technical Univerity Budapest to set up a special training program for talented students with interdisciplinary curriculum on the field of integrated manufacturing strait from the first year on. The Hungarian Ministry for Machine Building realized the impact of this education program on the future of the Hungarian production industry and donated an education CIM system for teaching and research purposes.

In other countries similar reseach centers have been established around leading faculties of technical universities so the education has embarked on the road of training new type of experts.

Statement by J.Paiuk

From the viewpoint of an industrial engineer the awareness and consciousness of the importance of economical implications is very important in this field. Especially in the field of factory automation the engineer has to keep in mind the industry's needs in relation to cost/economical implications:

  - on time delivery of projects according to expected objectives

  - training in both senses for technicians, operators as well as the upper management

By means of automation new technologies can be incorporated in the quickest possible way to any process. Thus professionals in our field have to be convinced of their roles as agents of changes, as educators.

Statement by J.Scrimgeour

Education and Flexible Automation -

An Examination of Need and the IFAC Role

The question before us, this evening, is on the subject of education, and I belief also on the role of IFAC. Perhaps even more specifically we might consider the role of the INCOM seminars as part of the education process.

When we speak of "education", as related to automation, we tend often to limit our thoughts and discussion to the design of curricula at the university level. An attempt will be made to indicate that a much broader view can be taken, which adds another dimension and which adds an even more vital meaning to the concept of education for advanced manufacturing technology.

In his very thoughtful opening remarks for the INCOM '89 seminar, L. Nemes drew attention to the three main elements of the name INCOM. They are information, control and manufacturing. He stressed that information is seldom organized or complete; that control includes management of the process as well as just control; and that the application area for INCOM is manufacturing. In our discussion we can come back to these three areas in a few moments.

First, let us come to the subject of education. We need to remind ourselves that there are at least three education streams for education in computer integrated manufacturing or flexible automation.

There are:

1.University education;

2.Professional personnel employed in the manufacturing industry (engineering and technical personnel); and

3.Management and executive training.

These three education streams represent three distinctly different groups in terms of their membership. They differ greatly in age, background, interests, motivation and outlook.

They also differ substantially in terms of their timing and contribution to the transformation process of industrial automation (as a requirement) and in order of their subsequent contribution, their order is the reverse of the above sequence. Management awareness, education and understanding of CIM and advanced manufactoring technologies must precede any large scale adoption. Professional training, and re-education of the engineering work force already employed and experienced in industry is the second key to a rapid transformation and adoption of new technology. Only then is a demand created for new technical personnel and a work force trained in the new technologies.

One can also ask if IFAC and the INCOM conferences are fulfilling their potential role in this. Is IFAC really involved? If my observation is correct, we do not have a large industrial population attending IFAC conferences for educational purposes.

Is this an information and control problem in manufacturing?

It is a most difficult task, at the chief executive officer level of a company, to develop a deep sense of awareness of CIM or advanced manufacturing technology, its impact on the company and its impact on the organization structure.

Is creating this awareness an information and control problem in manufacturing?

It is also a difficult task for those within a company to identify the economic justification for CIM, to recognize how it affects the total organization and to develop the strategic planning required.

Is this an information and control problem in manufacturing?

We know that to properly control a process it is necessary to fully understand that process. We also know that information about the behaviour of a physical process must be wrestled from nature. For INCOM the physical processes are the processes used in manufacturing and the list is long. Not only machining, but also welding, annealing, soldering, moulding, casting, forming, extrusion assembling, and many others. Within IFAC there is an apparent tendency to devote attention to the design of the controller, especially in its mathematical aspects, and in manufacturing to focus attention on the mathematics of sheduling, while ignoring the information and control aspects of the many other processes. There seems to be a tendency to assume that someone else will provide the information that is needed about the process, of what is to be controlled, and more importantly, how the process will respond when controlled. If we continue to do so, the outside world may assume that we in IFAC have taken a narrow view of what is meant by information and control problems in manufacturing.

There is also a too easily developed tendency to adopt the view that those in research or education are somehow in the lead, ahead of industry. Industry is regarded as lagging behind, but this view may not be totally accurate. It is fair to ask two questions:

*If somehow IFAC ceased to exist, would the adoption of automation in industry continue? Obviously the answer is yes.

*If somehow industry ceased to exist, would IFAC and the development of advanced automation continue? Sometimes, one fears that the answer to this question is also "yes".

These potential answers indicate a sense of isolation between IFAC and industry, and a gap between theory and practice. It is easy to berate industry for seeming to ignore technology, but we in the R&D community need to educate ourselves with regard to the real gaps in manufacturing technology and to identify and address the real knowledge needs of industry. Not just the ones we assume, find easy and likeable. Maybe we are even missing some big opportunities. This means talking to the user in industry who is the real costumer. It is only by exercizing this dialogue that theory and practice can stay together.

Summary by P.Kopacek

At the symposium a lot of papers dealing with factory automation and especially with control problems were presented by industrial as well as scientifically oriented people. For this discussion I tried to unite engineers and scientists.

Education in this field cannot only concentrate on Engineering schools and Universities. Education must also take place in factories, research institutes and universities in a postgraduate way. Especially small and medium sized factories have an increasing demand for skilled employees in factory automation. Therefore modular educational concepts will be necessary in the future. The members of the EDCOM Working Group W.G.7.2 " Education in Robotics and Manufacturing Automation " are strongly involved in this task.

INDUSTRY AUTOMATION

Panelists: Dr. Ch. Bühler. Institut für Roboterforschung.
University of Dortmund. FRG.

Dr. A. El Mhamedi. Laboratoire d´Automatique. Grenoble.
France.

Prof. J.J. Rowland. Dept. of Computer Science. University
College of Wales. Aberystwyth. UK.

Mr. A. J. Wells. Division of Manufacturing Tech. CSIRO.
Woodville. Australia.

Organizer and Chairman: Prof. P. Albertos. Dept. of Systems Eng.,
Computers, and Control. UPV. Spain.

INTRODUCTION.

Industry Automation is a too broad topic to be
covered in a round table session, but it
is also so challenging that, in the
framework of this meeting -INCOM 89-, it has
been considered the opportunity to
discuss the actual Difficulties in
the Implementation of the Industry Automation.
In order to analize this problem, industry
activities can be considered at different
levels. In the first level, production and
manufacturing are directly involved with
the operation of processes and machines,
interacting with operators. In upper
levels a variety of activities can be
considered, mainly in the processing of
information. As shown in fig. 1, some of them
are:

- Monitoring and surveillance of the
  production process.

- Product quality control

- Production planning

- Control of input of raw materials and
  components and output of products

- Finantial management.

All these activities are carried out
directly or by the supervision of
different people, all them sharing information
and interacting with machines.

Dealing with the automation of these
activities,some questions arise:

- What activities should be automatized?

- How this automation must be carried out?

- In what extent these activities will be
  automatized?

- How these automated processes will
  interact with the users?

Of course, many other questions, technical and
social, remain open. Some of these issues
arepresented by our invited speakers and some
others appeared in the discussion period.
The last section sumarizes the conclusions
of this Round Table.

MODULAR DESIGN OF AUTOMATION SYSTEMS. (J.J. Rowland).

If automation is to become more widespread it
must become easier and cheaper to integrate
all the various components of the system
during re-configuration.

The various automation components must fit
together in different combinations to form
systems for different applicatios,relatively
easily, in a cost efective way.

Thus we require **modules** that perform the
various functions. The modules must be
designed so that they can connect to, and
operate in conjuction with, other appropriate
modules. The modules must be re-usable, off-
the-shelf items.

Computational components can be designed as
functional modules:

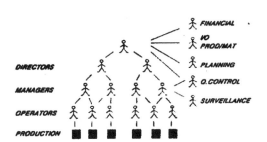

Fig. 1. Industry Activities

- each with a single, identifiable function

- with uniform **information** interfaces

- with uniform electrical interfaces

MAP can provide a basis for interconnection at both these levels (and at the intermediate levels).

Is functional modularity applicable to automation? Let us consider an example:

Sensors are important for monitoring processes and for inspecting products. **Flexible** automation implies greater reliance on sensors and, more generally, on systems that are easy to configure and reconfigure.

Currently:

The variety of available sensors allows sensing of most physical properties relevant to manufacturing industry.

Work on computational methods to extract specific types of sensory information from combinations of sensors is well advanced. This is oftem referred to as **sensor data fusion**.

But...methods that allow different sensing devices and sensor fusion techniques to coexist and cooperate in different areas of automation are not generally available. Consequently, integration of sensing into automation is difficult, and sensing is not as widely used as the fundamental techniques permit.

Some current research can produce easily reconfigurable industrial sensing systems.

Could functional modularity be used for the design of entire automation systems for mechanical as well as computational modules?

In designing and installing modular systems:

- Account must be taken of the limitations imposed by the need to incorporate existing equipment.

- Reliability needs to be considered in the cases of :

  * failure of the automation system itself.

  * failure of the **process flow** in a way that was not foreseen when the automation system was designed.

- Modularity can improve system maintainability. A faulty module can be replaced.

  * automatically ("redundancy")

  * manually

In all cases diagnostic facilities are important.

- One obstacle is that at present it is difficult even for systems integrators to integrate additional sensors into propietary robot controllers.

SENSOR INTEGRATION. (Ch. Bühler).

Sensor integration has to be performed on different levels of the automation system. Every level can be interpreted as a feedback loop, where the sensors act to feed back the information of the actual current situation to the controller of the relevant loop.

The basic loop is the classical direct feedback control. A next level generates the inputs for the direct feedback control level. This can be cascaded in succession with quality control, flow and throughput control, security control, task replanning level, learning components and overall control and monitoring. At the latest on the monitor level man is again in the loop.

Therefore sensors are needed at every level of a CIM-hierarchy to close the different loops.

The integration aspects are especially stressed by the real time constraints of the corresponding loop.

While at low levels quite simple informations, such as displacements, forces, ect. are needed in short cycles, at higher levels preprocessed and interpreted data are necessary in larger time intervalls.

State of the art:

- A broad range of sensor technology for automation purposes is available.

- A lot of single sensors are already used in the systems.

- The interpretation of the sensor data is process dedicated.

- Mostly the use of sensor information is local (decentral).

- A large number of single solutions exists.

Evolution influences:

- Miniaturizing sensors and electronics.

- Advances in computer hardware

- Advances in the software design methods.

Research and development activities:

- Integration of the already installed sensors into several automation levels.

- Smart sensors (sensor system integrated with signal conditioning, analog to digital conversion).

- Multisensor systems (smart sensors in a combination, with internal data preprocessing).

- Parallel computing to solve the real time constraints.

- Standard interfaces (digital serial or parallel) integrated in the sensor system.

- Standard interfaces for sensors in the controllers.

- Standard communication protocolls between sensor system and controller.

- Multisensor integration. Software to integrate and interpret different kinds of sensor information. Pattern recognition problems. Adequate environmental descriptions (world model).

- Improvement of the control methods and strategies to make full use of the accessible sensor informations.

INFORMATION INTEGRATION (A. WELLS).

With respect to this issue, the objective of making ALL RELEVANT INFORMATION ACCESSIBLE AT ANY ACCESS POINT, should be considered.

Examine the scope of this objective:

A. RELEVANT: Likely to influence a decision about resources.

- decisions cover a range of time horizons (now; today; this month)

- resources are actual or conceptual objects we can confortably think about.

B. INFORMATION: Has many origins, and may be:

- current (now; present status)

- historical (totals, averages, statistics, exceptions)

- future (final target, plans, schedules)

C. ACCESIBLE: Any user will need:

- various levels of abstraction

- segmentation at each level

- navigation by specification or incrementally

D. ACCESS POINT: Appropriate functionality is:

- Passive (read) anywhere

- Active (Enter/Modify) restricted

Items A and B relate to the DATA MANAGEMENT SYSTEM that serves the INFORMATION PRESENTATION SYSTEM covering items C and D

The DATA MANAGEMENT SYSTEM may be:

- a central system, receiving all raw data for processing and storage

- a hierarchical system, where some local processing is done, with data transfer at the request of a supervisory level.

The INFORMATION PRESENTATION SYSTEM should provide for:

- multiple simultaneous views

- navigation both vertically and horizontally

- guided assessment of proposed actions.

COMMENTS by J.J. ROWLAND

The demand from industry for multisensor robots is not high. This is partly because the difficulty of integrating sensors makes multisensor robots expensive. It is also because industry in general is not aware of the potential of this technology. Multi sensor robots are a"solution looking for a problem" in the way that, for example, lasers and microprocessors were in the last decade.

HUMAN-MACHINE INTERACTION (A. EL MHAMEDI).

With automated manufacturing, such as in the fully automated factory, the roles of the human operators are becoming less and less important. As a result, interaction or cooperation is entering upon new age.

What is Human-Machine Interaction?

Human Machine Interaction may be characterized as:

- a branch of human factors

- a branch of cognitive sciencies

- a branch of software Engineering

- a set of know-hows

- an emerging technique

It is very difficult to develop all of these branchs. For a discussion, four points of view are developed

1. Risks to let human operator in production system.
   Several risks to let human operator may be mentionned:

   - Human makes several errors due to:

     - bad representation of system the control

     - personality and responsability

     - sensitivity ,... etc.

     Several accidents are caused only by human errors.

   - In full automation, the human operator intervens rarely, so his competence is decreased.

   - A rapid evolution of automation and lack of human motivation lead to reject the installation.

   - The environment and the physical ambiances rarely protect the human health.

So the human operators must be relieved of dirty, undesirable ,and monotonons tasks. Their health and safety must be protected.

2. Why to let human operator in Production Systems?
   The human intervention is necessary for several reasons such as:

   - Taylor organization is generally inefficient

   - Human operator disposes a know-how and may adapt to several situations.

So a high flexibility may be obtained by human operator teams. It is necessary to integrate human operator in automated production system.

3. How to integrate human operator?
    Large dynamic systems, such as FMS, are
    becoming increasingly complex, and so as
    consequence human and computers interact
    and share the decision making
    responsabilities to control these systems.

    If man and computer share the
    resonsabilities for controlling a FMS,
    this bring up the question of how the
    decision-making responsabilities should be
    allocated between them. It seems
    reasonable to propose that the optimal
    solution is a symbiosis of the human and
    computer activities rather than to slow
    the function of either two alone.

    So man and machine must cooperate to
    control and monitor a system.

4. Actual tendencies?
    For a cooperation between man and machine,
    two possibilities exists:

    - a static cooperation, where each
      decision maker -human or computer-
      would be allocated by a subset of the
      total tasks available to the system

    - a dynamic cooperation, where the set
      of tasks is allocated as a function
      of each decision maker
      characteristics.

A.I. techniques are generaly used for man-
machine cooperation.

A.I. programs share and also extend the
informatic processing responsabilies and
capabilities of the human operator. Thus some
of the human intelligent which is needed to
make decisions will be transferred to the
computer machinery and a new hybrid
intelligence will be emerge from the combined
capabilities of the human and the
computer.This hybrid intelligence will now
supervise and monitor the FMS.

In conclusion, some remarks may be mentioned:

    - An improvement of whole Production
      Systems should be taken into account the
      human, technical, cultural, and
      organizationals aspects. So multi-
      disciplinary tools and methods must be
      developed for the analysis and evaluation
      of production system.

    - For a real integration of human operator
      some techniques and tools must be also
      developed:
      - ergonomics tools such as tasks and
      activities analysis - knowledge
      acquisition - programming languages -
      explication and learning techniques -
      distributed artificial intelligence - etc

3. Human tasks generally are diagnostic,
    detection, and compensatio.n So these tasks
    must be also analyzed and developed for
    perhaps an auto-diagnosys in the future
    discussion.

4. An integrated automation and
    multidisciplinary groups of research must be
    encouraged.

CONCLUSIONS

After the discussion period, it was clear
that there are many common problems at
any level of automation and a general
methodology allowing to deal with a
particular problem will be desirable. Some
of the issues emphasized in this session are
the followings:

    - Modularity and Standardization are key
      issues in the automation
      implementation.

    - Automation should not imply loose of
      flexibility.

    - Sensors are always in the feedback
      path. They are critical. Sensor
      integration is a basic goal in further
      steps in automation.

    - Automation must assure reliability to
      provide confidence to the management.

    - Some problems require continuous
      improvement:M/M interaction, Information
      integration and automation cost.

    - Industry and Academia must maintain close
      contact in all areas of CIM research. In
      this way academia will keep ensure that
      its work is relevant to industry, and
      industry will be more aware of
      advances made in academic research.

        P. Albertos

CONTRIBUTION OF EUREKA PROJECTS TO INNOVATION IN MANUFACTURING TECHNOLOGY

F. Jovane

Politecnico di Milano, Italy

Competitiveness of Industry is of great importance to establish and strengthen the European Market. The new "success model" for industry is based on innovation, which in turn is more and more depending on R&D and Technology Transfer.

National and supranational institutions have been launching Research Programmes concerned with Research from pre-competitive level to technological applications.
In the area of Manufacturing at pre-competitive level a great effort has been produced by the European Community through such as ESPRIT, BRITE-EURAM Programmes.
Technological projects to develop innovative products, manufacturing systems and services ready for the market have been launched by EUREKA. If national Research and Innovation Projects are also added, Europe looks very active in the area of Manufacturing Technology.

In this paper, following a presentation of EUREKA goals and "criteria", projects in the area of Manufacturing, whether belonging to the EUREKA FAMOS Umbrella Project or running as single EUREKA projects, will be considered to show in which final production areas and related enabling technologies the current research and innovative activities are concentrated. Such a picture may be helpful on one side to foster cooperation between research institutions and ongoing projects, on the other side to propose new projects by industry and, as some ongoing projects show, by research institutions whose creative activities may promote and match industrial involvement. EUREKA is industry led, but, as some projects show, new proposals may come from advanced research institutions whose activity may push industry to innovation through experiencing new applications in advanced technologies.

Finally the survey presented shows an impressive Research and Innovation community which may be considered as a source of technological knowledge and expertise and a "testing facility" for research people developing new tools, system configurations and technologies for manufacturing.

1.   THE EUREKA INITIATIVE

The EUREKA Initiative was launched by Seventeen European Countries and the CEE in the Ministerial Conference held in Paris in July 1985.

The Declaration of Principles, approved at the Hannover Ministerial Conference, set as Eureka goals:

- to raise, through close cooperation among enterprises and research institutes in the field of advanced technologies, the productivity and competitiveness of Europe's industries and national economies on the world market, fig. 1;
- to develop products, process and services based on advanced technologies with a world wide potential;
- to ensure a stable technological, economic and social position of European countries and industries;
- to make easy the exchange of technologies between European enterprises and institutes, that is a prerequisit for a high technological standard of European industry.

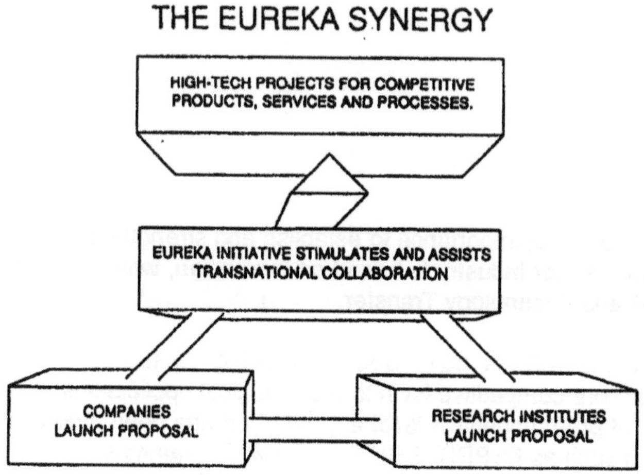

THE EUREKA SYNERGY

fig. 1 Cooperation of Companies and Research Institutes to
strengthen Europe's Industry and Economy

The above goals lead to the following EUREKA project criteria and pre-requisites:

- project relates to products, processes and services in significant areas of advanced technologies.
- use of advanced technologies
- project cooperation between partners in more than one EUREKA member country
- identified benefit from pursuing the project on a cooperative basis
- appropriately qualified participants (technologically and managerially)
- adequate financial commitment by participants
- performed within and exploited to the benefit of the EUREKA countries.

The Eureka Status of a project is granted by Governments of member countries to which enterprises and research institutes, that participate to the project, belong. It is granted after an analysis of the project suitability with EUREKA pre-requisites and criteria.

The technological areas considered are:
- Information technology
- Telecommunication
- Robotics and Production Automation
- New materials
- Energy
- Biotechnologies
- Laser
- Environment
- Transport

The countries participating to the Eureka initiative are:

- AUSTRIA
- BELGIUM
- DENMARK
- FINLAND
- FRANCE
- GREECE
- IRELAND
- ICELAND
- ITALY
- LUXEMBURG

- NORWAY
- NETHERLANDS
- PORTUGAL
- UNITED KINGDOM
- GERMANY
- SPAIN
- SWEDEN
- SWISS
- TURKEY
- CEC

2.   EUREKA PROJECTS

Following the Vienna Ministerial Conference, June 1989, EUREKA may be described as follows:
19 countries involved plus the European Commission; more than 1500 participating organisations, more than 6000 million ECUs planned investments in 297 different high technology projects oriented to development of products, systems and services ready for market.

The projects distribution among the nine sectors previously introduced is shown in fig. 2.

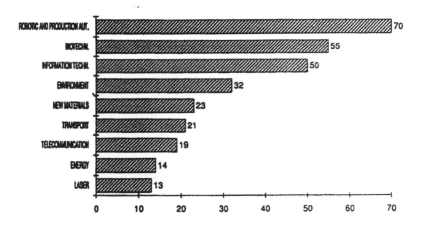

fig. 2 Eureka projects: number per area

The amount of money invested is reported in fig. 3. It is worth noticing that Robotics and Production Automation is leading in project number and shows the second highest investment. If we consider together the three areas mostly related to manufacturing:i.e. Laser, New Materials, Robotics and Production Automation, we find that the related projects represent 36% of total projects and the investement is 27% of total investment.

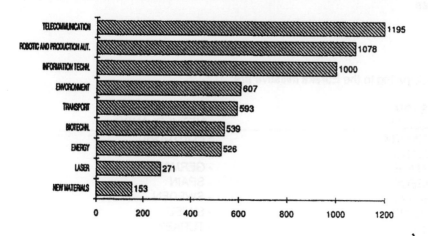

fig. 3 Eureka projects: total cost per area (MECU)

The size of projects is related to the sector involved as shown in fig. 4. The Robotics and Production Automation sector shows a concentration of projects in the low-medium range.

PROJECT COST (MECU)	TOTAL PROJ. NUM. *	NUMBER OF PROJECTS PER SECTOR								
		BIO	COM	ENE	AMB	INF	MAT	ROB	LAS	TRA
< = 5	139	30	5	4	22	24	16	24	4	10
> 5 < = 10	48	8	2	1	4	7	3	20	.	3
> 10 < = 20	46	9	3	2	1	11	3	10	4	3
> 20 < = 40	29	5	2	4	2	4	.	9	2	1
> 40	35	3	7	3	3	4	1	7	3	4
* 44 projects are still in definition phase										

fig. 4 Financial consistence of the projects (MECU)

The duration distribution of all projects is reported in fig. 5.
Middle (25-48 months) and long term (more than 49 months) projects prevail.
The participation of the 19 countries is shown in fig. 6.

PROJECT DURATION (MONTHS)	PROJECTS TOTAL NUMBER *
< = 24 (SHORT-TERM)	44
25 - 48 (MIDDLE-TERM)	136
> = 49 (LONG-TERM)	117
* 44 projects are still in definition phase	

fig. 5 Scheduled duration of the Eureka projects

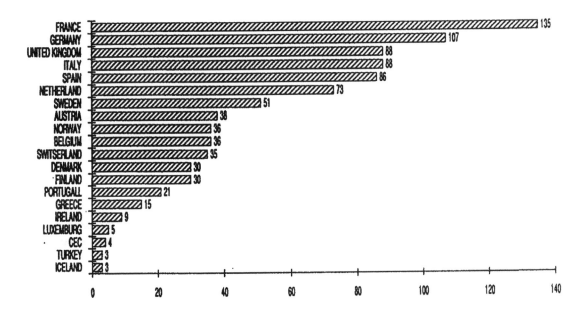

fig. 6 Eureka - projects participation per member State

3.   EUREKA PROJECTS RELATED TO MANUFACTURING

The analysis of the 297 ongoing projects shows that 106 are related to Manufacturing. They include projects belonging to the Robotic and Production Automation area, but also to laser and new materials areas (the investiment is 1500 MECU).
Of the aforementioned 106 projects, 78 projects have been generated following the typical bottom-up Eureka approach; 28 projects, mainly related to Factory Automation and Assembly, have been conceived through the FAMOS Umbrella Project.
All the projects may be classified as shown in fig. 7. In this general classification Famos projects show to be mostly related to the areas of manufacturing systems and technologies.

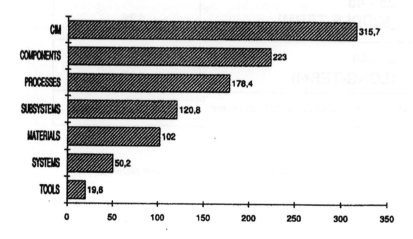

fig. 7 Eureka - Financial resources for projects in
manufacturing production (MECU)

The interaction between the bottom-up approach and the guiding action of the Famos Working Group has lead to an appropiate distribution of projects between sectors and technologies. An apport which paves the way towards the automated factory.

4.   THE FAMOS UMBRELLA PROJECT AND ITS INDIVIDUAL GENERATED PROJECTS

The Famos Umbrella Project was proposed by seven Countries in June 1986 following a German initiative. After an appropriate preliminary phase it was granted the Eureka status (EU72) on november 17th 1987 during the Ministerial Conference held in Stockholm.

The rational behind Famos is:
  - production and jobs are drifting away from Europe to high-tech and to low-labour-cost countries;
  - assembly has become the bottleneck of production in some areas; it accounts for up to 40% of costs; it is less automated in Europe: 40% of robots in Japan versus 10% in Europe.

Hence:
Europe must develop and use advanced technologies in flexible automation of assembly and open the potential for a significant increase in:

- productivity
- flexibility
- quality
- reliability of production

Famos projects are meant to provide novel solutions to assembly problems, according to the following aims:
- less jobs moving to non-European countries
- products and production once lost are brought back
- power for competition is recovered in lost and changed areas in respect of USA and Japan
- European strength in strategically important sectors

Seventeen countries and the EEC are supporting the Famos Umbrella Project. This is coordinated by a Working Group composed of: a National Coordinator and a National Project Leader for each country; a member from the Eureka Secretariat. The Famos Secretariat helps in keeping the continuity and the relationships among the various partners.
The process of launching Famos projects follows the typical bottom-up approach combined with an active interaction with potential proposers and partners. It refecs to the Assembly Technologies reference model fig. 9 and to field assessments of industrial sectors where assembly problems are strategically crucial. The Working Group activity is fundamental to help establishing projects partnerships, beside avoiding duplications and leaving out unexplored areas of interest.

Typical general FAMOS project funding priorities are derived from the EUREKA criteria:
- projects have to cover Flexible Automated Assembly
- be led by industry
- product based
- support/commitment by company board
- benefit from European approach
- balance of contributions
- use of enabling technologies
- long term objectives

As stated before 28 projects have been conceived and launched during the 3 years of activity of the Famos Working Group. The total investment is 360 MECU.
They came out through a four stage process which has involved in its first step some hundred proposals.
Each project, which is industry led, leads to a pilot plant and advanced tools/technologies: the partners are technology suppliers, end users, research institutions.

The industrial sectors covered are:

- light engineering	1 - vehicles components	4
- engines	3 - electrical and electric components	4
- household appliances	2 - shoe	1
- machine tools	1 - aerospace	1
- mechanics	1 - leather	1
- metal working	1 - toys	1
- food	1 - wood	1
- Instrumentation	1	

Four of them are tools that are not related to specific industrial sector.

The successful development and implementation of the results of EUREKA projects may require certain "enabling conditions" to be met which are beyond the capabilities and influence of the project participants themselves. Fulfilling these conditions may therefore require action from FAMOS and EUREKA bodies. This is the concept of "supportive measures", as an added value to the projects.

The FAMOS Umbrella offers a help to identify such requests of the individual projects and to channel them to the appropriate bodies.

This is of particular importance in the standarization area on advanced manufacturing technologies.

5.   A PRELIMINARY ANALYSIS OF FAMOS PROJECTS

Starting from the avalaible data, if we analyse the different projects in terms of products manufactured, their volume and variety, i.e. the production main performances of a pilot plant (which is one of the products of a research project), we may classify the various projects as shown in fig. 8. The dimension of cicles in the table is related to the project dimension measured by financial effort requested (the semi-circles are due to a project used by two different end-users that belong to different sector typologies).

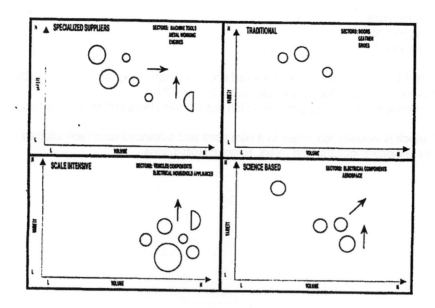

fig. 8 Classification, following Pavitt's,
of Famos pilot plants productions

The concerned sectors fall in a wide range of categories; following the Pavitt's classification, the projects may be classified into four sectorial typologies: traditional and specialised suppliers, scale intensive, science-based sectors. According to the leading technological trends in each sector, the goals of each innovative activity in the field of assembly are determined by different factors, in each sector taken into consideration.

The "traditional sectors", typology characterized by innovative trajectories, where the acquisition of new technoloy is obtained through capital goods, appears in a few projects. In some of them we see that the "technology-push" driving force prevails, i.e. the technology suppliers recognize the possibility of extending some automative applications to labour intensive sectors, such as the traditional sectors.

The "scale intensive" sectors (i.e. cars and household appliances) are broadly present, confirming thus that automation is a potential solution to the problem of increasing flexibility. Such sectors, characterized by high volume production and by dominant scale economies are experiencing a phase of basical change in the demand that leads to a high differentiation in the product. From the high volume - low variety position, automation, especially in final assembly phase, is a tool that enables these sectors to manage a wider range of products at low costs (the arrows show which are the trajectories that would be followed through the project).
The "specialised suppliers" use the customisation of production, as strategic weapon, but search to reach scale economies through the standardization of components and flexible assembly.

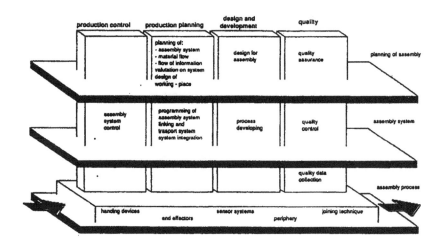

fig. 9 FAMOS reference model

At last among the "science-based" sectors there are relevant experiences in consumer electronics and electric-electronic components, where the resourse to flexible automation is a solution to high product variety for small series production; plants allow the assembly of non-standard products, increasing the production scale processed by a single system.

The enabling technologies involved in the various projects are those shown by the reference model, fig. 9.

If we analyse the various projects in terms of enabling technologies, as shown in fig. 9, we find the classification shown in fig. 10.

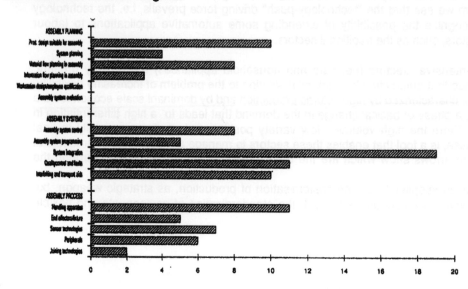

fig. 10 FAMOS: number of projects per assembly technological area

The most traditional manufacturers and, anyway, the sectors at medium or low technological intensity search new solutions to production problem through new design techniques and not through new processes. The research for new processes is the most appropriate innovation factor regarding the R&D high intensity sectors, because the rapid product innovation and the lack of consolidated processes open many opportunities in process innovation.

The systems are the technological area of particular interest for the high intensity scale fields besides the introduction of new managerial methodologies regarding the material flow.

New design techniques are the most important innovative factors in the traditional sectors.

As FAMOS is an "industry led" project, it follows that industries are the leaders of the projects: it is interesting to observe how besides an underlined prevalence of end users of the pilot plant as main partner of the projects, there is an important percentage of projects led by technology suppliers that are sometimes also research institutes or research centres, fig. 11.

Besides the implementation of the pilot plants, there are projects that have the tools production as goal, where starting from the enabling technologies, the conditions for the development of new solutions are created. The latter can then be applied to various sectors with similar characteristic of production.

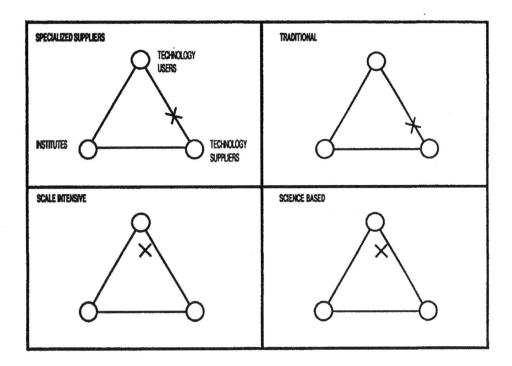

fig. 11 Projects Participant compositions as a function of Pavitt's classification of production sectors

6.   CONCLUSIONS

Some 100 projects, a third of EUREKA ongoing projects, are related to manufacturing.
The financial involvement is as high as 1500 MECU. If CEE and National Research Programmes are also considered it looks that manufacturing in Europe has been receiving great attention.
The projects falling in the area of Flexible Assembly Automation have been conceived and launched through the EUREKA FAMOS Umbrella Project. They cover various sectors. Pilot plants and new technologies are among expected results.
The number of launched EUREKA FAMOS projects and the related involvement of resources and expected results called for an appropiate technological and economical analysis.
A preliminary study has been carried out and the results are reported here.
An extensive analysis has been proposed by the Famos Secretariat and the final results will be available by the next Ministerial Conference in Rome (May 1990).
In this preliminary analysis we consider the final performances of pilot plants, the enabling technologies involved and the partners involved in Famos Projects already launched to see how projects relate to innovation in industrial sectors, in production technologies and to involvement of companies and research institutions.

Having classified pilot plants production performances following Pavitt's sectors classification, a preliminary analysis shows their contribution to move the companies involved along the best technological trajectory.

As this implies the use of advanced enabling technologies a strategic the market advantage may be obtained.

The analysis of enabling technologies (from software tools to processes, systems) shows that a wide range of them is being considered. As new advanced "capital goods", they will be produced and tested in a real advanced industrial environment for a high technology market.

A survey of the partners involved shows that the technology producers and end users behave following a push-pull scheme which is related to Pavitt's classification.

Although projects are industry led, Research Institutions are very active as knowledge developers and technological culture centers. In some cases their innovative ideas and methods helped in starting and launching projects.

Famos people and its research and innovation activities, represent an important, growing European "Technological Space" which may be integrated, more and more, with CEE and National Research Programmes. Such a space may be seen as a "Laboratory" for research and innovation in manufacturing, as a source of knowledge and advanced industrial culture, as a reservoir of relevant technological expertise, to be used at its best to strengthen European Industry, to help education of new European Technologists.

The diffusion of automation, at a slower pace than expected, calls for some thoughts about effectiveness of Research and Innovation.

The lack of field experience directly related to the research process may account for some of the barriers encountered by automation. EUREKA FAMOS projects in Robotics and Automated Production for their specific approach may represent a great contribution to effective implementation of advanced automation in most industrial sectors taking advantage of research Institutions. The game is open. New projects may be started.

ACKNOWLEDGEMENTS

The kind help of the EUREKA Secretariat, Bruxelles, Ing. Vittorio Chiesa (CNR-PFTM, Milano) and Ing. Antonio Diterlizzi (CNR-PFTM, Milano) is gratefully acknowledged.

GRAPHICAL SIMULATION OF COMPLIANT
MOTION ROBOT TASKS

P. Simkens, J. de Schutter and H. Van Brussel

*Department of Mechanical Engineering, Katholieke Universiteit Leuven,
Celestijnenlaan, Heverlee, Belgium*

Abstract. A graphical robot simulation, including a hybrid force/velocity
control strategy, is necessary to verify the off-line specification of
compliant motion robot tasks, before they can be downloaded to a robot.
In this paper an approach is presented to incorporate a 3D force control
strategy as a closed loop around the position control modules of a normal
3D off-line programming and simulation system. In this approach, input
data for the force control module comes from an emulated six-dimensional
force/torque sensor. This emulated sensor calculates the contact forces
from the available geometrical data about the free contacts with objects
in the environment. Output velocities from the force control modules are
integrated yielding desired positions which are transferred to the
position control modules. With this 3D force control simulation system,
compliant motion primitives based on orthogonal task frames with force
and velocity directions are executed and visualised. The applicability of
the methodology for the visual verification of compliant motion robot
tasks is illustrated by means of a 2D insertion task.

Keywords. Robots; force control; sensors; computer simulation.

INTRODUCTION

In order to program robots and manipulators without the use of actual workcell
equipment, robot off-line programming and
simulation systems have been developed.
Such systems include a graphic simulation
capability, which allows motions of robot
and peripheral equipment to be displayed
through computer graphical animation. This
allows robot programs to be developed and
verified as if the actual workcell devices
are available.

A central feature of such off-line programming and simulation systems is the
geometrical model of the workcell that is
maintained internally in the computer.
Problems arise when this internal model
does not match the external reality. In
addition to the geometrical uncertainty of
the environment, industrial robots themselves have a limited degree of absolute
accuracy and repeatability. Feedback from
vision, proximity and tactile or force
sensing can be used to refine the positional information and to make the execution
of the program more robust with respect to
geometrical uncertainties in the workcell
environment and the limited degree of accuracy of the robots.

Other applications, such as deburring or
grinding or "compliant" robot tasks in
general require a controlled contact between workpiece or tool in the robot gripper and some object in the environment.
These applications require appropriate
force control strategies, which are by
necessity an integrated part of the manipulator control system. The interface to
these force control capabilities comes
through special off-line language statements, further called compliant motion
primitives, which allow the user to specify the force control strategy. According
to Mason's theory (1981) about the hybrid
control functional specification, further
extended by De Schutter (1986, 1988a,
1988c), every compliant task environment
can be characterized by an orthogonal task
frame formed by force and position (or velocity) controlled directions.

However, no commercially available off-
line programming and simulation system includes possibilities to develop robot programs based on force control strategies.
In this paper an approach is presented for
emulating a force sensor in order to generate the necessary input for a force control strategy. This force control strategy
is then integrated into a normal off-line
programming and simulation system, which
includes a position control strategy.

The main objective for implementing a simulation for compliant motion is to test
the validity of the compliant motion primitives. The objective is NOT to test the
detailed dynamic behaviour of a real robot, its position or force controllers and
its interaction with the environment.

A COMPLIANT MOTION SIMULATION
SYSTEM: MAIN COMPONENTS

A compliant motion simulation package includes five major parts: a contact force model, a contact force transformation from sensor frame to task frame, an (idealised) model of the force control behaviour and a velocity integration, all to be built around an existing graphic robot simulation system based on velocity or position control (Fig. 1).

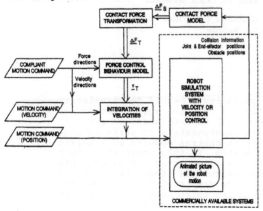

Fig. 1. Compliant motion simulation system

A Conventional Graphic Off-line Programming and Simulation System Based on Position Control

This component of the compliant motion simulation system converts a desired destination position, defined in a task frame, into a sequence of joint positions. This sequence of joint positions is calculated by an inverse kinematics module and is displayed as an animated picture on a computer display through some 3D-visualisation modules.

We suppose that this graphic simulation system also gives information about detected contacts or collisions between the manipulator and its environment. This collision information is available from a collision detection module, with the end effector position and the obstacle positions as input. This graphic simulation system, based on position control, is an already existing component.

A Contact Force Model

This component calculates the contact force from the collision information, the position of the robot end effector and the position of the objects in the environment (also called obstacles). Contact forces are calculated in the sensor frame, which has a fixed location between the last robot arm segment and the end effector.

A Contact Force Transformation

After the contact forces have been calculated in the sensor frame, they are transformed to the task frame, specified in the compliant motion primitive.

A Model of the Force Control Behaviour

This component converts the force error (in the task frame), calculated from the

difference between the desired forces and the emulated contact forces, into velocity information.

Velocity Integration

Calculated velocities are integrated into an incremental position input. This gives the immediate input for the position controlled simulation system. With this component, the force control loop is closed around the position control loop.

Contact force model, contact force transformation, force control model and velocity integration are explained in detail in the next sections.

CONTACT FORCE CALCULATION

Kankaanranta (1988) and Cai (1988) calculate the resulting contact forces in the sensor frame from motion equations, derived from D'Alembert's principle. However, this requires a complete dynamic description of the robot and a dynamic model for the contacts between rigid bodies. Moreover, the solution of the dynamics is very demanding for computer time.

In our approach, contact forces are calculated from the deformation of a flexible element. However, in contrast with Merlet (1987), where a flexibility is introduced in the object to be manipulated by the robot, we calculate the contact forces from the deformation of a flexible element, which is located between the last robot arm segment and the robot end effector.

The flexible element emulates a force / torque sensor. As long as the robot is moving in free space, the emulated force sensor is in its reference position. However, when the emulated force sensor collides with an object in the robot environment, the contact force model calculates the required deformation of this flexible element in order not to penetrate the object (Fig. 2). The required deformations are calculated according to the specified force directions of the compliant motion specification. Contact forces are calculated directly from these deformations: deformations are multiplied by the stiffness matrix of the emulated force sensor (De Schutter, 1988c).

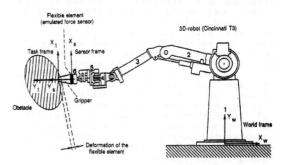

Fig. 2. Calculation of contact forces from the deformation of a flexible element.

This approach of calculating contact forces from the sensor deformations is very realistic: in a force control implementation on a real robot controller, contact forces are also obtained from the deformations of a force/torque sensor, measured with strain gauges (Van Brussel, 1985). The approach of concentrating the flexibility in the force/ torque sensor is valid when the stiffness of the emulated force/ torque sensor is lower than the robot stiffness.

The input of the contact force model consists of: end effector position, collision information (coordinates of a point, in contact with a plane) and geometrical information about objects, all available from the conventional off-line programming and simulation system.

Sensor deformations and contact forces are calculated using five different types of equations:

A. a 3D geometrical transformation;
B. free contact geometrical constraints;
C. frictionless contacts;
D. force and moment equilibrium;
E. relation between deformation and force.

In full 3D, where we have six degrees-of-freedom, two different cases are distinguished.

First case: the number of detected contact points is less than six: the complete set of equations of type A through E is required to calculate the deformations. Information about the individual contact forces in each contact point (equations of type C and D) is necessary to solve the under-determined geometrical problem.

Second case: the number of detected contact points is greater than six: a set of equations of type A, B and E is defined for six arbitrary chosen contact points, making contact with at least three pairwise non-parallel planes. Indeed, the sensor deformation is physically completely defined by the geometrical constraints for six contact points.

Geometrical Transformation

A first set of equations expresses the transformation of the contact point position due to the deformation of the emulated force/torque sensor. The equations, for small values ψ, θ and ϕ of the Roll-Pitch-Yaw angles, are (Simons, 1983):

$$
\begin{bmatrix} x_i' \\ y_i' \\ z_i' \\ \hline 1 \end{bmatrix} = \left[\begin{array}{ccc|c} 1 & -\phi & \theta & \Delta x \\ \phi & 1 & \psi & \Delta y \\ -\theta & \psi & 1 & \Delta z \\ \hline 0 & 0 & 0 & 1 \end{array}\right] \cdot \begin{bmatrix} x_i \\ y_i \\ z_i \\ \hline 1 \end{bmatrix}
\quad
\begin{array}{c}(1)\\(2)\\(3)\end{array}
$$

where:

Δx, Δy, Δz: translational deformations;
ψ, θ, ϕ : rotational deformations around x-, y- and z-axis;
x_i', y_i', z_i': contact point coordinates after transformation;
x_i, y_i, z_i: contact point coordinates before transformation;

All variables are expressed in the sensor frame.

Geometrical constraints for a free contact

For a free contact, i.e. a contact point which is not influenced by any kinematical constraints like rotational or prismatic joints with the environment, the contact point makes contact with a surface. The contact point is lying on that surface, after transformation according to the deformation of the emulated force / torque sensor. For objects, modelled as polyhedrals, we obtain the following equation:

$$A_i x_i' + B_i y_i' + C_i z_i' + D_i = 0 \qquad (4)$$

where A_i, B_i, C_i, D_i:

coefficients of the contact plane of the polyhedral in contact with the i-th contact point, expressed in the sensor frame.

The assumption of working with polyhedrals is not a limitation: most graphical computer systems internally work with polyhedrals. On the other hand, when working with free form surfaces, the equations of the surfaces are to be linearised at the contact point.

The specified equations (1) through (4) are valid when the contact point is a point of the grasped object or end-effector, and when the surface is a boundary of a polyhedral in the environment. For the opposite case slight modifications are necessary.

Frictionless contacts

For an idealized case we suppose there is no friction between the two objects in contact. This means that the contact force lies in the direction of the normal to the contact surface. In a completely free contact there are no contact moments. Mathematically:

$$F_{xi} = k_i \cdot A_i \qquad (5)$$
$$F_{yi} = k_i \cdot B_i \qquad (6)$$
$$F_{zi} = k_i \cdot C_i \qquad (7)$$
$$M_{xi} = 0 \qquad (8)$$
$$M_{yi} = 0 \qquad (9)$$
$$M_{zi} = 0 \qquad (10)$$

where:

k_i : ratio between normal vector and force vector.
F_{xi}, F_{yi}, F_{zi} : contact forces in the i-th contact point.
M_{xi}, M_{yi}, M_{zi} : contact moments in the i-th contact point.

Forces and moments are expressed in the sensor frame.

If friction is taken into account, the contact force does not necessarily lie in the direction of the normal to the surface, but lies in the friction cone.

Force and moment equilibrium

The total forces and moments expressed in the sensor frame are calculated from the individual components of all the contact forces and moments:

$$F_{xs} = \Sigma \ F_{xi} \qquad\qquad (11)$$

$$F_{ys} = \Sigma \ F_{yi} \qquad\qquad (12)$$

$$F_{zs} = \Sigma \ F_{zi} \qquad\qquad (13)$$

$$M_{xs} = \Sigma \ (-z_i \ F_{yi} + y_i \ F_{zi}) + \Sigma \ M_{xi} \qquad (14)$$

$$M_{ys} = \Sigma \ (\ z_i \ F_{xi} - x_i \ F_{zi}) + \Sigma \ M_{yi} \qquad (15)$$

$$M_{zs} = \Sigma \ (-y_i \ F_{xi} + x_i \ F_{yi}) + \Sigma \ M_{zi} \qquad (16)$$

where:

F_{xs}, F_{ys}, F_{zs} : total forces in the sensor frame;

M_{xs}, M_{ys}, M_{zs} : total moments in the sensor frame.

Relations between forces and deformations

Forces and moments are related to the deformations of the emulated force / torque sensor by its stiffness, defined by a 6x6 stiffness matrix. Mathematically:

$$\begin{bmatrix} F_{xs} \\ F_{ys} \\ F_{zs} \\ M_{xs} \\ M_{ys} \\ M_{zs} \end{bmatrix} = \underline{K}_s \cdot \begin{bmatrix} \Delta x \\ \Delta y \\ \Delta z \\ \psi \\ \theta \\ \phi \end{bmatrix} \qquad \begin{matrix} (17) \\ (18) \\ (19) \\ (20) \\ (21) \\ (22) \end{matrix}$$

where $\underline{K}_s$: stiffness matrix in the sensor frame.

The stiffness matrix is a general 6x6 matrix. However, in the case of a well designed real force/torque sensor, all forces and moments are decoupled (Van Brussel, 1985). This means that there is no relation between a force or a moment in one direction and a deformation in another direction. In this case the stiffness matrix is a diagonal matrix.

Number of equations and unknowns

First case: the number of detected contact points is less than the number of degrees-of-freedom. Suppose the number of contact points is n (n < 6 in 3D).

Number of equations:
- 3D geometrical transformation: 3n
- constraints for free contacts: n
- frictionless contacts: 6n
- force and moment equilibrium: 6
- relation forces/deformations: 6
                                     _____
                                     10n + 12

Number of unknowns:
- $\Delta x, \Delta y, \Delta z, \psi, \theta, \phi$:   6
- x'_i, y'_i, z'_i :   3n
- k_i :   n
- $F_{xi}, F_{yi}, F_{zi}, M_{xi}, M_{yi}, M_{zi}$:   6n
- $F_{xs}, F_{ys}, F_{zs}, M_{xs}, M_{ys}, M_{zs}$:   6
                                                       _____
                                                       10n + 12

Second case: the number of detected contact points is greater than the number of

degrees-of-freedom. In this case the number of contact points to work with is reduced to six in 3D.

Number of equations:
- 3D geometrical transformation: 18
- constraints for free contacts: 6
- relation forces/deformations: 6
                                     _____
                                             30

Number of unknowns:
- $\Delta x, \Delta y, \Delta z, \psi, \theta, \phi$:   6
- x'_i, y'_i, z'_i :   18
- $F_{xs}, F_{ys}, F_{zs}, M_{xs}, M_{ys}, M_{zs}$:   6
                                                       _____
                                                               30

CONTACT FORCE TRANSFORMATION

The relation between forces and torques in sensor frame and task frame is expressed by a force transformation matrix, which is a Jacobian. Moreover, since both sensor frame and task frame are Cartesian frames, the Jacobian has a simplified form. According to De Schutter (1986) the relationship is:

$$\begin{bmatrix} \underline{F}_{ct} \\ \hline \underline{M}_{ct} \end{bmatrix} = \left({}^s\underline{J}_t \right)^T \cdot \begin{bmatrix} \underline{F}_s \\ \hline \underline{M}_s \end{bmatrix} \qquad (23)$$

where:

$\underline{F}_{ct}, \underline{M}_{ct}$: calculated force and moment vector in the task frame.

$\underline{F}_s, \underline{M}_s$: force and moment vector in the sensor frame.

${}^s\underline{J}_t$: Jacobian from task frame to sensor frame.

FORCE CONTROL

From the difference between the calculated contact forces and the desired contact forces, both expressed in the task frame, the force error vector is calculated. This force error vector is converted into a velocity vector by multiplying the force error vector with the matrix of the force control feedback constants and with the compliance matrix, also expressed in the task frame.

The compliance matrix in the task frame is (De Schutter, 1986):

$$\underline{C}_t = \left(\underline{K}_t \right)^{-1} = {}^t\underline{J}_s \cdot \left(\underline{K}_s \right)^{-1} \cdot \left({}^t\underline{J}_s \right)^T \qquad (24)$$

where:

$\underline{C}_t$: compliance matrix in the task frame;

$\underline{K}_t$: stiffness matrix in the task frame;

$\underline{K}_s$: stiffness matrix in the sensor frame;

${}^t\underline{J}_s$: Jacobian from sensor to task frame.

After this conversion of the force error vector into a velocity vector, only the resulting velocities for the force controlled directions are retained by multiplying with the compliance selection matrix, where the diagonal elements s_{ii} are 1 if the i-th direction is force controlled, all other elements being 0.

This yields the following global scheme for the force control behaviour model (Fig. 3):

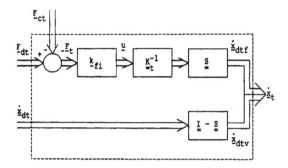

Fig. 3. Force control behaviour model

where:

$\underline{F}_{ct}$: calculated forces in the task frame;

$\underline{F}_{dt}$: desired forces in the task frame;

$\underline{\Delta F}_t$: force errors in the task frame;

$\underline{k}_{fi}$: matrix of feedback constants;

$\underline{S}$: compliance selection matrix;

$\underline{\dot{x}}_t$: velocity output in the task frame;

$\underline{\dot{x}}_{dt}$: desired velocities in the task frame;

$\underline{\dot{x}}_{dtf}$: desired velocities in the task frame (force directions);

$\underline{\dot{x}}_{dtv}$: desired velocities in the task frame (velocity directions).

VELOCITY INTEGRATION

The desired velocity vector in the task frame is converted to the world frame. Then this velocity is integrated, yielding incremental displacements, by multiplying by the sampling time T_s:

$$\begin{bmatrix} d_x \\ d_y \\ d_z \\ \delta_x \\ \delta_y \\ \delta_z \end{bmatrix} = T_s \cdot \begin{bmatrix} v_{xw} \\ v_{yw} \\ v_{zw} \\ \omega_{xw} \\ \omega_{yw} \\ \omega_{zw} \end{bmatrix} = T_s \cdot {}^w\underline{J}_t \cdot \underline{\dot{x}}_t \quad (25)$$

The updated position of the robot end effector in the world frame at moment j, expressed by the homogeneous transformation matrix $[A]_j$, is (Paul, 1983):

$$[A]_j = \begin{bmatrix} 1 & -\delta_z & \delta_y & d_x \\ \delta_z & 1 & -\delta_x & d_y \\ -\delta_y & \delta_x & 1 & d_z \\ 0 & 0 & 0 & 1 \end{bmatrix} \cdot [A]_{j-1} \quad (26)$$

EXAMPLE: INSERTION IN 2D

The described simulation strategy for force controlled robot tasks has been implemented in 2D on an Apollo DN 580 workstation. Geometrical data about collisions

are given by a simple collision detection module: it is tested whether any line segment of the gripper or the grasped object intersect the boundaries of any of the obstacles (Sun, 1987). With these data the force control modules are executed, and the resulting position in the world frame is converted into joint values with an inverse kinematics module.

The task to be executed is (Fig. 4): insert a conical tap in a hole. In this example there are an initial misalignment and translational error to be corrected. The insertion motion in the X_t-direction is obtained by specifying a desired contact force, greater than zero, in X_t. The misalignment and translational error in the Y_t-direction are corrected by specifying respectively a desired zero-torque for the rotation and a zero-force for the translation in Y_t.

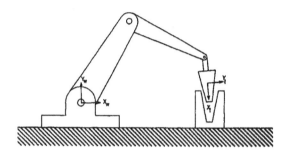

Fig. 4. Compliant motion in 2d: example.

Compliant motion specification:

```
Move compliantly
    with task frame defined fixed at the
        center of mass of the tap

    with task frame directions
        x_t : force :   50 N
        y_t : force :    0 N
        α_t : force :    0 Nm

    until all desired forces and moments
    reached.
```

Fig.5 shows the calculated results for the emulated forces and moment. Contact is detected after 0.2 seconds. The initial misalignment results in a contact moment, until the orientational error is completely corrected after 2.2 seconds. Then also the Y_t-component of the contact force raises, until the desired value for the X_t-component is reached (t = 2.5 sec). Stiffnesses are 10000 N/m and the force control feedback constants are chosen such that the product $k_{fi} * T_s \leq 1$ for a stable control behaviour.

CONCLUSION

In this paper an approach has been presented to include a hybrid force/torque control strategy in a graphical off-line programmation and simulation system. The main component of such an implementation is the force sensor emulation module. In this module, contact forces are calculated from geometrical constraints, constraints for a

free and frictionless contact, relationships between deformations and forces, and
force and moment equilibriums.

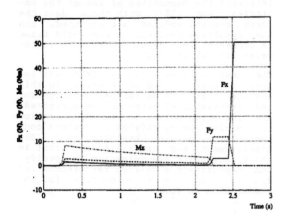

Fig. 5. Simulated contact forces.

REFERENCES

Cai, L., and A.A. Goldenberg (1988). General dynamic model for analysis and
simulation of robots in contact tasks.
In U. Remboldt (Ed.), Proc. SYROCO'88
(Karlsruhe), Pergamon Press, Oxford.

De Schutter, J. (1986). Compliant Robot
Motion: Task Formulation and Control.
Ph.D. Thesis, Faculty of Applied
Sciences, K.U.Leuven.

De Schutter, J., and H. Van Brussel
(1988a). Compliant robot motion. I. A
formalism for specifying compliant
motion tasks. Int. J. of Robotics Research, 7/4, 3-17.

De Schutter, J., and H. Van Brussel
(1988b). Compliant robot motion. II. A
control approach based on external
control loops", Int. J. of Robotics
Research, 7/4, 18-33.

De Schutter, J., and P. Simkens (1988c).
CAD-based Verification and Refinement
of high level Compliant Motion Primitives. In B. Ravani (Ed.), CAD-based
programming for Sensory Robots, NATO
ASI Series F 50, Springer-Verlag, 203-
222.

Kankaanranta, R.K., and H.N. Koivo (1988).
Dynamics and simulation of compliant
motion of a manipulator. IEEE Journal
of Robotics and Automation, 4/2, 163-
173.

Mason, M.T. (1981). Compliance and force
control for computer controlled manipulators. IEEE Transactions on Systems, Man and Cybernetics, SMC-11/6,
418-432.

Merlet, J.P. (1987). Programming tools for
force-feedback commands of robots. In
U. Remboldt (Ed.), Languages for Sensor-based Control in Robotics, NATO
ASI Series F 29, Springer-Verlag, 69-
81.

Paul, R.P. (1983). Robot Manipulators. Ma-

thematics, Programming and Control,
M.I.T. Press.

Simons, J. (1983). Mathematica, Programma-
tie en Controle van Industriele Robots. Summer Course, Faculty of
Applied Sciences, K.U.Leuven.

Sun, K., and V. Lumelski (1987). Computer
simulation of sensor-based robot collision avoidance in an unknown environment. Robotica, 5, 291-302.

Van Brussel, H., H. Belien and H. Thiele-
mans (1895). Force sensing for advanced robot control. In N.J. Zimmerman
(Ed.), Proc. 5th Int. Conf. on Robot
Vision and Sensory Control (Amsterdam), IFS, Kempston, 59-68.

A STUDY ON DISTRIBUTED SIMULATION
FOR FLEXIBLE MANUFACTURING SYSTEMS

S. Fujii, H. Sandoh, H. Matsuda and M. Tasaka

*Department of Systems Engineering, Faculty of Engineering, Kobe University, Kobe,
Japan*

Abstract.   This paper proposes a distributed simulation model for FMS. In the model an FMS
system is described as a network where the nodes represent the components of FMS and the links
correspond to the flows of the objects processed by FMS. For the realization of the model on a
distributed computer system this paper presents a synchronization mechanism by message passing
which is implemented on a distributed system. Effectiveness of the proposed method is demonstrated
by an example.

Keywords.   Distributed processing; Simulation; Flexible manufacturing system; Synchronization;
Microprocessors.

INTRODUCTION

Computer simulation is recognized as one of the key techniques for the effective design of a flexible manufacturing system (FMS). As FMS becomes complicated, the physical configuration and the relation in FMS become difficult to describe and the simulation model tends to become large and complex. This paper proposes a *distributed simulation* for FMS using a network of processors. The distributed simulation is defined as follows (Wagner, 1988):

- Simulation which runs on a distributed system
- The timing information in simulation is spread out among a group of independent processes

The distributed simulation described in this paper makes the components in FMS correspond to the processor elements in a distributed system. It provides a view for natural understanding of the physical configuration of FMS.

This paper also describes how to realize the distributed simulation in a real distributed system. The processor elements do not have any shared memory and all information is exchanged by message passing.

In a distributed system, a synchronization mechanism is required to keep the consistency of the entire system. Several methods for this mechanism are reported in the past, e.g., TEXS (Chandy, 1979), the virtual ring synchronization algorithm (Peacock, 1979), the link time algorithm (Peacock, 1979), and the time warp mechanism (Jefferson, 1985). The algorithm of the synchronization mechanism realized in this study extends the link time algorithm by allowing communication from one processor to some other processors.

DISTRIBUTED SIMULATION MODEL

Component Model

In the simulation model three components, i.e., *facility*, *transaction* and *cart* are introduced. A facility is a fixed component in FMS (e.g., machining center, material warehouse, inspection equipment, etc.) and a transaction is an object which moves among facilities (e.g., workpieces, tools, jigs, etc,). But transportation equipment which moves by itself (e.g., automated guided vehicle) is modelled as a *cart* and distinguished from the other moving objects since it has special functions such as path planning. Behavior of the other equipment which moves objects on a fixed path (e.g., conveyer, etc.) is presented by the movement of transactions. A facility, a transaction and a cart have a *facility type*, a *transaction type* and a *cart type* respectively. These types indicate their functions in the FMS. For example, a facility type presents the processing function of the facility. A transaction type decides the behavior of the transaction, e.g., processing steps of the transaction and processing times in facilities where the transaction arrives. A cart type denotes the capability of the cart, e.g., the transfer speed of the cart and the number of the objects which the cart can carry at the same time. A transaction also has a *transaction flow* as the attribute. It indicates processing steps of the transaction in the FMS and is denoted as a sequence of facility types.

As the timing information, *simulation clock* and two types of *events* are introduced. Simulation clock is the abstraction of real clock and is discretely advanced by events. As the types of events an *arrival event* is processed when a transaction reaches a facility and a *departure event* is processed when a transaction leaves a facility. Scheduling time of event can be given by random numbers, if necessary, which are decided corresponding to facility, transaction and cart types. It is to be noted that the simulation clock is distributed on all facilities and is called *local clock*.

It will be shown later how to synchronize the distributed simulation clock.

Network Description

The connection among facilities is described as a *network*. A *node* of the network represents a facility and an arc, called *link* in the following, denotes a path of transactions moved by transportation equipment and is decided by a transaction flow. A link toward a node and a link which goes out of a node are called the *input link* of the node and the *output link* of the node respectively.

Figures 1 and 2 show a simple example of FMS. In Fig. 2 facility i-j denotes the j-th facility of which facility type is i. As mentioned before, transaction flows are described as the sequences of facility types. In network description all possible links should be given for all possible transaction flows.

We introduce a hierarchical structure called *subsystem* in the network description. A subsystem is also described as a node. Figure 2 can be shown as in Fig. 3 by introducing two subsystems. By introducing subsystems, a distributed simulation model can be constructed as a hierarchically structured model. The efficiency of synchronization mechanism is improved by collecting all facilities into a few subsystems.

A subsystem has a network description which denotes the connection of facilities collected in the subsystem. This network description requires an input node and an output node which play the roles of interfaces among subsystems. In a subsystem these nodes play roles as *deputies* of the facilities in the other subsystems (Linton, 1987).

Message Passing

The control of transaction and cart is carried out by passing *control messages* between a transmitting node and a receiving node. The messages in the simulation model are listed as follows and are also used for synchronization as mentioned later:

- Transaction message: A message transferring the information of a transaction
- Request message: A request to transmit a transaction message
- Reply message: A message replying to a request message

- Cancel message: A message canceling a request message
- Cart request message: A request for a cart which can transfer a transaction
- Cart reply message: A message replying to a cart request message

Transaction messages which are transferred to the next node are scheduled as output waiting transactions in a *waiting list*. The waiting lists are placed in nodes as FIFO output buffers but the placing methods are different corresponding to the types of the nodes. In subsystem nodes waiting lists exist on each output link. In facility nodes waiting lists are placed corresponding to the facility types of the nodes where transactions are transferred at the next step since transaction flows are given as sequences of facility types.

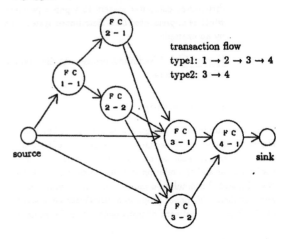

Fig. 2. Network description of the FMS in Fig. 1.

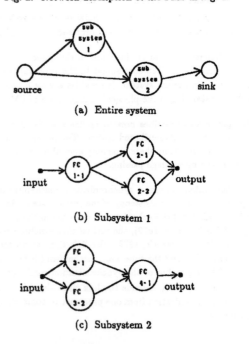

(a) Entire system

(b) Subsystem 1

(c) Subsystem 2

Fig. 3. Network description of the FMS in Fig. 1 with subsystems.

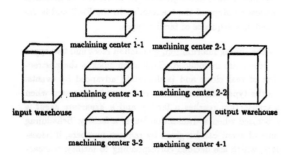

Fig. 1. Facilities in an FMS.

A cart request message and a cart reply message are required for the preparation to transfer a transaction using a cart. The information of all carts (e.g., the place, the state, etc.) is collected into a *cart monitor*. A node which transfers a transaction using a cart transmits a cart request message to the cart monitor. When the cart monitor receives the message, the monitor selects an available cart and replies to the node by transmitting a cart reply message. More details on the cart control is described later.

SYSTEM CONFIGURATION

In the simulation model a distributed computer system is assumed as a system which has several processors communicating by message passing through a network. The number of processors is assumed not to be less than the number of the nodes in the network description mentioned in the previous section. The network is not assumed to have any specific configuration. Only assumption on the network is that messages are transmitted and received without errors. The distributed computer system used for the development of a simulation system is mentioned in the later section.

Figure 4 shows an abstract image of the simulation system realized on a distributed computer system where its model is shown in Fig. 3. In Fig. 4 *monitors* supervise the simulation clock of their lower-level layers. In addition a subsystem monitor and a cart monitor control transaction and cart transfer between subsystems respectively. As the lower-level layer of a subsystem monitor, *facility processes* corresponding to facilities in a subsystem are placed under the monitor. A facility process controls event processing and transaction transfer between facilities. An *entire system monitor* is a special subsystem corresponding to input and output storages in FMS. It supervises the situation of transaction generation and termination and detects the end of transaction generation or the end of simulation.

In the distributed computer system each facility process is allocated on a processor with one to one mapping. Each subsystem monitor is also allocated on a processor. The subsystem monitors in Fig. 4 correspond to the input and output nodes in Fig. 3. The links between subsystem monitors and facility processes in Fig. 4 correspond to the links between the input, output and facility nodes.

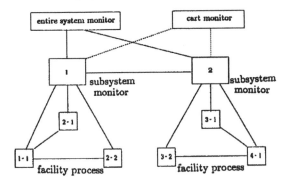

Fig. 4. An abstract image of simulation system.

SYNCHRONIZATION

Each node does not recognize the processing in the other nodes. The only information that a node should take is whether the node can transmit a transaction to the other nodes or not. The control of transaction transfer is carried out based on this information. It is realized by the synchronization mechanism presented in this section.

Definition

A synchronization problem in distributed simulation is fundamentally to realize the arrival sequence of transactions at a node in a consistent order. This means that each arrival time in the transactions does not decrease compared with the preceding arrival time.

An objective system for the simulation study can be modelled by a network description as mentioned in the previous section. Each node in the network description has input links and output links except the source node and the sink node of transactions. A node which is the start node of a link is called *departure node* of the link and a node which is the goal node of a link is called *target node*. Each node also has waiting lists for storing transactions to be transmitted.

The state of links is denoted as follows:

(1) occupied: A transaction waiting for transmitting exists on the link.

(2) empty: No transaction exists on the link.

In the case (1) the number of the transactions on a link should be only one. Another transaction is placed in a waiting list and will be moved on the link as soon as the previous transaction is removed from the link.

Each transaction has timing information called *SAT (Scheduled Arrival Time)*. SAT means the scheduled time when the transaction will arrive at a target node and will be scheduled only when the transaction is at a departure node (i.e., the node before transmitting to the current node). The real arrival time may be greater than SAT depending on the situation of the target node. SAT, therefore, denotes the lower bound value of an arrival event time.

In each transaction, the value of SAT is given as follows:

$$SAT = t_{DEP} + t_{WCT} + t_{MOV} \qquad (1)$$

where t_{DEP}, t_{WCT} and t_{MOV} mean the time of the departure event scheduled by the transaction left the departure node, the waiting time for arriving a cart to the departure node and the moving time of the transaction from the departure node to the target node respectively. If no cart is used in the transaction transfer, t_{WCT} is set to zero.

A node which is a target node for some transactions has a timing information called *RLS (node release time)*. It presents the time when the node becomes in the condition to accept a transaction. When a transaction arrives at the node, SAT of the transaction is compared with the node release time in the node. Since the node release time and SAT never decrease, the following properties holds:

$$t_{ARV} = max(RLS, SAT) \qquad (2)$$

where t_{ARV} denotes the time of an arrival event.

Lower Bound Time

LBT (Lower Bound Time) means the lower bound value on SATs of all possible transactions which will be arrive at a node through an input link. A request message has LBT and the state of a link, i.e., occupied or empty. The information is used for link selection and target selection to be discussed later.

The value of LBT is decided as follows:

- If the input link is occupied,

$$LBT = SAT \qquad (3)$$

- Otherwise and if transactions exist in the departure node, i.e., the node before the current node,

$$LBT = t_{NXT} \qquad (4)$$

where t_{NXT} denotes the time of the next event which will be processed in the departure node.

- Otherwise,

$$LBT = \max(t_{RCT}, t_{MIN}) \qquad (5)$$

where t_{RCT} and t_{MIN} denote the time of the arrival event which was processed most recently and the minimum value among LBTs of all input links of the departure node respectively.

Link Selection

If a node becomes to be able to accept transactions, the node gets a transaction from one of the input links as shown in Fig. 5. If some input links exist in a node as shown in Fig. 5, the node selects an input link from the links. The selection is called *link selection*.

Link selection is performed by the following algorithm called *R/R algorithm*. The R/R stands for "request and reply" where the control is carried out by transmitting request messages and receiving reply messages. In the following, t_{OCP} and L_{OCP} denote the minimum value of LBTs on oc-

cupied input links and the occupied link which has the minimum value respectively. t_{EMP} and L_{EMP} are same as t_{OCP} and L_{OCP} except the input links are empty.

(1) If all input links of a node are occupied, the node gets a transaction from L_{OCP}.

(2) Otherwise,

(2-1) If t_{OCP} is not greater than t_{EMP}, the nodes gets a transaction from L_{OCP}.

(2-2) Otherwise, the node cannot accept transactions since a transaction which has SAT that is less than t_{OCP} may appear on L_{EMP}.

After all transactions transferred from an input link, LBT of the input link is set to infinite. t_{EMP} becomes, therefore, greater than t_{OCP} by the end of transaction generation.

LBT is spread out by transmitting request messages. Each node transmits request messages at the following time:

- the time before a transaction transmits,
- the time when the node receives a request message.

A node receives request messages from input links, selects the link from which the node gets a transaction and notifies the result of selection to the departure node connected to the link by transmitting a reply message. The reply message has RLS of the node as timing information. The information is used for target selection described in the next section.

If the target node transmits a reply message, the node turns to unacceptable state which does not accept any transactions from the other links and waits for transmitting a transaction from the link.

Target Selection

In the distributed simulation model, a transaction can select a target node from some nodes as mentioned before. The selection is called *target selection*. Figure 6 shows this situation where a node has some output links to the nodes which have the same facility type. Since transaction flows indicate only facility types, transactions in the departure node can go to any node among the target nodes.

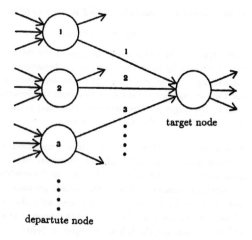

Fig. 5. Link selection.

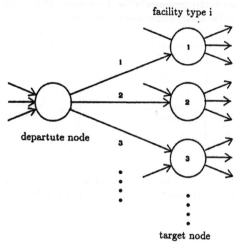

Fig. 6. Target selection.

Target selection is carried out as follows:

(1) A transaction occupies all output links to the target nodes, that is, same request messages are transmitted to the nodes simultaneously and all of the links are occupied.

(2) Each target node performs link selection and transmits a reply message to the departure node which includes RLS of the target node.

(3) The departure node compares all RLSs in the reply messages and selects the target node which has the minimum RLS.

(4) The departure node transmits a transaction to the selected target node and transmits cancel messages to the other target nodes.

The R/R algorithm in the previous section should be extended to apply the algorithm with target selection. The extended algorithm called *R/R* algorithm* includes target selection and the modified R/R algorithm. The modification is made so that the target node performs link selection again if the node receives a cancel message instead of a transaction.

The R/R algorithm cannot be applied to target selection. However the R/R algorithm is more efficient than the R/R* algorithm since the number of message passing is less. The R/R* algorithm is, therefore, only chosen in the nodes where the algorithm is required. It can decide which algorithm is to be selected on each node before simulation starts based on the network description and the transaction flows of the simulated system.

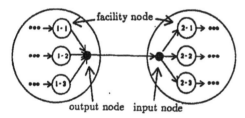

(a) Network description with subsystems.

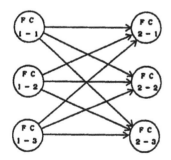

(b) Network description without subsystems.

Fig. 7. Different network description.

Communication among Subsystems

As discussed in system configuration, the communication between two subsystems is carried out by considering the output node of a subsystem as the departure node and the input node of another subsystem as the target node. The communication algorithm described in the previous section can be applied to input and output nodes.

By introducing subsystems into the network description of a simulated system, the number of nodes increases but the number of links and the number of message passing can be decreased in complex systems. Figure 7 shows two different network description of a same simulated system where subsystems are introduced in (a) and not introduced in (b). In (a) the input and output nodes are added on (b). The number of the links from the facility nodes 1-i (i = 1, 2 and 3) to the facility nodes 2-j (j=1, 2 and 3) is 7 in (a) and 9 in (b). In general the number of the links from m facility nodes to n facility nodes is changed from m × n to m + n + 1 by introducing subsystems.

On message passing, the numbers of link selection and target selection affect the number of messages. In Fig. 7 (a) the number of link selection is only 1 in the output node and the number of target selection is also only 1 regardless of the number of facility nodes. However in Fig. 7 (b) the number of link selection is 3 in the facility nodes 2-j (j=1, 2 and 3) and the number of target selection is 3 in the facility nodes 1-i (i = 1, 2 and 3). The total number of each selection is, therefore, both 9. By our experience the number of message passing affects the efficiency of simulation extremely and thus we conclude the introduction of subsystems is effective to improve the efficiency of simulation.

CART CONTROL

When a transaction is transferred using a cart, a cart monitor supervises the cart control as described before. The processing on the cart monitor is as follows:

- Selection of a request for cart transfer
- Assignment of the selected request to a cart
- Path planning for the cart
- Reply to the request

A request for cart transfer is presented by transmitting a cart request message from a node which has or will have a transaction to transfer using a cart. The message includes timing information same to the request message. Selection of a cart request is carried out by a method like target selection.

After assignment of a cart and its path-planning, the cart monitor transmits a cart reply message to the node. The reply message also has timing information for cart transfer. Based on the information, the transaction is transferred with the cart. Details of cart control will be reported in another paper.

AN EXAMPLE

Effectiveness of the proposed method is demonstrated by applying the algorithm to a relatively simple manufacturing system shown in Fig. 8. In this example there are 3 subsystems and 12 facilities where a subsystem has 4 facilities in which two facilities have a same type.

To execute this example 16 processors are required in total, i.e., 1 for the entire system monitor, 3 for subsystem monitors and 12 for facility processes.

The computer system used for the development of a prototype simulation system is composed of fifty 16-bit microprocessors (IBM 5540 personal computers, the performance is same as IBM PC/AT) connected with a mainframe computer (IBM 3083) by serial lines (IBM 3270 lines). The communication among the processors is carried out through the serial lines.

The execution times required for various numbers of transactions are given in Fig. 9 which are measured by elapsed times from the start of simulation to the end of simulation. In Fig. 9 the execution times denoted by PSS-FMS mean the times by sequential simulation on the same microprocessor. PSS-FMS (Fujii, 1986) is a simulation system for FMS developed by the authors. As shown in Fig. 9, the execution time by sequential simulation rapidly increases as the number of transactions increases. But the execution time by distributed simulation increases only in proportion to the number of the transactions. The method proposed in this paper is, therefore, concluded to become more effective when the size of the simulated system becomes larger.

CONCLUSION

We proposed a distributed simulation model for FMS in this study and developed a prototype system for simulation on a distributed computer system. In the model an FMS system is described as a network. The nodes of the network correspond to the components of FMS and the links of the network correspond to the flows of the workpieces. In the implementation the nodes are allocated on processors in the distributed computer system. We also proposed synchronization mechanism for the distributed simulation. We presented two algorithms for the synchronization mechanism. Effectiveness of the proposed method is shown by applying the algorithms to an example, although the computational experiences on a distributed computer system showed some necessity to improve efficiency. The improvement of computational efficiency are further problems to be studied in future.

REFERENCES

Chandy, K. M., V. Holmes, and J. Mistra (1979). Distributed simulation of networks. *Computer Networks*, 3-1, 105-113.

Fujii, S., M. Tanaka, H. Sandoh and others (1986). Flexible manufacturing system simulator with graphic modeler. In B. Wahlström and K. Leiviskä (Ed.), *Modelling and Simulation in Engineering*. Elsevier Science Publishers, North-Holland. pp.319-324.

Jefferson, D. R. (1985). Virtual time. *ACM Trans. Programming Language and Systems*, 7-3, 404-425.

Linton, M. A. (1987). Distributed management of a software database. *IEEE Software*, 4-6, 70-76.

Peacock, J. K., J. W. Wong, and E. G. Manning (1979). Distributed simulation using a network of processors. *Computer Networks*, 3, 44-56.

Wagner, D. (1988). Thoughts on Distributed Simulation. *Simulation*, 50-6, 225.

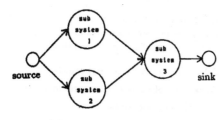

(a) Entire system

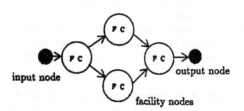

(b) Subsystems

Fig. 8. An example.

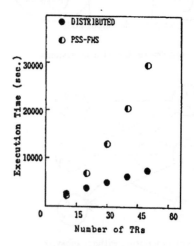

Fig. 9. Execution times.

Copyright © IFAC Information Control Problems in
Manufacturing Technology, Madrid, Spain 1989

AN APPROACH TO THE FOOTWEAR
ASSEMBLY AUTOMATIC SYSTEM

R. Aracil, C. Balaguer, A. Barrientos and A. Yela

Departamento Automática, E.T.S.I. Industriales (DISAM), Madrid, Spain

Abstract. This paper presents an automatic system for footwear assembly (SAMCA), which is being developing in the EUREKA-FAMOS EU-285 project. The main feature of the proposed assembly system, in contrast to current methods, is that parts remain fixed on pallets during the process, which are placed, by means of a transport system, into different machines and robots, in order to perform automatically several operations, which presently are carried out manually. Furthermore, a simulation model of the system has been built, in order to calculate correctly the requirements of the system, and to quantify the effects of various technological and operational factors that affect its performance.

Keywords. Assembling. Discrete event simulation. Manufacturing processes. Operations research. Robots.

INTRODUCTION

This paper presents some results of the definition and evaluation phase of the EUREKA-FAMOS EU-285 project "SAMCA: Footware Assembly Automatic System". The project is presently being developed by a French-Spanish partnership. The Spanish partners are the research institutes INESCOP and DISAM, and the companies IMEPIEL and MERQUINSA. Together, they are responsable for the definition of the subsystem intented to attach the soles and heels and carry out all the complementary operations.

In the definition of the SAMCA system the most important considerations have been the size of the factories and present structure of the footwear industry. The defined system must be modular, in order to be installed in different stages in medium and small sized factories. Also it must be intelligent enough to allow the input of components made by different auxiliary makers, in which there are appreciable desviations from the standard. The following sections describe the system definition and its simulation.

SYSTEM SPECIFICATIONS

Several footwear manufacturers have been consulted in order to select the specifications of the system. They refer to the types of shoes to produce, required productivity, conditions on the system inputs, and type and necessities of future customers.

The selected shoe types are:

A) Men's shoes with one piece sole-heel, and flat base.
B) Men's shoes with one piece sole-heel, and base with welt.
C) Women's shoes with separated sole and heel (cuban heel shoe).
D) Women's shoes with heel covered by sole (built heel shoe).

Shoes are classified in three kinds, according to their size ranges (European units):

S. Children   : 18 to 30.
M. Youth   : 31 to 37.
L. Men-Women : 38 to 45.

All selected shoes have the remaining leather of its top part folded on the shoe lining. This type of shoe is called "top line folded shoe".

In the case of shoes with separated sole and heel, it is necessary to attach the heel to the shoe bottom. This joint is performed by fusion of a plastic pivot included in the bottom, previously inserted into a heel cavity.

The required productivity is about 100 pairs per hour. This productivity must be evaluated at steady state, after a warm-up period. Continuous changes of shoe types to manufacture are not considered in this type of production. However, the developed system permits this flexibility, and the productivity of the system has been evaluated under these conditions.

In order to guarantee the correct operation of the system, it is necessary to make sure of the homogeneity of shoe components: sole, heel, upper, etc. These components arrive at the system in the following conditions:

1. Shoe upper and bottom are put together on the last.
2. All the lasts have a standard fixture system.
3. All the lasts and soles have an identifier code, to determine the product type.

The system is designed to be used by medium and small sized shoe factories. For this reason, it must be modular and not very expensive. In this way, small shoe factories will be able to use only some of the system cells, while the rest of the production steps may be performed with traditional machinery.

DESCRIPTION OF THE SYSTEM

The SAMCA system is formed by four flexible cells which have the following objectives:

i) Preparation of pallets (**Preparation Cell**).
ii) Roughing and cementing the sole and shoe upper (**Roughing and Cementing Cell**).
iii) Attaching the sole and heel to the shoe upper (**Assembly Cell**).
iv) Completion of the shoe (**Finishing Cell**).

The link between these cells is done by means of a special conveyor belt. The system works with only one worker and one supervisor.

The easy reprogramming of the whole system makes it useful to manufacture simultaneously different kinds of shoes. Nevertheless, the productivity of the system is medium-high. On the other hand, the system is modular, which makes its application in the medium and small sized industry easy. Therefore, the SAMCA system has the main characteristics of CIM: high flexibility with medium productivity, and modularity.

The concept of the SAMCA transport system is based on the fact that there is no separation of the last and sole from the pallet during the whole process. This is a great advantage, due to the constant positioning of the last and sole respect to the pallet. The pallet is universal for all kinds and sizes of shoes. Moreover, the same pallet is used for lasts and soles. In the case of the last, it is situated on the pallet through special holes. The sole is placed on the pallet using a sole carrier attached in the same holes, as shown in Fig. 1. The various kind of soles have different ways of being attached to the sole's carrier. In fact, there are two types of attaching systems: using a gripper or a suction pad (vent).

The first cell is the Preparation Cell. Its mission is to prepare the pallets and to verify the correspondence between soles and lasts. The inputs of the system are the raw materials (a pair of lasts and soles) and two empty pallets: one for the lasts and the other for the corresponding soles. Each pallet has its own magnetic programmable identification card in order to store the type and size of the shoe, the type and number of operations, etc. The cell is made up by different stores (soles, carriers, etc.), a manipulator to remove the soles carriers, and several verification machines, and it is the only one which needs a human operator.

The next fully automatic cell is the Roughing and Cementing Cell (Fig. 2). This cell performs the operations of roughing and cementing lots of soles and shoe uppers, using a 6 DOF robot. The lots of pallets are stored in two special rotating tables (lasts and soles). Each table has two independent zones, one of them is used for the input and output of lots, and the other one for robot operations. The number of pallets in each table depends on the simulation results, which will be described in the next section.

The Roughing and Cementing Cell has some intermediate stores and specialized machines. One of them is the marking machine for shoe uppers. This is why the transport system has a bifurcation in order to treat the soles and shoe uppers separately.

The output pallets from the Roughing and Cementing cell are the input pallets to the Assembly Cell. In the first place the pallets of lasts and soles enter in the reactivation furnace to reactivate the adhesives. Later, a 6 DOF robot attaches the sole and the heel to the shoe upper. This is one of the most delicate operations, due to the necessity of positioning the sole with a high level of precision. A vision system helps to obtain the geometrical center of the sole in order to position it over the shoe upper.

Finally, the pallet with the assembled shoe (last, shoe upper, sole and heel) and a free pallet go into the Finishing Cell. In this cell a robot performs most of the completion operations, such as polishing, lacquering, edge painting, etc.

SYSTEM SIMULATION

In this section, the construction of a model to simulate the performance of the Footwear Assembly Automatic System, is first presented, and then, it is shown how this model is used to quantify the effects of several operational control strategies on the performance of the system.

Simulation has proved to be a powerful tool in the system design phase. The two main objectives of the simulation in this phase were, in the first place, to size up correctly the system, for example, processing times, storage sizes, number of pallets, etc., in order to achieve the required productivity; and after, to show how various operational factors, such as lot-sizing, input sequence of parts or number of shoe types, affect the output of the system.

The simulation model was built by means of the simulation environment SIMFACTORY on a PC computer. This tool provides an easy way to develop, with a menu based programming method, the model and the animation graphics, and reports valuable graphical and statistical outcomes to evaluate the behaviour of the system.

Description of the Model

At this point, the developed simulation model is explained. It may be divided into the following cells:

- **Preparation Cell.** It is made up of two stores of raw materials (lasts and soles) and empty pallets, and a worker who performs several operations to place lasts and soles on the pallets.

- **Roughing and Cementing Cell.** This cells consists of a marking machine, and a roughing and cementing robot. A setup is made between these operations. Furthermore, the two rotating tables and loading and unloading of lots of lasts and soles pallets are considered.

- **Assembly Cell.** It is formed by the adhesive reactivation furnace, a robot for sole and heel assembly, a heel attaching machine, and a pressing machine. The processing times and the order of these operations depend on the shoe type.

- **Finishing Cell.** This cell contains the polishing tunnel and the finishing robot. When work is completed, empty pallets return to the store.

The union between these cells is made by queues with unlimited sizes, that will be replaced afterwards with a material handling system (i.e. conveyor belts). In Fig. 3, a schematic diagram of the developed model is presented.

Model Simulation. Outcomes Analysis

The first objective of simulation was to size up the whole system to satisfy a required productivity of one hundred pairs of shoes each hour. Since the times of the different tasks to be executed by the system cells are not yet completely defined, as they depend on many technological factors, the first step was to obtain the maximum value of these operation times, which should be verified on the real machines and robots.

With this objective in mind, several simulations have been realized to evaluate the sizes of the stores and union queues (conveyors capacity), and the number of pallets needed. Moreover, other system performance parameters, such as stations utilization or flow time, are obtained. As an example, Table 1 presents the input parameters and the numerical outcomes of two model simulations, corresponding to lot sizes of 10 and 5 pairs, respectively.

Once the system parameters were sized up, the second objective of simulation was to quantify the effects of several operational factors on the behaviour of the system. In Table 2, the results of ten simulations are shown, in which some changes on the operation times, lot size, input sequence of parts, and types of shoes, are considered. All other input parameters have the same values of the preceding simulations.

Based on the analysis of the numerical outcomes of these ten simulations, the following conclusions are obtained:

- **Operation times.** Comparing simulations 1) and 2), on the one hand, with 4) and 5), on the other, it's seen that, in this case, reducing soles assembly time increases the productivity of the system. The utilization levels of the robots show how a higher reduction would not increase anymore that productivity. Similarly, in simulations 6) and 7), in which roughing time is reduced, the productivity increases, and a further reduction would increase it even more. From the values of the flow time, it follows that the system becomes faster as operation times decrease.

- **Raw materials input.** From the values of the eight first simulations, it can be affirmed that, when the input of parts exceed maximum productivity, the levels of the queues and the number of required pallets increase substantialy. Therefore, the difference in these two output parameters due only to other factors (such as lot size), should be observed under the condition that the input of parts matches up to the obtained productivity, as happens in the two last simulations.

- **Roughing and cementing lot size.** We conclude that the productivity of the system decreases as the lot size is reduced, except in the two first simulations, because of the complete utilization of the assembly robot. Furthermore, the system becomes faster, and less pallets are required, by decreasing the size of the lot. In any case, the outcomes corresponding to lot sizes of 10 and 5 pairs result to be quite similar.

- **Three shoe types manufacturing.** In case 8), manufacturing three types of shoes is considered. Comparing with 6), it can be observed that it does not affect significantly to the output parameters of the system, except in the assembly cell, as type A) does not require heel attaching.

Further Applications

The applications of simulation of the system in the future are various. Some problems to be considered could be the following:

- Replace the union queues between cells by conveyor belts, and quantify the effects of their speeds and capacities on the output of the system.

- Check whether assigning priorities to the different types of shoes affects the performance of the system.

- Determine the optimum input sequence of raw materials to minimize the required capacity of the conveyors.

- Consider interruptions in the operation of the machines and conveyors, as breakdowns or maintenances.

- Model manufacturing of defective parts and the actions to be performed on them.

- Consider operation times as probabilistic functions, in tasks in which it is suitable, because of their nature (i.e. worker operation time).

CONCLUSION

In this paper, a system that performs automatically all the required operations for footwear assembly, has been presented. Its main features are: flexibility and modularity, so that its utilization in footwear factories, usually medium or small sized, is feasible. The construction of a simulation model of the system has been shown, too. This model is used to evaluate the dimensions of all system parameters, and to study the sensitivity of the system to several technological and operational factors that affect its performance, quantifying their effects.

The robot operation times used in the simulation of the system have been verified in a prototype cell, which carry out soles attaching, and shoe upper roughing and cementing, using the ASEA's IRB-2000 robot (Fig. 4).

ACKNOWLEDGEMENTS

The SAMCA's is a EUREKA-FAMOS project (EU-285), and it is supported by a grant of the Ministerio de Industría y Energía of the Spanish Government.

The authors acknowledge the colaboration of managers, technicians and whole staff of the project partners INESCOP, IMEPIEL and MERQUINSA in the realization of this article.

REFERENCES

N.T. Burton. E.J. Lansdown.
Implementation of a Flexible Manufacturing System to Make Parts in a Small Batch Environment.
Planning for Automated Manufacturing. pp. 89-96. 1986.

H.I. Nisanci, R.J. Sury.
Production Analysis by Simulation in a Shoe Manufacturing Factory.
International Journal of Production Research. Vol. 18, No. 1, pp. 31-41. 1980.

L. Pardo, T. Valdés.
Simulación. Aplicaciones Prácticas en la Empresa.
Ed. Díaz de Santos, Madrid. 1987.

B. Pourbabai.
Effects of Control Strategies on the Transient Performance of an Integrated Manufacturing System.
Robotics & Computer-Integrated Manufacturing. Vol. 5, No. 1, pp. 83-89. 1989.

A.A.B. Pritsker.
Introduction to Simulation and SLAM II.
Halsted Press, New York. 1986.

P. Ránky.
Computer Integrated Manufacturing.
Prentice Hall Int., UK. 1986.

U. Rembold, C. Blume, R. Dillmann.
Computer-Integrated Manufacturing Technology and Systems.
Dekker Inc., New York. 1985.

P.M. Taylor, G.E.Taylor, I. Gibson.
A Multisensory Approach to Shoe Sole Assembly.
6th International Conference on Robot Vision and Sensory Control. pp. 117-126. 1986.

/../
SIMFACTORY v1.6. User's and Reference Manual.
CACI Products Company. La Jolla, California. 1988.

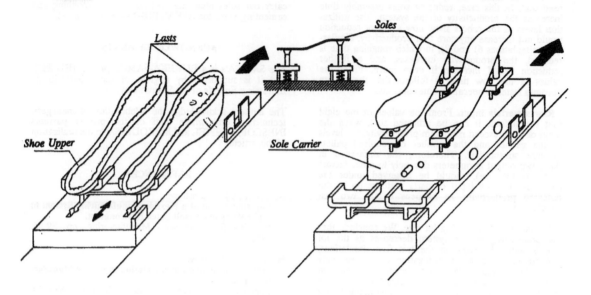

Fig. 1. Pallet for lasts and soles

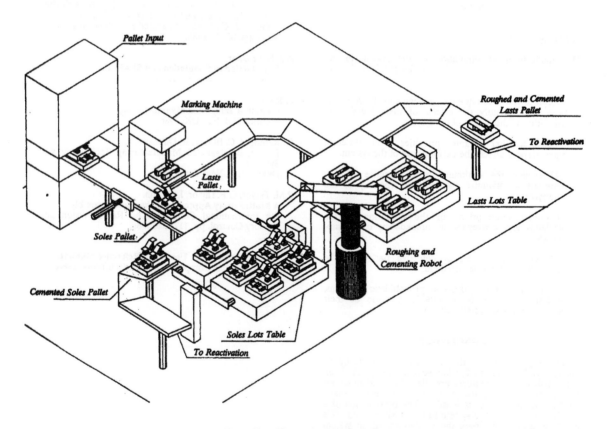

Fig. 2. Roughing and cementing cell

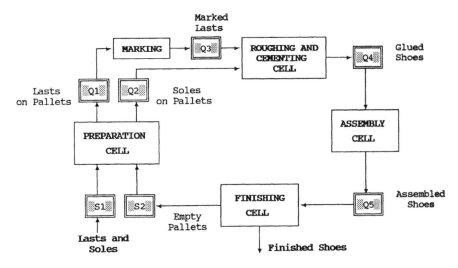

Fig. 3. Schematic diagram of the simulation model

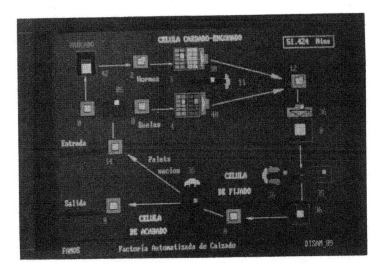

Fig. 4. Prototype cell

Fig. 5. Simulation execution. Layout Animation Graphics

Table 1 System Parameters Estimation

INPUTS

- Operation Times.
  - Pallets preparation .. 30 s/pair

  - Marking .... 25 s/pair
  - Roughing .... 14 s/pair
  - Cementing .... 10 s/pair
  - Setup .... 10 s/pair

  - Cement reactivation .. 15 s/pair
  - Soles assembly .... 26 s/pair
  - Heels assembly .... 10 s/pair
  - Heels attaching ...... 5 s/pair
  - Pressing .... 12 s/pair

  - Finishing .... 36 s/pair
  - Polishing .... 10 s/pair

- Raw Materials Input.
  - 50 shoe pairs each 30 minutes.

- Roughing and Cementing Lot Size.
  - 10 and 5 pairs.

- Shoe Types.
  - Type C), only.

- Simulation lenght.
  - 4 hours. Warm-up of 30 minutes.

OUTPUTS

- R & G Lot Size	10	5
- Productivity (pairs/hour)	100	96
- Average Flow Time (min) .	17	11
- Machine Utilization.		
- Worker	83%	80%
- Marking machine	67%	65%
- R & G robot	100%	100%
- Furnace	42%	40%
- Assembly robot	100%	94%
- Heel attaching machine	14%	13%
- Pressing machine	33%	32%
- Finishing robot	100%	94%
- Polishing tunnel	28%	26%
- Queues Size. (pallets)		
- Q1 (lasts on pallets) .	1	1
- Q2 (soles on pallets) .	1	5
- Q3 (marked lasts)	4	8
- Q4 (glued shoes)	22	9
- Q5 (assembled shoes) ..	2	2
- Used Pallets	59	44

Table 2 Performance Simulation Results

INPUTS

Simulation	(1)	(2)	(3)	(4)	(5)	(6)	(7)	(8)	(9)	(0)
- Roughing and Cementing Lot Size	10	5	1	10	5	10	5	10	10	5
- Soles Assembly Time (s/pair)	40	40	40	30	30	30	30	30	30	30
- Lasts Roughing Time (s/pair)	20	20	20	20	20	15	15	15	15	15
- Lasts and Soles Input (pairs each 30')	50	50	50	50	50	50	50	50	38	37
- Types of Shoes to Manufacture	C	C	C	C	C	C	C	ACD	C	C

OUTPUTS

Simulation	(1)	(2)	(3)	(4)	(5)	(6)	(7)	(8)	(9)	(0)
- Productivity (pairs/h).	65	65	49	69	66	76	73	76	76	73
- Average Flow Time (min)	67	61	80	61	61	51	49	50	24	19
- Machine Utilization (%)										
- R&G robot	100	100	95	100	100	100	100	100	99	100
- Assembly robot	100	100	75	86	83	96	92	88	95	92
- Finishing robot	73	73	55	77	74	85	82	85	84	82
- Queues Levels (pallets)										
- Q2	130	150	231	130	150	100	120	100	6	14
- Q3	139	151	231	139	151	106	122	106	11	16
- Q4	54	19	1	20	9	20	9	20	20	9
- Used Pallets	356	342	470	322	330	260	271	261	71	60

AN INTELLIGENT PLANNER FOR ASSEMBLY PROCESS PLANNING

A. L. Dowd and Y. P. Cheung

University of Warwick, Department of Engineering, Coventry, UK

Abstract. There has been an urgent need to integrate design with
manufacturing following the increase in competition among manufacturers and the change in market demands. It is widely recognised today that
integration of design and manufacture is the key to reducing production
costs. One answer to lowering costs even further lies in providing the
vital link between CAD and CAM i.e. Computer Aided Process Planning.

Where conventional methods have failed to provide a solution,
designers have sought to apply alternative techniques using Artificial
Intelligence. This seems to be the next logical step towards the next generation of Computer Aided Engineering systems as demands for more
flexibility, and intelligence increase. Such systems help remove the tedious repetition of engineering thus allowing more time and consideration
for the creative and innovative aspects of engineering.

This paper outlines the development of a prototype process planner
for assembly. Two examples were taken from a local car manufacturing
company. The basic concepts of the planner were adapted from Artificial
Intelligence Planners because of similarities between Artificial Intelligence planning and process planning. Finally, the techniques adopted and
difficulties encountered are also discussed.

Keywords. Artificial intelligence; assembly; computer aided process planning;
frame problem; qualification problem; ramification problem; temporal logic.

1. AI PLANNING AND PROCESS PLANNING

1.1. AI Planning

Many AI planning systems have evolved over the
years as a result of the need to provide general problem
solvers using AI techniques. The planning problem in AI
terminology can be formalised using the *State-Change*
model (Genesereth, 1987). In this model, a plan is the
result of proving that a goal state is achievable from an initial state or vice versa. An assumption of this model is
that, a state remains unchanged until an action is done to
change it, hence converting it into another (intermediate or
goal) state.

The State-Change model can be described using a formal language such as *Predicate Calculus* (PC). Sentences
in PC have the basic structure :

relation(term1,term2).

PC allows the use of logical connectives such as: and,
or, negation and implication to form more complex sentences from simple ones. Using a logic programming
language based on PC, the State-Change model can be
computerised. This idea of using PC to describe the State-
Change model was first formulated by Greens and Kowal-
ski (Nilsson, 1980). This paper describes a planner for
component assembly which is based on the principles
described above.

Some of the problems associated with this formulation
are the so called *frame problem, qualification problem* and
ramification problem.

The problem of having to represent facts that change
as well as those that do not is known as the frame problem
(Rich, 1983). Consider the robot planning problem, where
a robot has only one arm and it is able to stack or unstack
one block at a time. By starting with an initial state
description and applying rules to make changes to the
states, the robot problem can be solved. However if during
the search there is a need to backtrack, there are no indications of what need to be undone unless this has been explicitly defined at each state description. It seemed that this
problem could be solved by recording all the possible
changes at all the states but in some problems this may not
be enough. For example, it may be essential to record
every instance of the position of the robot's arm. The
frame problem becomes increasingly important as the complexity of the problem increases. Even in the simple robot
planning problem, the number of frame axioms could be
immense since a description of what remains unchanged
may include facts such as : the robot is still at its original
position after the operation, the locations of other objects
remain unchanged, the house that the robot is in is still
intact, etc. The number of frame axioms needed is proportional to the product of the number of relations to the
number of actions described in the model. Potentially,
there can be a very large number of frame axioms and
managing these efficiently is termed the frame problem.

The qualification problem is that there can be a large
number of pre-conditions in the model (McCarthy, 1977).
Many things can prevent the robot from performing its task
: the work surface may not be strong enough to withstand
the weight of the block, the blocks might crush under the
robot's gripper, there may be a fire in the vicinity, etc.
Computationally, it is impossible to consider all possibilities. However in real world and real time applications, the
qualification problem remains an obstacle.

Similarly, for any given action there are essentially an infinite number of possible consequences depending on the complexity of the situation in the State-Change model. The ramification problem is the difficulty of recording all the possible consequences for a given action (Finger, 1971). Computationally, it is not feasible to consider all possible consequences. Hence, in specific applications, only the most likely consequences are taken into consideration.

There are various approaches to the above problems. The classical monotonic approach involves explicitly defining frame axioms for the model. As shown earlier, the number of frame axioms is also immense. This approach can be cumbersome in that each time a new action or relation is added, more frame axioms will also need to be added. By making the frame axioms implicit as in the STRIPS (Fikes, 1971) approach, there is no need to define frame axioms. Here, actions are described in terms of pre-conditions, add lists and delete lists. Maintaining a consistent database using this approach may be difficult. However, this approach is viable when the facts used to describe the world can be guaranteed to come from some predetermined set (Lifschitz, 1986). In the possible worlds approach (Ginsberg, Smith, 1988), the result of the action in the *nearest world* is taken. For example, the *nearest world* to the current world in the robot planning example, would be the block that is being stacked is at its new location while everything else remain the same.

1.2. Process Planning

Process planning is the stage between design and manufacture. Plans involving instructions on the manufacture of a part are drawn up from the information obtained from the drawings and the experience of the planners. Very often, it is an iterative process, involving much consultations between design and the production engineering before a final plan is accepted. This stage also involves making an estimation of the cost of manufacture based on the process plans drawn up with cost estimates being made against each operation in the process plan.

Inital attempts at automating this stage in the manufacturing process have produced variant-type systems where much effort have been spent on documenting old process plans. New process plans are then obtained by retrieving the existing plans and edited where necessary to cater for a new product. Currently, the trend of research in this field have been aimed at generating plans without the need to consult existing plans and AI techniques have been used in this field. An example of a generative process planning system for machined components is GARI (Descotte, Latombe, 1981) which is still under research at the moment. By using this approach, it may be possible to integrate the process planning system to other systems in an organisation, where the process planning system is able to share its information with other parts of the manufacturing system. Most of the process planning systems developed so far are for machining processes where most of the data on machining are well documented. Hence producing such process plans are more straight forward than assembly process plans.

It is believed that an analogy can be drawn between AI planning and process planning in manufacturing. By formulating the process planning problem using the State-Change model it is possible to implement such a system on a machine. For example, a sentence to represent the fact that a bearing is pressed onto the shaft is : *press(bearing,shaft)*. In this way, generative process planning systems can be implemented. The generative approach to process planning involves synthesizing plans from the logic incorporated in the system.

Below is a description of a prototype for assembly process plans. Initially much effort was spent on linearising the goal list i.e. ordering a partial ordering of the goal list which can be a list of goals in any given order.

2. DEVELOPMENT OF THE SYSTEM

The examples used in the development of the prototype are the sub-assemblies of the piston-connecting rod and the primary shaft. The piston-connecting rod subassembly consists of the piston, the connecting rod and the gudgeon pin as illustrated in Fig. 1 (Cheung, Dowd, 1987).

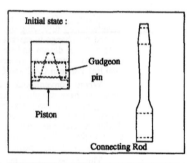

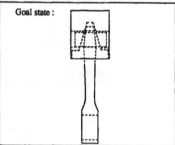

Fig. 1 An initial state and goal state of the
piston-connecting rod sub-assembly

During the assembly process, the operator has to remove the gudgeon pin, insert the small end of the connecting rod into the centre hole of the piston and then secure it with the gudgeon pin. The sub-assembly of the primary shaft involves pressing the bearings onto the two ends of the primary shaft (see Fig. 2).

The constraint of this sub-assembly is that one of the bearings has to be pressed first because the reaction face for performing this action is also the location for the other bearing. Hence the order of pressing the bearings is important.

The piston-connecting rod sub-assembly is used as an example because of the geometry of the parts, i.e. there is a cross-section or intersection of the areas to be filled in by the components (see Fig. 1) In the primary shaft sub-assembly, reaction faces on the components are used for pressing bearings. It was found that these features are common among many sub-assemblies of the car. Having obtained these two examples, the next stage of the development involves formalising the information and converting it to the State-Change model. The pre- and post- conditions of the various states of the assembly problem have to be defined. By maintaining a world model which is updated each time an action is performed as in the STRIPS planner, the state and consistency of the database can be maintained.

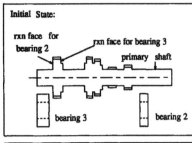

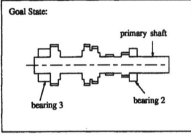

Fig. 2 An initial state and goal state of the primary shaft
sub-assembly

The development was divided into two main modules:

i)   The Top-level Planner.

ii)   The Low-level Planner.

2.1. The Top-level Planner.

The main objectives of this module is to reason about
the input goal list and produce a linearised goal list from a
partial ordering of the goal list. Currently, for the primary
shaft example, the input to the system is a goal list such as:

assemble(bear3,whole,shaft,face3,shaft,face1).
assemble(bear2,whole,shaft,face2,shaft,face3).

The above goal state is the final state of the assembly
of the two bearings, bear2 and bear3 onto the respective
positions of the shaft. The term, *whole* indicates that the
whole of the bearing is assembled onto the shaft. The third
and fourth terms are the names of the parts where the bearings are pressed and the position on the part respectively.
The last two terms indicates the part and the reaction face
for the action as pressing bearings would normally require
reaction faces. Similarly, the goal state of the piston-
connecting rod sub-assembly is :

assemble(pin,whole,piston,transverse_hole,nil,nil).
assemble(rod,whole,piston,axial_hole,nil,nil).

where the last two terms, *nil, nil* suggest that no reaction
face is required , the first goal suggests that the whole of
the pin is assembled into the transverse hole of the piston.
Similarly for the latter goal.

The following tests are incorporated as rules in the
planner:

i)   Redundancy check.
      Here any redundant goals, i.e. repeating
      goals and goals that are achieved as a result
      of another goal in the list are deleted form
      the goal list.

ii)   Addition.
      A goal may not be achievable because its
      pre-condition is not true in the current
      model. Hence it may be necessary to add
      its pre-condition(s) to the goal list. Put
      these pre-conditions into another list for the
      time being.

iii) Sorting.
      Here the goal list is sorted by checking the
      pre- and post-conditions if the goals. Basi-
      cally, if the post-condition of a goal, X is
      the same as the pre-condition of another
      goal, Y then it is possible to place goal X
      before Y.

iv)   Merging.
      Finally, merge the ordered goal list with the
      list of added pre-conditions.

Some of the above tests were adapted from AI
planners such as NOAH (Genesereth, Nilsson, 1987) and
TWEAK (Chapman, 1987). The nonlinear planner, NOAH
first attempts to order a partially ordered set of actions
using some of the above rules or critics. The research in AI
planning seems to be concerned with developing a minimal
set of rules for a complete and generalised planner.
Recently, the TWEAK planner suggested a formulation for
a complete planner.

However, most of the AI planners are conjunctive
planners, i.e. an attempt is made to put all the actions into
a sequence assuming that they all interact with each other.
For real applications, there is normally a mixture of both
conjunctive and disjunctive goals (i.e. goals that do not
interact with each other). For example, the assembly of the
primary shaft and the piston-connecting rod do not interact
with each other and could be done in parallel. Hence a formalisation catering for both conjunctive and disjunctive
goals is necessary. This was first suggested by Taylor
(1988). The representations for parallel goals and sequential goals are illustrated below:

par(goal_1,goal_2,....goal_n).
seq(goal_1,goal_2,...,goal_n).

Thus the final goal list of the assembly of the shaft is:

seq(assemble(bear2,whole,shaft,face2,shaft,face3),
assemble(bear3,whole,shaft,face3,shaft,face1).

Similarly, the goal list for both assemblies is :

par(seq(assemble(bear2,whole,shaft,face2,shaft,face3),
assemble(bear3,whole,shaft,face3,shaft,face1)),
seq(assemble(rod,whole,piston,axial_hole,nil,nil),
assemble(pin,whole,piston,t_hole,nil,nil))).

Only the higher level temporal logic of the problem is
dealt with here, since the actual timing (start and end
times) of the actions is not very significant at this level.
This is because the only concern is the production of an
ordered goal list for the next stage of the planner.

2.2. The Low-level Planner

This part of the system is concerned with decoding the goal list from the previous stage. The reasoning here is straight forward planning, i.e. proving that the goal states are achievable from the given state of the model. After the success of each proof, the sequence of operations is produced.

The pre- and post conditions of each state is defined accordingly. A definition of the state, *assemble* is:

state(assemble(Obj1,Face1,Obj2,Face2,Rxnobj,Rxnface)):-

result(press(Obj1,Face1,Obj2,Face2,Rxnobj,Rxnface)).

and

state(assemble(Obj1,Face1,Obj2,Face2,Rxnobj,Rxnface)):-

*findall(W,recorded(world,W,_),W1), (member
  (assemble(Obj1,Face1,Obj2,Face2,Rxnobj,Rxnface),W1);
  member(assemble(Obj2,Face2,Obj1,
    Face1,Rxnobj,Rxnface),W1)).*

where the first definition can be read as: to achieve the assembled state given the various terms, it is the result of the press action using the given terms. The six terms in the *assemble* and *press* predicate are as described in 2.1. The latter definition simply suggests that the assemble state is already true in the current model. The *findall* predicate finds all the facts recorded in the database called *world*. The *member* predicate checks if *assemble(Obj2,....)* is a member of the list, *W1* which should be the list containing all facts in the current *world* database.

As can be seen from above, formalisation of the problem is critical. Much time and effort has to be spent on gathering and seiving through the enormous amount of information on how and why parts are assembled in practice.

3. INTERFACE WITH DESIGN

A description language can be used to describe a design to the system which will then need to be decoded. However this method can be cumbersome as words used for describing the design may have to be known by the system. It is recommended that an interface to a CAD model be developed so that information from say, a solid modeller can be down loaded onto the planning system. In this way, cumbersome input descriptions can be avoided. Investigations are being carried out at the moment regarding such a possible interface to the planning system.

4. COMMENTS ON THE APPROACH

The system was originally developed by adapting techniques of AI planning which has been going on for many decades but has only been recently applied to engineering and other fields.

Although the planner is rather naive in the sense that it only knows about the connecting rod and primary shaft sub-assemblies, it has been able to produce consistent plans and maintain the database. This suggests that there is much potential for using techniques of Artificial Intelligence in process planning as well as manufacturing as a whole.

The blackboard approach was first used in the development of the HEARSAY-II speech-understanding project (Rich, 1983). It can be used to organise large AI programs. The system consists of separate modules or knowledge sources (KS) that contain domain-specific knowledge. The blackboard is simply a shared data structure through which all the knowledge sources communicate. By using this, the process planning system could be considered as a sub-section of an Integrated Manufacturing system. Other parts of the system, would be able to communicate via the blackboard which in a way, behaves like a master controlling its slaves (knowledge sources such as the process planning system, design system, management information system, etc). The architecture of the system is illustrated in Fig. 3.

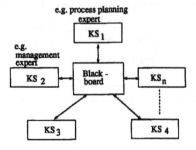

Fig. 3   Integration of the
         system with the overall
         configuration

5. CONCLUSIONS

The planner has been written in Edinburgh Prolog (Version 1.5) using the SUN workstation. Further work on the system includes enhancing its *knowledge* so that it knows about more assemblies, incorporating a CAD interface, user interface and perhaps adding a blackboard system as well. Other constraints such as resources and low-level timing of each operation would also need to be considered in a useful planner.

The planner has shown that there is an analogy between AI planning and process planning (perhaps mainly the derivation of the sequence of operations) and that it is possible to employ AI techniques in manufacturing.

Finally, the next generation of manufacturing systems would perhaps employ more techniques of Artificial Intelligence and new manufacturing systems would be designed using the blackboard approach.

REFERENCES

Chapman, D. (1987). Planning For Conjunctive Goals. Artificial Intelligence 32, 333-377.

Cheung, Y.P. and Dowd, A.L. (1988). Artificial intelligence in process planning. Computer-Aided Engineering Journal, August 153-156.

Descotte, Y., and Latombe, J-C. (1981) GARI: A problem solver that plans how to machine mechanical parts. Int. Joint Conf. On AI. 7, August, Vancouver, 766-772.

Fikes, R.E. and Nilsson N.J. (1971). STRIPS: A new approach to the application of theorem proving to problem solving, Artificial Intelligence., 2, 189-208.

Finger, J.J. (1987) Exploiting Constraints In Design Synthesis. Ph. D. thesis, Stanford University, Stanford, CA.

Genereseth, M.R., and Nilsson, N.J. (1987). Logical Foundations Of Artificial Intelligence. Morgan Kaufmann, USA.

Ginsberg, M.L. and Smith, D.E. (1988). Reasoning about Action I: A Possible Worlds Approach. Artificial Intelligence., 35, 165-195.

Lifschitz, V. (1986). On the semantics of STRIPS. Proceedings Of The 1986 Workshop On Planning And Reasoning About Actions. Timberline, OR.

McCarthy, J. (1977). Epistemological problems of artificial intelligence, Proceedings Of The Int. Joint Conf. On AI. Cambridge, MA, 1038-1044.

Nilsson, N.J. (1980). Principles Of Artificial Intelligence.' Tioga Pub. Co.

Rich, E. (1983). Artificial Intelligence. McGraw-Hill.

Sacerdoti, E.D. (1977). A Structure For Plans And Behaviour. Elsevier-North, Holland.

Taylor, P.W. (1988). Application Of Temporal Logic To Programmable Assembly - Micro Scheduling Within An Assembly Cell. Internal Publications, GEC Research Ltd.

ARTIFICIAL INTELLIGENCE TECHNIQUES ON A DISTRIBUTED CONTROL ENVIRONMENT

J. Borda, R. González and M. Insunza

Datalde S.A., Bilbao, Spain

Abstract. The distributed control systems are becoming, by several reasons, the process-control architecture for the factory of the future.

The established hierarchy allows also a distribution of the "intelligence" and for this reason facilitates the construction of relatively simple and specialist Expert Systems which perform their role at the different control levels.

The paper defines the different areas for the application of A.I. in process-control, and shows the main features of the involved Expert Systems and the necessary multi-level information flow.

Keywords. Adaptive control; Artificial Intelligence; Control engineering applications of computers; Distributed control; Expert systems; Fuzzy control; Linear optimal control; Pattern recognition; PID control; Supervisory control.

INTRODUCTION

The distributed control hierarchy tries to set, from the hardware and software point of view, a functional oriented architecture.

Thus, the execution of algorithms and the idea of making them intelligent can also be distributed and simplified, and this represents a real opportunity to develop suitable expert systems.

Fig. 1 shows the possible utilization of expert systems around six functional levels of distributed process control. Also see that a predeterminated-time communication interface is required to link the two present hardware levels.

Why the A.I. application to process-control?

In a CIM environment, process repeatibility is perhaps the most basical requirement, and has an important role in lead-time reduction, avoiding rework and scrap.

Consequently, the A.I. application is expected to get:

* Parameters repeatibility under closer tolerances.
* Predictive quality control.
* Lower operational costs.
* Intensification of the R&D process.

On-line SPC and predictive maintenance spectrum-based diagnosis are other subsystems which must be integrated in the supervisory level in order to have under integration the control, process, quality and machine condition.

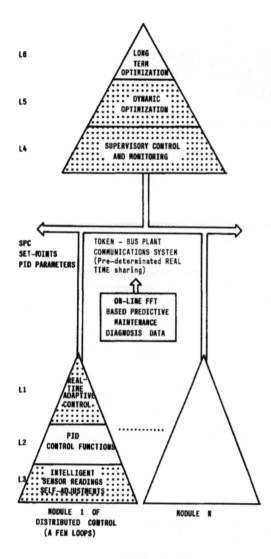

Fig. 1. The hierarchy of distributed control.
Long term optimization can be arranged
by conventional operation - research
techniques (e.g.: simplex)

[::::] A.I. possible areas

THE A.I. APPROACH IN DIFFERENT FUNCTIONAL LEVELS
OF THE DISTRIBUTED PROCESS CONTROL (DPC)

Fig. 2 represents the approach of A.I. in DPC.
In this section we will comment more deeply each
level.

LEVEL	A.I. APPROACH	REMARKS
L1	RULE-BASED ADJUSTMENT FROM DIFFERENT LOOPS READINGS CONFIGURATION.	SPECIAL PURPOSE DATABASE.
L3	PATTERN MATCHING OF TRANSITORY DISTURBANCE SHAPES.	SPECIAL PURPOSE LOCAL DATABASE.
L4	RULE-BASED SYSTEM FOR DATA AND ALARM SEQUENCING PRESENTATION ON A "SUFFICIENT EVENT" FRAMEWORK.	- -
L5	RULE-BASED SYSTEM USING SIMULTANEOUS INFORMATION FROM LOOPS OPERATING VS/ STANDARD SET-POINTS, PID PARAMETERS, ETC, AND PROCESS-MODEL PARAMETERS.	ACCURATE ENOUGH MATHEMATICAL MODEL OF EACH LOOP IS REQUIRED
L6	THE O.R. SIMPLEX-MODEL CAN BE NORMALY USED.	PLANT FLOW MODEL IS REQUIRED.

Fig. 2. A.I. Approach in the functional levels
of distributed process control.

LEVEL L1. Consider: LiRt = Reading at the time
t for the i-loop.

For better integration, it is convenient to have
under the same hardware the loops that are going
to be computed together with the L1 expert system.

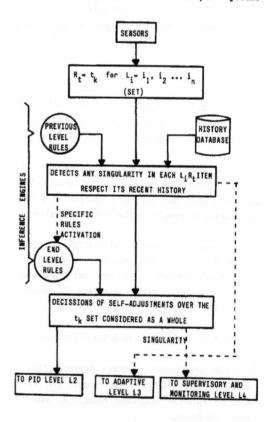

Fig. 3. L1 Expert System. Block diagram.

LEVEL L2. At this level a typical PID algorithm is performed. PID constants Kr, Ti and Td are going to be considered in levels L3, L4 and L5 for adaptation.

$$\delta Y = Kr \left[e + \frac{1}{Ti} \int_{t}^{t+Ti} e.dt + Td \ \frac{de}{dt} \right] \quad (1)$$

being:

δY = PID generation

e   = Error

Kr = General and proporcional constant

Ti = Integral time

Td = Derivative time

LEVEL L3. Good practical results can be achieved by the dynamic adaptation of the Kr, Ti and Td values for each loop, taking into account the - resulting shapes of the disturbance waves after the setting of a significative PID actuation.

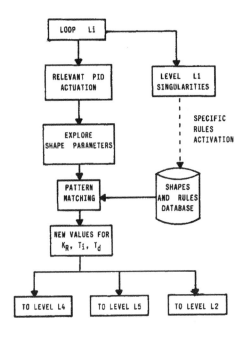

Fig. 4. L3 Expert System. Block diagram.

Thus, just as an expert control-room operator, the system would analyse shape parameters as:

τ (delay time)

To   (extinction time) /   (extinction factor)

Xo   (amplitude)

ωn (frecuency)

and would also do a pattern-matching process which assign to the real wave the most equivalent database existing shape, associated with a determinated change in the set of PID parameters Kr, Ti and Td.

LEVEL L4. The "sufficient event" framework is given by the information from L1 and L3 levels. In these cases, a particular rule-based expert system takes a significant role as operator's guide monitor, presenting and sequencing the most -- relevant information in an effective and friendly way.

This level is going to be the general purpose "system window" with the user.

LEVEL L5. We have here the first general optimization level.

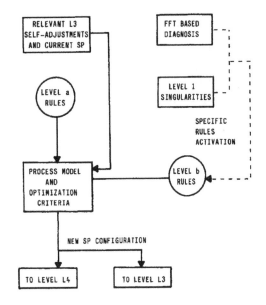

Fig. 5. L5 Expert System. Block diagram.

As shown in Fig. 5, the input data are:

* Current set-point and PID relevant   self-adjustments from L3.

* Relevant self-adjustments of readings from L1.
* Selected on-line FFT-based equipment condition diagnosis.

Mathematical model of each loop is required. This model can generally be based on order-2 differential equations after accurate simplification supposing operations around a set-point tolerance.

Order-2 equations provide a good mathematical - tool to define parameters as τ, ωn, ε, etc, that must be taken into account at this stage.

Let us remember briefly the order-2 differential equations applied to process-control:

$$J.\ddot{x} + c.\dot{x} + k.x = F(t) \text{ , being} \qquad (2)$$
$$J.\ddot{x} + c.\dot{x} + k.x = 0 \qquad (3)$$

This equation represents the free oscillation of the system after a perturbation. F(t) also represents any PID action.

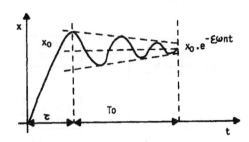

The solution of (3) is done by:

$$x = x_0.e^{-\varepsilon\omega nt} . \cos(\omega t - \varphi) \qquad (4)$$

being:

$$\omega n = \sqrt{\frac{k}{J}}$$

$$\varepsilon = \frac{c}{2J\omega n}$$

$$\omega = \omega n \sqrt{1 - \varepsilon^2}$$

$$\varphi = \tan^{-1} \frac{\varepsilon}{\sqrt{1 - \varepsilon^2}}$$

$$\tau = \frac{J}{c}$$

$$x_0 = \frac{F}{k} . \sqrt{1 + \frac{\varepsilon^2 \omega n^2}{\omega^2}}$$

The error equation e(t) is given by:

$$e(t) = \text{set point} - x(t) \qquad (5)$$

Optimization results in this level L5 are new set points variation bands communicated to the level L3.

Two optimization criteria are selected to be used by the expert system,

$$\text{MINIMIZE} \int_0^t e(t)^2 \, dt \qquad (6)$$

$$\text{MINIMIZE} \int_0^t t.e(t) \, dt \qquad (7)$$

and also two different levels of rules are set up pursuing this goal:

. Level a: The framework is defined by the - established relations among loops.

. Level b: The framework is defined by the relevant self-adjustments and current set points from Level L3. This level is also affected by information from L1 and FFT diagnosis in the way of some specific rules activation - (see fig. 5), and structures the process-model around the optimization criteria done by (5) or (6).

CONCLUSIONS

Significant simplification of process control intelligent automation can be achieved on a distributed process control framework.

The paper suggests four levels of expert systems which can be implemented under a down-top methodology, and only the latest level requires an approximate mathematical model of the process.

The whole can be considered as a modular software package, to be developped around a distributed control philosophy.

REFERENCES

Borda, J. (88). "Arquitectura y modelos computacionales CIM en inyección de termoplásticos". Doctoral Thesis p. 324, paragraph 4.2. (Escuela Superior de Ingenieros Industriales y de Telecomunicaciones de Bilbao).

Borda, J. (87). "Inteligencia artificial en el control de procesos industriales". Revista Química 2000, Enero 87, p. 4 in Ediciones Arcadia.

Borda, J. - Gutierrez, A. (87). "Aplicaciones de la inteligencia artificial en mantenimiento" Revista de Mantenimiento, Marzo - Abril 87, p. 10 in Puntex (Barcelona).

Darius, J.H. (87). "Rule based control: a new - tool". Proceedings of the 2nd international conference on machine control systems in Birmingham (UK), p. 9 in IFS, Bedford (UK).

DATALDE S.A. (87). "Prisma user handbook". pp 100 - 256, p. 578. (Predictive Maintenance Techniques).

Holland, R.C. (83). "Microcomputers for process control". pp. 108 - 124, in Pergamon Press, -- Oxford.

Kuo, B.C. (75). "Sistemas automáticos de control" pp. 577 - 578 in Cecsa, Mexico D.F.

Pau, L.F. (85). "Failure detection processes by - pattern recognition and expert systems". Proceedings on artificial intelligence in maintenance. Working group, Denver-Colorado. p. 10 in Noyes, New Jersey.

AN ABSTRACT DATA TYPE FOR FAULT TOLERANT CONTROL ALGORITHMS IN MANUFACTURING SYSTEMS

S. A. Andreasson, T. Andreasson and C. Carlsson

*Department of Computer Science, Chalmers University of Technology, Göteborg,
Sweden*

Abstract

Fault tolerance will in the future become an important issue in manufacturing systems. This paper addresses the issues of supporting fault tolerance control in such systems. To achieve fault tolerance, there must be more than one way to configure the manufacturing system. We describe algorithms which dynamically distribute the work among the units. As a usable concept for the algorithms we introduce a data structure called a Mission Pool (MP).

A manufacturing system can be viewed as a hierarchically structured system. We use a method which we call a General Recursive System (GRS), to model the manufacturing system. GRS gives a flexible and uniform way to model all the levels of the manufacturing system hierarchy.

Keywords: Fault Tolerant Systems, Distributed Computer Systems, Factory Automation, Data Structures, Algorithms.

1. Introduction.

Computer Integrated Manufacturing (CIM) will be an important issue in future manufacturing. All different uses of computers in a factory will be incorporated into a single system and thus form a "totally" automated fabrication. Fault tolerance will become an important issue in these manufacturing systems. The enlarging complexity of the interconnected system increases the demands for maintenance and fault handling. The probability that components may fail increases with the number of components included in the system. Failures, stopping the production, are very expensive and situations where failure of one component disables the entire system is unacceptable. This paper addresses the issues of supporting fault tolerance control in manufacturing systems.

A manufacturing system is often viewed as a hierarchically structured system. To model such a manufacturing system we introduce a method called a General Recursive System (GRS). GRS gives a flexible and uniform way to model all the levels of the manufacturing system hierarchy. The GRS model reflects that different levels of a manufacturing system have different requirements, which influence the choice of appropriate control methods.

To achieve fault tolerance in manufacturing systems, there must be more than one way to configure the system and an appropriate control of redundant resources. The work at each level in the structure must be distributed among the different units within the level. This is a configuration of the system. The work is described in a *mission*, which is the output of a CAD/CAM system. In this paper we describe algorithms to dynamically distribute missions among the units. As a usable concept for the algorithms we introduce a data structure called a *mission pool*. A mission pool is a container of missions from which suitable fractions of the missions can be bound to

the different units. Fault tolerance is achieved by these algorithms, since faulty units are avoided during the work distribution. When a unit fails, the algorithms also can be used for reconfiguring of the system. We describe algorithms both for centralized and distributed control.

When a mission is distributed among the units there must be a control among the units during the fabrication. This is real time control, which can be conveniently described by using the General Recursive System.

2. Background.

Factory automation can be viewed as a hierarchically structured system. We identify the following five hierarchical levels:

- Device Level
  This is the lowest level in the hierarchy, and consists of the individual resources, e.g. robots, machine tools, sensors and material handling systems.

- Manufacturing Cell Level
  A Manufacturing Cell is an assembly of devices, which are controlled by a common computer.

- Assembly Line Level
  An Assembly Line consists of a number of interconnected Manufacturing Cells.

- Plant Level
  This level manages the coordination of resources and production in the plant.

- Corporate Level
  This level includes global process planning, production management, financial functions and administrative functions.

The levels in this model are in accordance with the model in (Mohideen, 1987). Other subdivision of the hierarchy is possible. The ISO reference model for factory automation, e.g. in (McGuffin, 1988), identifies a Station Level between the Device Level and the Manufacturing Cell Level. The model in (Chintamaneni, 1988) uses four levels, the Assembly Line Level is not identified.

Information flows in both directions of the hierarchy. In order to execute the production - planning, commands and schedules are passed downwards through the levels of the hierarchy. Operational data, e.g. production status information and exception indicators, are passed in the reverse direction upwards through the levels.

3. General Recursive System.

We now present the General Recursive System, which will serve as the base for our manufacturing system model. The General Recursive System (GRS) is a simple and powerful method to model fault tolerance in distributed systems.

A GRS consists of a *set of entities* $\{e_1, e_2, ..., e_n\}$ glued together with a GRS internal *data network* N_D and a GRS internal *material network* N_M. The GRS is connected to the surrounding world via two *in-ports* and two *out-ports*. One in-port and one out-port correspond to each network. When using the GRS-model to describe a real system, some of the ports of a GRS may be absent if not needed.

An entity is either a GRS itself or an *atomic entity*. An entity which is a GRS can be expanded further into more elementary entities, connected with their own data network and material network. An atomic entity can not be expanded further. An atomic entity is connected to each internal GRS network via one in-port and one out-port, $N_{D_{in}}$ and $N_{D_{out}}$ respectively $N_{M_{in}}$ and $N_{M_{out}}$. An atomic entity AE is associated with a *set of offered operations* O_{AE}.

More formally the concept of the GRS is defined as:

$$GRS = <E, N_D, N_M, N_{D_{in}}, N_{D_{out}}, N_{M_{in}}, N_{M_{out}}>$$

$$E = \{e_1, e_2, ..., e_n\}$$

$$e = GRS \mid AE$$

$$AE = <black\_box, N_{D_{in}}, N_{D_{out}}, N_{M_{in}}, N_{M_{out}}, O>$$

A General Recursive System could be graphically described as in fig. 1. The concept of GRS is more thoroughly described in (Adlemo, 1989).

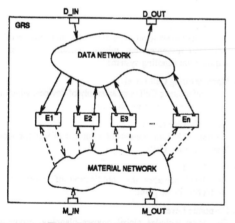

Fig. 1. GRS described graphically.

A GRS can be used in a top down fashion or in a bottom up fashion to model a real system. This gives a flexible and

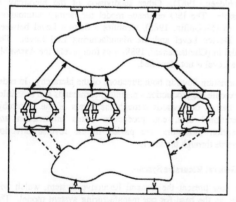

Fig. 2. Expanded General Recursive System.

simple way of describing complex systems in a uniform manner. Figure 2 shows how a system is expanded into two GRS levels.

A GRS is controlled by loading missions. In the missions it is described what is going to be performed by the GRS. A *mission M* is a partially ordered set of *entity missions EM*, that together forms the mission. For each *EM* there is a set of requirements R_{EM}. An entity mission can be either a mission or an *atomic mission AM*. An atomic mission is a mission containing operations that can be performed by a corresponding atomic entity. Subsequently, we have the following definition of a mission M:

$$M = \{ EM_1, EM_2, ..., EM_{n-1}, EM_n \}$$
$$EM = M \mid AM.$$

3.1 General Recursive System Control.

The control of the General Recursive System is divided into three levels.

- System Control Level.
  At the highest level of control, the System Control Level, it is described what is going to be produced by the entire system. We call this description a *mission*. The mission can be viewed as a program for the GRS.

- System Internal Control Level.
  The task described in the mission is divided into proper subtasks called *entity missions,* which are distributed among the entities. The work to be executed by the entity is specified by the corresponding entity mission. The System Internal Control Level is responsible for the coordination of the work produced by the different entities.

- Entity Control Level.
  The lowest control level in the GRS, the Entity Control Level, is responsible for the control in the single entities. Each entity mission that is loaded into an entity will form a mission for this entity. Together the different entity missions, that are loaded into an entity, will form a *mission pool* for that entity. If the entity is expanded to a new GRS, its mission pool corresponds to the mission pool of the expanded GRS.

Fig. 3. Levels of the System Control.

3.2 Modeling Manufacturing using the GRS.

We now give an example how to model a manufacturing system using the General Recursive System. We treat the units at the Device Level, e.g. robots and sensors, as atomic entities. The Manufacturing Cell is an assembly of devices which are controlled by a common computer. Figure 4 shows an example of a manufacturing cell described as a General Recursive System. The cell is composed of two robots, one sensor, one database and one computer.

A number of manufacturing cells can be connected to form an Assembly Line. This system, too, can be described using a GRS. An example of an assembly line is given in fig. 5.

Similarly, assembly lines can be put together to form plants, which in turn can be put together to form corporates. The GRS can in this way be used to model manufacturing systems in a uniform way.

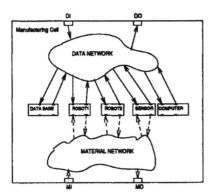

Fig. 4. Example of a Manufacturing Cell.

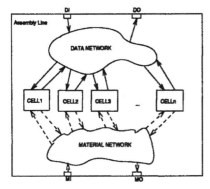

Fig. 5. Example of an Assembly Line.

3.3 Fault Tolerance Issues.

In a system modeled as a General Recursive System, a failure is caused by a fault in one or more entities and/or in a network. There are mainly two ways to cope with these problems to achieve a dependable system.

- Fault Avoidance.

   One way to increase the dependability in the GRS is to increase the dependability of the entity that may cause a failure. In this case we can choose two different ways

   — The entity can be treated as an atomic entity. Higher reliability is obtained by choosing an entity of higher quality, although at a higher price.

   — If the entity is not treated as an atomic entity, it is instead described as another GRS (recursively). Increased dependability for the entire system will be achieved by increasing the dependability in this new GRS.

- Fault Tolerance.

   The other way to increase the dependability in the GRS is by replication of entities. To achieve fault tolerance, the replicated entities must be handled by appropriate algorithms. In many cases, it is essential to get automatic reconfiguration of the system in order to get high availability.

4. Definition of the Mission Pool Data Structure.

A *mission pool MP* is a data structure where the intended missions are stored. The missions are inserted into the mission pool in a FIFO order. When a mission is inserted into the mission pool, each of its entity missions is put behind the entity missions of the missions already inserted into the mission pool. This is done according to the partial order.

MP = queue of missions M_i
        = set of partly ordered entity missions EM_{j_i}

The mission pool MP can be centralized or distributed among a set of entities. A mission M is loaded into the MP of the GRS through the data in-port $N_{D_{in}}$. Normally, the mission pool is used by entering missions into it and then pick entity missions from it following the partial order among the entity missions.

To describe what an entity E_i can perform, the entity is associated with a set of offered operations O_{E_i}. The set O_{E_i} describes what equipment and manipulations the entity E_i offers.

O_{E_i} = set of operations offered by E_i

For the entity E_i to be able to perform entity mission EM_j it must hold that E_i can perform all operations in EM_j, i.e. $R_{EM_j} \subseteq O_{E_i}$.

In fig. 6 three missions $M1$, $M2$ and $M3$ are described with the partial ordering of their entity missions. Furthermore, a mission pool MP containing the three missions is described. The missions are inserted in the order $M2$, $M3$, $M1$. First the MP is shown on mission level. Then the MP is shown on entity mission level with the subsequent partial order among the entity missions.

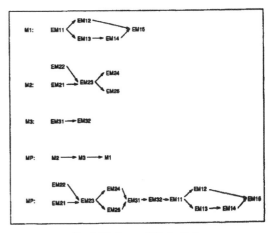

Fig. 6. Missions and Mission Pool.

4.1 Operations on the Mission Pool.

We now proceed to define the mission pool more formally. A mission pool has the following constructors, selectors and predicates.

Constructors:

$$make() \rightarrow MP$$
$$insert(M, MP) \rightarrow MP$$

Selectors:

$$get(MP) \rightarrow EM$$
$$pick(O_{E_i}, MP) \rightarrow EM_j \quad \text{where} \quad R_{EM_j} \subseteq O_{E_i}$$
$$delete(M, MP) \rightarrow MP$$

Predicate:

$$isempty(MP) \rightarrow boolean$$

The operation $get(MP)$ returns an entity mission EM_j from the mission pool MP, such that EM_j is not preceded by any other entity mission in MP.

The operation $pick(O_{E_i}, MP)$ returns an entity mission EM_j from the mission pool MP such that:

- $R_{EM_j} \subseteq O_{E_i}$,

    i.e. the requirements R_{EM_j} of the entity mission EM_j is a subset of what can be offered by the entity E_i,

- $\forall\, EM_k$ such that $EM_k \rightarrow EM_j : \neg\, (R_{EM_k} \subseteq O_{E_i})$,

    i.e. no entity mission EM_k preceding EM_j in MP has a requirement set that is a subset of O_{E_i}.

    The notation $EM_k \rightarrow EM_j$ denote the partial order on the entity missions.

5. Control Algorithms.

The System Control is divided into two main parts, Work Configuring and Work Control. Work Configuring is the division of work among the entities, while Work Control is the control of the actual execution of the planned work.

5.1 Work Configuring.

A mission that should be executed is loaded into the GRS. The work described in the mission must be divided among the different entities. It must be decided which entity should execute which entity mission. Our work configuring algorithms use the mission pool of the GRS (MP_{GRS}) to link a suitable entity to each entity mission. The control algorithms work recursively on each GRS level and each level has a mission pool. For each entity mission EM_j in MP_{GRS} the work configuring algorithms select an appropriate entity E_i. The entity mission is loaded into E_i's own mission pool MP_{E_i}. This selecting procedure is done recursively on each level until atomic missions are reached. Then also corresponding atomic entities should be reached.

In order to apply the missions to the General Recursive System we define the following two operations on a GRS:

$$load\,(M\,,GRS) \rightarrow boolean$$
$$select\,(R_{EM}\,,GRS) \rightarrow E$$

The operation $load\,(M\,,GRS)$ loads the mission M into the GRS. This will lead to an insertion into the GRS's mission pool MP_{GRS}, i.e. $load\,(M\,,GRS)$ implies $insert\,(M\,,MP_{GRS})$.

The operation $select\,(R_{EM}\,,E)$ selects an entity E within the GRS such that $R_{EM} \subseteq O_E$.

Work Configuring can be done in a centralized or decentralized manner.

- Centralized Work Configuring.
    One of the entities is selected to execute the Work Configuring. The selected entity is called the Central Controller. The Central Controller manages the MP_{GRS} and choses an entity within the GRS for each entity mission in MP_{GRS}. When new missions are loaded into the GRS they are put in the MP_{GRS}. The centralized configuring algorithm is described by the process $CentralPlanner$:

```
Process CentralPlanner;
  while true do
  begin
    get entity mission;
    select an appropriate entity;
    load entity mission into
        the selected entity;
  end while;
```

- Entity Distributed Work Configuring
    Instead of having a Central Controller all the entities pick adequate entity missions by themselves. The GRS Mission Pool can be situated at one special entity. All the entities execute the process $EntityPlanner$:

```
Process EntityPlanner;
  while true do
  begin
(1)   pick an adequate entity mission
          from the GRS mission pool;
      insert the entity mission into
          the entity mission pool;
  end while;
```

A more fault tolerant solution can be achieved by distributing the Mission Pool among several entities. In this case more than one entity will hold a copy of the MP_{GRS}. When a mission is loaded into the GRS, the mission must be distributed to all these entities. The case when all entities have a copy of the MP_{GRS} is showed in fig. 7. When an entity picks an entity mission, the entity must inform all the other entities in order to get a consistent system. To achieve this, there will be a need for a reliable broadcast within the GRS, i.e. a guarantee that all entities get all broadcast messages in the same order.

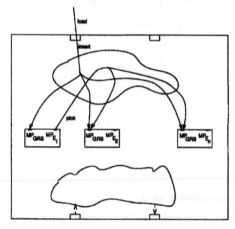

Fig. 7. Entity distributed work configuring.

5.2 Fault Tolerance.

To allow work reconfiguring it must be possible for more than one entity to perform a certain entity mission. If the entity mission EM_k should be able to configure dynamically in more than one way, there must exist at least two entities E_i and E_j for which it is valid that $R_{EM_k} \subseteq O_{E_i}$ respectively $R_{EM_k} \subseteq O_{E_j}$. Subsequently, there must be some kind of redundancy among the entities, i.e. $O_{E_i} \cap O_{E_j} \neq \emptyset$.

To achieve a fault tolerant distributed system, atomic multicast is a useful concept. How this can be done is discussed in (Birman, 1987; Cristian, 1988; Stumm, 1988). When using atomic multicast it is possible to let all the processes, that together forms a distributed application, observe the same order of events concerning the application.

6. Further Research.

To introduce further fault tolerance into the system it is of importance to study how the mission pool can be distributed among the different entities.

It would also be necessary to investigate different algorithms for Work Configuring in order to get a deeper understanding of the configuring problem.

Another essential issue of research is to analyze which effects the choice of Work Control algorithms have on the Work Configuring algorithms and vice versa.

7. Conclusions.

In this paper we have introduced an abstract data type, called a mission pool, which is a useful concept when designing fault tolerance algorithms for factory automation systems. Examples of algorithms using the mission pool are given. We have also introduced a model for describing manufacturing automation, called a General Recursive System. This model gives a simple, flexible and uniform way to describe all the levels of the manufacturing system hierarchy.

8. References.

Adlemo, A., S. A. Andreasson, T. Andreasson, C. Carlsson, A. Dahlberg and G. Lindberg (1989). A recursive model for analyzing fault tolerance in manufacturing systems. Will appear in *the 12th International Conference on Fault-Tolerant Systems and Diagnostics*, Prague September 1989.

Birman, K.P. and T. A. Joseph (1987). Exploiting virtual synchrony in distributed systems. In *Proceedings of the 11th ACM Symposium on Operating Systems Principles*, November 1987.

Chintamaneni, P. R., P. Jalote, Y. B. Shieh and S. K. Tripathi (1988). On fault tolerance in manufacturing systems. In *IEEE Network*, May 1988.

Cristian, F. (1988). Reaching agreement on processor group membership in synchronous distributed systems. *IBM Research report RJ 5964 (59426) 3/22/88.*

McGuffin, L. J., L. O. Reid and S. R. Sparks (1988). MAP/TOP in CIM distributed computing. In *IEEE Network*, May 1988.

Mohideen, M. F. and Y. J. Kang (1987). Application of CHORUS to factory automation. *TENCON 87*, Seoul 1987.

Stumm, M. (1988). Strategies for decentralized resource management. In *Frontiers in Computer Communications Technology*, ACM 1988.

APPENDIX: Examples using the GRS and the Mission Pool in Prolog.

We will describe a simple manufacturing system, which consists of four types of primitive objects: a set of *primitive parts P*, a set of *atomic entities AE*, a set of data networks *ND* and a set of *material networks NM*. Each object has the following elements:

P = {a, b, c, d}
AE = {robot_1, robot_2, robot_3, robot_4, robot_5}
ND = {ethernet}
NM = {conveyor}

GRS.

An atomic entity is a black box which offers one operation and one resulting composition of two parts. The parts are either primitive parts or composed parts. The primitive parts a, b, c, and d have the following appearance:

The operations of the five atomic entities are given below. In addition a Prolog implementation is given for each atomic entity.

Robot_1 offers the operation [ass,a,b] -> [a,b].

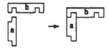

```
atomic_entity(robot_1,Operation,Complex_part):-
    Operation = [ass,a,b],
    Complex_part = [a, b].
```

Robot_2 offers the operation [ass,[a,b],c] -> [[a,b],c].

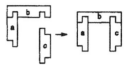

```
atomic_entity(robot_2,Operation,Complex_part):-
    Operation = [ass, [a,b], c],
    Complex_part = [[a,b],c].
```

Robot_3 offers the operation [ass,[[a,b],c],d] -> aquepart.

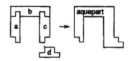

```
atomic_entity(robot_3,Operation,Complex_part):-
    Operation = [ass, [[a,b],c], d],
    Complex_part = aquepart.
```

Robot_4 offers the operation [ass,aquepart,d] -> triumphal.

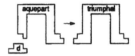

```
atomic_entity(robot_4,Operation,Complex_part):-
    Operation = [ass, aquepart, d],
    Complex_part = triumphal.
```

Robot_5 offers the operation [ass,triumphal,aquepart] -> aqueduct.

```
atomic_entity(robot_5,Operation,Complex_part):-
    Operation = [ass, triumphal, aquepart],
    Complex_part = aqueduct.
```

A General Recursive System has two in-ports and two out-ports, to connect the system to the surrounding world. In this example the data network and material network is modeled like:

```
data_network(ethernet).
material_network(conveyor).
```

A General Recursive System consists of a set of entities glued together with a data network an a material network. An entity is either a General Recursive System or an atomic entity. In Prolog this is implemented as below:

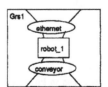
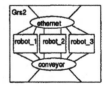

```
Grs1 = [[robot_1], ethernet, conveyor]
Grs2 = [[robot_1,robot_2,robot_3], ethernet,
        conveyor]
```

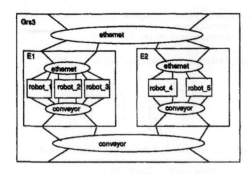

```
E1     = [[robot_1,robot_2,robot_3],ethernet,
          conveyor],
E2     = [[robot_4,robot_5],ethernet,conveyor],
Grs3   = [[E1,E2],ethernet,conveyor],
```

Each GRS offers a set of operations *O*, which in Prolog can be calculated as follows:

```
offers([E,ND,NM],O) :-
        set_of_entities(E,[[],O]),
        data_network(ND),
        material_network(NM).

set_of_entities([E|[]],[Offers,O]) :-
        entity(E,[Offers,O]).
set_of_entities([E|L],[Offers,O]) :-
        entity(E,[Offers,Tmp]),
        set_of_entities(L,[Tmp,O]).

entity(BB,[Offers,[Operation|Offers]]) :-
        atomic_entity(BB,Operation,_).
entity(E,[Offers,O]) :-
        offers(E,Operations),
        union(Operations,Offers,O).
```

We can now ask the Prolog system what operations Grs1, Grs2 and Grs3 offer by running the following questions:

```
| ?- offers(Grs1,O).
O = [[ass,a,b]]

| ?- offers(Grs2,O).
O = [[ass,[[a,b],c],d],[ass,[a,b],c],[ass,a,b]]

| ?- offers(Grs3,O).
O = [[ass,triumphal,aquepart],[ass,aquepart,d},
     [ass,[[a,b],c],d],[ass,[a,b],c],[ass,a,b]]
```

Mission.

A General Recursive System is controlled by loading missions. In a mission it is described what is going to be performed. A mission *M* is a partially ordered set of entity missions. An entity mission *EM* can be either a mission or an atom mission *AM*. An *AM* is a mission containing operations that can be performed by an atomic entity. Using Prolog the missions can be described as:

```
Aquepart   = [[ass,a,b], [ass,[a,b],c],
              [ass,[[a,b],c],d]],
Triumphal  = [Aquepart, [ass,aquepart,d]],
Aqueduct   = [Triumphal, Aquepart,
              [ass,triumphal,aquepart]],

Mission_1  = [[ass,a,b],[ass,[a,b],c]],
Mission_2  = [Aqueduct],
Mission_3  = [Triumphal],
Mission_4  = [Triumphal,Triumphal,Aqueduct,
              Aqueduct],
```

A mission *M* demands a set of requirements *R* in order to be executed by an GRS. The requirements *R* can in Prolog be calculated as follows:

```
requirements(M,R) :-
        set_of_entity_requirements(M,[[],R]).
```

```
set_of_entity_requirements([],[R,R]).
set_of_entity_requirements([EM|L],
                [Requirements,R]) :-
        entity_requirements(EM,[Requirements,Tmp]),
        set_of_entity_requirements(L,[Tmp,R]).

entity_requirements(EM,[Requirements,
                [EM|Requirements]]) :-
        atom_requirements(EM).
entity_requirements(EM,[Requirements,R]) :-
        requirements(EM,Operations),
        union(Operations, Requirements, R).

atom_requirements(AM) :- atomic_entity(_, AM, _).
```

We can now ask the Prolog system what requirements Mission_1, Mission_2, Mission_3 and Mission_4 demand by running the following questions:

```
| ?- requirements(Mission_1,R).
R = [[ass,[a,b],c],[ass,a,b]]

| ?- requirements(Mission_2,R).
R = [[ass,triumphal,aquepart],[ass,aquepart,d],
     [ass,[[a,b],c],d],[ass,[a,b],c],[ass,a,b]]

| ?- requirements(Mission_3,R).
R = [[ass,aquepart,d],[ass,[[a,b],c],d],
     [ass,[a,b],c],[ass,a,b]]

| ?- requirements(Mission_4,R).
R = [[ass,triumphal,aquepart],[ass,aquepart,d],
     [ass,[[a,b],c],d],[ass,[a,b],c],[ass,a,b]]
```

Mission Pool.

A mission pool *MP* is a data structure where the missions are stored. This is a structure where it is described what is to be performed by the General Recursive System.

The missions are inserted into the mission pool in a FIFO order. Each mission is put behind the missions already inserted into the mission pool. The operations on the mission pool are in Prolog implemented as:

```
make([]).

insert(Mission, [], [Mission]).
insert(Mission, [M|MP], [M|Res]) :-
        insert(Mission, MP, Res).

delete(M, [M|MP],MP).
delete(M, [X|MP1],[X|MP2]) :- delete(M,MP1,MP2).
delete(_, [],[]).

isempty([]).

get([EM|MP], EM).

pick(O, [EM|MP],EM) :-
        requirements([EM], R),
        subset(R, O).
pick(O, [X|MP], EM) :-
        pick(O, MP, EM).
```

Assume that the mission pool contains the following entity missions:

```
Mission_pool = [[ass,triumphal,aquepart],
                [ass,a,b],[ass,[a,b],c]]
```

Then we can ask the following questions:

```
| ?- get(Mission_pool,EM).
EM = [ass,triumphal,aquepart]

| ?- offers(E1,O),
pick(O,Mission_pool,EM).
EM = [ass,a,b]

| ?- offers(E2),
pick(O,Mission_pool,EM).
EM = [ass,triumphal,aquepart]
```

HOW TO USE THE MANUFACTURING INFORMATION SYSTEMS AS A COMPETITIVE WEAPON

J. Borda

Datalde S.A., Bilbao, Spain

Abstract. CIM is, basically, an advanced strategical and organizational concept focused to get a good position in the competition race.

Self-integrated with the "new automation technologies", the information and -- organization technologies have to be considered as an active and useful "know how" to improve the competitive position.

The paper explains how the CFS (critical factors for sucess) method can help to design a SDSS (strategical oriented decission support system) which will contribute to establish some kind of competitive barrier.

Keywords. Computer organization; Computer software; Database management systems; Decision theory; Distributive data processing; Information theory; Management systems; System theory.

INTRODUCTION

If we have to define the key-factor to increase the competitiveness in the 80's this is "flexibility".

Flexibility and lead-time reduction are, obviously, two visions of the same thing: low inventory manufacturing (or synchronized manufacturing).

Let us divide the lead-time into the next sections:

. Engineering
. Purchasing and raw materials stock
. Manufacturing
. Delivery and finish products stock

The first synergycal effect in flexibility is - obtained through the integrated simultaneous reduction of engineering and manufacturing lead-times, which affects directly the other components of the lead-time.

There are well-known factors for this: Group technology and appropriate production strategy, scheduling and control, set up reduction, SPC, preventive/predictive maintenance, lay-out configuration, and organizational integration between product and process engineering. In this paper we are going to focuse the influence of the information systems (I.S.) and information technology (I.T.) creating a competitive barrier through - their role in the lead-time reduction.

THE CFS APPROACH

In the CFS approach, information is largely "customized" to face with the strategic development areas. In our case, the goal is lead-time reduction.

What does "customization" mean?

. Self-generation of desired organizational attitudes by the use of I.S.

. Emphasys on key-information around the goal, and at different levels.

. Use the I.T. as a technological weapon to approach the most effective working of both.

An intention to achieve organizational changes using I.S. must take into account, basically, two rules:

Rule 1. Establish appropriate interdepartmental information interfaces. In this way, some departments which have to work synchronized, share common information concerning matters that must be solved in a team.

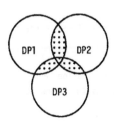

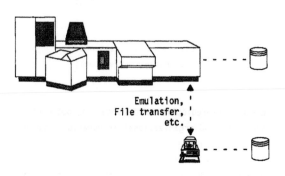

Emulation,
File transfer,
etc.

Fig. 1. Inter-departmental information interfaces. The dotted area represents shared information, probably allocated in a common data-bank.

The rest (undotted) can be supported by local-microcomputer-based databases.

Rule 2. The I.S. Organizational Model (IS-OM) must have key-connection points with the existing organizational model (EOM). Otherwise, the IS-OM will not have enough energy to conduct the desired change.

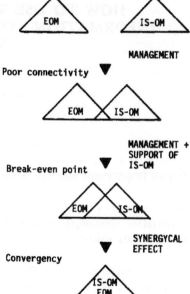

Fig. 2. Key-connection points between the IS-OM and the existing organizational model EOM. See the evolution and the break-even effect.

These are, obviously, two necessary but insufficient conditions, because management thrust and non-significant human negative attitudes are essential.

In the other hand, multilevel adaptive and close loop management around the key goals and concerning information is a generator of organizational synergya, as shown in Fig. 3.

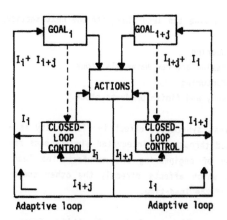

Fig. 3. Multilevel adaptive and closed loop management and information system. A synergya engine.

Being:

. $(I_i + I_{i+j})$ the results obtained by the system without adaptive control,

. $(I_i + I_{ij})_{ac}$ the results obtained by the system with adaptive control,

. δ an arbitrary period of time,

the synergya can be symbolically formulated as follows:

$$S = \sum_{t=t_i}^{t=t_{i+}} (I_i + I_{i+j})_{ac} - (I_i + I_{i+j}) \qquad (1)$$

Is quite easy to see that, if I.T. is pursuing the lead-time reduction through SDSS (and this is the strategic goal), I.T. must be considered as an investment oriented to create a competitive barrier. We have there an unusual approach for investment in computers, systems engineering or information technology in general.

Fig. 4 shows the relative position of these concepts.

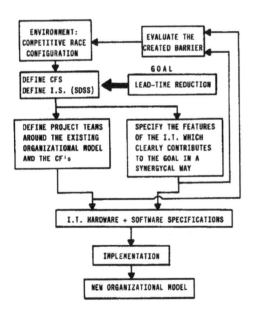

Fig. 4. The CFS approach to use CIM as a competitive weapon.

The approach has other important effects in the always difficult CIM implementations:

. Gives a meaningful sense to integration.

. Avoids many organizational obstacles.

. Prepare positive human attitudes.

FEATURES OF THE I.T. ON A LEAD-TIME REDUCTION ENVIRONMENT

As introduced before, main factors with influence in the lead-time reduction are:

1. Product-process quality, and maintenance engineering integration.

2. A real-time closed-loop production scheduling and control.

3. SPC, process control and predictive maintenance.

4. Flow-shop oriented lay-out.

5. Set-up reduction.

6. Multi-level project-teams "flywheel"

Fig. 5 defines the mentioned influence of these factors. The different nature of the paths talks about the complexity of the lead-time reduction goal.

FACTOR NB. (AREA)	PATH OF INFLUENCE
1	Interactive design for manufacturing in a preventive process quality framework.
2	Dynamic optimal decissions.
3	Maintain the process under control and repeatibility.
4	Eliminate lead-time unnecessary components.
5	Physical flexibility.
6	Synergical "continious improvement flywheel".

Fig. 5. Influence of factors in lead-time reduction.

In this context, the barrier is going to be set with an agressive utilization of the I.T. in each one of the six before mentioned factors. Let's have a look at it.

* Product-process quality:

I.T. agressive factors
. CAD-CAM and CAD-MRP integration.
. Relational real-time manufacturing DBMS.

Special difficulties to be solved in each case
. Release internal resources for training and DBMS completion.
. Investment on a "fuzzy" pay-back environment.
. Customized software development.

* Real-time closed-loop PSC:

I.T. Agressive factors
. Real-time plant data capture.
. Expert systems for MRP-2 simulation and short time finite capacity scheduling (OPT frame work).

Special difficulties
. Instrumentation lay-out.
. HW lay-out, communications and architecture.
. Plant model construction.
. Customized software development.
. Fuzzy pay-back environment.

* SPC, process control and predictive maintenance:

I.T. agressive factos
. On-line automatic SPC.
. Expert systems for adaptive control.
. Expert systems for diagnosis.
. Real-time unnormal conditions reporting.

Special difficulties
. Process knowledge level.
. Instrumentation lay-out.
. Direct labour training and "minding".
. Maintenance labour training.
. DBMS completion and availability of technical documentation.
. Fuzzy pay-back environment.
. Customized software development.

* Flow-shop oriented lay-out:

I.T. agressive factors
. Group technology and relational real-time manufacturing DBMS.
. Real-time plant data capture and buffer control.

Special difficulties
. Permanent integrated engineering and manufacturing taskforce.
. Dynamyc bottleneck based production strategy (OPT).

* Set-up reduction:

I.T. agressive factors
. Process-data capture around the set up.
. Integrated manufacturing DBMS.
. Real-time multilevel scheduling decissions.

Special difficulties
. HW lay-out, communications and architecture.
. Customized software development.

* Multilevel project teams "flywheel":

I.T. agressive factors
. Relational distributed DBMS.
. Integrated manufacturing DBMS.

Special difficulties
. Leadership.
. Integrity of distributed data.
. Real-time availability.
. HW lay-out, communications and architecture.

It must be said that the effective resolution in each case of the special difficulties is what creates the "difference" and contributes to competitive advantage.

CONCLUSIONS

The information technology (I.T.) must be considered as a generator of competitive advantage in the framework of CIM.

Main difficulties concern with internal resources proper utilization, investment on "fuzzy" pay-back environment, Hardware and Software architecture and special developments, and also with an effective leadership, but the resolution of special difficulties on I.T. implementation around the goal (lead-time reduction) is what creates an added-value "barrier" which must be used to get a better position in the competitiveness race.

REFERENCES

Bessant, J - Lamming, R. (87). "Organizational integration and advanced manufacturing techno- logy". Proceedings of the 4th european confe- rence on automated manufacturing in Birmingham UK. p.12. In IFS, Bedford, UK.

Bonczek, H - Holsapple, C.W. - Whinston, D.B. - (81). "The evolution from MIS to DSS: from - data management to model management". Procee- dings of the N.Y. Simposium on DSS, New York, p. 18. In North Holland Publishing Co., Ams- terdam.

Borda, J. (86). "Teoría General de Sistemas: - Substrato de las nuevas tecnologías y mode- los de gestión" In Boletín de Estudios Econó- micos de la Universidad Comercial de Deusto, Bilbao - Spain. "La empresa del futuro p.15 (Universidad de Deusto, Bilbao).

Borda, J. (88). "Arquitectura y modelos computa- cionales CIM en inyección de termoplásticos". Tesis doctoral p.331 (Escuela Superior de In- genieros Industriales y de Telecomunicaciones de Bilbao).

Borda, J. (88). "Optimización de modelos organi- zacionales". MADS VØ2 DATALDE S.A., p. 6. (DA- TALDE S.A., Bilbao - Spain).

Borda, J. (86). "CIM: Una aproximación tecnológi- ca y organizativa a la fábrica del futuro". Speech held in APD, CVT and IMI Program, p. 95 (DATALDE, Bilbao - Spain).

Borda, J. - Gonzalez, R. - Gutierrez, A. (88). "Estrategia de producción en entornos flexi- bles utilizando OPT". Congress on Machine-tool research, design and use, p. 13 (INVEMA, San Sebastian - Spain).

DATALDE S.A. (87). MADS-VØ2. Methodology for - systems analysis and development, p. 578 -- (DATALDE S.A., Bilbao - Spain).

Goldratt, E. (86). "The race" p. 179 in North - River Press, New York.

Gomez Pallete, F. (84). "Estructuras organizati- vas e información en la empresa", p. 403 in APD, Madrid - Spain.

Johansen, O. (82). "Introducción a la Teoría General de Sistemas", p. 164 in LIMUSA, México.

Peterson, D. (87) "G.T.: A foundation for CIM in- formation management". CIM Review, Winter 87 p. 8. In Auerbach, New Jersey, USA).

Von Bertalanffy, L. (81). "Teoría General de los Sistemas", p. 308 in FCE, Madrid - Spain.

Von Bertalanffy (84). "Tendencias en la Teoría General de Sistemas", p. 323 in Alianza Edito- rial, Madrid - Spain.

Weldon, J.L. (86) "Database integration". CIM - Review. Spring 86, p. 4 in Auerbach, New Jer- sey, USA.

HANDLING OBJECTS OF UNKNOWN CHARACTERISTICS

P. Adl, Z. A. Memon and R. T. Rakowski

Manufacturing & Engineering Systems Department, Brunel University, Uxbridge, UK

Abstract. Tasks such as grasping objects of unknown characteristic or handling items of varying batch sizes and compliance calls for intelligent manipulators. Excessive gripping forces could produce severe object damage. This paper outlines the details of a tactile sensor using magneto-resistive technology. The sensor not only has to make readings of normal, tangential and rotational forces but also be intergrated to a real time control system which copes with characteristic changes occuring within the body of the compliant object. Such a control strategy is described.

Keywords. Robots ; Sensors ; Hierarchical intelligent control ; Manufacturing processes ;Real time computer systems .

INTRODUCTION

No matter how robust and rigid a robot and how good its control system, a robot that is to be used in interaction with its environment needs sensory devices that can aid it in performing in these situations.Human beings achieve this through employing the visual and tactile sensation and a learning capability.

Achieving human capabilities in terms of perception and flexibility are not yet realised in the field of robotics. Although machine vision is well advanced the sense of touch remains a constant area of activity (Dario P., et al, 1987). Since most tasks in the area of gripping and placing can be achieved by responding to contact and force, it is desirable to instrument the gripper with force and touch sensors (Harmon,L., 1987).

This paper outlines a method of distinguishing between normal, tangential, and rotational force. Control of the optimum grip force through the use of these sensors in an adaptive real-time manner ensures damage free handling of compliant objects

MAGNETORESISTIVE SENSORS

Feromagnetic thin film magnetoresistive (MR) sensors are solid state magnetic sensors which can be employed in a wide variety of sensing and measuring applications. They have traditionally been used in read only heads for tape and credit card reading applications. In its simplest form the sensor operation relies on the fact that when a magnetic field which is oriented in the plane of the MR element and perpendicular to its length is placed near the sensor, a change in resistivity (resistance) of the element occurs. The magnitude of the change in resistance of the sensor is a function of the applied field amplitude.

Simple and Multi-element devices may be manufactured economically and sensors may be operated in DC fields, or AC fields of up to 5MHz and withstand working temperatures of greater than $200^\circ C$ .The sensor characteristic is essentially a cosine function and hence nonlinear, however, the sensors may be operated in their quasi-linear region provided that they are not saturated.

MAGNETORESISTORS IN TACTILE SENSING

A number of MR tactile skins have been proposed by Vranish (1984),Hackwood et al (1983),and Nelson et al (1985).The sensors proposed utilise varying techniques of signal aquisition and use DC, pulse magnetic fields or fixed magnetic dipoles mounted over a compliant medium. Tactile information is extracted in two distinct methods, i)Tactile imaging   ii)Tactile force and shear monitoring at the point of object/gripper interface.The sensors used in this work comprise the basic normal and shear force sensors , disscused later, and combination of the two provide information to be used in a pick and place problem.

FORCE AND SLIP DETECTION

In order to translate physical movement due to force and slip into electronic signals using a magnetoresistive sensor it is necessary to introduce a magnetic field whose characteristics change only through mechanical forces applied and this is achieved by introducing a compliant medium between the sensors and a magnetic field generated by a current carrying conductor directly above the sensor.

The basic force sensor comprises of a single element (Fig. 1) and any normal displacement in the overlay conductor causes a change in the voltage across the sensor due to the change of magnetic field strength at the sensor. Compression of the compliant medium will produce a change in the magnetic field intensity δH, and hence change in

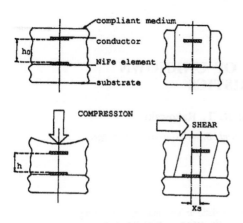

Fig. 1. Normal and Shear Force Detection

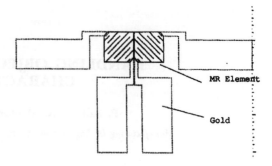

Fig. 2. Serpentine-path Half-bridge Sensor

resistance δR. By using an AC magnetic field any external electrical noise can be filtered out using band-pass filters (say, at 10kHz) which will improve the signal to noise-to-noise ratio and further improve the sensitivity. In future sensor systems cross-talk between normal force sensor elements can be eliminated by applying different signal frequencies to adjacent conductors and using band-pass filters. Hysterisis is also overcome by using AC fields by reversing the magnetic field to its initial state.

The slip sensor employs a similar principal but here two adjacent elements are used as zero centred potentiometers and any lateral displacement of the conductor shows as a differential change across the sensors. The elements are set up in bridge configuration. This reduces thermal drift and minimises the effect of common mode enviromental signals, such as the earths magnetic field.

At present a pair of shear and force sensors is used at each gripper finger. Since the spacial resolution of the sensors is in the order of microns and given the wide bandwith of such sensors complex arrays of such sensors seem quite feasible.

SENSOR CHARACTERISTICS AND PERFORMANCE

The change in resistance due to MR property can be small compared with inductive coupling between the sensor and overlay conductor. Such swamping of the MR sensor signal would produce a response which could be highly non-linear. By folding the element path back onto itself (Fig. 2) the inductive pick-up between adjacent elements is cancelled out. Additional benefits of the serpentine configuration include (i) increase in element resistance allows higher bridge voltages to be used improving sensitivity, (ii) current flow in one sensor pad, produces some bias field for the adjacent sensor pad ensuring operation in the linear region, (iii) the MR anisotropic axis for adjacent pads are set at 90° forming a complementary pair eliminating the steady component of magnetic field and making it immune to the temperature changes of thin film resistance, (iv) the sensors offer a linearity of ± 1%, far superior to the linearity of a single element.

The graph of Figure 3 shows typical performance of slip sensors in separations of 0.2 up to 2mm. The slip sensor output remains independent of the gap separation for up to ± 2mm displacement. The graphs diverge after this point. This displacement when applied to rubber in shear enables a wide range of forces to be detected.

Natural rubber is used as the compliant medium for its excellent mechanical characteristics and stability. In the compressive mode, the rubber response is linear up to 30% compression set and there are no problems with shear force detection as long as the bonded faces of the rubber to glass substrate do not fail. To this end, the dynamic range of the sensors depends largely on the type of compliant medium used and a dynamic range of 20dB is achievable.

The overlay conductor is constructed by etching copper patterns on a Mylar backing. Mylar is used for its flexibility and compliance enabling a range of grasping actions to be detected. The field outside the flat conductors is governed by

$$H = (\tan^{-1}(b + S/2h) - \tan^{-1}(b - S/2h))I/2$$

where I = Current through conductor (A)

    S = Conductor width (m)

    h= sensor element separation (m)

    b = horizontal displacement of the conductor (m)

The magnitude of H depends on the size of the current and width of the conductor directly so that the narrower the flat conductor, the sharper the resolution of the magnetic field and hence less cross-talk between adjacent sensors.

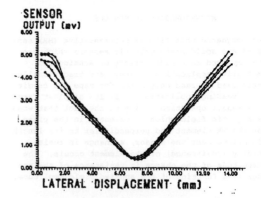

Fig. 3. Shear Force Sensor Characteristic

The load-deflection curves for rubber in tension and compression are approximately linear for strains of the order of several percent. As these curves remain linear through the origin, the value of Young's modulus remains the same during compression and tension. The bulk modulus of rubber is many times greater than its Young's modulus and this means that the rubber hardly changes its volume even under high loads, so that for most types of deformation there must be space into which the rubber can deform. The more restriction that is made on its freedom to expand the stiffer it will become, therefore allowance had to be made in the design of sensors to overcome this problem. The coefficient of friction (μ) for most dry rubber surfaces is generally unity, but for design purposes it is assumed that slip between object and sensor due to shear force will not occur if the ratio of maximum shear force to minimum compressive force is less than 0.3. If water (a lubricant for rubber) is present it will normally be squeezed out under load.

The dynamic range of the sensors, therefore, depends on the rubber compliant medium, the strength of magnetic field and the output amplification of sensor signals.

To build into the sensor a value for the mechanical force the simplified model of sensor geometry shown in Fig. 4 will be used. The displacement in the vertical direction is given by:

$$x = h - h_0$$
$$x = \frac{4FR^3}{E_o I}\left[\left(\frac{I}{AR^2}-1\right)^2 \frac{1}{2\pi} + \frac{\left(\frac{I}{AR^2}-1\right)}{\pi} + \frac{\pi}{16}\right]$$

for which the symbols are as defined in Fig. 4, and E_o = Young's modulus for the particular rubber, $R = (D+d)/4$, $I = L(D-d)^3/96$, $A = L(D-d)/2$.

Considering the case of no bending, then for rubber of IRHD 30:

$$x = (4.0473 \times 10^{-2})F \quad \text{(mm)}$$

From this expression for a maximum force of 50 Newtons,

$$x = 2.02 \text{ mm}$$

Therefore, for a dynamic range of 20dB, the minimum detectable value x_{min} should be 0.0202mm which relates to a force of 0.49 Newtons. This relationship holds if the ratio of the outer diameter D to the inner diameter d does not exceed 4.5.

The minimum detectable change in the magnetic field is $\pm 8 \times 10^{-2}$ Am^{-1} when the skin is uncompressed. The magnetic field at the sensors for I = 20mA, S = 500μm, h = 3mm is H = 60.65 Am^{-1}. If the rubber were pressed to the threshold point the field strength would become: H = 60.73 Am^{-1}, representing a new height of h_0 = 2.997mm well within the 20dB requirement for x_{min}.

SIGNAL ACQUISITION AND PROCESSING

Analog Information

The two sources of force and shear information are processed independantly using conventional signal aquisition techniques. All the lead outs from the edge of the substrates are screened to maximise the signal-to-noise ratio. The MR elements are powered using constant current sources and every sensor is activated using AC fields. In this way rows of sensory data can be processed using filters and differential amplifiers as close to the sensors as possible to eliminate ambient noise.

Hierarchical Control

As robots gain more intelligence and the tasks they carry out become more complex, to conserve valuable processing time of the controller and to keep functionality simple the need of a modular approach to design becomes more apparent. Tasks should be decomposed in an upside down approach and decisions should be made locally as far as possible with different levels interacting by simple commands providing natural break points for software and ease of debugging (Fig. 5a). The overall control strategy is based on the NBS Hierarchical Robot Control system (Barbera., et al, 1982) modified to allow real-time adaptive gripping between sensor-gripper-object.

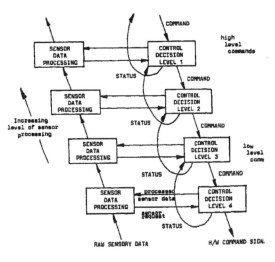

Fig. 5a. NBS Control Model

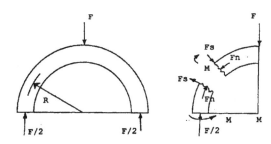

Fig. 4. Simplified Cross-section of Compliant Medium

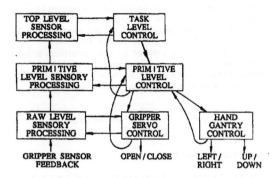

Fig. 5. Hierarchical Control for a Basic Adaptive
Gripper Application

System Realisation

The situation is somewhat simplified by the fact
that the gripper controller is dealing with the
lowest levels of control in the "Control Decision
Hierarchy", that is the joint and gripper coordination, up as far as the primitive control level.
The control block of Fig.5 represents a plug compatible module communicating with the higher
level controller through simple commands, issuing
status reports and in turn supplying commands for
levels below. At every control level each input
generates a definite output or a combination of
outputs according to the information organised in
the "State Table" at that level (Table 1). Hence
each level cycles around INPUT PRE-PROCESS-
ING; STATE TABLE IMPLEMENTATION; OUTPUT POST
PROCESSING. The pre and post processing are only
data preparation tasks and the bulk of processing
is carried out by state-table implementation.

Hardware and Software Implementation

To design the hardware in accordance with NBS
strategy a multi processor approach was chosen
with each processor dedicated to one control
task, such as gripper control, data aquisitaion,
etc.. **Figure 6** illustrates the hardware implementation of the Gripper control level. The state
table for each level is in the form of a 4kBytes
multiported common memory board. To eliminate
race conditions and prevent loss of data the communication bus was based on a 'First come -first
serve 'principle. Here, as soon as one CPU tries
to talk to the common memory the other processors
are blocked out immediately. As a result time
wastage associated with sequential polling is
avoided.

The Control software is based on the modified
state table implementation (Table 1) at each
level. Input commands are generated at the
highest level by the operator and passed down
through to lower levels. Every command then
starts a sequence of events that take the form of
procedures to which parameters are passed and
received .Since the sensory processes are continuous processes various process levels lend
themselves naturally to modular design. This
provides for ease of modification and trouble
shooting. Software is written in a high level language and compiled into machine code and loaded
into onboard EPROMS.

TABLE 1   Modified State Table Implementation

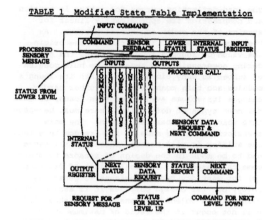

State Tables for Primitive level and Gripper Control level are given in Tables 2 and 3. To illustrate implementation of State Tables,
execution of the OPEN command through Primitive
Control level (Table 2) is explained as follows:
OPEN   This command is issued by the higher
level and initiates a check of the Internal
status for the Pick and Place operation. If this
is incomplete OPEN will return an appropriate
error signal and the process is terminated, otherwise the process will continue through the table
until OPEN COMMAND is decomposed to the lower
level or if the gripper were open at the time the
command was issued the OPEN CALL is cleared.
Other commands from higher level follow a similar
discription and every command initiates a chain
of IF THEN ELSE procedures.Other commands at
this level and at the Gripper Control level are
executed in much the same way.

Utility routines for various tasks of data
preparation, pre- and post- processing and state
table implementation were written using PL9 which
is a 6809 development tool and compiled into
machine code. The use of this development system
was purely due to availability and to this end
any other commercially available packages may
have been used. Since every "executing owner" has
a definite procedure, modifications can be easily
carried out and CPU EPROMs updated.

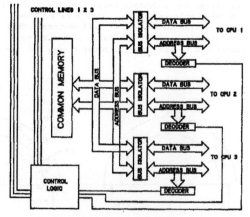

Fig. 6.   Hardware Implementation

ADAPTIVE GRIPPING

The task for tactile sensors in handling objects of unknown characteristics is to determine an optimum grasp level without losing or damaging the object. Here the important criterion is the state of the interface between the gripper fingers and the object. There are three possibilities: whether the object is in contact with the fingers; it is gripped   and or if the object is slipping. When a pick command is received, control is decomposed down to primary command level and subsequently to Gripper Controller. At this stage the slip and force informations are used to control the gripping process.

Adaptive control of objects of unknown weight and dimensions are achieved by aiming to attain an optimum gripping force from the Gripper Servo. To do this slip is translated as a judder or a rapidly changing shear signal due to an insufficient Grip Force. Normal force is then incremented until such time that slip is within ±5% of the previous cycle. **Figure 7** shows the output from the gripper drive circuit as applied to a simple pick problem utilising a servo-driven parallel jaw gripper. The first part of the diagram shows a maximum negative voltage (maximum motor torque) during the 'open gripper' operation. The second part of the diagram shows the gripper operating in 'adaptive gripping' mode. Firstly the gripper closes with a maximum velocity. As soon as contact is detected by the force sensors a short negative pulse stops the gripper motor dead. A minimum grip level is applied to ensure friction contact between the fingers and the object being picked up. The grip force level is then incremented during the lift operation until no change in slip signal is detected. This point would then indicate the optimum gripping force and the Lift/Place operation may then commence. The optimum force level proved somewhat difficult to achieve at first in that it caused chatter at the gripper fingers. This was overcome by introducing software hysterises at the servo-gripper driving module. Overall cycle time for gripping operation was 0.4ms for normal force operation but introduction of shear raises this value to over 10ms. Current work with single chip microprocessors indicates that cycle times of 470μs are achievable.

CONCLUSIONS

Work to date has indicated that Magnetoresistive technology can lead to devices sufficiently robust to be used in a manufacturing environment. Close packing of elements in arrays does not present a problem and the sensors may be employed in tailor-made applications such as pattern recognition, contour examination, gap detection and, of course, force and slip detection.

The use of hierarchical control strategy for adaptive gripping has shown much promise and current research is exploring performance limits of such a system. It is also possible to modify the gripper control system to realise an adaptive assembly gripper based on shear and normal force signals recorded during peg-in-hole assembly.

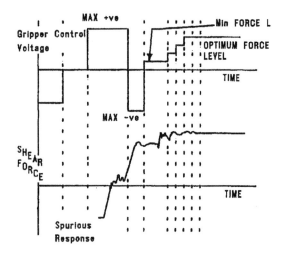

Fig. 7. Adaptive Gripping Profile

Furthermore, the software and hardware may be modified to include a bank of data on handling objects through a learn mode so that objects may be identified according to tactile sensory patterns.

ACKNOWLEDGEMENTS

The authors are indebted to the U.K. Science and Engineering Research Council for the support of the work.

REFERENCES

Adl, P., and Rakowski, R.T.(1988), Tactile Sensor for Robot System. *U.K. Patent Application*, 88-20889.7.

Barbera, A.J., Fitzgerald, M.L., and Albus, J.S.,(1982),Concepts for a real-time sensory interactive control system architecture. *Proceedings of the14th South Eastern Symposium on System Theory.*

Dario,P., et al,(1987), Multiple sensing fingetip for robotic active touch. *Proc. of the3rd Int. Conf. on Advanced Robotics ICAR'87*,IFS Publications, Springen Verlag.

Hackwood,S.,Beni,G., and Nelson,T.J.(1983), Torque sensitive tactile array for robotics, *Proc.ROVISEC 3.*

Harmon,L.(1987), Automated tactile sensing,*Int. J. of Robotics Research*, **1**, No. 2.

Nelson,T.J.,Van Dover,R.B.,Jin,S., Hackwood,S., Beni,G.(1985), Magnetoresistive tactile sensor for robots, *Proc. of Int. Conf. on Materials in Computers, Robotics and Comunication Idustry*,Monterey, Canada.

Vranish,J.M(1984)., Magnetoresistive skin for robots, *Proc. ROVISEC 4.*

TABLE 2   Primitive Control Level State Table

COMMAND	LIFT / PLACE	VERTICAL CONTACT	GRIPPER COMPLETE	HAND COMPLETE	GRIPPER STATUS	OUTPUT PROCEDURE
OPEN	COMPLETE	X	YES	X	CLOSED	CALL OPEN COMMAND
OPEN	COMPLETE	X	NO	X	OPEN	CLEAR OPEN COMMAND
CLOSE	COMPLETE	X	YES	X	OPEN	CALL CLOSE COMMAND
CLOSE	COMPLETE	X	NO	X	CLOSED	CLEAR CLOSE COMMAND
MOVE	COMPLETE	NO	X	YES	X	CALL MOVE COMMAND
MOVE	COMPLETE	NO	X	NO	X	CLEAR MOVE COMMAND
MOVE	X	YES	X	X	X	REPORT ERROR
LIFT	COMPLETE	X	YES	YES	CLOSED	CALL GRIPPING METHOD
LIFT	LIFTING	X	X	X	X	CALL MOVE UP
PLACE	COMPLETE	NO	X	YES	CLOSED	CALL MOVE DOWN
PLACE	PLACING	YES	X	X	X	CALL OPEN GRIPPER
SUB-COMMAND	INTERNAL STATUS	SENSORY INPUT	STATE FEEDBACK FROM LOWER LEVELS			OUTPUT EXECUTING OWNER

X - DON'T CARE STATES

TABLE 3   Gripper Control Level State Table

COMMAND	GRIPPER STATUS	CONTACT FOUND	HARDWARE SLIP	SOFTWARE SLIP	OUTPUT PROCEDURE
OPEN	CLOSED	X	X	X	OPEN GRIPPER FULLY
CLOSE	OPEN	NO	X	X	CLOSE GRIPPER QUICKLY
CLOSE	OPEN	YES	X	X	STOP GRIPPER QUICKLY
APPLY FORCE	X	NO	X	X	REPORT ERROR
APPLY FORCE	X	YES	X	X	INCREMENT TO FORCE
USE FRICTION	X	NO	X	X	REPORT ERROR
USE FRICTION	X	YES	X	X	MONITOR SHEAR FORCE
TEST FOR SLIP	X	NO	X	X	REPORT ERROR
TEST FOR SLIP	X	YES	NO	X	NO SLIP CONDITION
TEST FOR SLIP	X	YES	X	NO	NO SLIP CONDITION
TEST FOR SLIP	X	YES	YES	X	SLIP CONDITION
TEST FOR SLIP	X	YES	X	YES	SLIP CONDITION
SUB-COMMAND	INTERNAL STATUS	SENSORY INPUT			OUTPUT EXECUTING OWNER

X - DON'T CARE STATES

EXPERIMENTAL RESULTS ON IR SENSOR
SIMULATION

J. Ilari

Departamento d'Enginyeria de Sistemes, Automàtica i Informàtica Industrial,
Institut de Cibernètica, Barcelona, Spain

Abstract : The paper describes the work undertaken at the Institut de Cibernètica on the field of infrared (IR) sensor simulation. This work has led to the proposal of a simple IR sensor model as well as to the design and implementation of a simulation package allowing to emulate the behaviour of different IR sensors in a variety of working circumstances. The program permits independently defining the position and the beam patterns of both the IR emitter and receiver as well as dealing with different types of surfaces. The experimental results so far obtained confirm the validity of the simple model proposed for the IR sensor as well as the adequacy of the simulation approach introduced in the paper.

Keywords : finite element method, infrared sensors, modeling, robots, simulation

1 Introduction

Remote Range Sensors (RRS) have an outstanding interest for robotic applications, particularly in the field of on-line collision avoidance and parts grasping. On the one hand, they provide non-contact measures of the distance to a given target. This permits anticipating collisions and planning adequate evasive strategies. On the other hand, they are devices which give information that require neither a lengthy nor a sophisticated processing (as opposed to vision-based systems) by reducing the measurement scope to a small area of the robot workspace.

Several RRS are currently being used by the robotic community to provide robot manipulators with an adequate knowledge of their distance to close and middle-range obstacles (Idesawa and Kinoshita, 1986; Masuda, 1986). Ultrasonic, infrared and laser-based sensors are probably the best known of these devices (André, 1983).

In spite of the simplicity of the data they provide –a scalar value from which an estimate of the distance to the target is to be computed– range sensors convey a type of information which is seldom exploited : the trace of measurements supplied as the sensor changes its position in relation to a target. In fact, it is rather usual that these sensors be used on a *on-off* basis, ignoring the benefits that could be obtained from the use of more sophisticated techniques (e.g. firm analysis). To a large extent, this can be attributed to the little knowledge about the behaviour of this kind of sensors in a real environment.

To gain insight into the behaviour of infrared (IR) sensors, a software simulation package allowing to predict the output of such sensors under a variety of circumstances has been developed. As a simulation tool, this software permits experimenting with IR sensors without having to resort to real tests. This permits acquiring a good knowledge of the way IR sensors behave and, through this knowledge, to foresee the behaviour of real IR sensors.

This paper describes the mathematical model which has been proposed for IR sensors as well as the simulation testbed which has been implemented to carry out simulation experiments based on this model. Both have been successfully used in the development of a technique that permits locating edges in a set of obstacles by applying firm-analysis techniques to the trace of measurements provided by IR sensors as they approach the obstacles.

2 IR Sensor Model

An IR sensor consists of an *emitter* (E) and a *receiver* (R). Its working principle is simple : E emits a radiation, a portion of which reaches a given surface S. The radiation reaching S is partly absorbed and the rest is reemitted. Part of the radiation reemitted by S reaches R, which generates a signal that is a function of both the radiant power and the direction of the incoming radiation (Figure 1).

Emitter E (typically an infrared-emitting diode) is characterized by its *radiant flux function* $\varepsilon(\vec{u_e})$ –the power it radiates per unit of solid angle as a function of the unit vector $\vec{u_e}$ which defines the direction of outcoming radiation–. Receiver R (typically a phototransistor) is characterized by its *radiant sensitivity function* $\rho(\vec{u_r})$ –the signal it generates per unit of radiation power received as a function of the unit vector $\vec{u_r}$ which defines the direction of incoming radiation–. Finally, the surface S is characterized at every point p by its *absorption coefficient* $\alpha(p)$ –the proportion of radiation reaching p absorbed by S– as well as by its *reflectance function* $\sigma(p, \vec{u_e}, \vec{u_r})$ –the radiation it reflects as a function of the relative position of S and the directions of incoming and outgoing radiation–. E, R and S all have a well defined position and orientation in space.

Let us consider a differential of surface dS with sides dx and dy and a point $p = (x, y)$ such that with $p \in dS$. The

*This work has been supported by the Fundación Ramón Areces under the project SEPETER.

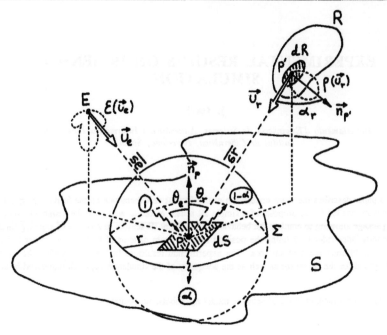

Figure 1: *Generic IR sensor facing an heterogeneous planar surface.*

radiant intensity (radiant power per surface unit) at p will be :

$$\gamma_{in}^{E}(p) = \varepsilon(\vec{u_e}).\frac{\vec{u_e} \bullet \vec{n_p}}{d_{se}^{2}} \quad , \quad (d_{se} = |\,\overline{se}\,|)$$

where $\vec{n_p}$ stands for the unit vector normal to S at p and $\bullet$ denotes scalar product. Thus, the amount of radiant power received by dS will be :

$$\pi_{in}^{E}(dS) = \gamma_{in}^{E}(p).dS$$

Part of $\pi_{in}^{E}(dS)$ will be absorbed by dS $(\alpha_S(p).\pi_{in}^{E}(dS))$ and the rest $(\pi_{out}^{E}(dS))$ will be reemitted :

$$\pi_{out}^{E}(dS) = (1 - \alpha_S(p)).\pi_{in}^{E}(dS)$$

A radiant flux appears as a consequence of this secondary radiation. $\phi^{dS}(p')$, the radiant flux in $p' \in dR \in R$ originated by the radiation reemitted by dS, can be computed as :

$$\phi^{dS}(p') = \pi_{out}^{E}(dS).k.\frac{\sigma(p, \vec{u_s}, \vec{u_r})}{d_{sr}^{2}} \quad , \quad (d_{sr} = |\,\overline{sr}\,|)$$

The value of constant k can be deduced by means of a radiation balance between radiation leaving dS and radiation leaving a sphere $\sum$ of radius r centered at p and containing dS :

$$\pi_{out}^{E}(dS) = \int_{\sum} \pi_{out}^{E}(dS).k.\frac{\sigma(p, \vec{u_s}, \vec{u_r})}{r^2} d\sum$$

But, since $\pi_{out}^{E}(dS)$, k and r are independent of the particular $d\sum$ selected,

$$k = \frac{r^2}{\int_{\sum} \sigma(p, \vec{u_s}, \vec{u_r}).d\sum}$$

The radiant power received by a differential of surface $dR \in R$ due to the radiant flux originated in dS will be :

$$\pi_{in}^{dS}(dR) = \phi^{dS}(p').dR.(\vec{u_r} \bullet \vec{n_{p'}}) \quad , \quad (p' \in dR)$$

where $\vec{n_{p'}}$ denotes the unit vector normal to dR in $\vec{p'}$. Thus, the differential of signal generated by dR as a result of the radiant flux originated by dS will be :

$$d\varsigma^{dS}(dR) = \pi_{in}^{dS}(dR).\rho(\vec{u_r})$$

and integrating all over the surface of receiver R, we get :

$$d\varsigma^{dS}(R) = \int_R \pi_{in}^{dS}(dR).\rho(\vec{u_r})$$

Finally, to compute the signal generated in R as a result of power radiated by the overall surface S, the expression above has to be integrated over S :

$$\varsigma^{S}(R) = \int_S \int_R \pi_{in}^{dS}(dR).\rho(\theta_r, \varphi_r)$$

that is,

$$\varsigma^{S}(R) = \quad\quad\quad (1)$$

$$\int_S \varepsilon(\vec{u_e}).\frac{\cos\theta_e}{d_{se}^{2}}.(1 - \alpha_S(p)).k.\int_R \frac{\sigma(p, \vec{u_e}, \vec{u_r})}{d_{sr}^{2}}.\cos\alpha_r.\rho(\vec{u_r}).dR.dS$$

where $\cos\theta_e = \vec{u_e} \bullet \vec{n_p}$ and $\cos\alpha_r = \vec{u_r} \bullet \vec{n_{p'}}$

3 Simplified IR Sensor Model

Equation (1) can be remarkably simplified if some assumptions are made about the environment of the IR sensor :

- If the absortion coefficient is assumed to have a constant value in all surface S, then the term $(1 - \alpha_S(p))$ can be assimilated to a constant k' [1].

- If the size of R is small as compared to d_{sr}, then $d\varsigma^{dS}(R) = \phi^{dS}(p').R.\cos\alpha_r.\rho(\vec{u_r})$ and the integration over R can be eliminated.

- If receiver R is assumed to have a spherical shape, then its effective surface will be constant when viewed from S. This means that the term $R.\cos\alpha_r$ can be assimilated to a constant (k'').

[1] In a steady state all radiation reaching S is reemitted, that is, $\alpha_S(p) = 0$ and, thus, $k' = 1$.

In this case we get the following expression for the signal generated in R as a result power radiated by surface S :

$$\varsigma^S(R) = k.k'.k''. \int_x \int_y \varepsilon(\vec{u_e}). \frac{\cos \theta_e}{d_{se}^2}. \frac{\sigma(p, \vec{u_e}, \vec{u_r})}{d_{sr}^2}. \rho(\vec{u_r}).dx.dy$$

(2)

In the particular case of S being Lambertian, that is, perfectly matte and reflecting light equally in all directions, $\sigma(p, \vec{u_e}, \vec{u_r}) = \vec{u_r} \bullet \vec{n_p} = \cos \theta_r$ and, thus :

$$\varsigma^S(R) = k.k'.k''. \int_x \int_y \varepsilon(\vec{u_e}). \frac{\cos \theta_e}{d_{se}^2}. \frac{\cos \theta_r}{d_{sr}^2}. \rho(\vec{u_r}).dx.dy$$

4 Modelling E, S and R

To tackle the simulation of IR sensors, the functions describing the behaviour of the emitter E, surface S and receiver R must be defined. In practice, these functions will depend to a large extent on the particular type of IR sensors and surfaces involved in the simulation. However, it is possible to derive expressions for them that are general enough to accommodate a large number of particular cases.

Experimental results have shown that both the radiant flux of the emitter ($\varepsilon(\vec{u_e})$) and the radiant sensitivity of the receiver ($\rho(\vec{u_r})$) have a set of common features :

- They exhibit a marked symmetry around a particular axis.

- They have a set of lobes (very frequently a single one) that can be qualitatively characterized by a gaussian function.

To reflect these features, we have modelled $\varepsilon(\vec{u_e})$ and $\rho(\vec{u_r})$ by means of the following rational functions (see Figure 2) :

$$\varepsilon(\vec{u_e}) = \sum_m (b_m + \frac{k_m - b_m}{1 + ((\theta_e - \alpha_m)/a_m)^2})$$

(3)

$$\rho(\vec{u_r}) = \sum_n (b_n + \frac{k_n - b_n}{1 + ((\theta_r - \alpha_n)/a_n)^2})$$

(4)

where :

- m and n are the number of lobes in $\varepsilon(\vec{u_e})$ and $\rho(\vec{u_r})$, respectively.

- b_m and b_n are the minimum values of lobes m and n in $\varepsilon(\vec{u_e})$ and $\rho(\vec{u_r})$.

- k_m and k_n are the peak values of lobes m and n in $\varepsilon(\vec{u_e})$ and $\rho(\vec{u_r})$.

- a_m and a_n are the parameters defining the shape of lobes m and n in $\varepsilon(\vec{u_e})$ and $\rho(\vec{u_r})$ (the lower their value, the narrower the corresponding lobe).

- α_m and α_n are the angles defined by lobes m and n and the main axis of E and R.

Other functions have been proposed to model the radiant flux of the emitter and the radiant sensitivity of the receiver in an IR sensor (Espiau, 1982). The function proposed in this paper has the advantage of not involving trascendent functions –it has a rational expression– while adequately modelling the behaviour of both the emitter and the receiver. This allows to

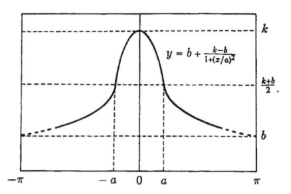

Figure 2: *Function used to model a single lobe of the emitter (receiver) of an IR sensor.*

significantly speed up the simulation of the behaviour of both elements.

Several functions have been proposed to model the radiation pattern of a surface. Probably the simplest one is the Lambertian model which states that this radiation pattern is a function of only the angle defined by the normal to the surface and the outgoing direction considered (see Section 3). Since Lambertian surfaces are perfectly matte and have no glossiness at all, they represent only ideal surfaces. To adequately represent real surfaces, other models have been proposed to complement the Lambertian behaviour with a specular component (Shafer, 1985). Based on them, a surface radiation model that includes near-Lambertian and near-specular components, while avoiding the use of trascendent functions (see Figure 3), is used :

$$\sigma(p, \vec{u_e}, \vec{u_r}) = \frac{k_l}{1 + (\theta_r/a_l)^2} + \frac{k_s}{1 + (\theta_{er}/a_s)^2}$$

(5)

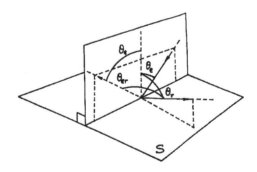

Figure 3: *Relative position in space of the incoming and outgoing radiation beams in S.*

According to this model, the pseudo-Lambertian component of σ exhibits a radial symmetry around an axis normal to

S. The pseudo-specular component exhibits a radial symmetry around the axis defined by the reflected beam. The values of k_l, k_r, a_l and a_r have a meaning which is analogous to their counterparts in Equations (3) and (4).

Expression (5) captures the idea of the radiant flux originated by dS being stronger :

- When θ_r is small (Lambertian behaviour). This corresponds to the first term of the right side of expression (5).

- When the direction of the outgoing beam is close to the direction defined by the incoming beam reflected in S (specular behaviour). This corresponds to the second term of the right side of expression (5).

5   Simulation Approach

To check the correctness of the IR sensor model proposed, a simulation environment has been built to evaluate the behaviour of IR sensors when facing a *planar heterogeneous surface S*.

To compute the value of $\varsigma^S(R)$ in (2), finite element techniques are used. This allows to tackle the problem of computing IR sensor reading numerically, avoiding the problems derived from an analytical approach when the function to integrate is complex. Surface S is modelled, thus, as a *lattice of homogeneous surface elements* and the integral of Equation (2) is performed through a summation over all surface elements :

$$\varsigma^S(R) = \sum_m \sum_n \varsigma^{dS}(R) \qquad (6)$$

Now, the contribution of each one of these elements to the overall signal generated by R has to be computed. This introduces a new integration problem but, since σ can be assumed to be constant within all the surface element area, integration is simplified. Even if in this case an analytical solution could be reasonably developed, to keep the program as general as possible, we have performed integration through the *Gauss integration formula* (Zienkiewicz, 1980). This formula allows to compute a definite integral from the knowledge of the value of the function to be integrated in a set of well defined points within the integration interval :

$$\int_a^b \int_p^q f(x,y).dx.dy = \sum_i \sum_j c_{ij}.f(x_i, y_i)$$

Moreover, it guarantees a null error if the function to be integrated is a polynomial of degree less than n, where n is a function of the number of points selected to compute the integral [Zienkiewicz 80]. Using the Gauss integration formula to compute $\varsigma^{dS}(R)$ we get :

$$\varsigma^{dS}(R) = \sum_i \sum_j c_{ij}.\varepsilon(\vec{u_e}).\frac{\cos \theta_e}{d_{se}^2}.\frac{\sigma(p, \vec{u_e}, \vec{u_r})}{d_{sr}^2}.\rho(\vec{u_r}) \qquad (7)$$

where c_{ij} stands for the Gauss coefficient associated with point (i, j) within the integration interval.

In the simulation package developed at the IC, up to 36 (6 × 6) points can be selected to compute the sought integral in each one of the surface elements. In this case, the Gauss integration formula can be guaranteed to be exact for polynomial functions up to degree 11. Even if the function to be integrated is not a polynomial, the Gauss integration formula gives a rather good approximation to the value of the integral (the greater the number of points used, the better the approximation). This will be true as long as the function to be integrated could be approximated by a polynomial within the integration interval. In our particular case, since the function to integrate is smooth enough within the integration interval to be accurately approximated by a polynomial of degree 5, (3 × 3) integration points have been used to implement the Gauss integration algorithm.

Through the finite element approach and the use of the Gauss integration formula to compute the contribution of each surface element to the overall signal generated by the IR receiver, the integral in (2) can be reduced to a summation (over the surface lattice) of summations (over the surface element) :

$$\varsigma^S(R) = \sum_m \sum_n (\sum_i \sum_j c_{ij}.\varepsilon(\vec{u_e}).\frac{\cos \theta_e}{d_{se}^2}.\frac{\sigma(p, \vec{u_e}, \vec{u_r})}{d_{sr}^2}.\rho(\vec{u_r}))$$

This makes this simulation approach simple and general enough to be effectively used to simulate the behaviour of IR sensors when facing heterogeneous planar surfaces when the radiant flux function of the emitter, the radiant sensitivity of the receiver and the reflectance function of the surface are known.

6   Implementation Details

Both the IR sensor model and the finite element simulation software have been programmed in C (aprox. 700 lines of code). They were originally implemented on a VAX 785 under VMS 4.7. Afterwards they were migrated to a SUN 4/260 under SunOS 4.0.1 (UNIX). Graphic postprocessing routines to help visualize IR behaviour were written in VAX Fortran.

The simulation package we have implemented allows to independently define the position, the orientation and the radiant functions of both the IR emitter and the IR receiver. Reflectance features of the surface to be dealt with are defined through the reflectance function associated with the several pixels within the surface lattice (see Equations (5), (6) and (7)).

7   Experimental results

To emulate the behaviour of the actual IR sensors available at the IC, some of them were selected, their radiant functions were obtained experimentally and the several simulation parameters (see Equations (3) and (4)) were set accordingly. Two surfaces with rather different radiant features were also selected and the parameters of the respective models were set according to experimental data. Then, two experiences were simulated :

- An IR sensor going away from an homogeneous surface, starting from a position very close to the surface.

- An IR sensor executing successive sweeps at increasing heights over the boundary between two half-planes with rather different radiant features.

From the comparison of the simulation results obtained with the readings of real IR sensors under similar circumstances, the validity of both the simplified IR sensor model and the simulation approach proposed in the paper can be concluded (see Figures 4 and 5). It should be noted that the model correctly gives rise to the bimodal curve of real IR sensors,

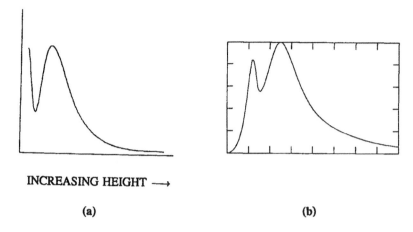

INCREASING HEIGHT $\longrightarrow$

(a)   (b)

Figure 4: *(a) Experimental and (b) simulated IR readings obtained from an IR sensor going away from an homogeneous surface.*

correctly reflecting their behaviour in the short to long distance ranges. As far as we know, existing IR models are valid only in the mid to long distance ranges.

8   Computational considerations

The finite element technique on which the computation of the IR sensor reading when facing a planar surface is based, determines this process to have an $O(n_x.n_y.n_i.n_j)$ complexity, n_x and n_y being the X and Y resolution of the lattice describing surface S, and n_i and n_j being the number of points selected in the X and Y direction to implement the Gauss integration formula. Obviously, this fact will slow down the simulation process when high resolutions are needed.

The efficiency of the program could be improved by reducing the surface area contributing to the IR reading. In the current implementation all the surface is considered to contribute to this reading. However, preliminary tests have provided rather similar results at a much lower computational cost, if the active surface elements are limited to those falling within a given influence cone defined from the transmitter and the receiver beam patterns. In relation to applications where speed is mandatory it should be noted that the finite element method adopted to tackle the simulation process is especially suitable to be implemented in parallel architectures.

9   Conclusions and Future Work

The IR sensor model proposed is simple and does not involve complex computations. It makes use of a reduced set of parameters, each making a well defined contribution to the overall model expression. In spite of the simplicity of the proposed model, this has been seen to correctly emulate the behaviour of real IR sensors. A high resolution in the description of the surface has been shown to slow down the simulation task. In this case, some speed up techniques are available, being also possible to tackle the simulation computation with parallel architectures.

An automatic procedure to set the several parameters involved in the model has still to be designed. Both to evaluate the shortcomings of the proposed approach and to further refine the model, more accurate analysis tools should be developed.

In a real environment it is possible that IR sensors have to face a 3D scene, not a 2D one. The simulation techniques we have described can be easily adapted to the 3D case by computing from a CAD database an adequate digitized image of the environment as seen by the IR sensor. In this image each pixel will convey information concerning both the *depth* and the *orientation* of the surface element it represents as well as its *radiant characteristics*. Once the image has been computed, the simulation techniques proposed in this paper can be employed to correctly predict IR sensor readings.

10   Acknowledgements

I wish to thank *Dr. Federico Thomas*, who designed the IR sensing hardware at the IC, for his helpful comments all along this work. I wish to thank also *Prof. Luis Pérez Vidal* from the *Dpt. de Llenguatges i Sistemes Informàtics (UPC)* for his suggestions and valuable hints on the subject of finite element integration methods.

11   References

André G. (1983) Conception et modélisation de systèmes de perception proximétrique. Application a la commande en téléopération. *Phd. Thesis, Université de Rennes I, October 1983*

Espiau B. (1982) Prise en compte de l'environnement local dans la commande des robots manipulateurs. *Phd. Thesis, Université de Rennes I, June 1982*

Idesawa M., Kinoshita G. (1986) New type of miniaturized optical range-sensing methods RORS and RORST. *Journal of Robotic Systems, 3(2), pp. 165-181*

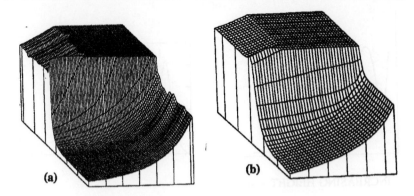

DECREASING HEIGHT ⟶

Figure 5: *(a) Experimental and (b) simulated IR readings obtained when executing an IR sensor successive sweeps at increasing heights over the boundary between two half-planes with rather different radiant features.*

Masuda R. (1986) Multifunctional optical proximity sensor using phase modulation. *Journal of Robotic Systems, 3(2), pp. 183-190.*

Shafer S.A. (1985) Using color to separate reflection components. *Color research and application, Vol. 10, Nr. 4, Winter 1985, pp. 210-218*

Zienkiewicz O.C. (1980) El método de los elementos finitos. *Editorial Reverté, Spain* (translation of the original published by Mc. Graw-Hill)

AN EXPERT SYSTEM APPLICATION IN MANUFACTURING

E. Di Chio

*Department of Industrial Design & Production, School of Engineering,
University of Bari, Bari, Italy*

Abstract.The use of Artificial Intelligence tools for Manufacturing and Expert Systems
in Computer Integrated Manufacturing is an important factor for advanced applications.
This function is verified in the present work by an industrial case of welding
systems.
The Expert System development in welding systems is interesting,for a quality execu
tion it is necessary,as seen in the examined case,the choise of such parameters
and their coordination.
It is also possible to obtain the objectives capable to value Expert System introduc
tion in a work cell control (system definition and control hypothesis) and to extend
control even to cell maintanance problems (data-base,definition of main problems,safe
ty functions and timing activity).

Keywords.Artificial Intelligence tools;Expert System of consultable type;work cell
control;Welding System;Knowledge Engineering.

INTRODUCTION

Expert System function may consent either the data
organization concerning various welding process
in a data bank,obtainable through recording data
by means of appropriate devices used for interacti
ve talk with operator,also in reference to the
needs of the manufacturing industry,or an excel
lent tool for optimization of various welding pro
cedures through preselected programs guided by
welding parameters.
Expert Systems are computer programs for advanced
applications,that use the knowledge underlying
human expertise to solve difficult problems.
Manufacturing Process Expert System in welding
industrial application allow also the automatic
correlation of process parameters,the introduction
of automatic quality control and the execution of
automatic diagnostics of plant sections.
The use of Expert System in welding has been par
ticularly applied before as preweld expert
system.
The application concern the parameters choise and
coordination,but it is also possible to include
welding process selection,welding procedure de
termination,material selection,expert emulation
of joint design,preweld inspection and the simu
lation for timing activity.

EXPERT SYSTEM USE

Expert System use is developed by three levels
for two types of knowledge:
theoric data or public,as documented definitions,
facts and theories;
heuristic or private,as knowledges undocumented
and based on individual experience.

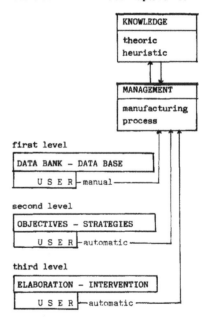

Fig. 1. Expert System Utilization Levels.

The Expert System is able to give also the most
appropriate answers for choising technological
and control parameters of process under examina
tion in real time,once solved dimension and incom
patibility problems for computer control structu

re.

The specific area analysis concern <u>acquisition</u>,
<u>representation</u>,but essentially
<u>investigation</u>:define all desirable result by
                    information about procedures
                    and work modalities;
<u>advices</u>:useful advices for the best definition
           of manufacturing process;
<u>control</u>:to fix work priorities among different
           selection.
The appropriate development of Expert Systems
Shell for manufacturing process needs to fix the
purposes to be gained,the variants to be consi<u>de</u>
red,the rules and the excellent strategies for
achivieng the objective.

WELDING EXPERT SYSTEM ROLE

The expert system application in welding is impo<u>r</u>
tant for a quality execution it is necessary,as
seen in the industrial examinated case,the choise
of such parameters and their coordination.
The choise of such parameters depends upon joint
design;it is identified essentially by material,
geometry,thickness and type of preparation.
Actually in the examinated and considered case of
industrial application,when it is necessary to
weld a new type of joint,parameters search and
coordination take place by attempts;that is,sta<u>r</u>
ting values are fixed and then,they are adjusted
in accordance with joint deficiencies.
This procedure may result enough long and theref<u>o</u>
re expensive and it is also necessary that wel<u></u>
ding operator have remarkable experience.
As already noted,however,it is necessary not only
to be able to increase parameters,variable and
disturbances knowledge but also,above all,to value
their connexions and interferences in order to be
able to stating efficiently a correct control / :
system.
Analysis have been developed for inox steel wel<u></u>
ding about which significant data are available,
also for comparison between arc and laser welding.
For this process laser at solid state or at CO_2
have been considered in impulsed or continual m<u>o</u>
dality with different powers which for the same
process result changeable in function of 20 W
until 20.000 W according to the range of applic<u>a</u>
tion.Particularly,Nd-Yag lasers is utilized for
brass welding,spot welding,above all in precision
mechanics.
The cordon depth obtainable by sources of this
type is limited to 1.5mm typically "conductivity"
welding shapes are obtainable.
Laser at CO_2 is utilized continually passing from
conductivity to deep penetration welding with po<u>s</u>
sibility to weld by one passage steel 15-17mm thick
with 15 kW power.
Indeed,to these quantities other variables are also
obviously connected,light beam power being equal,
different welding,form and type of assistance gas,
of focus position compared with the surface of the
piece to be welded,the lenght of the focal lens,
the light power,the energy density,etc.
As one can see,for speaking of automatic welding
it is necessary to have a good knowledge, a not<u>a</u>
bly high control and management capacity of it;
therefore the need for automation to consider an
expert system supporting user's options.

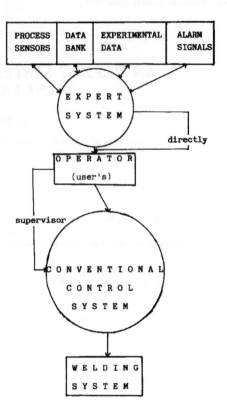

Fig. 2. Welding Expert System.

As we can see,the volume of knowledge for exec<u>u</u>
ting a correct welding,becames very large and it
seems almost necessary to apply to knowledge
system efficiently advanced.

ROBOTIZED WORK CELL

Furthermore,greater advantages derive from the
fact that robotized system allows the configur<u>a</u>
tion,the efficient services,and then also greater
possibilities of reference to different material
and process.
The welding methods for stainless steel are either
by impulsion or continuos.
An impulsed laser welding characteristics depend
upon average outgoing power and a single impulse
energy.The impulsed energy is referred to the me<u>l</u>
ted zone dimension and penetration whereas the <u>a</u>
verage power determines the frequency repetition
in case of spot welding or the cordon maximum ex<u>e</u>
cution speed in the continuous case.
Other variables may be considered,as the cordon
width in function of penetration,type of burner,
polarization,type of assistance gas,pressure and
flow rate.
The theoric study about some applications of
Expert System to the welding robotized work-cell
cross the focalization of a determinated method<u>o</u>
logy based essentially on there phases:modelling,
defining main problems and building suitable m<u>o</u>
dels;description,finding and evaluating the obta<u>i</u>
nable goals by other parameters too.
Near this theoric study,a research,restricted only
to a few significant parameters but very inte<u>re</u>

sting for the possible implications,could be deve
loped by just experienced methodology of Expert
System.
The model is supported by elements of Artificial
Intelligence,as database or Expert System,and this
lets the users to program correctly the activities
of workcell in perfect syntony with purposes of
production management and flexible manufacturing.
There is no doubt about the basic xonsideration
of workcell control inside the strategies that aim
to Flexible Manufacturing System management.
In fact it knows some case of workcell,consisting
of a numerically controlled machine tool,an in
dustrial robot,an automated fixturing system and
a material handling system,for which a specific
control hierarchy is supposed.
Moreover there are some approaches to the Artifi
cial Intelligence,used essentially into automatic
planning systems and for the control of robot cell
(especially in assembly)with error recover capabi
lity.
About the control and the management of robotized
work cells,many applications have been developed,
particularly about "gantry" robot.
It must let know the realization of laser-robot
workcell for jet engine component rework for which
a database is just incorporate into the software
to improve the materials and processes management.
The next phase,also in the examined case of
nuclear power plant components manufacturing,is
to study what an Expert System may give as a sup
port about the activities of production management
for welding robotized work cell,that is essential
ly a Flexible Manufacturing System.

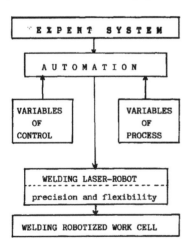

Fig. 3. Robotized Work Cell Configuration.

The consequences refer to construction of a pattern
through the use of Expert System,also for laser-ro
bot cell control and timing activity for production
management applications are the optimization in
work cell management under using profile of the
performances,manufacturing cycle and materials
choise.
It is also possible to obtain the objectives ca
pable to value Expert System introduction in a work
cell control(system definition and control hypothe
sis) and to extend control even to cell maintanance
problems(database,definition of main problems and
Expert Systems possibilities for safety).

CONCLUSIONS

The objectives in the expert systems for examina
ted welding process,as it is developed in the fol
lowing program listing,are essentially to "guide"
operator parameter selection.
Artificial Intelligence technology has much to of
fer the welding system automation.
The use of Expert Systems,the interesting develo
ped branch of Artificial Intelligence tools,offers
a set of behavioral characteristics for a welding
control system that can increase the quality and
reliability of welding activities and of all manu
facturing process scheduling and planning.
The Expert System Generative Process Planning
uses part features as input,correlates features to
manufacturing operations and to part operation se
quences and generates total process plan.
The solution methodology theoretical study on
system possibilities through verification of other
parameters,results including the simulation with
Expert System example for robotized work cell
timing activity.

REFERENCES

Davies,B.J. (1985).The integration of process plan
    with CAD CAM including the use of expert
    systems.CIRP Annals,pp.35-37
Cornu,J. (1988).Welding Automation,Vol. 2 & 3,
    IFS Pubblications,Springer-Verlag
Lane,J. (1987).International Trends in Manufactu
    ring Technology Robotic Welding,IFS Pubblica
    tions,Springer-Verlag
Bernold,T. (1987).Artificial Intelligence in Manu
    facturing,North-Holland
Aström,K.J.,J.J.Anton and K.-E.Arzén (1986).Automa-
    tica,Vol.22,No.3,pp.277-286

PROGRAM LISTING

```
1   model laser_welding
2   version "3.0,1989"
3   intro
4   "!this expert system is able for the"
5   "!estimation of welding speed for"
6   "!stainless steel,the investigation"
7   "!is restricted only to the following"
8   "!data:power=400w,thickness= 4 - 12mm"
9   firstarea control
10  constant limit=4.0
11  scale reporting_scale:"Reference"
12  "scale for welding speed"
13  from
14  "3000 mm/min"    1*limit
15  "1500 mm/min"    2*limit
16  " 400 mm/min"    3*limit
17  up to
18  "indeterminable"
19  area investigation:"welding speed"
20  action Joint_char:"characteristic"
21  consider welding_speed
22  also
23  advise"the recommended speed is"
24  assertion welding_speed:"calculation"
25  "of optimal welding speed"
26  using reporting_scale
27  rule wel_spe_cal:"This rule considers"
28  "the sheet thickness to determine"
29  "the optimal speed"
30  welding_speed depends on
31  mm 4           affirms3.0 denies 0.0,
32  mm 8           affirms8.0 denies 0.0,
33  mm12           affirms12.0 denies 0.0
34  assertion mm 4:"the thickness is 4mm"
35  default 0.0
36  question ask_mm4:"the sheet thick4mm"
37  obtain yesno mm4
38  assertion mm8:"the thickness  is 8mm"
39  default 0.0
40  question ask_mm8:"the sheet thick8mm"
41  obtain yesno mm8
42  assertion mm12:"the thickness is12mm"
43  default 0.0
44  question ask_mm12:"the sheet thi12mm"
45  obtain yesno 12mm
46  //***********advice*****************
47  area advice :"speed depending on"
48  "sheet thickness"
49  action mm4_advice:"speed for 4 mm"
50  advise"!N the speed is 3000 mm/min"
51  provided (mm4=4)
52  action mm8_advice:"speed for 8 mm"
53  advise"!N the speed is 1500 mm/min"
54  provided (mm8=8)
55  action mm12_advice:"speed for 12 mm"
56  advise"!N the speed is 400 mm/min"
57  provided (mm12=12)
58  //***********control****************
59  area control:"welding speed control"
60  action sequence:"speed sequence"
61  consider investigation,advice
    * * * quit
```

```
Current model is "laser_welding"
Version 1989
this expert system is able to estima-
tion of welding speed for stainless
steel
     --more(y/n) ? yes
the investigation is restricted only
to the following data:
power=400w
sheet thickness from 4 to 12 mm
the sheet thickness is 4 MM?
(you may answer yes or no):no
the sheet thickness is 8 mm?
(you may answer yes or no):yes
the sheet thickness is 12 mm?
(you may answer yes or no):no
the reccomended is 1500 mm/min
* * * That conclude the consultation
Current model is "laser_welding"
Version 1989
this expert system is able to estima-
tion of welding speed for stainless
steel
     --more(y/n) ? yes
the investigation is restricted only
to the following data:
power=100w
sheet thickness from 4 to 12 mm
the sheet thickness is 4 mm?
(you may answer yes or no):yes
the reccomended speed is 200 mm/min
* * * That conclude the consultation
Current model is laser_welding
Version 1989
this expert system is able to estima-
tio of welding speed  or focus posi-
tion for stainless steel
     --more(y/n) ? yes
the investigation is restricted only
to the following data:
power=100w to 10000w
focal lenght=63.5mm
sheet thickness from 1 to 10 mm
the sheet thickness is 2.5mm(depth)
the welding speed is constant?
(you may answer yes or no):yes
the reccomanded speed in continuous
welding   at 2000w is 2540 mm/min
the focus position is -1.3 mm
at 6000w is 10160 mm/min
the focus position is -2.5 mm
* * * That conclude the consultation
Current model is laser_spot_welding
Version 1989
this expert system is able to estima-
tion of  frequency or pulsed energy
the investigation is restricted only
to the following data power from
100w to 400w
     --more(y/n) ? yes
the power is 100w
the pulsed enrgy is 10J? (you..)yes
frequency is  10 Hz
* * * That conclude the consultation
```

PILOT EXPERT SYSTEM (ROBEX) FOR CONCEPTUAL DESIGN OF INDUSTRIAL MANUFACTURING TASK-BASED ROBOTS

V. R. Milacic, N. D. Covic and I. Z. Race

*Mechanical Engineering Faculty, Department of Production Engineering,
University of Beograd, Beograd, Yugoslavia*

Abstract. The paper presents the principal elements of a developed pilot expert system for knowledge engineering based conceptual design of robots - ROBEXP. The ROBEXP expert system suggests to the designer the solutions for construction variants and corresponding functional subsystems of industrial robots (IR), i. e. their components. The suggested solutions are explicated by ES at the demand of the user, while the final choise is made by the user, and the conceptual designing develops further "step by step".

Keywords. Robots; artificial intelligence; expert systems; conceptual designing.

INTRODUCTION

During the designing process of robots, the designer must be acquainted with different components of robots, as well as the dynamics and kinematics of the robotical structural parts. One of frequent approaches to robot designing is to seek the solution to problems through the comparison of robot specifications in the catalogues of their manufacturers. Such an approach, obviously, is a complicated and long procedure which does not guarantee successful results. This is due to the fact that in robot designing the geometrical properties of robots are not of primary importance, but some subjective decisions related to their application.

The initial demand put forward during the building of the ROBEXP expert system was to suggest to the man-designer the solutions for certain components of robots based on tasks to be executed by them. The main task of the ROBEXP system is to define the functional subsystems of manufacturing task-based robots. The building of the pilot ROBEXP expert system is founded on real data of existing solutions and manufacturing tasks for which robots will be used.

THE STRUCTURE OF ROBEXP SYSTEM

In the building of the ROBEXP expert system the natural procedure was to follow up the way in thinking of conceptual designing domain-experts, that is the meta-graph for conceptual designing of industrial robots as illustrated in Figure 1.

Each of the illustrated meta-graph nodes represents an entity, i. e. the set of entities in the IR conceptual designing process. The meta-graph nodes as non-terminate elements represent the knowledge base classes of the domain-expert, and thus of the ROBEXP pilot expert system. The nodes are connested with terminate elements representing operational structures passing from one to the other node. The knowledge symbolically represented with nodes, corresponds to the series of rules and facts manipulated by the expert for achieving desired solutions.

The representation of knowledge in the ROBEXP pilot expert system is based on production systems, so that the knowledge is defined as:
- declarative (represented by overall data base describing the problem under consideration-the facts in data base),
-- procedural (represented by production rules used for declarative knowledge manipulation-with the knowledge base rules), and
- control (containing control strategy for the solution of the problem under consideration).

The knowledge base obtained through structuring of corresponding data bases (Čović, Milačić, 1987a, 1987b) (containing the existing knowledge) about the technologies and methods and achieved IR solutions and methods thus forms the overall existing data base. Therefore in the ROBEXP expert system the knowledge required for IR conceptual designing is organised as:
- manufacturing task knowledge - IR processes,
- the knowledge related to IR functional subsystems based on developed solutions.

With such organisation of the knowledge the main designing entities and attributes required for IR conceptual designing are defined in the form of several hundreds of production rules and facts defining the knowledge about:
- IR type tasks,
- IR end effectors (gripper, tools-mini machines),
- IR kinematic subsystems (supporing structure),
- IR driving subsystem (drive+gear),
- IR measuring subsystem,
- IR sensor subsystem,
- IR control subsystem.

Starting from the IR conceptual designing meta-graph (Figure 1), the complex graph (Čović, 1988), Figure 2 , was made, the basis of which is the recognition, which is one of the fundamental knowledge blocks (Milačić, 1987).

Complying with the basic engineering designing activities the following general structure of ROBEXP system was made up of the following modules, Figure 3, (Milačić, 1987):
- the module for the recognition of type task-based designing requirements,
- the optimisation module,
- technologies and methods data base,
- the knowledge base, and
- the inference engine.

IR MANUFACTURING TASKS

The problem encountered in the beginning of the IR

conceptual designing is the identification of manufacturing task elements, i. e. the decomposition
of manufacturing task into type tasks. IR manufacturing tasks within the ROBEXP system represent
two sets of a series of type tasks decomposed to
the level of type tasks (Milutinović, 1980) as illustrated in Figure 4.

The relations in the tree describing this decomposition, Figure 5, are given in the form of facts
subordinated in the PROLOG language (in the examp
le of the decomposition of machines served by robot
for machining treatment, applications 1. 2. 3 in
Figure 5:
  subordinated ("Machines for machining treatment",
  1, "Lathes").
  subordinated ("Machines for machining treatment",
  2, "Machining centers")/
  subordinated ("Machines for machining treatment",
  3, "Boring machines"),
  subordinated ("Machines for machining treatment",
  4, "Machines for manufacture of gears and screw
  threading").
  subordinated ("Machines for machining treatment",
  5, "Grinding machines"),
  subordinated ("Machines for machining treatment",
  6, "Sheet metal cutting machines").

Type tasks are directly coupled with manufacturing
task sequences out of which only those of interest
to the IR conception system are considered. These
sequences represent IR designing requirements based
on the manufacturing task to be performed by the
robot, Figure 5. Project designing requirements are
established on the basis of the extensive recommendations from available literature. In PROLOG-language this connection in the case of the above machine for machining treatment is represented by the
following facts:

table ([1,2,3,1], [[ps,s,ps],[s,ps,s,s,ns], [ns,s,s],
  [s,s,ns]]).
table ([1,2,3,2], [[ns,s,ps],[s,s,s,s,ns], [ns,s,ns],
  [s,s,ns]]).
table ([1,2,3,3], [[ps,s,ns],[ns,ps,s,ns,ns],
  [ns,s,ns].[s,ps,ns]]).
table ([1,2,3,4], [[ns,s,ns],[ns,s,ps,ps,ns],[ns,s,
  ns,s,ns],[ns,s,ns]]).
table ([1,2,3,5], [[ns,s,ns],[ns,s,s,ps,ns],
  [ns,s,ns],[s,ps,ns]]).
table ([1,2,3,6], [[ns,s,ns],[ns,s,ps,ps,ns],
  [ns,s,ns], [ns,s,ns]]).

where the first element of the table structure is
the number of a corresponding type task, and the
other element of this structure the list with four
sublists containing corresponding designing requirements determined by type task in the following
order: the degree of freedom, robot configuration,
maximum speed and the positioning accuracy. The
symbols ns, s and ds in the list stand for non-
-standard, standard and partially standard designing requirement.

In the beginning of work of the ROBEXP system the
user is continuously asked about the details of the
type task for which the IR conceptual designing is
to be made. Each answer represents "a step forward"
on the basis of which a new question is put forward
until the type task in question is fully defined.
For the type task defined in this way IR designing
requirements are recognised which are then recommended to the user of the ROBEXP system.

On the basis of designing requirements defined by
the user the corresponding construction IR variant
is recognised and recommended. At the request of
the user each recommended configuration may be elaborated.

THE CALCULATION OF ORIENTATION MASSES AND TORQUES

For the selected robot configuration the orienta-

tion segment robot masses (5) are calculated, which
are regularly distributed along the IR supporting
structure, for the purpose of torque compensations
caused by gravitation forces. The calculation of
robot segment orientation masses is based on the
optimal structure of robot segments from its osci-
lation aspect (Čović, 1988, Nnaji, 1983, Demaurex,
Gerelle, 1979). The oscillation segment shapes directly depend on segment configurations, damping
and loading characteristics. In the calculation of
segment masses the real configuration of the robot
approximates its model, Figure 6. The points of concentrated masses are located in the model, while
joint rigidity are exchanged with springs of corresponding rigidity. Approximative speeds and accelerations, and positioning freedom degrees are
calculated on the basis of the maximum speed of the
robot recommended for a given manufacturing task.
With known maximum speed and the length of segments
maximum angle speeds of the positioning freedom
degrees are obtained. The user states desired acceleration time for positioning freedom degree axes,
and with the knowledge of the speed the acceleration is easily obtained. Approximative torques in
joint axes are calculated with Lagrange second order equations for extreme robot positions
For the system illustrated in Figure 7 the kinetic
and potential energy is calculated through the definition of the most unfavourable IR position thro-
ugh $\Theta 1$, $\Theta 2$ and $\Theta 3$ angles. The torque in i-th joint
is calculated in the following way:

$$M_i = M_{gi} + M_{i_i} + M_{c_i} \tag{1}$$

where is:

$$M_{gi} = \frac{\partial P}{\partial \Theta_i} \tag{2}$$

and P-the potential energy of the system;

$$M_{i_i} = \frac{d^2}{dt^2} \Theta_j \text{ , inertion forces torque;} \tag{3}$$

$$M_{ci} = C_{ijk}(\Theta) \dot{\Theta}_j \dot{\Theta}_k, \text{ Coriolis acceleration torque} \tag{4}$$

(Coriolis force includes centripetal force).

In order to calculate the tourqe, particular torques for each joint-connestion have to be added o.e.:

$$M_{i_n} = \sum_{j=1}^{n} M_{ij} \tag{5}$$

where M_{ij} is the torque in the link (axis)
due to segment j

The torques calculated in this way represent one of
designing requirements in the selection of robot
driving groups. The calcularion of masses, speeds,
accelerations and torques is an algorhytmic part of
the RCBEXP expert system.

THE SELECTION OF IR FUNCTIONAL SUBSYSTEM COMPONENTS

After the definition of project recuirements it is
necessary to decompose the same to corresponding IR
subsystems, followed by their composition and cop-
marison with IR subsystem solutions already existing in the world. With the aid of designing requirements the decomposition of the gripper/tool is
first undertaken, to define accurately those which
satisfy proposed requirements. After that follows
the composition of subsystems in order to find the
final solution of the gripper/tool.

DESIR-I (Race, 1988) was used as shell general diagnostic expert system for the selection of IR subsystems (drive, measurement, sensor and control,
that is their components). The selection of individual components is becoming particularly complex as
it is necessary to "reconciliate" different, frequently contradictory requirements. It is also neces-

sary to bear in mind all additional requirements established during the previous part of the robot conceptual designing process. In the ROBEXP expert system, the choise of individual solutions are structured within each subsystem in the form of a tree (Nnaji, 1983). Figure 8 shows the decision tree for the choise of the drive subsystem type.

To each node in the tree a corresponding table is attached, Table 1, which contains the types of motors with frequency percentages (P(A)). These frequencies represent the percentage of the application of corresponding drive types in industrial robots. The frequency percentages (P(B)) ask for one characteristic out of the total sum of user's demands relating to specific robot characteristics. As during the inference process the probability share of speific drive types are known, i. e. the probability facts, it is necessary to calculate specific conclusion probabilities on their basis, and thus the solution probabilities as well. The apparatus for the operation with such formulas is Bayes conditional probability formula. With it is inferred that in addition to a priori specific choise probability two conditional probabilities should be known as well, i.e.: $P(B/A)$ and $P(B/\neg A)$. The relationship between these probabilities is called sufficiency factor-S. N-factor, or the necessity factor is the probabilitiy relationships $P(\neg B/A)$ and $P(\neg B/\neg A)$. With high S factor B-event becomes sufficient condition for the occurrence of A-event. With low N-factor B-event becomes necessary condition for the accurence of A-event. Sufficiency factor should corespond with requirement and choise connection quoted in the previous table. On the basis of these tables, the existing data, and the acceptance of certain information provided by the user, ROBEXP system proposes to the user one of the choices (one tree branch linking the node under observation with its sons). In such cases ROBEXP is solving the following problems:
- finding of the most selective question in the given situation at that moment,
- the acceptance of the answer provided by the user elaborating requirements which may be of different intensity,
- finding of a solution in case of incomplete data,
- the adoption of a solution after being "sufficiently" convinced into its correctness, without l listing of the entire repertory of questions,
- omitting of a specific question if the answer is already known,
- providing of an adequate explanation at the request of the user,

In this way hierarchical method of thinking of the human expert in the selection process is simulated.

The rules connecting specific request and corresponding solution have the following form in the ROBEXP system:

IF 〈choise title〉 THEN〈P1〉 〈P2〉 ARE〈request title〉 ;

The solution facts relating to the choise of driving subsystem have the following form:

FOR SOLUTION 〈request title〉 THE PROBABILITY IS〈P〉

FOR REQUEST 〈request title〉 THE QUESTION IS 〈question relating to request〉 ;

These rules in the ROBEXP system are translated into corresponding PROLOG-language notation, and may be modified when necessary. These rules are then one after the other translated into facts in the language of the PROLOG form:

request_choise(〈choise title〉,〈request title〉, 〈P1〉,〈P2〉).
the_solution(〈choise title〉, 〈P〉),
the_question(〈request title〉,〈request-based question〉).

With this solution the use of the so called "hard"

knowledge is avoided, which will impose the repeating of the same procedure regardless of the user's answers. During its work the system attempts to follow up the through process of the expert for conceptual robot designing, ant to simulate his reasoning activity. In this way the "building up" of a corresponding process of through is achieved, during which some answers provided by the user are in favour and some against the established hypothesis.

The solution of the selection problem develops in cycles. The question relating to the circle of final choice candidates at each given moment, is being selected until the final criteria (there remains only one candidate for the final choice or there are no more questions) is satisfied. The selected question is put forward and the answer to it is whether the request under consideration, in the given case, is important for the selection considered.After the acceptance of the answer, which may be differently ranked, the choise probabilities in the circle of candidates for the final answer-based selection are up-dated. This condition reflects the "reasoning" about the suitability of specific choices at a given moment. At the end of the cycle, out of the final choice candidates are excluded the selections without prospects to be included into the user's answer-based final selection relating to tasks to be executed by the robot.

In this way the system is enriched with positive properties of the DESIR-I shell, which thus provided a good counselor for choices made in individual nodes. Together with retaining of the general strategy of the robot conceptual designing system, the system which may serve as an "intelligent" advisor in the selection of robot components was obtained, because it is able to elaborate its standpoint for each recommended choice. Normally, the role of the expert system in this case is advisory, because the final decision at each node is made by the man (who may decline the recommended choice made by the system).

Possible dialogue (the form of the dialogue depends on user's answers) between the ROBEXP system and the user during the selection of the driving subsystem may have the following form (the answers of the user during interactive work in the system are underlined):

Drive subsystem

Are small dimensions of the drive required?: yes
Does operation takes place in explosive environment?: why.

THE EXPLANATION OF THE QUESTION

The question - Does operation takes place in explosive environment?

was put because
- my opinion at the moment is that the selection - electrical - according to the available data until present was almost sure and the probability of positive answer to the question
if thus choise is selected is 0
while the probability of a positive answer to the same question
if the above has not been selected is 0.12

Repeating the question:
Does operation takes place in explosive environment?: no
Is the price a decisive limiting factor?: ignore
Does work takes place in dirty environment?: -2

Suggested choice(s):
electrical is very probable choise.

Is explanation of the answer required?: yes

1 - hydraulic
2 - pneumatic
3 - electrical   recommended

For which choice is the explanation required?: 3

EXPLANATION
SUGGESTED

Choice - electrical - a very probable choice

I think this because:

- The question - Are small dimensions of the drive
                 required?
                 was answered with yes

- The question - Does work takes place in very ex-
                 plosive environment?
                 was answered with no

- The question - Does work takes place in dirty
                 environment?
                 was answered with - 2

Against such opinion are the facts that:
- The question - Required carrying capacity for the
                 task?
                 was answered with 35

No effect on the reasoning about this choice had
- Is the price a decisive limiting factor?
     the answer   ignore it

Is explanation of the answer required?: no

Decide for one of the proposed solutions:

1 - hydraulic
2 - pneumatic
3 - electrical   recommended

Your choice is: 13

Decide for one of the proposed solutions:

1 - hydraulic
2 - pneumatic
3 - electrical   recommended

Your choice is: 3

Type of electrical motor.

CONCLUSION

The inference engine developed in the ROBEXP system takes over some of the designer's activities during the IR conceptual designing. The rules in the knowledge base may be enlarged, modified, added to and eliminated. These rules are defined in the form to the PROLOG-language which may automatically examine already passed procedure, which facilitates the explanatory process of decisions reached. As the structure of ROBEXP is modular, and represented in the form of the rules in PROLOG-language, it has greater flexibility is easy to use and accelerates the type-task IR designing process.

REFERENCES

Demaurex, M.D., and Gerelle, E.G.R. (1979) Can I Build this Robot, 9th International Symposium on Industrial Robots, Washington.

Milutinović, D., (1980), Systems of Industrial Robots and Manipulators for Machine Tools, M.Sc. thesis, Beograd (in Serbo-Croatian).

Nnaji, B.O., (1983), Computer-Aided Design of Robots. Ph.D. Thesis, Virginia Polytechnic Institute and State University, Blackburg, Virginia.

Milačić, V., (1987). Manufacturing Systems Design Theory, Production Systems III, Mechanical Engineering Faculty-Beograd University, JUPITER Association, Beograd (in Serbo-Croatian)

Čović, Dj.N., and Milačić, V., (1987a). Robot-Selector, 13th JUPITER Conference, Cavtat (in Serbo-Croatian), 299-314.

Čović, Dj. N., and Čović, Dj.,N., (1987b). Selection of Robot DC Servo Motors by Extreme Position Method, 13th JUPITER Conference, Cavtat (in Serbo-Croatian), 315-322.

Čović, Dj. N., (1988). The Development of the Pilot Expert System for Knowledge Engineering-Based Conceptual Design of Industrial Robots-ROBEXP, M. Sc. thesis, Mechanical Engineering Faculty, Beograd (in Serbo-Croatian)

Race, I.Z., (1988). Program Implementation of Diagnosis Expert System in PROLOG Language, M.Sc. thesis, Faculty of Natural Sciences and Mathematics, Beograd (in Serbo-Croatian)

TABLE 1 Drive System types

Characteristics	Type P	Hydraulic 30	Pneumatic 10	Electrical 60
Load	60	large	medium-large	medium-small
Dirty environment	40	no effect	no effect	avoid
Explosive environment	40	no effect	no effect	at no cost
Dimensions	50	large	small	small
Exploit. level	65	small	medium	high
Price	30	high	high	low
Noise	20	high	medium	low
Sealing loss	20	yes	low	no
Clean operation	40	no	yes	yes

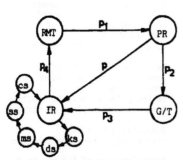

RMT - robot manufacturing task
PR - project design
      requirements
G   - gripper
T   - tools
IR - industrial robots
      ks-kinematic subsystem
      ds-drive subsystem
      ms-measuring subsystem
      ss-sensor subsystem
      cs-control subsystem

Fig. 1. Meta-graph for conceptual designing of task based
industrial robots (Čović, 1988).

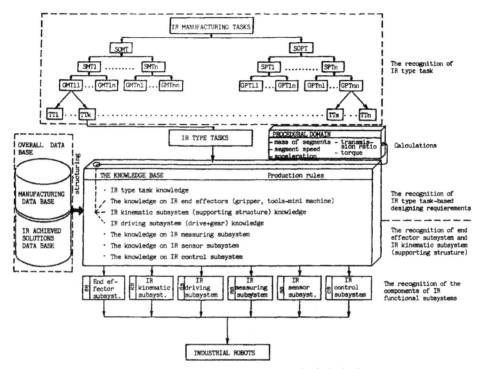

Fig. 2. Complex graph of IR conceptual designing

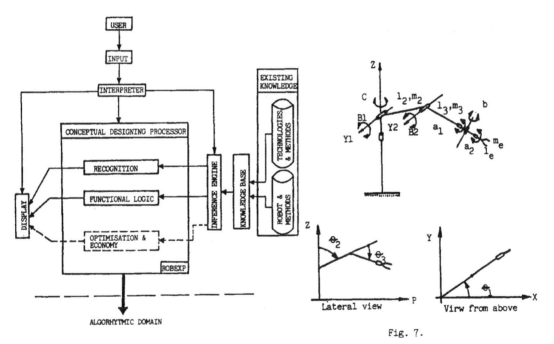

Fig. 3. The structure of ROBEXP system.

Fig. 7.

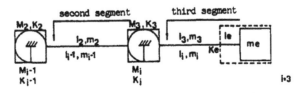

Fig. 6. The model of the robotical arm with two segments

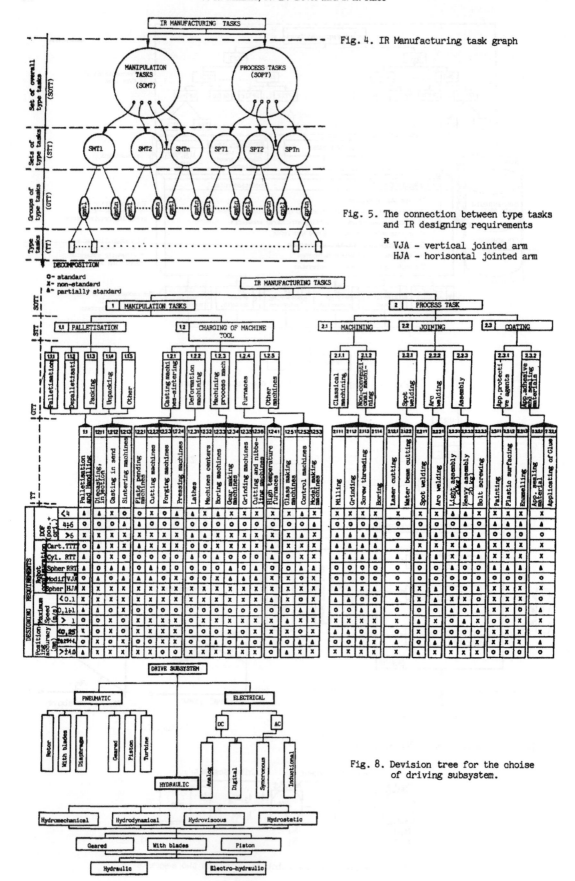

Fig. 4. IR Manufacturing task graph

Fig. 5. The connection between type tasks and IR designing requirements

X VJA – vertical jointed arm
  HJA – horisontal jointed arm

Fig. 8. Devision tree for the choise of driving subsystem.

DESIGN OF DECISION-MAKING ENGINE IN KNOWLEDGE ASSISTED PROCESS PLANNING SYSTEM (KAPPS)

K. Iwata* and Y. Fukuda**

*Department of Mechanical Engineering for Computer-controlled Machinery,
Osaka University, Osaka, Japan
**Technical Research Institute, Japan Society for the Promotion of Machine Industry,
Tokyo, Japan

Abstract. Process Planning is recognized as the critical problem between design and manufacturing in the Computer Integrated Manufacturing (CIM) System. Recently many types of Computer Aided Process Planning (CAPP) systems using expert system technique have been developed.
However, most of the systems developed so far lack an ability to handle a knowledge and know-how of the experienced process planners. To solve this problem, the new decision-making engine of process planning is designed through the interviewed result and its analysis of thinking process and procedures of the experienced process planners. This engine is consisted of three main problem solveres, such as the pattern matching, constraints searching and sequencing.
The engine can be written by using common-lisp and is installed at Knowledge Assisted Process Planning System (KAPPS) which was developed by our laboratory.
It is shown, through the case studies, that the engine designed here is effective to determine the process plan in machining.

Keywords. Knowledge engineering; Computer-aided process planning; Machining; CIM; Expert systems;

INTRODUCTION

In the discrete parts manufacturers, much attention has been paid to constitution of a Computer Integrated Manufacturing (CIM) system. In realizing the CIM system, the critical problems to be solved are summarized as follows:

(a) Development of integrated system of Computer Aided Design(CAD), Computer Aided Manufacuturing (CAM) and Computer Aided Production Management.

(b) Development of Computer Aided Process Planning (CAPP) which can adequately handle the know-how and knowledge of experienced process planners,

(c) Development of a data-base which can effectively store and retrieve the dynamic status data and the know-ledge in CIM,

and

(d) Development of effective local area networks and communication systems which can integrate data and inter-face among design, production, and production management.

In order to solve the items (a) and (b) above described, a number of process planning systems have been developed throughout the world in this decade (Evershiem 1985). However, most of these systems could not sufficiently processed the know-how and knowledge on which experienced process planners have.

Recently, several process planning systems have been proposed, which can consider the know-how and knowledge accumulated by process planners (Matushima, 1984; Descotte, 1985; Iwata, 1985). However, utilization of know-how and knowledge in these systems were rather limited, since it was described by simple production rules.

In order to realize the practical process planning system, system should be included the following abilities, such as the representation of knowledge which process planners have, the knowledge base system, and the engine for re-utilizing process planners procedures.

The objectives of the research work is to design a decision-making engine which can handle experienced process planners' know-how and knowledge in determining of process plans. This paper defines the role of knowledge in process planning, proposes the system configuration and basic procedure of decision making engine, and evaluates the feasibility of the proposed system through case studies.

THE ROLE OF KNOWLEDGE AND PROPOSED KNOWLEDGE ASSISTED PROCESS PLANNING SYSTEM

A great deal of knowledge and know-how is required to carry out the process planning. The knowledge and know-how is considered to be composed of the experience and intellect in production engineering. The knowledge and know-how belongs essentially to the brains of experienced process planners and the skilled workers.

The process planning process has been investigated through interview with the experienced process planners from fifteen Japanese major discrete manufacturers. It has been made clear that the experienced knowledge and know-how plays an important role in process planning . It is also recognized through the analysis that the know-how and knowledge must be considered at least of eight decision-making problems in process planning (Iwata, 1986).

These decision-making problems are as follows;
(1) Recognition of machining surface,
(2) Recognition of rough shape of machining parts,
(3) Selection of reference surface for machining,
(4) Determination of preference relation for machining,
(5) Selection of machine tools,
(6) Selection of cutting tools,
(7) Determination of cutting conditions,
(8) Selection of jig and fixtures

522 kinds of know-how and knowledge were collected from the interviews. From the analysis of data on know-how and knowledge, it has been recommended that the combination method of frame and production rule is comparatively useful as method of knowlwdge representation in process planning (Iwata 1988).

Based on the analysis, a new type KAPPS (Knowledge and Know-how Assisted Production Planning System) has been proposed, and the some detailed configuration of knowledge base was discussed in separate paper (Iwata, 1987). KAPPS consists of the know-how base

in applying both frame and production rule, the CAD interface which connects the solid modeling system for machining parts, the know-how acquisition system which can acquire the knowledge from the experienced planners by using the dialogue technique, and effective decision-making engine for process planning. Figure 1 shows the system configuration of KAPPS.

DECISION-MAKING ENGINE IN KAPPS

One of the important items on developing the KAPPS is to design the decision making engine which can re-utilize the decision process of experienced process planners exactly.

The decision process of experienced process planners has generally some specific features. For examples;
(1) The experienced process planners can consider both the knowledge and know-how on both the procedures and the status of planning process.
(2) The experienced process planners change the process procedure to fit the dynamic status of parts/products and shop floor.

In order to satisfy the above features, decision making engine were newly developed. The engine consists of kernel, eight solvers and three modules which were programmed by LISP basic functions. The system configuration of decision making engine is shown in Fig.2.

The kernel directs the decision process based on the status of the temporary file and knowledge of control. Knowledge of control is stored in know-how base at KAPPS. The kernel has the ability to determine the best sequence of decision processes based on planning process, the engine can randomly process the procedure of process planning.

In the kernel, the following steps are executed.
Step 1 Generate Temporary File.
Step 2 Define the decision-making problem based on the status of the Temporary File.
Step 3 Determine the procedure to solve a problem, and select the solver. The selected solver make a solution and return to kernel.
Step 4 Up-to-date the temporary file according to the solution.

The temporary file (TPF) is a file which includes parts shape data, production information of planned parts and the intermediate result. This file plays an important role in the decision-making procedure of KAPPS.

Eight solveres can execute each decision-making problems and make an adequate solution by applying the knowledge and know-how in the given knowledge base.

BASIC PROCEDURE IN DECISION MAKING ENGINE

In order to realize the proposed engine, the analysis of decision-making in process planning were carried out. The eight decision-making problems described in the second section are classified into the following three categories from the standpoint of the characteristic of the problems, which are ;
(1) Pattern matching,
(2) Selection,
(3) Sequencing.

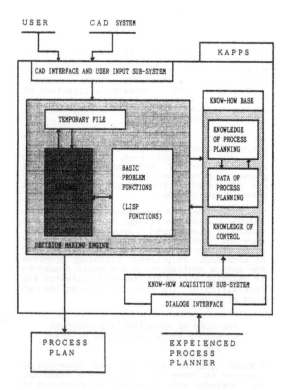

Fig. 1 System configuration of KAPPS

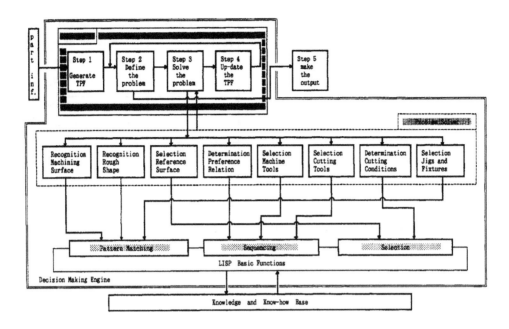

Fig. 2 Configuration of decision-making engine

Pattern matching is a problem that whether or not the predetermined pattern is included in the pattern of part to be machined. The predetermined patterns are useally formed by the knowledge of the experienced process planner and are stored in the knowledge base by using the frame representation method in this system..

This problem can be solved by the following steps, as shown in Fig. 3.
- Step 1 Generate the standard pattern from the predetermined pattern stored in the knowledge base, considering the charastristics of process planning problems to be solved.
- Step 2 Extract the attributes set from the standard pattern.
- Step 3 Extract the attribute set from the part data.
- Step 4 Compare between attribute values of standard pattern and the part data.
  If both attribute value is matched, return "true" to top-level, but if not, return "nil" to top-level.

This procedure can be applied to the problems for the determination of the recognition of machining surface, the recognition of rough shape, and the selection of jig and fixture in the process planning.

The problem of selection is to reflect the sets which can be satisfied the requirements and constraints. The requirements and costraints are stored as the data of experienced knowledge in the knowledge base. Each requirement and constraint is represented by form of frame, referring to the environmental conditions.

This problem can be solved by useing the algorithm that is similar to the constraints search (Baykan 1987). The flow chart of this procedures is shown in Fig 4. Main procedures steps are as follows;

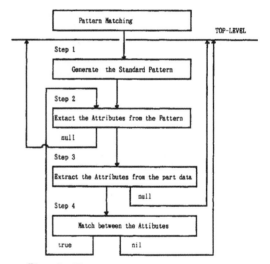

Fig. 3   Flow chart of the pattern matching module

- Step 1 Generate a conditions set (Ks) from the knowledge base according to the environment of the problem.
- Step 2 Select a condition (Ks) from the set (Ks) based on the priority value.
- Step 3 Make a sub-set (Sn) based on the selected condition. Iteration from step 2 to 3 is continued until the set (Ks) is null.
- Step 4 Make a sub-set (Q) by using the product set. In case that Ks has n conditions, the iteration times is n. In case that sub-set (Q) is null, go to step 5. But, if not, return subset (Q) to top-level.

Step 5   Each of the following two modes is selected based on the knowledge of control.

Mode 1
Mode 1 is used when the condition cannot be relaxed. In this case, return to top-level the "null" and the list name when Q is null.

Mode 2
Mode 2 is used when the condition is able to relax and the conditions have a priority for relaxation.
According the priorities for relaxation, conditions are relaxed until the solution be found.

This procedure can be applied to the problem for the determination of the the selection of reference surface and determination of cutting conditions in process planning.

The problem of sequencing is to transform the given set to the well-ordered set. To be solved the problem, the precedence relation between attributes are needed. The experienced process planners have much knowledge of the precedence relations between attributes and it's restrictions. In this system, the precedence relations and it's restrictions are formed by using the frame representation method and are stored in the knowledge base.

The flow chart of the procedure to be solved this problem is shown in Fig. 5. Maine procedured steps are as follows;

Step 1   Extract the attributes to be ordered from the part data to be machined. And the extracted attributes are listed in the working list.
Step 2   Make a set of paired attributes in the working list.
Step 3   Decide the precedence relation between the paired attributes by

Table 1 The Correlation between basic problems and process planning problems

Basic Problem	Process Planning Item
Pattern Matching	Recognition of Machining Surface Recognition of Rough Shape Selection of Jig and Fixture
Selection	Selection of Reference Surface Determination of Cutting Condition
Sequencing	Determonation of Preference Relations Selection of Machine Tools Selection of Cutting Tools

using the knowledge in the knowledge base.
Iteration between step 2 and 3 is continured until all precedence relation between the paired attributes are determined

Step 4   Make a well-ordered set of attributes in the part by using the precedence relations between the paired attributes. And return to top-level the working list which includes the well-ordered set.

This procedure can be applied to the problems for the determination of preference rerations, the selection of machine tools, and selection of cutting tools in the process planning.

The correlation between the three basic modules in the decision making engine and the eight decision-making problems in the process planning is summarized in Table 1.

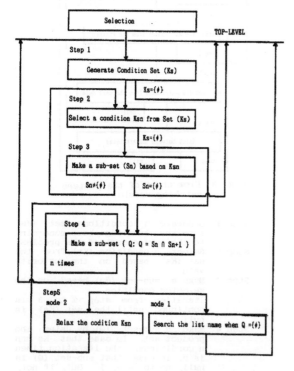

Fig. 4   Flow chart of the selection module

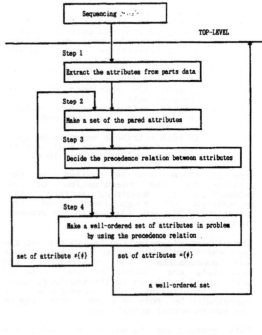

Fig. 5   Flow chart of the sequencing module

Three basic modules were developed as the general purpose algorithm by using the common-LISP language, and were put into the decision making engine of the process planning system.

CASE STUDIES

In order to evaluate the effectiveness of the proposed engine, the engine is installed in the KAPPS and typical cases in process planning were studied.

The KAPPS is run on Tektronix 4405 (CPU; 68020 32 bit processor, 5Mbyte), the knowledge base is constructed on the external memory (40Mbyte, Fixed Disk Drive), and 522 kinds of collected knowledge were stored in the knowledge base. The engine and know- ledge base are programmed by common-LISP, the man-machine interface in KAPPS is programmed by smalltalk-80.

In the first case study, the rotational part

Table 2   The result of the second Case Study (prismatic Part)

	Sequential	Random
01	Rough Shape	Rough Shape
02	Machining Surface	Machining Surface
03	Reference Surface	Reference Surface
04	Pref. Relation	Pref. Relation
05	Machine Tools	Machine Tools
06	Cutting Tools	Machining Surface
07	Cutting Condition	Pref. Relation
08	Jig and Fixture	Cutting Tools
09		Cutting Condition
10		Jig and Fixture
11		Machine Tools
12		Cutting Tools
13		Cutting Condition
14		Jig and Fixture
A	183 sec.	473 sec.
B	187 frame	248 frame

A: processing time (except input time)
B: number of knowledge frame

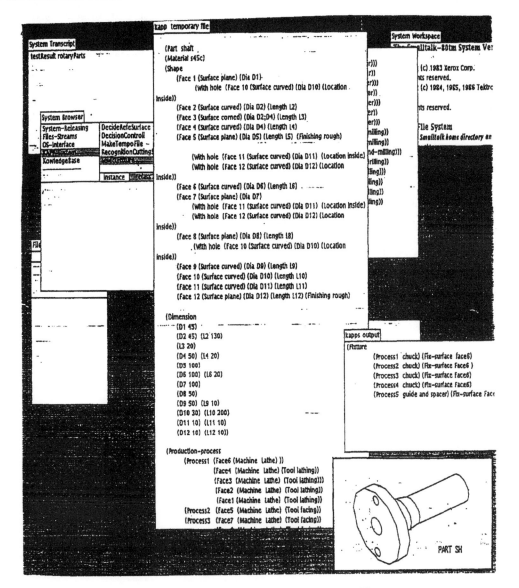

Fig. 6   Case study output (rotational part)

and the prismatic part were used in order to
evaluate the effectiveness of the decision-
making engine, and all problem of process
planning in both parts were solved. One of
the case studies outputs is shown in Fig. 8.
These results are almost the same as those
determined by the experienced process
planner. In the prismatic part, the
processing time was about 3 minutes, and 187
kinds of knowledge were used. In the
rotational part (Fig. 8), the processing time
was about 2 minutes, and 102 kinds of
knowledge were used.

In order to re-utilize the procedure of
experienced process planner, the second case
study was executed. In this case study, the
prismatic part was used, and the usage
machine of partial processes were
predetermined. The KAPPS and the engine
could determine the process planning on
considering the status of temporary file.
The process of determination is shown in
Table 2. In this table, "Sequential
Procedure" means a normal procedure of
KAPPS, "Random Procedure" means the
procedure for re-utilizing the procedure of
experienced process planner.

The result is almost the same as decision-
making procedures of the experienced process
planner.

According to the results of two case
studies, the engine and KAPPS can replay the
process planner's decision making procedures
and they can apply to a practice problems.

CONCLUSIONS

The following conclusions can be obtained;
(1) It has been found through the analysis
of decision process of experienced
process planner that the process
planning contains the three basic
modules, which are the pattern
matching, the selection, and the
sequencing.
(2) The decision-making engine in process
planning was developed by using the
basic problem functions, which was
programmed by the common-Lisp
language.
(3) The knowledge based process planning
system in which the engine proposed
is installed based the engine has been
developed and the effectiveness of the
proposed system was shown through the
case studies.

REFERENCES

Baykan C.A., and M.S.Fox, An Investigation
of Opportunistic Constraint Satisfaction
in Space Planning, Proc. of IJCAI-87,
1035-1038.
Descotte,Y., and J.C.Latombe (1985), Making
Compromises among Antagonist Constraints
in a planner, Artificial Intelligence,
27, 187-217.
Eversheim,W., and J.Schulz (1985), Survey of
Computer Aided Process Planning Systems,
Proc. of 1st CIRP Working Seminar on
CAPP, 1-7.
Iwata,K., and N.Sugimura (1985), An Inte-
grated CAD/CAPP System with Know-hows on
Machining Accuracies of Parts, Proc. of
the ASME Winter Annual Meeting, PED-19.
Iwata,K., and Y.Fukuda (1986), Representa-
tion Know-how and Its Application of
Machining Reference Surface in Computer
Aided Process Planning, Annals of the
CIRP, 35-1, 321-324.
Iwata,K., and Y.Fukuda (1987), KAPPS: Know-
How and Knowledge Assisted Production
Planning System in the Machinig Shop,
Proc. of 19th CIRP International Seminar
on Manufacturing Systems, 287-294.
Iwata,K., and Y.Fukuda (1988), An Approach
of dynamic Process Planning for Very
Small Batch Production in Machine Shops,
Proc. of 3rd International Conference on
Computer Aided Production Engineering,
66-75.
Matushima,K., N.Okada, and T.Sata (1984),
The Integration of CAD and CAM by
application of Artificial Intelligence
Techniques, Annals of the CIRP, 31-1,
329-332.

SCLES: STRATEGIC CONTROL LEVEL
EXPERT SYSTEM FOR INDUSTRIAL ROBOTS

V. Devedzic

"Mihailo Pupin" Institute, Beograd, Yugoslavia

Abstract. - Expert system approach to the strategic control level synthesis for industrial robots is described. A short introduction to the hierarchical robot control and the concept of the strategic control level are presented, and a particular expert system implementation is discussed. The main task of the proposed expert system is to select the appropriate model and algorithm to be used to execute the given robot task at the strategic control level. The system has two modes of operation: off-line and on-line. An example of the proposed system application is also presented.

Keywords. Expert systems; robots; on-line operation; hierarchical systems; strategic control level.

INTRODUCTION

Robot control in industrial applications is usually organized hierarchically, in several control levels. An example of hierarchical robot control system organization can be found in Albus et al. (1983). Simple robot tasks, like moving the robot hand from one point to another, part grasping, compliant motion, simple assembly tasks, etc., are executed under the control of *the tactical control level* (TCL). On the other hand, more abstract and more complex robot tasks can be controlled by *the robot control level* (RCL). In general, more complex tasks, which are controlled by higher control levels, can be decomposed in sequences of simpler tasks, which are then executed by lower control levels. For example, relatively complex task at the RCL, like: "Move the part from machine 1 to machine 2", is decomposed by the RCL in a sequence of elemental robot moves needed to complete the given task. The corresponding sequence of simple commands is then issued to the next lower control level (possibly the TCL), which is responsible for further execution of each simple task. For the RCL command mentioned above, this sequence might look like: "Move the hand from position (x1,y1,z1) to position (x2,y2,z2)"; "Grasp the part"; "Move the part to position (x3,y3,z3)"; "Release the part"; etc.

A number of control algorithms have been proposed for various simple robot tasks at the TCL (see Stokić and Vukobratović, 1986). All of these algorithms are based on certain models of robots and their workspace, and each model/algorithm has both good and bad features. For example, the results from applying a certain model/algorithm are achieved with high precision and uncertainties are reduced to a large extent. However, it takes too long to compute the solution, or vice versa. In general, no model/algorithm exists which could be qualified as the best one for the majority of cases, and from every point of view. Applications themselves, and particular conditions under which robot tasks are executed, can influence a great deal the judgement on what control algorithm in a whole class of algorithms would be the most convenient to apply to execute every single task.

Stokić and Vukobratović (1986) have proposed an expert system (ES) approach to solve the problem of selecting the appropriate model/algorithm to control the execution of simple tasks at the TCL. In their approach, this problem, as well as several other robot control problems, can be solved by another control level, which is inserted between the RCL and the TCL - *the strategic control level* (SCL). An overview of this approach is presented in the following section.

This paper describes an expert system SCLES, which has been implemented to synthesize a part of the SCL. SCLES architecture and the application example are presented in the subsequent sections.

STRATEGIC CONTROL LEVEL

The main idea introduced by Stokić and Vukobratović (1986) is to insert the SCL between the RCL and the TCL, and to combine several techniques, i.e. several models and algorithms in solving every simple robot task, in order to solve these tasks more efficiently, taking real application requirements into account. The idea is to have a whole data base describing various techniques, approaches, models and algorithms, and to have an expert system that makes decisions on which model and algorithm to use in every particular case, according to the particular application, current process status, and current robot task at the higher control level. This approach is particularly suitable for synthesizing the SCL in robot applications requiring high degree of flexibility, like when using robots as parts of flexible manufacturing cells (FMCs).

The block diagram of the SCL using this approach, previously discussed by Stokić and Vukobratović (1986), is presented in Fig. 1. According to this concept of the SCL synthesis, the SCL performs its function by:

* Selecting the optimal models of the robot and the workspace, as well as the optimal algorithm to execute the given task in every particular case. Selections are made from the specific models and algorithms data base (see Fig. 1). The selections are optimal in the sense that more accurate results can be achieved with the selected model and algorithm *in the estimated time for solution computation and task execution*, than with any other model and algorithm suitable for the same task. Depending on the particular task and particular algorithm, solution computation and task execution using that solution may be separated.

* Decomposing the higher level command into the sequence of simpler commands, through the process of trajectory planning; the sequence of simpler commands is passed on to the tactical control level for further execution.

* Identifying the actual system parameters during the system operation, and improving models and algorithms in the models/algorithms data base with the actual parameter values.

In general, models and algorithms that are contained in the SCL models/algorithms data base are procedural programs and the corresponding records which describe their features. When more then one model/algorithm could be applied, SCL automatically selects the best solution at the moment, to satisfy the requirements of the given task.

SCL should also be capable of improving particular models during the system operation, thus introducing some basic concepts of learning in the control system. There is another control domain at the SCL when learning is also used. If the system has to solve a task for the first time, usually some simple algorithm is applied, in order get to a satisfactory solution as fast as possible. The solution is memorized, and if the task is repeated later, which is often the case when robots are applied in actual FMCs, the system needs not to compute the solution again, it can directly apply the already known solution. However, the system can make an effort to achieve a better solution, knowing the previous one and then applying some more complex algorithm. If a better

solution can be computed during the estimated time allowed for the solution computation, the new solution replaces the old one in memory, and the next time the task is repeated, the system will already have an improved solution. This may gradually lead to obtaining the best solution possible with the models and algorithms existing in the data base.

The way the proposed system works can be described using Fig. 1. When it gets a higher level command, the first SCL sublevel is activated. It is implemented as an ES (a knowledge based system (KBS)). The ES identifies the type of the particular task, estimates the time that can be spent to compute the solution and to execute the task, and selects the optimal model and algorithm to complete the task, according to the task type and the estimated time. KBS performs these tasks using the knowledge base, which contains decision rules describing problem solving heuristics for the particular robot application (for example, as a part of a particular FMC). The solution is transferred to the second sublevel, where the trajectory planning is performed, according to the selected model and algorithm. The solution is then sent to the lower (tactical) control level. The computation time at the second sublevel can be significant. Therefore, solutions for particular task classes are memorized at this level. If the task is repeated, the memorized solution can be transferred directly to the tactical level by the request of the first sublevel (ES), or the solution can be recomputed using more accurate model/algorithm, if the time constraints are satisfied. In this way the system can gradually learn the best solution possible. This is the objective of the system's second learning function, which is performed by ES.

The solution from the second sublevel is transferred to the third sublevel as well, which also receives the feedback information from the tactical level. Comparing the requirements from the first two sublevels to the actions actually performed by the robot, the third sublevel can improve the model/algorithm parameters. This is another objective of the system's learning function.

SCL is used to solve tasks in real time. It is necessary to provide the SCL with the permanent information about possible changes of the specific features and parameters of the system and the workspace, and about the status of the FMC process. Therefore, the sensor system information is continuously updated and examined at every SCL sublevel.

In this concept of the SCL synthesis, the point is to achieve the most convenient tradeoff between the accuracy and the precision of the task solution required at this control level, and the computation times and the computer system capacity.

SCLES DESCRIPTION

One particular program implementation of the first SCL sublevel - KBS - is described in this section. It is called SCLES (Strategic Control Level Expert

System), and it has two basic modes of operation. In *development and simulation mode (off-line mode)*, the user can define or change (in an interactive procedure)the descriptions of the objects in the system and the rules in the knowledge base. The user can also perform simulations of various SCL tasks in this mode. During these simulation experiments, SCLES may may require additional information from the user if the solution can't be found. The user may require the system to explain its reasoning during the inference process.

In *real time operation mode (on-line mode)*, all the processes featuring the off-line mode that slow down the system operation are eliminated. In this mode of operation, the quantity of knowledge built in the knowledge base should be enough for the system to solve all the tasks that could appear in the particular application, without asking the user to supply additional information.

If some unexpected conditions arise in the FMC, the system immediately sends the information about it to the higher control level (or to the operator). The communication with the lower control levels and with the sensor system is completely automated.

The block diagram of the implemented system is shown in Fig. 2. The figure represents the largest version of the system, in the development and simulation mode of operation. In the real time operation mode, some blocks from Fig. 2. are out of use, and what remains is the reduced version of the system.

SCLES knowledge base consists of: rules in block A; various data about the robots and the other objects at the workspace in block B; and the models and algorithms data base block H. The user can define, update and review the rules and the other data in the knowledge base by running the program from block C, which is built in the SCLES shell. All of these update actions are performed off-line. They are followed by the reorganization of the new or updated parameters and rules, performed automatically, to get the internal record of the entire knowledge base, block D, which is used by the system's inference engine during the task execution.

Rules in SCLES knowledge base represent the human expert knowledge about the problem being solved by SCLES. This knowledge includes the general expert judgement and heuristic strategies on how the SCL tasks should be solved, and the application dependent heuristics. An example rule, extracted from SCLES knowledge base developed for the application described in the following Section, is:

Rule 22

    IF   RobotState is Wait AND
          Sensor_S1_val greater_than 12.3
    THEN GripperLoaded is 1

stating that "If the robot is in the wait state, and if the value from the sensor S1 is greater than 10.8 [N], then the robot gripper is loaded by a part". Every

parameter used in rule clauses, like RobotState, Sensor_S1_val and GripperLoaded, must be described by a specific record in SCLES knowledge base.

There is also a user defined set of records in SCLES data base describing relevant objects and facts about objects in the workspace (see block B in Fig. 2). For example, to describe a part type, the following record might be inserted into SCLES data base:

                part type name : IC_6
                part type code : 2
                material code : 3
                weight :   3.7
                shape code :   4

User defined model/algorithm records in block H contain codes for the model/algorithm itself, the SCL task type that can be solved and the accuracy of the solution achieved by applying the model/algorithm, as well as the estimated computing time required to apply the model/algorithm, and the short comment on the model/algorithm. The example of such a model/algorithm record in SCLES models and algorithms data base is the following :

model/algorithm code : 11
task type :   1
accuracy :   3
computing time :   220
comment :   move in space with
                        obstacles with
                        collision checking
                        for all segments

In implementing SCLES, no commercially available shells or other ES development packages have been used. The reason for this is the specific system purpose, and the requirement to incorporate the expert system into the more complex hierarchical control system, along with the other already implemented systems for lower level robot control. Therefore, the entire ES shell has been newly developed, including the inference engine and the user interface, blocks F, C, and E.

Since this is a relatively simple system, compared to other expert systems, well known from the literature, (see Hayes-Roth, Waterman and Lenath 1983, and Winston 1984), the shell developed resembles the examples given by Duda and Gasching (1981), Thompson and Thompson (1985), and Winston (1984), but it is enlarged and enriched with various appendices, due to the specific system purposes. There are two main levels in SCLES shell: Knowledge level and Task level. When working in Knowledge level, the user can develop and modify rules, parameters, object records, model/algorithm records etc. Task level allows for particular task specifications and simulations. The user can specify constraints under which the task should be performed, and certain sensor system information that may influence the task execution.

When solving robot tasks at the SCL, the inference engine searches the rule base backward and forward. In the off-line mode of operation, the solution is presented to the user by the appropriate user interface, block E. When solving a

particular task, temporary data are
generated, block G, deduced from already
known facts about the problem being
solved, and from applying the rules from
the rule base.

In real time operation mode, only blocks
D,E,F,G,H and I are active. In this case,
block E represents the higher control
level (FMC), and the solution is
transferred directly to the tactical
level. The knowledge base should contain
enough knowledge for the system to be
capable of solving all of the robot tasks
that could appear during cell operation.

AN APPLICATION EXAMPLE

SCLES has been developed on VAX 11-750, in
Pascal, as part of a simulator program for
a particular FMC. The entire SCLES code
occupies about 430K. Although it has been
integrated with other modules and packages
in a complex program for simulating the
FMC operation, it can be used as a
standalone ES as well.

The schematic representation of the FMC is
shown in Fig. 3. It is composed of two
robot manipulators (R1 and R2), three NC
machines (NC1 - a milling machine, NC2 - a
lathe, and NC3 - a press), two pallets,
(P1 and P2), two conveyors, (T1 and T2), a
buffer for rejected parts (RP), and a
sensor system, simply represented by a
camera (TV).

The simulated process in the FMC includes
the machining of three part types (a
shaft, a flange, and a ball). Each
unprocessed part, coming into the FMC by a
conveyor belt T1, is recognized by the
camera. Robot R1 moves the part from
conveyor T1 to the first NC machine, NC1
(the milling machine), which processes the
part by executing the program related to
the part type. All three NC machines in
the cell are capable of processing parts
in this way, i.e. by executing the
appropriate programs, according to the
manufacturing process.

When the process at the milling machine is
finished, robot R1 moves the part to
pallet P1. Robot R2 then takes the part
from the pallet, and passes it to NC2
machine (the lathe), and subsequently to
the press, NC3, via part pallet P2.
Finally robot R2 moves the part to
conveyor T2, which removes it from the
cell after the machining is completed. If
the part is not processed properly, it is
moved from the press and rejected into the
buffer RP.

It is assumed that the time required to
process a part on each NC machine depends
on the part type and on the machine
itself. The arrival of a new part to the
cell is always sensed by the camera. The
system is equipped with other sensors as
well, like touch sensors in robot
grippers, force sensors in NC machines,
position sensors in various places in the
cell, et cetera. Most of these sensors are
simulated with binary values only. The
overall control system (box CS in Fig. 3),
which monitors the entire process, is
organized according hierarchically,
according to the ideas from Albus et al.
(1983).

In such an application, SCLES was applied
for controlling both robot R1 and robot R2
at the SCL. Independent knowledge bases
were developed for each robot. The
knowledge base for robot R1 contains about
70 rules, and the knowledge base for robot
R2 contains about 120 rules. Five typical
SCL tasks were examined : move, stop,
grasp, release, and a constraint motion.
20 various models/algorithms were inserted
into the models/algorithms data base. For
example, models/algorithms that are used
for for move task include point-to-point
move algorithm using decoupled model with
world coordinates, continuous path move
algorithm within the specified time,
obstacles avoiding move algorithm using
configuration space model, etc. (for more
details see Stokić and Vukobratović 1986).

Robot commands are generated by the FMC
control system, and decomposed to SCL
commands by the robot control level. SCLES
executes all of these SCL commands using
rules and facts from the appropriate
knowledge base.

Several experiments with SCLES have been
conducted while developing the knowledge
bases for this application. The results
from these experiments showed that:

* SCLES can be successfuly applied as a
  standalone ES for the SCL synthesis and
  for off-line application development
  and experiments.

* When integrated with other control
  modules, SCLES can be used for on-line
  SCL control synthesis for robots. Since
  FMC applications require more
  flexibility in SCL control than other
  robot applications, ES approach to the
  SCL synthesis appears to be very
  suitable particularly for FMC
  applications.

* SCLES knowledge bases for typical FMC
  applications include about 100 rules,
  which makes SCLES a relatively small
  KBS. Moreover, these rules are very
  simple in structure. Since SCLES
  inference engine applies simple and
  efficient search and conflict
  resolution algorithms (see Devedžić,
  1988), the system operates fast enough
  to satisfy real time constraints for
  typical applications.

* In the majority of cases when SCL tasks
  are repeated, the system can achieve
  the best solution possible with models
  and algorithms existing in the data
  base, by gradually improving previous
  solutions. This is the way the
  convenient tradeoff is achieved between
  the accuracy of task solutions and the
  computation times and the computer
  system capacity.

CONCLUSIONS

The main purpose of the ES described in
the paper is the SCL synthesis for robots,
in complex hierarchical control systems,
such as FMC controllers. There is an
intention to apply SCLES for other
manufacturing control tasks as well (with
certain modifications), but further
research is needed. The main advantage of
the proposed system is its flexibility. It
can be easily adjusted to various robot

applications. This flexibility is achieved by exploiting the tradeoff between the precision of task solutions and computation times required by various models and algorithms. Another advantage is that the system can be applied off-line, as a standalone ES, for development and simulation purposes, as well as on-line, in real-time robot applications.

REFERENCES

Albus, J.S., McLean, C.R., Barbera, A.J., Fitzgerald, M.J. (1983). Hierarchical control for robots in an automated factory. Proc. 13th ISIR / Robots 7, Vol.1, Chicago, IL, April 17-21., 1983., pp. 13.29-13.43.

Devedžić, V.B. (1988). An expert system for robot control in flexible manufacturing cells. Proc. of 15. Symp. on Operational Research, Brioni, Yugoslavia, Oct. 11-14., 1988., pp. 15-18. (in Yugoslavian).

Duda, R.O., Gasching, J.G. (1981). Knowledge-based systems come of age. BYTE, Sept. 1981., pp. 238-281.

Hayes-Roth, F., Waterman, D.A., Lenat, D.B. (eds.), (1983). Building Expert Systems. Addison-Wesley, Reading, MA.

Stokić, D., Vukobratović, M. (1986). A model-based expert system for strategical control level of manipulation robots. Proc. of 6th CISM - IFToMM Symposium on Theory and Practice of Robots and Manipulators, Ro.man.sy, '86., Cracow, Poland, Sept. 9-12., 1986.

Thompson, B.A., Thompson, W.A. (1985). Inside an expert system. BYTE, Vol.10, No.4, pp. 315-330.

Winston, P.H. (1984). Artificial Intelligence. Addison-Wesley, Reading, MA,

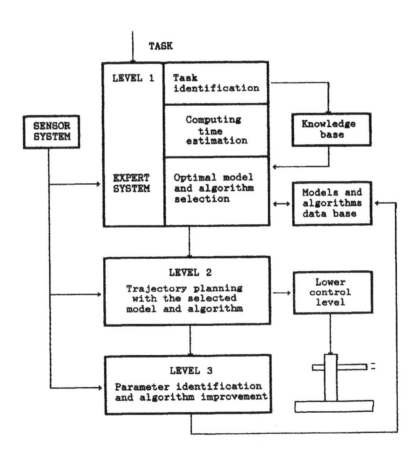

Fig. 1. - SCL block diagram

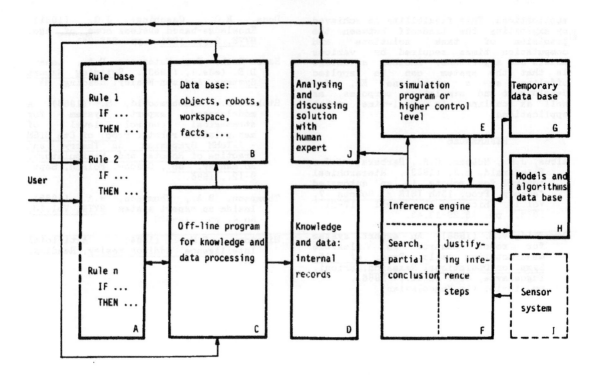

Fig. 2. - SCLES Block Diagram

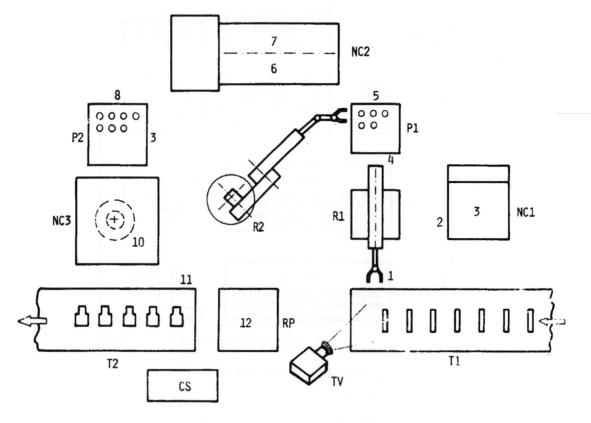

Fig. 3. - Simulated FMC Process

MEASUREMENT OF INFORMATION SYSTEMS INTEGRATION

G. Brace, A. L. Dowd and P. N. Johnson

Manufacturing Systems Engineering, University of Warwick, Coventry, UK

Abstract. It has long been advocated, by Universities and others, that integration of information systems is a 'good thing'. However, it is very difficult to find any quantitative evidence for this in the literature. As part of a wider ranging research project it has been necessary to formulate a measure of integration and subsequently arrive at a conclusion as to whether the lack of integration is economically significant. This is to be followed by a cost benefit analysis of the organisation undertaking further integration. This paper looks at the first part of the project by examining structured systems analysis methodologies, in particular SSADM, for recording the information system and then gives consideration to the process of measuring integration and some of the difficulties found in practice. The research involves companies both with and without computer systems. However only integration within the confines of manual intervention is considered - questions of internal computer processing efficiency are ignored. Within this context, a definition of integration is offered which is believed to provide the opportunity of undertaking a realistic measurement of the level of integration.

Keywords. System Analysis; integration measurement; computer integration; information system integration; data reduction; management systems.

INTRODUCTION

The research project[1] which gave rise to this paper is involved in examining the information systems of three manufacturing companies in the UK. It is one of three related projects that were established by the Centre for Manufacturing Renewal, which is part of the Manufacturing Systems Engineering Group (MSEG) at the University of Warwick. These were started in 1986 following discussions with J G Waterlow, the ACME technical consultant who later wrote (1986, on page 1):

> The degree of integration ... which can be achieved in practice depends not only on technical factors but on the extent to which the organisation can adapt its operational and organisational practices.

The overall aim of the research project is to establish whether there is a causal link between the style and structure of management on the one hand and the degree of information system integration on the other. This paper addresses the problem of establishing objective criteria by which the relative states of "Integration" of respective information systems may be judged.

The three companies selected for the project were similar in size and in industrial sector. However, each was at a different level of computerisation, one almost totally computerised, one without a computer and the third in between. Initially this variation was seen as a means of clarifying measures of disparity be ensuring a wide variation in values, but in the event, we found more disadvantages in this design feature than advantages.

STRUCTURED SYSTEMS ANALYSIS

When the research was started it was obvious that some formal methodology would be required to document the information flow of the companies. Such a methodology would have to cater both for purely manual systems and computer systems with manual interfaces. The initial intention was to use IDEF-0 (Ross, 1977), the Data Flow Diagram element of IDEF, as the structured systems analysis methodology. This was the then de facto standard of the MSEG at the University of Warwick, however it quickly became apparent that this was not an appropriate methodology for this project. The major reason being that IDEF-0 does not have a symbol for data store, also it records little organisational detail. Therefore, a survey of methodologies was undertaken.

From an initial survey of the structured systems analysis methodologies available in the public domain, it quickly became apparent that all of them made some use of one or more of three elements:

---

[1] We are grateful to acknowledge the assistance of the ACME Directorate of the UK Science and Engineering Research Council in providing funds and to individual members of SERC for providing advice and guidance on the project.

	IDEF-0	SSADM DFD
Is intended for documenting	Manufacturing systems	Information systems
Position in project life cycle	Business / functional analysis	Software analysis and design

Fig. 1: Summary of reasons for selecting SSADM

<u>Data Flow Diagram (DFD)</u>. Used for the system description, to portray what the system does, with symbols to represent: the path data travels and the transformation and storage of data. This is normally constructed as a hierarchy of diagrams, with each succeeding layer showing greater detail. IDEF-0 is a format for data flow diagrams

<u>Data or Entity Diagram</u> (possibly supported by a data dictionary). Used to record related groups of data and to show the relationships between them. IDEF-1 is a format for entity diagrams.

<u>Data or Entity Life History Diagram</u>. Used to show the chronology of events on the data within the system. IDEF-2 is a discipline for depicting such a chronology.

However, of the methodologies examined in more detail only two, SSADM and IDEF, made use of all three of the elements and did not aspire to be something special - such as a design tool for Real-Time control systems. Although at that early stage of the project it was only intended to use DFD's, it was felt that the availability of the other two elements may be of some use in the future.

Following the survey, SSADM (Structured Systems Analysis and Design Methodology) was selected. SSADM provided a number of benefits to the research:

  i) it does have a symbol for data store.

 ii) it catered equally for both manual and computer systems

iii) it was sufficiently versatile for use in recording data relating to some features of organisation, in addition to the flow of the information system.

A more detailed reasoning for the selection of SSADM rather than IDEF-0 is given in a previous paper by one of the authors (Wood and Johnson, 1989), however the conclusions are summarised in Fig. 1.

While introducing the notions of SSADM this paper makes no attempt at a full description of its construction and use. For a more detailed explanation of SSADM see Cutts (1987) or Downs (1988).

<u>SSADM</u>

SSADM was originally created in Britain by Learmonth and Burchett Management Systems in conjunction with the Central Computer and Telecommunications Agency (CCTA), a UK government agency. First introduced in 1981, the methodology has progressed through several versions. In 1983, it became a mandatory requirement for UK government commercial computing system developments. Developments in the methodology are continuing with moves to create a British standard, to be followed by a European standard and with further enhancements to cater for distributed computer systems. (Computer Weekly, 1988)

<u>Concepts of SSADM</u>

The concepts of SSADM which are summarised in Fig. 2 and discussed here are probably applicable to all structured analysis methodologies.

- Graphic
- Concise
- Communication
- Review
- Model

Fig. 2: Concepts of SSADM

The first concept is that the technique is graphic; using the tenet that a picture is worth a thousand words. Following the same argument, the method is concise with much information being recorded in a relatively small space. The obvious purpose of producing such diagrams is to aid communication: between computing professionals, between groups of users and between the professionals and users. Such communication is ideally contemporaneous to allow a process of recording and review to take place. For example, is the system being offered by the computer professional the same as that required by the potential user? Perhaps a less obvious purpose is to communicate through time. Certainly when the system is on a computer there will be a future need to change the system. Such changes include both the need to 'fix' the software to remove bugs and to enhance the software to meet the growing needs of the users and the changing circumstances of the business. For both of these purposes the diagrams provide a major source of information on how the system is supposed to perform and why it was designed that way. Lack of precision in communication was perhaps the most common cause of failure in early computer applications which implemented the deficiencies long accepted and hardly noticed in manual system design. A similar discipline for use with manual systems is needed. Its advent will be another example of the more advanced technology passing improvement back to its parent technology.

Finally, the results of the structured analysis is a model of the system being analysed or being designed.

Principles of SSADM

The principles which underlie SSADM are considered here only in relation to data flow diagrams, although the integrated nature of the methodology makes the underlying principles constant throughout all aspects of analysis. Figure 3 summarises these principles.

- **Top — down hierarchical approach**

- **Process is iterative**

- **Users actively involved**

- **Data structures determines system structure**

- **Separation of logical and physical**

Fig. 3: Principles of SSADM DFD's

They are:

1. Analysis requires a Top Down Hierarchical approach to make the complexity manageable

2. Analysing a system is an iterative process. This not only assists comprehensive records, but is an insurance against diversion along a 'branch line' activity.

3. Effective systems analysis demands the input of system users.

4. Data structure is the key to function, rather than the other way round.

5. A system cannot always be completely described by a logic model, the physical method of fulfilling the logic may be an essential parameter.

Although the first three points are universally recognises the last two are not. The endurance of data structure is important. The heart of any information system is its files and databases which may be a major capital asset in the host organisation. Operating methods and administrative procedures change, but the underlying data structure is remarkably constant. A design approach which fails to recognise this priority will yield an incapable system, which may become a serious liability.

The second point is on the need for both physical and logical models. When analysing an existing system, some of the activities recorded will not be essential to the task in hand. To take a simple example within a manual system: consider the task of putting incoming invoices into a filing cabinet which stores them in alphabetical order by company name. The person who performs this task takes a group of invoices, sorts them initially into alphabetical order on the desk and then places the invoices in the filing cabinet. The desk sort is a part of the physical system and should be recorded as such, but it is not part of the logical system which says 'transfer invoices to the file'.

Within this project DFD's are being produced at a fairly high level to show sufficient of the information system as is necessary. The data structure is being ignored. There is further comment on the question of data analysis later.

Data flow diagrams

The symbols used within SSADM for the production of DFD's are few in number and are shown in Fig. 4. Even within this limited set there is some variation in the symbol used for the External Entity. However, this always seems to follow the pattern of a 'squashed circle'.

When considering other DFD methodologies, apart from a variation in symbols, the construction method is very similar for all. About the only major difference is in the different methods of dealing with transformations where the entry / exit position of flows is of significance only within IDEF-0.

One problem which arose in the early stages of DFD construction in this project, was the definition of a database or data store. Within a computer system such a concept is easily understood, but it is rather more difficult in a purely manual system. At what point does a piece of paper stop becoming a trigger mechanism and start becoming a data store? To overcome this difficulty, each of these terms were defined:

If the paper is used to pass data on to the next function, then it is a database.

If the piece of paper acts merely as a trigger to initiate the next function then it is a mechanism and not a database. (A 'Kanban' is a mechanism whether it is a tote pan or a piece of paper.)

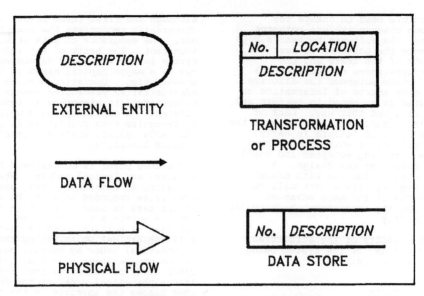

Fig. 4: DFD symbols for SSADM

The use of a hierarchy in the production of DFD's avoids the need to put all the detail on one very large piece of paper. By starting with an overview diagram and then decomposing each transformation into its own DFD, a hierarchy effect is created. This process is repeated until the requisite level of detail is obtained.

Ideally, the information system for the whole company should be contained in a single hierarchical structure. However in practice this is very difficult to achieve. Data collection normally takes place on an ad hoc basis as individual forms are chased around the system and as people can spare the time to discuss their work. This leaves a major task of consolidating the various mini-hierarchies gathered into a single picture. However, the consolidation process is both a very time consuming and an error prone process. This difficulty could be avoided to some extent if:

i)   The researcher were a more fully integrated member of the company's staff.

ii) A basic DFD hierarchy were already in existence (eg from a similar company) on which the details of the company information flows could be constructed.

INTEGRATION

Even when a DFD hierarchy has been produced, the measurement of integration is not easy. After much deliberation it was decided that the most straightforward approach was to define the converse. It is easier to perceive 'duplication' than 'integration' since the one is tangible and the other a concept. More formally:

> Integration is the availability of data for more than one purpose without conflicting records or duplication of effort.

However this definition does not provide a complete answer. Lack of integration may have consequences of "communication failure" in addition to, or instead of, duplication. The situation is far more complex than might be readily apparent.

When DFD's are examined, some of the duplication found may be considered essential. For example duplication may be required, quite reasonably, by the auditors as protection against fraud. This normally takes the form of splitting responsibilities and even performing similar work in parallel. We have ignored such duplication.

Examples of contentious issues which specifically relate to manual systems are:

i)   The practicalities of running a manual system. If two people need the same piece of information to perform their jobs, it is necessary to have the information duplicated in two data stores, unless it is feasible to co-locate the workers.

ii) The use of indices, for example a drawing register acting as an index to a set of manually produced drawings. Could such an index be considered essential to the operation of a manual system, even though a separate index need not be maintained in a computerised system?

Should these be considered unnecessary duplication? Rather than taking a stance on necessary or unnecessary duplication the more pragmatic approach is followed; that duplication is duplication. Inherently, duplication is neither a vice or a virtue, it is just a fact. There is also the question of duplication between companies which has been explicitly excluded from this project.

The search for integration has concentrated on manual systems. For the two companies with significant computer systems, no attempt was made to judge the internal efficiency of those computer systems. However, any need to transfer data manually between separate computer systems or to carry out unnecessary manual processing was included for consideration as to whether it is duplication. A watch was also kept for the duplication of data in computer databases.

In order to make sense of the duplication found, it is considered misleading to assign a 'simple number of occurrences'. Some indication of scale is required. Therefore, a further stage in the research is to put a valuation on the duplication found. At present it is intended to calculate initially the manpower effort required to support the duplication and then to apply an average marginal cost of employment to provide the valuation. Such factors can be easily measured using such techniques as work study. These we have termed the Direct cost of lacking integration.

Other costs which it is felt need to be considered are connected with the penalties of having multiple copies of data or data stores. These take the form of data (un)reliability due to discrepancies between nominally identical copies. Costs arise due to the need to periodically reconcile the data stores to identify and remove differences in the content. A more major cost (but less easily quantified) can be associated with making incorrect decisions because the data store used to supply the information, on which the decision was based, was not fully up to date and correct. Both these circumstances add to the unnecessary costs of the business, but are much more difficult to measure. These, more nebulas costs have been termed the Indirect costs of lacking integration.

One major problem within the project is the need to check manually the DFD's produced. Such checking is required for two purposes. First to ensure the accuracy of the DFD's produced and second to identify the occurrences of duplication. Without the use of automated checking, it is necessary for the researcher to retain an overall picture of the DFD hierarchy in his mind since data or effort may be duplicated on diagrams far removed from each other in the hierarchies.

USE OF CASE TOOLS

As part of the research, two Computer Aided Software Engineering tools were reviewed. Both packages support the three main elements of analysis: DFD, Data (Entity) Diagram and Life History and both had further elements for use in system design and to allow code generation. However, these latter points are outside the interest of this project.

Of the packages selected, one was used to receive data gathered from the participating companies, while the second package was reviewed when problems arose with the first. The major problems found during the project were:

i)   The rudimentary nature of the diagram validation, especially where changes were made after the initial creation, for example during later discussions with company staff.

ii) The inability of the software to identify duplication either directly or indirectly.

Due to the lack of useful output, the use of CASE tools was therefore dropped from the project.

It should be appreciated that this use of CASE tools was outside their designed purpose and hence it is perhaps not too surprising that they were not of assistance. This is a pilot study and in subsequent work it will be recommended that a 'CASE' technique be suitably adjusted to be of use. The need to generate a single DFD hierarchy has already been mentioned. In addition, it is felt that the production of Entity Diagrams and a basic Data Dictionary would assist in the identification of duplication and perhaps allow a CASE tool to play a more productive role in the work.

DISCUSSION OF RESULTS

An initial inspection of the DFD's for the three companies has led to some tentative conclusions although, at the time of writing, it is not possible to fully quantify them. Duplication has undoubtedly been found in all three companies. The amount found would seem to be dependent on the degree of computerisation within the company with a relationship similar to that shown in Fig. 5.

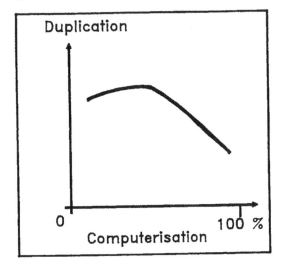

Fig. 5: Outline of initial findings

This observation is in line with a comment by Aleksander (1987) on the findings of Paul Strassmann, to the effect that the more successful companies either leave computing alone or else they develop a bold strategy - they do not tinker with it. However this initial observation does not address other important, although not necessarily definitive, factors such as the size of the company, the geographic spread of the operations and the organisation type.

For the first company, Company A, the direct costs of lacking integration have been calculated. For a turnover of £10.5m per year, the Direct cost of duplication amounts to only £32.6k (= 0.31 %). Of this it is anticipated that over half (£18k) will be eliminated by the introduction of a shop floor data collection system and the subsequent automation of the processing of production, quality and time keeping records. However, we would have expected this company to have the highest figure of duplication because it uses a mix of both manual systems and a variety of computer systems. Thus it would fall in the middle of the horizontal axis of Fig. 5.

In addition to this direct cost, two other figures have been identified. The first is for internal processes that are not carried out due to the constraints of manpower and time. Should these tasks be performed, the direct cost of duplication would increase by some £5.6k pa. The second figure is related to the external integration of information systems. Company A has CAD equipment which is capable of producing NC tapes for the production of tooling. However, as a matter of policy such tooling is always produced by sub-contractors - some of whom have NC equipment. As Company A never supplies NC programs, the sub-contractors are obliged to part-program the tooling, which adds some £6.7k to the overall costs of the supply chain. There are a number of reasons for not taking advantage of such possible savings. There are two major problem areas. The first relates to the legal problems of ownership when drawings are transferred by electronic means. The second problem is on the physical difficulties of either transferring a 3D model or of ownership of the NC post processor.

CONCLUSION

In principle SSADM does provide a method of studying the level of integration within information systems and using the definitions given, especially that for integration:

> The availability of data for more than one purpose without conflicting record or duplication of effort.

it is believed that accurate and useful results will be obtained from this research. This definition of integration may be considered fundamental to the whole basis of the research.

The costs of lacking integration have been split into two:

i) the direct costs, which can be directly measured by using, for example, work study type measures.

ii) the indirect costs which include the more nebulas factors. This would include the costs of reconciling differences between two data stores that are supposed to have identical content. Alternatively, the cost of making incorrect decisions because the data store used to supply the information on which the decision was based was not fully up to date and correct, due to duplication of data stores.

At the time of writing, the direct costs have been calculated for the first company. These amount to less than 0.5 % of the annual turnover. However, the indirect costs, which we feel are potentially much greater, have yet to be evaluated.

The project has made use of only the data flow diagram element of SSADM. However, two factors need to be reviewed if commercially available CASE tools are to be able to assist in the identification of the lack of integration:

i) the generation of a single DFD hierarchy.

ii) the use of entity diagrams, possibly supported by a data dictionary.

The preliminary results available at the time of writing appear to confirm the work of Strassmann. However, when the full results are available, it is hoped to confirm (or not) a link between the degree of integration and the organisational structure.

REFERENCES

Aleksander, I. (1987). The Culture of Failure. Business Computing and Communications, April, 13.

Computer Weekly (Journal). (1988). (No Author). Government wants method in Europe. 6 October, 10.

Cutts, G. (1987). Structured Systems Analysis and Design Method. Peradign, London.

Downs, E., P. Clare and I Coe. (1988). Structured Systems Analysis and Design Method: Applications and Context. Prentice Hall, Hemel Hempstead, UK.

Ross, D.T. (1977). Structured Analysis (SA): A language for communicating ideas. IEEE Transactions on Software Engineering SE-3 No 1, 16 - 34.

Waterlow, J.G. (1986). ACME Initiatives in CAPM Research. A paper presented to the ACME grantholders' conference, September, 1.

Wood, P.J. and P.N. Johnson (1989). A review of the use of SSADM and IDEF at the University of Warwick. SAMT'89, Sunderland Polytechnic, UK. March.

CONCEPTS FOR THE REALIZATION OF DISTRIBUTED, FAULT TOLERANT CIM STRUCTURES

R. Bernhardt

*Frunhofer Institute for Production Systems and Design Technology,
Department of Automation, Berlin, FRG*

Abstract.   Today industrial enterprises are often decentrally organized. The
objective of this is to concentrate resources and capacities for the elaboration of
specific tasks and the effective solution of problems. On the other hand, an effective
information exchange is required between the various departments which are
located at different sites. Existing networks and communication systems are not
suitable to fulfil the necessary tasks. This paper reports on R&D activities which
aim at overcoming the impediments to the integration of decentrally stuctured
enterprises. The basis for the integration is a broadband communication system
based on an optical fiber network which has already been installed in Berlin by the
Deutsche Bundespost. The BERCIM project has been launched with the purpose of a
conception and realization of distributed, fault tolerant CIM structures using such
a communication system. The main tasks of the projekt are the conception and
prototypical realization of distributed CIM applications and their informational
linkage via this communication system. Therefor the communication kernel,
transport system and a management information system have to be realized. For
the communication system and the distributed CIM applications local and network
overlapping fault tolerant procedures are to be considered.

Keywords.   CIM, BERKOM, Communication System, ISDN-B, Wide Area
Networks, Information System

INTRODUCTION

The increasing change from a supplier-domina-
ted to a buyer-oriented market forces companies
more and more to meet the diverse demands of
their customers. Costs and overall processing
time have to be reduced while at the same time
productivity, quality and flexibility must be increased. Industrial companies face the task of
implementing features of service companies
within their existing structure. This leads to a
growing importance of a linkage by information
technology of the areas within a company which
are concerned with the production, as well as
with suppliers, customers and external service
rendering companies. Such a linkage facilitates
the optimization of not only the production process but also all other processes within a company and external sites which are concerned. Consequently, structure, organization and management processes often have to be changed, too. The
introduction of such computer-aided production
methods is still at its very beginning, and it is
mostly hampered by existing structures comprising heterogenous, partly autonomous systems,
some of which by now have become difficult or
impossible to link up /1/.

---

*)   The author reports about a project conducted by researchers of different institutions
mentioned in the acknowledgement

Therefore an integrated information system has to be realized which is capable of supplying all necessary data to connected systems in an up-to-date, complete and fail-safe way. Within todays decentralized industrial companies a high-performance telecommunication system capable of linking up the different areas is an essential basis. In this context standardization aspects are of major importance. The high demands for system reliability which result from CIM-applications necessitate the integration of advanced fault-tolerance concepts for telecommunication and information systems as well as for the distributed CIM structures.

The realization of such a concept results in:

- increased transparency within the company by application of information systems which at any time are capable of providing information in necessary detail;

- reduction of overall processing time from product idea to manufacturing and from order to delivery by reducing hold-ups due to idleness and passing-down, by switching from successive to simultaneous processing, and by reducing repeated, manual data reprocessing and transfer, and the

- stimulation of synergetic effects by making use of information bases which can be applied to different functions and work areas, such as by improving the cooperation of research and production engineers who work at different sites.

By taking into account these aims it becomes evident that the main function of CIM lies in supplying information at hand to all concerned sides completely and in an up-to-date form, and to ascertain that every authorized user can gain access to all necessary information at any time. Thus the basis of CIM is an integrated open information system. As modern companies are decentrally organized, a high-performance communication system for linking up the various information bases is required to fulfil this task.

To put this approach into concrete terms the requirements of the information and communication system for distributed CIM structures have to be analized. As a first step into this direction, a functional analysis of industrial production systems has been carried out.

FUNCTIONAL ANALYSIS AND MODELLING OF THE INDUSTRIAL PRODUCTION

The result of the analysis is the functional model of the industrial production system which contains the functions and their interrelations, their informational linkage and mechanisms. This model acts as a reference allowing to reflect CIM structures from different points of view (e.g. communication system, information access / exchange). The model is further used

- to assign to each function systems or components which fulfil the function and manage their interaction,

- to identify information sources and drains,

- to classify the information and

- to define data rates.

In this context, the functional model is the basis for defining requirement of the necessary communication mechanisms related to type, features and quality. It is also a basis to identify the integration potential and advantages coming from a powerful communication system allowing data rates up to 140 Mbit/s.

For the representation of the functional model the semi-graphical notation of Structured Analysis and Design Technique (SADT) /2/ has been used. Thereby the production system is hierarchically decomposed (fig. 1). At the top level (parent diagram) the information relations of the production system to the outside world are specified globally. At the lower levels (child diagram) the structures and interrelations are more and more detailed. Thereby one box in a parent diagram is completely replaced by the child diagram.

Using SADT a system can be modeled by stressing functions (activities) or data. Both representations are dual. For modeling the industrial production the activity representation has been chosen. In fig. 2 a model element and its interrelations to other elements are shown. Beside the activity the

- data flow structure (input/output) specifying the produced and consumed data by the activity,

- control flow structure specifying data controlling the activity, and

- mechanisms specifying constraints and realization aspects have to be analized and defined.

Mechanisms are realizing or supporting the activity or they are additional informations. Due the realization aspects are not directly included in the model structure they can be brought in by the mechanisms.

This technique has been applied to an industrial production system. Regarding industrial enterprises a functional separation can be identified which mirrors the division of labour. In the first decomposition step the main activities are separated in

- sales and services,
- production program planning,
- design and development,
- manufacturing planning,
- component manufacturing,
- manufacturing program planning,
- manufacturing control and supervision.

In fig. 3 these main activites are presented showing only the data flow structure. The control flow and the mechanisms have been neglected in this article for better comprehensiblility. Starting with this level a further decomposition has been elaborated down to the forth level.

The principle result of this analysis is a functional model of an industrial production system showing also the information flow structure. It is further assumed that all functions or activities can be locally distributed if a powerful communication and information system is available. This system has to support the different activites and to bridge the heterogeneity as well as the local distribution. This means that the communication and information system has to lead to an informational integration of the differernt functions of the industrial production system even if they are locally distributed. Furthermore an access is required to data bases, powerful computers, expensive peripherals and the use of complex software for solving specific problems, e.g. optimization problems. Also video conferences dealing with e.g. design drawings, visualization of process simulation or real production processes

should be enabled.

DERIVATION OF REQUIREMENTS OF THE COMMUNICATION SYSTEM

Subsequently the principle procedure applied for the derivation of requirements of the communication system is presented. As a first step, different general goals for the realization of CIM as well as views in different enterprises and industrial areas are analized. Out of this the characteristic features and trends can be summarized as follows:

- Integration of existing and operating automation islands which are composed out of heterogenous components often characterized by centralized, closed structures.

- Adaption to hierarchical and existing organizational structures as well as the different requirement concerning performance, reliability, safety and security.

- Transparency and open system structures are required for all components in all hierarchical levels to realize cooperation, communication and information integration.

In the next step the different information views from the functions of the model of the industrial production have been analized. Basic multi media document (MMD) information types and some application specific information types are relevant for the CIM area /3/. These information is used not only for design, manufacturing system planning, off-line programming and simulation, manufacturing process execution and supervision, but also in areas like sales and services. In these different areas also different views do exist concerning the access, exchange and representation of information. Some of these viewes are listed below:

- Model data (incl. technical descriptions, technological, geometrical kinematic data in different representations) are used for design, system planning and programming.

- Access to company internal and external data bases containing e.g. formalized descriptions of standardized parts and components is required for design and planning.

- Access to and derivation of specific data for off-line programming and simulation must be made possible.

- Derivation of supervision and diagnosis information is required for manufacturing control and quality ensurance.

- Visualization of products, real manufacturing processes and their simulation for acquisition purposes and public relations.

Beside these views, the specific interest and future developments in the frames of BERCIM and BERKOM have been taken into account where those structures and applications are of utmost interest which stresses the aspects:

- product and manufacturing information processing considering multi media documents and communication structures,

- integration of technical administrative and commercial areas,

- information linkage and cooperation between enterprises,

- fault tolerant concepts which can handle in a distributed environment the failure of components and to ensure the security of information transmission.

Within the project these views and specific interests have been analized in detail and requirements for the communication system have been derived. Some examples are given subsequently.

- Simultaneous access to distributed information bases which implies specific requirements related to e.g. safety, allowance, synchronization and speed.

- Visualizations of processes for remote control require high data rates and minimum time delays for interaction.

- Access to model data within or company overlapping require high speed due to inter-activity, synchronization of the various information types and check of access rights.

These general requirements for the communication system are to be used for the derivation of requirements for communication services and their qualities which are related e.g. to data rates, safety and fault tolerance. Therfor also different CIM scenarios are defined and partly realized which are based on the functional model of the industrial production. Thereby the structuring of distributed applications, the identification of closely related functions and their coordination as well as the selection of communication services of specific qualities will lead to an architecture of distributed CIM systems. In this context two aspects are of major importance.

The first is to reach an information integration within the production systems, e.g. design, manufacturing planning, manufacturing control and supervision.

The second is to demonstrate the possibilities of broadband communication mechanisms related to interactivity, high data rates and fault tolerance for distributed CIM structures.

BERKOM

Today applied communication services and protocols are designed having in mind data rates and unreliable networks (max 64 kbit/s for wide area and 10 Mbit/s for local area networks).

For that reason, today proposals are discussed about new protocols which consider à priori the features of wide area networks with data rates greater than 100 Mbit/s and error rates less than 10^{-14}. Furthermore, protocols and services of the transport system used today do not support broadcast and multicast functions which are required from distributed applications. To overcome these disadvantages, the Berlin Communication System Project (BERKOM) has been launched.

The general goal of BERKOM is the advancement of the development of communication services, distributed applications and end systems based on a wide area network (WAN) with the above mentioned features. The result will be a universal Integrated Services Digital Network for broadband communication (ISDN-B) in the public domain.

Beside the development of end system and the related architectures for a direct connection to the WAN also private communication structures are considered to enable an indirect connection of end systems. Such private structures are e.g. standardized networks (ISO 880Z) or to be standardized broadband local area networks (FDDI, FDDI II) which should be connected to the public WAN. The required transition systems are also developped in the frame of BERKOM.

The BERKOM reference model /4/ structures the functions for open system communication in layers in agreement with the ISO reference model /5/. In fig. 4 the BERKOM and in fig. 5 the ISO reference model is presented. It should support the integration of existing standardized communication services and also the development of new communication services and end systems which use intensively the inherent possibilities of broadband communication. Beside this horizontal structuring of the communication in layers the BERKOM reference model will be enlarged by a vertical structuring related do distributed applications. This allows a better description of aspects concerning distributed processing, storage and retrieval of distributed data and the management of application instances in a distributed environment.

The BERKOM reference model can be decomposed into five functional blocks:

-	applications (e.g. office area, information system, CIM area),

	information types, communication services and management of distributed applications,

-	communication control supporting synchronization of application instances,

-	transport services supporting a transparent information exchange between application instances, and the

	test network.

On the basis of 30.000 km already installed optical fibers a broadband network for the public domain is realized and connections to this network are available to the BERKOM project participants.

Within the BERCIM project this network will be used for the prototypical realization of distributed CIM structures.

Based on the functional model of the production system different scenarios for distributed CIM applications have been elaborated. These will be discussed subsequently.

DISTRIBUTED CIM STRUCTURES

The BERCIM infrastructure for development and test of distributed, fault tolerant communications and applications within the CIM area is realized in two steps. In the first step the project partners build up local islands which will be linked in the second step via the BERKOM test network. These islands are based on standardized local area networks (LAN) for the administrative, technical and manufacturing area allowing data rates up to 10 Mbit/s.

These islands do also reflect the fuctional model of the industrial production whereby the areas manufacturing planning and manufacturing control and supervision are considered. In fig. 6 the according SADT diagram is shown. Therby the function manufacturing planning is decomposed into the subfunctions layout and task planning and robot off-line programming which contains also a simulation system for test purposes. The different functions are allocated to the islands of the project partners which are locally distributed. In fig. 7 the information exchanged is shown in more detail. Due to the realization state of the gateways which are also developed within the frame of BERKOM, the LAN´s can only be connected via routers allowing data rates of 64 Kbits/s. This means that layout and task data, robot programs and control commands (DNC) are to be transferred by this data rate. Additionally a 140 Mbit/s channel is available for the transmission of video data. Therefor a specific coder/decoder (CODEC) is required which has already been installed by the different project partners. In fig. 8 a detailed installation plan of the CIM-islands linked via the BERKOM test network is shown.

CONCLUSION

A first approach showing the advantages of broadband communication for distributed CIM structures is the linkage of end systems based on standardized LAN´s and protocols. Due to the lack of suitable gateways this can only be done via components available on the market allowing data rates up to 64 Kbit/s. In the near future gateways will be available allowing the connection of LAN's with data rates of up to 100 Mbit/s which will lead to an increase of the performance of the system necessary for practical applications. In areas which require the transmission of video data (conferences, process monitoring etc.) the coder/decoder for the 140 Mbit/s channel is already available on the market and also used for the first realization.

In the further devlopment of the BERCIM project distributed CIM scenarios will be realized which utilize the possibilities provided by a broadband communication architecture with high data transfer rates and of novel transport services. These scenarios will be the frame for the design, test and demonstration of innovative protocols, communication services and for fault-management procedures. Moreover they will be the basis for a practical evaluation and examination of the feasibility of novel CIM solutions which in the final step will be the starting point for a reevaluation of the structures appropriate for the manufacturing plant of the future.

In order to show the application of broadband communication with the connection of local area networks with a data rates up to 140 Mbit/s, the BERCIM structure will be built up step by step. During the conception phase, the basis for distributed CIM scenarios was laid using transit systems available on the market. This arrangement is already suitable to show potential users the advantages of a connection to the BERKOM network.

REFERENCES

1. Der Bundesminister für Forschung und Technologie:
Fertigungstechnik, Programm 1988-1992.
Bonn 1988.

2. D.T. Ross: Structured Analysis (SA) - A Language for Communicating Ideas.
TOSE, Vol. SE-3, No. 1, January 1977, S. 16-34.

3. MMD-Documents in ISDN-B:
Anforderungsanalyse V 3.0 DETE KON,
Projektleitung BERKOM, Berlin January 1989

4. BERKOM-Referenzmodell, Version 1.0,
DETECON,
Deutsche Telepost Consulting GmbH,
Projektleitung BERKOM, June 1987.

5. A.S. Tannenbaum: Computer Networks,
Prentice-Hall Inc., Eaglewood Cliffs, New Jersey, 1981

ACKNOWLEDGEMENT

The Deutsche Bundespost and the Senat of Berlin have placed an order to their associated company Deutsche Telepost Consulting GmbH (DETECON) to launch a so-called BERKOM project (Berlin Communication System). As a result the DETECON installed a BERKOM Project Management in 1986 seated in Berlin. The task of BERKOM is to sponsor the development of applications, services, and various end systems for ISDN-B (Integrated Services Digital Network-Broadband) in parallel to the installation of the optical fiber network. The Deutsche Bundespost finances the project and also makes available the ISDN-B optical fiber network for technical tests.

The R&D tasks within the project are taken over by industrial companies and research institutions. In the meantime a broad spectrum of projects has been kicked-off reaching from general information systems, office communication, medical information systems up to CIM applications.

In the CIM area the BERCIM project has been launched, which is based on a cooperation between

- AEG Aktiengesellschaft, Research Institute Berlin (Dr. Merker,G. Heiner, F. Michel, D. Krämer, F. Kuehl)

- GMD FOKUS, Gesellschaft für Mathematik und Datenverarbeitung mbH, Research Center for Open Communication Systems (V. Tschammer, L. Henckel, U.W. Brandenburg, K.-P. Eckert, J. Hall, D. Strick)

- IPK, Fraunhofer-Institut für Produktionsanlagen und Konstruktionstechnik Berlin (Dr. R. Bernhardt, H. Linnemann, C.-L. Hartke, S. Fang, H. Fredrich, L. Menevidis, M. Timmermann)

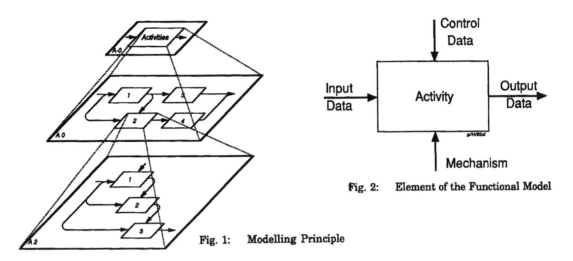

Fig. 2:   Element of the Functional Model

Fig. 1:   Modelling Principle

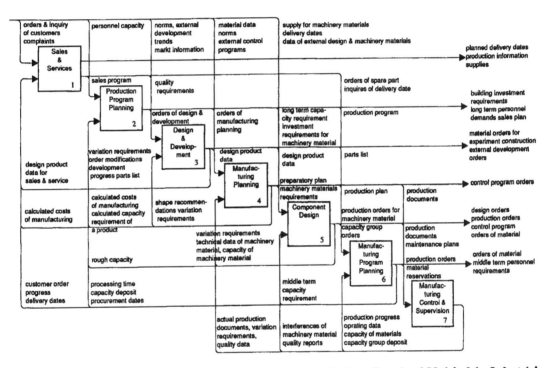

Fig 3.   Functional Model of the Industrial
Production

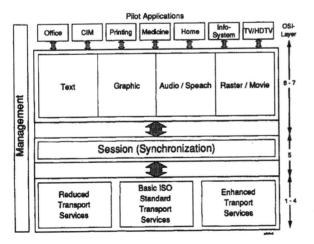

Fig. 4:   BERKOM Reference Model

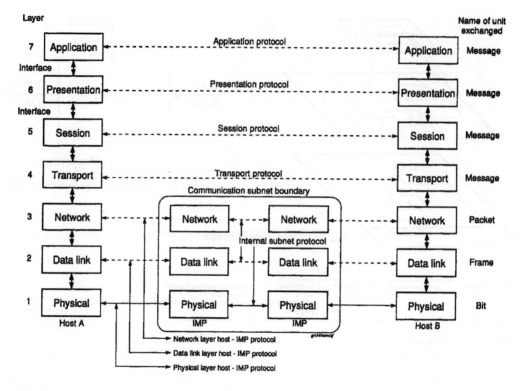

Fig. 5:   ISO-OSI Reference Model

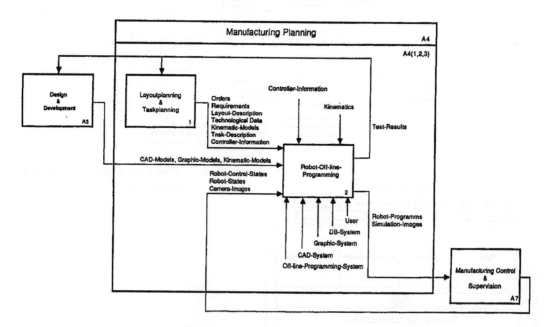

Fig. 6:   Part of the Functional Model Chosen
for Prototypical Realization

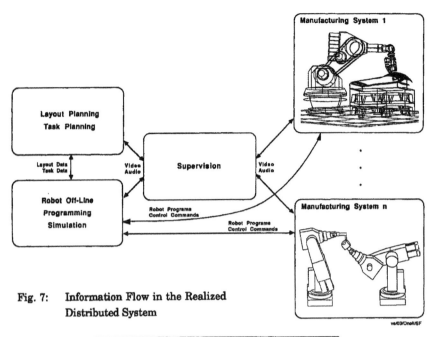

Fig. 7:   Information Flow in the Realized
           Distributed System

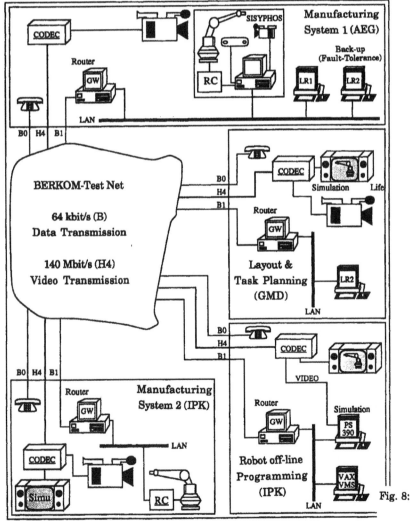

Fig. 8:   Installation Plan of the
           Distributed System

MANUFACTURING AUTOMATION AND PROTOTYPING FOR PRINTED WIRING BOARDS

J. A. Kirk, D. K. Anand and J. D. Watts

*Department of Mechanical Engineering & Systems Research Center,
University of Maryland, College Park, USA*

Abstract. This paper presents an improved method of manufacturing automation and prototyping for Printed Wiring Boards. The protocol requires that three data sources be present [circuit specifications, component specifications, and a circuit layout database], in computer interpretable form. The specifications for these three data stores is presented and a generic workcell, suitable for low volume high part mix prototyping. is discussed. In order for the protocol to control the generic workcell, the subtask orders which are required for proper function are also presented and discussed.

Keywords. Assembling, automation, computer hardware, logic circuits, printed wiring board.

INTRODUCTION

Work is currently under way at The University of Maryland is into the development of a protocol for rapid automated assembly of Printed Wiring Boards. The aim of the protocol is to develop a methodology for controlling the placement, soldering, and inspection of both, plated through hole and surface mount components in a high part mix/low lot size production environment. Additionally, the results of this work must be integratable into an existing manufacturing environment.

Figure 1 shows a diagram of the typical present design and manual assembly operation for printed wiring boards. Here the design engineer develops an electrical schematic which is then used to generate a computer database of the required circuit layout. Taken by hand from the design database is the component parts list, procurement requirements, assembly work instructions, electrical test parameters, and quality inspection and acceptance criteria.

Although some of the data interchange between functional groups in the process is automated, typically it is still administered manually. For instance, parts lists, which are created on a CAD system, are manually transferred into the material control/purchasing computer system through the re-keying of the data. Additionally, tracking the transition from design to layout to manufacturing is done either on paper (manually), or verbally. The final database Configuration Control Drawing Database in Figure 1 is then utilized to generate a circuit board drawing package (i.e., photomaster) which is used to fabricate a bare circuit board. Once the bare circuit board is fabricated, it is returned to an assembly area where it is kitted with an assembly manual and the required parts. The assembly manual includes a work order (a detailed set of assembly instructions), the required assembly drawings, and the control documentation (which are used to trace the assembly/inspection process and all subsequent rework of the board).

After the assembly kit is complete, the board is assembled by hand and then subjected to an iterative cycle of inspection and rework. After passing assembly, inspection and rework, the board is electrically tested (with appropriate reworking and inspection), conformal coated, visually inspected and electrically retested. The assembly is then given a final over-all visual inspection and placed in stock were it is made available for delivery to the customer or use in a higher level assembly.

Figure 2 is shows schematic diagram for rapid prototype assembly of printed wiring boards. In rapid prototype assembly, the design information is used to both produce a drawing package containing all the circuit layout drawings, and to directly control an automated assembly workcell. A main component of the rapid prototype system is the assembly protocol. This protocol provides the important link to establish the methodology, via the use of existing industrial standards, or by developing in-house standards, which allows electronic circuit assembly design information to be directly used in controlling a generic assembly workcell.

The rapid prototyping assembly protocol provides interface definition and data links for the operation of an automated production system for the manufacture of electronic circuit assemblies. Additionally, the protocol identifies the required design information which must be available in order to produce a functional assembly. By identifying the information which is needed in electronic assembly production, and then configuring an appropriate generic assembly workcell, it is possible to utilize the research presented here in a wide variety of existing commercial or military production facilities.

BACKGROUND

The Advanced Design and Manufacturing Laboratory (ADML) at The University of Maryland has been actively involved with a local electronic systems contractor in developing a protocol for the automated assembly of printed wiring boards. This particular organization is currently utilizing manual assembly and soldering methods in their printed wiring board manufacturing process. They are, however, in the process of developing products that will require the use of automated placement, soldering, and inspection processes. These new designs will require the integration of both, plated through hole (PTH) and surface mount devices (SMD's).

The manual production of electronic assemblies diagram was developed for this work and is abbreviated in Fig. 3. The diagram shows a abbreviated picture of the design and assembly process, while remaining unencumbered by incorporation of the particular nuances of each process as performed by any specific entity or manufacturer.

The flowchart shows the development of the circuit design, the generation of the assembly instructions, the purchasing/ inspection paths for the bare boards and parts, and finally, the assembling and inspection of the completed hardware. In short, the diagram expands upon the information presented in Figure 1.

The goal for this project is to develop a rapid prototyping assembly system which links an automated factory for electronic assembly into the existing manual environment. Figure 4 shows the flow of the system and how it is integrated into this existing environment. The system, consisting of both hardware and software, is discussed further in the following section.

AUTOMATED PWB ASSEMBLY

The rapid prototyping assembly system, must allow for short assembly workcell setup times and rapid downloading of CAD generated placement data. Additionally, the protocol must be capable of dealing with the placement, soldering, and inspection of both surface mount devices (SMD) and plated through hole (PTH) components. A schematic diagram for the rapid prototyping assembly system is shown in Fig. 5. From Fig. 5, it can be seen that the system inputs will consist of standard design information, inventory information, cell status information, and system queuing, all of which must be presented in a defined standard format. In turn, the protocol will produce the cell control commands (standard output driver codes) to operate the assembly workcell and provide status feedback to the designers and the inventory system.

To identify the locations of key data necessary for protocol operation, the data flow requirements of a typical design/assembly process were analyzed. A Ganes and Sarson Dataflow format was used in this analysis and has identified the design and manufacturing data parameter requirements for production of PWB's. This information also includes the format and location of the design data for access by the assembly protocol. Charting the dataflow of the current assembly process was mandatory to identify the final source of information used in the assembly process. The decision to define and chart the present design/assembly information paths was made to ensure that the flexible assembly protocol would utilize existing data in an efficient manner by accessing the needed information automatically at the source. Accessing this data at the source will help eliminate errors of using outdated data, and it will help avoid the inefficiencies of creating and maintaining multiple data stores.

Since the rapid prototyping assembly protocol will generate the process plan for the automated

assembly workcell [showing all the available inputs and outputs from each software module] it is desireable to constrain the inputs and outputs to coincide with information commonly available in a typical manual assembly process. Following this line, each process in the design and manufacturing process flowchart is depicted as a operation module with its required inputs and developed outputs. For such modules, where automation techniques are applicable, software drivers are then needed to control the actual assembly equipment. For those modules where human involvement is a necessity, the guidelines and task definitions must provide sufficient detail to control the operation in a production environment. Furthermore, once computer terminals and screens are introduced to the assembly floor, the needed operator inputs to the system should not require typing text for data entry. Bar codes and voice recognition systems would be the preferred data entry mechanisms.

Development of rapid prototyping assembly protocol in this manner supplies additional benefits in that it reflects exactly what the requirements of the software are and what functions each software module must perform. This serves as a safeguard against the protocol becoming too dependent upon commercial or poorly designed software which, although adequately performing the designated operations, requires cumbersome communications and information preparation modules to be generated so as to satisfactorily integrate it into the system. Such a dependency on commercial or poorly developed software drastically reduces the flexibility of the automated system, which was one of the initial driving factors in the development of the rapid prototyping assembly protocol.

For the purposes of this research, it is assumed that the following restrictions apply:

1. The information available to the protocol is in a standardized format.

2. The choice of components for use in new electronic assembly designs is limited to those on a preferred parts list and the designer be made aware of these limitations prior to the initial design.

3. The workcell is capable of placing all components on the preferred parts list onto the unpopulated substrate.

4. An inventory system is available to supply the workcell without increasing down times or reducing production through puts. The supply of components into the workcell can be fully automated via an automated material handling system, or can be manually loaded into the workcell magazines and matrix trays prior to a production run.

Interface drivers are required to convert the proprietary design information into standardized formats, and then additional drivers are required to convert the standardized protocol outputs into "machine specific" driver codes. The restriction that inputs and outputs of the rapid prototyping assembly system be standardized has an additional benefit in that each driver is modularized and becomes a distinct piece of software, while still an integral part of the overall software plan. Since the proper operation of these software modules is conditional only upon the presence of the correct inputs (as it is the module itself which generates the outputs), each module becomes interchangeable with any other module requiring the same inputs and producing similar outputs. As a collection of dedicated hardware drivers, the rapid prototyping assembly system allows for the introduction of a wide assortment of equipment and easy system expansion. So long as the interface definition of the protocol is maintained, it may be adapted to a variety of specialized tasks within the production environment.

Figure 6 shows a simplified flowchart of the design and assembly flow for Printed Wiring Board assemblies. This figure further defines the process flow diagram shown in Figure 4 and its primary use here is to illustrate the required information (called the specifications stores) needed to develop the rapid prototyping assembly system. At the present time, it is proposed that three data stores should be available to the assembly system. The first specifications store, CIRCUIT SPECIFICATIONS, is a data store containing all information pertaining to the operation of the completed assembly. This includes design requirements, restrictions on component types, component reliability, temperature requirements, currents and voltages at various test points, error analysis, margin of error, worst case analysis, reliability calculations, test validations and diagnostic tree.

The second data store, COMPONENT SPECIFICATIONS, is an integral part of the inventory database kept by the manufacturer, and must contain inventory part number, manufacturer part number, manufacturer identification, part function, part reliability, part temperature considerations, part package type, number of pins, Phase dimensions, origin (pin 1, center, other), placement offsets, tool number to handle part, tool point offsets, location of tool, feeder type, location of feeder and relative rotation of feeder axes from assembly axes.

A third datastore is the CIRCUIT LAYOUT DATABASE. This database contains the necessary information related to the assembly such as component rotation/orientation, X coordinate, Y coordinate, Z coordinate, logical placement order, placement pressure, component lead configuration/length, feeder location, tool needed, tool location, and board based vision landmark.

In order for the rapid prototyping assembly system to become operational, the following functions must be in place:

1. Standard format information be declared and made available to the protocol concerning component placement on the assembly substrate.

2. A link to the inventory system be made which provides parts availability information.

3. Confirmation be given when parts are loaded into the workcell.

4. The workcell be made controllable by the rapid prototyping assembly system for the placement of the parts onto the substrate.

The first task, standard format information, stems from the requirement standardization. This task demands the installation of the proper translation modules, or interface drivers, between the protocol and the specific design system. At present these standards are under review, but the initial work has involved both IGES and PDES standards. The second task, inventory control, requires that the components available to the

designer appear on a preferred parts list. There is no limitation on the parts individual manufacturers place on their preferred parts lists, just that the final workcell be configured such that it is capable of handling and placing all the components on the preferred parts list .

This stipulation that the parts the designer uses must appear on the manufacturer's preferred parts list and further, that the part's specifications and dimensions be entered into the components specifications datastore, introduces an additional benefit. Currently, the costs involved in the entering new part information into the computer database is a hidden cost. With the use of the rapid prototyping assembly system, however, an actual cost can be derived for this process. This cost can then be assigned as a direct cost of not using parts already on the preferred parts list.

The rapid prototyping assembly systems approach is consistent with just in time (JIT) inventory control. As the required design information is in computer intelligible form, purchase orders for the required components for a scheduled production run can be produced automatically at a set time period prior to the run initiation date. In this way, components are brought in-house on a job basis, thereby reducing inventory storage and maintenance costs.

In the final task, that of workcell control, it is important to remember that the protocol accepts design information in a specified standardized format, and produces cell control information, in a standardized output format. Hence, if the manufacturer is configuring a new workcell, it is desireable to purchase equipment that accepts information in this format. If the protocol is to control a currently existing cell, interface drivers that convert the protocol outputs to the machine specific driver codes are required.

If a new cell is being configured, it is important to accurately estimate what the production demands will be. Demands for low part mixes with little or no dependence on mixed technology applications will warrant the configuration of a workcell which differs significantly from a cell configured by the demands for high part mixes with a higher dependence on mixed technology applications.

In the present work, the production demands imposed were high part mixes and small production runs. Based on these demands, the generic workcell shown in Fig. 7 was configured. The choice to utilize SCARA robots was based on the premise that short production run setup times are more important than the shorter production run times attainable through the use of dedicated placement machines. Additionally, as the cell will be installing PTH components, one of the SCARA robots can easily be programed to pre-tin all PTH components prior to the production run.

For lower volume production, which is the emphasis in the rapid prototyping assembly system development, flexible pick and place machines, such as SCARA robots, are preferred. These machines are usually robotic manipulators which can perform a multitude of different tasks and can be reprogrammed quickly. Typically, these machines are not taught component placement sites. Instead, the component placement locations, dictated by the designer's CAD drawing, in conjunction with component databases, are used to drive the movement of the manipulator. Additionally, since these systems usually incorporate vision systems, they are capable of correcting for variations in board to board layouts through board based bindmarks.

The choice of utilizing printing type solder
dispensing equipment is made based on the wide
acceptance these systems have achieved in
industry. As no reprogramming is needed to
institute a new production run, the systems are
highly flexible. Additionally, as the system down
time is limited only by the loading of solder
paste and the changing of solder screens,
production through put is sufficient to meet the
production volume requirements set previously.

The pneumatic syringe indicated in the post reflow
robot envelop serves as a post reflow (post
soldering) or rework device, allowing for the
deposition of solder when screening or stenciling
is not possible. Its primary purpose is to apply
solder paste to the pads of SMD's which cannot
undergo the conventional reflow operations due to
the risks of thermal shock.

Solder reflow is completed through a conveyor fed
in-line vapor phase reflow system (VPS). Although
VPS has attained the highest recognition and use
in the industry, it is not a truly flexible nor
economic system to use. Since the saturation
temperature of the working fluid is a constant,
the only variables available for customizing the
soldering process are variations in the conveyor
feed speeds and in changing the working fluid. As
the working fluids commonly used produce
relatively high operating costs ($600 per gallon),
changing the fluid unnecessarily quickly proves to
be too costly. VPS was used inspite of this
drawback since it eliminates the need to format
the placement of components on the board on the
basis of component sizes and colors, a major
drawback in the use of near infrared reflow
systems.

For those components which cannot undergo the VPS
operation, a heater bar or laser soldering system
are two viable options for the post reflow robot
envelop. Although the components can be soldered
manually, a regulated reflow operation promotes
better solder joint quality. The PTH components
inserted at this step, however, will need to be
soldered manually as no reliable system was
identified for the soldering of PTH components on
a mixed technology board with the exception of
wave soldering, which is not being considered due
to the thermal stress it places on the SMD's and
the produced solder joint quality in fine pitched
SMD's.

CONCLUSIONS

The manufacturing automation needs of a typical
manual Printed Wiring Board assembly process have
been evaluated. A protocol for the rapid assembly
of Printed Wiring Boards (PWB's) has been
developed for a high part mix low volume
operation. The resulting protocol has all
attributes necessary for the rapid manufacture of
various PWB prototypes. In this protocol three
data stores; circuit specifications, component
specifications, and circuit layout, have been
identified as required for automated control of a
PWB workcell. A generic PWB assembly workcell has
been configured and the suggested drivers for the
workcell have been linked to the protocol data
stores.

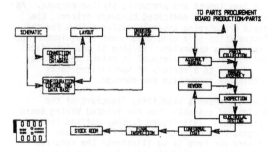

FIGURE 1   PRESENT ASSEMBLY PROCEDURE

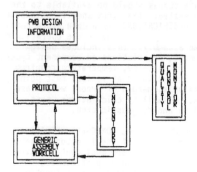

Figure 2   Schematic Diagram for
Feedback and Control Paths in
Automated PWB Assembly

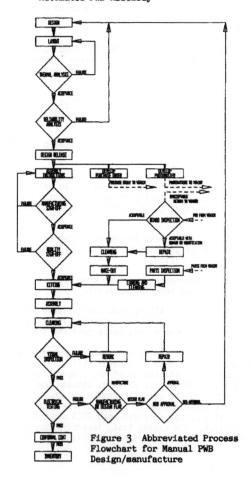

Figure 3   Abbreviated Process
Flowchart for Manual PWB
Design/manufacture

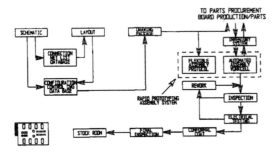

Figure 4   Proposed Assembly Proceedure

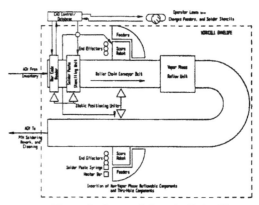

Figure 7   Proposed Generic Workcell

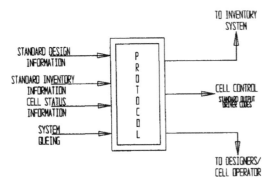

Figure 5   Rapid Prototyping Assembly System

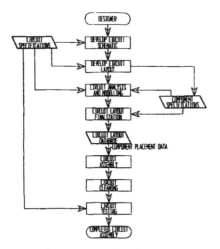

Figure 6   Data Store Requirements
for a Rapid Prototyping Assembly System

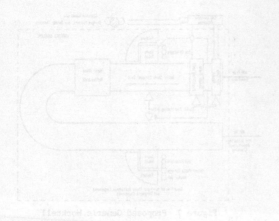

Figure 4 Proposed Assembly Procedure

Figure 7 Proposed Generic Approach

Figure 5 Rapid Prototyping Assembly System

Figure 6 Case Store Requirements
for a Rapid Prototyping Assembly System

AUTOMATED WAREHOUSING AND MANUFACTURING PILOT PLANT: HARDWARE AND SOFTWARE

J. A. Sirgo, A. M. Florez and G. Ojea

*Department of Electrical Engineering, University of Oviedo, E.T.S. Ing. Industriales,
Gijón, Spain*

Abstract. In today's factory, totally automated warehouse has become increasingly integrated with the manufacturing process in order to reduce cost and attain high degrees of flexibility in storage. A prototipe of a totally automatic warehousing and manufacturing plant is being constructed in the "Departamento de Ingeniería Eléctrica, Electrónica, de Computadores y de Sistemas" (DIEECS) in the School of Engineering at the University of Oviedo. The whole warehousing and transport system has being designed at this Department. It consist of an automated pallet storage by stacker crane and an automated guided vehicle system controlled by the same host computer. This paper will be a brief description of the automated warehouse and, in detail, of the Automated Guided Vehicle System (AGVS) and the inteligence distribution through the whole system.

Keywords. Automation; warehouse automation; guidance systems; materials handling; flexible manufacturing.

INTRODUCTION

At present, the School of Engineering at the University of Oviedo is dealing with notable Spanish Companies to collaborate in the development of technologies in which these Companies are interested.

Actually, we are about to accomplish the development of an Automated Guided Vehicle System (AGVS), entrusted by Duro Felguera S.A.

Previous to this development, an automated pallet storage by stacker crane was designed and constructed collaborating with the same Company. Both projects have been financed by the "Fundación para el Fomento en Asturias de la Investigación Científica Aplicada y la Tecnología" (FICYT).

These projects are part of a totally automatic manufacturing plant. A prototype of this is being constructed in the "Departamento de Ingeniería Eléctrica, Electrónica, de Computadores y de Sistemas" (DIEECS) in the School of Engineering at the University of Oviedo.

In this paper we will describe this pilot plant, with more details in the case of the most recent works: the AGVS and the software structure and distribution through the plant.

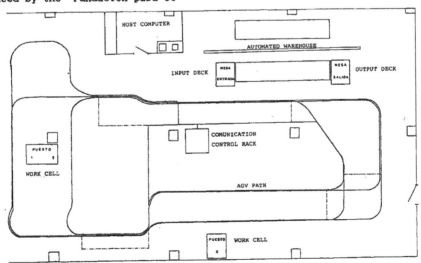

Fig. 1. Automated manufacturing pilot plant.

GENERAL DESCRIPTION

The most notable elements in this plant, which are shown in Fig. 1, are:

- Automated warehouse
- Automated Guided Vehicle
- Work cells
- Comunications control rack
- Host computer (MicroVAX 2000)

All this items may be described as follows.

Automated Warehouse

The warehouse installation for palletized goods are constituted for one or more rows of good cells. Between these rows there are one or more aisles with cells on both sides. This allows us to place there a mobile system responsible for good displacement.

The mobile system (stacker crane) can give service at one or more aisles simultaneously and there is frecuently to be found in one warehouse, various cranes working in co-ordination.

The developed prototype is a small-size installation of 5 m. heigth and 10 m. length.

It consists of one aisle and two rows. Each row has 6 sections of 1 m. These sections have 3 or 4 levels at 500 Kg. weigth per level. Two different good-cell sizes are implemented, symmetrically distributed in both rows (Fig. 2).

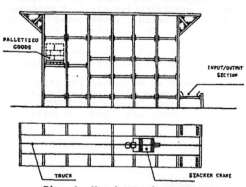

Fig. 2. Warehouse lay-out.

Material handling is achived by a stacker crane running on rails, through the aisle, picking up goods by an automatic fork, dispatching or retrieving pallets and allowing internal material movements. The input/output section consists of two roller conveyor decks where pallets are placed and picked up by the stacker crane.

The proposed solution can be easily applied to large scale warehouse systems with several aisles, rows, stacker cranes, etc.

The stacker crane has a mechanical structure as shows Fig. 3. The crane allows three movements: horizontal, vertical and fork displacement.

Displacement speeds in the three axes are respectively:

Horizontal :   60 m/min.
Vertical :   20 m/min.
Fork :   6-12 m/min.

This speeds can be increased in a large scale warehouse system. There has been utilised for both horizontal and vertical movements an asynchronous squirrel-cage motor-brake.

In both axes a slip-frequency control strategy is selected. The speed is measured by digital encoders and each axis control algorithm is implemented in a different microprocessor.

For the velocity control of the induction motor there has been utilised PWM inverters, which maintain constant the torque to a nominal speed, over a power constant up to 2.4 nominal velocity.

To diminish the time between each operation, a simultaneity of movements has been introduced. Therefore the total time is reduced considerably and will be equal to the maximun time between translation and elevation. In this way, the system works so that the horizontal and vertical movements are simultaneously controlled. An effect of this is that the system is more complex. The number of intelligent elements and the communications between them are necessarily increased. It is vital that they are perfectly synchronized.

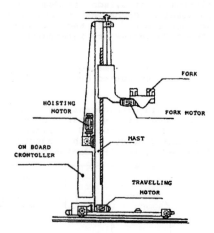

Fig. 3. Stacker crane.

Automated Guided Vehicle

The construction of an Automated Guided Vehicle (AGV) was undertaken as the next step in the design of an automated manufacturing plant. Among the different avilable guidance chances, the inductive guidance was selected as a suitable choice in the objective.

The designed inductive guidance AGV incorporates an on-board microprocessor in order to control all the vehicle task and sensors, as shown in Fig. 4. The foremost sensors are:

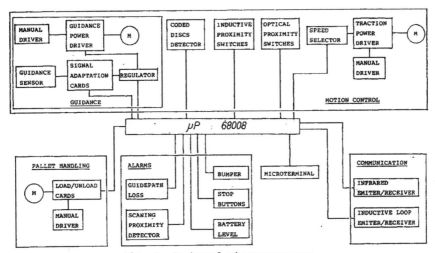

Fig. 4. On-board microprocessor.

- Position controls
  Inductive Proximity Switches
  Optical Proximity Switches
  Coded Discs Detector

- Collision Prevention
  Warning Lamps and Beeper
  Scanning Proximity Detector
  (two levls: slown down and stop)

- Emergency Stop
  Full Safety Bumper
  Strategically Placed Stop Buttons

Another remarkable sensor, not included in the former schedule, is the one which allows the vehicle guidance.

Vehicle guidance. To allow the vehicle guidance forward and backwards, two guidance sensors have been placed in the vehicle. One of them is assembled in front of the steering wheel and turns with it. The other is at the back, behind the rear wheels, to manage a reliable backward guidance.

The inductive drive system, shown in Fig. 5, is well-known for people who are famiNlarized with AGVS. It consist of a high frequency current flowing throught a wire embedded in the floor which is detected by two coils with magnetic cores. The difference between the signal amplitude from the two coils shows the sensor deviation from the guidance path.

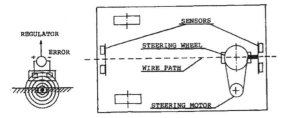

Fig. 5. Inductive guidance system.

It is easy to notice that since the vehicle has only three wheels, the guidance forward control structure will be

diferent from backwards.

Both guidance control block diagrams are shown in Fig. 6 and 7. When the vehicule is guided backwards, an aditional feed-back must be added to the control diagram, and regulators are more complex and hard to adjust. Therefore, the vehicle will move more slowly backwards than forward. Nevertheless this is not a trouble as the backwards guidance is only to be used in maneuvers.

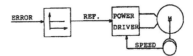

Fig. 6. Forward guidance.

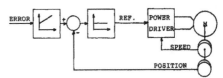

Fig. 7. Backward guidance.

AGV handling modes. Depending on the position of a modeswitch, the vehicle may work on the following three different modes.

- Manual mode. The vehicle guidance is to be done by a set of push-buttons. This is an aditional mode which is only used to control the vehicle out of the wire-guides.

- Semiatomatic mode. The vehicle tasks are controlled by a microterminal located in its back. The microterminal supplies commands to the microprocessor of the vehicle. When all the required commands have been compilled, the microprocessor executes each command automaticly. This mode affords an automated work of the vehicle without a required communication network.

- Automatic mode. In this mode, a communication network is required to

TABLE 1 Vehicle Specifications

```
Guidance..............Inductive Guidance

Guidepath.............Wire-loops embedded in the floor
                      (three frecuencies generators)

Guidepath sensor......Two coils with magnetic cores

Drive system..........Front-Wheel Steering and
                      Front-WheelDrive (Motor in Wheel)

Type of steering......Manual/Automatic

Travel direction......Forward and Backward

Travelling speeds.....0.1, 0.25, 0.5 and 1 m/s

Minimun turn radius...1 m

Capacity..............1000 Kg

Load/Unload...........Roller conveyor deck for
                      palletized goods

Data transmission.....Infrared Data Transmission and
                      Inductive Loop Data Transmission
```

Fig. 8. Load operation.

Fig. 9. Designed AGV.

supply commands to the vehicle from a remote host computer (MicroVAX 2000), which controls the whole warehouse and the AGVS. The execution of these commands is similar to that in semiautomatic mode, but workers are not to control the vehicles. Hence, this is the work mode to be used in an automatic warehouse, though semiautomatic and manual mode might also be useful occasionally.

Further information about vehicle characteristics are supplied in the vehicle specifications table.

Work cells

Two roller conveyor decks, equipped with keyboards allow us to simulate work cells in the plant. The AGV supplies them raw materials and takes the manufacturated products which will be stored by the stacker crane.

Keyboards at work cells are connected in Daisy-Chain to supply error-free the requirements of every cell to the host computer through a single RS-232C plug. The keyboard message includes a work cell code and a function code to inform the host computer about the requested function and the requesting cell. A reply from the host computer will turn on a lamp in the requested function is being processed.

Comunication Control Rack

There are two microprocessor boards placed into the communication control rack. They may be an interface between host computer and low level devices communication to release the computer from some tasks.

One of them allows on-route communication through an inductive loop between an AGV and host computer. The other one takes the operations requested from the work-position keyboards and controls infrared communication.

Host Computer

The host computer is responsible for controlling the AGVS and the automated warehouse. Moreover, it may be an operator terminal, which supplies additional information about warehouse stocks, manufacturing plant status and AGVs traffic.

Therefore, it must be fully informed of each item status and requierements through a communication network. This involves a complex software structure with a distribution of microprocessors along the whole plant (Fig. 10) in order to hand out system tasks, and thus, to optimize information fluxes and plant processes.

SOFTWARE DISTRIBUTION

Most software is spread among several microprocessors to release the host computer from routinary tasks (Fig. 10). The software distribution can be described by the functions that each element provides.

Host Computer

The foremost software programs are integrated into the host computer to accomplish the following functions:

Management of the warehouse. It allows the user to choose among different standars of input/output (FIFO, LIFO, ...).

Management of the AGVS. Decide the task that each vehicle must carry out.

AGVS traffic control. Avoid AGVs collisions and detect problems that the vehicles might come across, by means of the inductive loop microprocessor board information.

Display information of the system. AGV locations, warehouse inventory and movements, goods locations, etc.

Attend work-position needs. by means of the microprocessor board for work cells.

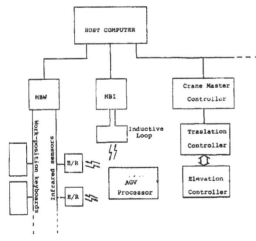

Fig. 10. Communications.

Microprocessor Board for Work cells (MBW)

The main functions of this microprocessor are:

- Acquire the operations requested from the work-position keyboards.

- Control the work cell roller conveyor decks.

- Information interchanges with vehicles through infrared sensors (communication physical link and protocol).

- Syncronice work cell load/unload.

Microprocessor Board for Inductive Loop (MBI)

This microprocessor interchanges information by means of the inductive loop (communication physical link and protocol) to manage an on-route control of each vehicle, and send it to the host computer.

Both microprocessor boards release the master computer from a considerable amount of work, specially considering that pooling is to be done to initiate information interchanges between vehicles and the computers. Thus, the host computer fluxes of information are optimized to keep it fully informed with scarcely wasted time.

On the other hand, both microprocessor boards are responsible for allowing vehicles to use physical links. As shown in Fig. 10, all the infrared sensors are connected to a single RS-232C port . Thus, a reliable method to control the port must be addopted to avoid errors which would come up if several vehicles undertakes communication simultaneously.

Polling is the solution given to this problem. The MBW addresses one of the vehicles of the system sending a message through infrared port. As a result of the infrared-sensors connection, this message is detected by vehicles at any work position. If the addressed vehicle is ready to undertake communication, the MBW receives a reply. A vehicle code in the reply message confirms to the MBW whether this reply comes from the addressed vehicle or not.

Even though this solution is highly reliable, in large systems, vehicles might be waiting for instructions at work cells for long time. Therefore, if the number of work cells is high enough, installing several MBW is advisable in order to improve efficiency. Each MBW, whose structure would be equal to that shown in Fig. 10, is to control a set of work cells. Then several operations may be carried out simultaneously and vehicle waitings at work cells drop down. Hence, traffic fluency and speed are remarkably increased.

The MBI incorporates the same method to control its RS-232C port than that incorporated by the MBW. All the above-mentioned MBW characteristics and considerations are also applicable to the MBI.

AGV On-board Microprocessor

Its main functions are:

- Control the vehicle sensor and tasks that had been already mentioned.

- Choose the path that the vehicle is to take to its destination.

- It has to support links with the host computer by means of infrared sensors at work cells and inductive loop when the vehicle is on-route.

Stacker Crane Microprocessors

Control has been carried out with three microprocessors. The firs, Crane Master Controller, attends to the following functions:

- Communications with the management computer.

- Communication with Translation Microprocessor of each crane.

- Manual/automatic control of exchange.

- Height control of palletized goods.

The Translation Microprocessor attends to:

- Communications with the Crane Master Controller.

- Communications with Elevation Microprocessor.

- Control the horizontal position.

- Driving translation induction motor.

- The alarm control.

Finally, the Elevation Microprocessor attends to:

- Communications with Translation Microprocessor.

- Weight control of palletized goods.

- Control of vertical and fork position.

- The alarm control.

CONCLUSIONS

An automated manufacturing pilot plant has been developed: an automated warehouse and an Automated Guided Vehicles System has been desingned and verified in a small scale installation located in the laboratory of the "Departamento de Ingeniería Eléctrica, Electrónica, de Computadores y de Sistemas" in the School of Engineering at the University of Oviedo.

All the communication equipment, such as work cell keyboards, infrared sensors and inductive loop cards, have been designed and constructed for this pilot plant in this Deparment.

An accurately management and control software has been implemented along the pilot plant to optimize communication fluxes and plant processes.

In the future, this pilot will allow us to study industrial plant problems, such as goods distribution and fluxes, storage and transport time optimization, etc, on a real flexible manufacturing plant without simulation computer aid.

Copyright © IFAC Information Control Problems in
Manufacturing Technology, Madrid, Spain 1989

FACTORY AUTOMATION: PRODUCTION
SUBSYSTEM MANAGEMENT

M. D. del Castillo, A. Alique and F. Cano

*Instituto de Automática Industrial, Consejo Superior de Investigaciones Científicas,
Madrid, Spain*

Abstract.   Factory automation is a job that intends to map all the cooperative
and competitive relationships that happen among the elements existing in a
factory into the software domain. The goal of this paper is to show an important
phase when a software system of this kind is to be built. This is the design
phase. We have chosen an object oriented methodology. We think several concepts
belonging to this methodology are applicable on the industrial environment to be
controlled. We think it is very important to build a conceptual model of the
system which will control the factory. The final software system will depend on
the consistency of this model. The main feature of the model is to consider the
factory as a set of different classes of components. The components belonging to
each class have the same behavior.

Keywords. Object oriented design, OOD, factory automation.

INTRODUCTION

The structural and functional features of every
manufacturing system are based on the features
of the elements which constitute it and on the
interconnection between those elements.

The majority of the factories are nowadays
composed of very highly automated elements which
realize a wide spectrum of tasks.

We can think of them as components of the whole
manufacturing system, which transport, process
and store physical objects.

The existence of those different elements in a
factory can show us a very high degree of
automation in the different areas of the whole
production process, but this fact does not mean
that the factory is automated.

So, for achieving an automated factory we must
automate this information flow, that is, we can
be able to transport, store and process
information objects (Naylor, 1987).

A flexible manufacturing system must be
conceived through the integration of the
majority of the factory components. This
integration means communication between the
different devices. In this way, the factory will
be anything more that the ensemble of automation
islands (Naylor, 1987). The integration of all
the devices has to be solved in two ways: the
physical and functional one.

We are studying the functional one and so we try
to build a software system which makes easy the
information acquisition, storage and circulation
in such a way that different areas of the
factory could access to it at the just time and
in the most appropiate manner.

THE SOFTWARE SYSTEM MODEL

It is necessary to emphasize the first steps of
the software system life cycle when we are
performing a software system to support an
automated factory : requirements collection,
analysis and design.

Our factory model will be formed by abstract
entities that have a direct association with
physical elements (machines, stores, parts,
tools, etc) and with no physical elements (CNC
programs, productions orders, etc). Both of them
are implied in the whole manufacturing process.

Then, our model of the real world will be
composed by entities. As we have said till the
moment those entities should only show the
behavior of every factory elements in an
independent manner. However, we are trying to
design a software system which support the whole
manufacturing system integration, that is, the
relations between elements to get, in a
cooperative manner, a factory model where
machines and automated centers do the required
task at the required time in order to obtain the
final product.

These entities include structural and behavior
properties belonging to those elements in the
factory they state. But the entities, still in
the modelling stage, must be able of
communicating with each other in order to work
in a cooperative manner. This necessity of
communication is due to each entity stores and
processes only the information associated to it.
When an entity requires some unknown information
it establishes a relation with the entity that
has got it.

Therefore, it is necessary that the model also should include those relations. This is possible by building entities which process their own information and besides they can communicate between them by sending and receiving information from other entities. The final software product depends on the completeness and consistency of the model (Sommerville, 1985).

SOFTWARE SYSTEM DESIGN

The model of the real world is described as a set of entities and relations between them. With this basis, we choose an object-oriented design methodology, where every object is an entity in the way we have described at the modelling level.

An object shows a behavior which is fixed by the actions that it does and the actions that it needs from other objects (Booch, 1986), (Pascoe, 1986).

An object or entity is characterized by an state, which is defined by the value that its data structure has got. Every feature that configures the state of an object is called an attribute of this object. Two objects communicate between them only if someone of them needs an operation from the other.

At time to approach a complex problem the first step to do is assembling those elements of the problem, which have an analogous behavior, in classes.

When we have the problem of automating a factory, that is, designing and implementing a software system which controls the information flowing through it, it is difficult at first sight to find all the objects that will conform the definitive design. But it is possible to establish different abstraction levels into the model. Each of them could be associated to the different subsystems existing into the factory. The objects into a subsystem can communicate with all the objects within their subsystem and they only know certain objects from others.

We would like that the design process of the problem should be as closed as possible to the real world. In a manufacturing environment the communications occur in an asynchronous way. Therefore, the communications between the objects also must be of the same kind. We mean that the object sends messages whenever it likes irrespective of the state of the receiver object (Yonezava, 1986). In the implementation stage we must study what means this asynchronous communication between objects of the software system.

AUTOMATING A FACTORY. EXAMPLE

In this paper we show the problems associated to the job of automating a certain production environment.

We show an example where the factory automation is based on the automation of the information flow associated to the cutting tools. So, through this information flow, the manufacturer may control the elements staying in the factory. This project is being performed in NASA (Logroño, Spain) in collaboration with Duñaiturria y Estancona, S.A.

Such an environment is a factory working on the manufacturing of big and expensive parts. The number of cutting tools used in this class of factory is very large and due to the cost of these tools the manufacturer has to think of the neccesity for achieving an exhaustive and fiable control over them.

Because there exist flexible manufacturing cells in the factory able of carrying out different sorts of jobs and so able of wanting a great diversity of cutting tools, depending on the job, it is needed a centralized control system devoted to supervise the tool locations in all the manufacturing proccess.

After studying the real performance of the factory we can distinguish several stages along the tool file cycle: within a magazine and without a certain job assigned, within a magazine and ready for a later job, working in a machining center, in a unloading center, in a measuring center, etc..

The tool cycle usually starts when a part is being designed and a tool is required for it. This job is performed in the design center. Once all the necessary tools for making the part are known, the next step will be to perform the production order. These productions orders contain, according to a certain production philosophy of the enterprise, the different stages where the part has to carry over in order to achieve its full machining. Every stage includes the suitable tools, the numerical control programs, etc.

One of the reasons why makes this software system so complex is the great number of relationships that happen in the real world, since that the system has to supervise all the elements in which the tool could stay and all the information associated to the flow tool.

Due to the continuous advance of the technology, a factory could be seen like a dynamic environment, where new centers will be added or the existing centers will be improved, by automating or joining new operations. Therefore, it is necessary to make a software system ables of taking into account these changes without causing expensive reorganizations.

We establish a conceptual system model, that should be complete and consistent and that this model leads us to a maintainable design. This criterion is the most important for the final user. A maintenaible design means that the cost of the system changes is minimized, that is, changes should be local. Due to the great degree of relatedness among the elements in the factory it is very easy to make a system highly coupled forgetting the maintenance in the sense described above.

An schematic view of the model of our particular system is shown in Fig.1. This model is the result of studying the real system. At first, we divided the full system into several subsystems. This division , schematically shown in the Fig., is consequence of studying all the tasks carried out in the factory. We have found that it is

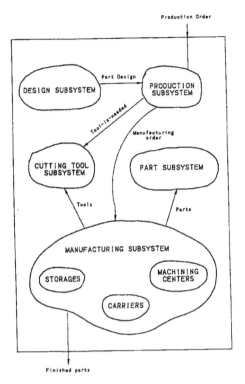

Fig. 1

The Production Subsystem

In this section we shall mainly concentrate on a certain subsystem : the production one.

First, we must be concerned with the production subsystem performance in order to obtain the final design of it.

The production subsystem receives all the production orders of parts which have already been designed in the design center. These production orders hold the features of the different tools manufacturing the part, besides the machine-tool dedicated to machine the part, the machining date, the part, etc...

Several production orders can be run simultaneously by the production manager according to the enterprise job planning. At this level the manager does not know which is the real burden of the machining centers.

So, once the production subsystem has received the production orders schedules them by producing real working plans for the manufacturing centers belonging to the factory.

The scheduling is necessary due to different production orders may require the same machining center at the same time. In order to perform all the production orders, the production subsystem must allow an interleaving between them, building real working plans associated to each machining center.

These real working plans are composed of a part, a machining center and a set of cutting-tools devoted to work in that center. The plans must be ordered by the date and this date depends on the real burden of the machining center.

In order to perform these working plans the production subsystem needs to communicate with the manufacturing and the tool subsystems. It requires to know if the tools associated to the wornking plan are free in the fixed date. If it is true the tools can be selected and in other case the production subsystem could warn to the production manager. The production system also needs to know when the machining centers are to be free of work or in what state are these centers.

According to the selected design method we divide this subsystem into different objects. We found these objects after studying the steps that the production order follows, in this particular example, from it has been elaborated to it has been changed into working plans for the machining centers.

We show part of the design of the production subsystem in Fig.2. Every object has got some attributes and methods. The lines joining different objects mean a necessity of communication between them.

In the production subsystem there are objects of the model which have not a correspondance with physical objects of the world, but also with logic objects. Within this kind of objects we include lists of production orders, logic carriers of cutting tools, etc.

possible to distinguish several entities holding data and actions. Some entities may arrange a set in which they often communicate with each other or they perform similar jobs. Every set is a subsystem in the model. No hierarchical communication order between objects exist in the subsystems as well as between different subsystems.

As an example we briefly explain the tool subsystem. This subsystem is composed of all the tool objects found in the factory, besides other objects that represent accessory elements of the tool. Those elements can be cones, nippers and nuts.

The difference between tools is in the tool shape. There are as many tool objects as different tool shapes exist. Every tool object have got different attributes and methods, both of them associated to the features of the tool.

This subsystem receives requests from the production one when the tools are selected to make a part and from the manufacturing one when for example a tool breaks in a machining center and a substitute tool is needed.

As we said before there exists no hierarchical order in the subsystems, but in this particular one we have considered to create a tool object that is the communication interface with other subsystems. This object receives a tool code and an action and it is devoted to find out which particular tool object has to carry out the action and after that it transfers the control to it.

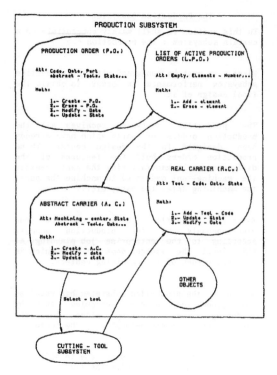

Fig. 2

On the other hand this design approach adapts oneself very well to the problem of the natural concurrency underlying in a factory, helping for an easy implementation in distributed devices (Klittich, 1988).

REFERENCES

Booch, G.(1986). Object-Oriented Development. IEEE Transactions on Software Engineering, Vol 12, No 2.
Klittich M.(1988). CIM-OSA: The Implementation Viewpoint.In E. Puente and P.McConaill(Ed) Proceedings of the 4th CIM Europe Conference.
Naylor, A.W., R.A.Volz(1987) Design of Integrated Manufacturing System Control Software. IEEE Transactions on Systems, Man and Cybernetics, Vol.17, N 6.
Pascoe, G.A.(1986).Elements of Object-Oriented Programming.BYTE.
Sommerville, I.(1985).In A. Wesley (Ed) Software Engineering.
Yonezava, A., J.P. Briot, and E.Shibayama(1986). Object-Oriented Concurrent Programming in ABCL1. OOPSLA'86 Proceedings.

IMPLEMENTATION

As we said till now our objective is to build a software system able to control a factory where parts are mechanized. We have emphasized in the design stage with a fit methodology to do an abstraction of that industrial environment.

In this implementation stage there exist several conditions fixed beforehand. One of them is the programming language. The enterprise requires a programming language sufficiently tested. With this language we have tried to adapt the design stage to the features of the industrial environment.

The object attributes are records of a relational database and the object methods are carried out by executing blocks that perform operations on the database. On the other hand, every object has got a buffer where the messages coming from other objects are stored. Every message invokes one executing block belonging to the receiver object. The messages are attended according to the arrival order and to a priority degree. The message buffer is also a relation in the database.

CONCLUSIONS

We have intended to show the design stage like the main stage when a complex problem of this kind is to be treated. The result of this stage gives the fit view in order to achieve the implementation stage in a structured way.

So, the final product is more maintanaible, the changes are locals, the integration of new entities or of new operations to existing entities is fast and easy.

MANUFACTURING MODELLING AND MULTI-LEVELS EVALUATION INTEGRATED HUMAN ASPECTS

A. El Mhamedi and Z. Binder

Laboratoire d'Automatique, Grenoble, France

Abstract. The human and organizational aspects in production system evaluation implies multi-disciplinary research and development of new methodology. Three evaluation levels are studied:
- The intuitive level which generates organizational structures.
- The qualitative level which analyzes the productivity and flexibility tendencies according to changes in means of production and/or their structures.
- And finally, the quantitative level which realises the manufacturing system simulation, using Colored Timed Petri Nets.
This paper highlights some human aspects in production systems, presents briefly our methodology and discusses qualitative evaluation.

Keywords. Flexible manufacturing; Human factors; Man-machine systems; organizational structure; Modeling; Simulation; Optimization; Artificial intelligence.

INTRODUCTION

A higher degree of flexibility and productivity are major slogans among production system engineers nowadays. Variety-seeking, demands from customers, fierce competition among firms, frequent changes in demands, and production technologies have been demanding more sophisticated production with effective management. To be adapted to this environnement, a great effort has been made on technical, social systems, as well as on interaction between them.

In the developing computer aided manufacturing systems, the human no longer directly controls the system processes. Instead, the operator interacts primarily with a computer which directly controls the processes. The human is becoming less involved in the manual control of the inner loops of the manufacturing system and is more concerned with the supervisory control from the outer loops of the system. In terms of tasks performed, in computer aided manufacuring systems the operator deals with information processing and decision making more than materials handling and manipulation.

It is of the utmost importance that the production system shall be understood as a whole, both during its design and when it is put into operation. Methods and tools for a quantitative and rational analysis concerning the behaviour of automated production systems are therefore required.

Recently, numerous attempts to model these systems have appeared from independant sources in the literature (Buzacott 1984, Bel 1985, Inoue 1985), including simulation methodologies and from the use of artificial intelligence techniques. Unfortuneatly, however, the human and organizational aspects are hardly taken into consideration in this work.

The aim of this paper is to integrate human and

organizational aspects into the modelling and the evaluation of production systems. The first section highlights human and organizational aspects in modelling and evaluation of production systems. The second section introduces a multi-level modelling and an evaluation methodology, and the third presents a qualitative evaluation level.

PRODUCTION SYSTEM EVALUATION AND HUMAN ASPECTS

Production system modelling and evaluation

The understanding of the manufacturing process is a difficult problem, its evaluation appears to be also very difficult. This evaluation tries to :

- understand the production status.
- improve the production management.
- provide a stepping stone for further improvement.
- have a common understanding of the current status, and
- share the satisfaction of achievements.

It is very important, when improving production systems, to see that the production system is not a technical system, but one with human, organizational, social and cultural aspects. On the basis of such recognition, the system has to be developed and improved through a wider view analysis to achieve problem-solving.

It is now generally admitted that it is necessary to have formal models of integrated, automated manufacturing systems. These models can be used to predict performance and to address key design issues such as the provision of storage, the degree of flexibility and the appropriate structure of control, i.e. which decisions should be made at what level in the hierarchical control structure and what information should be used for the decision.

Because of the inherent complexity of integrated automated systems the development of adequate formal models is by no means easy. The state of the art in model developement is given in (Buzacott 1984). The human and organizational aspects are generally neglected in this work. We try to take into account some one of these aspects.

Human aspects and flexibility

Flexibility in production systems is defined as the ability to adjust the systems to exernal/internal changes. A long developoment about human aspects and a flexibility in production system is presented in (Inoue 1985). The consideration of human, organizational, social and cultural aspects is indispensable for acheving flexibilities in production systems (Inoue 85).

For example, human and organizational problems in manufacturing plants may be due to : man - machine affectation, allocation of tasks, command and control of operators, ... These problems are very difficult.

The allocation of tasks and responsibilities between humans and computers in manufaturing plants are developed in (Barfield 1986). Man / machine design; principles and tools are discussed in (Rouse 1988). A model of human operators in problem solving tasks is presented in (Rasmussen 1986).

To integrate human and organizational aspects, we suggest a multi-level evaluation and an analysis of the production system. We are going to develope here a multi-level evaluation.

As an illustration, the problem of man / machine affectation in manufacturing plants is presented in this paper : Assuming that we have N operators and K machines, we look for one optimal structure $ST^* = \{$ (operator, machine) $\}$, with social, human, economical, and technical criteria. This is a complex problem.

Example.

If we have one workshop composed of five operators; OP_1, OP_2, OP_3, OP_4, OP_5, and five machines M_1, M_2, M_3, M_4, M_5, which manufacture a variety of products P_1, P_2,....P_k (k>1).

Our problem is to find an "optimal" structure with different criterion (social, human, technical, economical). For this we proposed a decomposition of the problem, and a multi-level resolution.

We use a Coloured Petri Nets (Alla 87) to model a structure (see Appendix). The result of the simulation is a Gantt diagram, the production costs and the rates of committment.

QUALITATIVE EVALUATION

Problem statement and modelling.

The intuitive evaluation gives for any group of machines, a team of operators concerned :

$$I_1 = \{ M_1, M_2, M_3 \} \text{ for } T_1 = \{ OP_2, OP_3, OP_4 \}$$
$$I_2 = \{ M_4, M_5 \} \quad \text{ for } T_2 = \{ OP_1, OP_5 \}.$$

The qualitative level propose above shows the man / machine affectation in the group of machines and in the correspondent team : For (T_1, I_1) and (T_2, I_2), we look for an "optimal" structure. For this, we have a biparty graph (Fig. 3). We define human and machines characteristics, and the optimal criterion.

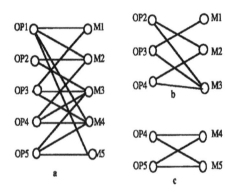

Figure 3 : A biparty graph of man / machine affectation;
a : complete graph, b : (T1, I1), c : (T2, I2)

Using research algorithms of a maximum matching in one biparty graph (Pillou 89), we obtain a set of structures concerning a team T_i and a group of machines I_j.

We study only the case where $|T_i| = |I_j|$
($|X|$: cardinal of X).
For (T1, I1) ($|T1| = |I1| = 3$), (Fig. 3b), we have two realisables structures :

$$ST_{11} = \{ (OP_2, M_3), (OP_3, M_1), (OP_4, M_2) \}$$
and
$$ST_{12} = \{ (OP_2, M_2), (OP_3, M_1), (OP_4, M_3) \}.$$

Where ST_{ij} is a structure number j associated to the biparty graph G_i.
Given that ST_i is a set of structures associated to a biparty graph G_i :

$$ST_i = \{ ST_{i1}, ST_{i2}, ST_{i3},...,ST_{iN} \}, \text{ and}$$

$$J_k = J(ST_{ik})$$

the performance of a structure ST_{ik} for a given criteria J.
Problem : Find one structure $ST_{i.}^{*}$ in STi that optimises a criteria J, with the constraint given by a graph.

Note that a set of realisables structures is finished, for k operators and k machines, the maximal number of structures is k! ($|ST_i| <= k!$). Next sub-sections presents the data of our problems, and algorithms approach to solve it.

Data model

To describe data, the Entity-Relationship model (Chen 76) is used. We introduce three abstraction mechanisms :

i. Entities : An entity describes a set of objects (entity occurrences) sharing common characteristics.

ii. Relationships : A relationship R among the entities E_1, E_2,..., describes the set of relations (relationship occurences or links) relating to the objects e_1 in E_1, e_2 in E_2,..., and (possibly) sharing common characteristics.

iii. Attributes : An attribute describes a characteristic or property of an entity or a relationship, and takes its values from a set or domain.

As an example of attributes for a human operators, we define *"competence"*. This is shown by an extended task for the operator in work (Eyraud 84), which takes it values in :

{ loading/unloading, regulation, resumption, adjustment, correction, testing, programming }

We note that, the elements of this set is hierarchized, i.e. an operator who can test a program, can correct it, adjust it, so we define an order relation ">" like :

MULTI-LEVEL EVALUATION APPROACH

Presentation.

The multi-level evaluation (Fig. 1) methodology proposed is motivated by the integration of some characteristics of human and organizational aspects in production system evaluation.

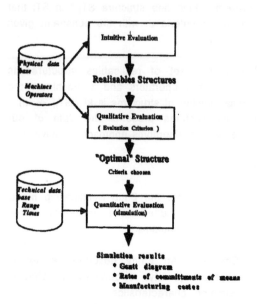

Figure1:Multi-level evaluation methodology

This section presents briefly the intuitive and quantative levels. The qualitative level is to be discussed in the next section.

Intuitive evaluation.

We suppose that our workshop is split into sub-systems, where we have two groups of machines;

$$I_1 = \{ M_1, M_2, M_3 \} \text{ and } I_2 = \{ M_4, M_5 \}.$$

I_j is a group of machines number j.
The methods and tools for decomposition of manufacturing systems are known as group technology (GT) and is surveyed in (Kusiak 1988).

To improve human relations, we introduce a group of human operators or a team. A team is an organization in which there is a single goal or pay off common to all the members (Ho 1972).

We want to choose a group of people to form a team for a given group of machines.

we look for :

$$T_1 = \{ OP_{i1}, OP_{i2}, OP_{i3} \} \text{ and }$$
$$T_2 = \{ OP_{i4}, OP_{15} \}$$

with i_k in $\{1,2,3,4,5\}$,

. T_i (Team) is a group of operators number i. T_i is affected by a group of machines I_i.

A weighted graph is used to present human relation. An heuristic to decompose one weigthed graph into some sub-graphs is proposed and developed in (El Mhamedi 1989).

For our example (Fig. 2), we obtain :

$$T_1 = \{ OP_2, OP_3, OP_4 \} \text{ and } T_2 = \{ OP_1, OP_5 \}$$

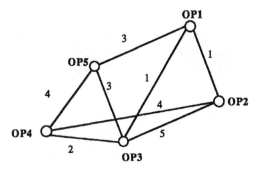

Figure 2 : Operators relations graph

In other levels, we are intereseted in (T_1, I_1) and (T_2, I_2), specially man/machine affectation problems in a team and a group of machines.

Quantitative evaluation

The production system is generally attributed to a multi-purpose multi-variable problem and analytical methods can hardly be used. Simulation is a common and effective technique, especially for the problem of evaluation (Inoue 84).

In this level, we are interested in the functionning of the structure. We suggest one model for product flow simulation.

A ">" B $\Leftrightarrow$ given an operator op,

    if op is in A then op is in B.

(i.e. An operator characterised by A is always characterised by B).

Application :

programming ">" testing ">" correction ">" adjustement ">" resumption ">" regulation ">" loading/unloading.

The "*experience* " is defined and takes its values in two distinct sets, because we have a "*general experience* " takes its values in the following set :

{0-2m, 2-6m, 6-8m, 8-12m, 1-3y, 3-5y, 5-10y, 10y+ } and

a "*professional experience* " in the following :

{0-6m, 6-12m, 1-3y, 3-5y, 5y+ }.

The values of these attributes are also hierarchized.

Machines, products objects and their attributes are given in (Bel 85).

Qualitative evaluation algorithms

In the last sub-section, we have defined the characteristics of men and machines. We have one static representation of manufacturing represented by the objects and their attributes (i.e. Physical data base in Fig. 1). We intend to give some performance evaluation of the system using only a data model. The functionning model is used in simulation.

The data model, that we have is symbolic, vague and imprecise, so it is difficult to use it. To simplify, we suppose that we have an object operator characterized by k attributes; o_1, o_2, ..., o_k : operator (op#, o_1, o_2, ..., o_k) and an object machine : machine (m#, m_1, m_2,..., m_n). One structure is a set of (op#, m#). The difference between two structures in ST_i is the man-machine affectation. To find the best structure of ST_i, we give two methods :

method 1 : qualitative simulation

Qualitative simulation is a key inference process in qualitative causal reasoning (Kuipers

1986). We give one structure ST_{i0} in ST_i, and we assumed to have known the performance of this structure :

$$Perf0 = J(STi0),$$

We define one relation "$\Re$" between two structure performances, given two structures ST_{*i} and ST_{*j} :

$$J(ST_{*i}) \, \Re \, J(ST_{*j}) \Leftrightarrow \{$$

    if (OP_k, M_l) in ST_i and

        (OP_k, M_l) in STj

    then o_i^l ">" o_i^h

    and if (OP_l, M_l) in ST_i and

        (OP_l, M_h) in STj

    then m_i^l ">" m_i^h

    }

(i.e. ST_{*i} is better than ST_{*j})

The relations ">" between human characteristics are defined in the last sub-sections.

For one set ST_i of structures and a given ST_{i0}, the application of this relation ($\Re$) gives a best structures.

This method analyzises the performance tendencies according to changes in production means and/or their structures.

method 2 : optimization

We define one variable x_{ij} in {0, 1} like :

$$x_{ij} = \begin{cases} 1 \text{ if } Op_i \text{ is affected to a machine } M_j. \\ \\ 0 \text{ else.} \end{cases}$$

Contraints of our problem :

$$\begin{cases} \sum_{j=1}^{k} x_{ij} = 1 \quad i = 1, ..., k \\ \Rightarrow \text{ for one machine, we have one operator.} \\ \\ \sum_{i=1}^{k} x_{ij} = 1 \quad j = 1, ..., k \\ \Rightarrow \text{ for one operator, we have one machine.} \end{cases}$$

Criteria :

$$Max (Z) = \sum_{i,j=1}^{k} (c_{ij} \, x_{ij})$$

Where c_{ij} is a performance of the sub-system (OP_i , M_j).

A Simplex algorithm gives an optimal solution of our problem.

Conclusion

Human characteristics define are symbolic, vague and imprecise, and its difficult to used them in computer systems.
The influence of these charachteristics on production system evaluation is difficult to determine.

Two methods are proposed to evaluate a system. The method implemented uses the simple criteria that : " an experienced man works with an unreliable machine".

- The intuitive level generates a team of operators for a given group of machines.
- The qualitative level defines an optimal structure for any given group of machines. This level uses artificial intelligence techniques.
- And the quantitative one suggests a plants simulation using Tempered Coloured Petri Nets.

The qualitative level is developed in this paper, but the other levels are only briefly discussed.

The result of this research is a hybrid system prototype SAMEAH (System Assisting the Modelling and Evaluation of Plants Integrating Human aspects) implemented in PROLOG and C languages on IBM-PC-AT.
Proposed human characteristics are valid in real cases. Our actual work is oriented to the validation of this system in industrial cases.

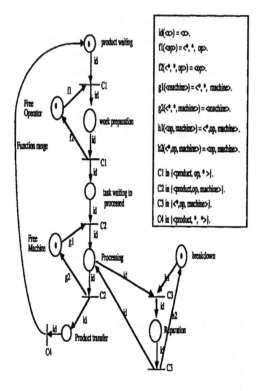

Figue 4 : Coloured Petri Nets model of structure.

CONCLUSION

The object of the paper is to highlight human and organizational aspects in the modelling and evaluation of production systems. To integrate these aspects, a multi-level evaluation methodology is proposed. which consists of three levels.

ACKNOWLEDGEMENTS

The research reported here is supported in part by the Centre National de la Recherche Scientifique (C.N.R.S) : GSIP-G and GR Automatique - A.I.

REFERENCES

Alla, H. A., (1987), " Réseaux de Petri Colorés et Réseaux de Petri Continus : applications à l'étude des systèmes à évenement discrets", Thèse d'etat Grenoble Juin 1987.

Barfield, W. and all., (1986) , "Human - Computer Supervising Performance in the operation and control of Flexible Manufacturing Systems : Methods and Studies", in : A. Kusiak (Ed.), Modelling and Design of Flexible Manufacturing Systems, Elsevier Science, Amsterdam 1986.

Bel, G. and Dubois, D., (1985) " Modélisation et Simulation des Systèmes automatisés de production", APII-1985, 19, 3, 43, Systèmes de production.

Buzacott, J. A.,(1984) " Modelling Manufacturing Systems", in Proc. IFAC, Budapest Hungary, July 1984.

Chen, P. P., (1976), "The Entity Realtionship model : Toward a Unified View of Data", ACM Trans. Database Syst. 1(1), 9-36 (1976).

El Mhamedi, A. and Binder, Z., (1989) , " On the evaluation of organizational structures" IFAC Int. Workshop on Decisional Structures in Automated Manufacturing, Genova, Itay, 18-21, Sept, 1989.

Eyraud, F., Maurice, M., D'Iribane, A. and Rychener, F. (1984) ," Développement des qualifications et apprentissage pour les entreprises des nouvelles technologies : le cas des M.O.C.N dans l'industrie mécanique" , Revue Sociologie de travail No 4-1984.

Ho, Y. C. and Chu, K. Ch., (1972) : " Team Decision Theory and Information Structures in Optimal Control Problems" , IEEE trans. on Automatic Control, Vol. ac-17, No. 1, Feb 1972.

Inoue, I. and Managaki, M. (1984), "A Human-Computer Interactive Evaluation Method in Variety Production Systems", Advances in Production Management Systems, pp.115-124, North-Holland, 1984

Inoue, I. and Yamade. Y., (1985) : "A total evaluation model/ methodology of production systems with the consideration of socio-cultural aspects", in: P. Falster and R. P. Mazumder (Ed.), Modelling Production Management Systems, Elsevier Science Publishers B. V. (North-Holland) pp. 123-134.

Kuipers, B. (1986) , "Qualitative Simulation", Artificial Intelligence 29 (1986) , pp 289 - 338.

Kusiak, A. and Chow, W. S. , (1988) , "Decomposition of Manufacturing Systems", IEEE Jou. of Robotics and Automation, Vol. 4, No. 5, October 1988.

Pillou, C. and Rech, C. (1989), "A simple algebraic algorithm for the determination of the generic-rank of structured systems", Int. journal of Control.

Rasmussen J. (1986) , " Information processing and human-machine interaction", North-Holland Series in System sciences and Engineering. Andrew-P., Sage (Ed.)

Rouse, W. B. and Cody W. J. (1988) , "On the Design of Man - Machine Systems : Principles, Practices and Prospects", Automatica Vol. 24, No. 2, pp 227 - 238, 1988.

DEVELOPING INDUSTRIAL SYSTEMS ACCORDING TO THE PROCESS INTERACTION APPROACH

R. Overwater and J. E. Rooda

Department of Mechanical Engineering, Eindhoven University of Technology, Eindhoven, The Netherlands

Abstract.

When developing automated industrial systems, models and modelling play a prominent part. Models are built according to a world view and it is felt that many of today's problems and difficulties encountered with industrial automation can be removed by chosing a different approach to the modelling of industrial systems. This paper presents the process interaction approach as an alternative world view for modelling industrial systems. The approach yields two major benefits: its concepts can be used to model all kinds of activity throughout the system, varying from machines control to factory control, and they can be used throughout the entire development process, i.e. the same concepts apply to the specification, the validation and the implementation of industrial systems.

Key words.

Industrial systems, modelling, simulation, control.

INTRODUCTION

The last decade was witness to enormous growth in the application of computer-based solutions to complex industrial problems. Unfortunately, this continually increasing complexity exacerbates new problems and difficulties that have become increasingly apparent as time passed.
When developing automated industrial systems, models and modelling play a prominent part; since, it is not possible to formalize just any part of a system without forming a picture of its operations.

Although systems development cannot be defined as a detailed step-by-step procedure, it is possible to give some global indication of the progression of a system under development. In this perspective, three phases can be recognized: the specification, the validation and the implementation of the system-to-be. The transitions between these phases are not readily apparent. Rather, the distinction between the phases will be suggested by the nature of the activities, more than by the recognition of the attainable milestones during the development process. In fact each phase includes something of the other two, because in each phase the system may undergo modifications, which, in their turn, will require specification, validation and implementation. Still, the nature of the activities gradually

changes: from abstract and qualitative during specification, to exact and quantitative during validation and, finally, to concrete and realistic during implementation.
All three phases involve a lot of modelling. During the specification, a model is built to reflect the system's purpose and its essential properties, and to enable ideas and views on the system to be discussed. Next, validation requires a model of the system that can be examined in order to gain an insight into the system's performance. And finally, the difference between the system and its model fades away entirely when the model becomes incorporated in the system during its implementation.

Models are built according to a world view and it is felt that many of today's problems and difficulties encountered with industrial automation can be removed by chosing a different approach to the modelling of industrial systems. This paper presents the process interaction approach as an alternative world view for modelling industrial systems.

PROCESSES AND INTERACTIONS

It is hard to give a precise definition of the

notion 'industrial system'. Factories, distribution centers and harbours are typical examples of
it, but manufacturing systems and conveyor
systems can be regarded as industrial systems
too. The operations in an industrial system can
be described as the manufacture, handling and
transfer of products. These operations are
performed by three kinds of processes, namely
production processes, transport processes and
control processes. All processes take place
simultaneously and relate with each other through
the exchange of products and information. This
understanding reveals the two basic concepts of
the process interaction approach to the modelling
of industrial systems: processes and interactions
[Overwater, 1987]. Through these concepts, it is
possible to describe the sequential, the parallel, as well as the non-deterministic aspects of
an industrial system in one and the same model.
The idea of parallel processes, on itself, is not
new: examples of this world view can be found in
the fields of discrete event simulation and
computing science. Applying it to the development
of industrial systems, however, is indeed new. In
this context a process can be defined as a set of
changes of state in a system produced by an
active element. Active elements are those parts
of the system that can perform actions. Examples
of active element are machines, robots, conveyors, controllers and, off course, men. Passive
elements, on the other hand, are those parts of a
system that can only undergo actions. Examples
are raw materials, work pieces, pallets and
boxes, but also, data or signals.
An interaction is defined as the transfer of a
passive element (or its contents) in order to
achieve cooperation between active elements.
Cooperation has two different aspects, synchronization and communication. Therefore, two fundamental transfer mechanisms can be discerned: a
discrete and a continuous mechanism. The discrete
mechanism emphasizes the synchronizing quality of
interaction, while, the continuous uses the
communicative quality. Both mechanisms provide a
'send-type' operation (S-action) and a 'receive-
type' operation (R-action) through which interaction can be accomplished.

THE SPECIFICATION OF SYSTEMS

This section describes a language for expressing
models developed according to the process
interaction approach. The language is supported
by the software package D86 [Munneke, Overwater,
1986] and it combines graphic notations with a
restricted use of natural language. The graphical
part is formed by the Process Interaction Diagram
(PRIND), the textual part by the Diagram Documentation (DIDOC). A Process Interaction Diagram is
a graphic representation of a system showing the
processes taking place and the interactions
between them. Each Process Interaction Diagram is
associated with a Diagram Documentation in which
these processes and interactions are documented. Models represented in this

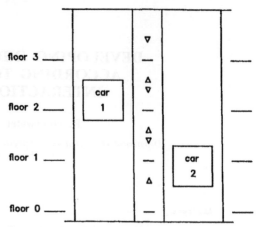

Figure 1. The Elevator System

language must obviously consist of at least one
PRIND and one DIDOC, but may be expressed as
structures of PRINDs and DIDOCs as well. In this
way the language can represent parallel processes
as sequential routines.The creation of PRINDs and
DIDOCs is subject to a number of conventions and
rules which are explained here by means of a
small case study. The object of the case study is
to develop a control structure for an elevator
system consisting of two cars (see figure 1). An
important requirement is that the cars must
operate under mutual exclusion, since both of
them may serve each floor.

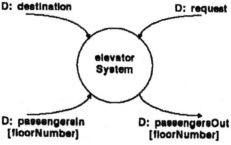

Figure 2a. PRIND Context

One of the first issues to be considered is how
to recognize the system's interactions with its
environment. At first sight, this seems to be
rather trivial. Passengers are transported by the
system and, thus, can be considered as passive
elements. From this point of view, the system
exchanges passengers with the environment.
Passengers, on the other hand, can also be
considered as active elements, as they make
demands on the elevator system and choose their
destinations. This duality can be solved by
assuming that the requests and destinations are
generated outside the system's boundaries. Now,
the PRIND Context can be created. As follows from
figure 2a, a process is represented by a 'bubble'
and an interaction by means of an arrow. The
PRIND shows how passengers, as well as, their
requests and destinations are modelled as passive
elements and that they are exchanged in a
discrete manner as indicated by the character

'D'. The DIDOC Context (figure 2b) contains the
documentation associated with the PRIND Context.
A DIDOC can be divided into three parts: an
object definition part, an interaction definition
part and a process definition part.

```
*** Objects ***
numberOfFloors  @ cardinal;
callDir         @ up | down;
floorNumber     @ [0..numberOfFloors-1];
numberOfCars    @ cardinal;
carNumber       @ [1..numberOfCars];
*** Interactions ***
Request       @ dis / object
                         floor @ floorNumber;
                         dir   @ callDir
                         end;
Destination   @ dis / object
                         carNr @ carNumber;
                         floor @ floorNumber
                         end;

PassengersIn  @ dis / array [floorNumber] of
physical;

PassengersOut @ dis / array [floorNumber] of
physical;

*** Processes ***

elevatorSystem @ expanded;
```

Figure 2b. DIDOC Context

The first part contains definitions of passive
elements that can serve as a reference in other
definitions. In these definitions the symbol '@'
should be read as 'appears as'. Several structuring operators are available to build up a new
definition. The second part defines the interactions that were new introduced in the corresponding PRIND. Interaction definitions show both the
transfer mechanism ('dis' or 'con') and the kind
of passive element that is transferred. And the
last part shows the process descriptions.
Processes can be defined using a Modula2-like
pseudo-code. This pseudo-code permits informal
texts to be used as long as their mood is
imparative and, also, it provides a notation for
describing S(end)- and R(eceive)-actions.
The process 'elevatorSystem' still contains
parallel processes and, therefore, it can be
expanded into a new PRIND. This is indicated by
the desription 'expanded'. Through this concept
of deepening hierarchies of PRINDs can be
created. The PRIND ElevatorSystem (figure 3a)
contains only two bubbles, one representing the
control process ('elevatorControl') and the other
representing the car processes involved ('car
[carNumber]'). The pattern of interaction between
these processes, and the car movements resulting
from it, can be explained as follows: each time a
car approaches a floor, it sends a 'commandRequest' to the 'elevatorControl'. The 'elevator-
Control' replies with a 'carCommand' indicating

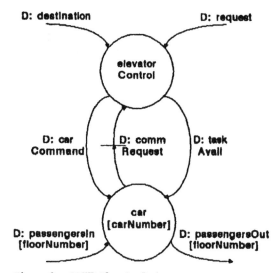

Figure 3a. PRIND ElevatorSystem

whether the car should continue moving 'up' or
'down', or should 'stop' at this floor. As soon
as it has stopped and its doors are opened, it
sends a new command request; the next command can
be a command to continue in either direction, or
a second stop command - when a request for the
opposite direction has been assigned - or, the
command 'none' which means that no task is
available yet. In the latter case, the car will
wait stationary for a signal ('taskAvail') from
the control process to say that a new task may be
available. In response to this signal, the car
will release a new 'commandRequest'. Figure 3b
shows the part of the DIDOC ElevatorSystem which
contains the description of the car processes.

```
car @ array carNumber of
process
  curCarMode @ mode;
begin
  curCarMode:= none;
  loop
    wait for approach signal;
    commRequest:= carNumber;
    give CommRequest;
    take CarCommand;
    case carCommand of
      stop: begin
              if curCarMode # stop
              then
                slow motor down;
                wait for position signal;
                stop motor;
                open the doors;
              end;
              make passengers enter and
              leave
            end;
      up  : if curCarMode # up
            then shut doors and start motor
            upwards
            else keep moving upwards end;
```

```
     down: if curCarMode # down
           then shut doors and start motor
                downwards
           else keep moving downwards end;
     none: loop
              take DestAvail[carNumber]
              within 0;
              if destAvail[carNumber] #
                 timeOut
              then exit end;
              take RequestAvail[carNumber]
              within 0;
              if requestAvail[carNumber] #
                 timeOut
              then exit end
           end
     end;
     curCarMode:= carCommand
  end
end;
```

Figure 3b. The process 'car'

The process 'elevatorControl' governs the way how
the different requests are assigned to cars. A
review of various control strategies can be found
in Barney and Dos Santos (1977). The strategy
chosen in this case is known as the 'longest
route strategy' which aims to keep each car
travelling in the same direction as far as
possible. To this end, the control process
includes a data structure which retains all the
current tasks of the system. Each car is associated with one set of tasks – composed of destinations and requests of passengers – that have been
assigned to that car. Each time the control
process receives a 'commandRequest' from a car,
it updates the position of that car in the data
structure and searches a new task for it. After
the search has been completed, a new 'carCommand'
is released.

THE VALIDATION OF SYSTEMS

Models built with D86 serve the purpose of
specification and provide a means to talk and
think about industrial systems, as well as,
revealing the processes and interactions in the
system in a structured way. The processes in the
model not only relate to the activities on
production level, but also to all control
activities needed to achieve this production.
Although these models yield a great insight into
the functioning of an industrial system, they can
never guarantee that the system will work as
intended. For this purpose the model must be
validated.
Validation of models is a common practice in
engineering and usually, it is achieved with
mathematics. This, however, is not always
possible for industrial systems. Many relations
in industrial systems have a stochastic nature
and cannot be modelled mathematically so that
they can be solved analytically or numerically.
One solution for this problem is the application
of discrete simulation. Discrete simulation is a

validation technique which can be applied to
discrete systems. The characteristic feature of
discrete systems is that their behaviour can be
described as a sequence of events. At each event,
the state of the system alters; whereas, in the
time between two successive events, the state
remains unchanged.
Simulation experiments require a model of the
system based on the system's layout and control
strategies. It should be possible to describe the
model in terms of a computer program, thus,
offering an opportunity to run the experiment on
a computer. The computer program is then referred
to as 'simulation program'.
A simulation model is built according to a
certain world view [Nance, 1981]. The simulation
package S84 [Rooda et al., 1984] uses the process
interaction approach and, thus, it would appear
to be appropriate for the validation of models
built with D86. The package provides facilities
that describe processes and interactions in much
the same way as those presented in the previous
section. Other facilities support the random
generation of input, the collection of data on
the system's performance, the tracing of events
and the generation of graphical output. Figure 4
gives an impression of this latter facility.
Animation of this diagram, makes it possible to
detect deficiencies in the strategy.
In case of the elevator system, the animation
revealed that the initial implementation of the
control strategy could not prevent 'bunching'.
'Bunching' is the phenomenon of two or more cars
travelling in phase over a certain period. As a
result, the control strategy was modified.
An important aspect of simulation is that a
simulation model should include a model of the
real-world environment in which the simulated
system has to function, too. The environment of
the elevator system was modelled by means of
'floorPanel' processes simulating the generation
of 'requests', and 'carPanel' processes simulating the generation of 'destinations'. In spite
of these extensions, the basic structure of the
model remained unchanged.

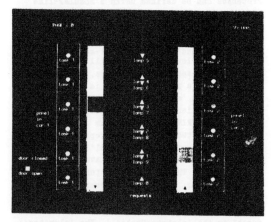

Figure 4. Graphical simulation of the Elevator
          System

THE IMPLEMENTATION OF SYSTEMS

Control forms an integral part of a system and determines the system's overall behaviour in association with the processes to be controlled. Control processes are often referred to as internal processes, because they usually remain within the limits of a computer. In a similar way, the processes to be controlled can be considered as external processes.

Control strategies are designed with the aid of a model of the external processes plus a model of the control processes. Both models have been defined in the specification of the system. In the previous section, simulation models have been introduced as a major tool for the validation of these specifications. Due to their purpose, these models must contain the required control strategies in a more or less explicit way. According to Rooda and Boot (1983), an interesting opportunity has arisen with respect to the implementation of control. Fundamental to their idea is that the simulation model, as well as, the necessary control system are computer programs, provided that a computer is used as the controller. From this perspective they suggest that, if the simulation program is detailed enough, part of it is exchangeable with the control program. In order to implement this strategy, the simulation model should be divided into three parts: the model of the control system, the model of the 'real world' and the synchronization between these two. Ideally, the simulation model could be converted into a control program as easy as replacing the 'real world' processes by interfaces with this 'real world'. Signals would now be generated by the hardware rather than by the simulation processes. The rest of the model should remain unchanged.

The software package ROSKIT [Rossingh, Rooda, 1984] has been developed in order to meet the real-time requirements mentioned above and it provides control facilities based on the process interaction approach.

Figure 5. The scale model of the Elevator System

The simulation program mentioned in the previous section was converted into a ROSKIT program to control a scale model of the elevator system (figure 5). The program contains the same processes as the simulation program. Now, however, the

processes 'car', 'floorPanel' and 'carPanel' form the interface with the 'real world'.

Unfortunately, the conversion is not as simple as suggested above. When building control programs the problem of parallelism in real-time operations is encountered to its full extent for the first time. In case that the control program is executed on a single processor machine, the concurrency found within the computer does not correspond to the parallelism in the 'real world'. In fact, the control program just feigns parallelism. As a consequence, operations may require mutual exclusion, and timing problems must be solved by assigning priorities to processes. The use of priorities, on the other hand, must be regarded as an 'escape' solution, as – strictly speaking – they conflict with the idea of parallelism. Too many priorities will interfere with fair scheduling and may cause the starvation of processes, quite easily.

It should be noted that specification models do not suffer this discrepancy, as they simply assume that each process is executed by its own processor. In simulation modelling the problem is less apparent too, since a simulation model contains its own 'real world' and, consequently, all processes are based on the same kind of parallelism.

CONCLUSION

This paper introduced the concepts of process and interaction as an approach to the development of industrial systems. The approach has been applied successfully to a variety of cases. A few of them will be mentioned for further illustration.

Some major examples, in the field of production and machines control, were the development of a transport and storage system for a bicycle-tyre factory [Arentsen, 1981], the simulation of a modular internal transport system [Miert, 1984; Pont, 1984; Bergkamp, 1985], and the simulation and control of a power-and-free conveyor system [Toet, 1986]. Another case concerned the control of an automated machine for building up and breaking down a stack of pallets [Rooda, Boot, 1986].

In the area of distribution and transportation, some characteristic examples have included an examination of the operational characteristics of closed-loop conveyors [Rooda, Nawijn, 1980; Nawijn, Rooda, 1982], a strategic study concerning coal transshipment [Rooda, Schilden, 1982], an analysis and simulation of a flower auction [Rooda et al., 1982], and the development of a planning system for various combinations of public transport services [Stiphout, 1983]. Other examples have included the development of an architecture of loosely-coupled processors [Vuurboom, 1984] and the simulation of a local area network in a highly automated production environment [Welmer, 1985].

Some recent (unpublished) case studies include the development of an order picking system for a dairy factory, a feasibility study for a conveyor

and storage system for fibre products, and the
design of a control system for the production of
so-called 'wafers' in a chip factory.

The considerable experience gained so far,
demonstrated that the process interaction
approach did indeed provide a suitable world view
for the development of industrial systems. Its
power springs from the distinction between the
cyclic, reversible phenomena (the processes) and
the non-cyclic, irreversible phenomena (the
interactions) in an industrial system. A system
no longer needs to be modelled as a single
sequence of events, but it can be divided into a
number of simultaneously operating work cycles
which can be considered independently. As a
result, two major advantages could be attained.
Firstly, it became possible to establish an
integration between all kinds of activity
throughout the system, varying from machine
operations to production control. And secondly,
it permitted a far going integration of the three
phases of systems development: the same concepts
applied to the specification, the validation and
the implementation of industrial systems.

REFERENCES

Arentsen, J.H.A. (1981). Ontwerp van een trans-
port- en opslagsysteem t.b.v. de fabricage
van fietsbuitenbanden (in Dutch), MSc.
thesis, Fac. of Mech. Eng., Twente
University, Enschede.

Barney, G.C., and S.M. Dos Santos (1977). Lift
Traffic Analysis, Design and Control, Peter
Peregrinus, London.

Bergkamp, W.M. (1985). Een simulatieprogramma
voor een intern transportsysteem (in Dutch),
MSc. thesis, Fac. of Mech. Eng., Twente
University, Enschede.

Miert, G.J.C. van (1984). Simulatie en besturing
van een modulair accumulerend transportsysteem met het simulatiepakket SOLE (in
Dutch), MSc. thesis, Fac. of Mech. Eng.,
Delft University of Technology, Delft.

Munneke, B., and R. Overwater (1986). Design '86
(in Dutch), Manual, Fac. of Mech. Eng.,
Eindhoven University of Technology, Eindhoven.

Nance, R.E. (1981). The time and state relationship in simulation modelling, Comm. ACM, 24
(4), 173-179.

Nawijn W.M., and J.E. Rooda (1982). Some further
results for closed-loop continuous conveyors, Mechanical Communications, Twente
University, Enschede.

Overwater, R. (1987). Processes and Interactions,
An approach to the modelling of industrial
systems, Dissertation, Eindhoven University
of Technology, Eindhoven.

Pont, M.J.H. de (1984). A simulation model for
handling and routing of product carriers in
a modular internal transport system, MSc.
thesis, Fac. of Mech. Eng., Twente University, Enschede.

Rooda, J.E., and W.C. Boot (1983). A combined
approach for the design and control of
logistics systems, Proc. 4th Int. Logistics
Congress, Dortmund.

Rooda, J.E., and W.C. Boot (1986). Processcompu-
ters, Processen en interacties (in Dutch),
Memo.nr. WPA 252, Fac. of Mech. Eng.,
Eindhoven University of Technology, Eindhov-
en.

Rooda, J.E., Jansen, P.F., and M.E.A. Striekwold
(1982). Analyse des Verteiler-Systems einer
Auktionszentrale fuer Blumen (in German),
Foerdern und Heben, 32 (10), 774-777.

Rooda, J.E., S.M.M. Joosten, T.J. Rosssingh and
R. Smedinga (1984). Simulation in S84,
Manual, Fac. of Mech. Eng., Twente Universi-
ty, Enschede.

Rooda, J.E., and W.H. Nawijn (1980). The
analysis of operation characteristics of
loop continuing belt conveyors using
simulation and queueing approximations,
Mechanical Communications, Twente University, Enschede.

Rooda, J.E., and N. van der Schilden (1982).
Simulation of maritime transport and
distribution by sea-going barges: an
application of multiple regression analysis
and factor sreening, Bulk Solids Handling,
2, 813-824.

Rossingh, T.J., and J.E. Rooda (1984). RealTime
Operating System KIT, Manual, Fac. of Mech.
Eng., Twente University, Enschede.

Stiphout, J.W.P.J. van (1983). Mobiele Service
Systemen, modellering en simulatie (in
Dutch), MSc. thesis, Fac. of Applied
Mathematics, Twente University, Enschede.

Toet, E.M. (1986). Simulatie en besturing van een
power-and-free transportsysteem (in Dutch),
MSc. thesis, Fac. of Mech. Eng., Eindhoven
University of Technology, Eindhoven.

Vuurboom, R. (1984). Communication in a loosely-
coupled multi-microprocessor system, Its
architecture and implementation, MSc.
thesis, Fac. of Computing Science, Twente
University, Enschede.

Welmer, H.J. (1985). Ontwerp van een computer-
netwerk werkend onder een MUPOS of ROSKIT
Modulair Pascal beheerssysteem (in Dutch),
MSc. thesis, Fac. of Computing Science,
Twente University, Enschede.

KINEMATICS OF A THREE DEGREE-OF-FREEDOM, TWO LINKS LIGHTWEIGHT FLEXIBLE ARM

V. Feliú*, K. S. Rattan** and H. Benjamín Brown, Jr.*

*Robotics Institute, Carnegie Mellon University, Pittsburgh, Pennsylvania, USA
**Department of Electrical Engineering, Wright State University, Dayton, Ohio, USA

Abstract. The objective of this paper is to develop the kinematic transformations of a three degree-of-freedom, two-link flexible arm. These transformations are needed to carry out real-time control of robots. For rigid arms, these relations can be easily calculated, but in flexible arm they turn out to be nonlinear and coupled. Several algorithms to calculate these transformations are proposed in this paper, and a comparative study is performed. It was founnd that a noniterative algorithm to accurately estimate these kinematic transformations exists and is well suited for the control of flexible arms.

Keywords. Flexible arm; kinematics; robotic manipulators; compliance.

1 INTRODUCTION

Traditionally, robot manipulators have been modelled and controlled as if they were rigid structures whose states could be completely and precisely described in terms of joint coordinates[1-2]. This assumption substantially simplifies the kinematic and dynamic equations, and minimizes the amount of sensing required. The position and orientation of the end effector could be determined directly from the joint angles, independent of the forces on the system. While ignoring flexibility in a robot arm does simplify its modelling, it places severe limits on its design and performance. For the rigid model to be reasonable, all joint and links of a manipulator must be designed with adequate stiffness so that the deflection at the end effector is negligible under normally encountered loads, acceleration, external disturbances, etc. Thus, we find that a typical robot arm weighs many times its load-bearing capacity. The mass of the arm dominates, and actuators, sized accordingly, tend to become heavy and slow. Allowing robot arms to be flexible is advantageous in many ways. Flexible arms can be lighter, less expensive to build, faster, consume less energy, and can be easily transported. The last two features are critical in outer space applications and in mobile robots. Flexibility may appear both in the joints and in the links, but the term *flexible arm* has generally been used for manipulators with flexible links. In this paper, the term flexible arm is associated with flexible links. Flexibility in the links is studied because of two reasons:

1. It is an undesired phenomenon that dominates the dynamics of the arm at high frequencies (this also happens with joint flexibility [3]). Thus, controllers have to be designed to compensate for undesired oscillations that appear when performing fast movements. Undesired link flexibility typically appears in large robots.

2. There are a lot of potential applications where flexible links may be advantageous. In this case, flexibility is viewed as a feature to be exploited in the mechanical and control design, rather than a problem to be studied in isolation. Two features that may be exploited in flexible links are their light weight and compliance.

While flexibility may be exploited to improve the mechanics of the arm, their control becomes significantly more complex, and more sensors are needed. The goal of our research is to build an industrial type three degree-of-freedom, two-link flexible arm (see figure 1, where the convention $l_1 = 0$ has been adopted). Most of the existing work in flexible arms is oriented to one-link one-joint planar arms that exhibit several vibrational modes [4-7]. Little work has been done on two degree-of-freedom planar arms moving either in the horizontal plane [8] or vertical plane [9-10]. All of the existing work has focused on the dynamic control of these robots. Henrichfreise [11] controlled a three degree-of-freedom, two-link flexible arm. In this paper, the kinematic relations of a two-link three-jointed flexible arm is carried out. We show that efficient algorithms may be used to compute the kinematics of flexible arms, facilitating their control.

Our arm is structurally compliant in all directions, with links exhibiting flexibility in the plane of rotation as well as in the plane perpendicular to it. Torsional deformation as well as compression-extension of the links are also considered. The arm is designed in such a way that the masses are concentrated on the tip and in the joints, using mass-less links. This is a reasonable approximation in many other flexible arms. The first important problem that appears when dealing with flexible arms, whose links are compliant in all directions, is in the definition of the kinematic equations. As was noted in [12], kinematic equations not only depend on the position values but also on the weights and external forces/moments applied to the arm. The kinematics for the case of two rigid masses connected by a chain of massless beams having an arbitrary number of mass-less rotational joints has been developed by Book [12]. In this paper, we will first generalize this method to the case of mass-less links, with masses concentrated in the joints and later extend it to the two-link case.

The kinematics of flexible arms are defined by transformation between positions and coordinates axis, as in rigid arms. However, in the case of flexible arms, a new kinematic parameter, *the compliance matrix*, that represents the changes in position and orientation of the tip of the arm as a function of the applied external forces/moments, also needs to be considered. Section 2 extends Book's method to the case of concentrated masses in the joints. Section 3 extends this method to our arm configuration. Section 4 proposes a numerical method to compute direct kinematics, and conclusions are drawn in Section 5.

---

[1]Visiting Prof., Dpto Ingenieria Electrica, Electronica y Control, UNED, Ciudad Universitaria, Madrid-28040, Spain.

2   Kinematic Equations

In order to generalize the analysis, we will assume that the masses are concentrated at the neighbourhood of the joints, instead of at the joints. We assume that all joints are rotational. Using the method and notation described in [12] (see figures 2, 3), the coordinate transformation matrices are given by

$$\left[\begin{array}{c} 1 \\ \bar{X}_{i,i-1} \end{array} \right] = A_i E_i \left[\begin{array}{c} 1 \\ \bar{X}_{i,i} \end{array} \right] = A_i E_i \left[\begin{array}{c} 1 \\ \bar{0} \end{array} \right] \qquad (1)$$

where

$\bar{X}_{i,i-1}$ = the position of the origin of coordinate system i in terms of coordinate system $i-1$,

A_i = transformation matrix with no deflection,

E_i = transformation matrix due to deflection,

$\bar{0}$ = a 3×1 vector whose elements are zero,

$\bar{X}_{i,i}$ = location of point P in the ith coordinate system. In this case, P is the origin of coordinate system i.

Matrix E_i is a function of the forces/moments applied to P, and is of the form

$$E_i = \left[\begin{array}{cccc} 1 & 0 & 0 & 0 \\ \Delta X_i & 1 & -s\theta_{Zi} & s\theta_{Yi} \\ \Delta Y_i & s\theta_{Zi} & 1 & -s\theta_{Xi} \\ \Delta Z_i & -s\theta_{Yi} & s\theta_{Xi} & 1 \end{array} \right] \qquad (2)$$

where $s\theta = sin(\theta)$ and $c\theta = cos(\theta)$. For small deflections, E_i may be expressed as

$$E_i = \left[\begin{array}{cccc} 1 & 0 & 0 & 0 \\ \Delta X_i & 1 & -\theta_{Zi} & \theta_{Yi} \\ \Delta Y_i & \theta_{Zi} & 1 & -\theta_{Xi} \\ \Delta Z_i & -\theta_{Yi} & \theta_{Xi} & 1 \end{array} \right] \qquad (3)$$

where θ_{Xi}, θ_{Yi}, and θ_{Zi} are the angles of rotation about the X_i, Y_i, and Z_i axes, respectively. The elements of this matrix are given by

$\Delta X_i = \alpha_{ci} \cdot F_{Xii}$

$\Delta Y_i = \alpha_{F_y i} \cdot F_{Yii} + \alpha_{M_z i} \cdot M_{Zii}$

$\Delta Z_i = \alpha_{F_z i} \cdot F_{Zii} - \alpha_{M_y i} \cdot M_{Yii}$

$\theta_{Xi} = \hat{\alpha}_{M_z i} \cdot M_{Xii}$

$\theta_{Yi} = -\hat{\alpha}_{F_z i} \cdot F_{Zii} + \hat{\alpha}_{M_y i} \cdot M_{Yii}$

$\theta_{Zi} = \hat{\alpha}_{F_y i} \cdot F_{Yii} + \hat{\alpha}_{M_z i} \cdot M_{Zii}$

where, in general,

$F_{Xij}, F_{Yij}, F_{Zij}$ = forces at the end of link i, in terms of coordinate system j and $F_{ij}^T = \left(\begin{array}{ccc} F_{Xij} & F_{Yij} & F_{Zij} \end{array} \right)$

$M_{Xij}, M_{Yij}, M_{Zij}$ = moments at the end of link i, in terms of coordinate system j and $M_{ij}^T = \left(\begin{array}{ccc} M_{Xij} & M_{Yij} & M_{Zij} \end{array} \right)$

α_{ci} = coefficient of compression, (displacement/unit force)

$\alpha_{F_y i}$ = coefficient of bending in Y direction, (displacement/unit force)

$\alpha_{F_z i}$ = coefficient of bending in Z direction, (displacement/unit force)

$\alpha_{M_z i}$ = coefficient of bending in Y direction due to M_Z, (displacement/unit moment)

$\alpha_{M_y i}$ = coefficient of bending in Z direction due to M_Y, (displacement/unit moment)

$\hat{\alpha}_{M_z i}$ = coefficient of torsion, (angle/unit moment)

$\hat{\alpha}_{F_y i}$ = coefficient of bending in θ_z rotation due to F_Y, (angle/unit force)

$\hat{\alpha}_{F_z i}$ = coefficient of bending in θ_y rotation due to F_Z, (angle/unit force)

$\hat{\alpha}_{M_y i}$ = coefficient of bending in θ_y rotation due to M_Y, (angle/unit moment)

$\hat{\alpha}_{M_z i}$ = coefficient of bending in θ_z rotation due to M_Z, (angle/unit moment)

Forces and moments $\bar{F}_{ii}, \bar{M}_{ii}$ at the end of the link i, in coordinates system i, are needed to compute E_i. In the case of massless chains, the only stresses supported by the structure are produced by the force/moment at the tip of the arm ($\bar{F}_{N0}, \bar{M}_{N0}$). Then we have

$$\left[\begin{array}{c} \bar{F}_{ii} \\ \bar{M}_{ii} \end{array} \right] = \left[\begin{array}{cc} R_{0i} & 0 \\ \bar{r}_{iN} \times R_{0i} & R_{0i} \end{array} \right] \left[\begin{array}{c} \bar{F}_{N0} \\ \bar{M}_{N0} \end{array} \right] \qquad (4)$$

where R_{ij} is the rotation matrix from system i to system j, $\bar{r}_{ij}$[2] is the distance vector from the origin of system i to the origin of system j in terms of coordinates i, and $\times$ indicates a vector cross product.

This equation is easily generalized to our case, in which we include in the model a mass concentrated in the neighbourhood of each joint. Using the notation introduced in figure 4, and assuming that the forces and moments $\hat{F}_{i0}, \hat{M}_{i0}$ produced by mass m_{i+1} (which is related to joint $i+1$ and is located at a point $\bar{p}_{i+1}$ in coordinates i) are applied at the point of intersection of links i and $i+1$, the above equation becomes

$$\left[\begin{array}{c} \bar{F}_{ii} \\ \bar{M}_{ii} \end{array} \right] = \sum_{j=i}^{N-1} \left[\begin{array}{cc} R_{0i} & 0 \\ \bar{r}_{ij} \times R_{0i} & R_{0i} \end{array} \right] \left[\begin{array}{c} \hat{F}_{j0} \\ \hat{M}_{j0} \end{array} \right] +$$

$$\left[\begin{array}{cc} R_{0i} & 0 \\ \bar{r}_{iN} \times R_{0i} & R_{0i} \end{array} \right] \left[\begin{array}{c} \bar{F}_{N0} \\ \bar{M}_{N0} \end{array} \right] \quad \forall i < N \qquad (5)$$

where $\hat{F}_{j0}^T = \left(\begin{array}{ccc} -m_{j+1} \cdot g & 0 & 0 \end{array} \right)$, and $\hat{M}_{j0}^T = (R_{j0} \cdot \bar{p}_{j+1}) \times \bar{F}_{j0}$.

Compliance Matrix

The kinematic model developed above allows us to express the tip position and orientation as a function of the joint angles, and the forces and moments applied to the tip ($m_i, \bar{p}_i$ are structural parameters) as

$$\Delta X = \underbrace{A_1 \cdot E_1 \cdot A_2 \cdot E_2 \cdots A_N \cdot E_N}_{T_{N0}} \cdot \left(\begin{array}{c} 1 \\ \bar{0} \end{array} \right) \qquad (6)$$

Assuming that the joint angles remain fixed, the compliance matrix of a flexible arm is defined as the differential relation that describes the changes produced in tip position and orientation caused by changes in tip forces and moments. The compliance matrix C is thus given by

$$C^T = \left(\begin{array}{c} \frac{\partial}{\partial \bar{F}_{N0}} \\ \frac{\partial}{\partial \bar{M}_{N0}} \end{array} \right) \left(\begin{array}{cc} \bar{X}_{N0}^T & \bar{\theta}_{N0}^T \end{array} \right) \qquad (7)$$

where $\bar{\theta}_{N0}^T = \left(\begin{array}{ccc} \theta_{N01} & \theta_{N02} & \theta_{N03} \end{array} \right)$ represents the rotation of coordinates system N relative to the base coordinate system. Differentiating $\bar{X}_{N0}$ with respect to F_{XN0}, we get

$$\frac{\partial \bar{X}_{N0}}{\partial F_{XN0}} = [\sum_{i=1}^{N} A_1 E_1 \cdots A_i \frac{\partial E_i}{\partial F_{XN0}} A_{i+1} \cdots A_N E_N] \left[\begin{array}{c} 1 \\ \bar{0} \end{array} \right] \qquad (8)$$

Similar expressions are obtained when $\bar{X}_{N0}$ is differentiated with respect to $F_{YN0}, F_{ZN0}, M_{XN0}, M_{YN0}, M_{ZN0}$, and

$$\frac{\partial}{\partial F_{XN0}} \left[\begin{array}{cccc} 1 & 0 & 0 & 0 \\ (X_0)_{N0} & 1 & -\theta_{ZN0} & \theta_{YN0} \\ (Y_0)_{N0} & \theta_{ZN0} & 1 & -\theta_{XN0} \\ (Z_0)_{N0} & -\theta_{YN0} & \theta_{XN0} & 1 \end{array} \right]$$

---

[2] notation of [12] has been adapted here to include our case

$$= \sum_{i=1}^{N} A_1 E_1 \cdots A_i \frac{\partial E_i}{\partial \bar{F}_{XN0}} A_{i+1} \cdots A_N E_N \qquad (9)$$

Similar expressions are obtained when we differentiate with respect to $F_{YN0}, F_{ZN0}, M_{XN0}, M_{YN0}, M_{ZN0}$. Details may be found in [12].

Note that E_i is explicitly a function of $\bar{F}_{ii}, \bar{M}_{ii}$, but we want the derivatives with respect to $\bar{F}_{N0}, \bar{M}_{N0}$. These derivatives can be calculated by using (5). Assuming that R_{0i} and $\bar{r}_{ii} \times R_{0i}$ are independent of $\bar{F}_{N0}, \bar{M}_{N0}$, which is true to the first order, the form of the differential relations among these variables may be easily calculated. It may be shown that they are explicitly independent of the joint masses. There exists an implicit dependence on these masses because the derivatives are calculated at points of static equilibrium of the arm, and are dependent on all the loads in the structure.

3 Application to the 3-d.o.f. 2-links Case

3.1 Arm description

Consider the mechanical configuration of figure 1 with rotational joints and two flexible links. The links are made of fiberglass tube with a 2 inch outside diameter and 1.84 inch inner diameter. Both links are identical, and are 30 inches long. Each link weighs 1 pound, and joint 3 weighs 6 pounds (motor included). The properties of the material are: $E = 3 \times 10^6 psi$, $G = 1.5 \times 10^6 psi$ and maximum flexural stress $= 30 \times 10^3 psi$. From these coefficients and the geometry of the link, we get

$$\alpha_c = 0$$
$$\alpha_{F_Y} = \alpha_{F_z} = 13.45 \times 10^{-3} in./lb$$
$$\alpha_{M_x} = \alpha_{M_y} = 0.224 \times 10^{-3} lb^{-1}$$
$$\hat{\alpha}_{M_x} = 0.0449 \times 10^{-3} in.^{-1} lb^{-1}$$
$$\hat{\alpha}_{F_y} = \hat{\alpha}_{F_z} = 0.673 x 10^{-3} in.^{-1} lb^{-1}$$
$$\hat{\alpha}_{M_y} = \hat{\alpha}_{M_z} = 0.0244 x 10^{-3} in.^{-1} lb^{-1}.$$

The mass of the links is small, and we assume that it is concentrated half on each extreme of the link. We also assume that the center of mass of joint 3 is placed on the tip of link 1. Then, we have

- There is a mass m_3 concentrated at the joint 3 of 7 lbs ($\bar{p}_3 = \bar{0}$).

- There is a concentrated mass of 0.5 $lbs.$ at the tip. This will be added to the external force applied to the tip.

- The concentrated mass in the neighbourhood of joint 2 does not affect the static and kinematic analysis because its resulting force and moment are applied to the base of the arm.

3.2 Arm Equations

The kinematic equations of our arm are

$$T_{N0} = A_1 A_2 E_2 A_3 E_3. \qquad (10)$$

Matrices A_i correspond to the rigid links case and are given by

$$A_1 = \begin{bmatrix} 1 & 0 & 0 & 0 \\ 0 & 1 & 0 & 0 \\ 0 & 0 & c1 & -s1 \\ 0 & 0 & s1 & c1 \end{bmatrix}, \quad A_2 = \begin{bmatrix} 1 & 0 & 0 & 0 \\ l_2 c2 & c2 & -s2 & 0 \\ l_2 s2 & s2 & c2 & 0 \\ 0 & 0 & 0 & 1 \end{bmatrix},$$

$$A_3 = \begin{bmatrix} 1 & 0 & 0 & 0 \\ l_3 c3 & c3 & -s3 & 0 \\ l_3 s3 & s3 & c3 & 0 \\ 0 & 0 & 0 & 1 \end{bmatrix}.$$

E_2 and E_3 are of the form (3), and are given by

$$\begin{bmatrix} \bar{F}_{33} \\ \bar{M}_{33} \end{bmatrix} = \begin{bmatrix} R_{03} & 0 \\ \bar{r}_{33} \times R_{03} & R_{03} \end{bmatrix} \begin{bmatrix} \bar{F}_{30} \\ \bar{M}_{30} \end{bmatrix} \Rightarrow$$

$$\bar{F}_{33} = R_{03} \bar{F}_{30}, \quad \bar{M}_{33} = R_{03} \bar{M}_{30} \qquad (11)$$

$$\begin{bmatrix} \bar{F}_{22} \\ \bar{M}_{22} \end{bmatrix} = \begin{bmatrix} R_{02} & 0 \\ \bar{r}_{22} \times R_{02} & R_{02} \end{bmatrix} \begin{pmatrix} \hat{F}_{20} \\ \hat{M}_{20} \end{pmatrix} +$$

$$+ \begin{bmatrix} R_{02} & 0 \\ \bar{r}_{2N} \times R_{02} & R_{02} \end{bmatrix} \begin{pmatrix} \bar{F}_{30} \\ \bar{M}_{30} \end{pmatrix} \Rightarrow$$

$$\bar{F}_{22} = R_{02}(\hat{F}_{20} + \bar{F}_{30}), \quad \bar{M}_{22} = R_{02}\bar{M}_{30} + \bar{r}_{23} \times R_{02}\bar{F}_{30}, \quad (12)$$

where we have taken into account that $\hat{M}_{20} = \bar{0}$ because $\bar{p}_2 = \bar{0}$.

Additional relations are given by

$$R_{02} = (A_1 \cdot A_2)^T \qquad (13)$$

$$R_{03} = (A_1 \cdot A_2 \cdot E_2 \cdot A_3)^T \qquad (14)$$

$$\begin{pmatrix} 1 \\ \bar{r}_{23} \end{pmatrix} = E_2 \cdot A_3 \cdot E_3 \cdot \begin{pmatrix} 1 \\ \bar{0} \end{pmatrix} \qquad (15)$$

4 Computation Methods

4.1 Methods

Several simplifications have been made in this paper to obtain the kinematic model of the previous section. They are based on the assumptions that the deflections are small and the links are massless. Even after these simplifications, the solution of the system of equations (1-5) for a generic arm is complicated because they are nonlinear, making necessary an iterative procedure in order to calculate the tip position and orientation. Iterative procedures are slow, and may make the use of this transform impractical for real time control of flexible arms.

We explore several simplified algorithms in this section, and show that, for arms with mechanical configuration similar to that discussed in the previous section, iterative procedures may be avoided, provided that the deflections are reasonable (deflection at the extremity of the link is less than 10 % of the length of the link).

The kinematic model T_{N0} is obtained here using these simplifications, and the results are compared. The following case is considered: the arm in the horizontal position and completely extended ($\theta_1 = 0°$, $\theta_2 = 90°$, $\theta_3 = 0°$). We apply a force $\bar{F}_{30} = \begin{pmatrix} -78.842, & 78.842, & 0.0 \end{pmatrix}$ at the tip. It is easy to show that this force generates the maximum permissible flexural stress at the base of the arm (joint 2). This is really an extreme case, where maximum deflections are expected. No external momentum is applied at the tip. The methods compared here are:

1. Rigid arm: $E_2 = E_3 = I$.

2. Flexible arm under the assumption of small deformation, with the simplification that R_{02}, R_{03}, and $\bar{r}_{23}$ are independent of the load in equations (13-15). This assumption allows us to compute matrices E_i in a direct way, without the need of iterations.

3. Flexible arm, assuming now that R_{02} and R_{03} change with the tip load according to equations (13-14). We still assume that $\bar{r}_{23}$ is fixed. We show that a direct procedure may be used.

Notice that R_{02} is independent of the applied forces/moments. E_2 may now be calculated. Once we know E_2, we can calculate R_{03}, from which we can get E_3. This procedure is general and may be applied to a flexible arm of N links and N joints.

According to (5): $F_{N0}, M_{N0}, \hat{F}_{j0}, \hat{M}_{j0}$ $(j = i, N-1), R_{0,i} \Rightarrow$
F_{ii}, M_{ii}.
According to (3): $F_{ii}, M_{ii} \Rightarrow E_i$.
Homogeneous transform: $E_i \Rightarrow R_{0,i+1}$.
This is repeated for $i = 1$ to N

4. Flexible arm, assuming that r_{23} also changes. An iterative procedure must now be used. However, we will show that it converges quickly.

5. Flexible arm, assuming that R_{03} and r_{23} change with the load, and the deflections are not small. We use expression (2) instead of (3).

6. Both expressions (2) and (3) approximate $\cos(\epsilon)$ by 1 if ϵ is small. The real transformation, calculating all the trigonometric functions involved, is of the form

$$E = \begin{bmatrix} 1 & 0 & 0 & 0 \\ \Delta x & c\theta_y c\theta_z & -c\theta_y s\theta_z & s\theta_y \\ \Delta y & c\theta_z s\theta_z + s\theta_z s\theta_y c\theta_z & c\theta_z c\theta_z - s\theta_z s\theta_y s\theta_z & -c\theta_y s\theta_z \\ \Delta z & s\theta_z s\theta_z - c\theta_z c\theta_z s\theta_y & s\theta_z c\theta_z + c\theta_z s\theta_y s\theta_z & c\theta_z c\theta_y \end{bmatrix} \tag{16}$$

We also assume here that R_{03} and r_{23} change with the load.

4.2 Results

Method 1 (rigid arm):

$$T_{N0} = \begin{bmatrix} 1 & 0 & 0 & 0 \\ 0 & 0 & -1 & 0 \\ 60 & 1 & 0 & 0 \\ 0 & 0 & 0 & 1 \end{bmatrix} \tag{17}$$

Method 2:

$$T_{N0} = \begin{bmatrix} 1 & 0 & 0 & 0 \\ -6.104407 & -0.164823 & -0.99405 & 0.005912 \\ 59.768642 & 0.988423 & -0.164823 & -0.159103 \\ 5.831943 & 0.159103 & -0.005662 & 0.994373 \end{bmatrix} \tag{18}$$

$$E_2 = \begin{bmatrix} 1 & 0 & 0 & 0 \\ 0 & 1 & -0.111426 & -0.106042 \\ 1.694478 & 0.111426 & 1 & 0 \\ 1.590243 & 0.106042 & 0 & 1 \end{bmatrix} \tag{19}$$

$$E_3 = \begin{bmatrix} 1 & 0 & 0 & 0 \\ 0 & 1 & -0.053397 & -0.053061 \\ 1.06715 & 0.053397 & 1 & 0 \\ 1.060425 & 0.053061 & 0 & 1 \end{bmatrix} \tag{20}$$

Method 3:

$$T_{N0} = \begin{bmatrix} 1 & 0 & 0 & 0 \\ -6.104407 & -0.164823 & -0.99405 & 0.005912 \\ 59.768642 & 0.988423 & -0.164823 & -0.159103 \\ 5.831943 & 0.159103 & -0.005662 & 0.994373 \end{bmatrix} \tag{21}$$

$$E_2 = \begin{bmatrix} 1 & 0 & 0 & 0 \\ 0 & 1 & -0.111426 & -0.106042 \\ 1.694478 & 0.111426 & 1 & 0 \\ 1.590243 & 0.106042 & 0 & 1 \end{bmatrix} \tag{22}$$

$$E_3 = \begin{bmatrix} 1 & 0 & 0 & 0 \\ 0 & 1 & -0.053397 & -0.053061 \\ 1.06715 & 0.053397 & 1 & 0 \\ 1.060425 & 0.053061 & 0 & 1 \end{bmatrix} \tag{23}$$

In this particular case, the result of method 3 coincides with method 2.

Method 4:

After 3 iterations, the solution given by this method is:

$$T_{N0} = \begin{bmatrix} 1 & 0 & 0 & 0 \\ -6.08803 & -0.164414 & -0.994067 & 0.006713 \\ 59.769508 & 0.988467 & -0.164458 & -0.158652 \\ 5.815655 & 0.158696 & -0.004818 & 0.9944 \end{bmatrix} \tag{24}$$

$$E_2 = \begin{bmatrix} 1 & 0 & 0 & 0 \\ 0 & 1 & -0.111016 & -0.105635 \\ 1.690382 & 0.111016 & 1 & -0.000827 \\ 1.586173 & 0.105635 & 0.000827 & 1 \end{bmatrix} \tag{25}$$

$$E_3 = \begin{bmatrix} 1 & 0 & 0 & 0 \\ 0 & 1 & -0.053441 & -0.053016 \\ 1.068034 & 0.053441 & 1 & 0 \\ 1.059535 & 0.053016 & 0 & 1 \end{bmatrix} \tag{26}$$

Method 5:

After 3 iterations, the solution given by this method is:

$$T_{N0} = \begin{bmatrix} 1 & 0 & 0 & 0 \\ -6.081224 & -0.164161 & -0.994082 & 0.006695 \\ 59.769958 & 0.988495 & -0.164205 & -0.158431 \\ 5.809796 & 0.158475 & -0.004808 & 0.994413 \end{bmatrix} \tag{27}$$

$$E_2 = \begin{bmatrix} 1 & 0 & 0 & 0 \\ 0 & 1 & -0.110789 & -0.10544 \\ 1.69039 & 0.110789 & 1 & -0.000824 \\ 1.586181 & 0.10544 & 0.000824 & 1 \end{bmatrix} \tag{28}$$

$$E_3 = \begin{bmatrix} 1 & 0 & 0 & 0 \\ 0 & 1 & -0.053416 & -0.052992 \\ 1.06803 & 0.053416 & 1 & 0 \\ 1.059539 & 0.052992 & 0 & 1 \end{bmatrix} \tag{29}$$

Method 6:

After 4 iterations, the solution given by this method is:

$$T_{N0} = \begin{bmatrix} 1 & 0 & 0 & 0 \\ -6.029069 & -0.161903 & -0.986784 & 0.00669 \\ 59.427574 & 0.974432 & -0.160939 & -0.15678 \\ 5.739968 & 0.155593 & -0.018854 & 0.987621 \end{bmatrix} \tag{30}$$

$$E_2 = \begin{bmatrix} 1 & 0 & 0 & 0 \\ 0 & 0.988434 & -0.109577 & -0.104839 \\ 1.684305 & 0.110091 & 0.993921 & -0.00089 \\ 1.580134 & 0.104106 & -0.010662 & 0.994489 \end{bmatrix} \tag{31}$$

$$E_3 = \begin{bmatrix} 1 & 0 & 0 & 0 \\ 0 & 0.997234 & -0.052409 & -0.052697 \\ 1.049347 & 0.052482 & 0.998622 & 0 \\ 1.05364 & 0.052624 & -0.002766 & 0.998611 \end{bmatrix} \tag{32}$$

5 Conclusions

Flexibility in mechanical manipulators has previously been considered as a problem. We consider flexibility as a feature that can be exploited in many applications like assembly, etc. For this reason, we are building a 3 degree-of-freedom 2-links flexible arm, whose links exhibit flexibility in all directions.

This extended kinematic model is applied later to our flexible arm. Some methods are compared in order to calculate this model in the most computationally efficient way. This is a critical aspect to consider if we want to build flexible robots for industrial applications, since these models have to be calculated in real time in most cases.

The results of the simulations for our arm may be summarized as follows

- The effects of flexibility are quite noticeable. There are differences larger than 10 % of the total length of the arm (60 *in.*) between the tip position predicted by the rigid arm model and the position predicted by the flexible arm models.

- Methods 3 and 4 give similar results. The difference in the tip position predicted by these methods is about 0.5 *mm.* (0.04 % of the total length). This suggests that iterative methods can be avoided by using Method 3.

- Differences between methods 4 and 5 are negligible (0.2 *mm.*), but the differences with method 6 are relatively important (9 *mm.*, about 0.6 % of the total length).

Method 6 is most exact, but is iterative and requires the calculation of some trigonometric functions (16). The simulation results described in this paper showed a noticeable difference between the results given by method 6 and the others. However, these results were obtained for the extreme case where maximum deflections were produced. Simulations performed for other arm configurations, and with other forces/moments applied to the tip, showed that method 3 gives a very good estimation of the transformation matrix, T_{N0}, within a normal range of working conditions. This method is not iterative and only common trigonometric functions are required to calculate the rigid links homogeneous transforms $A_1 \ldots A_N$. Thus, it is well suited for the real time control of flexible arms.

REFERENCES

[1] Paul, R., P. *"Robot Manipulators-Mathematics, Programming, and Control."* The MIT Press Series in Artificial Intelligence. Cambridge MA. 1982.

[2] Craig, J., J. *"Introduction to Robotics: Mechanics and Control."* Addison-Wesley Publishing Company, 1985.

[3] Guillén A. "Estudio del Comportamiento en Frecuencia de Robots Industriales." Efecto de la Torsión. *Masters Thesis in Control Engineering. E.T.S.I. Industriales Universidad Politécnica of Madrid.* 1986.

[4] Cannon, R. H., and Schmitz, E. "Precise Control of Flexible Manipulators." *Robotics Research #1.* MIT Press. Cambridge MA. 1985.

[5] Matsuno, F., Fukushima, S., and coworkers. "Feedback Control of a Flexible Manipulator with a Parallel Drive Mechanism." *International Journal of Robotics Research.* Vol. 6, No. 4, Winter 1987.

[6] Kotnick, T., Yurkovich, S., and Ozguner, U. "Acceleration Feedback for Control of a Flexible Manipulator Arm." *Journal of Robotic Systems.* Vol. 5, n-3, June 1988.

[7] Feliu, V., Rattan, K.S., and Brown, H.B. "Adaptive Control of a Single-Link Lightweight Flexible Manipulator in the Presence of Joint Friction and Load Changes." *1989 IEEE International Conference on Robotics and Automation.* Scottsdale, Arizona. (USA). May 1989.

[8] Daniel, R.W., Irving, M.A., Fraser, A.R., and Lambert, M. "The Control of Compliant Manipulator Arms." *Robotics Research #4.* MIT Press. Cambridge MA. 1988.

[9] Hennessey, M.P., Priebe, J.A., Huang, P.C., and Grommes, R.J. "Design of a Lightweight Robotic arm and Controller." *Proc. of 1987 IEEE International Conference on Robotics and Automation.* Raleigh NC. April 1987.

[10] Gebler, B. "Feed-Forward Control Strategy for an Industrial Robot with Elastic Links and Joints." *Proc. of 1987 IEEE International Conference on Robotics and Automation.* Raleigh NC. April 1987.

[11] Henrichfreise, H. "The Control of an Elastic Manipulation Device Using DSP." *1988 Automatic Control Conference.* Atlanta GE. 1988.

[12] Book, W.J. "Analysis of Massless Elastic Chains with Servo Controlled Joints." *ASME Journal of Dynamic Systems, Measurement, and Control.* Vol. 101. September 1979.

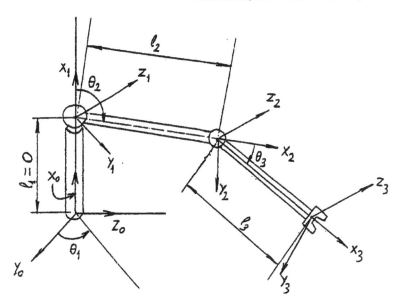

Figure 1. Flexible Arm Configuration.

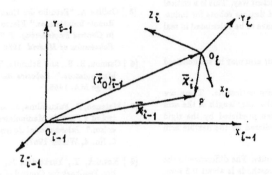

Figure 2. Transformations between Coordinates in Link i.

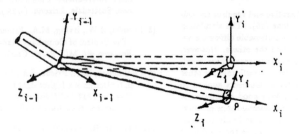

Figure 3. Transformations between Deformed and Undeformed ith Link.

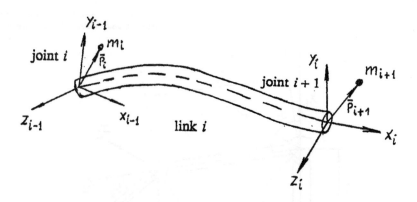

Figure 4. Generalization including Concentrated Masses Supported by the Joints.

A CRITERION FOR THE OPTIMAL
PLACEMENT OF ROBOTIC MANIPULATORS

J. A. Pámanes-García

*Centro de Graduados e Investigación, Instituto Tecnológico la Laguna, Torreón,
Mexico*

Abstract. The relative motion between adjacents links of robotic
manipulators are usually limited. Thus, in order to avoid the
interference between links when some task is carried out, it is
necessary to specify a suitable position of the robot frame. In
this paper a criterion is presented for the determination of such
a position. The proposed criterion is such that it keeps the links
away from their limit positions as much as possible. Thus, the
possibility of interference between links is minimized. An illustrative example shows the advantages of the criterion.

Keywords. Robots; optimization; nonlinear programming; placement;
working volume.

INTRODUCTION

In general, for each joint of a robotic
manipulator, a range of permissible
relative displacements between links
connected is defined which depends on the
links and joint geometries. Thus, the
reachable volume of the terminal organ of
the robot is determined by the limit
positions of the links.

On the other hand, it is evident that the
magnitudes of relative displacements of
the links, corresponding to a given position of the terminal organ, depends on the
location of the robot frame. Therefore, a
position and orientation of this frame
must be chosen so that the limit positions
of all the links are respected when some
task is carried out. In this paper, the
problem dealing with such positioning and
orientation of the robot frame will be called placement problem.

In the case of tasks to be carried out
which require only a small working volume
of the terminal organ, the placement problem may be solved by a trial and error
process "in situ", which may be accomplished in a short time. A large working volume and/or a great number of orientation
changes of the terminal organ result in a
complicated problem. Thus, a formal procedure may be required.

A criterion for a formal solution of the
placement problem is based on the insertion of the working volume of the terminal
organ into the reachable volume (Salisbury
and Craig, 1985). Nevertheless, since in
the reachable volume are not considered
orientation changes of the terminal organ,
the violation of the limit position of some links is possible. Therefore, this criterion is not reliable.

In the present work an optimization approach to solve the placement problem is
applied. Thus, such problem is considered
as a nonlinear programming problem, which

can be solved by using some classical
algorithm.

FORMULATION

Figure 1 shows any two adjacents coplanar
links of a robotic manipulator, these links
being connected by the i th joint, which
without loss of generality is considered as
one as revolute.

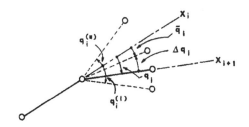

Fig. 1. Relative position between
two adjacents links.

For the relative position between links, in
Fig. 1 also is shown the upper limit $q_i^{(u)}$,
the lower limit $q_i^{(l)}$, the middle $\bar{q}_i$, the
current q_i, and the difference Δq_i defined
as

$$\Delta q_i = \bar{q}_i - q_i \qquad (1)$$

In order to avoid the interference between
links, the condition (2) must be satisfied
for all the points in the trajectory of the
terminal organ:

$$q_i^{(l)} \leq q_i \leq q_i^{(u)} \qquad 1, 2, \ldots, n \qquad (2)$$

where n is the number of joints. However,
since it is convenient to minimize the possibility of interference between links, a
solution is desirable such that the values
of q_i are kept as close as possible to

those of $\bar{q}_i$ during the execution of the task. Such a solution could be considered an optimum solution.

In order to minimize the possibility of interference, we define the ratio

$$k_{ij} = \left(\frac{\Delta q_{ij}}{\Delta q_{i\,max}} \right)^2 \qquad \begin{matrix} i=1,2,\ldots,n \\ j=1,2,\ldots,m \end{matrix} \quad (3)$$

where

$$\Delta q_{i\,max} = \bar{q}_i - q_i^{(1)} \qquad (4)$$

In Eq. (3), m is a certain number of suitable points chosen of the trajectory to be described.

Now, since the value of k_{ij} concern to the proximity of q_{ij} to $\bar{q}_i$, the optimum placement of the robot frame must to make all the $m \times n$ values of k_{ij} as little as possible. Therefore, we define the function

$$f = \bar{k} + z k_\theta \qquad (5)$$

as the function to be minimized, $\bar{k}$ being the mean, k_θ the standard deviation of the $m \times n$ values of k_{ij}, and z is the standard variable corresponding to a certain great value of k_{ij} in a normal distribution.

In order to complete the formulation we consider one coordinate system Σ_c attached to a robot[1] as shown in Fig. 2.

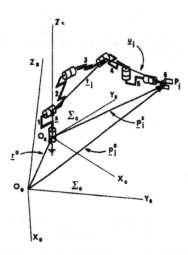

Fig. 2. Coordinate systems of the manipulator considered.

Axis z_c of the system Σ_c is oriented so that coincides with the rotation axis of the link connected to the frame. The position and orientation of the terminal organ corresponding to some point P_j in the trajectory, are specified by the vector p_j^c and the matrix a_j^c with respect to a system fixed Σ_0. The position of the origin of the system Σ_c is defined with respect to Σ_0

---
[1] For this formulation, a robot with architecture of revolute coordinates and 6 degrees of freedom is considered. However, it does not substract generality to the development.

by the vector

$$\underline{r}^\circ = [\, r_x, \, r_y, \, r_z\,]^T \qquad (6)$$

and the orientation of the system Σ_c referred to Σ_0 is determined through

$$a_b = \begin{bmatrix} c\mu c\upsilon & -c\mu s\upsilon & s\mu \\ s\lambda s\mu c\upsilon + c\lambda s\upsilon & -s\lambda s\mu s\upsilon + c\lambda c\upsilon & -s\lambda c\mu \\ -c\lambda s\mu c\upsilon + s\lambda s\upsilon & c\lambda s\mu s\upsilon + s\lambda c\upsilon & c\lambda c\mu \end{bmatrix} \qquad (7)$$

$$s = \sin \qquad c = \cos$$

where λ, μ and υ are the Bryant angles (Gorla and Renaud, 1984) corresponding to sucesives rotations which allows coincide the system Σ_0 with the Σ_c. Through a_b, any vector refered to system Σ_c can be refered to Σ_0. In addition, by multipling of a_b and a_j^0 is obtained the orientation matrix a_j^c of the terminal organ with respect to the system Σ_c:

$$a_j^c = a_b \, a_j^0 \qquad (8)$$

Now, can be noted that the independent variables to be determined in order to minimize the function (5) are the components of $\underline{r}^\circ$ and the Bryant angles. The explicit constraints for these variables, depends on the available space for the placement of the robot at its work station. These constraints are expressed by

$$r_x^{(1)} \leq r_x \leq r_x^{(u)} \qquad (9a)$$

$$r_y^{(1)} \leq r_y \leq r_y^{(u)} \qquad (9b)$$

$$r_z^{(1)} \leq r_z \leq r_z^{(u)} \qquad (9c)$$

$$\lambda^{(1)} \leq \lambda \leq \lambda^{(u)} \qquad (9d)$$

$$\mu^{(1)} \leq \mu \leq \mu^{(u)} \qquad (9e)$$

$$\upsilon^{(1)} \leq \upsilon \leq \upsilon^{(u)} \qquad (9f)$$

It must be noted that, if the robot frame is constrained to remain at some specified plane, the number of independent variables is reduced to 4. In this case, is convenient to define the system Σ_0 so that the specified plane is parallel to the $x_0 y_0$; thus, λ and μ are zero.

One elemental condition must be satisfied by any set of independ variables being feasible in the minimization problem: every point of the trajectory must be reachable. This condition is represented by

$$\| \underline{t}_j \| \leq \ell \qquad (10)$$

where ℓ is a parameter depending on the geometry of the links which determines the reach of the manipulator (for the architecture of Fig. 2, ℓ is equal to the sum of lengths of the links 2 and 3), and the vector $\underline{t}_j$ is the part of the position vector asociated to these links. It is observed in Fig. 2 that

$$\underline{t}_j = \underline{p}_j^c - \underline{y}_j - \underline{s} \qquad (11)$$

It is assumed that in the work station of the robot there are not obstacles which must be specified as constraints in the placement problem. Literature deals with the obstacle avoidance problem (Fu, González and Lee, 1987) and it is out of objective in this work.

It is interesting to observe that is not necessary to include the condition (2) as constraint of independent variables, because such condition is implicit in the function (5) through k_θ. However, if the condition (2) is not satisfied by the optimal placement, then not any placement will satisfy it; in this case the assigned task is inadequate for the manipulator considered. Consequently, after obtain the optimal placement, must be realized an additional test to verify if the condition (2) is satisfied.

SOLUTION TO THE OPTIMAL
PLACEMENT PROBLEM

In agreement to the formulation developed at previous section, the optimal placement problem is stated in the following terms: to determine the vector $\underline{r}^s$ and the Bryant angles which minimizes the function (5), subject to constraints (9) and (10). This problem can be solved by using some method of nonlinear programming; however since neither the objective function and constraint (10) are explicitly expressed in terms of independent variables, the use of some method which require the derivation of the objective function makes the problem enough complicated. Thus, some huristic search method result appropiated (for example Box,1965; Rosenbrock,1960). In this work, the Box method is applied through the OPTIM program (Evans, 1975).

In OPTIM program at each iteration, the evaluation of the objective function and the implicites constraints is required. Such evaluations are carried out in a subroutine called MODEL, which must be prepared by the user.

In order to evaluate the objective function, we must determine the generalized coordinates of the robot, corresponding to m positions of the trajectory. Thus, the operational coordinates a_j^e and p_j^e will be specified, and the inverse geometric model (Paul, 1981) will be solved. It is noted that, since changes of $\underline{r}^s$ and a_b occurs in each iteration, the operational coordinates must be updated by the equations (8) and (12).

$$\underline{p}_j^c = a_b(\underline{p}_j^e - \underline{r}^s) \qquad (12)$$

The vector $\underline{t}_j$ of condition (10) is obtained by using the Eq. (11). Right hand vectors of this Eq. are determined in the following way: $\underline{p}_j^c$ is obtained by Eq. (12), $\underline{u}_j$ result from the solution of the inverse geometric model, and $\underline{t}$ is defined from the geometric parameters of the link connected to the robot frame.

Finally, the MODEL subroutine is structured as shown in Fig. 3.

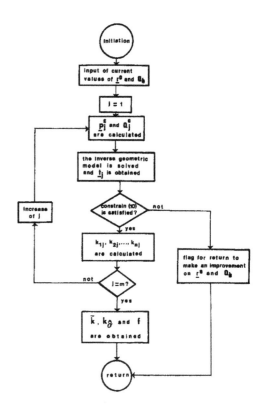

Fig. 3. MODEL subroutine.

EXAMPLE

Determine the placement of the robot manipulator SALVIATI TL-1 (Pámanes, 1986) which allows the transfer of objects from a feeder to a transportation band describing the trajectory shown in Fig. 4.

The coordinates of the points indicated in Fig. 4, refered to the system Σ_0, are given in the Table 1. The eight points presented in such table are the m points considered in the Eq. (3).

TABLE 1   Coordinates of Points in the Trajectory

j	POINT	X_0 cm	Y_0 cm	Z_0 cm
1	A	48	0	80
2	B	80	0	40
3	C	80	-20	20
4	D	80	-20	0
5	E	84	-20	0
6	F	84	-20	20
7	G	80	20	60
8	H	84	20	80

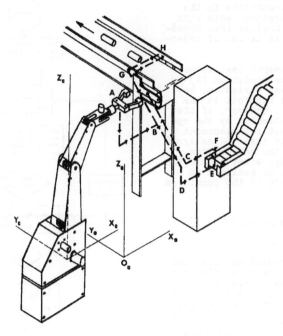

Fig. 4. Task of the robot.

The matrices of orientation of the terminal organ, with respect to system Σ_e, for the initial position (at A), for transfer and clamping (from B to G), and for deliver (at H) are

$$\mathbf{Q}_j = \begin{bmatrix} 0 & 1 & 0 \\ 1 & 0 & 0 \\ 0 & 0 & -1 \end{bmatrix} \qquad \mathbf{Q}_j = \begin{bmatrix} 0 & 1 & 0 \\ 0 & 0 & 1 \\ 1 & 0 & 0 \end{bmatrix} \qquad \mathbf{Q}_j = \begin{bmatrix} 0 & 1 & 0 \\ \frac{1}{\sqrt{2}} & 0 & \frac{1}{\sqrt{2}} \\ \frac{1}{\sqrt{2}} & 0 & -\frac{1}{\sqrt{2}} \end{bmatrix}$$

$$j = 1 \qquad\qquad j = 2,3,\dots,7 \qquad\qquad j = 8$$

The parameters corresponding to the relatives positions of each joint of the manipulator are given in Table 2.

The initial values of the independent variables for the optimization process are displayed in Table 3, in which also are presented the limits established for such variables, and the values of the objective function and of the dependent variable $\|\underline{t}_j\|$ (this corresponding to the critical position).

The solution was found after 158 iterations. In Table 3 the solution is given. In Fig. 5, for comparative purpose, the values of $\sqrt{k_{ij}}$ concerning to both the optimal and the initial placements are plotted.

CONCLUSIONS

The criterion proposed in this paper let to arrive in fast and safe way to the solution of the placement problem. In addition, all the links are kept away from their limit positions as much as possible.

It is important to observe that through the function (5) the criterion minimize the mean and the standard deviation of all the k_{ij}. Therefore, the reduction of some k_{ij} initially greats may to implicate

TABLE 2 Parameters of Relative Displacements

JOINT i	$q_i^{(1)}$ deg	$q_i^{(u)}$ deg	$\bar{q}_i$ deg	$\Delta q_{i_{max}}$ deg
1	-135	135	0	135
2	-20	120	50	70
3	-140	45	-47,5	92,5
4	-120	90	-15	105
5	0	180	90	90
6	-360	360	0	360

TABLE 3 Data of Independent Variables

	r_x cm	r_y cm	r_z cm	λ deg	μ deg	ν deg	$\|\underline{t}_j\|$ cm	f
LOWER BOUND	-50.0	-50.0	-50.0	-360.0	-360.0	-360.0	0.0	—
UPPER BOUND	50.0	50.0	50.0	360.0	360.0	360.0	70.4	—
INITIAL VALUE	0.0	0.0	0.0	0.0	0.0	0.0	65.6	1.14
SOLUTION	-3.5	6.5	-4.1	13.1	-1.2	18.4	69.9	0.86

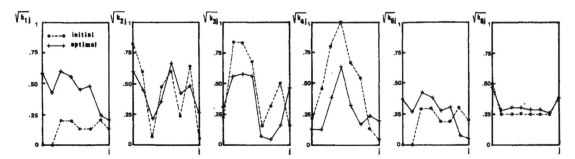

Fig. 5. Behavior of the $\sqrt{k_{ij}}$ of each joint corresponding
to the initial and the optimal placements.

that increases others k_{ij} initially smalls.
Nevertheless, these increases will be as
little as possible. So, in Fig. 5 can be
noted that after of the optimization the
maximum $\sqrt{k_{ij}}$ of the joints 1 and 5 increases, while those of the remainder joints
are reduced.

The criterion proposed is useful, of course, for the placement of sedentary robots;
however, since the process of solution is
automatic, its use is ideal for the placement of movil robots by incorporing such
criterion to the algorithm of control.

In this work not any rule of selection of
the m points to be considered has been established, although it is evident that
critical points must be chosen.

ACKNOWLEDGMENTS

The project which takes part this work is
supported by the Consejo Nacional de Edu-
cación Tecnológica.

REFERENCES

Box, M.J. (1965). A new method of constrained optimization and a comparison
with other methods. Computer J., 8,
42-52.

Evans, L.B. (1975). OPTIM: Un programa de
optimización. In M.A. Murray-Lasso,
E. Chicurel, L.B. Evans, and colleagues, Aplicaciones de la Computación
a la Ingeniería, Ap. 4. Limusa, México. pp. 331-342.

Fu, K.S., R.C. Gonzalez, and C.S.G. Lee
(1987). Robot intelligence and task
planning. In Robotics. Control, Sensing, Vision and Intelligence, Chap.
10. Mc Graw-Hill, New York. p. 512.

Gorla, B., and M. Renaud (1984). Modèle
géométrique direct d'un robot manipu-
lateur. In Modèles des Robots Manipulateurs. Application a leur Commande,
Chap. 4. Cepadues, Toulouse. p. 94.

Pámanes-García, J.A. (1986). Aspectos de
diseño y manufactura de un robot ma-
nipulador. Gestión Tecnológica, 5,
9-18.

Paul, R.P. (1981). Kinematic equations. In
Robot Manipulators. Mathematics, Pro-
gramming and Control, Chap. 2. MIT
Press, Cambridge. pp. 41-63.

Rosenbrock, H.H. (1960). An automatic method for finding the greatest or least
value of a function. Computer J., 3,
175-184.

Salisbury, J.K., and J.J. Craig (1985).
Articulated hands: force control and
kinematic issues. In M.T. Mason, and
J.K. Salisbury, Robot Hands and the
Mechanics of Manipulation. MIT Press,
Cambridge. p. 116.

Fig. 5. Behavior of the q_i of each joint corresponding to the initial and the optimal placements.

OFF-LINE PROGRAMMING ENVIRONMENT
FOR ROBOTIC APPLICATIONS

E. Hagg and K. Fischer

Institut für Informatik, Technische Universität München, München, FRG

Abstract. Programming major applications for robots demands a comfortable programming environment. This paper presents the high-level robot language MRL for Siemens (RCM3) and Unimation (VAL II) robot control systems. MRL facilitates structured programming by supplying typed and locally defined variables, procedures (with parameters), complex movement statements, processes and modules. A robot terminal emulator under VAX/VMS provides full access to the Unimation control and integrates an upload and download link between both systems with efficient file transfer capabilities. In robotic systems with low memory resources, only the currently needed parts of a large program are kept in the robot's memory. The multi-user access to the robot given by the emulator consists in transmission of commands, dynamic loading and execution of programs in parallel.

Keywords. Programming environment; high-level language; industrial robots; robot interfaces.

INTRODUCTION

A short review of programming systems and languages used in applications for the control of robots shows two different approaches: First there is one type of language, as in (Mujtaba, 1981), (Blume 1985) or (IBM Corporation, 1982), with all the features of a high-level language. These languages are either extensions of general-purpose languages or have been developed to solve robotic problems. But they lack sufficient propagation, there being few installations outside research institutes which use them, and most of them have been developed for special robots or robot controls.

Then there are languages found in robots on the factory floor, for an example see (Unimation Incorporated, 1986) and Siemens AG, 1986), which have been developed out of languages for numerically controlled machines. If we compare them to general-purpose languages, they belong to assembler dialects or to the BASIC family. They were designed for teaching the robot to find locations, for point-to-point movements or programmed trajectories and, therefore, need only simple control structures and data types.

In a present-day programming environment single-user mode, limited storage resources in the robot control, insufficient backup facilities and primitive editors do not hamper programming by teaching for simple manipulation tasks, but handicap efficient software production. Program development of the near future will - in part - rest on graphic-based simulation tools. These tools supply operations to define and manipulate objects in the robot's environment producing program code the execution of which can be simulated in the graphic system itself. The real robot is only needed for a last, short testing. Therefore, one robot can be used by several programming teams, provided that there are links to a development system for program uploading and downloading, combined with a robot terminal emulator to transmit commands and start the execution of programs. These simulation tools simplify the day-by-day robot programming, but there will still be a need for sophisticated

robot programs with complex control structures. Examples can be found in connecting the robot to its environment, i.e. programming reactions to external events, in sensor integration, on-line modification of robot trajectories and so on. In order to solve these problems, a programmer needs a high-level programming language with features for structured programming, which enable him to build a more general problem solution out of smaller elements, and application-specific constructs like parallel processes and interrupts.

These applications also tend to grow to a size which is difficult to handle, but can be structured and simplified by modules. Because of the frequently small memory of robot control systems, even middle-sized programs cannot be kept completely in it. Thus, dynamic loading of currently used parts of a program system has to be provided by the robot runtime system and not - as happens today - by an operator.

MRL

At the Institut für Informatik within the Technische Universität of Munich we use in general two types of robot controls: Siemens's RCM and Unimation's VAL II. Our attempt to develop the language for our programming environment consisted in selecting and combining the best qualities of the given languages and completing this subset with important characteristics. In order to avoid any redesign of the robot control software, only those extensions were included which we knew could be implemented on both of them. Out of this evolved MRL, the Modular Robot Language.

MRL does not only offer 'reals' and 'locations' as simple data types. We also included 'boolean', 'integer' and 'joint coords' (the last meaning the definition of a robot location by the angles of the joints). Like Pascal or other high-level languages, MRL forces the programmer to declare variables by assigning data types to freely chosen identifiers. The basic type 'char' is also useful for robot control applications, but is

155

not provided by both of the underlying control systems and is, thus, not included in MRL. Basic types can be grouped to form one-dimansional arrays, but general compound data types, i.e. records, can only be implemented with some effort and low efficiency, because there is no byte-adressing feature. Therefore, a later version of MRL will have to supply these data types.

Procedures, functions and processes with parameters and local variables are the most desirable and usually lacking features in common robot languages. MRL, therefore, has been endowed with them. Parameters are transmitted using a call-by-value or call-by-reference mechanism giving much more flexibility in program structuring. The implementation of this feature was hampered by limitations in RCM causing some inefficiency in the generated SRCL code. The locally fixed validity of variables in MRL excludes access to them from outside a procedure or function, and guards cooperating programming teams against errors due to the use of identically named variables. Global variables for synchronization or communication are provided, too.

Nearly all of the control structures, like 'for...', 'if...then...else...', 'case...', and so on stem from VAL II, with an only slightly modified syntax.

A MRL statement for the control of the robot´s movements, for example

'move <loc1> linear, via <loc2>, via..., speed 100'

includes the destination and optionally the locations between starting point and destination which the robot should pass. Velocity and type of movement (linear instead of interpolating the robot´s joint angles) can be specified within the statement, too.

The formal construct of I/O (input/output) channels helps in structuring digital, bcd and analogue I/O lines to one-dimensional or simple-type units.

Each of the languages on which we rely - SRCL and VAL II - provides specific elements which cannot be implemented using the other one. MRL includes all of the important ones. They are included in specific runtime libraries available for each control type. For an example, the valuable MRL feature of monitoring external signals which asynchronously activate interrupt procedures rests on the same property in VAL II but cannot be implemented using SRCL. It was the main reason why we dropped, for some definitions, the idea of one language for all robot systems and created one language version for each supported control type.

Another group of language structures which can only be implemented in a restricted manner cannot be excluded from a language which is expected to fulfil the requirements outlined above. In this light, MRL contains the general, non-restricted feature; the necessary checking of the limitations in the implementation is done by the compiler and the runtime system. As an example: MRL allows parallel activities; processes can be started and stopped, and we provide basic functions for synchronization. But in the generation of VAL II code we are restricted to one foreground and one background process, and in SRCL we find only some type of asynchronously executed subprogram with runtime limitations.

The robot control systems which are supported by our

programming environment do not provide any language interface below SRCL or VAL. Thus, two MRL compilers produce either SRCL code or VAL II code, respectively. The compilers differ only in the code-generating back end. The common front end, a recursive-descent parser, is based on the same attributed grammar defining MRL.

Modules

Modularity obviously is the most valuable characteristic of structured programs. Complex problems are solved by breaking them down into smaller parts. The locally fixed validity of variables and procedures or functions (in the discussion which follows the syntactic term 'function' stands for procedures, functions and processes) helps to hide implementation details and to concentrate on the functional specification of a problem.

MRL offers modules; every module declares local variables and functions which can be used within the module. Global functions are exported to all of the other modules.

Figure 1 shows the components of an MRL program and their relations. One special module, the definition module, describes the modular structure of an MRL program. It contains a list of modules belonging to the program. For every item in this list a programmer specifies a corresponding list of global functions, submodules and files. At compile time each module is built out of the code in the files listed under the module description in the definition module. By inserting a file in two separate module descriptions, it may be used for different modules without making it global (see also the 'file' paths in Fig. 1). The submodule relation defines a tree structure on the modules (see the 'submodules' path in Fig. 1). The definition module designates the root of the tree, the main module. This module is the only one containing the main program which will be executed at the start of an MRL program. Global constants, variables and I/O channels have to be declared in the definition module, too.

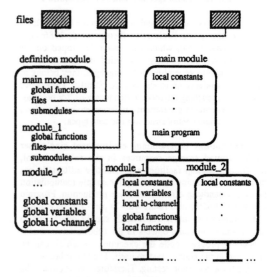

Fig. 1. Modular structure in MRL

Compiling a program starts by analysing the definition module and evaluating the information it carries about the program´s structure. It is possible to check the syntactic correctness and the semantics of imports and exports at compile time, because all of the exports are listed in the definition module. Even the correctness of external procedure calls, the limitation of which is described later, can be checked statically.

Overlays

Applications for robots tend to need more storage resources than robot controls usually are provided with, and most of the robot systems do not supply any program-controlled access to mass storage media for swapping or paging purposes. In order to cope with low memory, the whole of a program system has to be stored on a host computer, which normally will be the program development system, and only currently executed program components are to be kept within the robot control. Then, a mechanism for the dynamic loading of program parts has to be provided.

MRL uses the tree structure on a program's moduls to decide which parts of code have to be fixed in the robot control's memory. MRL modules are considered as indivisible parts of an overlay structure belonging to every MRL program. The main module is the only one which is always kept in memory. Every module which is executed and, therefore, stays in memory can either call a function in a higher modul, in which case the module is already loaded, or in a son module. Calling a function of the latter type activates an 'overlay manager' on the host system which loads the module's code into the robot's memory.

PROGRAMMING ENVIRONMENT

Programs for robots should not be developed on a robot control system, but on a standard workstation. We chose a VAX with VMS operating system as development and host system because most of our robot programming tools use this environment. A comfortable editor and a graphic simulation system, which is described in (Milberg, 1988), run on these workstations and are valuable tools for an off-line way of producing and testing robot programs. An action planner, which is outlined in detail in (Fischer, 1988), generates programs in MRL or VAL at runtime. In addition, MRL precompilers, which generate either SRCL or VAL code, allow structured programming for robots. This off-line environment enables us to use the robot rarely during program development and to share it by several programming teams.

The interface program to the robot's control is an emulation of the robot's system terminal and supplies it's full functionality. It can be activated by several users and allows them in addition to access the host computer's file system and to download robot programs. Parallel requests to the robot are synchronized.

Uploading and downloading of program systems from VMS to the VAL system and vice versa is only possible for VAL code. Therefore, MRL programs have to be compiled before any downloading.

Both of the possible formats are supported: ASCII-format for files which are generated by an editor or other program development tools, and the VAL internal format, which can be transmitted more quickly. The VAL backup facility is also able to communicate with the host system and allows us to

use the VAX as VAL file server. The multi-user access to the robot allows the transmission of programs even during program execution.

These transfers can be done statically, that means, the whole program is loaded into the robot control before execution, or dynamically. The former is necessary for programs written in VAL II. Dynamic loading, which is only implemented for MRL, uses the overlay manager, which downloads only requested MRL modules into the robot control.

A user who links up to the interface program does not lose the VMS functionality. At any time he can call any VMS service, for instance the editor or the MRL compiler.

Some functions in the robot control require exclusive access to the robot. For instance, a parallel execution of separate robot program systems is not possible. Thus only one user, the superuser of the robot link, is granted all of the robot control's functionality. All other users are not allowed to use commands and functions which would affect the correct execution of the superuser's activities. Any user may attain the superuser status by declaration and hold it until he leaves the status explicitly. Requests from other users for superuser status are queued by the system.

Link to the Robot Control

The Unimation robot control offers basic services to implement an interface to a supervisory computer. In defining it, the VAL system software is grouped into several 'logical units', which are parallel processes with I/O functions (see Fig. 2). One logical unit deals with input and output of user programs, one with asynchronously sent system messages, one with the VAL command interpreter and one represents the VAL backup facility. Input and output of the logical units can be directed from the robot terminal to a serial line which is connected to a supervisory or host station, which is in our case a VMS workstation. The host system communicates in our implementation with these logical units by means of 'servers'. These are VMS processes which correspond to each of the logical units.

At a higher level, the VMS emulator of the robot terminal permits downloading of programs and transmission of robot commands in parallel. Therefore, an additional VMS process, the master process, became necessary to synchronize parallel activities. The master process accepts requests from VMS users by various mailboxes and receives the robot control's responses from the servers (see Fig. 2). It buffers messages and identifies the receiver.

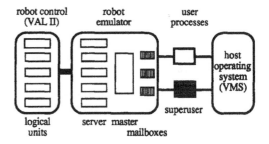

Fig. 2. Structure of the supervisory interface

CONCLUSION

A comfortable programming environment for the development of robot control applications was implemented. It consists of the new high-level language MRL, which up to now is restricted to Siemens's RCM3 and Unimation's VAL II, and compilers which generate either SRCL or VAL code. MRL is a basis for the structured programming of robots. Its concept of moduls and the dynamic downloading of MRL modules into a robot control facilitate the development of major robot applications. A supervisor link between VAX workstations under VMS and VAL II provides full access to both VMS and VAL by supporting a comfortable robot emulator with efficient file transfer capabilities.

REFERENCES

Aho, A. and J. Ullman (1980). Principles of Compiler Design. Addison-Wesley, Reading.

Blume, C. and W. Jakob W. (1985). PASRO - Pascal for Robots. Springer, Berlin.

Blume, C. and W. Jakob (1986). Programming Languages for Industrial Robots. Springer, Berlin.

Fischer, K. (1988). Regelbasierte Synchronisation von Roboter und Maschinen in der Fertigung (Rule-based synchronization of robots and machines). Report TUM I8816, Institut für Informatik, TU München.

Gruver, W., B. Soroka, J. Craig and T. Turner (1984). Industrial robot programming languages: a comparative evaluation. IEEE Transactions on Systems, Man & Cybernetics.

IBM Corporation (1982). A Manufacturing Language - Reference. Boca Raton, Florida.

Matzeder, A. (1983). Untersuchung der Realisierbarkeit von AL-Sprachelementen in den Systemimplementierungs-sprachen ADA und Modula-2 (A study on how to implement AL constructs in ADA and Modula-2). Diplomarbeit (MS Thesis), Institut für Informatik, TU München.

Mujtaba, S.and R. Goldman (1981). AL Users's Manual, 3rd edition. Computer Science Department, Stanford University.

Schmidt, C. (1988). Entwurf einer für die beiden Robotersysteme VAL II und RCM geeigneten höheren Programmiersprache und Implementierung der Abbildung auf das RCM-System (Design of a robot programming language for the VAL II and RCM control and implementation of the SRCL compiler). Diplomarbeit (MS Thesis), Institut für Informatik, TU München.

Siemens AG (1986). Sirotec RCM 2 und RCM 3 - Programmieranleitung. Nürnberg-Moorenbrunn

Milberg J., N. Schrüfer and A.Tauber (1988). Requirements of Advanced Graphic Robot Programming Systems. Symposium on Robot Control, Karlsruhe.

Unimation Incorporated (1986). User's Guide to VAL II - Programming Manual, Version 2.0. Danbury, Connecticut.

COMPUTER AIDED LAYOUT PLANNING FOR ROBOT ASSEMBLY APPLICATIONS

M. Huck

*Institute for Real-time Computer Control Systems and Robotics,
Faculty for Informatics, University of Karlsruhe, Karlsruhe, FRG*

<u>Abstract</u>. This paper presents *ROSI-Layout*, a computer aided planning system for the generation of suitable workcell layouts for robot assembly applications. The idea is to derive first the spatial transformations for the product parts, which lead to the assembly of the final product. A global database provides the product data and the precedence graph. This information comes from product design and assembly planning. The design of spatial transformations is derived from constraints given by the geometric shape of the product parts and from rules that define how assembly with robots should be accomplished. This is followed by the generation of a suitable layout. A layout can be developed that is specially designed for the product to be assembled. During the entire planning session, the database manages all planning data and ensures consisteny. ROSI-Layout is an extension of the RObot SImulation system ROSI.

<u>Keywords</u>. Computer-aided planning, workcell design, layout planning, robot assembly, robot simulation, assembly planning, work frame, data model.

INTRODUCTION

In the wide area of manufacturing layout planning means the development of a well organized workcell sized to the requirements given by the application. Layout planning includes both the selection of equipment, in robot based manufacturing these are robots, grippers and peripherals, and their geometric arrangement to a workcell such that reasonable assembly is possible. Together with the programming of the cell devices, substantially robots, workcell design is the main task in the planning of a robot application.

A great advantage of industrial robots is their versatility (Lozano-Perez, 1983). Robots are enabled to handle parts almost without restrictions in their three-dimensional working space. This flexibility implies additional complexity for the design of a robotic workcell. For this reason no methods and tools are available to generate a optimal workcell automatically. However, a lot of planning systems have been developed to support the human planner with the aid of the computer. Computer aided planning systems related to robot based assembly applications can be classified into three categories:

1. assembly planning systems,
2. automatic robot planning systems,
3. interactive graphical robot simulation systems.

Assembly planning systems mainly support the optimization of the assembly sequence as well as the selection of reliable assembly strategies. Usually the user determines the layout of the assembly cell applying simple layout facilities of CAD-systems (Levi, 1988).

The purpose of *automatic robot planning systems* is that they automatically produce robot level programs derived from a task oriented description of the robot application. Such systems are still in the stage of research prototypes. Nevertheless for specific problems, e.g. planning of collision-free trajectories, they have already yielded some remarkable solutions. From the known systems very few consider the layout problem as part of the planning process. However, even these systems only deal with the conceptual aspects of cell layout planning. (Lozano-Perez, Brooks, 1985).

Interactive graphical robot simulation systems represent computer aided planning and programming systems specialized for robot applications. These systems are applied sucessfully even in industrial applications. They support the operator to design a robot workcell and to develop the respective programs. The graphic system displays the robot workcell as a precise three dimensional model and visualizes all movements in the workcell (Sjolund, Donath, 1983; Dillmann, Hornung, Huck, 1986). However, layout planning is performed in an empirical style. With the aid of graphical functions the user first produces some gross layouts by arranging the equipment and parts in the cell. In a second step he then refines the layouts by applying reachability tests for the robot and known work points as well as cycle time analysis.

REQUIREMENTS TO LAYOUT PLANNING FROM ASSEMBLY

Automatic assembly of technical products using robots is a multi discipline task. The performance of the assembly execution depends on how good the assembly process fulfills different requirements. For some requirements there are mutual dependencies with the layout of the assembly cell. Consequently these requirements should be included into layout planning. Requirements regarding the layout are:

- The correct assembly of the product. The structure of the product is represented by the geometric shape of the parts as well as the location of the parts in the final product.
- The suitability of specific assembly strategies that must be applied to execute assembly tasks using robots.
- The efficient use of the equipment, especially robots according to their functional chracteristics.

ROSI-Layout uses a structured approach to incorporate these demands into layout planning. Central objective of the planning is the structure of the product and how the individual parts must be manipulated to match the requirements above. The method is based on the central idea that the product represents the task. There is a lot of experience in the field of assembly even with the use of robots. Evaluating the features of the product and the knowledge about assembly, spatial transformations of the part can be generated. These transformations represent constraints for the layout planning process. Robots and other equipment are merely tools to perform the assembly task. So it is of great advantage to know task specific constraints. This facilitates the selection of adequate equipment and the creation of an efficient layout.

The planning method consequently determines constraints to refine the assembly process. With the help of constraint propagation the derived constraints influence the subsequent planning steps. Selection of specific equipment can be performed if necessary constraints are known. The inclusion of selected equipment generates further constraints.

PRODUCT ORIENTED APPROACH FOR PLANNING AN ASSEMBLY CELL

Task oriented work points for the robot are the most important information needed for the design of a robot workcell. For process oriented robot applications, for example welding, the work points on a workpiece may be defined during the design of the workpiece with the help of a CAD-system. For assembly operations the work points depend on the spatial transformations necessary to move each part from its start position to its end position in the product. ROSI-Layout aims to ascertain these spatial transformations so that they guarantee a correct and efficient assembly process. Such a spatial transformation is specified by a start and an end point, which are work points for the robot. A work point specifies the position and the orientation of the part at a discrete situation during its assembly. To describe work points the *frame* concept is used. Product oriented assembly information is derived from the geometric constraints imposed by the parts of the product and from knowledge about how assembly with robots should be accomplished.

This concept requires the integration of the robot specific planning tasks into the comprehensive production planning as it is shown in fig. 1.
Product design using a CAD-system provides the geometric models of the parts.

From *assembly planning* a precedence graph results (see fig. 7). A node of the graph outlines an assembly operation (AO<i>). Each assembly operation embodies the task to add a part to the final product. The edges between the nodes show all valid order sequences for the execution of the assembly operations (Frommherz, Hornberger, 1988).
The geometric models and the assembly graph provide the information for the robot specific planning phases.
During *assembly operation planning* each assembly operation is analysed to derive the spatial transformations for the assembly of the corresponding part. If there exists more than one possible transformation for the part this results in additional degrees of freedom for the work points. To represent work points with degrees of freedom in their position and orientation the frame concept is enhanced in the so called *work frame* concept. Work frames allow to propagate product oriented transformations as constraints for the design of the workcell. This planning step is accomplished independently from a specific robot. The part transformations and work frames are stored as planning data in the database.
Considering the product oriented assembly information *gross layout planning* supports the generation of workcell layouts. This includes the selection of equipment and the computation of possible sites.
Detail layout planning incorporates the evaluation of the layout variants according to certain criteria and the optimization of equipment selection, sites, work frames and the assembly sequence.
In the phase of *program development* the robot program can be explicitly coded and tested graphically. The work points to specify the robot motion are given by the work frames. The degrees of freedom in position or orientation of the work frame can be evaluated to optimize the assembly operations performed by the robots.

ASSEMBLY OPERATION PLANNING

Goal of this planning task is the derivation of work points that determine the assembly process. A work point incorporates

possible degrees of freedom in position and orientation. At this time merely the assembly operation without any robot is considered. Planning is performed on the level of assembly operations.

Decomposition of an assembly operation

For the analysis of constraints each assembly operation is decomposed into several phases. The example in fig. 2 illustrates three phases to transform a peg from its start location (start frame) in the part holder to its end location (end frame) in the final product. In this example the fetch phase requires only two transformations. First the peg is removed from the start frame to the contact frame. The contact frame describes the geometric arrangement of the peg in reference to the part holder, where the last contact between both parts happens. Then the peg is moved to a depart frame that ensures a safe transformation. The frames (start, contact, depart) and transformations are affixed to ´Part holder´. For the assembly phase the approach and contact frame define one transformation of the peg from a safe approach location to the location of the first contact with ´Sideplate´. The other transformation represents the insertion of the peg. These frames are affixed to ´Sideplate´. As the end frame of a part results from assembly planning, the analysis for both the fetch phase and the assembly phase starts from end frame or start frame. The transfer phase has no influence on the product oriented planning. It is considered in gross layout planning.

For both phases the analysis of the transformations is logically devided into two steps. In the *contact area* the transformations are derived from the geometry of the part. The reasoning process for the *near-by area* is mainly influenced by the knowledge about assembly.

Structure of a work frame

The product oriented constraints are modelled by work frames. Such a work frame describes the geometric location of an object relativ to a reference object including existing degrees of freedom in its position and orientation. Figure 3a shows the structure of a work frame. An ordinary frame (´Location frame´) records the actual location of the fixed coordinate system of ´Peg´ (´PC´) in reference to the coordinate system of ´Block´ (´BC´). For the description of the degrees of freedom in the arrangement of the peg inside the hole of the block a *variant frame* (´V-Fr1´) is generated. ´V-Fr1´ locates an adequate transformation coordinate system (´TC´) that serves as reference for the degrees of freedom in the position of the peg expressed by lower and upper bounds for the x, y and z axis of ´TC´. Now the peg may be rearranged inside the patterned face as it is shown in fig. 3b. In this case the parameters x_a, y_a, z_a indicate the actual position of ´TC´. After a rearrangement of the peg its new ´Location frame´ (see fig. 3b) can be computed.

The variant frame also allows to describe degrees of freedom in orientation. However with one variant frame degrees of freedom either in position or in orientation can be modelled. Combinations of degrees of freedom require chaining of variant frames.

Geometric reasoning and assembly specific knowledge

Geometric reasoning is a process of deriving potential movements of an object from the geometric shapes of the object itself and its environment. This technique is applied to derive potential directions for fitting or removing parts in the contact area and to find free space for transformations of parts in the near-by area. For a successful reasoning the geometry of the parts is modelled using a boundary representation scheme with additional normals to indicate material.

Especially for the near-by area the transformations can not be generated by geometric reasoning only. For this case the planning procedures are influenced by rules expressing assembly specific knowledge. These rules are classified into strong rules (SR) and weak rules (WR). If a strong rule is not obeyed then the assembly will fail. However weak rules only give recommendations for a "good" assembly. In the planning procedures the following rules are incorporated:

SR1   move the part and the gripper on a collision free path,

SR2   avoid grasping the part where it touches other parts in start or end situation,

SR3   a grasp of a part must be stable (center of mass of the part located in the tool center point of the gripper),

WR1   insert or remove parts in vertical direction,

WR2   approach and withdraw the part along same directions as for insertion or removal,

WR3   the distance to approach and to withdraw the part should be small,

WR4   avoid changing the orientation of the part during approach and depart.

Analysis of the contact area

Analysis of the contact area starts with the location of the part at its end position in the product. To generate the transformation of the part including corresponding work frames the part will be removed from its end location. Geometric constraints between parts determine how the part can be removed. They exist as long as the assembled part is in contact with other parts. Figure 4 demonstrates two different "peg-in-a-hole" operations. In both examples the pegs can only be moved along a straight line in the positive z-axis relative to its end location. After the analysis is finished the work frames ´End frame´ and ´Contact frame´ specify the transformation necessary to move the peg inside the hole. For the symmetrical peg the orientation along z-axis is free. Thus, the orientation along the z-axis in both work frames is specified to an interval from 0° to 360°. In the second example an additional constraint results from the slot and feather. This constraint limits the orientation along the z-axis to a unique value and consequently the z-orientation of both work frames to 0°.

Analysis of the near-by area

In the near-by area the parts must be moved between a contact frame and a depart or approach frame. There are no immediate contacts between the part and any environmental objects. Usually the bounds for the transformations of the part result from objects in the environment and from side effects regarding the transformations in the contact area. Potential transformations can be derived from the already mentioned rules. Important rules with relevance to the near-by area are SR1, WR2, WR3, WR4. With the help of the rules moving directions, planes for the location of approach/depart frames and free spaces as potential places for the part are defined. This results in work frames with additional degrees of freedom. Moreover the geometric elements *moving direction, approach plane* and *free space* can be handled to support the interactive definition of the approach/depart frame. The lower right picture in fig. 7 shows the use of a moving direction and a depart plane to define the corresponding depart frame graphically. For such cases the system provides the user with functions to define relations between geometric elements, for example "align parallel" the bottom face of the peg with the depart plane. Based on this input the system computes the numerical arrangement for the peg. According to the rules the depart frame in fig. 7 allows:

-    to move the peg away from the depart frame (transfer phase) without a collision,

-    to remove the peg along the moving direction that results from the contact transformation,

-    to remove the peg on a minimal path avoiding collisions.

Grasp planning

With grasp planning the assembly operation planning will be finished. Grasp planning means on one side the choice of an adequate gripper and on the other side the planning of the geometric relations between gripper and parts. Since that task is merely effected by the fetch phases and the assembly phases, grasp planning is located to the part oriented planning process. It is mainly influenced by:

-    the geometric situation in the fetch phase, where the part is located at its start frame (grasp part and move away),

-    the situation in the assembly phase, where the part is located in its end location in the final product (move down and release part),

-    the rules SR1, SR2 and SR3.

The geometric relation (grasp frame) between the reference coordinate system of the gripper and the coordinate system of the object is described using the work frame concept. This allows to incorporate many potential grasps into the grasp frame. For grasp planning graphical functions support the operator to define potential grasps for each assembly operation in three steps:

1.   With the part alone and the gripper a grasp frame with all potential grasps may be specified.

2.   Then the grasps valid for the part will be restricted considering the gripper and both the start situation in the fetch phase and the end situation in the assembly phase. Now the grasp frame specifies one or possibly more grasps valid for the context of the part given by the assembly operation.

3.   In the last step the part specific transformations will be refined with consideration of the gripper holding and moving the part. To avoid collisions between the moving objects (gripper and part) and the environmental objects the transformations must probably be restricted further. The degrees of freedom of work frames and the grasp frame are reduced respectively.

Figure 5 gives an example for the restriction of work frames in grasp planning. The left side demonstrates the start situation with the part specific grasp frame, that allows to grasp the part with a free z-orientation. Due to geometric constraints the part can not be grasped with arbitrary orientation of the gripper. This requires to reduce the degrees of freedom of the end frame, the contact frame as well as the grasp frame. The right side shows the remaining arrangements (0°,180°) of the gripper in reference to the part holder.

GROSS LAYOUT PLANNING

The start situation for gross layout planning is demonstrated on the part ´Sideplate´ in fig. 7 All the sketched work frames are affixed to ´Sideplate´. They represent the work points for the robot to assemble each part on its end location. These work frames ensure the correct assembly of the parts without any collision for the part and the gripper holding the part. With an arrangement of e.g. ´Sideplate´ or an equipment carrying ´Sideplate´, all work frames affixed remain unchanged relative to ´Sideplate´. For each part holder or peripheral device carrying parts such a group of affixed work frames exists.

With the constraints specified through work frames, in gross layout planning variants of layouts are developed with the following planning steps:

-    selection of robots and additional equipment,

-    arrangement of the units consisting of parts, peripherals and work frames,

-    assigning work frames to a robot if more than one robot are selected to execute the assembly,

-    computation of potential locations for each robot according to fixed work frames.

The layout variants ensure for each work frame the reachability by the respective robot. These gross layouts are refined in the next planning phase.

DETAIL LAYOUT PLANNING

The refinement of layouts is accomplished by the interrelating processes of evaluation and optimization. To evaluate a layout means to test how good a robot can do the assembly. Criterions for evaluation are:

-    the arm configuration of the robot for distinct work frames,

-    the travelling distance and the travelling time for the entire task,

-    potential singularities during motion.

Characteristic numbers for these criterions can be computed calling the robot emulator facilities of the robot simulation subsystem ROSI.

In reference to the evaluation results gross layouts can be optimized. The system supports some optimizations like

- the rearrangement of robots or peripherals including affixed work frames,
- the modification of an individual work frame inside its assigned degrees of freedom,
- the exchange of equipment against one of the same type with a slightly varied size.

PLANNING DATA MODEL

To support the robot specific planning tasks a submodel to manage the planning data was developed. Figure 6 shows this planning data model in a very simplified scheme. It outlines the most important abstract data objects and the relations between them. The model itself is further devided into the product oriented assembly information and the part that describes the layout of the workcell. This separation allows to manage several alternatives of workcell layouts for one assembly task during the planning process.

The precedence graph takes the role as an entry point for the planning information. If the product consists of many parts then several assembly plans can be set up. Each assembly plan represents one inspected sequence of assembly operations and points to the first item of this sequence. An assembly operation refers to the part to be assembled and to a list of transformations derived for the part. Each transformation then points to its work frames. A transformation has a reference to the device that feeds the corresponding part. In reference to fig. 6 transformation 'PT1' can be accomplished by the feeding device 'Feeder_A' to feed part 'Sideplate1' to the work frame 'WF2'. Device 'Robot_A' then executes all subsequent transformations to finish assembly operation 'AO1'. These generic objects allow to represent the transformations of the part which lead to the assembly process.

Relations between assembly operations with parts and between part transformations with devices establish the link between assembly specific data and the corresponding workcell. Each physical object (e.g. part or device) holds a frame that defines where the object is placed. This frame gives the position in relation to another object ('Part frame') or to an abstract cell coordinate system ('World frame'). Additional relations of the type 'is_above', 'is_below', 'connected_with' or 'feed' map topological links between different objects. Figure 6 represents a state when device 'Feeder_A' actually feeds part 'Sideplate1'.

Transforming this rough data scheme into a database requires the development of a conceptual data model, that describes all objects, their attributes and all the relations that exist between the objects. For ROSI-Layout the data model is actually implemented in a relational database.

REALIZATION

The concept is realized as an extension of an existing graphical robot simulation system (Dillmann, Huck, 1986). It is the purpose of the system to support the user as much as possible. Depending on the complexity of the current task the system is able to solve parts of the design automatically. However to make the concept practical to a wide range of applications ROSI-Layout is on purpose designed as a supporting tool for the engineer. He controls the planning process using graphical interactive functions and is responsible for the more difficult decisions. The graphical model represents the current planning state.

Automatic support from the computer mainly concerns:

- geometric oriented computations to derive transformations of parts mainly for the contact area and to detect collisions,
- computation of object positions implicitly specified by relations between elements of the geometry of the objects,
- automatic computation of potential locations for the robots,
- emulation of the functionality of robots and peripherals to evaluate layouts using robot motion simulation,

reachability tests for robots, singularity tests, estimation of assembly time, etc.

Figure 7 demonstrates with an example the graphical support of ROSI-Layout. During the planning session the precedence graph together with the state of the workcell is visualized on the graphics display. The user selects the appropriate assembly operation. With support of the system he then defines adequate part transformations and work frames. To construct the depart frame of the example the system automatically computes the direction for removal (visualized as an arrow). Using a graphical input device like a dial the user moves the peg along this direction. With the aid of a depart plane he constructs the minimal work frame for departure. Using a picking device he picks the top face of a neighbouring face as depart plane. Then he picks the bottom face of the peg being removed and commands the system to move the peg along the removal direction until the bottom face of the peg aligns parallel with the departure plane. As mentioned before a structured database takes care of the data handling.

CONCLUSIONS

A concept has been presented for designing workcells for robot assembly applications. The approach aims to adapt the layout of a workcell stronger to the restrictions imposed by the structure of the product. This is achieved by the investigation of the transformations necessary to assemble an individual part to the final product. Influenced by specific knowledge about assembly with robots the transformations are derived from the geometric models. This planning task provides work frames including potential degrees of freedom that guarantee the correct assembly. The work frames represent the product oriented constraints for the layout generation. For derived work frames layout planning supports the arrangement of peripherals and the automatic computation of potential locations for robots. In a second step gross layouts may be evaluated and optimized. The implementation of ROSI-Layout combines automatic planning functions with the interactive graphical planning to open the system for a wide range of applications.

For off-line programming the work frames represent the motion frames for the robot. Thus the programming effort can be significantly reduced. What remains to be done is the definition of the program's control flow, the synchronisation of simultaneous actions as well as the definition of technological parameters for the program.

REFERENCES

Dillmann R., Huck, M. (1986). A Software System for the Simulation of Robot Based Manufacturing Processes. Robotics Vol. 2 No. 1.

Dillmann, R., Hornung, B., Huck, M. (1986). Interactive Programming of Robots Using Textual Programming and Simulation Techniques. Proc of 16th ISIR, Brussel.

Frommherz, B., Hornberger, J. (1988). Automatic Generation of Precedence Graphs. Proc of 18th ISIR Lausanne.

Levi, P. (1988). TOPAS: A Task Oriented Planner for Optimized Assembly Sequences. ECAI: European Conf. on Artificial Intelligence.

Lozano-Perez, T. (1983). Robot Programming. Proc. IEEE. Vol. 71. No. 7.

Lozano-Pérez, T., Brooks, R.A. (1985). An Approach to Automatic Robot Programming; MIT, A.I. Memo No. 842.

Sjolund, R., Donath, M. (1983). Robot Task Planning: Programming Using Interactive Computer Graphic. Proc of 13th ISIR.

ACKNOWLEDGEMENT

This research work was performed at the Institute for Real-Time Computer Control Systems and Robotics, Prof. Dr.-Ing. Ulrich Rembold and Prof. Dr.-Ing. Rüdiger Dillmann, Faculty for Informatics, University of Karlsruhe, 7500 Karlsruhe, Federal Republic of Germany. This work was sponsored by the government of Baden-Württemberg.

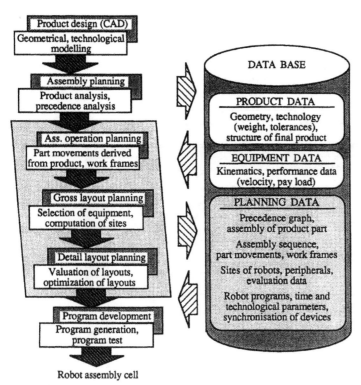

Fig. 1. Integration of the robot planning tasks into production planning

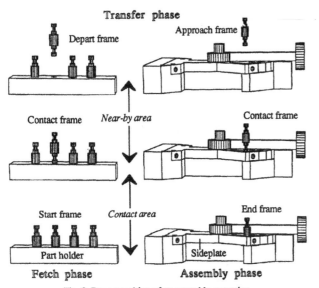

Fig. 2. Decomposition of an assembly operation

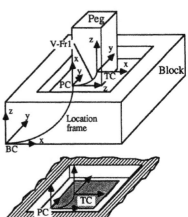

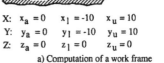

X: $x_a = 0$ $x_l = -10$ $x_u = 10$
Y: $y_a = 0$ $y_l = -10$ $y_u = 10$
Z: $z_a = 0$ $z_l = 0$ $z_u = 0$

a) Computation of a work frame

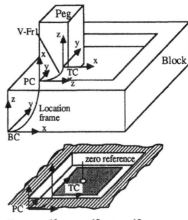

X: $x_a = -10$ $x_l = -10$ $x_u = 10$
Y: $y_a = -10$ $y_l = -10$ $y_u = 10$
Z: $z_a = 0$ $z_l = 0$ $z_u = 0$

b) Rearrangement of an object

Fig. 3. Characteristics of a work frame

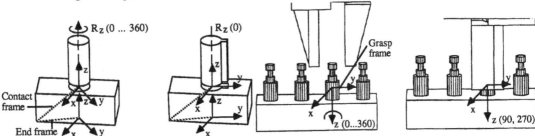

Fig. 4. Geometric based constraints for the contact area

Fig. 5. Limitation of work frames during grasp planning

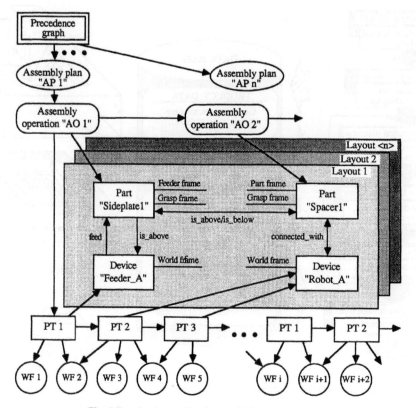

Fig. 6. Rough scheme of the data model for planning data

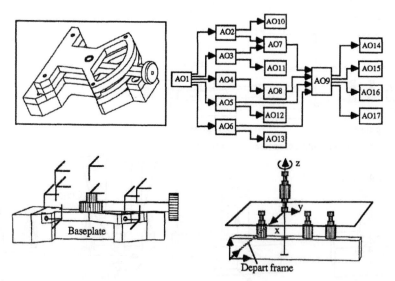

Fig. 7. A scene of a planning session

UPON ONE CONTROL PROBLEM FOR
SEMIAUTOMATED PRODUCTION SYSTEMS

D. Golenko-Ginzburg, Z. Sinuany-Stern and L. Friedman

*Department of Industrial Engineering and Management,
Ben Gurion University of the Negev, Beer Sheva, Israel*

Abstract. A production system includes a deterministic assembly-line and a supplementary
production unit to process parts for the line. There are several possible production speeds
to process these parts given in the form of stationary stochastic processes. Given the
routine control point, the actual accumulated production observed at that point and the
deterministic rate of demand, the decision-maker determines both the speed to be
introduced and the timing of the next control point. The problem is applied to
semiautomated production processes where the advancement of the process cannot be
measured or viewed continuously, and the process has to be controlled in discrete points by
the decision-maker. Since the cost of performing a single control is relatively high, the
control should be carried out as rarely as possible but has to ensure a preset confidence
probability of achieving production output no less than that required. Formulae for
determining the next control point for an arbitrary distribution function of the stationary
process with a certain autocorrelation function are presented. They depend on the status of
the system (shortage or surplus), the relation between the rate of demand and the mean
value of the speed, the variance of the speed, and on the confidence level $1 - \alpha$.

Keywords: Industrial production systems, production control, constraint theory, on-line
operation, simulation.

INTRODUCTION

The traditional optimal control models (see Lefkowitz
[1977], Nagasawa et al. [1983], Maxwell et al. [1989] deal
with fully automated systems where the output is
continuously measured on line. However, in semiautomated
production systems under random disturbances, e.g., in
textile industry, building projects, mining, agriculture, R&D
projects, system analysis and design projects, the output can
be measured only in preset control points as it is impossible
or too costly to measure it continuously.

At every control point the decision-maker observes the
amount produced and has to determine the timing of the next
control moment. For the sake of reducing the production
expenses the control should be carried out as rarely as
possible but under a chance constraint which with a pregiven
confidence probability $1 - \alpha$ ensures achieving production
output not less than that required.

In many production control models, including "just-in-time"
systems (see Bitran and Chang [1987]), the problem is to
control the supplement of an assembly-line with various
parts. In the case of semiautomated production systems the
assembly-line is usually supplied by parts produced by a

production unit under random disturbances and with several
possible speeds to process the parts.

Golenko-Ginzburg et al. [1989c] have examined the case of
an assembly-line which operates at a constant rate, which in
turn imposes a given deterministic rate of demand for the
parts at any stated moment until the end of the planning
horizon. The supplementary production unit possesses
several possible speeds, which may be given either in the
form of a random variable, or a stochastic process. Given
the probability distributions of the production speeds, the

planning horizon, the confidence probability not to be in on-
line shortage of the parts, and the deterministic rate for the
parts to supply the assembly-line, the management has to
control the unit's work periodically. Namely, on the basis
of the actual amount of parts processed at each control point
we have to determine both the next control moment and the
proper speed to be introduced. There are two possible
decision-makings:

a) Either if there is a shortage a higher speed will be used in
order to close the gap. It is assumed that a high speed is
more expensive, may cause a log-jam of parts in the line and
is undesirable. Therefore, in this case we want to minimize
the time span in which this speed is used.

b) On the other hand, if we are in surplus a lower speed can
be used and we want to maximize the time span of using this
lower and cheaper speed.

The general problem boils down to a multiple solution of the
subproblem where the interval between two control points is
maximized or minimized under a chance constraint that
insures that at any point the actual production will not fall
below the planned production at a given confidence level.

This paper deals with a production unit that operates at a
speed which is a stationary stochastic process with an
arbitrary distribution function, but with an autocorrelation of
a white noise. This means that at every moment from the
routine control point, the speed varies randomly and
independently from the moment before. Given the routine
control point, the actual accumulated production observed at
that point, the rate of demand, the pregiven probability and
the parameters of the stationary process, the next control
point is determined.

Similar mathematical formulations have been treated in the literature (Zangwill [1969], Wilde and Beightler [1967]). This paper is a further development of our previous articles (Friedman et al. [1989a], Golenko-Ginzburg et al. [1989c]) where the general problem has been considered and the subproblem has been solved for the case of production speed being a random variable only.

THE GENERAL PROBLEM'S FORMULATION

Let us define the following terms:

T^p - the end of the planning horizon;

$V^f(t)$ - actual amount of parts produced up to moment t;

t_i - the i-th control point;

W - the deterministic rate of demand for parts;

$V^*(t) = W \cdot t$ - the planned amount of production at moment t;

$B_i = V^f(t_i) - W \cdot t_i$ - the difference between the actual and planned amounts at moment t_i;

$v_j(t)$ - the j-th production speed, $j = 1, 2, ..., m$, as a stationary stochastic process in the wide sense with mean value μ_j and an autocorrelation function $\sigma_j^2 \cdot \delta(t_k - t_l)$, where

$$\delta(t_k - t_l) = \begin{cases} 1 & \text{if } t_k = t_l \\ 0 & \text{otherwise} \end{cases} ; \qquad (1)$$

$1 - \alpha$ - pregiven confidence probability;

N - the number of control points;

m - the number of possible speeds;

k_i - the index of the speed introduced at the control point t_i, $1 \le k_i \le m$;

d - the minimal time span between two consecutive control points t_i and t_{i+1}.

The problem is to minimize the average actual amount

$$F_1 = \min_{t_0, ..., t_{N-1}, k_0, ..., k_{N-1}} E \left\{ \sum_{i=0}^{N-1} \int_{t_i}^{t_{i+1}} v_{k_i}(t) \, dt \right\} \qquad (2)$$

and the number of control points

$$F_2 = \min N \qquad (3)$$

s.t.

$$\Pr [V^f(t) \ge W \cdot t] \ge 1 - \alpha \qquad \forall \, 0 \le t \le T^p$$

and control point restrictions

$$t_0 = 0, \qquad (4)$$

$$t_N = T^p, \qquad (5)$$

$$t_i < t_{i+1} \le T^p, \quad i = 0, ..., N-1 \qquad (6)$$

$$t_{i+1} - t_i \ge d, \quad i = 0, ..., N-2. \qquad (7)$$

THE SUBPROBLEM

Since the problem is too difficult to obtain a precise or even an approximate solution, a heuristic approach seems to be the most effective one. Golenko-Ginzburg et al. [1989c] have presented an algorithm based on a combination of several heuristic rules and a multiple solution of the subproblem:

For a fixed speed $v = v_j$ at point t_i determine the next control point, t_{i+1}, to maximize (in the case of surplus), or minimize (in the case of shortage) the objective function

$$\text{Max (Min)} \, \{T = t_{i+1} - t_i\} \qquad (8)$$

subject to

$$\Pr \{V^f(t_{i+1}) \ge W \cdot t_{i+1}\} \ge 1 - \alpha. \qquad (9)$$

Two alternative assumptions may be imbedded in the problem's formulation:

a) the speed is a random variable, namely, it varies at the routine control point and remains constant until the next control point;

b) the speed is a stochastic process, i.e., it varies at each point between two adjacent control points.

The case of the first assumption has been examined by Friedman et al. [1989a]. In this paper we present the general solution of problem (8-9) for the second case.

THE SUBPROBLEM'S SOLUTION

The problem is to determine the next control point t_{i+1} in order to maximize or minimize the objective function

$$\text{Max (Min)} \, \{T = t_{i+1} - t_i\} \qquad (10)$$

subject to the constraint

$$\Pr \left\{ V^f(t_{i+1}) \ge V^*(t_{i+1}) \right\} \ge 1 - \alpha. \qquad (11)$$

Using

$$V^f(t_{i+1}) = V^f(t_i) + \int_{t_i}^{t_{i+1}} v(t) \, dt, \qquad (12)$$

we obtain constraint (11) in the form:

$$\Pr \left\{ \int_{t_i}^{t_{i+1}} v(t) \, dt < W \cdot T - [V^f(t_i) - W \cdot t_i] \right\} \le \alpha. \qquad (13)$$

Let us examine the stochastic integral $\int_{t_i}^{t_{i+1}} v(t) \, dt$ in (13). It is known (see Parzen [1962]) that if $v(t)$ is a stochastic process of white noise, the stochastic integral $\int_{t_i}^{t_{i+1}} v(t) \, dt$ is a random variable with mean value $\mu \cdot T$ and variance $\sigma^2 \cdot T$. Moreover (see Zayezdny et al. [1989]), this integral is approximately a normal random variable for any distribution function of the stochastic process under consideration. This follows from the fact that any stochasic integral $\int_a^b v(t) \, dt$ can be written as a Riemann sum $\sum_{i=0}^{n-1} v(t_i)(t_{i+1} - t_i)$, for any partition

$$a = t_0 < t_1 < ... < t_i < ... < t_n = b.$$

According to the Central Limit Theorem, whatever be the distribution of n independent random variables, the distribution of the sum is approximately normal for large value of n.

If $v(t)$ is a stationary stochastic process of white noise, where each variable $v(t_k)$ at time t_k is independent of the variable at the moment before, as it is given in (1), the Central Limit Theorem can be applied.

Denote the constraint in (13) by g(T).

$$g(T): (W-\mu) T + Z_{1-\alpha} \sigma \sqrt{T} - (V^f(t_i) - W \cdot t_i) \le 0 \quad (14)$$

where $Z_{1-\alpha}$ is the $(1-\alpha)$-th quantile of the standard normal distribution. The solution for (10) under the constraint g(T) is obtained by solving the quadratic equation in (14). It is given in the Appendix.

Summary of the results is given in Table 1 (based on the Appendix). The solution differs for surplus (B > 0) versus shortage (B $\le$ 0) and for various relations between the production speed (mean value μ) and the rate of demand (W).

Table 1: The Subproblem's Analytical Solution

W \ B	B > 0	B ≤ 0
W > μ	Min T = 0	
	Max T = T₁	No solution
W = μ	Min T = 0	
	Max T = T₃	
μ - ε ≤ W < μ	Min T = 0 or T₂	Min T = T₂
	Max T = T₁ or ∞	Max T = ∞
	Min T = 0	
W ≤ μ - ε		
	Max T = ∞	

$$T_{1,2} = \frac{B}{W-\mu} + \frac{1}{2}\left(\frac{Z_{1-\alpha}\sigma}{W-\mu}\right)^2 \left[1 \mp \sqrt{1 + \frac{4(W-\mu)B}{Z_{1-\alpha}^2 \sigma^2}}\right],$$

$$T_3 = \left(\frac{B}{Z_{1-\alpha}\sigma}\right)^2, \quad \varepsilon = \left|\frac{Z_{1-\alpha}^2 \sigma^2}{4B}\right|,$$

$$\int_{-\infty}^{\alpha} \frac{1}{\sqrt{2\pi}} e^{-z^2/2} \, dz = \alpha,$$

$$Z_\alpha = -Z_{1-\alpha}, \quad \alpha < 0.5.$$

In order to evaluate the efficiency for various distribution laws v(t) extensive simulation has been performed. The production process was simulated and the actual probability achieved was compared with the confidence probability given, 1 - α. In the simulation the stochastic process has been discreticized. The results of the simulation for various distribution laws are described by Friedman et al. [1989b]. It can be well-recognized from the study that the analytical

solution is independent of the distribution of v and fits the simulation results.

APPENDIX

Solving Max (Min T) With Chance Constraint

The necessary and sufficient conditions for an optimal solution T* of the non-linear programming problem:
$$\max f(T)$$
$$s.t.$$
$$g(T) \le 0,$$

where the functions are differentiable, are the Kuhn-Tucker conditions (see Zangwill [1969]):

(1) T* is feasible.

(2) There exists a multiplier $\lambda \ge 0$ such that $\lambda \cdot g(T^*) = 0$.

(3) $\dfrac{\partial f(T^*)}{\partial T} - \lambda \dfrac{\partial g(T^*)}{\partial T} = 0$.

The conditions apply to the minimization case as well, with the exception that λ must be nonpositive.

These conditions can be applied to the objective function in (10), f(T) = T, subject to the constraint:

$$g(T): (W-\mu)T + Z_{1-\alpha} \sigma \sqrt{T} - B \le 0 \quad (A.1)$$

Since (3) in the Kuhn-Tucker condition is:

$$1 - \lambda \cdot \frac{\partial g(T)}{\partial T} = 0,$$

$$\lambda = \frac{1}{\frac{\partial g(T)}{\partial T}} \ne 0,$$

the optimal value T* is the solution of (A.1).

If $W \ne \mu$, g(T) has a quadratic form, and

$$\lambda^{-1} \frac{\partial g(T)}{\partial T} = (W-\mu) + \frac{Z_{1-\alpha}\sigma}{2\sqrt{T}}.$$

For positive λ the maximum is obtained, and the minimum is for negative λ.

Denote $\sqrt{T} = y$; substituting y in (A.1) we obtain

$$g(y): (W-\mu) y^2 + Z_{1-\alpha} \sigma y - B \le 0, \quad y \ge 0. \quad (A.2)$$

For all y's satisfying (A.2), value T = y² satisfies (A.1). The solutions of the quadratic equation

$$(W-\mu) \cdot y^2 + Z_{1-\alpha} \sigma y - B = 0 \quad (A.3)$$

are y₁ and y₂:

$$y_1 = \frac{-Z_{1-\alpha}\sigma + \sqrt{Z_{1-\alpha}^2 \sigma^2 + 4(W-\mu)B}}{2(W-\mu)}$$

$$y_2 = \frac{-Z_{1-\alpha}\sigma - \sqrt{Z_{1-\alpha}^2 \sigma^2 + 4(W-\mu)B}}{2(W-\mu)}$$

The corresponding solutions for T of (A.1) are T₁ and T₂ (T₁ < T₂):

$$T_1 = y_1^2 = \frac{B}{W-\mu} + \frac{1}{2}\left(\frac{Z_{1-\alpha}\sigma}{W-\mu}\right)^2 \left[1 - \sqrt{1 + \frac{4(W-\mu)B}{Z_{1-\alpha}^2 \sigma^2}}\right] \quad (A.4)$$

$$T_2 = y_2^2 = \frac{B}{W-\mu} + \frac{1}{2}\left(\frac{Z_{1-\alpha}\sigma}{W-\mu}\right)^2 \left[1 + \sqrt{1 + \frac{4(W-\mu)B}{Z_{1-\alpha}^2 \sigma^2}}\right] \quad (A.5)$$

We consider the following situations of the system under examination:

I.   If the system is in surplus at t_i, i.e., $B = V^f(t_i) - W \cdot t_i > 0$, we consider 4 different cases;

Ia. $W > \mu$;

Ib. $W = \mu$;

Ic. $\mu - \varepsilon \leq W - \mu,\ \varepsilon = \dfrac{Z_{1-\alpha}^2 \sigma^2}{4B}$;

Id. $W < \mu - \varepsilon$.

II.   If the system is in shortage, i.e., $B < 0$, we consider two different cases:

IIa. $W \geq \mu$;

IIb. $W < \mu$.

III.   In the case of equality ($B = 0$) we also consider two different cases:

IIIa. $W \geq \mu$;

IIIb. $W < \mu$.

Ia.   There are two solutions of (A.3) with opposite signs, and the center of the convex parabola is on the negative side. Therefore among all the positive y's between y_2 and y_1 ($y_2 < y_1$), where $g(y) \leq 0$, the minimal is $y = 0$, (or $T = 0$), and the maximal is y_1, $T_1 = y_1^2$ in (A.4) being the solution. The $\partial g(T)/\partial T$ for this maximum solution is:

$$(W-\mu) + \frac{Z_{1-\alpha}\sigma}{2\sqrt{T_1}} = (W-\mu) + \frac{Z_{1-\alpha}\sigma}{2y_1} =$$

$$= (W-\mu) \left\{ \frac{\sqrt{1 + \dfrac{4(W-\mu)B}{Z_{1-\alpha}^2 \sigma^2}}}{-1 + \sqrt{1 + \dfrac{4(W-\mu)B}{Z_{1-\alpha}^2 \sigma^2}}} \right\}$$

which in case Ia is always positive. Therefore a maximum solution is obtained at T_1.

Ib.   Chance constraint (A.1) becomes a linear constraint $Z_{1-\alpha}\sigma\sqrt{T} - B \leq 0$. All the T's that satisfy the constraint are between 0 and $\dfrac{B^2}{Z_{1-\alpha}^2 \sigma^2}$. Their minimum is at $T = 0$ and their maximum is at T_3, where

$$T_3 = \frac{B^2}{Z_{1-\alpha}^2 \sigma^2}.$$

Another way to obtain this case is as a limit of case Ia. By using L'hospital rule twice, the limit solution for T_1 is obtained as T_3.

Ic.   There are two positive solutions to the concave parabola. In this case two sets of y's satisfy $g(y) \leq 0$, namely, the first one being $0 \leq y \leq y_1$, and the second one $y_2 \leq y < \infty$. For the first set the minimal solution for T is $T = 0$ and the maximal is $T_1 = y_1^2$. For the second one, the minimal is $T_2 = y_2^2$ and the maximal is $T = \infty$.

Id.   There are no real roots to (A.3). It means that all the points of the concave parabola are below the time axis. Therefore each T on the positive time axis between 0 and ∞ satisfy (A.1). The minimum solution is $T = 0$ and the maximal is $T = \infty$.

IIa.   If $W > \mu$ either
        - there are no real roots to (A.3); that means that all the points of the convex parabola are above the time axis, therefore no T's satisfy (A.1); or
        - there are two negative solutions for y and no positive solution for T.
        If $W = \mu$, (A.1) becomes a linear constraint $Z_{1-\alpha}\sigma\sqrt{T} - B \leq 0$, and also in this case no positive T satisfies this constraint.

IIb.   There are two roots for (A.3) with opposite signs, and the center of the concave parabola is on the positive side. Among all the positive y's satisfying $g(y) \leq 0$ which are between y_2 and ∞ (here y_2 is greater than y_1), the minimal is y_2 and the maximal is $y = \infty$. Therefore the minimal solution for T is $T_2 = y_2^2$ and the maximal is $T = \infty$.

        Since relations

$$\lambda^{-1} = \frac{\partial g(T)}{\partial T} = (W-\mu) + \frac{Z_{1-\alpha}\sigma}{2y_2} =$$

$$= (W-\mu)\left\{ \frac{\sqrt{1 + \dfrac{4(W-\mu)B}{Z_{1-\alpha}^2 \sigma^2}}}{1 + \sqrt{1 + \dfrac{4(W-\mu)B}{Z_{1-\alpha}^2 \sigma^2}}} \right\}$$

and $W-\mu < 0$ hold, $\partial g(T)/\partial T$ is always negative. This means a minimum solution at T_2.

IIIa.   There are two roots to (A.3), one always being $y=0$ and the other a negative one. Therefore there is no positive solution for T. If $W = \mu$, the constraint (A.1) is $Z_{1-\alpha}\sigma\sqrt{T} \leq 0$, and again, there is no positive T which satisfies the constraint.

IIIb.   There are two solutions of (A.3). One is always $y=0$, the other, y_2, is a positive one, and the center of the concave parabola is on the positive side. Among all the positive y's satisfying $g(y) \leq 0$, which are between y_2 and ∞, the minimal one is y_2 and the maximal is $y = \infty$. Therefore the minimal solution is $T_2 = y_2^2$, and the maximal is $T = \infty$.

        Substituting $B = 0$ in T_2, one obtains $T_2 = \left(\dfrac{Z_{1-\alpha}\sigma}{W-\mu} \right)^2$

as the minimal solution in this case.

REFERENCES

Bitran, G.R. and Chang, Li. (1987). A mathematical programming approach to a deterministic kanban system. Management Science, 33, No. 4, 427-441.

Friedman, L., Golenko-Ginzburg, D. and Sinuany-Stern, Z. (1989a). Determining control points of a production system under a chance constraint. Engineering Costs and Production Economics (in press).

Friedman, L., Golenko-Ginzburg, D. and Sinuany-Stern, Z. (1989b). Semiautomated production systems under disturbances. Tech. Rep. No. 89-18, Dept. of Ind. Eng. & Mgmt., Ben Gurion University, Beer Sheva, Israel.

Golenko-Ginzburg, D., Sinuany-Stern, Z. and Friedman, L. (1989c). Feeding parts with random production speed to an assembly line. Engineering Costs and Production Economics (in press).

Lefkowitz, I. (1977). Integrated control of industrial systems. Phil. Trans. R. Soc. Lond. A. 287, 443-465.

Maxwell, W., Muckstadt, J., Thomas, J., Vander Eecken, J. (1983). A Modeling framework for planning and control of production in discrete parts. Manufacturing and assembly systems interfaces 13 (6), 92-104.

Nagasawa, I. and Kobayashi, T. (1973). Computerized integrated production control system. Third Interregional Symposium on the Iron and Steel Industry, UNIDO.

Parzen, E. (1962). Stochastic Processes, Holden-Day, Inc. San Francisco.

Wilde, D. and Beightler, C. (1967). Foundations of Optimization, Prentice-Hall, Englewood Cliffs, N.J.

Zangwill, T. (1969). Nonlinear Programming, Prentice Hall, Englewood Cliffs, N.J.

Zayezdny, A.M., Tabak, D., Wulich, D. (1989). Engineering Applications of Stochastic Processes Theory, Problems and Solutions, RSP.

THOR, A CAPP SYSTEM FOR TURNING WITH A HIGH GRADE OF INTERACTIVE IMPLEMENTATION

A. Bengoa and I. Yañez

Department of Production Engineering, Asociación de Investigación Tekniker, Eibar, Spain

Abstract. THOR is a generative process planning system for rotational parts. Given the geometry of the part, from any CAD system or from its own graphics package, the system determines the lathe, set-ups, operations sequence, cutting tools, cutting conditions, time and costs, route sheet and the CLDATA file. The main characteristics of the THOR system can be described as flexibility and integrability.

Keywords. CAPP, CIM, CAM, CAD, CNC.

INTRODUCTION

We are deeply involved in the so-called communications age where up-to-date data processing is a production tool as important as time and money. Present companies need for its good management to have available a sound information, sufficient and on time. On the other hand and due to today market evolution, companies have found themselves in a situation where they have to react as soon as possible when facing market demands and consequently they have to reduce the lead time on new products.

Computer Integrated Manufacturing (CIM) consists of a integrated data processing system of all company data (management, technical and production) in order to know in each of its deparments the latest state of bussines going on that moment and take the right steps in order to achieve the best of results.

CIM strategy is going to allow companies to prepare and handle quickly and properly the information flow in such a way that a quick reaction to market demands imposed on them and reduction of lead time can be achieve.

At each company department back-ground information is prepared, and when this is carried out with the help of computers, they become what is know as islands of automation. These data should be shared, creating a data flow among departments, in order to be able to work in real time.

Working within a CIM environment requires a great amount of speed when a step has to be taken, this has to be done on time when facing different events that alter the production process. This reaction capacity becomes more critical when working is based on JIT (Just In Time) philosophy, since any incident related to production process increase risks by the time order delivery terms have to be met and there is not enough time to achieve this.

Because all this, a CIM component characteristics could be expressed as the flexibility degree to cope with diverse situations where it has to be used and the system integrability or its capacity for sharing information with others CIM components.

One of the most important steps when establishing a CIM system for the industry is the computarisation, automation and integration of the process planning function. Computer aided process planning (CAPP) is the critical link between desing (CAD) and manufacturing (CAM) and fills the technology blank that exists between them. CIM advantages will never be reached without an effective integration and computarisation of the process planning function.

Because of this and due to CAPP strategic situation within a CIM environment (See fig. 1) is the reason why special attention and efforts are paid to CAPP system developing in both, the academic and the industrial comunities.

CAPP can be defined as the systematic determination of the methods (how) and the means (machines, fixtures and tools) by which a product is to be manufactured economically and competitively. Therefore, it specifies the variety and amount of needed resources and the sequence and manner they should be used. A CAPP system should be able to produce all these data in a short period of time.

Therefore, a CAPP system easily allows to face production problems and to provide, in due time, alternative ways or other changes in the means, in order that the new process can match with the availability of these means at any time.

A CAPP system integrated within a CIM environment is the source of manufacturing technical data and allows for a flexible use of production means. CAPP together with scheduling allows for a quick answer matched to means available as well as to reach a good degree of their use.

THE BASIS OF THE THOR SYSTEM

THOR (TecHnological ORiented modules for turning) is a generative process planning system that produces a complete process plan in the turning field. It is now under development at a very advanced state.

Given the geometry of the component and the general input data, the system chooses the most suitable machine tool, the work-holding parameters, determines the operation sequence, chooses the best cutting tool for each operation, calculates the optimized cutting parameters for each pass of every operation, works-out the time spent, outputs a process sheet and a CLDATA file and monitors the process. (See fig. 2)

In order to be able to perform the above tasks, and to fulfil the basic requirements a CIM component should have, i.e. flexibility and integratibility, the THOR system is based on the following:

- A generative approach,
- High interactive implementation,
- Modularity,
- Graphic facilities,
- Technological algorithms,
- Word processing,
- Artificial intelligence techniques,
- A technological database,
- An IGES interface,
- A CLDATA file.

The system has been developed under the generative approach, i.e. it generates a complete process plan for every particular piece. This approach gives the system more flexibility than only the variant approach. The system also uses the variant approach to solve some functions like the machine selection and the time calculation. In order to give flexibility to the system and to make it able to produce alternative process plans the system needs to be complete,i.e. to be able to solve all the process planning functions since all of them are close related to each other.

The system has been provided with a high degree of interactivity in all its stages to give a greater decision taking capacity to the operator when dealing with process planning. This interactivity has several facilities that allow and speed-up a great deal the man-machine interface. Thanks to this interactivity a larger system flexibility is attained. Moreover the planner has also the capacity to validate the process in accordance with the manufacturing conditions.

Since the system has to carry out many particular functions of very different nature, like calculations, decision-making, graphics and so on, it has been used for each function the most suitable informatic tool for software developtment, like conventional algorithms, word processing, graphics facilities and artificial intelligence techniques.

Due to CAPP assumed complexity and the great amount of different working environments where it has to work, it is not possible to reach an universal solution for all those environments. This implies a new flexibility in the system in order to be able to cope with, without meaning a special implementation dificulty, the different working environments where there will be a particular experience and specific working philosophy.

In order to achieve this kind of flexibility, expert systems development in CAPP is of great importance. On one side due to how knowledge is expressed in an E.S. and on the other because E.S. allow to add, delete and change quite easily the rules. They allow to collect the expertise gathered within a specific company and its later implementation. They are easily 'tuned' to fit the particular logic of those different environments where the system is going to be operating.

System integrability is proved by the integration of system input/output. Data input is made up by the piece geometry and its parameters as well as general data. Geometric and manufacturing requirements integration is obtained thanks to a graphics interface between CAD systems and the THOR system, using the IGES standard (Initial Graphics Exchange Specification) for geometric data interchange. Integration of the general data input/output and technological data output generated by the system is attained by the use of the SQL standard in the technological system database. CAPP-CAM integration is given directly thanks to the CLDATA neutral file made by THOR. It contains all process data which are needed by a post-processor in order to prepare the part program for a specific combination of machine-tool and control equipment. (See fig. 3).

MAN-MACHINE INTERFACE

The THOR system has a powerful menu driven man-machine interface for the input task and the later process modifications. It has been designed to be very simple to use with several facilities in order to give the planner the chance to contribute with his/her experience to improve the process planning.

The user interface has three main parts. A graphics package, a graphics editor and a process editor.

Graphics Package

This input module has been developed to give

independence to the system from existing CAD systems. This package enables the user to define the shape of the component and the blank. It is menu driven an very easy to use.

The user chooses sequentialy the geometric elements, introduces the known parameters of those elements and the system does the rest. There is no need for the user to calculate the intersections or tangency points. In the cases of various shape posibilities the package displays them and the user selects the wanted one. (See fig. 4).

Once the complete profiles are stored, the systems asks the user to introduce all the manufacturing requirements, such as the surface roughness, size tolerances, concentricities, parallelism, perpendicularities, run outs and so on.

Graphics Editor

This editor permits the user to perform many modifications and checks during and at the end of the geometry input. Geometric elements can be added, deleted or modified. Machining requirements can be changed as well. It can be checked the distances between the part points, points and lines, between lines, check the slope of any line and the radius of any arch. This option is very useful to verify if the whole part geometry has been properly introduced.

This editor is also very useful to complete and to correct the data that has been introduced from a CAD system, since all the required manufacturing information (requirements, footnotes,...) can not always be captured from the drawing.

Process Editor

The process editor enables the user to carry out a complete operation sequence choosing from the menu between the basic turning operations and the portion of the part that has to be machined by that particular operation. This process is carry out until the whole process planning is finished. This operation sequence can be done fully automatic by the system itself.

Once the whole process planning has been made this editor allows the planner to realize many different changes in the operations (to change the cutting tool, cutting parameters, make all passes with the same depth of cut...) and in the whole process (renumber the operations sequence, change the setups, define new operations...).

In this way, the planner can define more than one process for the same part. In order to compare those processes, to modify or create new ones, some inquiring facilities are provided by this editor. The user can ask the system for information about all tools employed in each process, process general

data and a complete information on each operations of every process. A simulation of all current processes is also available.

All of the above tasks can be performed thanks to a file called PTFILE (Process Technology File). The system has been storing in this file all the original process information and the later (if any) alternative processes data the technological processors have been generating. I.e., selected machine, fixtures, the operations that constitute a particular process plan, selected tools for every operation, part geometry involved in the operation, cutting conditions and total time and cost.

At the end of the process, and if there is more than one process planning done, the user has to choose the best one in his/her opinion. Then the system retrieves all the information about that process from the PTFILE, it puts it in the process post-processor and the later generates rapidly the route sheet, the CLDATA and, on the planner's request, it displays the selected process monitorization. At this point some of the data has to be recalculated, like the final total time and cost. (See fig. 5).

DESCRIPTION OF THE THOR MODULES

The following modules perform all the CAPP functions and constitute the technological processor of the system. They feed PTFILE with all the information the process post-processor will need.

Machine Selection

The choice of the most suitable lathe from the workshop is a very important task. The characteristics of the chosen machine will greatly influence over the rest of the process planning. This machine selection involves a lot of decision-making, like in the other selection tasks in the THOR system. Artificial Intelligence has been proved to be the most suitable informatic tool in order to eficiently achieve the desired solution in issues of this nature. The most succesful A. I. application, the Expert System, permits to express the expert's knowledge with a more suitable syntax. E. S. rules can be added, deleted or modified quite easily which makes this kind of systems very apt to be adapted to particular environments and different working philosophies. The use of expert systems adds flexibility to the system and makes it more efficient.

Due to all previous reasons, an expert system has been developed to solve the machine selection task, at a prototype level.

Some of the rules deal with the physical characteristics of the machine and the workpiece (sizes, weights). Some others with the technological aspects involved in the turning operations (Power, feed and speed ranges, metal removal rate, numerically controlled or not). Others deal with the availability of the machine through the time, and the rest with general production and finacial aspects (batch size, complexity of the part, machine cost rates...). (See fig. 6).

Work holding configuration

A particular work-holding configuration will directly affect the operation sequence and the cutting conditions, and must fulfil the technological requirements placed on the component such as surface roughness, concentricity and run outs. For these reasons the determination of the work-holding configuration is important.

The system has to take into consideration the following:

- Slenderness of the part.
- Technological requirements placed on the part.
- Different clamping system devices available on the machine and their characteristics.
- Geometry of the part.

The system must determine:

- Work-holding configuration (chuck,chuck and center, between centers) and its elements.
- Minimun number of setups to complete the job.
- The different clamping surfaces on the blank and the part.

Machining methods. Operation sequence determination

Due to the large amount of possible solutions that can be reached to make the operation sequence and due to the complexity of this task, two operation modes have been developed. An automatic mode and a manual-interactive operating mode.

In the automatic mode the system itself determines the complete operation sequence. It takes into consideration the different setups and compares the component profile against the blank profile. It is based on a series of logic rules and the preceding relationships between some of the basic turning operations. (e.g. drilling before boring). The operation sequence obtained this way is quite an optimal one and fulfils all the manufacturing requirements placed on the part.

In the manual mode, the user enters the process editor and chooses sequentialy the operations that he/she wants to be done and then chooses the part of the workpiece that has to be machined by that operation. The system does the rest. It determines the toolpath, chooses the cutting tool and calculates the cutting conditions. This editor mode is menu driven to choose the desired operation and based in graphic facilities to select the portion of the part to be machined. (e.g. Rubber band).

Figure 7 shows an example of how a part can be manufactured using these two methods.

Cutting tool selection

The system in this module chooses automatically the best couple toolholder-insert from those in the database for each particular operation taking into account the part material.

This task is quite a complicate one due to the large amount of different kinds of toolholders, inserts, chipbreakers and insert grades.

In an automatic process a good chip control is of vital importance. If the chip is not properly broken that may lead to unwanted and even dangerous situations.

Selecting a good cutting geometry is basic because of its influence in the cutting parameters which strongly influence the economy of the operation. Selecting the most suitable carbide grade for an operation and a part material will lead to a longer tool life which means a high percentage of the total cost of the operations.

The cutting tool selection module works with the ISO code for turning tools. It is based on:

- The type of operation. (Roughing or finishing).
- The turning operation. (Facing, boring, drilling, threading...).
- Working conditions. (Slenderness, vibrations, machine rigidity and so on).
- Physical characteristics of the piece material.

It has to be noted that in turning there are many cutting tools that can perform many different operations and others just one. This tool selection module optimizes the number of chosen tools by substituting when posible the actual chosen tool for other already chosen (nearly equivalents form a technological point of view) or vice versa in order to optimize the number of tool changes and not to exceed the maximum number of tools that the turret can hold.

Determination of the cutting parameters

As pointed before, and since all processes are becoming more automatic every day, the cutting time has a strong influence over an operation economic side. This THOR module can work with two different optimizing criteria. Minimum cost or maximum production.

This module is based on the Taylor's tool life formula and takes into account many different constrains, and calculates for each pass the depth of cut, feedrate and cutting speed.

Since all these calculations require a large amount of data not always available, it is not always possible to apply this optimizing method. In such cases this module uses a different method based on recommendations. In the database are stored cutting conditions for all turning operations and for hundreds of working materials. This recommendation values are used as starting points, checked against all the constrains and accordingly modified in case those constrains are not satisfied. (See fig. 8).

Time calculations

Total time spent in manufacturing a part is a very important data for scheduling and capacity planning.

There are two different time calculating methods depending on the type of machine being used for the process. Numerically controlled or conventional. In the first case and since all the speeds and feeds as well as all the toolpath and tool changes are known by the system, a more exact time calculation is done. In the second case, group technology techniques are used to estimate the total time based on statistical and empiric formulae for each part family.

Results Output

The system provides the user with three different results output: - Monitoring.
                          - Route sheet.
                          - CLDATA file.

Monitoring is a very important tool to check if the process is properly run by the system and for checking if there is not any collition risk between the part and the tools. The system screens the blank and the component as well as all tool movements, using plain lines to indicate the tool is removing material, feed rate, and dotted lines when moving at rapid rates. Tool changes are also monitored.

This monitoring enables the user to easily interact with the system in all its process editing facilities.

At the end of the planning session, and when a process is chosen as the final one (if there are more that one), the system, post-processing the information stored in the PTFILE, generates a complete route sheet of the chosen process. (See table 1).

This route sheet is very useful to prepare the NC machine and should be used as a general guide line if a conventional lathe is to be used.

The third output facility, CLDATA file, is the most important one from the integration point of view. This file, which has been made through all the process determination, is a standard file that contains the cutter location data for a particular part without reference to the machines on which it may be made. A NC post-procesor adapts this output to produce a machine program for the production of the part on a particular combination of machine and control equipment. Obviously, if the selected machine is a conventional one the post-processor doesn't generate this output.

All the results of the system are stored in the database for a later use.

TECHNOLOGICAL DATABASE

THOR is provided with a technological database in order to feed the system with all the technological data it requires. In this relational database there are stored general data and particular data environment dependent.

The general data comprises recommended cutting conditions for all turning operations and hundreds of workpiece materials, recommended tool geometries and grades as well as cutting fluids recommendations. There is a lot of data that depends on the environment the system will be operating and is related with the available turning resources a particular firm has. Machine tool parameters, fixtures, available cutting tools and so on. There is a specific user interface to easily store, retrieve or modify any particular data in the database.

FINAL REMARKS

The main characteristics of the THOR system are its flexibility and its integrability. The system is very flexible because it can simply be adapted to different working environments, it can work automatically or driven by the user and features many modifications and inquiring facilities. THOR can be easily integrated because it could get the geometrical information from a CAD system through the IGES standard and because the system generates a standard CLDATA output file.

The use of this system will lead to drastic cuts in the process making time. It can rapidly adapt itself to new workshop situations and it can quickly make alternative process for different machine tools. The processes this system generates will lead to better machining quality and machining conditions because it takes into account all manufacturing requirements and the cutting conditions fulfils all the process constrains.

It has to be noted that it has been used the most suitable tool for software development for each task the system has to perform, conventional algorithms for the calculation tasks and expert systems for some decition-making ones, which makes the THOR system very efficient.

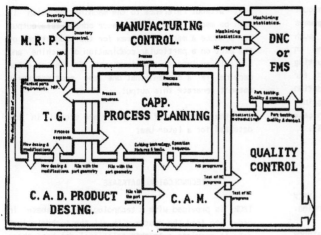

Figure 1

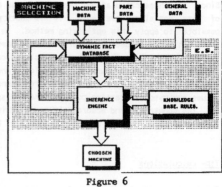

Figure 6

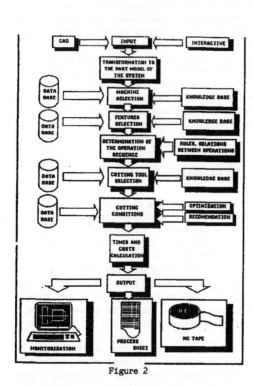

Figure 2

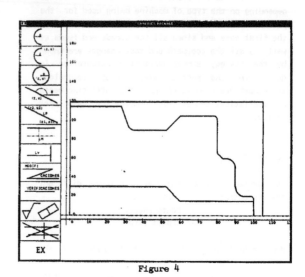

Figure 4

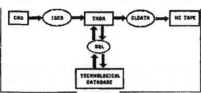

Figure 3

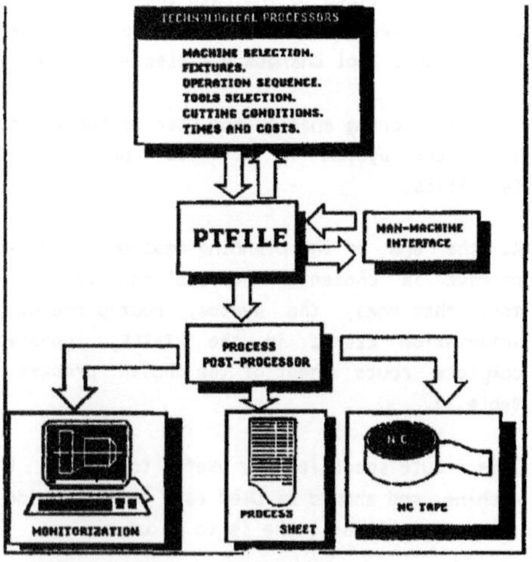

Figure 5

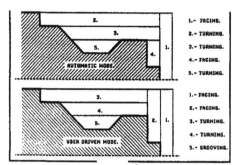

Figure 7

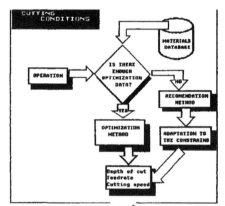

Figure 8

Oper.	Operation	Tool	Machining Conditions							Time s
1	Rough Facing	1	Diam.1 mm	Diam.2 mm	Depth mm	Feed mm/rev	Speed m/min	N rpm	Power kw	
	1		144.00	0.00	4.00	0.38	178	316.6	9.60	25.7
	1	PTFNR 1212F16	TNMG 16 04 12 15							
	2	PCBNR 2525M12	CNMG 12 04 12 15							

PART OF THE ROUTE SHEET

Table 1

REFERENCES

- "TECHTURN: a technologically oriented system for turned components", S.Hinduja, BE, MSc, PhD and G.Barrow, DipTech (Eng), MSc Tech, Ph University of Manchester Institute of Science and Technology.

- "Fixes, a System for Automatic Selection of Set_ups and Desing of Fixtures". J.R.Boerma, H.J.J.Kals. University of Twente. Annals of the CIRP Vol. 37/1/1988.

- "XPLANE, A Generative computer Aided Process Planning System for Part Manufacturing", A.H. van't Erve, H.J.J.Kals; Twente University Of Technology, Netherlands. Annals of the Cirp Vol. 35/1/1986.

- "Process Planning in the New Information Age", Joseph Tulkoff. Lockheed Georgia Co. Cim Technology, August 1987.

- "TOJICAP - A System of Computer Aided Process Planning for Rotational Parts". S.Zhang, W.d.Gao. Tongji University, Shanghai. Annals of the Cirp Vol. 33/1/1984.

- "Strategy in Generative Planning of Turning Processes". M.van Houten, Twente University, Netherlands. Annals of the Cirp Vol. 35/1/1986.

- "Education for CIM", Prof John R.Crookal, Cranfield, UK. Keynote Paper. Annals of the Cirp Vol. 36/2/1987.

- "Computer-Aided Process Planning: The Present and the Future", Inyong Ham, The Pennsylvania State University/USA; Stephenc.-Y.Lu, University of Illinois at Urbana. Champaign/USA. Annals of the Cirp Vol. 37/2/1988.

- "Coats: An expert Module for Optimal Tool Selection". F.Giusti, M. Santochi, G. Dini; Istitut di Tecnologia Meccanica, Pisa University. Annals of the CIRP Vol. 35/1/1986.

- "Artificial Intelligence on factory floor: A succes story". Robert L. Moore and Rod Khanna. Lisp Machine Inc. Andover. Massachusetts. Proceedings of conference sessions. AMS'86.

- "A perspective on the evolving role of artificial intelligence technology in manufacturing". Yoh-Han Pao. Center for Automation and Intelligent Systems Research. Cleveland. Ohio. Proceedings of conference sessions. AMS'86.

- "The use of Expert Systems in Process-Planing", B.J. Davies and I.L. Darbyshire. Annals of the CIRP Vol./33/1/1984.

- "Expert Control", K.J. Astrom and J.J. Anton. IFAC World Congress, Budapest, Agosto 1984.

- "The evolution of an expert system for process planning". Proceeding IEEE Expert Systems in Government Symposium. Mclean, Virginia.

Figure 7

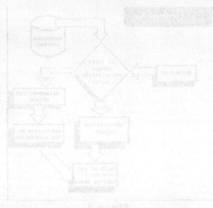

Figure 8

Table 1

SCHEDULING EDITOR FOR PRODUCTION MANAGEMENT WITH HUMAN-COMPUTER COOPERATIVE SYSTEMS

T. Fukuda, M. Tsukiyama and K. Mori

Industrial Systems Laboratory, Mitsubishi Electric Corporation, Hyogo, Japan

Abstract. The computer-aided scheduling environment of manufacturing system is described. The cooperation of the human and the computer is proposed, in which the complex simulation and data handling are executed by computer and the evaluation of schedule and the editing of schedule is indicated by the human planner. The cooperation of human and computer is essential in the system to edit the schedule as a planner desires. The simulation and the modelling of manufacturing systems as discrete event systems is based on the timed Petri net.

Keywords. Computer simulation; manufacturing processes; production control; modeling; Petri net; man-machine interface; scheduling.

INTRODUCTION

As a flexible manufacturing system is introduced, it tends to be more complex to manage the operations of facilities to fulfill the many varieties of requirements and goals. It is considered a difficult problem to construct an optimal scheduling plan to meet with many constraints. Scheduling problem can be theoretically solved by mathematical programming methodology, meanwhile it requires giant iterative manipulation, that is, combinatorial explosion. To circumvent these burdens of complex problems, there have been developed Artificial Intelligence approach(Fox, 1984; Fukuda 1986; Kanet 1987; Hillion 1987; Numao 1988; Shen 1986). The investigation of simulation technology has been examined by discrete event systems(Carlie 1985; Fukuda 1986; Gershwin 1987). The complexity of manufacturing system is typically found in Wafer fabrication system(Bitran 1988) and the decision rule is investigated to manage the system(Blackstone 1982). It is a substitutional but powerful methodology to use the human ability for these scheduling problems such as Numao(1989) than the automatic scheduling system .

Our paper proposes a scheduling editor which is a human-computer cooperative environment for planning an operation schedule. The scheduling editor is composed of four sessions of modelling, simulation, evaluation and schedule editing. The last function of editing is a completely new concept of the interactive system which offers to a planner to make and modify freely a desired schedule with satisfying the constraints of manufacturing systems. Simulation technologies of manufacturing system have been broadly developed, while there has not existed a computer editor with which a planner can organize a schedule with the aids of computer as a planner desires. Modelling used in this editor is based on the timed-Petri Net which makes it easy to recognize and analyze the behavior of complex manufacturing systems.

SCHEDULING PROBLEM OF MANUFACTURING SYSTEMS

There are many varieties of manufacturing systems in the real world such as flow shop type or job shop type. We treat in this paper the job shop type systems because the complexity of process makes it difficult to construct an optimal schedule.

The properties of manufacturing systems to be focused here are such that;
(1) Varieties of machines: The most simple machine is the single process machine which is processing one job at one time. The batch machine is processing multi jobs at once. The multi machine is a set of single or batch machine which is

processing jobs at any machine. The common machine is available to different process or task which select job to be done.
(2) Complex process route: We here define that a job needs a sequence of process on machine respectively, such that job Ji = Process< p1,p2,...pn>, where pj is a process. The sequence of process is different with variety of product. In the job shop type process, a product has a sequence of process to be executed at different machines and they are visiting machines not in a serial order but in a arbitrary order with process requirement. Here we treat the cyclic process which is visiting the same machine for different process. It makes complex to control the flow of a product in the manufacturing systems.
(3) Varieties of products: There exist many varieties of product each of which has different sequence of processes. The requirements for products are different with each other, for instance.

The performance of manufacturing systems ca be evaluated from many points of view. They are differing with products and the manufacturing systems. The objectives are classified into the following categories such that;
(a) job; Each job has the duration time to be accomplished.
(b) machine; The utility of a machine is desired as high as possible in the case that the cost of operation or equipment is high.
(c) system; The distribution of work in process should not be local in a system. The equal distribution is desired from the smooth flow in the system.

We have the constraints from the operations of manufacturing systems. They can be categorized into two aspects of strong and weak constraints such that;
(1) Strong constraints. They are not to be violated in any case. The physical constraints are such strong constraints that the number of jobs to be processed is restricted to the maximum capacity. The sequence of process for a job should be maintained. It is not possible for a job to be processed that violates the sequence. These strong constraints are quite important for the schedule planning.
(2) Weak constraints. This constraints is desired to be maintained but is not necessarily held. The duration time of a job to be accomplished is desired in a required time limit. The utilization of a machine is desired as high. The waiting time is desired to be short. The volume of work in process is small. These weak constraints are desired to be held, yet they are not strictly required to be maintained.

Here, it should be noted that our approach is not on the optimal solution, which implies that the objective function is not explicitly expressed in this scheduling system. The objective or performance of manufacturing system is not so simple that it is not described by the numerical objective function. Our approach is focused on the human-computer cooperation, so that the decision that a schedule is accepted or not is made by the human. The strong constraint is not to be violated from the physical reason. The weak constraint can be considered to represent the requirements which are usually described by the objective functions.

PETRI NET MODELLING

For the purposes of scheduling a manufacturing system, it is required to establish a model of manufacturing system. As the manufacturing system can be described by the discrete event system, the timed-Petri net model is suitable to represent it. The Petri net model PN is described by four items $PN=<P, T, B, M>$, where P representing the places, T representing transition, B representing connecting relation among places and transitions and M representing marking of token. The places is either active or inactive corresponding to existing tokens in it. The transition is firable when the connected places are all active and during the given time the transition is being fired.The token represents jobs and status of machines. The status of places is changing and the simulation is executed.

While there exist many varieties of machine in the manufacturing systems, it is easy to represent them by the Petri net model. The typical models of machine are shown in Fig. 1.

A single processing machine is represented by a job place, a machine place and a timed transition. Timed transition implies a process of a job by a machine with time duration. A token in a job place shows that a job is arrived and waiting for a process, a token in a machine place shows that the machine is idle and available for job processing. The transition is firable with full tokens in the input places and fired during the duration time.

A batch machine is processing more than two jobs simultaneously which is represented by more than two job places with a single machine. In the case that the full or less number of jobs to the batch capacity arrives at the machine, the machine transition is firable.

A set of multi machines are represented by more than two tokens in the machine places. Where there exist some process that is too busy to execute jobs, more than one machine is introduced. They represent the concurrent operating process.

A common machine is usable to the different kinds of processes which is represented by more than two connecting transitions to which a machine place is connected in the case of different job processes.

With the Petri net modelling it is easy to represent a complex production line and easy to simulate the behavior of system dynamics.

DUAL LOOP OF
SIMULATION AND EDITING

The scheduling task of manufacturing systems can be considered as the following three subtasks to be done.
(1) monitor the current status of facilities and jobs,
(2) make clear the requirements of the facilities and the jobs,
(3) make decision of the processing order of jobs.

It is clear that the difficult tasks are the last two subtasks. Theoretical approach such as mathematical programming of combinatorial optimization technique is not applicable because of their combinatorial explosion.

To circumvent the burden of the combinatorial explosion, we propose a human-computer cooperative approach with the aids of computer simulation and interactive interface. The principal architecture is shown in Figure.2. A scheduling system is composed of four sessions; modelling, simulation, evaluation and editing. The last function is a key to the human-computer cooperative systems which aims that a planner can make a desired schedule on the simulated schedule or modify them on the display.

The conventional scheduling technique is essentially based on the simulation technique. A simulation gives a processed pattern given initial work and machine condition with decision rule. So that the iterative try and error is needed for the desired schedule, that is, no way to specify the schedule apriori. On the other hand, for a simple manufacturing system, it is often observed for a schedule expert to manage the desired schedule on the blackboard considering with the many requirements, that is, editing task of the schedule. Yet it is too complex to manipulate the schedule the complicated manufacturing systems except for a simple system.

For a complex system scheduling, both techniques of simulation and editing are needed. Simulation is necessary to know the response of jobs and machines for complicated manufacturing systems while it is not true to find the desired schedule by such try and error simulation. On the other hand, it is simple to draw (that is, "editing") a schedule manually on the blackboard which is daily seen in the factory. Yet this drawing method is too simple to manage the complicated manufacturing system. However from these observations, it is found that the dual methodologies will present a quite powerful paradigm on the computer. So we here propose the dual loop scheduling methodology of the simulation and the editing.

Scheduling is to manage the job processing in the time-space. The dual paradigm of scheduling consists of simulation loop and editing loop the role of which are;
(a) simulation: to know the response of systems given condition.
(b) editing: to specify a schedule of job at the desired position in the time-space.
Deficits of simulation lies in that it is not true for the desired schedule, and the deficits of editing lies in that specified schedule is not necessarily feasible.

By combining the two methodologies, it is possible to edit the simulated schedule to the desired one and to simulate the desired schedule for testing and satisfying the feasibility. The schedule task is done by the iteration of both simulation and editing as shown in Fig.2.

The software architecture of scheduling system is shown in Fig.3. The procedures of scheduling are five steps as follows.
(1) With the Modelling Editor (ME), model the manufacturing system in the Petri net form with the aids of iconic interface.
(2) With the Discrete Event System Simulator (DESS), simulate the behavior of jobs and machines. The simulation results are displayed on the Schedule chart (process-wise) and Gantt chart (machine-wise) along the time horizon.
(3) Evaluate the performance of the systems ,with the display of performance exhibit.
(4) With the Scheduling Planning Editor (SPE), modify or reschedule the schedule obtained by simulation as desired.
(5) End if the schedule is satisfied, or return to the 3. until satisfied.
The decision rule is stored in the rule base management systems which can be modified. The properties of both jobs and machines are stored in the object management systems as a data-base.

SIMULATION

Simulation is executed by DESS based on timed-Petri net model. The procedure of simulation is as follows. First the tokens are marked for existing jobs and available idle machines. Secondly the firable transition is surveyed and the firable condition is investigated. There occurs the conflicting condition to fire the transition. For instance, if two jobs are waiting for a single machine, the selection of one job is needed. At the multimachines, which one of a set of machines is to be decided. These conflicting resolution is managed by the dispatching rules such as;
(1) priority of a job arrived first,
(2) priority of a job arrived last,
(3) priority of a job which has shortest number of processes left,
(4) priority of a job which has the longest process time at the machine,
and others.

These dispatching rule is considered as reflecting each real world operation. So, the general and numerical expression is not sufficient and expressed by rule-base. The another rule base are such as machine priority, job priority, process priority and product priority.

With these rule base the conflicting firable transitions are decided. The transition is fired in the duration time which is the processing time of the machine. The clock is advanced and surveyed the firability of transition and tokens of places is moved out and in. The simulation is proceeded in forward time. The status of machines and jobs can be seen on the Petri net model display with token and firing at any time horizon.

EVALUATION

The result of simulation is displayed as a Schedule chart or Gantt chart. The schedule chart is like a time diagram of processed jobs expressed by process-wise along the time and the Gantt chart is expressed by machine-wise along the time. (Note that if a machine executes only one process, both chart is the same.) With the two charts, it is possible for a planner to evaluate the job processing and machine utilization at a glance. The Petri net is also displayed at any time horizon which shows which machine is busy (that is, the bottleneck where many job tokens are waiting) and distribution of job tokens over all machines.

The evaluation is done from the many points of view. The scheduling system is available to show the statistical performance exhibits such as machine utilization, job completed time and other indexes. It is most important for a planner to observe a pattern of schedule at a glance. Pattern evaluation is the very human ability, so the graphic interactive interface is critical for a planner to organize the satisfied schedule solution.

EDITING FUNCTION
AND REALIZATION

After a simulation is executed given job, machine and decision rule base, it is desired to change the schedule; the order of jobs processing, the machine allocation and also decision rule base. Our purposes lie the stresses on this editing function which changes the processing order of jobs, job sequences, job priority, machine priority, machine availability and others to the desired schedule. The conventional scheduling system is just only simulation given initially. Our editing function is to modify and change the result of simulation to the desired schedule on the display. The editing interface is available on the schedule chart as well as Gantt chart by iconic interface.

The purposes of editing system lie in the pursuit of usable interactions between a man and a computer. Editing functions are the followings;
(1) visual display of Schedule chart or Gantt chart
(2) ability to move, delete, copy and insert the desired job on the schedule chart with constraint satisfaction
(3) data input by icon not by keyboard
(4) simulation in forward or backward
The fundamental editing function is shown in Table1.

Table 1 Fundamental Function of Schedule Editing

editing function	explanation
insert:	insert any job at any position in the chart
move:	move the job to the desired time
delete:	remove a job from the schedule
copy:	copy same pattern of job schedule
forward:	specify the starting time of process sequence
backward:	specify the ending time of process sequence
machine:	make it active or inactive
machine number:	add or delete the machine

The scheduling system is designed for a planner to move a job sequence, to change order, to insert another job into the existing schedule on the display of simulation. So the Gantt chart can be modified independently from the simulation results to the planner's desired schedule.

CONSTRAINT PROPAGATION

As it is clear from the above, it is not free to modify the schedule as desired. There exist varieties of constraints for a schedule to be satisfied. The hard constraints are not to be violated in any case such that a single machine is available for only one job at the same time or the order of processes for a job should be maintained. The constraint satisfaction is needed as mentioned before. A planner can draw and delete a schedule on the Schedule chart or Gantt chart. In the case of editing scheduling chart, these constraints are automatically checked and satisfied with the hard constraints. It is clear that the constraints are propagating to another job schedule in succession by a change of one job schedule. The function to control the constraint propagation is realized by the rule base.

Let us consider the case to insert the new jobs into the already made schedule. The constraints to be considered are the followings;
• constraint of machine capacity; which job among the waiting jobs should be processed first?
• constraint of process sequence; a preceding or succeeding process should not be violated, then a process is delayed or advanced to effect the processing time.
• constraint of waiting time; some process should be followed by the succeeding process in time such as chemical process.
So in the case to insert the new jobs, it is required to reschedule all jobs until they are satisfied with the constraints. One constraints satisfaction makes another conflicts and so on. This procedure is made automatically by referring to the rule base.

AN ILLUSTRATIVE EXAMPLE

An example of visual display is shown as Fig. 4 through Fig.6., which show a Schedule chart and Gantt chart on the display in which a job (shown as a sequence of boxes and a box shows one process by a machine) is processed with a sequence of continuing process along the time.

While a simulation is executed given machine and job assignment, it is required to control a job at the desired time or at the desired machine. Editing function realizes scheduling task as if a planner draws a chart on a display. For instance, a process sequence of one job can be drawn on the schedule chart at any desired starting point or finishing point of time. A job has a constraint of process order in which, for instance, the process (a1) should precede the process (a2), and so on. This sequence constraint is automatically satisfied. A change of process timing of one job makes another job's process timing to forward or backward in succession. While the schedule chart is obtained by a simulation and a planner desires to insert a high priority job newly into the full of existing jobs schedule chart, it is possible for this new job to be assigned to the desired point of time by moving the other existing jobs. On this procedure, there occurs many side effects to the other existing jobs to satisfy the constraints. This rearrangement can be automatically done by computer.

Let us consider three cases as shown in Fig.4 through Fig.6. In Fig.4, there is only one product(A) which requires eight processes on the eight kinds of machines. Product (A) has now four jobs to be processed. Schedule chart is represented process wise and Gantt chart is machine wise. On the other hand, in the case of product(B), drawing box in fill, existing alone, the process and machine utilization is shown in Fig.5. where no product(A) exist. Now consider the case to desire to process both products in a mixed pattern. As it is clear the conflicts of machine allocation and job priority occurs. Specification of starting time of each jobs is indicated by a planner to the editing system. The scheduling system investigates all the side effected constraints and moved the existing job in advance or putoff automatically. The constraint is managed by the scheduling system to resolve and to obtain the mixed process pattern as shown in Fig.6. Such insertion or specifying the completion time can be executed with this schedule editing system. As described here, it is convenient for a planner to obtain the desired schedule and to make the computer to check all the constraints to resolve the conflicts occurred by the modification of schedule.

CONCLUSION

The motivation of our paper lies in the development of human-computer cooperative problem solving environment for a complex manufacturing scheduling. The scheduling system does not aim the optimal solution but make it usable for a planner to arrange the schedule as desired. This is the concept of human-computer cooperative problem solving system to construct a schedule for a complex manufacturing system.

So far there have existed two methodologies of simulation and mathematical programming. The simulation gives a schedule as a result given starting job, but it is not able to control a job schedule to the desired timing position. On the other hand, mathematical programming gives the optimal solution but it has a problem of combinatorial explosion and difficulties of formulation to represent the non quantitative aspects which can be realized by an expert system.

Our target lays stresses on the explicit utilization of human ability to evaluate the schedule chart pattern at an intuitive observation and the easiness to organize the schedule on the computer display. A computer's role is to handle with the complex constraint propagation arose in editing the schedule. This human-computer cooperative environment consists of varieties of functions; interactive interface, simulation, evaluation, expert system and editing. Modelling methodology is based on the timed Petri net for representing discrete event system by which it is easy to recognize the system dynamics of a complex manufacturing systems. The graphic interface has been developed on the engineering workstation by which it is realized to execute the simulation instructively and rearrange the schedule as a planner desires to change. The further problem would be in the implementation in the real world for the real time scheduling.

REFERENCES

Bitran,G.R., and D. Tirupati (1988). Development and Implementation of a Scheduling System for a Wafer Fabrication Facility. Operations Research. 36. 377-395.

Blackstone,J.H. (1982). A State-of-the-Art Survey of Dispatching Rules for Manufacturing Job Shop Operations. Int.J.Prod.Res., 20, 27-45.

Carlier,J., P. Chretienne, and C. Girault (1985). Modelling Scheduling Problems with Timed Petri Nets. Lecture Notes in Computer Science, 188, Springer-Verlag, Berlin. 62-82.

Dayhoff,J.E., and R.W. Atherton (1987). A Model for Wafer Fabrication Dynamics in Integrated Circuit Manufacturing. IEEE Transactions on Systems,Man,and Cybernetics, 17,91-100.

Fox,M.S., and S.F. Smith (1984). ISIS - A Knowledge-Based System for Factory Scheduling. Expert Systems, 1, 25-49.

Fukuda,T., S. Takeda, and M. Hayashi (1986). Simulation of Dynamics and Operations Systems for the Evaluation and the Planning of Manufacturing Systems. Proceedings of JSST International Conference, 477-480.

Fukuda,T., S. Takeda, and M. Hayashi (1986). Distributed Expert Systems for Production Control. Proceedings of International Industrial Engineering Conference, 222-228.

Gershwin,S.B. (1987). A Hierarchical Framework for Discrete Event Scheduling in Manufacturing Systems. In M. Thoma, and A.Wyner (Ed.), Lecture Notes in Control and Information Sciences, 103. Springer-Verlag, Berlin. 197-216.

Hillion,H.P., J.M. Proth, and X.L. Xie (1987). A Heuristic Algorithm for the Periodic Scheduling and Sequencing Job-Shop Problem. Proceedings of the 26th Conference on Decision and Control, 612-617.

Kanet,J.J. (1987). Expert Systems in Production Scheduling. European J. of Operations Research, 29, 51-59.

Numao,M. and S. Morishita (1988). Scheplan - A Scheduling Expert for Steel-Making Process. Proceedings of International Workshop on Artificial Intelligence for Industrial Applications, 467-471.

Numao,M. and S.Morishita (1989). A Scheduling Environment for Steel-Making Processes. Proceedings of Conference on Artificial Intelligence Applications, 279-286.

Shen,S., and W.L. Chang (1986). An AI Approach to Schedule Generation in an Flexible Manufacturing System. Proceedings of 2nd ORSA/TIMS conference on Flexible Manufacturing Systems:Operations Research Models and Applications, 581-592.

Walter,C. (1989). Integrating Simulation and AI into a Production Scheduling System. In J. Browne (Ed.), Knowledge Based Production Management Systems, Elsevier Science Publishers B.V., North-Holland. 95-103.

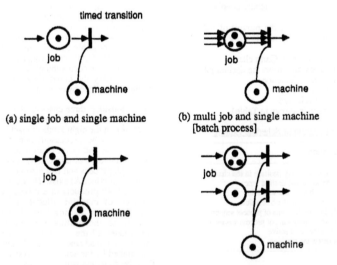

timed transition

(a) single job and single machine

(b) multi job and single machine [batch process]

(c) multi machine

(d) common machine for different process

Fig.1 Petri net representation of machine and job.

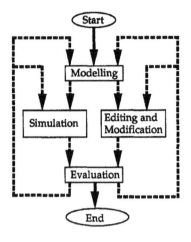

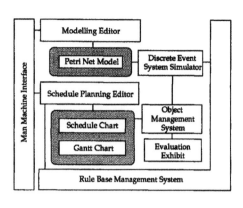

Fig. 2 Dual loop scheduling of simulation and editing.

Fig. 3 S/W architecture of Scheduling Editor.

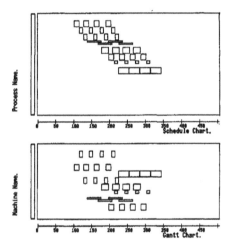

Fig.4 Schedule chart and Gantt chart for product(A) alone.

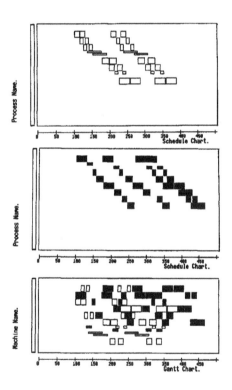

Fig.6   Schedule chart and Gantt chart for both product(A) and product (B). The individual existing schedule obtained for product(A) and product(B) are automatically modified to satisy the constraints.

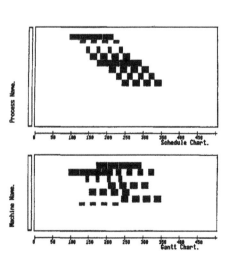

Fig.5 Schedule chart and Gantt chart for product(B) alone.

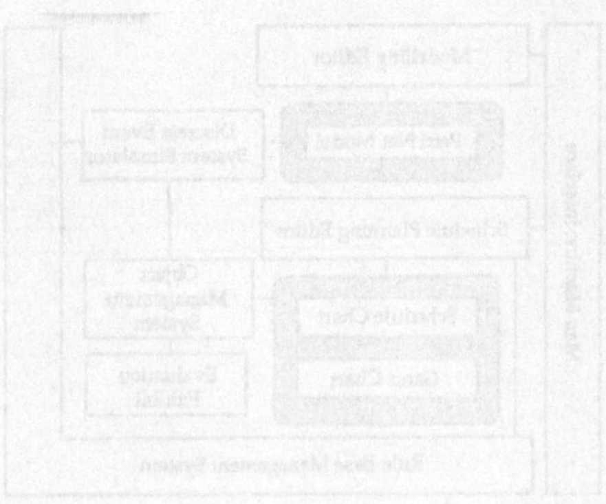

Fig. 3 S/W architecture of Scheduling Editor

Fig. 2 Dual loop scheduling of simulation and editing.

Fig.6 Schedule chart and Gantt chart for both products(A) and product(B). The individual existing schedule obtained for product(A) and product(B) are superimposed in action to satisfy the constraints.

Fig.4 Schedule chart and Gantt chart for product(A) alone.

Fig.5 Schedule chart and Gantt chart for product(B) alone.

GRID MODELISATION FOR AUTONOMOUS ROBOT

A. Pruski and V. Boschian

Laboratoire d'Automatique et d'Electronique Industrielles, University of Metz, Metz, France

Abstract. To allow the path determination, a mobile robot has in memory, a model of evolution environment. We propose an adaptative grid representation on which each cell is represented by a boolean variable. Since the research tree is not explicitly defined, we developpe an heuristic method to find the path based on logical combination between the grid and a mask.

Keywords. Autonomous robot; modeling; model reduction; path planning.

INTRODUCTION

The Problem

To allow an autonomous evolution, the mobile robot must know or discover the work area. In this way it memorises or creat along its evolution, a space representation used for the displacement strategy. A model must satisfy many criterions :
- representing faithfully a 2D or eventuelly a 3D environment with respect to predefined criterions
- allowing in a minimal time a path planning from a point to an other
- integrating robot sizes.

Different Models

A mobile robot environment depends of the type of task it performs. The inside robot (cleaning, building, transport...) meets all kind and all shape of obstacles (machinery, wall, table, hole...). The outside robot (farming, military...) moves in a different environment. The obstacles are not always present (example : a ploughed field) and its autonomy must be more important because the wideness of the task space. This article deals with a closed environment strewed with obstacles. Numerous works were performed to modelise an environment which defines a map of task space used by the mobile to establish its navigation plan. Generally the model lies on a space dividing in accessible, non accessible and eventuelly sensitive zones in which the penetration is allowed but may induce some problems. We shall cite some example of space dividing.
The quadtree dividing (Kambhampati, 1986) is performed in rectangular cells. If a cell which represent a portion of environment don't have an homogeneous content that is accessible or not accessible, it is divided again in four identical rectangles, divided again if their content is not homogeneous. In a second time the quadtree is transformed in a tree from which a leave node represents an homogeneous cell to which is associated a function defining the access possibility.
Brooks (1983) proposes a free space representation by generalized cones. Such a space is described with seven parameters representing the projection into a 2D space of cylinders and a cone combination. The analyse of intersection between the cones axis allows to establish the graph determining the links between the different cells.
Singh (1987) proposes a space dividing in 2n+1 horizontal and vertical strips. Each cell created by vertical and horizontal strips intersection has a binary value. An algorithm based on Quine and Mc Cluskey simplification method of swhiching fonctions allows to determine the intersection cells corresponding to the overlaped convex area, that is meaning the surfaces of which all points may be joint two and two by a line.
Chatilla (1981) divides the 2D space in convex polygonal cells. The cell set is represented by a graph with nodes and links representing respectively cells and allowed pathes. A cell grouping allows to creat a reduced graph (Laumond, 1983) which is developped by searching the cut point to accede to the group of considered cells.
The hypergraph model (Rueb, 1987) is an other convex polygon space representation method. The dividing is performed by projecting the segments which make up the vertices of the obstacles. Three types of regions are deduced : the obstacles, the free regions and the overlaped free space regions. A graph represents the free space which nodes and oriented arcs corresponding to the obstacles boundaries and adjacent boundaries. A graph dividing is realised to represent the set of vertices boundaries which are limits of free area. The graph and the dividing constitute an hypergraph. The path searche is performed by analysing the overlaped area of free regions.

Path Planning

The above described modelisation methods represents the environment with joining graph between elementary cells or a set of cells. Generally the A* algorithm determines the path between two cells. The path is not always optimum. This algorithm is a compromise betweenn swiftly path determining and associated computing cost.

MODELISATION

We propose a modelisation method associated to the path planning which takes a reduced memory space and which the path is swiftly computed.

The Grid Model

The grid modelisation consists on dividing the space in horizontal and vertical strips from which intersection forms elementary cells. To each cell is associated a function determining the access conditions, either a binary function (allowed access, not allowed access) or a more complex function including other elements (task, geometries...). The model precision depends on the mesh size. Commonly the grid is creat homogeneously depending on the desidered precision. This process manner needs a low mesh width if the desidered precision is high. That sets the problem of needing a large memory space. The path planning between two cells cannot be performed with an algorithm like A*. The search tree would be to large. An algorithm like LEE (1960) is more suitable but needs large memory space and long computing time.

The principle of the model we propose lies on whole space dividing by a non homogeneous grid which meshes having a width and a length on environment depending. The obstacles are represented in the 2D model, by polygons. The horizontal and vertical strips intersection forms the elementary cells which is associated a binary variable depending on wether a cell is a free space portion or an obstacle. The form of the object to model influences the mesh size. Let us define the model error as the biggest distance in a cell between the object to modelise and the boundaries of the cell. In the case of a space modelisation (2D) of a square with a width c, the grid modelising the square without error has a mesh width c and coincide with the boundaries. For whatever object the maximum error is $(1^2+L^2)^{1/2}$. with L and l the horizontal and vertical width of the meshes. For the mobile robot context it is not necessary, like in CAD, to modelise the environment with high precision. The precision criterion constitutes the navigation at the nearest of an obstacle. In close obstacle evolution, proximity sensor like US will help the lake of precision.

Let us consider two particular distances : the minimal and maximal distances approach between the center of a cell and the segment to modelise. The path planning is performed with respect to the center of cells. The model error depends on the parameter α that is the angle between the segment to modelise and the grid. (Fig.1)

"at the nearest"   "at the farest"

$$d0^2 = \frac{L^2}{4} + \frac{l^2}{4} \qquad d1^2 = \frac{L^2}{4} + \left(\frac{3\,l}{4}\right)^2$$

$$d\_min = d0 \sin(\alpha+\beta) \qquad d\_max = d1 \sin(\alpha+\Gamma)$$

with α the angle between the segment and the grid and

$$tg\,\beta = l/L \qquad tg\,\Gamma = 3l/L$$

d_min constitutes the minimal obstacle distance approach

d_max constitutes the maximal distance between the mobile and the obstacle at time of determining a path at the nearest of the environment. (Fig.2) In the case of the mobile displacement it is interesting to establish a minimal obstacle approach for security reasons.

We have

$$d\_mini = d0 \sin(\alpha+\beta)$$

This constant d_mini may be fit to robot size. For example 2 d_mini=K dim_robot with K a security coefficient. (Fig.3) In the particular case of L a finite value and l tends to zero (the mesh in infinitely small) :

$$\lim_{l \to 0} d\_mini = \frac{L}{2} \sin\alpha$$

d_mini tends to a finished value. In other hand if l is finished and L tends to zero :

$$\lim_{L \to 0} d\_mini = \frac{l}{2} \cos\alpha$$

we obtain :

$$L = \frac{2\,d\_mini}{\sin\alpha} \qquad l = \frac{2\,d\,mini}{\cos\alpha}$$

Let us take the case of a real obstacle described by a segment of a finite length LG and note m and n respectively the number of vertical and horizontal dividing. (Fig. 3)

We have :

$$m = \frac{LG\,\cos\alpha}{L} \qquad n = \frac{LG\,\sin\alpha}{l}$$

with

$$L = \frac{2\,d\_mini}{\sin\alpha} \qquad et \qquad l = \frac{2\,d\_mini}{\cos\alpha}$$

it comes $\quad m = \frac{LG\,\cos\alpha\,\sin\alpha}{2\,d\_mini} \qquad n = \frac{LG\,\cos\alpha\,\sin\alpha}{2\,d\_mini}$

simplifying

$$m = n = \frac{LG\,\sin(2\alpha)}{4\,d\_mini}$$

Procedure of Model Determining

Let us consider an environment which polygonal obstacles which are represented by sets of segments and an orthogonal referential. The model is performed in three stages :
- non homogeneous grid creation
- tranformation in homogeneous grid
- model creation.

Grid creation

1. The first stage consists to divide the space in lines parallel to and 2D referential axis and going through all vertices. Repeat the operation again with respect to the other axis of the referential. The obtained grid has m+1 horizontal and n+1 vertical strips with m and n the number of independant vertices to say that lies not on the same horizontal or vertical line. (Fig.4)

2. Creat a list including the coordonnates of the parallel lines according to the other.

$$A = \{\ A1, A2,\ \ldots\ ,\ An\ \}$$
$$B = \{\ B1, B2,\ \ldots\ ,\ Bm\ \}$$

3. Find for all vertical and horizontal strips the closest slope to 1, to say for α closest to $\pi/4$.

4. Computing the maximal dividing associated to each strip.

$$Pa = \{\ m1,\ m2,\ \ldots\ ,\ mn\ \}$$
$$Pb = \{\ n1,\ n2,\ \ldots\ ,\ nm\ \}$$

5. Perform the dividing according to mi and ni. (Fig.5)

Transformation in homogeneous grid and coordonnate transformation.

The following stage consists in transforming of the grid to be homogeneous. Each vertical or horizontal strip has the same size. This operation is expanded to the vertices. Given $Si = \{Ai,\ Bi\}$ the vertex coordonnates of vertex Si are

$$S^*i = \left\{\ \sum_{j=0}^{j=i} mj\ ,\ \sum_{j=0}^{j=i} nj\ \right\}$$

Propriety : the interest lies in the simplification of ulterior informations treatment to obtain the final model.

Model creating.

The final model (p,q) constitues a p words of q bits representing the state of the different cells by grid defined. The words represent the horizontal strips and each bit the state of cells in the strip. If the access is allowed, the bit is '1', in the other case it is forced to '0'. The model is generated in following manner. Let Dk be an obstacle line segment and $S^*k = \{\beta k, \Gamma k\}$ and $S^*k+1 = \{\beta k+1, \Gamma k+1\}$ the vertices of the reduced model of Dk, the slope is given by

$$\alpha = \frac{\Gamma_{k+1} - \Gamma_k}{\beta_{k+1} - \beta_k}$$

We have a discrete model where the cells coordonnates correspond to the bit position in the word.
Thus

$$\forall \beta, \Gamma : M(p,q)=1; \beta_1=\beta_k; \Gamma_1 =\Gamma_k;$$
FOR j=1 to m(k+1)
DO
$$\beta_j = \beta_j + \alpha_k; \Gamma_j = \Gamma_j + 1$$
$$M(INT(\beta_j), \Gamma_j) = 0$$

Propriety. In some configurations this model takes the robot size into account. Given two segments non parallel to the reference axis, the minimal distance to avoid in the reduced model the contact between them, must be such that at least (3+1)L distances separate them on the horizontal and (3+1)l on the vertical in order to avoid the cell including an obstacle to be contiguous.
Consequently

$$dis\_1 = 8 \ d\_mini$$

In order to realise a contact between segments, the distances must be less than 2L on the horizontal and less than 2l on the vertical.
Therefore

$$dis\_2 = 4 \ d\_mini$$

Between these two distances the contact between the model cells depends on the length of the segment and of the angle α.
If we take

$$d\_mini = k \ dim\_robot$$

with dim_robot the dimension of an homorphous robot.
We have

    dis > 8 k dim_robot   ---> possible path
    dis > 4 k dim_robot   ---> probable path
    dis < 4 k dim_robot   ---> non possible path.

With dis the normal distance between the segments.

Model Reducing

In a room the furnitures and accessories are often arranged in line and are parallel to boundaries. On other hand we remark the partition number depends on the angle α between the segment to modelise and a referential axis. The partition number tends to zero when α tends to zero. If we take an other basis (non obligatory orthogonal) such as α be minimal, we reduce the number of informations to memorise. (Fig.6)

Grid reduction procedure
1. Optimal angle research. This stage consists in finding the $\bar{\gamma}1$ and $\bar{\gamma}2$ rotation angle of the orthogonal axis in order to find the minimal dividing number.
1.1. Basis changing. We compute the vertices position in the new basis. Given $\bar{\gamma}1$ and $\bar{\gamma}2$ are the rotation of x an y axis and x' and y' on the axis after rotation.

$$\begin{bmatrix} x' \\ y' \end{bmatrix} = \frac{1}{\cos \varepsilon} \begin{bmatrix} \cos \bar{\gamma}2 & \sin \bar{\gamma}2 \\ -\sin \bar{\gamma}1 & \cos \bar{\gamma}1 \end{bmatrix} \begin{bmatrix} x \\ y \end{bmatrix}$$
while $\varepsilon = \bar{\gamma}2 - \bar{\gamma}1$

1.2. Minimal partition computing. Given $S*i \ \bar{\gamma}1, \bar{\gamma}2 = \{\beta_i, \Gamma_i\}$ the coordonnates in the [X'] and $\Sigma(\bar{\gamma}1, \bar{\gamma}2) = \{\Sigma\beta_i, \Sigma\Gamma_i\}$ the minimal dividing is given by

$$\bar{\gamma}1, \bar{\gamma}2, \text{such} \Sigma = Min \ \Sigma(\bar{\gamma}1, \bar{\gamma}2)$$

2. Model creating. The model is generated by the algorithm described above.

Comments. Actually the angle $\bar{\gamma}1, \bar{\gamma}2$ are performed in exhaustive manner what penalize the computing time of the model. If the environment doesn't have a privilegial orientation it's possible that the reduced model doesn't be better. A random arrangement of a great number of the segments gives a high probability to encounter a slope α_k close to 1.

PATH PLANNING

Principle

The path planning method bases on four elements : the M(p,q) model, a list $\pi(n)$ of the successive cells composing the path, a code C (n) allowing the displacement and a displacement priority function F. The grid model described by the vector M(p,q) represents the admissibility map of the environment. Each bit q of a vector component p represents a cell. All cells have a same dimensional and functional weight. The list π defines the set of successive cells from the source cell π (1) where is located the robot to the aim cell where is located the destination point. At each element in the list π is associated a code C(n) defining the movement possibility from a cell to the neighbour cells. Eight movements are coded considering always an orthogonal mesh grid :
- the base displacement up, down, left, right
- the associations of base displacements.
The code C(n) establishes the map of allowed displacements from cell n, part of the path. The function of priority evaluation defines in litigation cases the choice for which one has to opt. The function is by programmer defined considering the aim and the type of problem to solve.
Given M(i,j) the cell in which is the mobile is include and M(v,w) the cell destination. The M(i,j) following cell is defined according to the speedy approach criterion or to the difference between the indexes i,v and j,w. We obtain the following board:

	case :	:	$\pi(i,j)$	
i	: v>i	: t0=1 :		:
	: v<i	: t0=-1 :	i=i+t0	:
	: v=0	: t0=0 :		:
j	: w>j	: t1=1 :		:
	: w<j	: t1=-1 :	j=j +t1	:
	: w=j	: t1=0 :		:

The cell search is performed by boolean operations on the vector M(p,q) components. Let us take a concret case to explain the method. Given the following environment :

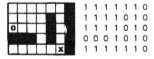

    1 1 1 1 1 1 0
    1 1 1 1 0 1 0
    1 1 1 1 0 1 0
    0 0 0 1 0 1 0
    1 1 1 1 1 1 0

We note the actual robot position in the third line, first column and the aim in the 6st column, fifth line.

    pos_robot : 1 0 0 0 0 0 0 line i
    pos_dest : 0 0 0 0 0 1 0 line v

Let us compute a mask from the robot position and from the calculation of t0.

    t0=1
    mask = pos_robot OR (pos robot/2)
    t0=-1
    mask = pos_robot OR (2*pos_robot)

The multiplying and dividing by two represent a
shift respectively left and right of the binary robot position word.
The example case gives mask=1 1 0 0 0 0 0
The following stage consists to perform the possibilities of movements on the actual line v and on
the line v+t. Therefore the following operation :

A = mask AND M(v) = 1 1 0 0 0 0 0

and

B = mask AND M(v+t) = 0 0 0 0 0 0 0

In our case the following cell choice is given by
A, the only possible since B=0. Generally more cases are present. Let us compute the three following
states.

SO = A AND NOT(pos_robot)
S1 = A AND B

S2 = S1 AND (S1>>1) si tO=1 (one bit right shift)
or S2 = S1 AND (S1<<1) si tO=-1(one bit left shift)

then

if (S2 != 0) pos_robot = S2
else
    if (S1 !=0) pos_robot = S1
    else
    if (SO!=0) pos_robot = SO
    else no possible displacement.

The SO and S1 test order may be imposed. It determines the priority movement horizontal of vertical
or inversely. Since the choice is realised the cell
containing the robot is prohibited in order to be
not a second time reachable.

SEARCHPATH Algorithm

The SEARCHPATH $(\pi(1),\pi(n))$ algorithm returning
the list of cells part of the trajectory from cell
source $\pi(1)$ to the cell destination $\pi(n)$ is given below :
1. To creat a list π of cells forming the path having one element $\pi(1)$, the actual robot position.
The code C(1) associated allows all displacement
around $\pi(1)$. Initialise i=1.
2. To do M(p,q)=O with p and q the actual robot position vector bit.
3. To compute the value tO an t1 determining the
possible way of displacement in $\pi(i)$ accordingly
to the possibility defined by C(i) and the priority
function F. If no possibility then jump in 10.
4. To modify the code C(i) with respect to the choice tO and t1.
5. Mask defining.
6. To compute the new robot position. If the new position don't exist, then jump in 3.
7. i=i+1. To Add to the list $\pi(i)$ the new robot position and the associated code C(i) allowing any
displacement around $\pi(i)$.
8. If robot position is equal to destination then
END.
9. To jump in 2.
10. i=i-1. If i=O then END : no displacement possible.
11. To jump in 3.

Path Optimisation

The SEARCHPATH $(\pi(1),\pi(n))$ results determine a
path that is not everytime interesting from point
of view the progressing to the aim. The speedy approach criterion used in SEARCHPATH induced, in some
cases, a path lengthening as shows the figure 7.
We propose a method derived from algorithm A* determining the path integrating the obstacle forecast.
The first stage consists of finding the path with
SEARCHPATH. Let us open, starting from the list π
two other lists. The first one Gi grouping the cells
from $\pi(1)$ to $\pi(i)$ and an other one Hi grouping the
cells from $\pi(i+1)$ to $\pi(n)$. Let us determine two
cost function Gg and Gh representing the cell number
of G and H respectively. The optimising algorithm

consists to compute for all i with $0<i<n$ the minimal cost $C^*=Cg + Ch$. We obtain the following algorithm.

```
i=s;
Gi=(π0,...,πi);
Hi=(πi+1, ..., πn);
cout=COST(GO);
WHILE  i<n
       {
       T=SEARCHPATH (π0,πi);
       if (COST(Gi) <= cout)
             Gi+s (T,πi+1,...,πi+s);
          else  Gi+s=(Gi,πi+1,...,πi+s);
       Hi+s=  (Pi+s+1, ..., Pn);
       i=i+1; cout=COST(Gi);
       }

   with COST (Gi)=Cg + Ch
```

s is a parameter varying from i to n allowing to
obtain a compromise between the search speed
(s big) and smalness (s small) of the path. In a
second stage let us begin again the same process
inverting the source and destination. The choice
of final path is performed by computing the cost of
the distance concerning real site. (Fig.7)

CONCLUSION

This article describes an adaptive grid based
environment modelisation method from a space defined by polygon objects. The method consist to divide the total space following vertices parallely to
a base referential. The formed strips are cut accordingly the minimal approach criterion between
the mobile and the obstacles. The path search between two points of the model is realised by an association of the LEE and A* algorithms by execution
of boolean mask operations. This method has the advantage to need a reduced memory space and to de-
terme swiftly the path.

REFERENCES

Brooks, R.A. (1983). Solving the find path problem
    by good representation of free space. IEEE
    Trans. on Syst. and Cyber. Vol SMC-13 march/ap-
    ril 83.
Chatilla, R. (1981). Système de navigation pour ro-
    bot mobile autonome : modélisation et processus
    de décision. Thèse de DDI. Toulouse 81.
Kambhampati, S. (1986) Multiresolution path plan-
    ning for mobile robots. IEEE Conf. San Francis-
    co 86.
Lee, C. Y. (1960).An algorithm for path connection
    and its applications.IRE Trans. on elec. comp.
    60.
Laumond, J.-P.(1983). Model structuring and concept
    recognition two aspects of learning for a mobile
    robot. 8 th IJCAI, Karlsruhe, 83.
Rueb, K.D., A.K.C. woreg, (1987). Structuring free
    space as an Hypergraph for roving robot path
    planning and navigation.
    IEEE Trans. on Pattern Analys. and Mach. Intel.
    March 87, Vol PAMI-9, N°2.
Singh, J. S., M.D. Wagh, (1987). Robot path planning
    using intersecting convex shapes : analysis and
    simulation. IEEE Journal of Robotics and Automa-
    tion. Vol RA83, N°2 April 87.

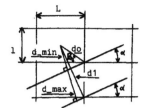

Fig. 1. The model error.

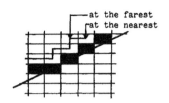

Fig. 2. The minimal and maximal distances approach.

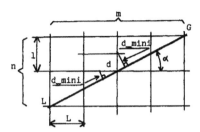

Fig. 3. Minimal distance and partition.

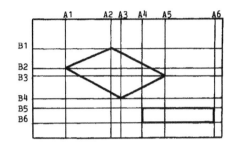

Fig. 4. Grid generation.

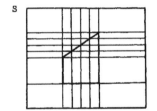

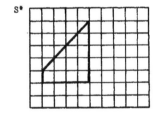

Fig. 5. Grid transformations.

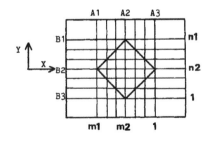

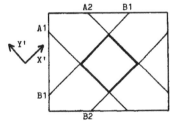

Fig. 6. Grid after basis changing.

o Source

x Destination

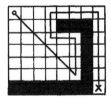

Fig. 7. Path planning and optimisation.

Fig. 2. The minimal and maximal distance operator.

Fig. 1. The model errors.

Fig. 4. Grid generation.

Fig. 3. Minimal distance and position.

Fig. 5. Grid transformation.

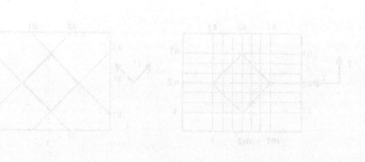

Fig. 6. Grid storm level changing.

Fig. 7. Path finding and utilisation.

MODELLING, CONTROL AND SIMULATION
OF FLEXIBLE ASSEMBLY SYSTEMS

M. Barbier and J. Delmas

*CERT/DERA, Dept. Automatique du Centre et de Recherches de Toulouse, Toulouse,
France*

Abstract: During the past ten years, the main research on real time control of flexible manufacturing shops
has been concerned with machining systems. Assembly systems are complicated by the need to control the
arrival times of required products at each assembly station.
This paper deals with control policies in the assembly environment where the components of the system have
been sufficiently well modelled and structured.

An outline of the real time control problem is first presented, which can be divided in two further problems:
"on time control" to meet due dates, and "at the same time control" to satisfy rendez-vous (simultaneous
arrival times).
In keeping with these requirements, we define a general control oriented model for assembly systems, with a
specific definition of cells. Control policies are studied at each level of the tree of decision points:
management of manufacturing orders in the system and stock control at shop, cell and station levels.
The simulation is used to validate the modelling and to test control policies: the software is written in LISP
Flavor and the main flavors and methods are given.
The theory is illustrated with an industrial example which has a shop structure.

Keywords: Assembling - Flexible manufacturing - Industrial production systems - Modelling - Simulation -
Process control - Stock control.

INTRODUCTION - ASSEMBLY ENVIRONMENT

Flexible manufacturing systems are either machining or
assembly systems (Margirier, 1987), but most research on the
control of these systems deals with the former (concept of
Flexible Manufacturing System (Blackstone, 1982; Browne,
1984; Smith, 1986), thorough studies on priority rules
(Falkner, 1986; Panwalkar, 1977), together with some
assembly research (Kusiak, 1986; Russel, 1985; Sculli, 1987)).

With regard to control problems, the main difference between
assembly and machining (Hall, 1986) is that the latter operation
involves only one part, whereas the former consists of the
control of the arrival of assembly parts (Fig. 1.).

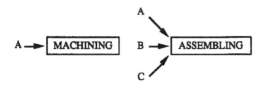

Fig. 1. The main difference between assembly and machining.

Not only do these parts have to arrive on time at the station (to
keep due dates) but also parts required for an assembly
operation should arrive at the same time (to avoid blocking and
waiting states).
New priority rules and new control policies therefore become
crucial to resolve these specific assembly problems.

CONTROL ORIENTED MODEL FOR ASSEMBLY
SYSTEMS

The control process acts at the short and very short range and is
then directly concerned with the layout of the shop you need to
regulate.
A large number of assembly systems have been set up in
industrial environments or investigated in research laboratories
(Correge, 1983; Gaillet, 1987; Miltenburg, 1987; Ranky,
1986). In order to solve the assembly problems arising from
each type of assembly layout, a generalized modelling approach
is then needed to include all the physical equipment possibilities
in the shop (machine-tools, storage places, transport systems),
taking into account the processing sequence of products.

Lots of terms are used in literature to characterize machining or
assembly systems. In the conventional hierarchy
shop/cell/workstation, cell is defined according to products, or
processing sequences, equipments, resources, transport,
operators ...

In this paper, cell is redefined considering assembly control
requirements. All assembly systems have in common the
control problem of managing the timely and simultaneous
arrival of products at the assembly workstations.
To meet due dates (on time control), one must manage
manufacturing orders in the process of production, whereas to
manage product rendez-vous (at the same time control), the
stocks upstream of the assembly stations must be controlled,
i.e. the departure of assembly parts from the buffer must be
managed.
The second problem shows the influence of buffers to allow for
the selection of required assembly parts, not only just upstream

the assembly stations but also at the cell level for anticipation. Using these control requirements, a cell surrounded with upstream and downstream buffers comprises a set of equipment - stations, storage and transfer systems - and behaves as a sphere of influence in which any action on an assembly part will have an effect on rendez-vous operations (delay, waiting or blocking state) (Fig. 2.).

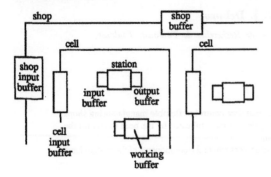

Fig. 2. Assembly system hierarchical model structure.

The buffers are at the shop level (supplying several cells), at the cell level (supplying several workstations) and at the station level (supplying the relevant station). Conveyor belts are modelled as buffer with a F.I.F.O. management whereas automated guided vehicles considered as resource required for transport are not modelled as buffer.

Studying various papers, we have identified eight patterns of assembly organizations, listed below and described in appendix 1.

  a) assembly station (a1: simple robot, a2: robots round a table).
  b) assembly transfer line (b1: conveyors, b2: pulling cables).
  c) loop conveyor belt.
  d) system with flexible transport (d1: transtockers, d2: rails vehicles, d3: Automated Guided Vehicles).

The organizations a) are cells; the other systems are either cell or shop depending on the position of buffers and stations.

THE "ON TIME" AND "AT THE SAME TIME" CONTROLS

For the "on time" control (to meet due dates), two algorithms will allow the management of manufacturing orders:

- the part input policy sequences the leaving time of raw parts from the shop input buffers. These launching dates are used as references to evaluate tardiness of manufacturing orders. The processing sequence and the real time data on waiting times in cell buffers give an evaluation of the time needed to process all the products of the new manufacturing order.

  A policy could be to launch the parts taking account of the slack between the operations given by the tree structure. Another policy is to launch at the same time the required parts for the first assembly operation; this policy is more relevant when all the raw materials have to be clamped on one machine.

- the assignment of priority numbers to the orders in process allows the comparison of their tardiness and a reaction by accelerating or slowing products. The sample time is long compared with the processing times of the orders because of the distant (short range) objective.

  The function can use the tardiness, the lateness or the greatest delay of the manufacturing orders in process. The evaluation

is made on each part of the product, on each product of the order and for each manufacturing orders.

For the "at the same time" control (to manage products rendez-vous), stock control policies must be defined taking into account assembly problems on subsequent stations and depending on the type of buffer (at shop, cell or station level). Figure 3. shows the different local policies (local because of their near objective) which relates to the general system.

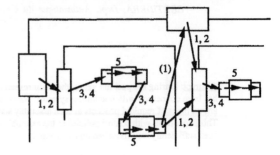

1 - Shop routing.
2 - Shop stock control.
3 - Cell routing.
4 - Cell stock control.
5 - Station stock control.

Fig. 3. Local policies for "at the same time" control.

The routing problem consists of the selection of the next destination (answering to the "where" request), and the stock control problem is the release of a part to the appropriate destination (answering to the "when" request).

For example, when a part arrives at a cell buffer, the policy 3 will select the next station and the policy 4 will authorize the routing to the selected station. If the authorization is not given, two others local policies will later allow the routing:

- the part type selection policy: a place on a station is released and the policy selects the next part(s) to be loaded,
- the "required job" policy: the stopped part is required by a leading part for an assembly operation.

The above scheme points out that the local policies have a hierarchical organization suited to the shop structure. Control policies are effected at the same level (e.g. cell/cell or station/station) or between two consecutive levels (e.g. shop/cell, cell/station or station/working buffer).

Control flexibility is related to the sizes of the buffers, and then directly to the concept of the cell.

Thus, the main purpose of the "at the same time" control is to avoid blocking states at the cell level (policies 3, 4 and 5) and to coordinate rendez-vous at the shop level (policies 1 and 2).

The policies 3, 4 and 5 are necessary in an assembly system. The corresponding rules are simple because the information required only involves the requested station and the states of work in process in the station.

The policies 1 and 2 are more complex due to the requirement of specific parts needed in the assembly operation. However the necessary information is only at the global level.

All policies are complementary. For instance, at its launching date (part input sequence) a part selects the destination cell (shop routing), is authorized to go into the cell (shop stock control), is put in the cell input buffer, selects the destination station (cell routing), is authorized to go into the input buffer of the station (cell stock control) and then into the working buffer (station stock control) to be processed subsequently.

Some local policies are proposed (Barbier, 1989):

- shop routing: the selection of the next cell for a part depends on the characteristics of the cells (see cell routing policy, where a cell is considered as a machine) and of the operations; in the second case, a policy could be to select the operation for which all the required parts are available at the shop level.

- shop stock control. A direct policy could be to authorize the release without restraint (in fact no policy); this policy have to be applied when the part have to free the working buffer (no output buffer at the station and cell levels). Another policy is to wait for all the required parts to be available at the shop level before releasing them to the cell. An advantage of a central buffer is the possibility of stocking works for cells; the policy could manage safety stocks for each of the sub-assemblies (these stocks could be use when a failure happens).

- cell routing. A first choice (when it occurs) is to leave the cell or not. The selection of the next station can otherwise use the classic priority rules studied on maching systems (Falkner, 1986) (availability of resources, utilization rate of equipments...).

- cell stock control. Some critical cases require special policy: when the station where is the part has no ouput buffer, the part must release the working buffer as soon as possible; when the selected station has no input buffer, the release must be as late as possible. A general policy is to authorize the release only for the leading part; another policy is to reserve the places of the input buffer for the required parts.

- station stock control: the authorization depends on the presence of the required parts in the input buffer of the station: the part could wait them or not.

VALIDATION OF MODEL STRUCTURE AND CONTROL POLICIES

Simulation tools have been chosen to validate the modelling and to test control policies.

A comparison has been made between object-oriented languages like LISP Flavor and simulation languages like SLAM and QNAP II. The first type of language has been chosen in this work due to its power and modular structure. Object-oriented languages are more suited to the varied nature of the tasks to be performed such as programming of the hierarchical model (objects shop, cell, station, buffer), simulation of the behaviour of product flow (objects product, part, manufacturing order, operation and methods linked by objects), the introduction of new control policies or priority rules (by means of new methods), coding of graphic display and statistical output report.

The purpose of the simulator ASPA is then to evaluate specific instances of control policies corresponding to particular assembly organizations in order to increase productivity.

The software, called ASPA - "Analyse des Systèmes de Production d'Assemblage" -, is developed in the CERT/DERA on a Texas Instruments Explorer machine in order to maintain continuity with a previous study on machining systems (Bel, 1988). The software can, however, be loaded on a Sun workstation, which facilitates its use in an industrial environment.

LISP is a programming language that is especially suited for manipulating structures and is widely used for aplications requiring symbolic processing. Flavor is an abstract structure that describes a whole class of similar objects. The flavors and the linked attributes used in ASPA are listed below:

- flavor shop: name, list-of-cells, list-of-input-buffers, list-of-central-buffers, list-of-output-buffers, part-input-sequence-algorithm, assignment-of-priority-numbers-algorithm, shop-routing-policy.

- flavor cell: name, list-of-stations, list-of-input-buffers, list-of-shop-buffers, list-of-output-buffers, shop-stock-control-policy, cell-routing-policy.

- flavor station: name, input-buffer, output-buffer, cell-stock-control-policy, station-stock-control-policy, part-type-selection-policy.

- flavor buffer: name, number-of-places, type-of-buffer.

- flavor transport (for transport modelled as resource): name, duration, quantity, transport-rule.

- flavor resource: name, quantity.

- flavor product: name, processing-sequence.

- flavor part: name, type-of-product, operation-in-process, position.

- flavor operation: name, required-products, duration, list-of-required-resources, list-of-stations, subsequent-operation.

- flavor manufacturing-order: name, product, quantity-of-products, due-date.

A specific object is an instance of a particular object. The appendix 2 gives some instances of flavors for the industrial application described in next section.

A method is attached to a flavor and define a generic operation that can be performed on any object of that flavor. The methods in the software ASPA authorize the simulation of the system: the operations concern the circulation of the parts, the utilization of the places... Methods are linked by messages or calls. ASPA is build around the following main methods:

- the method arrival-in-shop attached to the flavor manufacturing-order: call of the function that sequences the part input, and for each part: call of the method start-in-shop at the launching date.

- the method start-in-shop attached to the flavor part: if the buffer of raw materials is at the shop level, call of the method routing attached to the flavor shop, otherwise call of routing attached to cell.

- the method routing attached to shop: selection of the next cell (policy 1) and call of:

- the method stock-control attached to shop: authorization of departure (policy 2), if yes departure to an input buffer of the cell and call of:

- the method routing attached to cell-buffer: selection of the next station (policy 3), and call of:

- the method stock-control attached to cell: authorization of departure (policy 4), if yes departure to an input buffer of the station and call of:

- the method stock-control attached to station: authorization of departure (policy 5), if yes departure to the working buffer of the station and call of:

- the method ask-operation attached to station: if the required resources are available and if the required parts are also on the working buffer, call of:

- the method begin-operation attached to operation, and then, taking into account the duration of the operation, call of:

- the method end-operation attached to operation: release of resources, then, if the subsequent operation processes in an other cell: call of routing attached to shop, in another station of the cell: call of routing attache to cell, otherwise the operation is on the same station and call of ask-operation attached to station.

AN INDUSTRIAL APPLICATION

The physical system on which the modelling and control has been evaluated is build around a loop conveyor belt. The conveyor belt is linked with four derivative conveyors with each one feeding one or more workstations. Stations are either assembly machines or feeders; there are no duplicate machine, but each assembly operation can be processed on a manual machine.

This organization is of the type c2) and is then modelled as a shop with a central buffer feeding four cells; the two assembly cells have four (M1 to M4) and nine (M5 to M13) machines, each manual cell has one machine. Stations are simple: there are no input or output buffers.

The layout and the equivalent model of the system are given on Fig.4. and Fig. 5.

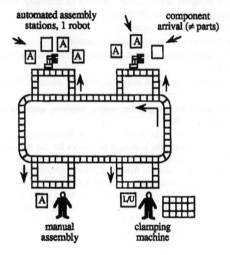

Fig. 4. Layout of the industrial application.

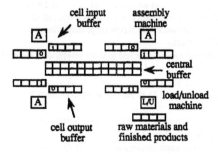

Fig. 5. Modelling of the industrial application.

Cells have either an advanced assembly robot or an operator which are resources required for each operation in the cell. Therefore, only one operation can be processed at one and the same time; this feature restrains the possibilities of part routing.

This shop manufactures one family of products, 10 types of raw materials are clamped on the load/unload machine. The tree structure of operations (Fig. 6.) has 8 assembly operations (an assembly operation concerns two or more parts from different pallets and not the components which are supposed to be always available). The life cycle of a product is about 180 minutes and a manufacturing order concerns 12 finished products.

When the shop has a set of products to manufacture, raw materials are clamped in the load/unload cell, and are placed on the main conveyor belt. The processing sequence gives the next operation and the part is subsequently routed to the next cell. A part proceeds on average through seven machines and is then unloaded from the shop.

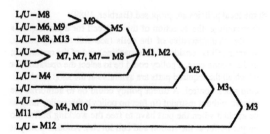

Fig. 6. The tree structure of operations.

Two types of events - an event is a decision point where a choice has to be made - can alter the work in process flow:
- arrival of manufacturing orders from the production level,
- end of processing for a set of pallets in a cell.

On receipt of a new manufacturing order characterized by the number of finished parts and their delivery date, the shop has to react to an additional work load: many raw materials have to be clamped and processed.

A relevant algorithm gives the ready date - i.e. the clamping date - for each introduced part. The main objective of this function is to meet expected deadlines, and we identify this as the part input sequencing.

When an operation is completed in a cell, the processed pallets are routed to the central buffer. The freed workspace can receive new pallets from the derivative conveyor or, if empty, from the main conveyor belt itself.

The shop control has then to select one or more sets of pallets to be processed in this cell, in order to avoid slowing down - if a required part is missing in the central buffer - and blocking - if there are bad parts in the F.I.F.O. queue of the derivative conveyor - of the cell.

This function is carried out by the control policy of the central stock in order to satisfy the assembly rendez-vous and one can easily recognize this as the shop stock control function.

Barbier (1988) relates in detail this application. After having tested different layouts and control policies with the simulator, the last stage of project development will be the implementation of relevant policies in the control module (using both software and hardware means) on the industrial system.

CONCLUSION

A generalized approach to industrial assembly system modelling is presented in this study. The modelling is based on the new definition of cell considering assembly control requirements.

Two algorithms which the objective is to meet due dates are introduced: the part input sequencing and the assignment of priority numbers to the manufacturing orders. In order to resolve the product rendez-vous problem, control policies are linked with the new hierarchical shop/cell/station structure: shop routing, shop stock control, cell routing, cell stock control and station stock control policies.

The modelling and the control system have been illustrated on an industrial application.

A general flexible simulator has been developed in LISP Flavor to validate the modelling and to test control policies; the main features of this tool are presented. Each type of assembly organization could thus be modelled and simulated using appropriate policies.

The simulator could be used in an aided design system in order to configure the layout and the control of any assembly industrial organization.

REFERENCES

BARBIER, M. (1989). Modélisation et conduite des systèmes de production d'assemblage. Rapport d'activité 1988 - CERT/DERA n°1/7661.

BARBIER, M., A. GAILLET, and C. LAMBERT (1988). Projet d'architecture de l'atelier CARMEN - fonctionnalités - moyens - gestion. Rapport intermédiaire - CERT/DERA n°5 et 6/7561.

BEL, G., and J.B. CAVAILLE (1988). Oasys: une nouvelle approche de la simulation des systèmes de production. Proceedings of USINICA, Paris.

BLACKSTONE, J.H and al (1982). A state-of-the-art survey of dispatching rules for manufacturing job shop operations. International Journal of Production Research, Vol 20 n°1.

BROWNE, J., and D. DUBOIS and al (1984). Classification of flexible manufacturing systems. The FMS Magazine.

CORREGE, M., and J.P. JUNG (1983). Automatisation d'un atelier de fabrication. Simulation et ordonnancement. Rapport final DRET - CERT/DERA n°4/7273.

FALKNER, C.H. (1986). Flexibility in manufacturing plants. Proceedings of the Second ORSA/TIMS Conference on Flexible Manufacturing Systems.

GAILLET, A., and C. LAMBERT (1987). Projet d'architecture de la cellule d'assemblage CARMEN. Rapport intermédiaire AICO - CERT/DERA n°168/87.

HALL, D.N., and K.E. STECKE (1986). Design problems of flexible assembly systems. Proceedings of the Second ORSA/TIMS Conference on Flexible Manufacturing Systems.

KUSIAK, A. (1986). Scheduling flexible machining and assembly systems. Proceedings of the Second ORSA/TIMS Conference on Flexible Manufacturing Systems.

MARGIRIER, G. (1987). Flexible automated machining in France: results of a survey. Journal of Manufacturing Systems, vol 6 n°4.

MILTENBURG, G.J., and J. KRINSKY (1987). Evaluating flexible manufacturing systems. IEE Transactions, Vol 19 n°2.

PANWALKAR, S., and W. ISKANDER (1977). A survey of scheduling rules. Operations Research, Vol 25 n°1.

RANKY, D.G. (1986). A program prospectus for the simulation, design and implementation of flexible assembly ans inspection cells. Proceedings of the Second ORSA/TIMS Conference on Flexible Manufacturing Systems.

RUSSEL, R.S., and B.W. TAYLOR (1985). An evaluation of scheduling policies in a dual resource constrained assembly shop. IEE Transactions, Vol 17 n°3.

SCULLI, D. (1987). Priority dispatching rules in an assembly shop. OMEGA, Vol 15 n°1.

SMITH, M.C., and al. (1986).Characteristics of US FMS: a survey. Proceedings of the Second ORSA/TIMS Conference on Flexible Manufacturing Systems.

APPENDIX 1

Description of the eight patterns of assembly organizations, using the following symbols:

(a1) assembly station with advanced robot

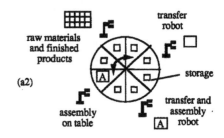

(a2) raw materials and finished products; transfer robot; storage; assembly on table; transfer and assembly robot

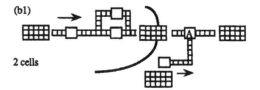

(b1) 2 cells

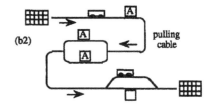

(b2) pulling cable

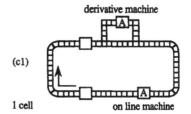

(c1) 1 cell; derivative machine; on line machine

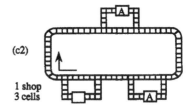

(c2) 1 shop 3 cells

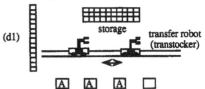

(d1) raw materials and finished products; storage; transfer robot (transtocker)

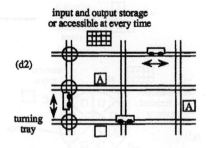

input and output storage
or accessible at every time

(d2)

turning
tray

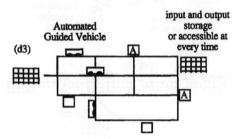

Automated
Guided Vehicle

(d3)

input and output
storage
or accessible at
every time

APPENDIX 2

Examples of instances for the layout of the industrial application.

- instance of the flavor shop: name = loop-conveyor-belt-shop, list-of-cells = (load-unload-cell, manual-assembly-cell, robot-assembly-cell-1, robot-assembly-cell-2), list-of-input-buffers = (raw-materials-buffer), list-of-central-buffers = (central-buffer), list-of-output-buffers = (finished-products-buffer).

- instance of cell: name = robot-assembly-cell-1, list-of-stations = (assembly-machine-1, assembly-machine-2, feeder-machine-3), list-of-input-buffers = (input-buffer-1), list-of-output-buffers = (output-buffer-1).

- instance of station: name = assembly-machine-1, input-buffer = nil, output-buffer = nil.

- instance of buffer: name = central-buffer, number-of-places = 24, type-of-buffer = shop-level.

- instance of resource: name = robot-1, quantity = 1.

AN INTERACTIVE SIMULATION MODEL FOR MULTIPARAMETER OPTIMIZATION OF CUTTING PROCESSES IN FMS

D. Polajnar, L. Lukic and V. Solaja

Lola Institut, Beograd, Yugoslavia

ABSTRACT. A multiparameter optimization model of cutting processes for a FMS production planning is presented. The model relies on a suitably designed technological database and a generation method with operation level optimization being the prime consideration. Technological knowledge and main machinability functions dependences are partly built into the logical database organization and data access paths. Model's output is the optimal technological parameters combination as well as tools assortment and quantity which satisfy the initially defined manufacturing conditions.

KEYWORDS. Computer simulation; cutting process; database; decomposition; machining centers; machining; optimization; software system; technological parameter; workpiece.

INTRODUCTION

The first stage of FMS production planning begins with a detailed technological study which includes an analysis of timing constraints of manufacturing cycle. In view of numerous specific requirements of various processes constituting a flexible manufacturing production system, the preparation of such a study may be a very complex task. Workpieces may involve hundreds of basic operations and require a similar amount of different tools. In practice, the search for a suitable combination of technological parameters is a laborious iterative process requiring a great deal of expertise and intuition. In order to facilitate this task, we have undertaken the development of an interactive simulation system, called LolaCut, that supports multiparameter optimisation of cutting processes. In the following section, we first outline the LolaCut software package specifically designed to support the optimisation procedure. The underlying model and its use in optimisation procedures is described in section 3. Finally, a brief conclusion is given in section 4.

LolaCut SOFTWARE PACKAGE

The LolaCut software package facilitates the chip removal manufacturing processes design for NC, CNC and other machine tools.

The main system components (Fig.1) are:

  o technological data and knowledge base (DB-manager),

  o supervisory module,

  o optimisation dialogue module,

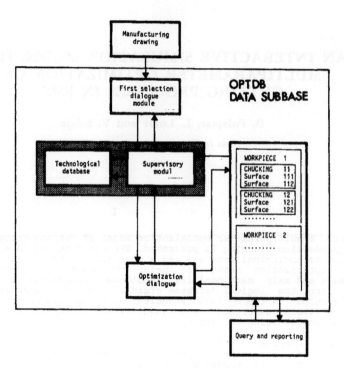

Fig. 1. LolaCut schema

o on-line query and report generation module.

The basis of the system is the technological database together with the supervisory module. The workpiece material table, tool tip material table, tool tip geometry tables, machinability table, analytical models tables and machine tool table form technological database. Technological knowledge and main machinability functions dependences are partly built into the logical database organisation and data access paths (Fig.2). DB manager enables complete database management and the creation of different queries and report generation.

Technological data in the database are a particularly valuable resource of the system, and their integrity must be adequately protected. Ordinary users are not permitted to update the database, even though they may use the full capabilities of the system. If some important data is missing in database, the user is guided by the system to supply necessary data to finish started procedure. These data stay temporary in the system and don't have any influence on the data in permanent database.

First selection dialogue is designed according to technical requirements defined in the workpiece machining drawing. The method used in LolaCut for flexible technology design is the generation method with operation level optimisation being the prime consideration. To implement this method, workpiece decomposition into workholdings, workholdings into surfaces and surfaces into operations is done (Fig.3).

Dialogue menues are hierarchically organized logically following the workpiece decomposition. Only reading of data from the database is allowed during

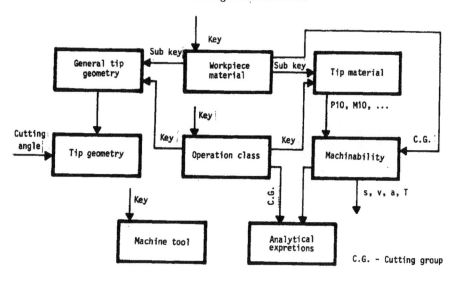

Fig. 2. Technological database schema

the dialogue process. As a result of this dialogue the first selection technological parameters data subbase is generated, OPTDB, as well as corresponding tool subbase DBT (Fig.5).

OPTIMISATION DIALOGUE

As mentioned earlier, the first optimisation level is performed during the first selection dialogue. The first selection dialogue is designed according to technological requirements which include assortment of workpieces, corresponding series, workpieces material and predicted production time (Fig.4). In some cases, the machining centers or/and FMS configuration may be defined.

All this information, together with decomposition scheme of workpieces into workholdings and surfaces present initial input into first level optimisation procedure.

The additional input data are supplied for each operation considered. With this data, some functional dependences in cutting processes at operation level are

established.

The operation is completely defined by the geometrical parameters (cutting diameter, length of cut, quality, precision), machining operation type (turning, drilling, milling) and tool (tool type, tool tip material type, tool tip basic geometry, tool tip cutting geometry, tool holders).

Geometrical parameters are found on the workpiece machining drawing, while the machining operation type is defined by user-technologist. These data actually present the input data at the operation level of the dialogue. The tool informations and machining parameters together with the machining time are output data from the optimisation procedure performed at the operation level.

The optimising procedure starts with the proper selection of machining operation type, because any of the three types of operation can be performed on the machining centers. Once the operation type was chosen the all valid (tool tip material,tool tip basic geometry)

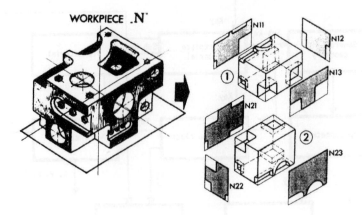

Fig. 3. Workpiece decomposition

combinations are presented to user. Any of these pairs satisfy the specified requirements for that operation. User selects one of this pairs. At this point the user also specifies the cutting angle. At the next step user selects one of a few (tool tip real geometry, depth of cut, feed interval, tool life) combinations. At this point, the proper tool tip is selected and checked against DBT in specified workholding. If such a tool tip already exists, its tool life is taken and no changes on its value are allowed. This way the tool reduction is done.

With this data, the cutting forces are calculated and compared to the machining centers corresponding characteristics. If calculated forces don't satisfy the required criteria, the new speed, feed data or tool is selected. Otherwise, the machining efects are calculated and presented through cutting time, which is then compared to the remainder of tool life. If cutting time is greater then tool life or the user wants to change it, the procedure is returned to the operation level. The change of cutting time may be achieved by changing, in hierarchical order:

o technological parameters (tool life, the depth of cut, the feed),

o tool tip material,

o machining operation type.

After attainding the "correct" value for cutting time, all informations about operation are recorded in OPTDB and about tool in DBT if such tool does not already exist in DBT (Fig.5). At this point the optimal technological parameters and tool are selected for that operation. This procedure is repeated for all operation in all workpieces of the given assortment. It may happen that some of the operations are not complited using this procedure. In such case, missing data are manually entered into OPTDB and DBT.

Next level of optimisation comes from interoperation relationship inside workholdings. The user is informed about the efects of the previous optimisation procedure throught different indicators such as:

o total machining time,

o machining path (how many workpieces, how many workholdings, etc),

o number of tools per assortment,

o total tool number.

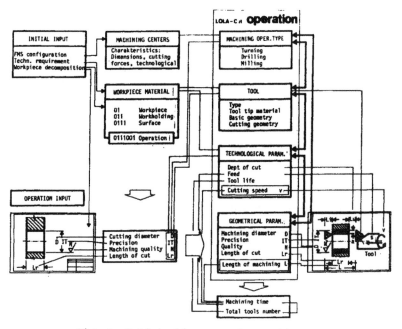

Fig. 4. Optimisation procedure schima

By analysing the results, user has an opportunity to optimise parameters in the following domains:

o tools (constant manufacturing conditions),

o manufacturing conditions (constant tools),

o both tools and manufacturing conditions.

Optimising procedures are repeated until satisfactory results are obtained. The results of the last optimising pass are permanently kept in OPTDB and DBT and later used for the generation of different reports needed in FMS production planning.

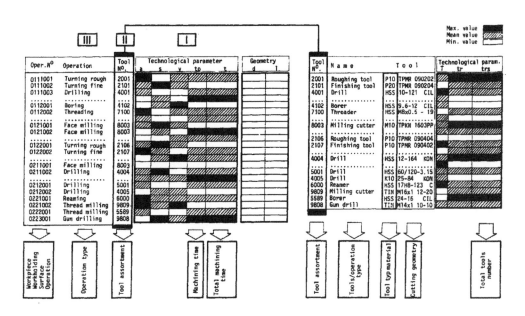

Fig. 5. OPTDB and DBT tables layout

CONCLUSION

The interactive simulation model for multiparameter optimisation of cutting processes presented in this paper enables the designer to proceed from a description of technological requirements to a specification of optimal cutting technology for flexible manufacturing. Technological requirements typically include the assortment of workpieces and corresponding series, workpiece materials, and total production time. In the current version of the LolaCut system, the designer must explicitly define workholding and operation sequencing. In the next version, these functions will be performed by an expert system module. LolaCut has been implemented in the programming language C, relying on the services of ORACLE RDBMS and runs on an IBM PC/AT under MS DOS.

REFERENCES

Hammer,M.,McLeod,D.: Database Description with SDM: A Semantic Database Model ACM Trans. on Database Systems, Vol 6,No 3, Sep.1981., p 351-386.

Lukic,Lj.,Polajnar,D.,Uzunovic,R.: Optimizacia tehnologii obrabotki rezaniem s pomosc'ju programmogo paketa LolaCut, V Mezdunarodnaja konferencija po gipkim proizvodstvenym sistemam (GPS) i voprosam CAD/CAM, g. Rydzyna 13.-19. nojabraja 1988.,(in russen)

Rosenberg,T.S.:HP-RL: An Expert Systems Language, HP Journal, Aug. 1988.

SIMULATION OF A FLEXIBLE MANUFACTURING SYSTEM (FMS) FOR FABRICATION OF BONDED STRUCTURES

R. M. A. Al-Yassin, M. Younis, S. M. Megahed and M. M. Sadek

*Mechanical Engineering Department, Faculty of Engineering & Petroleum,
Kuwait University, Kuwait*

Abstract

A flexible manufacturing system (FMS) is being
developed for the fabrication of bonded
structures. This flexible manufacturing system
consists of CNC-lathes, CNC-milling machines,
workpiece handling units and robots. The
cutting process has been optimized for cost
reduction. This paper presents a computer
package for the simulation of this FMS. The
synchronization of the various movements of the
components of the FMS is taken into consideration. The computer graphics capability is used
for the animation of such a system. This package
is written in BASIC language for its simplicity,
flexibility and polularity. This package can
also be used for any FMS production plant.

Introduction

Manufacturing simulation is a tool used to create
a computer model of a specific manufacturing,
storage, assembly or material and tool handling.
It is used to prove automated manufacturing
systems and test its various concepts with no
physical risk and at a low cost. Without the use
of simulation the potential interference of parts
in a transportation system has to be ignored
until the hardware is installed. It also
facilitates the optimization of all the
independant parameters and indentifies the
possible bottle-necks, examines the impact of
introducing novel oeprating proceedures and/or
variations in batch size and scheduling policies.
This computer model enables the analysis of
complex interactive processes to be performed.

Once the model is established, tests can be
applied to investigate the effect of batch size,
cycle time variation, set-up time... etc. on lead
time, work in progress, machine utilization etc.
This will result in the optimization of the plant
design, reduce the installation time, maximize
resource utilization, reduce leads time and
stocks.

Several ways of writing the simulation model have
been developed. A friendly simulation language
is "SIMAN" [1]. It is a Fortran based language,
process driven and can be quickly programmed by
non computer personell. Cinema has now been
introduced so that models written in SIMAN can be
linked with icons of resources, entities... etc.,
which move around against a static background in
real time.

Examples are given in [2] to show how
manufacturing simulation can be used to analyse
material's handling system. One of these
examples is the Bosh type accumulation conveyor
system where two products are manufactured on
·line [3]. The model is used to give information
on the effect of batch size and plant load. The
model is written so that other variables such as
maximum allowable queue lengths into each
operation, robot operational state and
operational times, could be changed before a
simulation run is started or during the run.
Another example was also given, [2] which deals
with powder mixing/pressing plant. These two
examples use just in time philosophy (JIT)
techniques developed by Hawker Siddeley [2].
This concentrates on improving door to door time,
reduce activities that do not add value such as
handling of parts between operations and set up
times between product types. It introduces
quality control at source.

A recnt paper [3] described how Fortran sub-
routines could be used to look at the results of
Simon simulation program. There are many
investigators who dealt with the use of expert
systems in simulation to help with the many
aspects of the operational and control of F.M.S.
[4, 5, 6, 7].

A graphical language has been developed [8] based
on operational specification and object oriented
programming. It is used for discrete event
simulation models and illustrate its application
to the simulation of a machining cell. It
provides production planning, detailed finite
scheduling and shop floor control. This
increasing manufacturing efficiency, reduce stock
levels and product lead times. A `what if'
facility is provided to investigate the effects
of controls and distrubances.

Simulation has been used in Jaguar Car Ltd., U.K.
[9] to evaluate a range of alternative operating
strategies taking into consideration, order
alteration batchings through the paint shops, and
need to balance labour loadings on the assembly
lines.

A flexible modelling technique is presented in
[10] for identifying resource utilization and
optimization and levels in the dynamic simulation
of assembly system, as applied on a robot
workcell using general purpose simulation
language [11,12]. This deals with idle time and
periods of resource activities.

The purpose of the F.M.S., presented in this
paper is to fabricate bonded structures.
Adhesive bonding is a joining technique that is
capable of replacing or supplementing
conventional methods such as bolting, rivets or
spot welding. The advantages of this technique
are listed in [13, 1ʌ]:

The technology of adhesive bonding for the manufacture of a casing of a lathe gear box has been developed and optimized [15], for which this simulation package has been developed. The cast gear box of a lathe shown in Fig.1(a) has been redesigned for fabrication by bonding. The philosophy of the design is to convert the commercial cast iron gear box into one to be fabricated by bonding, equivalent plates fitted in double containment joints and bushes. The side plates and bushes of the gear box are manufactured from mild steel. After the fabrication of the gear box components, it was assembled using two types of pre-designed jigs and fixtures. The first shown in Fig. 2(a) is for the assembly of corner joints, side joints and base plates, and the second presented in Fig. 2(b) is for the assembly of bushes with the plates. These jigs are designed so as to hold the parts in their net positions and to control the bond-line thickness. This paper presents a computer package for the simulation of bonded structures production plant. The synchronization of the various movements of FMS components (CNC- lathes, CNC-milling machines, workpiece handling system and robots) is taken into consideration. Computer graphics with anumation were chosen for this simulation. This package is written in BASIC for its versatility, flexibility and simplicity.

System Software

Computer simulation can be applied using either special simulation languages or general purpose programming languages. Special simulation languages such as GASP IV, SIMSCRIPT II.5, SIMULA and GPSS use event-scheduling or process interaction modelling [17]. Such languages provide most of the features required in system simulation resulting in reduction in user programming time.

A general purpose language such as BASIC or FORTRAN has many advantages compared to the special purpose languages. This can be summarized as follows:

- Most users already know a general purpose language, such as BASIC or FORTRAN which are available on all computers, contrary to a specific simulation language which may not be accessible in the available computer.

- BASIC or FORTRAN programs can be tailored to a particular applciation. On the other hand special simulation languages are designed to model a wide range of systems with one set of building blocks. This may lead to more execution time in the case of special languages.

- General purpose languages allow greater programming flexibility than special simulation languages. For example, the use of computer graphics is much easier.

In this work, BASIC is chosen as a programming language for its simplicity and flexibility in computer graphics. The graphics software is a collection of programs written to make it convenient for a user to operate the computer graphics system. It includes programs to generate images on the CRT screen, to manipulate these images, and to accomplish various types of interaction between the user and the system [18]. It also results in obtaining an overall system that can be specified in graphic mode and easily modified. This permits the examination and testing of specific system layouts prior to the actual use of the physical system [19].

Plant Layout

The arrangement of the existing entities within the production line depends on the production volume.

Process layout can be used also in quantity-type mass production. The production machines in this type are arranged into groups according to the manufacturing processes. The work-in-progress is moved from one work station to the next [18] by conveyor or similar means.

The plant layout for the presented FMS for fabricating bonded gear boxes, where both flexibility and efficiency are required, is a combination of process and product flow layouts. This work deals with the problem of the plant layout and how the existing machines are arranged around the conveyor.

FLOWCHART OF COMPUTER PACKAGE

The computer package presented in this work consists of a main program and four sub-programs linked together [18]. STEPO, being the main program simulates the production plant, and the other four sub-programs simulate the lathe (STEP1), turbing tools (STEP2), milling (MILR), and milling cutters (MILC). The interaction between these programs are shown in the block diagram presented in Fig.3.

The simulated plant consists of a closed track four-sided conveyer. One side of the conveyer is for loading/unloading material and finished work parts to the conveyer. Lathes, milling machines and robots will be located around the other three sides of the conveyer, for easy access of the machines for maintenance purposes. An interactive program is written to simulate the flow of parts within the system. The main operations of this program are:

a. **Data specification:**

- The dimensions of the working place of the production plant.
- The dimensions of the work range of each machine: lathe and milling machine etc.,
- The maximum and minimum reach of work range of the robot.
- Number of lathes on each side of the conveyer.
- Number of milling machines on each side of the conveyer.
- Direction of locating the machines around the conveyor (length or width).
- Order of location of machines and robots around the conveyor.

b. **Resultant operations**

Depending on the number of machines and their given dimensions, the program predicts the required work place required. If the given machines fit in the predicted work place, the program proceeds. If not, the program offers the user two choices: either to have an adequate recommended work place for that number of machines, or reduce the number of machines used. For the first choice, the program proceeds with the recommended dimensions. For the second choice, the program repeats the questions about the number and dimensions of machines:

The program will proceed by drawing the whole plant with the specified machines and then the pallet starts moving from loading station to the first work station. When the pallet reaches this station, the work piece will be picked up by a robot, and place the work part on the machine (lathe or milling machine). The corresponding simulation program is chained. After the process simulation is completed, the work part is unloaded and returned to the conveyor where it is transmitted to the next processing station.

The program predicts the time consumed in each step and synchronizes the movements of the various entities of the plant. The layout of this production plant is shown in Fig. 4.

SIMULATION OF THE LATHE AND THE MANIPULATING ARM

The operations performed on a lathe are centering, facing, turning, thread cutting and knurling. A numerically controlled lathe is used in the flexible manufacturing systems.

The flow chart of simulation of the NC-lathe used in this work is presented in Fig. 5. The cutting tools are stacked in a tool magazine mounted on the lathe, while a robot arm is used for loading and unloading tools. Cutting tools are coded as shown in Table 1. When the simulation program of the lathe is linked, a menu showing the codes of cutting tools appears first, as shown in Fig. 6.

Table 1: Lathe tools and their codes

TOOL	TOOL CODE
Centering Tool	CT
External Threading	ET
Parting Tool	PT
Right-cut Tool	RT
Left-cut Tool	LT
Internal Threading	IT
L-Cut side facing	LC
Curved Cutting	CC
End-Cutting	EC
Knurling Tool	KT
Forming Tool	FT

The required tool code, Diameter of the workpiece, Depth of cut, and Speed of the manipulating arm, is data which must be given by the user.

The output of the package:

The lathe and the manipulating arm are presented in Fig. 7. The speed and feed are optimized through a subroutine discussed in the Appendix. The arm starts moving to fasten the workpiece in the chuck and returns to normal position as shown.

According to the required tool, the tool magazine starts rotating till the required tool is in the loading position. The arm picks the tool and moves to fix it to the tool post, where the tool starts cutting at the optimum setting of speed and feed as shown as an example for the threading operation in Fig. 8. The animation of all these operations can be shown on a CRT terminal.

Simulation of the milling machine

The flow chart of the simulation program of the NC-milling machine is shown in Fig. 9. The cutters are in a carousel attached to the machine, with automatic cutter holder. This holder feeds the milling machine with the necessary tool and takes it off to the carousel as shown in Fig. 10. As in the case of the lathe each cutter is given a designated code shown. A menu showing each cutter and its corresponding code appear when the simulation program of the milling machine is linked.

Conclusion

This paper presents a computer package for the simulation of an FMS of a fabricated bonded structure consisting of NC-lathes, NC-milling machines, handling system and robots. This package is written in BASIC language for its simplicity, flexibility and popularity. Computer graphics is used t clarify various movements of entities of the manufacturing system taking into consideration their synchronization. The program can be used also for different layout of production plants, for different number of machines with any specified dimensions.

Acknowledgement

This work is sponsored by Kuwait University Research Council through project EM 023.

References

1. Pegden, C.D. 1982. "Introduction to "SIMAN", systems modeling corporation, state college. Pa,. USA.

2. Novels M.D. & Hackwell G.B., 1987 "Graphical Simulation of material handling system,". Proc. 4th European Conference Automated Manufacturing 507-528 (May) I.S.S.

3. Haddock, J. 1987. "An Expert System framework based on a Simulation Generator, "SIMULATION 48, No. 2 (Feb): 45-53.

4. Ben-Ariek, D 1986. "A knowledge based system for based on a Simulation and control of F.M.S. Generator". Simulation-Applications in Manufacturing, IFS (Publications) Ltd., U.K.

5. Shivnan,J, Browne, J. 1986. "AI-based Simulation of Advanced Manufacturing Systems" Simulation Applications in Manufacturing, IFS (Publications) Ltd., UK.

6. O'Keefe, R. 1986. "Simulation and Expert Systems - A taxonomy and some examples", SIMULATION 46, No. 1 (Jan): pp-10-16.

7. Shannon, Mayer and Adelsberger, 1985, "Expert Systems and Simulation', SIMULATION 44, No. 6 (June): 275-284.

8. Spooner P.D. 1985 "A Simulation Based Interactive Production Control System." Simulation in Manufacturing, I.F.S. Conference Limited, (March).

9. R.N. Ingrdm, R.N., "Strategic planning for C.I.M. using Simulation report of Jaguar Cars Ltd., U.K.

10. Rahueijat H., 1985 "Simulating for resource optimization in robot assisted automatic assembly," Proc. inst. Mech. Engineers Vol. 200/85-86 No. B. 3 pp 181-186.

11. Mejtsky, G.J. and Rahnejat H., 1985 "Introducing GPSL: a flexible manufacturing simulator" computer aided design (Jan).

12. Schriber, T., 1974 "Simulation using GPSL, John Wiley, New York.

13. Sadek, M.M. 1982, 1983 "Fabrication of structures using bonding by exposy resin adhesives". Part 1 & 2 Sheet Metal Industry, (Nov.) and (March).

14. Chang, H.C., Sadek, M.M., Tobias S.A. 1983 "Relative Assessment of the Dynamic and Cutting Performance of a bonded & Cast Iron horizontal milling machine. "A.S.M.E. Journal of Engineering for Industry vol., 105 No. 3 August 1983 pp 187-197.

15. Azayem, K.M. Sadek, M.M., Darwish, M., Design Philosophy of a Bonded Gear Box, M.Sc. Thesis of 1st Author, presented for publication.

16. Naylor, T.H., Balintfy, J.L., and others, 1968 Computer Simulation Techniques. John Wiley and Sons., New York.

17. Graybeal, W.J., Pooch, "U.W. Simulation, Principles and Methods", Winthrop Publishers, Cambridge, 1980.

18. Yassin, R.A.A., Younis, M.A., Sadek, M.M., "Simulation of Automated Production of a bonded Gear Box casing submitted for publication in Journal of Manufacturing, Institute of Mechanical Engineers London.

19. Younis, M.A, Sadek, M.M., "Simulation for Cutting parameters optimization for an Automated Machining process." in preparation.

20. Deprereux, W.R., "Determining the optimal cutting parameters for economical application of NC machine", Ph.D. Dissertation, Aachen, 1969.

APPENDIX: OPTIMIZATION OF CUTTING PARAMETERS

The parameters of a cutting operation of a workpiece are: the cutting speed v (mm/min), and the feed rate s (mm/revolution). The optimization of these parameters is based on the idea of a minimizing the cost of removing a unit volume of the workpiece material. The tool life T is given by Taylor's equation [20]:

$$T^n_v \, s^m = \text{constant} = k$$

The total cost per cubic millimeter of removed material is the sum of labor cost L (currency/hr), and the machine cost M (currency/hr). These costs depend on the dimensions of workpiece : length l (mm), diameter d (mm), and the required depth of cut a (mm).

The total volume removed volume $Rv = d\, l\, a$ (mm)^3. The removed volume needs the following periods of time in minutes: handling time t_h, setting time t_s, and changing tool time t_w. The costs of different cutting operations are given below:

$$\text{Setting costs} = \frac{(L + M)\, t_s}{60\, R_v} = A$$

$$\text{Handling costs} = \frac{(L + M)\, t_h}{60\, R_v} = B$$

$$\text{Machining costs} = \frac{L + M}{60\, a.s.v} = \frac{D}{s.v}$$

$$\text{Changing tool cost} = \frac{(L+M)\, t_w}{60 a.s.v.T} = \frac{E}{s.v.T}$$

$$\text{Tool costs} = \frac{C_{tool}}{a.s.v.T} = \frac{F}{s.v.T}$$

SO,

$$\text{Total cost/mm}^3 = C = (A+B) + \frac{D}{s.v} + \frac{E}{s.v.T} + \frac{F}{s.v.T}$$

$$= G + \frac{D}{s.v} + \frac{(E + F)}{s.v.T} \quad \ldots\ldots\ldots (1)$$

However, it is noticed from the logarithmic graph of the relationship of T-v, T-s Fig. 11 that the slopes τ_v, τ_s are dependent on v as s. By plotting $k(k = \tan \tau_s)$, $i (i = \tan \tau_v)$ versus cutting speed (v) and feed rate (s) respectively on a log-log graph, linear relations are obtained. These relations are represented by equations (2) and (3), where m and n are the slopes of linear equations.

$$k = \frac{\partial \text{Log } T}{\partial \text{Log } v} = K_v\, v^m \quad \ldots\ldots (2)$$

$$i = \frac{\partial \text{Log } T}{\partial \text{Log } S} = i_S\, s^n \quad \ldots\ldots (3)$$

Then,

$$-K_v v^m = [\frac{\partial \text{Log } T}{\partial T} \cdot \frac{dT}{dv} \cdot \frac{\partial v}{\partial \text{Log } v}]$$

$$[\frac{\partial T}{\partial v}]_s = -T.k_v.v^{m-1}, \text{ and } [\frac{\partial T}{\partial S}]_v = -T.i_s s^{n-1}$$

and since, $\quad \partial T = \frac{\partial T}{\partial v}\, dv + \frac{\partial T}{\partial S}\, ds$

$$\int \frac{dT}{T} = \int -k_v v^{m-1}\, dv + \int -i^{n-1} S\, ds$$

Then,

$$T = e^{-[\frac{K_v}{m} v^m - \frac{i_s}{n} S^n + C]} \quad \ldots (4)$$

Now, when substituting the value of T from eqn. (4) into eqn. (1) equating the partial differentiation to zero we get,

$$s^n = \frac{\text{In } \frac{D}{E+F} + c - \frac{k_v v^m}{m} - \text{In } (k_v v^m - 1)}{(i_s/n)}$$

$$v^n = \frac{\text{In } \frac{D}{E+F} + c - \frac{i_s}{n} s^n - \text{In } (i_s s^n - 1)}{K_v/m}$$

The optimum cutting speed (v_o), and the optimum feed rate (S_o) can be obtained by solving eqns. (5) and (6) simultaneously. The flowchart of the optimization subroutine is shown in Fig. 12.

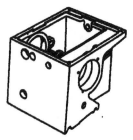

Fig.1a The M300 Harrison Cast Iron Lathe Gear Box (15)

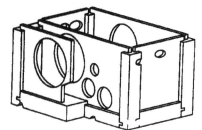

Fig.1b The Proposed Bonded Gear Box.(15)

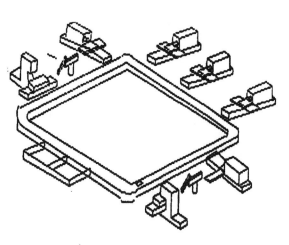

Fig.4 Production plant layout.

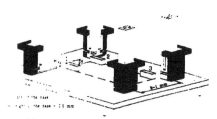

Fig.2a The Jig Used for Assembly of the Corner the side Joints and the Base Plate.(15)

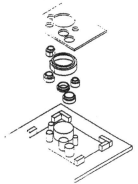

Fig.2b A Sample Jig Used for the Assembly of the Base Plate Along with its Bushes.(15)

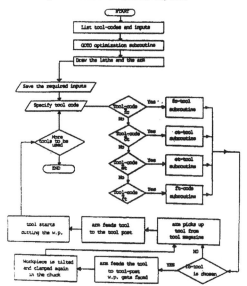

Fig.5 The flowchart of the simulation program of the lathe.

TOOL NAME	TOOL CODE	INPUT NAME	INPUT CODE
1.Centering Tool	CT	1.Tool code	TC
2.External Threading Tool	ET	2.Diameter of	
3.Parting Tool	PT	Work Piece	D
4.Right-cut Tool	RT	3.Speed of Arm	AS
5.Left-Cut Tool	LT	4.Feeding Depth	FD
6.Internal Threading Tool	IT		
7.L-cut Side-Facing Tool	LC		
8.Curved Cutting Tool	CC		
9.End-Cutting Tool	EC		
10.Knurling Tool	KT		
11.Form Tool	FT		

Do You Want To Run The Program (Y / N)?

Fig.6 Menu of the tool codes of the lathe.

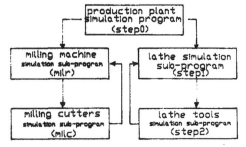

Fig.3 Block Diagram of programs constituting the package.

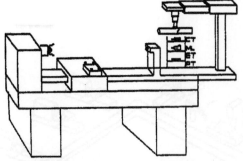

Fig.7 The lathe and manipulating arm.

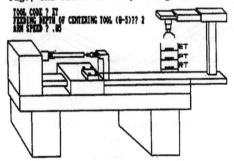

Fig.8 The arm is picking the next tool which
is external threading tool (ET)

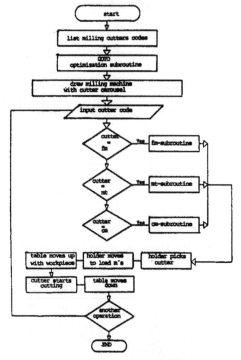

Fig.9 Flowchart of the simulation program of
milling machine.

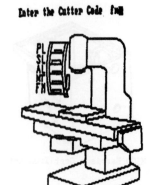

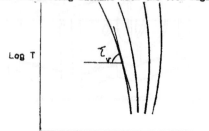

Fig.10 The milling machine with the tool magasine

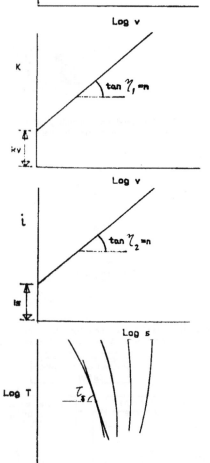

Fig.11

Optimization subroutine

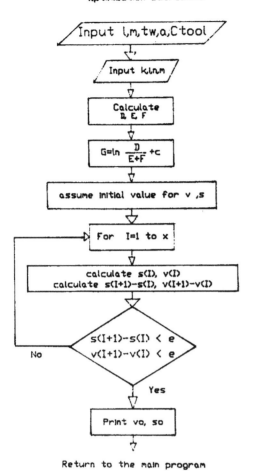

Fig.12 Flowchart of optimization subroutine

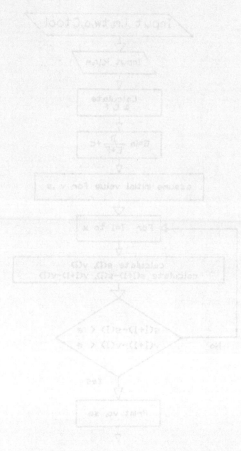

Fig.12 Flowchart of factorization subroutine

DEVELOPMENT OF UNMANNED MACHINE TOOLS IN CZECHOSLOVAKIA

I. Krsiak, P. Tomek and J. Kondr

Research Institute of Machine Tools and Machinery, Prague, Czechoslovakia

Abstract. The market requires more types and modifications of products and quicker response on placed orders. On the other hand people have wider choice in job selection and therefore fewer are willing to accept unpleasant working conditions in workshops or afternoon and night shifts. It is interesting, that the above mentioned trends seem to be alike with some alterations and some time delay in all political and social systems. It is no need to say that these trends are not reversible. All this brings us to the conslusion that unmanned mode of operation of machine tools will be more and more required in the near future. For this reason in 1985 a new state program for development of unmanned machine tools was started in Czechoslovakia. It will be finished in 1989 by introduction of four prototypes into production. Required features of unmanned machine tools and their components as well as trends of their development in Czechoslovakia are described in this paper. Basic information on control strategy is discussed.

Keywords. CAD, CAM, CNC, computer control, machine tools, manufacturing processes, minicomputers, programmable logic controller.

Unmanned Technology – Basic Terms and Scope of Applications

For better understanding of what we will consider as unmanned machine and what not, let's put down a basic definition:

"Unmanned machine tool does not require any human intervention for a period of time necessary for machining parts stored in integrated storage or parts delivered to the machine by automatic transport. There may be different parts in the storage (or brought to the machine by automatic transport). The machine setting for the new type of parts is done automatically".

As the scope of applications is considered – the flexible automation with unmanned machining is directed basically to the small batch production. But is does not need to be a general rule and exceptions are possible – e.g. in automotive industry unmanned machines are used in flexible transfer lines where the batch represents thousands of pieces.

As the workpiece properties are considered, the unmanned machining is applied mainly for complicated parts with long machining cycles. For prismatical parts it means machining from more sides in one set-up, for cylindrical parts besides turning operations boring and milling is included. All common materials are covered (steel, cast iron, light alloys). In most complex machining high precision must be guaranteed.

Required Features of Unmanned Machine Tools and Their Components

Unmanned machining constitutes new requirements in following fields: production planning, tooling, cutting tools, fixturing and gauging, machine desing, manipulation and transport, CNC controls.

Production plans need to be changed to reduce the number of parts set-ups, to enable measuring with touch probes on the machine and to machine the part completely.

Tooling concentrates on new progressive types of cutting heads, fixturing and gauging. The value of tooling may reach 25% of the total FMC investment.

Machine design is influenced by new precision and high cutting conditions requirements. Automatic parts and tool manipulation is integrated in the machine design. The whole concept of the machine tool enables stand alone or FMS operation.

Control. Let's discuss the strategy in more detail. The unmanned machining philosophy affects the CNC design a great deal calling for increase in both computing power and memory capacity.

One of the components of the state development program is also the desing of a new CNC system meeting the needs of unmanned production. Outlined below are basic features of the system.

The obvious requirement for the CNC is that the part program storage should be at least 128k byte. It has been calculated that an average program for milling taking an hour to run is about 400 blocks (12 000 characters) long. An 8 hour unmanned operation can thus require as much as 96 000 bytes of storage not including system data such as tool and zero offsets.

Auxiliary axes control is another feature of an unmanned machine CNC. Built-in manipulators require the control of as much as six axes that are quite independent of machining axes. These axes must be controlled while the main group of axes runs in interpolation. This problem is often being solved by "positioning units" receiving commands from the programmable logic controller (PLC). Another feasible method of manipulator control is a use of multi-channel NC, which is well suited to the task especially when operation cycles of the manipulator become more complicated. Then a need arises for more sophisticated method of programming the motion (use of subprograms, parameters etc.) usually not available on PLC's.

As an unmanned turning center should be able to perform some milling operations as well, the CNC must take it into account - e.g. it must be able to transform the tool path programmed in Cartesian coordinates into actual movement of a linear X axis and circular C axis on a lathe.

An implementation of measuring probe interface to the CNC places requirements not only to hardware of the system but it also affects the syntax of part progamming language. The language must provide the programmer with necessary tools to program measuring cycles and evaluation programs. The most straightforward way to do it is to implement a superset of common programming language giving access to NC-internal variables and a number of arithmetical and logical operands.
An automatic tool management system is also one of the crucial software subsystems making the unmanned operation possible. Our approach to the problem calls for a division of the task between the NC and PLC subsystems of the control. The NC keeps record of tool offsets pertaining to tools placed in the machine magazine. The PLC handles data related to actual wear of cutting edges and initiates a tool change when the usable life of a tool has been exceeded.

A reliable DNC communication becomes a necessity when integrating more machines into a production cell.

All the above mentioned features must be considered when designing a CNC for unmanned machines. In general it can be said that the more complicated the machine and its function become, the more tasks is being executed by the PLC which takes over in auxiliary axes control, tool and workpiece management, in-process gauging and other functions. Also the interface "window" between the NC and PLC must be able not only to communicate usual control and synchronization signals, but also to transfer data between the PLC and NC variables.

Development of Machine Tools in Czechoslovakia

As stated above new state program called "Unmanned Machines and System Solution in Machining Production" was started in 1985. This program comprises the development of four types of unmanned machines: three for prismatical parts on technological pallets 320 x 320 mm, 800 x 800, and 2 200 x 2 200 mm, one for cylindrical parts – machining features up to 200 mm diameter. Besides the machines, new CNC control for unmanned machining, tools and clamping devices and manipulators are developed.

The program will be finished in 1989 with production of developed machines, CNC systems and tooling. First prototypes have been already tested. Simultaneously, with two years delay, another 6 types of unmanned machines are developed to cover the whole range of machining technology including grinding.

Two typical machining cells developed in the state program are on the pictures: Fig. 1 illustrates two machines cell for turning flanges up to 200 mm diameter. Machines capable of turning, milling and boring from both sides (two spindle heads are used) are being served by one overhead gantry manipulator with 6 controled axes. New tooling system using small cutting blocks has been adopted. Tooling system on the machine comprises large tooling magazines for nearly 100 cutting blocks. Automatic change-over for a new part is used. Fig. 2 illustrates machining cell for prismatical patrs fixed on 800 mm technological pallet. Horizontal machining center is

equipped with large capacity tooling rack (for 144 tools) and chain-type pallet magazine (for 8 pallets). The cell may be integrated in-to automatic parts and tools flow in FMS.

Conclusion

The unmanned production area opens gradually. The necessary sophisticated modules are avialable already or they are developed intensively.

After a period of great FMS enthusiasm in all highly industrial countries – also in Czechoslovakia there are 6 FMS in operation – a cooling down period took place and much greater emphasis is being put on reliable flexible machining cells. The reasons are not only technical but namely economical and managerial. It is much easier to implement smaller units and build FMS gradually than to start with large complex FMS without previous experieces.

Czechoslovakia follows that pattern and through centralized financial sources invests money in the development of flexible automation. The state program for unmanned machines started in 1985, with its basic goals described in this paper, approves that.

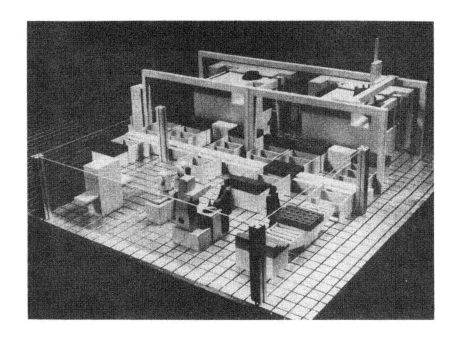

Fig. 1. Two machines cell for machining flanges up to 200 mm diameter.

Fig. 2. Machining cell for prismatical parts on 800 mm pallets.

Fig. 1.

Fig. 2.

PC's IN CAM EDUCATION

P. Kopacek, N. Girsule and R. Probst

Systems Engineering and Automation, University of Linz, Linz, Austria

Abstract. The paper starts with some considerations on CIM (computer integrated manufacturing) and especially CAM (computer aided manufacturing), which is the main topic of this paper. For efficient applications of CAM a well educated staff will be necessary. Therefore education in this field is very important. As a contribution to education in this field several software packages will be described. These packages illustrate the main fields of CAM namely the control of NC and CNC machines, machine feeding devices, transportation facilities, and storage devices.

Keywords. Education, CAM, simulation, digital computer applications

INTRODUCTION

The automation of discrete processes is one of the classical application fields of digital computers, in most cases microcomputers. In the sense of control manufacturing processes are of this kind. The automation of manufacturing processes started relatively early by controlling machine tools. First - approximately 20 years ago - for such purposes hardware controllers were used. But soon they were replaced by digital computers. Today the control of NC (numerical controlled) or CNC (computer numerical controlled) production machines is done by microcomputers. These machines might be the first step towards the so called "factory of the future" which will be characterized predominantly by total computer control.

CAM is defined as computeraided preparation of manufacturing including decisionmaking, process and operational planning, software design and artificial intelligence, and manufacturing with different types of automation (NC-, CNC-, DNC-machines and cells).

The main tasks of a CAM system are

- manufacturing
- assembly
- transportation
- storage

For the manufacturing process different types of machine tools as well as feeding devices for parts and tools including the changers for tools are necessary. In most cases today's assembly operations are carried out manually or by less flexible assembly machines. Flexible assembling requires the use of industrial robots (Kopacek, 1989). Transportation facilities are necessary for moving material and parts between the storage units and the production machines and assembly stations. Some examples are roller conveyors, shuttles, guided vehicles, shuttle cars and towlines. Various devices for storage of materials, products, and tools are available commercially today.

These devices could be assigned to these groups:

- potentially independent machine tools, assembly machines and inspection stations
- storage, transportation and orientation systems for parts and tooling
- an overall computer control system which coordinates all functions of the above

CAM is only a part of the "factory of the future". Other elements are CAD (computer aided design), CAP (computer aided planning), CAQ/CAT (computer aided quality control, computer aided testing) and PPS (production planning system). All these elements together form a complete CIM (computer integrated manufacturing) concept.

In general when a CIM system is installed the number of machine tool operators and setters goes down. But there is a need for people with new skills for example in maintenance and diagnostics.

EDUCATION FOR CAM

Because of the fact that a CAM system is complex a well educated staff is necessary for an efficient operation. Unfortunately education for CAM includes a lot of knowledge of the "classical" (Mechanical Engineering, Electrical Engineering...) as well as the "modern" (Electronics, Computer Sciences...) engineering fields. This might be illustrated by some related topics

- Kinematics
- Kinetics
- Control
- Hydraulics and Pneumatics
- Electrical Drives
- Expert Systems
- Artificial Intelligence
- Computer Hardware
- Programming and Programming Languages

215

The ways in which some of these topics should be introduced in educational programs need careful consideration. The Austrian government (Austrian Federal Ministry of Sciences, Austrian Federal Ministry of Education) for example starts with the development of an educational concept in automation in which the field of CIM is also included, too.

A short survey about effective methods for teaching basical knowledge of CAM will be given in the following.

EDUCATIONAL TOOLS

As pointed out earlier a very efficient tool in CAM education on all levels may be the PC. Generally speaking new abbrevations like

- CAT (Computer Aided Teaching)
- CBT (Computer Based Training)
- CML (Computer Managed Learning)
- CAL (Computer Assisted Learning)

have been generated lately. The software packages for such tasks are called "Teachware" or "Courseware". Especially for education in CAM and related fields two types of experiments are carried out. On one hand there is simulation by means of PC software, on the other hand there are control experiments using PCs. In this case the control computer is replaced by the PC.

Simulation

Nowadays simulation methods are increasingly used as a tool for planning manufacturing systems. For the design of the physical layout of machine arrangements simulation is necessary in particular. The user can study the material flow through the manufacturing process in an easy way and recognize e.g. bottlenecks. Modern simulation packages provide the generation of animated displays, giving users a

motion picture of their systems. Eventually, generalized interactive models will be available, permitting the user to stop model execution, incorporate changes, and continue model execution.

The instantaneous visual feedback appears to be an ideal method of training for the design, management control and operation of a flexible manufacturing system (FMS). For the simulation of discrete systems like manufacturing some software packages or simulation languages (e.g. GPSS, SIMSCRIPT, SIMAN, ...) are available today. For using such packages some experience is necessary. Therefore students of different fields (Mechanical Engineering, Computer Sciences, Industrial Engineering ...) need some preparatory training. For reduction of this time a special simulation package was developed at our institute.

This "Teachware"-package uses elements of SIMAN (Pedgen, 1986) with the animation package CINEMA - a commercially available software - as well as of PRODUL - a package developed at the Technical University of Ilmenau (GDR).

Disadvantages of SIMAN/CINEMA for educational purposes are

- complicated programming
- experience in simulation necessary
- too much features

Therefore in a first step the SIMAN software was extended by some additional modules (Angerer, 1989).

- SIMDES for system design'
- SIMGEN for generation of models and experiments
- SIMTST for testing and analysis

The connection of these new modules to SIMAN is shown in Figure 1.

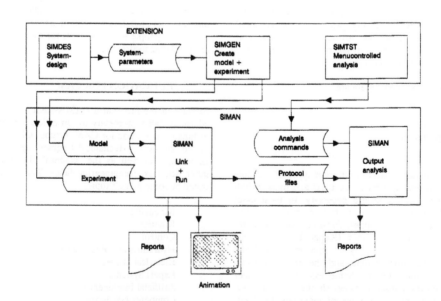

Figure 1: Structure of SIMAN with extended modules

In a next step the simulation package PRODUL was included in the modified SIMAN software. PRODUL offers some advantages especially in the theoretical field - the graphic and the animation are very simple.

The simulation system PRODUL is able to simulate a number of parallel sequences of tasks. These tasks can be done on different units. The workpieces in the simulated process must follow a well defined linear path through the different units. Each unit can process a defined number of workpieces at one time with a constant process period. Task of the simulation is to show flow of workpieces, filling of buffers, efficiency of the plant, pass through of workpieces, etc., and not dealing with changes of the workpiece itself.

Plants with fixed sequences of tasks give the best simulation results. It is possible to concentrate different tasks at one unit or to use different units for a distinct task. The sequence of tasks can be changed on one plant.

The program package includes edit functions to create and change files for the description of task sequences. The plant file editor supports plant files with a maximum range of 150 units and up to 300 tasks. At runtime up to four task files can be used. Simulation can be run in trace mode. Furthermore a protocol file is generated to use simulation data in external programs.

PRODUL uses the "Discrete-Event" model (Walther, 1975; Winkler, 1979; Kheir, 1988). The used algorithm simulates the transfer of one workpiece from one buffer to another executing one task. This action is named "elementary step". During simulation the program tries to execute elementary steps following distinct rules chosen by the user. Simulation is controlled by priorities for each defined task sequence.

The simulation of a simple transfer line consisting of three machine tools can be used as an example. The animation is shown in Figure 2.

In a very easy way it is possible to calculate various diagrams on manufacturing times.

Such examples have to be programmed by students of Computer Sciences as well as Mechanical Engineering in a laboratory course of CAD/CAM in a short time. This is only possible by using special educational simulation software.

Control

The control system for CAM in a CIM environment should be a highly flexible system satisfying the following conditions (Kochan, 1986)

- highly independent functions should be modularized
- modularized functions should be usable individually
- the total control system should take a hierarchical form made up of above mentioned modules
- the method of on-line and real-time control should be employed.

Usually control systems for CIM have a four rank hierarchy

- On the lowest level (machine level) control of different machine tools, handling devices (pick and place devices, robots, manipulators) as well as measuring or checking devices is necessary. These tasks are carried out by programmable controllers or small computers.
- On the next level (manufacturing cell level) controllers are necessary for supervision of machines, robots and measuring devices arranged in a cell.
- The next level includes controllers for the coordination of cells, complex assembly systems, conveyor systems and product inspection systems.
- The highest hierarchical control level is for coordination and optimization of the whole production (product management control level)

As a result of these facts mentioned above for education in the control field there are three main topics from the theoretical point of view:

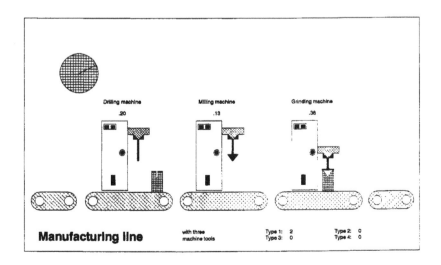

Figure 2: Example for SIMAN animation layout

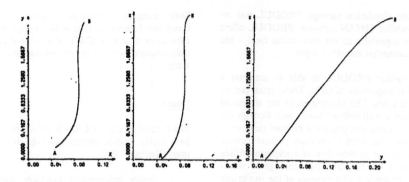

Figure 3: Robot motion between two points

- Simple control algorithms of motions (sequence control)
- Position control of axes of machine tools and robots
- Supervising control e.g. path planning of robots and transportation devices (robocars)

At our institute some software packages for control experiments using PCs are available. They were mainly developed by students during project works and diploma theses. As an example there will be a short description of two of them. The first example, described more detailed in Öhner (1989) is a simple sequence control for the model of the manufacturing system (Figure 2). A commercially available microcomputer controller was programmed by a PC.

The position control of a robot is a second example. Todays's robots are equipped with conventional PI-controllers. For robots of the future faster, digital control algorithms will be necessary. Therefore some software packages were developed in this direction. For robots with a kinematic structure RTT (rotational-translatorial-translatorial) and RRR simulation programs based on the equations of motion were created. By means of these programs analogue as well as digital position control algorithms can be tested (Kopacek, 1985). Figure 3 shows some results.

For control experiments on robots and transportation devices by means of PCs some equipments are available.

CONCLUSION

Starting with introductory considerations on CIM (computer integrated manufacturing) the necessity of a well educated staff is pointed out. Therefore some ideas for an efficient education are presented. Finally tools based on PCs are shortly described and discussed.

REFERENCES

Walther, U. (1975). *An Algorithm for Stochastic Simulation of Complex Processes.* Dissertation, TH Ilmenau (in German).

Winkler, U. (1979). *Stochastic Simulation of Complex Technological Processes and Relations to Control Devices.* Dissertation, TH Ilmenau (in German).

Kopacek, P., I. Troch and K. Desoyer (1985). Comparison of Digital Control Algorithms for Industrial Robots. *Preprints of the 7th IFAC/IFIP/IMACS Conference on "Digital Computer Applications to Process Control".* Vienna, pp. 291-294.

Kochan, D. (Ed.) (1986). *CAM, Developments in Computer-Integrated Manufacturing.* Springer, Berlin, Heidelberg.

Pedgen, C.D. (1986). *Introduction to SIMAN.* Systems Modelling Corporation, Calder Square, POB 10074, State College, Pennsylvania.

Kheir, N. A. (1988). *Systems Modelling and Computer Simulation.* Marcel Dekker, New York.

Angerer, G. (1989). *Simulation of Manufacturing Cells.* Diploma thesis, University of Linz (in German).

Kopacek, P. and K. Fronius (1989). CIM Concept for the Production of Welding Transformers. *Published in the same preprint volume.*

Öhner, S. (1989). *Control of a Manufacturing Line.* Diploma thesis, University of Linz (in German).

CIM: ON A NEW THEORETICAL APPROACH
OF INTEGRATION

T. Tóth and I. Detzky

*Department of Production Engineering, Technical University for Heavy Industry,
Miskolc, Hungary*

Abstract. It is expedient to study up-to-date theoretical means and practical resources of the automation of mechanical engineering through the
structure, functions, environment interrelations and the most important
features of Computer Integrated Manufacturing (CIM) systems because the
highest level of dual unity of automation for material- and data processing can be realized by them. Classification of CIM systems and their
subsystems is generally based on the automation levels and functional
roles, i.e. on the external features of the systems and subsystems. In
this paper the authors will prove that there is another classification
method well-established from the theoretical point of view, which determines the internal hierarchy of part manufacturing systems on the
basis of objective features. Starting from a uniformable model of these
systems the process optimization principles of discrete manufacturing
will also be generalized.

Keywords. Computer Integrated Manufacturing; automation; hierarchical
systems; computer aided process planning; CAD; CAM; optimization.

INTRODUCTION. THE CONCEPT OF CIM

According to a well-known interpretation
CIM (Computer Integrated Manufacturing) is
a high-technology approach to more efficient manufacturing. From the point of view
of practical realization CIM can be considered as a complex manufacturing system based
on intelligent electronics which is a proper
integration of manufacturing equipment, a
hierarchical information control system and
sophisticated operating softwares.

The following definitions (Williams,1985)
mainly emphasize the information processing
aspect of CIM:
- CIM is integration of systems for
  measurement and control of all production process operations with
  systems for managerial control of the
  factory and a wide range of corporate
  business function;
- CIM is a methodology for automating
  the gathering and sharing of information among computer systems to establish a closed loop in time feedback
  system for effective planning and
  control;
- CIM is the systematic implementation
  of computer technology within the
  company to achieve long term goals of
  maximizing efficiency, productivity
  and profitability.

In Fig.1. a general structure and information model of CIM systems are demonstrated.
The area with the legend "INTEGRATED
SYSTEMS ARCHITECTURE" being in the centre
of the Figure relates to a hardware-software
configuration of optional integration degree

including computer networks organized in a
hierarchical way. It means the concentrated
HW+SW resources of an information system
functioning as a main component of CIM and
a common data base belongs to it.The degree
of organization of integrated information
flow is the highest one between the integrated subsystems and the common data base.
The thick, hatched arrows relate to this.The
factory automation segment means the most
important tasks of integrated material
processing. Integrated data processing is
demonstrated by means of three further
segments, they are: design engineering,
manufacturing engineering, as well as manufacturing planning and control. The thick
black arrows refer to the information exchange between the neighbouring segments.
CIM, in its final extension,supposes such
a data structure that makes an optional
data flow possible towards any part of the
system. As it can be seen in Fig.1. the four
internal segments are surrounded by a nearer
and a farther information medium. Advancing
outwards in Fig.1. the degree of integration
is decreasing.

THE USUAL MAIN DIRECTIONS
OF INTEGRATION

The complexity of concept of CIM appears in
the fact that it includes integration of
three directions. They are as follows:

a) Fitting and joining the manufacturing
phases following successively in order
to maximize the product output (section in time, optimal manufacturing
program);

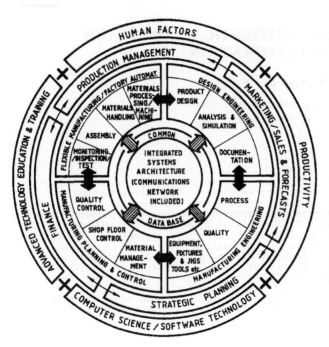

Fig.1. General structure and information
       model of CIM systems.

b) Integration of controlling levels
   superposed ("Organization pyramid");
c) Lateral integration of company functions.

In the "section" a) it can be examined how
the manufacturing phases following successively joint each other and how they can be
combined and concentrated. Here the most
important functions of CIM are to minimize
the stock being in the system and to maximize
the intensity of product output. To perform
these, external material transport coming
in time and sharply scheduled internal manufacturing are required. The elements of the
system (including the remain personnel) are
subordinated to the requirement of continuous
work (just-in-time=JIT).

In the "section" b) a well-organized, deeply-
structurized computer control hierarchy is
needed to the undisturbed manufacturing and
continuous motions of materials, as well as
semi-finished products. From the theoretical
analyses of competent publications (e.g.
Buzacott (1982), OTA Report (1984), Kusiak
(1985), Yeomans, Choudry and ten Hagen (1985),
Primrose and Leonard (1986), Scheer (1987),
Schuy (1987)) the inference can be drawn
that a CIM system can be dissected into 5
hierarchical levels. Advancing bottom-up
they are as follows:
   (1) The level of direct control of manu-
       facturing process (Process level);
   (2) Work Station Level;
   (3) The level of autonomous manufacturing
       units (Cell Level);
   (4) The level of manufacturing control
       subsystems (Center Level);
   (5) The level of company management (Top Level).

On the lowest level people were
ousted by automation in the industrially developed countries a
long ago. Programmable automation
has been expanding since appearance of microprocessors (intelligent controllers).

On the second level one or two
manufacturing equipment and a
quality controlling station create
a unit together. Typical controlling means are programmable part
units controlled by µP-s. For human
controlling and interrupting keyboard and display devices are used.
The scale of electronic processing
times and response times extends
from some seconds to some minutes.

As regards the third level it is to be
underlined that European and American
experts interpret the extension
of this level in different way but
there is a mutual understanding
in that the proper basic units of
CIM belong to this level. According
to the narrower European interpretation
a CIM unit of such level contains
generally more than two automatic
- mainly NC/CNC - machine tools,
special equipment, robots and
automatic material handling means.
The larger American interpretation
extend this level to workshops,
emphasizing that any of these units is
subordinated to the processing rythm of the
following unit. For computer controlling a
powerful professional microcomputer (e.g.
IBM PC AT) or a larger microcomputer (e.g.
microVAX) is required, the scale of response
times extends from some minutes to cca one
hour.

The fourth controlling level co-ordinates
all the manufacturing subsystems. If there
is such a controlling level in the given CIM
system, the access to the lower levels can
only be carried out through this center level
under normal operating circumstances. The
typical controlling equipment is a powerful
minicomputer (e.g. DEC VAX 11/7xx) or a main
frame computer with a simpler configuration.
Response time - in case of more sophisticated
decision sequences - can extend more hours.

The fifth level is, properly, the level of
production planning and control on the top
of such an "organization pyramid" in which
there are the previously described computer
hierarchy and manufacturing automation.The
typical equipment suitable for controlling
is a powerful large-size (main frame) computer with an appropriate configuration
(e.g. IBM 43xx). The decision and response
times can extend the intervals longer than
one day.

The "section" c) examines co-ordination of
the activities connecting to manufacturing:
it takes, really, possibilities of the lateral integration of company functions into
account. They are, among other things:CAE,
CAD, CAPP, CAM, CAST, CAQ and MRP. (The
interpretations of these abbreviations are
assumed to be known).

We attempt to demonstrate the three "sections" of CIM, in strongly simplified way,in Fig.2.

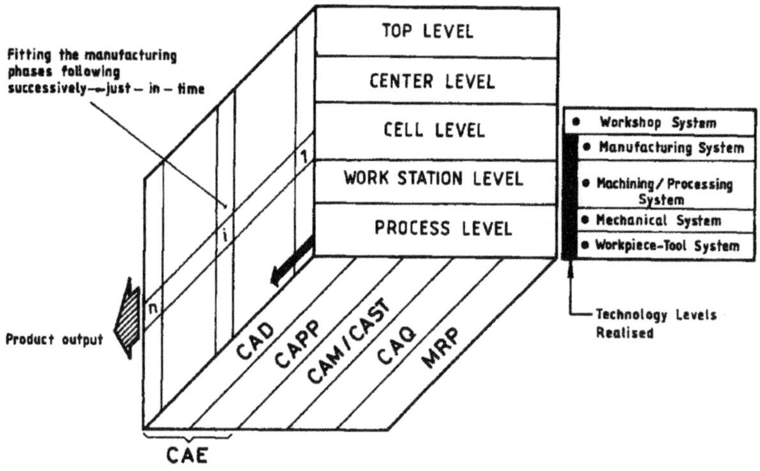

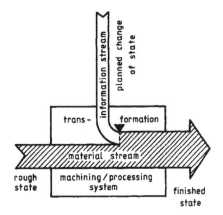

Fig.2. The three "sections" of CIM systems

SYSTEM-THEORETICAL APPROACH

Structure and process

Here if we speak about a CIM system we think of physically existing objects on all occasions. In this sense we also talk about reality of economical environment including manufacturing technology systems which is interwoven into an organization by three essential stream networks. They are as follows: material streams, information streams and value streams.

The functional substance of the manufacturing technology systems is that material transformation process which can automatically be executed along material streams.

The common substance of each such transformation is the objectivation on technological information related to the useful material properties and planned states (Fig.3). Here it has to be emphasized that rough material flowing in the system, independently of how it obtained its given state, in how many sorts of, and in what kinds of technological phases, resists a newer transformation as a "natural material" in all cases. Therefore a state of change always takes place according to objective laws of nature. Under such circumstances technology is able to influence the conditions system (constraints) only. Thus, in another approach, technology can be defined as the science of conditions systems of the laws of nature. In the last analysis all that what builds over the material processes in hierarchy is a function of possibilities and limits determined by objective laws and technical conditions, in a structural way too.

By means of this train of thought we wanted to verify that in the course of hierarchical

and structural decomposition of machining/ processing systems we have to start from those objective and more deep-seated physical conditions which determine the material processes of subsystems, i.e. the coming into existence of new,useful properties. Without these neither level of integration nor objective contents of flexibility can be explained.

Consequently, the CIM system hierarchy based on the automation levels and functional roles, i.e. on the external features of the systems and subsystems, has to be complemented with at least two further steps upwards from below in order to obtain an objective conception about adequate connections of functions and processes (c.f. with the right hand side part of Fig.2., proposed by Detzky (1988) and Tóth (1988)). This classification determines the internal hierarchy of part manufacturing systems.

Fig.3. The substance of material transformation process

Analyzing the technological processes

The authors demonstrate the analysis of discrete manufacturing technology processes through an example taken from the area of machining. This area represents high technology considering its specific applied means and it reflects the clearist system-models.

From the economical point of view processes of an autonomous technological system can be classified as follows:

- primary processes are those in which the object of work is present from beginning to end and some specific property or state of it changes in the expected way (cutting, heat treatment etc.);
- secondary processes are those in which the object of work is present but without changing any specific property or state of it and its relative position only is modified (part handling, storage measuring etc.);
- auxiliary processes are those in which the object of work is not present but the moments can be allocated identifiably to some object of work and they are necessary for generating the primary processes (preparing the means and information, machine-adjustment etc.);
- maintenance processes are those which cannot be allocated identifiably to a given object of work but they are directly needed for the predestined internal functioning of the technological system (energy supply, machine maintenance, etc.);
- environment processes are those which take place within the system but they are not reasonable for the object of work or the internal functioning of the system. Their necessity is reasonable for such aims which are determined by the superordinate environment (monitoring/controlling requirements, interventional constraints, etc.).

This division makes it possible to give an account of the costs of technological processes and to analyse them economically. Primary and secondary processes are together named basic processes. These have an important role in process planning and manufacturing programming.

In our example the quality of a certain machine construction is determined by the proper spatial relations and dimensions of the surfaces of parts in the overwhelming majority of cases. Thus, from the aspect of quality, a geometrical hierarchy of parts can be enforced in dissecting the processes in the first place. Accordingly, the following steps can be separated (see:Fig.4):

- operation group is that phase of the whole working process in which a certain part, starting from the rough piece, obtains its finished configuration suitable for assembling and to which some machine-group (manufacturing system) can be allocated in general;
- operation is that in which one of the characteristic configurations of the part having a common symmetry (shaped-group) is machined and to this, on the basis of symmetry, some kind of universal machine tool type (a homogeneous machine tool group) can be allocated;
- suboperation (partial operation) is that in which one of the shapes (surface groups) of a given configuration is machined and a typical set-up method and a clamping device set of common basis are attached to this;
- operation element is that in which only one surface of a certain shape is machined according to given dimension parameters and tolerances and to this some kind of tool type and tool path can be allocated.

- momentum is such a change of state which, in the geometrical sence, yields forming surface elements of a prescribed quality and cutting parameters are attached to this.

The enumerated hierarchical decomposition included only details of the different phases of the primary process but the realizing means allotable to them were also marked. The latters refer to the possible variety of secondary and auxiliary processes connecting with the different levels. These co-ordinations constitute the base of process planning.

Theoretical models

A significant part of practical and organizational requirements is formulated in the form of exact criteria for organizing the automated technologcal systems in a hierarchical way. The general requirements of the models valid for any level are as follows:

- Boundary surfaces of subsystems have to be defined in such a manner that the connections and streams through them should be expressed by means of mathematical (logical and arithmetical) variables.
- Structures of subsystems have to be synthetized in such a way that their common functioning can be described, on the base of independent disciplines if possible, by laws and constraints treatable mathematically.
- For the subsystems such optimum criteria should be interpreted which obtain a complete conditions system by means of optimization of the system above them.

Fulfilment and concretization of the requirements enumerated above depend on the adequate discipline applicable in the first place. In the part manufacturing exemplifying discontinuous processes these are as follows:
— operations research for manufacturing
— theory of operations sequences
— mechatronics for machine tools
— cutting theory.
Through the stepping we relate to a possible hierarchy of the system-models. We have to deal with the general requirement of optimization in more detail.

Technological optimization

Universality of the problem results from the fact that no special science can avoid optimum problems if it derives its products from aims of social origin. These have also long appeared in several methodological disciplines independently ("Mathematical Programming", "Operations Research", "Decisions Theory"). Technical problems can be divided into two large groups:

- Technological problems: all the possible processes of a given system existing physically have a subset which includs realizing possibilities of a product described in some kind of modul structure. From them we must select that process variant which is most suitable for prescribed aim exterior to the system.
- Design problems: all the possible models of a system of specific knowledge existing in thought have a subset which includes realizing possibilities of a function

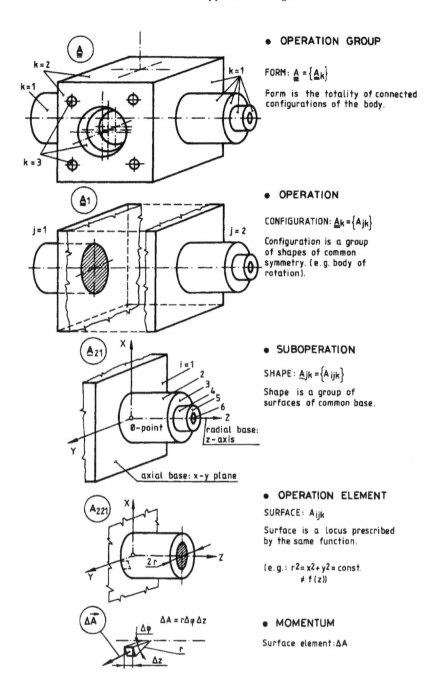

- **OPERATION GROUP**

FORM: $\underline{\underline{A}} = \left\{ \underline{\underline{A}}_k \right\}$

Form is the totality of connected configurations of the body.

- **OPERATION**

CONFIGURATION: $\underline{A}_k = \left\{ A_{jk} \right\}$

Configuration is a group of shapes of common symmetry. (e.g. body of rotation).

- **SUBOPERATION**

SHAPE: $\underline{A}_{jk} = \left\{ A_{ijk} \right\}$

Shape is a group of surfaces of common base.

- **OPERATION ELEMENT**

SURFACE: A_{ijk}

Surface is a locus prescribed by the same function.

(e.g.: $r^2 = x^2 + y^2 = const.$
$\neq f(z)$)

- **MOMENTUM**

Surface element: ΔA

$\Delta A = r \Delta \varphi \Delta z$

Fig.4. Relationship between process and geometry in case of machining

composed by some set of processes. From them we have to select that variant which is most suitable for a prescribed aim exterior to the field of specific knowledge.

The contents of the two categories, in spite of the symmetries of thought, are very different both theoretically and practically. Hence, there is a danger not to be underestimated that the methodologies built upon mathematical analogies confuse these categories in a perfunctory manner. Mentioning this in advance formal requirements of an elementary optimum problem will be outlined.

Thus, we need:
- variables (x), logical or arithmetical values of which are selectable and realizable independently of one another;
- functions or relations (M(x)) which are interpreted by means of the variables and given by the tools independently of the problem;
- functions or relations (A(x)) which are interpreted by means of the variables and given by the problem independently of the tools;

- **a function** (f(x)) which can be interpreted by means of the variables and it is determined by the **external aim** independently of both the tools and the problem.

Solution of the optimization problem, generalizing the symbols, can be formulated as follows:

1. X_I, totality of the technical field of knowledge has to be assumed, a vector of which, $x \in X_I$, will be the solution of technical problem.

2. $X_M \subset X_I$ is that subset which is needed for describing the technical tools disposable;

3. $X_A \subset X_I$ is that subset which is necessary for describing the technical problem;

4. $X_V = X_M \cap X_A \neq \emptyset$ is a subset of the solution variants possible technically;

5. $f(x \in X_V) \rightarrow opt. \rightarrow x_o$ is the optimum solution vector.

From the scheme it is easy to see that not the optimum-searching procedure itself, denoted by the last (5th) step, is primarily significant from the point of view of solution, but composing those sets, which have to be created in the previous steps.

In the Fig.5. the visual models of one-level and of multi-level optimization are demonstrated. (In the interest of simplicity feedback is not denoted.).

An example from cutting technology

The most up-to-date systems which represent high technology in production engineering are connected with cutting as a basic process. In Fig.6, as an example, we have summarized those characteristics by means of which the classes of technological systems can be identified. Along the secondary and auxiliary processes essentially similar kinds of levels can be separated on the basis of the models suitable for specific functions. Especially we emphasized importance of the additive objective function which comprehends the hiearchy of systems. For the sake of a uniform technological aspect we also propose to take into account the objective costs by additional equivalent times computing with the hourly costs of the direct environment.

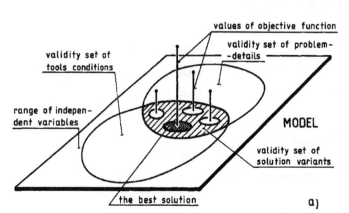

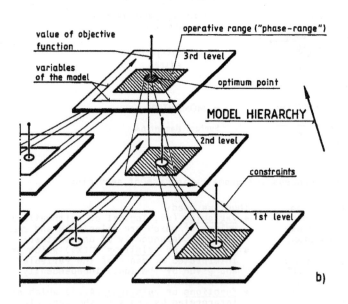

Fig.5. Visual models of optimization
a/ One-level model
b/ Multi-level model

REFERENCES

Williams, P.J.(1985). Computer Integrated Manufacturing (CIM). A Managament Perspective. Ryerson Polytechnical Institute, Toronto. (Draft papers).

Buzacott,J.A.(1982). The Fundamental Principles of Felexibility in Manufacturing Systems. Proceedings of the 1st International Conference on FMS, Brighton, United Kingdom.

Office of Technology Assessment (OTA) Report (1984). Computerized Manufacturing Automation. (Employment, Education and the Workplace). Congress of The United States, Washington.

Kusiak, A. (1985) Flexible Manufacturing Systems: a Structural Approach. International Journal of Production Research, Vol. 23, pp. 1057-1073.

Yeomans, R.W., Choudry,A., and ten Hagen, P.J.W. (1985). Design Rules for a CIM System. North-Holland, Amsterdam.

(cont.)

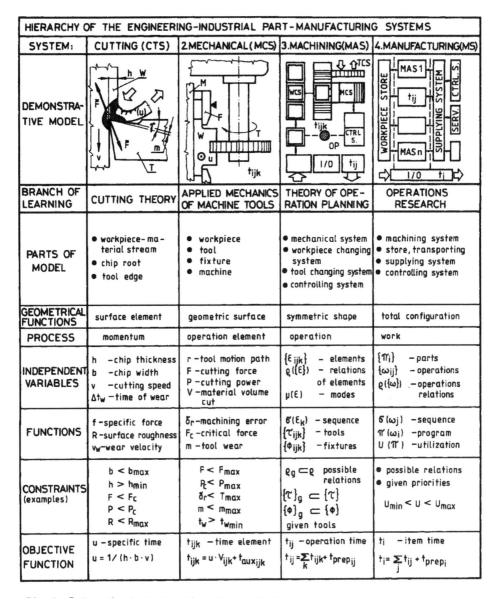

Fig.6. Internal hierarchy of part manufacturing systems, summarizing the most important characteristics in the case of the cutting processes

Primrose, P.L., and Leonard, R. (1986). Identifying the Flexibility of FMS Systems. Proceedings of the 26th International Machine Tool Design and Research Conference held in Manchester. Dept.of Mech. Engng., University of Manchester Institute of Science and Technology in Assoc. with MacMillan Publishers Ltd, pp.167-173.

Scheer,A.-W. (1987). CIM:Der computergesteuerte Industriebetrieb. Springer Verlag, Berlin-Heidelberg-New York-London-Paris-Tokyo.

Schuy,K.J.(1987). IBM Strategy and Directions. Mainz (Draft paper).

Detzky,I. (1988). Manufacturing Technology II. Theoretical Part 1. Educational Publisher,Budapest (in Hungarian).

Tóth,T. (1988). Computer Aided Process Planning in Manufacturing Technology. Doctoral Dissertation for Hungarian Academy of Sciences, Budapest (in Hungarian).

CONSTRUCTING PLANTWIDE MANAGEMENT AND INFORMATION SYSTEM

K. Yamashita

Computer Application Systems Department, Toshiba Co., Tokyo, Japan

Abstract The paper presents a conceptual overview, key technologies and experience in constructing plantwide management and information systems. The necessity of such a system is first described, followed by an outline of the system structure. Then, we present a millwide system, which has been in operation in one of the largest pulp and paper mill in Japan. This system employs key technologies, such as optimization algorithms, an expert system, relational database management, simulation techniques and local area networks. We also make several suggestions on constructing plantwide management and information systems based on our practical experiences in several plantwide systems. Finally, we discuss the future direction in which we expect these system to develop.

Keywords Plantwide system; millwide system; energy supply optimization; production schedule optimization; production planning expert system; relational database; large scale linear programming; LAN.

INTRODUCTION

Manufacturing and process industries in Japan are moving toward diversification of customer needs, increased varieties smaller lot and shorter lead time. Faced with this situation, industries must establish a flexible production system that minimizes the lead time for delivery and increase overall production efficiency. A plantwide management and information system (hereafter called plantwide system in short) can be an effective tool to cope with this situation.

The major problems concerning plantwide management, found in the above industies, includes the following.

(1) lack of commonly shared and accessible information on production, such as quality and quantity of product, material and product stock, status of related processes and schedule

(2) insufficient inter-departmental coordination in production activities

(3) lack of opimized operation

PLANTWIDE SYSTEM

Although there is no widely accepted definition of a plantwide system, our definition is as follows.

A plantwide system is one that supports management, supervision, and control tasks concerned with several production departments.

Functional configuration

In Fig. 1, such tasks are shown in a function configuration diagram. In an actual plantwide system, every function shown in Fig. 1 might not be realized, because of cost and time factors, as well as requirement and effectiveness which vary from plant to plant.

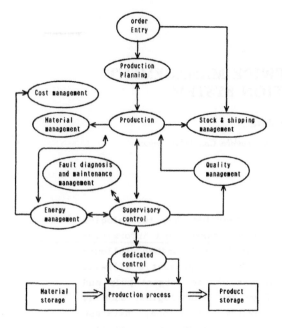

F i g. 1 Functional configuration of plantwide system

ESSENTIAL COMPONENTS

This section describes some of the essential components for implementing a plantwide system.

Hardware

Since the system concerns various plantwide activities and hence spreads throughout the plant, a local area network is indispensable to link the CRT displays, printers, PI/O and other peripherals of various departments. It also links computers installed in different departments.

Software

A commonly accessible database is a key component in plantwide systems. The database must be carefully designed to make access easy and quick.

Optimization algorithms are also necessary to improve productivity. For those problems, which are not mathematically expressible, an expert system approach is used. This approach is also effective in transfering human experiences and skills into the computer system. Man-machine interface, mainly through CRT displays plays an important role. A friendly and easy interface must be provided so that even older operators can readily adapt to

the system.

Humanware

Participation of the human being is indispensable to make a system successful. Our stance on plantwide system development is that the system can only be a support tool, to draw out the potential and intelligence of the human being. The information presented by the system stimulates the managers, staff and operators and to formulate methods to improve production efficiency.

REALISED MILLWIDE SYSTEM

Millwide system is a term used for plantwide systems in the pulp and paper industry. In this section, we present a millwide system, which has been developed in stages over the last decade.

Mill outline

The system was developed for Tomakomai mill of Oji Paper Co., Japan. This mill is the largest newsprint producer in the world. Fig. 2 gives a rough outline of the mill including power supply facilities. The outstanding features of the mill are given below.

1. Complex pulp mill, produces about 20 different types of pulp.
2. Extensive power generating facility has a steam power generating plant with 10 boilers and 15 turbine generators, and 10 hydraulic power stations with 19 turbine generators. The total power supply is 240 MW.
3. Ten paper making machines, producing 200 different types of paper products and totaling one million tons per year.

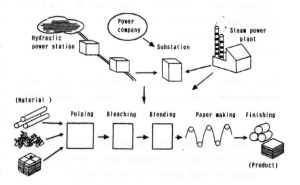

F i g. 2 Configuration of pulp andd paper mill

4. Total energy consumption is 250 MWH in electric power and 400 ton/H in process steam.

System objectives

The system is required to rearlize the following millwide tasks.

1. Optimization task.

(a) Pulp plant operation.

Fig.3 shows a simplified diagram of the pulping and blending processes. The objective here is to minimize pulp production cost which covers raw material, electric energy and bleaching chemicals. This involves optimal scheduling of pulp production and pulp blending by making the best use of buffer tank and optimizing bleaching chemical use.

(b) Energy plant operation.

Since the energy cost forms 20-30% of the production cost, substantial saving can be achieved by minimizing energy supply cost. Energy costs include electric company billings, boiler fuel and start-up and shut-down cost. Minimizing power supply cost involves steam power plant operation scheduling and optimal load dispatching to turbine generators.

2.Production planning.

The mill's operation is mainly based on customer orders. The planning department is required to provide monthly and weekly paper making schedule by taking into account various constraints. The task is so complicated, that in most cases, no plan is obtained that completely satisfies all the constraints, even by an experienced planner.

3. Millwide coordination.

This is one of the most important tasks of the system. The system has to provide functions that help operators know the status of the related processes and exchange operational information to each other, so that millwide coordination becomes easier.

4. Production cost calculation.

The continuous monitoring of production costs, calculated on various standpoints, provides timely information enabling effective measures to be taken to improve production efficiency.

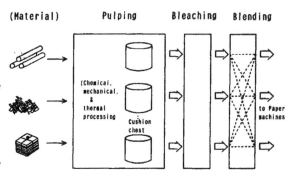

F i g . 3   Pulping,bleaching and blending processes

5. Production and quality database management.

The database is mainly to store historical data on production and quality and to provide easy access to the data.

Hardware system configuration

In Fig. 4, hardware system configuration is shown. The main elements of the system are five 32 bit minicomputers, one 16 bit minicomputer and two EWS's. The highspeed (10 Mbps) process LAN links 22 color graphic displays, printers and PI/O. In addition, there is an information LAN to which 32 character displays and printers are linked. These displays and printers are located at various production processes in the mill.

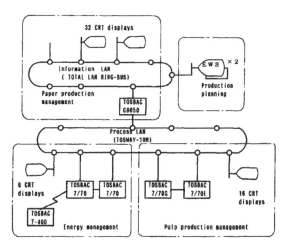

F i g . 4   Hardware system configuration

Table 1. Main functions of the system

Function group	Functions
Planning	Monthly paper making planning Weekly paper making scheduling
Optimization	Pulp production scheduling Pulp blending scheduling Pulp bleaching Power plant scheduling Steam turbine load dispatching Hydrauric turbine load dispatching
Production monitoring	Pulping progress monitoring Paper making progress monitoring Energy plant status monitoring
Cost caluculation	Pulp production efficiency calculation Energy consumption rate calculation
Supervision & control	Power demand supervision Turbine set point control Pulp quality supervision and control
Simulation	Pulping plant simulation Pulp blending optimization, case study Pulp production optimization, case study

System functions

In Table 1, system functions are described. It took eight years to develop the whole system. The construction started with the energy management block in 1980 followed by the pulp production management block in 1984, the paper production management block in 1986 and the production planning block in 1986. The system was completed at the end of 1988, although it might grow further.

KEY TECHNOLOGY

This section presents the main functions of the system and the technology applied to realize these functions.

Energy optimization(Yamashita,1983 & 1984)
Optimization functions are developed for steam power plant and hydraulic power plant operations.
The optimization problems are

(1) Boiler and turbine scheduling
(2) Steam Turbine load dispatching
(3) Hydraulic turbine load dispatching

These problems are formalized in a mathematical programming framework by using linear programming (LP), nonlinear programming (NLP) and dynamic programming (DP). For instance, problem (2) is formalized as follows :

$$\text{minimize} \quad f(x)$$
$$\text{s.t.} \quad h_i(x)=0, \quad i=1,\dots,m \qquad (1)$$
$$g_j(x) \leq 0, \quad j=1,\dots,l$$

where x_i is the independent variable vector, of which elements correspond to turbine steam load. $f(x)$ is the objective function consisting of fuel cost. $h_i(x)=0$, $i=1,\dots,m$, are the equality constraints, representing power balance, steam mass balance at each pressure level, and turbine and boiler mathematical models. $g_j(x) \leq 0$, $j=1,\dots,l$ are the inequality constraints representing various operational restrictions, mainly upper and lower limits.

The algorithm used is the multiplier method with conjugate gradient method. The optimal load dispatching is executed every five minutes and the optimal solution is sent to digital turbine controllers as set points.

Pulp production schedule optimization
(Yamashita, 1988 and Hara, 1988)
Fig. 3 shows a simplified flow diagram of a pulp mill. Under the condition that the paper production schedule for the paper making machines is given, it is required to provide a 24-hour pulp production schedule and pulp blending schedule that minimize cost. The problem formulation is done for each scheduling by using linear programming approach, i.e.

$$\text{minimize} \quad z=\sum_{i=1}^{p} c_i x_i \qquad (2)$$

subject to

$$\sum_{i=1}^{p} A_i x_i = b_0 \qquad (3)$$

$$B_i x_i = b_i, \quad x_i \geq 0, \quad \text{for } i=1,\dots,p \quad (4)$$

In the case of pulp production schedule, the variable vector x_i represents pulp production rate and cushion chest level at the i-th time segment, which is set to two hours. Consequently, p is equal to 12 .
The objective function (2) represents the production cost, including power cost and material cost. The constraint equation (3) represents the cushion chest models and

covers more than one time segment, which makes the size of the problem large. Conversely, each equation in (4) is contained in a single time segment. These constraints represents power consumption limits, production capacity limits and so on.

Even after careful simplification , the size of the problem became 2,400 variables and 6,002 constraints for pulp blending scheduling and 492 variables and 624 constraints for pulp production scheduling.

By considering the special strucure as in Eqs. (2)-(4), we applied Dantzig-Wolfe's decomposition principle(Dantzig, 1961). The scheduling problems are solved every 8 hours in real-time, for which several measures were needed to be devised and implemented.

Paper production planning

As in the mill outline, the mill has 10 paper making machines and produces more than 200 types of paper products each month. Therefore, mathematically, there are 10^{200} combinations for the production schdule. The production planning is done for the coming month at the end of every month. The major constraints in making the plan are the followings :

(1) production volume
(2) delivery date
(3) production order: product A must be/ must not be made just before/after product B on the same machine.
(4) production interval: product A must not be made within some specified period after product B is made on the same machine.
(5) product combination: product A and product B must not be made in parallel.
(6) pulp supply limit
(7) power supply limit

Note that (5)-(7) are cross-machine constraints. It is hardly possible to formalize this problem in a mathematical programming framework. We, therefore, took the following approach. First, we developed a planning support system on an engineering workstation (EWS), and provided planning support tools with a

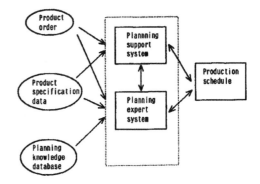

F i g. 5 Paper production planning system

frienndly man-machine interface. Then, in the second step, we developed an expert system that automatically formulates production plans, satisfying the constaints. These two systems can also interact each other to allow participation of the planners. Fig. 5 shows the planning system configuration. Main functions of the support system are :

(1) schedule display and modification
(2) production demand display and modification
(3) pulp mass balance calculation and display
(4) energy balance calculation and display
(5) expert system execution request

Fig. 6 shows a simplified format of resulting production plan, where the horizontal axis corresponds to time and vertical to machines. Paper products are distinguished by different shading.

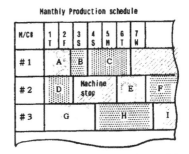

F i g. 6 Paper production schedule

Database management

In order to make database access easy and fast, we applied both an ordinary file management system and a relational database management system.

SYSTEM BENEFITS

We have delivered several plantwide systems to our domestic customers(Kishida, 1988). The following is a summary of the benefits of these plantwide systems.

(1) Plantwide coordination and cooperation became easier through commonly shared information and CRT displays.
(2) Production and energy costs are reduced by optimization measures.
(3) Flexibility to order changes is increased, i.e., production schedule changes became easier and faster.
(4) Quick and timely action can be taken to maintain higher productivity through production monitering and production cost calculation.
(5) Plant operators can take actions based on a wider scope, as more information becomes available.
(6) Managers can get more accurate production data and management decisions can be made easier and quicker.

KEY FACTORS FOR SUCCESS

(1) Organize a project team consisting of people from related sections, including operators.
(2) Make problems clear and set targets for the system.
(3) Make the system and functions as simple as possible.
(4) Consider reorganization of the plant and change process operations, if necessary.
(5) Design functions that allow operator intervention so as to make the system flexible and to make best use of human experience and ability and to motivate workers.

CONCLUSION

The paper presented several aspects of plantwide systems, such as system configuration, essential components, key technologies, system benefits and keys to successful construction. An actual millwide system has also been described.

Through our experiences in development of several plantwide systems, we believe we should keep in mind the concept 'Simple is best and beautiful'. Functional simplicity brings about easier operation, cheaper software maintenance cost and more flexibility in changes in plant operation and layout. The system provides abundant information on plant operation. Careful analysis of the accumulated data will improve return on investment. Optimization technology is a powerful tool for productivity improvement. The expert system we applied to production planning proved to be better than a skilled palnner.

The plantwide system is now moving toward including image database management to improve maintenance of the plant facility and equipment, and for less-paper plant management.

ACKNOWLEDGEMENT

The paper is based on several plantwide system projects. The author would like to express his sincere gratitude to the members who participated in the projects, in particular to Oji Paper Co..

REFERENCES

Dantzig, G.B. and P. Wolfe (1961). The Decomposition Algorithm for Linear Programming, Econometrica, Vol. 9, No. 4.
Hara, H., K. Yamashita and T. Watanabe (1987) On-Line Application of Large Scale Linear Programming to Pulp and Paper Mill by Mini-Computer, Proceedings of the 13th IFIP Conference on System Modelling and Optimization, 662-671
Kishida, M. and K. Yamashita (1988). Construction of Millwide System, Toshiba Review, Vol. 43, No.11, 852-856.
Yamashita, K. and T.Watanabe (1983). Real-Time Optimal Energy Management by Mathematical Programming in Industrial Plants, Proceedings of the 11th IFIP Conference on System Modelling and Optimization, 859-868.
Yamashita, K. and T. Watanabe (1984). Interactive Powerhouse Operation Planning Utilizing Optimization Methods, Proceedings of the IECON '84, Vol. 2, 691-695.
Yamashita, K. (1987). Real-Time Application of OR Methods in Industries Using Mini-Computers, The 11th Trienial Conference on Operations Research, Buenos Aires.

GROUPING PARTS TO REDUCE THE COMPLEXITY OF ASSEMBLY SEQUENCE PLANNING

N. Boneschanscher and C. J. M. Heemskerk

*Delft University of Technology, Laboratoire for Manufacturing Systems, Delft,
The Netherlands*

Abstract

A method of grouping parts to reduce the complexity of assembly sequence planning is presented. A
group, or cluster, consists of parts that belong together logically, because they have assembly features
in common, through which they can be assembled as a group. A key assumption is that the sequence of
assembly within a group may not be interrupted by parts from outside that group. Clusters introduce
a possibility to represent assembly sequence plans more abstractly, reducing the number of evaluations
that have to be made in the search for the optimal sequence. Three cluster types are defined: sorts,
stacks and layers. It is shown that a typical product contains several of these part groups, resulting in
a very simple sequence representation. A system for automatic detection of clusters was implemented
successfully.

Keywords

Assembling, Automatic Programming, Automation, Heuristic Programming, Industrial Robots.

1 INTRODUCTION

In the Laboratory for Manufacturing Systems of the Delft
University of Technology, a research project is underway
into automatic, off-line generation of programs for Flexible Assembly Systems [Heemskerk, 1987]. Part of this research focusses on the automatic generation of assembly
sequences.

Until now, assembly sequences are generated by hand. Because of the complexity of the problem, a human planner
will use heuristics and focus on one reasonable assembly
sequence. In our project, we want a sequence plan to
leave options open, so that a scheduling mechanism can
decide for the best sequence, depending on the actual circumstances (e.g. resource availability). A mechanism to
examine many sequence alternatives in parallel can thus
help to improve the quality of sequence plans, either as a
stand-alone automatic system, or -more likely- as an interactive tool to support the human planner.

In this paper a technique is introduced that reduces the
complexity of sequence planning by grouping parts into
clusters. Instead of having to deal with individual parts
all the time, clusters allow the use of a step-wise approach,
starting from a more abstract and less complex description.

In Section 2 the term 'cluster' is defined and some theoretical implications of the introduction of part groups are
discussed. Section 3 describes the three cluster types we
have discriminated so far, and how they can be detected
in a product. Section 4 presents an example of cluster detection in a semi-industrial product: the Cranfield Benchmark. The implementation of a prototype Cluster Recognition System (CRS) is described in section 5. The last

section presents some conclusions based on recent experiments with the CRS prototype.

An extensive description of the complete system discussed
in this paper is given in [Boneschanscher, 1988].

2 THEORETICAL EFFECTS OF CLUSTERING

2.1 Introduction

In determining the optimal assembly sequence for a given
product, a problem is the combinatorial nature of the actions involved in the decision process: even if the case is
restricted to assembly of only one part at a time, the number of alternatives increases drastically with the number
of parts involved. A product consisting of 10 parts can in
theory be assembled in 10! (over 3 million) different ways.
In practical cases, on one hand, the number of sequence
alternatives is even larger because simultaneous assembly
of some parts should be possible. On the other hand, the
structure of a product makes many of these sequences im-
probable, irrelevant or just impossible. Grouping parts into
clusters could help to focus the attention on the more probable sequences.

First we have to define a cluster:

*A cluster is a group of parts that belong together logically
because they have assembly features in common, through
which they can be assembled as a group. A key assumption
is that the sequence of assembly within a group may not be
interrupted by parts from outside that group. Depending on
the type of cluster, the sequence of assembly of the parts in
the product may be fixed, or can be determined at a later*

stage.

In the following paragraph some effects of clustering are illustrated using a representation called the Assembly State Transition Diagram (ASTD), and a related representation, the Layered Assembly State Transition Diagram (LAST-D) [Heemskerk, 1988]. In this representation, the *nodes* represent the static states of the parts in the assembly. The *arcs* represent the assembly processes to get from one state into the next. Figure 1 shows an example of an ASTD for a product with four parts. The parts are: Cap (A), Stick (B), Receptacle (C) and Handle (D). The example stems from [Homem de Mello, 1986].

In the remainder of the report, it is assumed, that parts are assembled one by one. This assumption causes the ASTD to be conveniently arranged, and will simplify the understanding of the concept of clustering. Please note however, that neither the concept of clustering nor the ASTD representation is restricted in this sense.

2.2   Advantages of Clustering.

Imagine a product with five parts: A, B, C, D and E. In theory there are 5! = 120 assembly sequences. Usually, many of these sequences are not relevant. Imagine for example that, for some arbitrary reason, the parts B, C and D have to be assembled right after another. The sequence B - C - A - D - E now is irrelevant, because part A interrupts the sequence of assembly of B, C and D. Instead of marking each sequence that contains such an interruption, we could group B, C and D into a cluster. In stead of 120, the number of sequences reduces to 36.

A convenient way to represent these sequences is to use a layered ASTD (LAST-D) as depicted in Fig. 2. In the top layer, a new part F is introduced that represents the assembly of the cluster. The composition of the cluster is 'hidden' in a lower layer ASTD. In the parent diagram, each transition with a double line represents the assembly of the fictive part F, acting as a *call* to the child diagram. During the call, the parts of the cluster are added to the already existing subassembly according to one of the assembly sequences depicted in the child diagram. Each assembly sequence represented in the parent diagram (each path from bottom node to the top node) contains exactly one call.

The first advantage of clustering is clear: It provides a tool to represent assembly sequence plans more abstractly. The complexity of evaluating assembly sequences depends directly on the number of nodes and transitions of a state representation (e.g. the accessability criterium checks transitions).

In some cases, the assembly of parts inside a cluster can only be done in *one* distinct sequence (e.g. because they are stacked on top of each other). The assembly of the cluster degenerates to one fixed sequence, but the abstraction is still worthwhile, because the cluster can occur many times. In other cases, the sequence of assembly inside the cluster is indifferent, i.e. *any* sequence will do (e.g. when putting similar components on a PCB).

Here is the second advantage of clustering: Clusters present a way to introduce specialized handling strategies for special product structures. Finding the exact sequence can even be postponed until real-time, while at the same time it is still possible to assure the overall strategy.

2.3   A Restriction

An important restriction of the clustering concept is illustrated in Fig. 3. Here, the Stick (B) and the Receptacle (C) have been grouped into a cluster E. The LAST-D shown in Fig. 3 is obviously less complex than the single level ASTD shown in Fig. 1. The first reason for this is that the layered representation better reflects the product structure (a good indication for this is that now we only need to mark three transitions as being technically infeasible, instead of ten transitions in the single level ASTD).

The second reason is that by clustering we have eliminated some technically unattractive, but theoretically possible sequences, because we assume the assembly sequence within a cluster to be noninterruptable. Thus, sequences B - A - C - D, B - D - C - A, C - A - B - D and C - D - B - A in Fig. 3 have been excluded. However, there may be situations where this exclusion is not optimal. The clustering concept should be regarded as a powerful but not infallible heuristic.

3   A FRAMEWORK FOR CLUSTER RECOGNITION

3.1   Introduction

This section discusses the integration of the theory of clusters into a complete system that can recognize clusters of various types. To be able to detect clusters, an appropriate product model is needed. Which information should be contained in this model is described in section 3.2. Section 3.3 presents the cluster types distinguished so far. This list is not extensive, but the complete system as discussed in section 3.4 has a modular structure, so recognition mechanisms for new clusters types can be implemented easily.

3.2   The Product Description

For the detection of clusters, an appropriate product model is required. The proposed model consists of a network of parts, and relations between parts. Figure 4 shows an example of a network, for the case of the Cranfield Benchmark [Rathmill and Collins, 1984]. The figure depicts a simplified version of the Cranfield Benchmark, consisting of nine parts. This product is also used in the following section, as an example to demonstrate the detection of clusters from an existing product.

The parts are listed below. The labels of the parts correspond to the names of the parts in the actual product depicted in Fig. 4, and to the names of the nodes in the relation network depicted in the same figure.

- SP1, SP2 = Side Plate
- BL = Spacer Block
- H = Handle
- PEG1 ... PEG4 = Spacer Peg
- SH = Shaft

Each relation in the network has an associated relation *type*. Currently two relation types have been defined:

- Peg / Hole contact
- Plane / Plane contact

This standard set of relations can be expanded very easily. The direction of assembly of a part relative to another part can be extracted from the relation type definition. From the exact positions of the parts involved and from the actual position and orientation of the relation, the absolute direction of assembly can be computed.

3.3 Some Important Criteria for Detecting Clusters

At the moment, three cluster types are defined:

- Sorts
- Stacks
- Layers

In a product, there usually are parts that have a similar or identical shape, often combined with similar functional specifications. Usually these parts will be handled with one specific gripper, which means that the parts ideally should be mounted one after another. Therefore, the sequence of assembly of these parts should not be interrupted by any other part in the product. These similar parts can be grouped into a *sort* cluster. However, to secure that the parts selected *can* actually be assembled one after another, the candidate parts are checked on similar relation properties. In summary:

- All parts in a sort cluster should have a similar or identical shape
- All parts in the cluster should have similar relations at similar relative positions

Often, the sequence of assembly of some parts in the product is fixed because the parts are stacked in some way (e.g. parts that are secured by bolts or other connective parts). At the moment, the detection criteria for *stack* clusters are:

- All parts related to *connective* parts are considered candidate stack cluster parts
- Each candidate part should at least have two directly related parts with relative assembly directions that contain an oppositely directed component

Occasionally, an assembly is characterized by a group of parts, that is sandwiched between two parts. The group in between is called a *layer* cluster. The parts in a layer cluster have the following properties:

- Each layer part has at the most two related parts by which it is sandwiched
- The direction of assembly of the two 'sandwich' parts relative to each of the candidate layer parts should have an oppositely directed component

3.4 The Complete Model

The mechanisms described in the previous paragraph form only a part of the complete framework for determining clusters in a product. The total system also has to account for the following situations:

- *Nested clusters:* It is possible that clusters form part of another cluster. An example is a stack that is secured by two bolts. These bolts are first detected as a sort cluster, after which they are treated as a single connective part.

- *Overlap:* There can be an overlap between different clusters. Some parts can be member of different clusters at the same time.

- *BOM analysis:* All sort clusters can be recognized from the bill of materials.

From the above properties, the complete framework can be defined. The sort clusters are detected directly from the bill of materials (BOM). The other cluster types are detected in a loop. The first action in each loop is to detect all clusters of different kinds in parallel. Then all clusters are checked for overlaps. If there are overlaps, the clusters are separated. The result is the forming of ideally large groups of clusters that do not overlap.

For each group of non-overlapping clusters, a new reduced relation network is produced from each of the existing relation networks. During reduction, parts that first appeared as separate entities now are eliminated, and replaced by one new, *imaginary* part (a cluster) with similar relations to 'outside' parts, as the parts that are member of the cluster. These new relations are not obvious, because they have to combine existing relations between more than one part inside and related parts outside the cluster.

After the reduction, the detection loop starts over again, for each reduced relation network, until no more clusters can be detected. The output is formed by the set of all reduced relation networks, augmented by the clusters detected.

4 AN EXAMPLE OF CLUSTER DETECTION

In this section the power of the clustering concept is illustrated using the simplified version of the Cranfield Benchmark (Fig. 4). The example will show that a typical product will contain more than one cluster and that the detection of all clusters will usually take more than one loop.

To make the example not too complex the simplified version of the Cranfield Benchmark is used, consisting of nine parts. This product has already been depicted in Fig. 4. The complete Benchmark contains eight more parts (small locking pins), that would make the reduction in the number of sequences more impressive, but would not illustrate any new basic ideas.

From the network and the description of the parts involved, it is clear that the first action is the detection of the sort clusters (SOCs):

- SOC1 = PEG1, PEG2
- SOC2 = PEG3, PEG4

The reduction of the relation network is done by eliminating the parts that occur in the clusters detected, and expanding the network with the new labels of the clusters, such that they can be treated as single parts. The relation network is transformed to the state depicted in Fig. 5.

Next, it can be seen that with the detection of two connective parts, in the form of clusters, there are two stack clusters, STC1 and STC2, that correspond with the definition for the direction of assembly of the parts in the clusters. For reasons of avoiding an overlap, the side plates are not considered part of the clusters. This has the advantage that the network can be reduced more effectively without introducing parallel reduced networks. At this point, there are no layers that can be detected. The result of the detection of stack clusters (STCs) is:

- STC1 = SOC1, BL

- STC2 = SH, H

The reduced network resulting from the detection is shown in Fig. 6.

After the reduction of the network by introducing the two stacks, it can be seen, that 'parts' STC1, STC2 and SOC2 correspond to the definition of a layer, according to the number of relations they have with other parts, and the direction of assembly relative to these parts. The result of the detection is:

- LAC = STC1, STC2, SOC2

The detection of the layer results in a reduced relation network, depicted in Fig. 7.

From the reduced relation network shown in Fig. 7, no other clusters can be detected, so that the reduction is complete.

Resuming, the following clusters are found:

- SOC1 = PEG1, PEG2

- SOC2 = PEG3, PEG4

- STC1 = SOC1, BL

- STC2 = SH, H

- LAC = STC1, STC2, SOC2

With the reduced relation network depicted in Fig. 7, which is the basis for the construction of the parent ASTD, and the clusters, which form the basis for the child ASTD's, Fig. 8, depicting the complete LAST-Diagram can be constructed.

Other examples have been tested too. From these tests it can be concluded that combinations of clusters of type sort, stack and layer occur frequently. Imagine the case of a gearbox, consisting of a house and a lid, that secure three gear shafts, by means of two bearings each. The three top bearings are pressed in the lid, the other three are pressed in the house. The house and lid are secured by four bolts. First the four bolts and the two groups of three bearings each are grouped as sort clusters. Then, the three shafts are only secured by the two bearing groups, which, taking into account the assembly directions, make the shafts a layer cluster. The fifteen parts are then reduced to only six (imaginary) parts. Using clusters, the number of relevant sequences reduces from $15! = 1.3 \times 10^{12}$ to 3.7×10^6.

5 A PROTOTYPE IMPLEMENTATION

A prototype Cluster Recognition System corresponding to the framework described in the previous paragraph is being implemented on a MicroVAX in the C language.

The most important part of this system is the framework for detecting clusters. Within this framework it is very easy to either implement more detection criteria for existing cluster types, and to implement criteria for new defined clusters.

Another advantage of the modular framework of the system is the fact that new cluster types can be evaluated on the basis of an interaction with the designer.

Three cluster types are being implemented at the moment: sorts, stacks and layers.

At this moment, part of the system still works interactively, so no computation times are available yet. The power of clustering can be expressed in numbers though: Theoretically, for the Benchmark consisting of nine parts, the number of sequences is $9! = 362,880$. The LAST-Diagram of Fig. 8 only holds 576 sequences.

6 CONCLUSIONS AND FUTURE WORK

6.1 Conclusions

The merits of determining clusters in a product are obvious: The number of sequences that are left for consideration is reduced drastically, and therefore the determination of all technically possible sequences is less time-consuming. Pruning the tree of possible sequences after determination of all clusters in a product will further reduce the number of relevant sequences.

The detection of a cluster should however be done very carefully. This is caused by the fact that the sequence of assembly of the parts in a cluster may not be interrupted. A wrong cluster therefore may introduce undesirable restrictions on the assembly sequence.

The model for determining clusters is modularly built. New criteria for existing cluster types or even new cluster types can be implemented easily in the existing framework. It is possible to evaluate the consequences of a new cluster type by defining an user-interaction with the model, before finally implementing the criterion on which to determine the new cluster type. The first version of the prototype was in fact built this way, and we found this feature very helpful in organising our thoughts.

6.2 Future Work

Although the framework for the model was implemented successfully, the number of detection criteria for a cluster is still relatively small. A more thorough investigation should be started for defining more cluster types and more criteria along real world products.

Some of the topics to be considered in this respect:

- The definition of more relation types

- The use of functional specifications of parts for cluster recognition, such as a frame or base function or the intentional degrees of freedom.

An interface to a CAD system will be built, including a mechanism for automatic feature recognition.

In order to avoid overlaps of clusters, that introduce parallel reduced relation networks, and therefore parallel ASTDs, it might be useful to split a part, that is a member of two different clusters, into two subparts that do not have an overlap. The implications of such an action are under consideration.

The CRS will be used in a much larger assembly planning system, that in turn is only a part of the Delft Intelligent

Assembly Cell (DIAC).

REFERENCES

Boneschanscher, N.: [1988] Grouping parts for the planning of assembly sequences. Technical Report no. WPS-88.026, Laboratory for Manufacturing Systems, Delft University of Technology, Delft, The Netherlands.

Heemskerk, C.J.M.: [1987] Programming an intelligent assembly cell. In Proceedings of the first European Symposium on Assembly Automation, Veldhoven, The Netherlands, March 1987.

Heemskerk, C.J.M.: [1988] The assembly state transition diagram, a representation for assembly sequences. Technical Report no. WPS-88.047, Laboratory for Manufacturing Systems, Delft University of Technology, Delft, The Netherlands.

Homem de Mello, L.S. and A.C. Sanderson: [1986] And/Or graph representation of assembly plans. Technical Report no. CMU-RI-TR-86-8, Robotics Institute, Carnegie Mellon University, Pittsburgh PA.

Rathmill, K and K. Collins: [1984] Development of an european benchmark for the comparison of assembly robot programming systems. In First Robotics Conference, Brussel, june 1984.

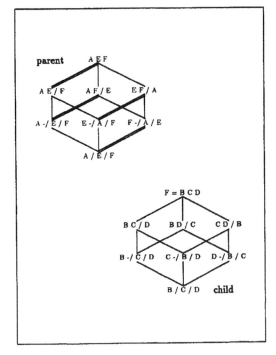

Figure 2: LAST-D Representation for a product consisting of five parts A B C D E, that has a cluster B C D.

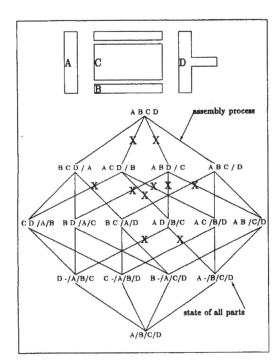

Figure 1: Example of an ASTD, for the case of the glue-dispenser, consisting of four parts. The assembly processes marked with a cross are not legal.

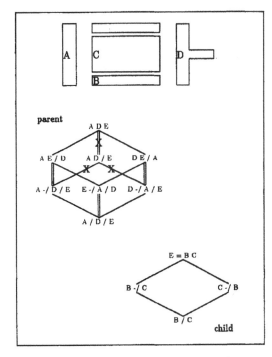

Figure 3: LAST-D representation for the glue-dispenser, that has a cluster B C. The assembly processes marked with a cross are not legal.

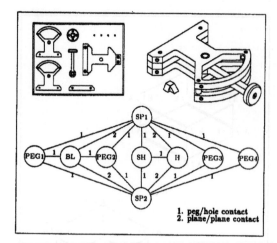

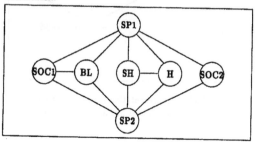

Figure 4: Example of a relation network for the Cranfield Benchmark.

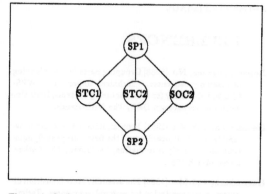

Figure 6: Reduced relation network after the detection of stacks.

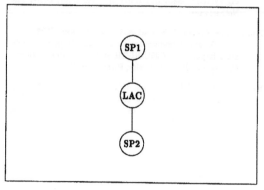

Figure 5: Reduced relation network after the detection of sort clusters.

Figure 7: Reduced relation network after the detection of layer clusters.

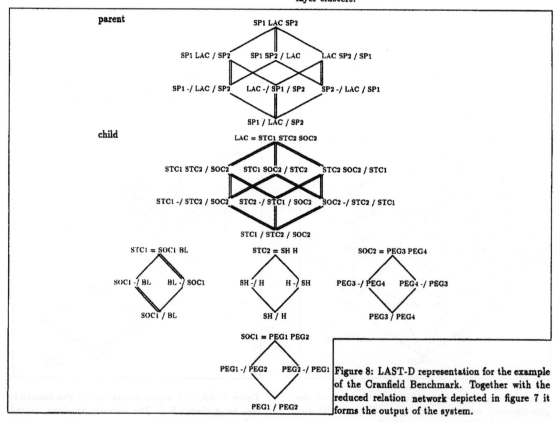

Figure 8: LAST-D representation for the example of the Cranfield Benchmark. Together with the reduced relation network depicted in figure 7 it forms the output of the system.

PARAMETER ADAPTIVE CONTROL STRATEGY FOR THE CYLINDRICAL SURFACE GRINDING PROCESS

A. Fuchs and R. Isermann

Institute of Control Engineering, Technical University, Darmstadt, FRG

Abstract.

This paper presents a new strategy of process control in cylindrical surface grinding. A self tuning control concept with a stop-measure and force cascade has been developped. Based on a mathematical model the on-line identification of grinding process parameters allows to design a control system, which adapts to the actual machining behavior. Based on data from a real process identification a simulation on a digital computer shows, that 25% of grinding time can be saved and the influence of abrasion and wrong process-adjustments can be eliminated.

Keywords. Adaptive control; Grinding; Machine tools; Modeling; Parameter estimation.

INTRODUCTION.

One of the major goals of CIM is to improve efficiency and quality of manufacturing through increasing computer control of production systems. Since 1960 research efforts on process control of grinding machines can be observed, Werner (1973). This leads to a more robust and quick-acting process behavior and an disturbance elimination. In the literature one distinguishes between

- ACC "Adaptive Control Constraint" and
- ACO "Adaptive Control Optimization"

ACC Systems control the grinding process to constant cutting forces or constant cutting power, see Saljé (1978). Mostly simple control loops with P-controller are applied, using the normal-force $F_N(t)$ as control-signal and the feed-rate $\dot{x}_F(t)$ as actuating signal. Tönshoff (1986) reports an adaptive force controller based on parameter estimation. The disadvantage of such ACC Systems is that workpiece roundness variances can't be eliminated by only force control. The reason is, that if the normal cutting-force is constant, the X-axis servo-system follows the profile of workpiece with a constant depth of cut.
ACO Systems are optimizing the production time or the production costs under the constraints quality and tool abrasion. Until now the actual process condition are adapted to a static abrasion- or quality-model using the information of a couple of grinding cycles, Bierlich (1976). Changes in the process during cutting are not considered.

The aim of the here presented method is to save production time during a grinding cycle by considering as well disturbances like workpiece surface variances as changes in process parameters, and to increase the tool-life. Changes in process parameters are caused by cutting wheel abrasion, varying process-adjustment like setting of wheel-speed v_S, wrong coolant flow but mostly by varying width of cut or workpiece-wheel combination.

Analogous to a servo-system with position and speed-control, a cascade control concept for workpiece-stop-measure and cutting-force, as shown in Fig. 1 is proposed.

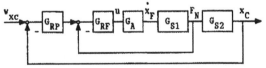

Fig. 1: Cascade control of grinding process

G_{RP} - contact-area position-controller

G_{RF} - normal cutting-force-controller

G_A - servo-drive X-axis

G_{S1}, G_{S2} - grinding process

The internal control loop controls the normal force F_N of the grinding process. The set-point for the normal force is given by the controller output of the external stop-measure control loop, which controls the output signal, the infeed position x_C.

For the controller design a theoretical dynamic model of the cutting process has to be developed in the form:

$$Y(t) = f[U(t), N(t), \varrho] \qquad (1)$$

In that mathematical description $U(t)$ and $Y(t)$ are the measurable input and output signals, ϱ the process parameters and $N(t)$ the unmeasurable disturbances.
Now a recursive least squares estimation method is used in order to obtain the process parameters ϱ. With these parameters the controllers for force and stop-measure control can be designed.

MODELING.

Fig. 2 shows the scheme of the kinematic relations of cylindrical surface grinding.

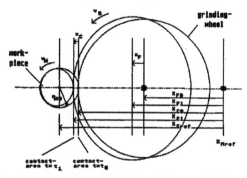

Fig. 2: Scheme of Cylindrical Surface Grinding

The X-axis drive presses the cutting wheel (peripheral speed $v_s \approx 45$m/sec) with the feed-rate $\dot{x}_F(t)$ ($\sim 10\ \mu$m/sec) against the rotating workpiece (peripheral speed $v_w \sim 0.7$ m/sec). The resulting cutting normal force $F_N(t)$ leads to a change in infeed cutting-velocity $\dot{x}_W(t)$.

Assuming a batch process, consisting of a point contact and a transport-lag during one workpiece revolution, the change in contact-area position $x_C(t)$ is given by:

$$x_C(t) = r_{W0}(t) - x_W(t-T_{tW}) \qquad (2a)$$

with the depth of cut a_p of one workpiece revolution

$$x_W(t) - x_W(t-T_{tW}) = a_p \qquad (2b)$$

$r_{W0}(t)$ is the variance of workpiece profile caused by the influence of workpiece base size and the roundness-error in chucking; T_{tW} is the transport-lag-time for one workpiece revolution, which is constant assuming a constant workpiece angular frequency.

Following Saljé (1978) and Fuchs (1988) a spring-mass-damper system describing the dynamics of motion in the contact area between wheel and workpiece can be stated:

- force balance in contact-area

$$c_C[x_F(t) - x_C(t)] + d_C[\dot{x}_F(t) - \dot{x}_C(t)] = F_N(t) + m_C\ddot{x}_C \qquad (3)$$

c_C – stiffness of wheel and workpiece in contact area
d_C – internal damping in contact area
m_C – deformed mass in contact area

- cutting force equation, see König and Werner /6/

$$\dot{x}_W(t) = \begin{cases} \left[\dfrac{v_s^{2\epsilon-1} \cdot v_W^{1-\epsilon} \cdot F_N(t)}{K_0 \cdot (r_{W00} - x_W(t))}\right]^{1/\epsilon} & F_N > 0 \\ \\ 0 & F_N < 0 \end{cases} \qquad (4)$$

r_{W00} – round profile start radius
K_0 – workpiece/wheel matter coefficient
ϵ – exponent describing workpiece- and grinding wheel characteristics

Combining eq. 2, 3 and 4 leads to the block diagramm, shown in fig.: 3.

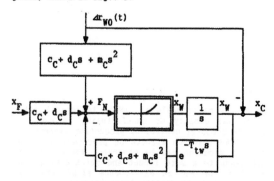

Fig. 3: Blockdiagramm of grinding process

In order to get a suitable process model one has to simplify some relations:

- $r_{W00} \gg x_C(t) \rightarrow (r_{W00} - x_C(t)) \approx r_{W00}$
- low acceleration and small mass m_C leads to $m_C\ddot{x}_C(t) \approx 0$
- linearizing the force-equation (4) in the operating point leads to $\dot{x}_W(t) = (1/\alpha)\cdot F_N(t)$, with

$$\alpha = \epsilon \cdot \bar{K}_0 \cdot v_{S0}^{1-2\epsilon} \cdot v_{W0}^{\epsilon-1} \cdot \dot{x}_{W0}^{\epsilon-1}$$

- neglecting internal damping terms, because $d_C \ll \alpha$

Thus the resulting transfer functions in the Laplace domain become:

$$G_{S1}(s) = \frac{F_N(s)}{x_F(s)} = \frac{\alpha / \alpha_0}{\frac{\alpha}{c}s + e^{-T_{tW}s}} \qquad (5)$$

and

$$G_{S2}(s) = \frac{x_C(s)}{F_N(s)} = \frac{1}{\frac{\alpha}{\alpha_0}s} = \frac{1}{T_{I}\alpha s} \qquad (6)$$

where α_0 is a scaling coefficient.

Discussing the dynamics of the grinding process the X-axis servo-drive has to be considered, see Fig. 4.

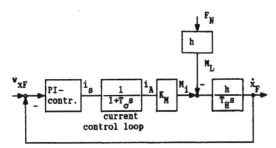

Fig. 4: Blockdiagramm of controlled servo-drive

Following Pfaff (1982) one gets the following transfer function of closed speed control loop with PI-controller.

$$G_A(s) = \frac{K_A}{1 + T_{Nn}s + \frac{T_{Nn}T_H}{K_M K_n h}s^2 + \frac{T_{Nn}T_H T_\sigma}{K_M K_n h}s^3} \qquad (7)$$

with T_H – run-up-time,

T_{Nn}, K_n – PI contr. integ. time and gain,

T_σ – equivalent current loop time constant

K_M – torque constant and

h – spindle lead

i_s, i_A – set-point- and armature current

Experimental identification results showed, that the time constants of the servo-drive are two decades smaller than the cutting process time constant $T_1 = \frac{\alpha}{c}$ and can be neglected for cutting-process parameter estimation and cutting-force-controller design.

So the transfer function for the servo-drive in cutting process becomes simply:

$$G_A(s) = \frac{\dot{x}_F(s)}{u(s)} = K_A \qquad (7a)$$

IDENTIFICATION OF PROCESS PARAMETERS.

Following Isermann (1988) a discrete time model

$$\begin{aligned}G_p(z) &= HG_A G_{S1}(z) \\ &= \frac{F_N(z)}{u(z)} = \frac{b_1 z^{-1} + b_2 z^{-2} + \ldots + b_m z^{-m}}{1 + a_1 z^{-1} + a_2 z^{-2} + \ldots + a_m z^{-m}}\end{aligned} \qquad (8)$$

is used for the identification of unknown parameters a_i and b_i, $i = 1, \ldots, m$.

Eq. 8 can shortly be written in a vector representation

$$y(k) = \underline{\psi}^T(k)\underline{\theta} + n(k) \qquad (9)$$

with the output signal

$$y(k) = F_N(k) ,$$

the measurement vector

$$\begin{aligned}\underline{\psi}^T(k) = [&-F_N(k-1), -F_N(k-2), \ldots, \\ &-F_N(k-m), u(k-1), \ldots, u(k-m)],\end{aligned}$$

$$k = T_0, 2T_0, \ldots, NT_0; \quad T_0 - \text{sampling interval,}$$

the parameter vector

$$\underline{\theta} = [a_1, a_2, \ldots, a_m, b_1, b_2, \ldots, b_m]$$

and an unmeasurable disturbance n(k).

With the generalized equation error

$$e(k) = y(k) - \underline{\psi}^T(k)\underline{\theta}$$

the solution for the unknown process parameters leads to a LS-problem and can be found by minimizing the lost function $V = \underline{e}^T\underline{e}$ for N time steps to

$$\hat{\underline{\theta}} = (\underline{\psi}^T\underline{\psi})^{-1}\underline{\psi}^T\underline{y} \qquad (10)$$

Eq. 10 can be obtained also in a recursive form (RLS) and realized with a numerical optimized recursive discrete square root filtering method (DSFI), see Isermann (1988) and Kofahl (1986)

CONTROL STRATEGY.

To control the workpiece stop-measure a cascade control with discrete PI - controller for normal force and P - controller for the stop-measure is used. An additional differential term in the force-controller (PID) does not improve the behavior of closed force-control-loop, but causes more actuator-motion and reduces the stability of closed loop, see Führer (1988).

● normal force control

If the transferfunction $G_p(z)$ is a second or higher order system, the force controller can be designed by a simple tuning method, following Kofahl (1985). If $G_p(z)$ is a first order system, the parameters of the PI - controller can be obtained by determining the behavior of closed force-control-loop $G_F(z)$ by a discrete first order system with the time constant T_F.

With the discrete PI controller transferfunction

$$G_{RF}(z) = \frac{q_0 + q_1 z^{-1}}{1 - z^{-1}} \qquad (11)$$

(Isermann, 1987) one gets by comparing the transfer function of closed force-control-loop $G_F(z)$ with the determined first order system

$$\begin{aligned}G_F(z) &= \frac{q_0 b_1 z^{-1}(1 + q_1/q_0 z^{-1})}{(1 + a_1 z^{-1})(1 - z^{-1}) + (1 + q_1/q_0 z^{-1})q_0 b_1 z^{-1}} \\ &= \frac{b_F z^{-1}}{1 + a_F z^{-1}}\end{aligned} \qquad (12)$$

the parameters

$$q_0 = \frac{b_F}{b_1} \quad \text{and} \quad q_1 = a_1 \cdot q_0. \quad (13)$$

with

$$b_F = (1 - \exp{-\frac{T_0}{T_F}}) \quad a_F = -\exp{-\frac{T_0}{T_F}}$$

where a_1, b_1 are the estimated process parameters of $G_p(z)$.

• **Workpiece stop-measure control**

For the design of the P - workpiece stop-measure controller we use the determined first order internal force control-loop behavior $G_F(z)$ to design the position control-loop. If we get a higher order system for the force control-loop, we can approximate it by an first order system with an equivalent time constant T_F, because the internal force control-loop is much faster as the stop-measure control loop.
The resulting control loop is shown in Fig. 5.

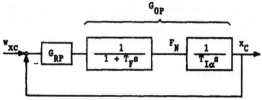

Fig. 5: workpiece stop-measure control loop.

We obtain a second order system for the closed control loop. But the dynamic behavior depends on the operating point, because the integration Time $T_{I\alpha}$ varies with the process gain α.
Thus the P-controller can be tuned determining a second order transfer function for the closed loop. With the transfer function of the discrete P-controller $G_{RP}(z) = q_0$ there are two possibilities to obtain the parameter q_0 by choosing either the resonance frequency ω_0 or damping D. To get a good workpiece quality we have to avoid an overshooting in the finishing phase, that leads to the demand $D = 1$ for the operating point "fine finishing".
The z-transformed transfer function of the open stop-measure loop $G_{OP}(s)$ is

$$G_{OP}(z) = \frac{b_1^M z^{-1} + b_2^M z^{-2}}{1 + a_1^M z^{-1} + a_2^M z^{-2}} \quad (14)$$

with $a_1^M = -(1 + \exp{-\frac{T_0}{T_F}}) \quad a_2^M = \exp{-\frac{T_0}{T_F}}$

$$b_1^M = [T_0 - T_F(1 - \exp{-\frac{T_0}{T_F}})]/T_{I\alpha}$$

$$b_2^M = [T_F(1 - \exp{-\frac{T_0}{T_F}}) - T_0\exp{-\frac{T_0}{T_F}}]/T_{I\alpha}$$

One gets with $D = 1$ the following condition for the closed-loop:

$$\left(\frac{q_0 b_1^M + a_1^M}{2} \right)^2 - (a_2^M + q_0 b_2^M) = 0 \quad (15)$$

Solving this quadratic equation for q_0 leads to:

$$q_{0_{1,2}} = -2 \cdot \left[\frac{a_1^M b_1^M - b_2^M}{b_1^{M\,2}} \right]$$

$$\pm \sqrt{\left[2 \cdot \frac{a_1^M b_1^M - b_2^M}{b_1^{M\,2}} \right]^2 - \frac{a_1^{M\,2} - 4a_2^M}{b_1^{M\,2}}}$$

$$(16)$$

$$q_{0_{1,2}} > 0$$

• **Choice of optimal set-point trajectory**

Conventional grinding cycles consist of three or more phases, like e.g. roughing, finishing and fine-finishing. A optimal set-point trajectory for the contact-area position has to perform the following tasks:

1. Building up the maximum cutting force F_N in a smooth transient to avoid force peaks (parabolic segment).

2. Roughing with maximum possible cutting force, limited by technology (linear segment).

3. Reducing the gradient of contact-area position to an minimum to guarantee a demanded workpiece quality (exponential segment).

4. Return traverse when stop-measure is reached.

EXPERIMENTAL IDENTIFICATION RESULTS.

With experimental data from a machine-tool SCHAUDT T3U a first order system for $G_{S1}(z)$ and a integral system for $G_{S2}(z)$ were estimated.

The process signals feed-motion x_F was measured by an incremental decoder and the contact-area position x_C with an in-process inductive displacement sensor.
Because one needs a reliable force sensor the cutting-force measurement was replaced by an electrical current measurement in the cutting-wheel-servodrive.
Fig. 6 a) and b) show the signals, used for parameter estimation. Table 1 shows the estimated parameters and the computed force- and stop-measure-controller parameters.
One can see, that the process parameters vary with the abrasion of grinding wheel. So also the parameters of the controllers have to be adjusted.

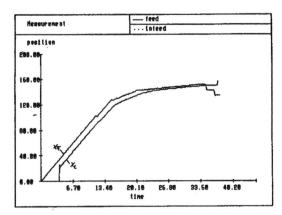

Fig. 6a) feed motion $x_F(t)$ and infeed-position x_C

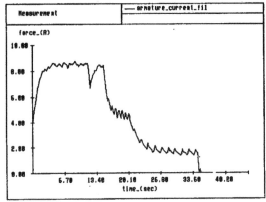

Fig. 6 b) armature current $i_A(t)$ of wheel-drive

Table 1: Estimated parameters

n	α/α_o	T_1	$T_{I\alpha}$	q_{OF}	q_{1F}	q_{OP}
1	1.076	3.405	1.155	9.106	−8.84	0.595
10	1.129	4.120	1.197	10.472	−10.47	0.626

$T_0 = 0.1$ sec, n = number of workpieces without drimming the grinding wheel, $T_F = 0.3$sec.

SIMULATION RESULTS.

With this adjustments grinding cycles were simulated on a digital computer considering as well non-linearities as the dynamics of X-axis servo-drive. Fig. 7a and b show the signals (x_F, x_C and F_N) of the uncontrolled grinding process in normal state, with blunt cutting wheel and reduced wheel-speed v_S performing a conventional 3 phase cycle. Depending on the operating point and the tool-workpiece characteristics the process gain and the time constant T_1 varies. Hence the production time can not be determined in advance and increases with bad process condition up to 30% of the nominal value.
Fig. 8a and b show the controlled output signal x_C, the setpoint v_{xc}, the actuating signal v_{FN} and the resulting cutting force F_N performing a conventional grinding cycle. Now the influence of process parameter changes is eliminated and a reduction of grinding time about 10% of nominal value can be achieved, because the normal cutting-force F_N is

built up quicker. Furtheron the controlled production cycle is more robust against varying process conditions and guarantees a constant workpiece quality. Fig. 9a and b show a production cycle with optimized set-point trajectory saving further 15% of production time, because the cutting-force can be reduced in an exponential transient, which guarantees a demanded workpiece quality.

Fig. 7: conventional grinding cycle (open loop)

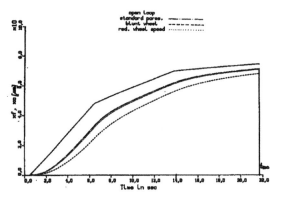

a) contact-area position $x_C(t)$ and X-feed motion $x_F(t)$.

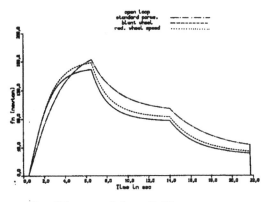

b) cutting normal force $F_N(t)$

Fig. 8: position controlled conventional grinding cycle

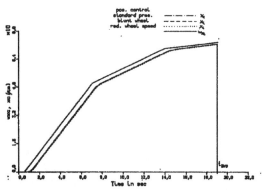

a) contact-area position $x_C(t)$ and set-point $v_{xc}(t)$.

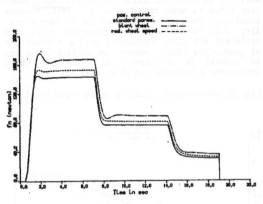

b) cutting normal force $F_N(t)$

Fig. 9: position controlled optimized grinding
cycle

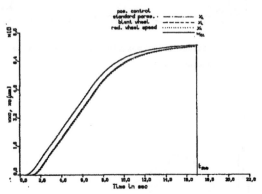

a) contact-area position $x_C(t)$ and set-point
$w_{xc}(t)$.

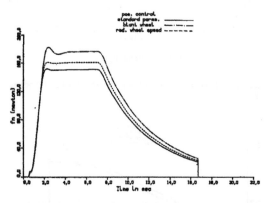

b) cutting normal force $F_N(t)$

CONCLUSIONS.

Starting with modeling and process parameter estimation at the real process the presented parameter
adaptive control concept has shown by simulations,
that production time up to 30% can be saved.
Furtheron the tool-life can be increased by using a
set-point trajectory with smooth transients , which
leads to a better workpiece quality.
Workpiece surface variances can be eliminated up to
the cut-off frequency of the closed contact-area-

position control-loop.
Now the concept will be realized on a real cylindrical surface grinding machine.

ACKNOWLEDGEMENTS.

This paper publishes results of a research project,
which is sponsored by the German Bundesministerium
für Forschung und Technologie (BMFT) with the project number 02FT45117.

REFERENCES.

Bierlich,R.: Technologische Voraussetzungen zum
    Aufbau eines adaptiven Regelungssystems beim
    Außenrundeinstechschleifen.   Dissertation   TH
    Aachen, 1976.

Fuchs, A.; Janik, W.; Isermann,R.: Model Based
    Supervision And Fault Diagnosis of The Cylindri-
    cal Surface Grinding Process. Prep. of XI Imeko
    World Congress, Houston, 1988.

Führer, J.: Untersuchung der Stabilität und dy-
    namischen Güte des Aussenrundschleifprozesses
    für verschiedene Regelstrategien. Diplomarbeit
    TH Darmstadt 1988

Isermann, R.: Digitale Regelsysteme Band I und II,
    Springer Verlag Berlin-Heidelberg, 1987.

Isermann, R.: Identifikation dynamischer Systeme
    Band I und II, Springer Verlag Berlin Heidel-
    berg, 1988.

Kofahl, R.; Isermann,R: A simple Method for
    Automatic Tuning of PID Controllers Based On
    Process Parameter Estimation. American Control
    Conference, Boston, 1985.

Kofahl, R.: Verfahren zur Vermeidung numerischer
    Fehler bei Parameterschätzung und Optimalfil-
    terung. Automatisierungstechnik at.11/34, 1986.

Pfaff, G.; Meier, C.: Regelung elektrischer
    Antriebe II, Oldenbourg Verlag München, 1982.

Saljé, E.; Pahlizsch, G.: Grinding Processes,
    Considered as Feedback Control System. Annals of
    the CIRP, 27, 1978.

Tönshoff, H.K.; Zinngrebe, M.: Optimization of
    internal Grinding by Microcomputer-Based Force
    Control. Annals of the CIRP, 35/1, 1986.

Werner, G.: Konzept und technologische Grundlagen
    zur adaptiven Prozeßoptimierung des Außen-
    rundschleifens. Habilitation TH Aachen, 1973.

Werner, G.; König, W.: Influence of Work Material
    on Grinding Forces. Annals of the CIRP, 27/1,
    1978.

REAL TIME QUALITY FEEDBACK IN A
FLEXIBLE MILLING CELL

D. Gien, L. Seince and R. Stepourjine

Ecole Nationale Supérieure de Mécanique et des Microtechniques, Besançon, France

Abstract. In this paper, an on-line adjustment system for a flexible milling center is
presented. Data from inspection are used for computation of tool offsets. The
measurement of a part allows some corrections for the machining of the next part. A
model taking tool wear and tool offset into consideration is proposed. Various
methods are given to solve the problem. An optimal approach aiming at automatic
control of the flexible cell is taken. Another approach, based on fuzzy sets, is
applied to the presentation of data to operator. Measurement filtering for the
estimation of part dimensions and for the prediction of tool wear is also outlined.

Keywords : Quality control, Machine tools, Adjustment, Flexible cell, Optimisation.

INTRODUCTION

Ever-increasing automation in manufacturing
systems leads to unmanned plants. Not only is
human work disappearing but also human
supervision. Inspection facilities are involved to
perform quality control tasks. The first goal is
to ensure that a part coming out an operation is
correct in order to be sent to the next operation.
The second goal is to monitor and to automatically
adjust the manufacturing process. Integrated
quality assurance is essential for flexible
manufacturing systems.

In a pilot flexible milling cell, the measurement
data are used to adjust the tool corrections.
Automatic computation is needed to obtain
error-free results within a short time. A real
time feedback allows cell optimal working. Each
part is measured and the computations are carried
out before a similar part is milled again. Keeping
track of apparent tool wear is made possible by
the measurement of the parts. Tool monitoring is
so easy and accurate.

In this paper, we first present the architecture
of a flexible milling cell including a coordinate
measuring machine. The proposed models are then
described and various adjustment tactics are
compared. Later on, we will deal with data
filtering and wear prediction. In the end, the
results from the simulation and from the real
manufacturing will be discussed.

PILOT FLEXIBLE MILLING CELL

The pilot cell includes a milling center, an
industial robot, a pallet conveyor and storage
devices for the parts, tools and fixtures. The
cell is designed to allow self-sufficiency for a
few hours. A computer is responsible for tasks
scheduling and equipment control. We take a
particular interest in measurement systems. The
parts are inspected as soon as possible after
machining. A measuring robot is used for this
purpose. It is a floor plant coordinate measuring

machine capable of giving a good idea of part
geometry within a short time. Only part washing
and cooling is needed before inspection. Real
time, i. e. with only delay of one part, is
preferred to maximum accuracy. The first goal of
measurement is to check the parts coming out of
the cell. Since measurement uncertainty is
significant, we have to take it into account. The
probability that a conformable part will be
obtained can be computed from the measurements :

$$P (\overset{\bullet}{\underline{x}} < \underline{x} < \overset{M}{\underline{x}}) = \int .. \int_{\overset{\bullet}{\underline{x}} - \tilde{\underline{x}}}^{\overset{M}{\underline{x}} - \tilde{\underline{x}}} f(\underline{x}).d\underline{x} \qquad (1)$$

where :

$\underline{x}$ actual dimensions of tooled parts (I-vector)

$\tilde{\underline{x}}$ mesured dimensions (I-vector)

$\overset{\bullet}{\underline{x}}$ lower tolerances (I-vector)

$\overset{M}{\underline{x}}$ upper tolerances (I-vector)

$f(\underline{x})$ probability density function of measurement
errors.

The obtained value is then used as an acceptance
test criterion :

Pa < P		=>	Acceptance
Ps < P < Pa		=>	Indecision
P < Ps		=>	Rejection

In doubt, the part is directed to a metrological
laboratory where the uncertain dimensions are
measured again.

Our aim consists in processing the raw measurement
data to adjust the milling center. At the start of
a batch or when a new tool is put into service, it
is necessary to measure the tool itself to preset
the milling machine. Hence, a tool inspection
station is still required.

245

A three-level control strategy is used to manage the cell. The first one, a reflex control level, is not our subject. The second level includes data filtering, estimation and adaptative control (figure 1). The optimal adjustment is computed for each manufactured part. The apparent tool size is calculated. The results are sent to the upper control level. The latter is responsible for decision making. Useful information comes not only from the inspection of the parts but also from various tool monitoring devices. The resulting action may be tool offset alteration, tool changing or process stopping. In this paper, we are mainly interested in the second control level and especially in feedback of measurement to the milling machine adjustment.

TOOL CORRECTIONS SETTING

In the machining process, the part programs are written using nominal dimensions. The real tool path is shifted from the material, a value, which one hopes, being equal to the tool radius or length. These offset quantities are named tool corrections. The initial tool offsets are set using the measured tool sizes. Then the tool corrections can be adjusted to balance tool wear. We, at first, adopt the following model to connect the part dimensions, the tool size and the tool corrections :

$$\underline{x} = \overline{\underline{x}} + Bc.\underline{c} + Bu.\underline{u} \qquad (2)$$

where :

$\underline{x}$ actual dimensions of tooled parts (I-vector)

$\overline{\underline{x}}$ nominal dimensions in part program
$\underline{c}$ tool radius and length corrections (K-vector)
$\underline{u}$ tool radii and lengths (J-vector)
Bc correction I.K-matrix
Bu wear I.J-matrix.

Bu matrix depends on the part geometry and the machining method. One can take :

$$Bc = - Bu \qquad (K = J)$$

To allow a more flexible adjustment, we may add redundant correction variables :

$$Bc = \begin{bmatrix} - Bu & Br \end{bmatrix} \qquad (K > J)$$

Br matrix represents the additional correction action on the part dimensions.

If we now consider the process time evolution we get :

$$\underline{x}_{n+1} = \underline{x}_n + Bc.\underline{\Delta c}_n + Bu.\underline{\Delta u}_n \qquad (3)$$

where :

n   time index (and, in our case, part number)

and:

$$\underline{\Delta c}_n = \underline{c}_{n+1} - \underline{c}_n$$

$$\underline{\Delta u}_n = \underline{u}_{n+1} - \underline{u}_n$$

The problem is to compute the tool correction changes in order to obtain the next part correctly made :

$$\overset{\blacksquare}{\underline{x}} < \underline{x}_{n+1} < \overset{M}{\underline{x}}$$

where :

$\overset{\blacksquare}{\underline{x}}$ lower tolerance

$\overset{M}{\underline{x}}$ upper tolerance

$<$ component by component comparison.

Assuming that we have estimates for the part dimensions :

$$\hat{\underline{x}}_n$$

and for the tool wear rates (prediction) :

$$\underline{\Delta u}^*_n$$

We must choose a correction Δc so that :

$$\overset{\blacksquare}{\underline{x}} < \hat{\underline{x}}_n + Bc.\underline{\Delta c}_n + Bu.\underline{u}^*_n < \overset{M}{\underline{x}} \qquad (4)$$

Generalized inverse

The first approach consists in searching for a setting point as close as possible to the nominal value. In order to obtain a consistent result, the nominal value must be the middle of the tolerance gap. The correction is given by :

$$\underline{\Delta c}_n = Bc^{\dagger}.(\overline{\underline{x}} - \hat{\underline{x}}_n - Bu.\underline{\Delta u}^*_n) \qquad (5)$$

where :

$Bc^{\dagger}$ Moore-Penrose generalized inverse.

That approach is very simple and does not need hard on-line computations. On the other hand, the method ignores tooling constraints. First, machining dispersion can change from one shape to another. Next, tolerances are more or less close. These specifications are not taken into account by a flat adjustment.

When machining the first part of a batch, the dimension and wear estimates are not yet available. We must compute the initial tool offsets from tool measurements or from previous tool wear prediction :

$$\underline{c}_o = - Bc^{\dagger}.Bu.\underline{u}_o$$

where :

$\underline{u}_o$ measured or estimated initial tool radii and lengths.

Optimal setting

To take manufacturing errors into consideration, we propose a second approach. We assume the statistical features of tooling deviations to be known. We can then compute the probability that a correct part will be obtained :

$$P (\overset{\blacksquare}{\underline{x}} < \hat{\underline{x}}_n + Bc.\underline{\Delta c}_n + Bu.\underline{\Delta u}^*_n < \overset{M}{\underline{x}}) =$$

$$= \int..\int_{\underline{x} - \hat{\underline{x}}_n - Bc.\underline{\Delta c}_n - Bu.\underline{\Delta u}^*_n}^{\overset{M}{\underline{x}} - \hat{\underline{x}}_n - Bc.\underline{\Delta c}_n - Bu.\underline{\Delta u}^*_n} f(\underline{x}).d\underline{x} \qquad (6)$$

where :

f($\underline{x}$) probability density function for machining errors.

Notice P depends upon the dimension estimates, the tool wear rate predictions and the tool corrections. The problem answer is the correction vector giving the maximum value for P. The approach is optimal in that if the dimension and wear estimates are equal to the real dimensions and wear rates, then the adjustment is the best one. Any other setting will give a lower probability that a good part will be made. However, that method requires that the statistical features be estimated. Computation complexity is significant but not actually deciding.

Fuzzy setting

Following the human thought process, we can classify adjustment quality according to manufacturing difficulty. The fuzzy set of easily feasible parts can be defined. A part with its membership function equal to one will surely be correct. A part with its membership function equal to zero will surely be wrong.

For each dimension i we define a fuzzy set:

$$Xi$$

with membership function :

$$\mu_{Xi} (x_i)$$

We choose a linear membership function (figure 2) quite convenient for our purpose. For the whole part we get :

$$X = \cap Xi$$

with membership function :

$$\mu_X (\underline{x}) = \min_i \{ \mu_{Xi} (x_i) \} \qquad (7)$$

Given the dimension and wear rate estimates, we obtain the membership of a part as a function of Δc :

$$\mu_X (\hat{\underline{x}}_n + Bc.\Delta c_n + bu.\Delta u^*_n) \qquad (8)$$

and, in that way, a correction choice criterion.

The best adjustment is then the correction for which the membership level is maximum. That criterion in fact evaluates the machining difficulty for the worst dimension. That allows an easy human interpretation and a feedback to the product design. If the process has a good capability, the solution is not a single point but a domain, in Δc space, limited by hyperplanes :

$$\overset{1}{\underline{x}} - (\hat{\underline{x}}_n + Bc.\Delta c_n + Bu.\Delta u^*_n) = 0$$
$$(9)$$
$$\overset{u}{\underline{x}} - (\hat{\underline{x}}_n + Bc.\Delta c_n + Bu.\Delta u^*_n) = 0$$

The adjustment point is then chosen as the center of the solution domain. The latter method allows a graphical data presentation using plane cuts (figure 3). The critical dimensions, i. e.

the domain limits given by equations (9), are clearly seen. The manufacture or/and design problems can thus be easily detected.

An example of fuzzy (o), optimal (+) and generalized inverse (x) corrections is shown (figure 4). The probability that a correct part will be obtained, i. e. optimal criterion, is plotted as a function of one component of the correction vector Δc. The involved tolerance limits are drawn (upper tolerances : ———, lower tolerances : ---). Thus a comparison of the methods is easy. Generally, the fuzzy set approach gives good results. In some cases, i.e loose tolerances, the generalized inverse method can also be employed.

DIMENSION AND WEAR ESTIMATION

The estimate values for the part dimensions and tool wear rates are needed in order to compute the tool offset corrections. At the same time, tool wear prediction is achieved. We propose a stochastic model derived from equation (3) :

$$\underline{x}_{n+1} = \underline{x}_n + Bc.\Delta c_n + Bu.\Delta u_n + \underline{v}_n \qquad (10)$$

where :

$\underline{v}_n$ random machining noise I-vector

with :

$$E\{\underline{v}_n\} = 0, \quad E\{\underline{v}_n .\underline{v}_m\} = 0 \text{ if } n \neq m, \quad E\{\underline{v}_n .\underline{v}_n\} = S_n$$

Measurement errors are taken into account by the expression :

$$\tilde{\underline{x}}_n = \underline{x}_n + \underline{w}_n \qquad (11)$$

where :

$\tilde{\underline{x}}_n$ measured dimensions

$\underline{w}_n$ random measuring noise I-vector

with :

$$E\{\underline{w}_n\} = 0, \quad E\{\underline{w}_n .\underline{w}_m\} = 0 \text{ if } n \neq m, \quad E\{\underline{w}_n .\underline{w}_n\} = R_n$$

We also introduce a model for tool wear :

$$\Delta u_{n+1} = \Delta u_n + \underline{t}_n \qquad (12)$$

where :

$\underline{t}_n$ random tooling noise I-vector

with :

$$E\{\underline{t}_n\} = 0, \quad E\{\underline{t}_n .\underline{t}_m\} = 0 \text{ if } n \neq m, \quad E\{\underline{t}_n .\underline{t}_n\} = Q_n$$

More $\underline{v}, \underline{w}$ and $\underline{t}$ are assumed to be independent.

We consider that, in chosen approach, wear is only given by an initial wear rate and by wear rate random deviations. Actually, we include wholly tool-dependent errors in the model.

We use a Kalman filter. For the dimension estimates, we can write :

$$\hat{x}_n = x^*_n + E_n . (\tilde{x}_n - x^*_n) \qquad (13)$$

$$x^*_n = \hat{x}_{n-1} + Bu.\Delta u^*_{n-1} + Bc.\Delta c_{n-1}$$

$$E_n = G_n .(R_n + G_n)^{-1}$$

$$G_{n+1} = (I - E_n).G_n + S_n .$$

For wear rate prediction we get :

$$\Delta u^*_n = \Delta u^*_{n-1} + F_n .(\Delta \tilde{x}_n - Bu.\Delta u^*_{n-1} - Bc.\Delta c_{n-1}) \qquad (14)$$

$$F_n = H_n .Bu'.(S + 2.R + Bu.H_n .Bu')^{-1}$$

$$H_{n+1} = (I - F_n .Bu).H_n + Q$$

with :

$$\Delta \tilde{x}_n = \tilde{x}_{n+1} - \tilde{x}_n .$$

Tool wear prediction is given by :

$$\hat{u}_n = \hat{u}_{n-1} + \Delta u^*_{n-1} .$$

This value is used by the upper control level to detect tool wear and to prevent tool breaking.

PART MODEL INSTANCE

We take a very simple part example consisting of a block with two pockets (figure 5). The part is milled by three tools, one for the outside and one for each pocket. We only consider the radius corrections and thus a two-dimensional part model. In fact, the length corrections are disconnected from the radius corrections. The part is completely defined by ten dimensions. A tool offset adjustment acts, at least, on four dimensions. The human choice of corrections is already complicated for that elementary example.

SIMULATION

Simulation is used for comparison of methods. According to equations (6, 7, 8) the measurement values of parts are produced. The tool initial wear rates can be chosen. Noises are normally distributed with adjustable variances. The machining process and the feedback loop are simultaneously simulated. The filtering law and the correction method may be thus easily selected. An instance of simulation sofware output is plotted (figure 5). Tool wear is always rapidly compensated at the cost of slightly increased random deviations.

EXPERIMENTATION

The experiments are carried out on the pilot flexible milling cell. First a part is milled with arbitrary tool offset. The part is measured by means of the measuring robot. The Raw measures are sent through a local area network to a microcomputer. The corrections are calculated using one of previously described methods. The results are sent back to the machining center controller. Then, the next part can be milled with the correct tool offsets. The operation is repeated until the tool is completely worn. The results of the machinig of sample parts agree with simulation. Conformable parts are manufactured except when a new tool is used. The tool measurement for initializing the process can partly offset that error. An alternative is to measure the part dimensions on the milling machine, using integrated gauging capabilities, before finishing machining.

CONCLUSION

The main contribution of the presented method is to allow the automatic adjustment of a milling center. The setting point can be evaluated by means of a criterion. The result is sent to an upper control level so that a decision may be taken to continue the process with updated corrections, to stop the process or to change a tool. The fuzzy set approach is especially interesting for human interface.
The proposed method can only compensate tool-dependent errors. The machine drifts are not part of that model. However, the model could be strainlessly extended.
For small batches, the statistical distributions must be estimated from the previous parts of the same kind. We also use a shape-oriented product definition for this purpose.
Finally, we point out a precaution one must take. The standard statistical quality control softwares cannot be used with the proposed feedback method. The statistical computations are an integral part of the correction software in order to obtain significant results. Hence the corrections can be taken into account.

REFERENCES

VAN DEN BERG B. (1987); Closed loop inspection of sculptured surfaces in a computer integrated environment; 8th Automated Inspection and Product Control; june 87; Chicago, U.S.A. .

DESCHANEL F. (1989); Pilotage d'une cellule flexible d'usinage; Thèse de Docteur de l'Université de Franche-Comté; Février 89; Université de Besançon; FRANCE.

DUPUIS G. (1985); Contribution à la surveillance de l'outil de fraisage; Thèse de Docteur-Ingénieur; Octobre 85; Université de Besançon; FRANCE.

GIEN D. (1987); Suivi du processus d'usinage sur cellule flexible automatisée; Robotics and automation; Juin 87; Lugano, SUISSE.

HILLYARD R. C. (1978); Dimensions and tolerances in shape design; Doctor of philosophy thesis; May 78; University of cambridge; U. K. .

NEVEL'SON M.S., MUSTAFAEV E.K., REVIS E. I.(1985); Current estimation of tool wear in unmanned machines; Meas. Tech. V6 N°28; June 85.

RICHARD J.(1985); Contrôle dimensionnel et suivi de production dans un îlot automatisé de fabrication de pièces mécaniques; Thèse de Docteur d'Etat; Novembre 85; Université de Nancy 1; FRANCE.

TREWIN E. T. (1985); Processing of measurement to control production; 7th Automated Inspection and Product Control; 26-28 March 85, Birmingham U.K.

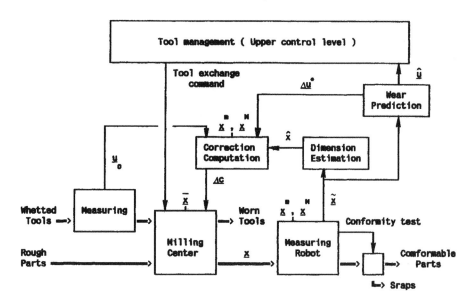

Fig. 1. Real time feedback in a flexible milling cell

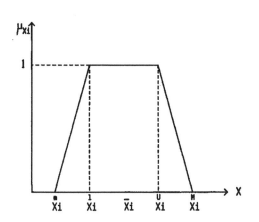

Fig. 2. Membership function

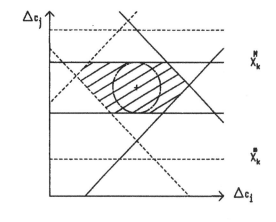

Fig. 3. Adjustment domain

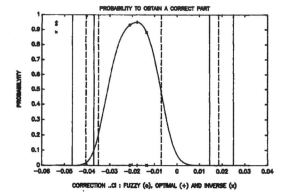

Fig. 4. Optimal criterion as a function of Δc_i

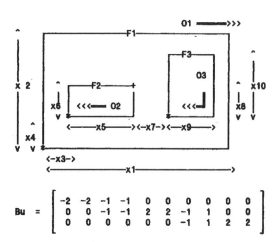

Fig. 5. Part example with associated Bu matrix.

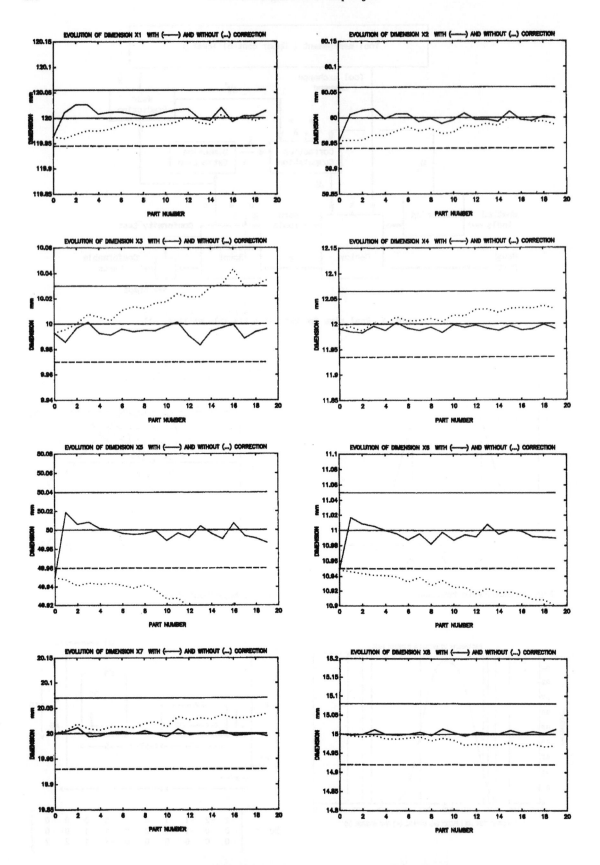

Fig 6. Simulation of machining with and without feedback

AN APPROACH TO THE SIMULATION FOR FMS DESIGN AND COST ANALYSIS

V. R. Milacic and B. R. Babic

Mechanical Engineering Faculty, University of Beograd, Beograd, Yugoslavia

Abstract. Rational decision making in the design of costly manufacturing structures,
such as flexible manufacturing systems, become increasingly important. Alternative production technologies, interdependence of the technology and equipment, the utilisation
of each element of the equipment, add up to the complexity of the investigation and
evaluation of such systems.

The paper describes the model for techno-economic assessment of FMS projects based on
the simulation and cost analysis. This model is used at the Mechanical Engineering Faculty in Beograd in the elaboration of preliminary design of new manufacturing systems.
The proposed method is illustrated with the example - a part of the project for a new
factory.

Keywords. Flexible Manufacturing, Simulation, Optimization, Modelling, Cost Analysis.

INTRODUCTION

Flexible manufacturing systems (FMS) offer significant advantages in comparison with less flexible
forms of manufacturing. Increased flexibility gains
in importance in small-scale manufacture with greater number of different parts. These advantages
are expressed in the form of smaller share of direct labour, lower stocks, greater utilisation of
machining stations, shorter leading times, improved product quality, etc. On the other hand these
advantages impose greater initial investment of
capital. more intensive utilisation of machining
stations, additional training of workers, organisational changes and new planning and control methods.

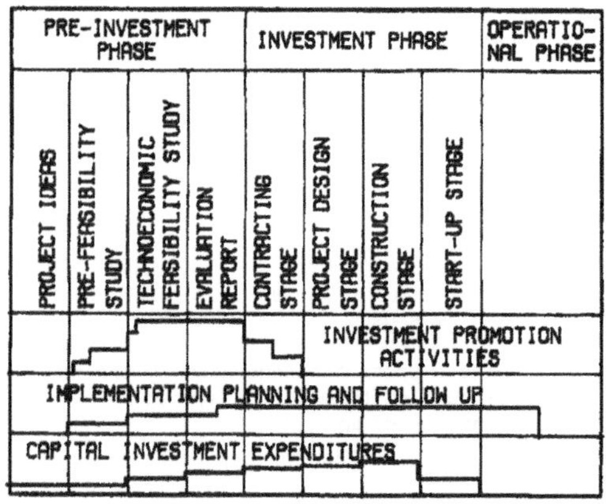

Fig. 1. Structure of industrial project evaluation

Rational decision making process during the pre-investment stage in these costly manufacturing structures becomes increasingly important. A simple model for the assessment of an industrial project may
be expressed with three main phases: pre-investment,
investment and operational phase (Fig. 1) (Milačić,
1981). The central effort is directed to the elaboration of the feasibility study, which defines

the technology required for a given project, evaluates manufacturing alternatives and selects the optimal combination of technical, financial and economic aspects of the project. In the metalworking
industry the selection of technology and equipment
are interdependent. Technological manufacturing
alternatives and the utilisation of each machining
station increase the complexity of the investigation. The paper describes a model for FMS design
projects assessment based on the simulation and
cost analysis.

THE BASIC CONCEPT OF THE FMS TECHNO-ECO-
NOMIC ANALYSIS SYSTEM

The basic concept of the system is illustrated in
Fig. 2. The inputs are technological procedures and
data relating to machines used for FMS layout design. System performances obtained by simulation
and cost analysis are used for the optimisation and
varying of the initial configurations for improving
of techno-economic properties of the initial system. The outputs represent the animation of simu-

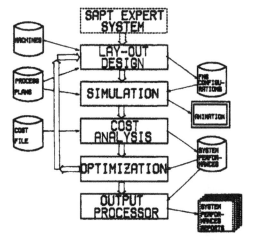

Fig. 2. Simulation based system for techno-economic
evaluation of FMS projects

lated system and the reports with techno-economic characteristics generated by output processor.

The system may be connected with automatic technology design system (SAPT-EXPERT system), developed as well in the Laboratory for Artificial Intelligence and Industrial Robots of the Mechanical Engineering Faculty in Beograd. More detailed description of individual blocks of the system is given below.

Layout Design Block

The graphical system for interactive layout design with the following functions was developed:
* definition of symbols for machining stations, transportation vehicles, buffers, etc.
* elaboration of the layout of equipment,
* definition of transport communications (paths)

On the basis of technological procedures of parts manufactured in a given FMS and available machines for which the symbols were made earlier (Fig. 3), the layout of the equipment is made and the transportation paths are entered (Fig. 4).

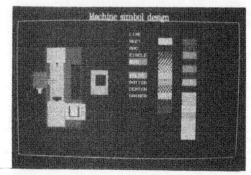

Fig. 3. An example of machine symbol making

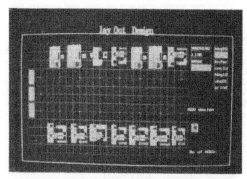

Fig. 4. An example of layout designing

Simulation Model

The simulation is the technique for the prediction of real system performances through the use of a model of such a system. For making of the model prior analysis of relevant activities determining the work of the system to be modelled and the carriers of these activities is necessary.

The main FMS components and their mutual relations are illustrated in Fig. 5. The activities to be incorporated into the model are as follows:

Machines
* workpieces are loaded on available machine
* finished workpieces are unloaded from machines
* machines break down in statistically determinate intervals

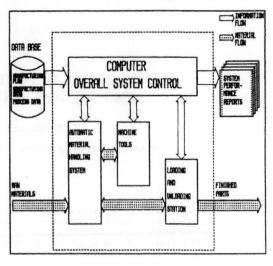

Fig. 5. FMS structure

Preparatory stations
* material is paletted
* repaletting of workpieces between operations
* depaletting of finished parts

Workpieces
* material is arriving from the store and the production is started
* workpieces have to "visit" machines according to prescribed routing and to remain fixed time on them (subject to technological procedure). For some workpieces there may exist alternative routings
* if the machine is not available waiting workpieces are placed into the buffer
* between some operations workpieces are transported to the preparatory station for repalletisation
* finished parts are transported into the store for finished parts

Material handling equipment
* workpieces are transported to assigned points
* exchange of workpieces on machines
* storage

Control
The management makes decisions, particularly in the area of introducing the work into the shop, and in the system of priority assigning to individual workpieces, and the like. The main activities linkked for the control will be specified as input data or built into the internal logic of the model. Simulation model must reproduce all the above mentioned activities. The basic logic of the model defined on the basis of the above stated consideration is illustrated in Fig. 6. In the making of this model the simulation package GPSS-F (GPSS Fortran version) was used.

The following outputs are generated by the model:
  - total running time of the system
  - utilisation of machines
  - slack time of machines
  - queue statistics
  - output rate (parts/hour)

One of outputs is the animation of simulated system (Fig. 7).

Costing Model

The comparison of two or several FMS projects may be achieved on the basis of the following criteria:
  1. Overall production costs
  2. Cost per product
  3. Output rate

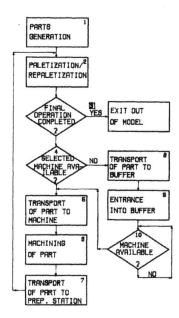

Fig. 6. Overwiev of the model

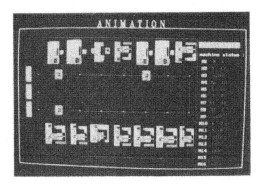

Fig. 7. Animation

These criteria are obtained from the costing model. The costing model was developed on the basis of the approach developed by B. N. Colding (1975, 1980). Certain additional alternations were made for the improvement of this model.

On the basis of the structure of the manufacturing system the overall production costs C_{ME} are expressed as:

$$C_{ME} = C_{SH} + C_{PR} \tag{1}$$

where:

C_{SH} - all costs pertaining to shop processing such as capital, wages, inspection, transportation, etc.

C_{PR} - all other costs, or preparatory costs, including cutting tools costs. These costs pertain to planning, scheduling, fixtures, cutting tools, etc.

Using analogy on metal cutting costs and productions costs, the equation (1) is defined as:

$$C_{ME} = C_{HP} * \mathcal{T}_p + C_o * \mathcal{T}_p/(\mathcal{T}_p - T_{sl}) \tag{2}$$

where:

C_{HP} - hourly shop rate
$\mathcal{T}_p$ - processing time
T_{sl} - slack time
C_o - preparatory cost

In order to extend the initial model the following equations are introduced:

$$M_{CY} = a \cdot K \tag{3}$$

$$L = men * Lch \tag{4}$$

where:

M_{CY} - machine cost/year
$K = \sum_1^n k_i$ - capital cost
$a = r * (1+I)$
r - rate of return
I - annual interest on invested funds
L - labour cost per hour
men - number of Labourers
L_{ch} - man/hour cost

On the basis of the equations (3) and (4) the hourly shop rate C_{HP} may be determined as:

$$C_{HP} = M_{CY}/(Ns * W_{HY}) \tag{5}$$

where:

Ns - number of shifts
W_{HY} - number of work hours per shift per annum

The following also may be determined:

Average cost/part:

$$T_C = C_{HP}/P_r \tag{6}$$

where:

P_r - output rate (part per hour)
and

utilisation of capital

$$\mathcal{7}_{cap} = (\sum_1^n util_i * k_i)/K \tag{7}$$

From the equations 1 to 7 the parameters for the comparison of FMS projects based on the above stated criteria are obtained.

Optimization

The optimization of the manufacturing system includes the improvement of performances of the system for the purpose of:
- cutting down of production costs,
- raising of the productivity of the system,
- increasing of the equipment utilisation level
attainable by:
- varying number of machining stations
- changing of technologies, and
- layout modifications.

On the basis of techno-economic indices obtained by the simulation and cost analysis, the initial system is varied in order to obtain an optimum solution.

Output Processor

The output processor generates the reports relating to performances of the simulated system. The following output data are obtained:
- utilisation level of machining stations,
- output rate (parts/year),
- total costs ($/year)
- average cost per product ($/piece),
- level of utilisation of invested capital.

THE EXAMPLE OF DEVELOPED MODEL`S APPLICATION

To demonstrate the proposed model an example is given of the FM cell for the production of couplings for pumps, representing a part of the mining industry pump factory, organised on CIM concept. This project is being realised in the Institute for Production Engineering and Computer Integrated Technologies of the Mechanical Engineering Faculty in Belgrade.

The product for the manufacturing of which FMS design was made consists of 12 components. Input data were taken from technological procedures including the specification of machine types (Table 1), and the sequence of operations with machining times for each part (Table 2).

TABLE 1   List of Machines

MACH NO.	SIGN	MACHINE TYPE
1	M1	MD 5 NC LATE
2	M2	MD 7 NC LATE
3	M3	C 12 NC GRINDING MACHINE
4	M4	VBG 50 VERTICAL MACHINING CENTER
5	M5	HBG 80 HORISONTAL MACHINING CENTER
6	M6	HBG 120 HORISONTAL MACHINING CENTER

TABLE 2   Operation Sequences

PART NO.	PART NAME	OP1	OP2	OP3	OP4	OP5	OP6	OP7
		OPERATION TIME [J]/MACHINE TYPE						
1	OUTLET SHAFT	7.5 / M1	2 / M4	1.5 / M3	1 / M1	0.5 / M6		
2	BEARING SUPPORT	1 / M1	2 / M6					
3	BEARING SUPPORT	2 / M1	1 / M6					
4	IMPELLER	17 / M2	30 / M6					
5	BEARING SUPPORT	3 / M1	1 / M6					
6	INLET SHAFT	5 / M1	1 / M6	2 / M1	2 / M4	1 / M6	1.5 / M3	
7	FRONT SUPPORT	2.5 / M1	1 / M4	1.5 / M6				
8	BACK SUPPORT	1 / M2	2.5 / M6	4.5 / M2	1 / M6	1 / M1	1 / M6	
9	BOTTOM HOUSING	1 / M1	20 / M6	2 / M4	3.5 / M6			
10	TOP HOUSING	1 / M2	1 / M4	20 / M6				
11	COUPLING HOUSING	5 / M1	4 / M6	7 / M6				
12	ELECTRICAL PUMP HOLDER	2.5 / M6	1 / M2	1 / M1	2.5 / M6	2.5 / M2	1 / M4	1.5 / M6

For the purpose of planning of capacities and establishing of the optimum structure of the system, it is necessary to determine the dependence of system's performances subject to output rate. On the basis of techno-economic indices obtained by simulation the "bottleneck station" is determined, which has critical for the limiting of output rate or the capacities of the system. With the addition of one or several stations of the bottleneck type, and repeated simulation new capacities are determined, together with other measurements of system's performances. Successive iterations of this procedure enables the generation of the relationship between the output rate and system's performances.

The results of the simulation and techno-economic analysis are presented in Table 3, and in graphical form in Fig. 8. The initial system has one machine of each type. The configuration of the system with optimum characteristics was obtained after eight iterations.

TABLE 3   Results of Techno-economic Analysis

SYSTEM ITEM NUMBER	M1	M2	M3	M4	M5	M6	AGV	OUTPUT RATE PRODUCTS /YEAR	TOTAL COST 10³ $/ YEAR	TOTAL COST 10³ $/ PRODUCT	UTILIZATION OF CAPITAL [%]
	NUMBER OF MACHINES IN SYSTEM										
	UTILIZATION OF MACHINES [%]										
1	1 40	1 40	1 4	1 13	1 96	1 80	1 30	30	67.3	2.27	42
2	1 42	1 42	1 5	1 14	2 52	1 85	1 33	31	78.5	2.50	39
3	1 62	1 62	1 7	1 20	2 75	2 62	1 46	46	86.0	1.91	50
4	1 75	1 85	1 8	1 26	2 95	2 67	2 23	58	90.4	1.55	61
5	1 83	1 83	1 9	1 27	3 87	2 84	2 25	61	99.1	1.61	58
6	2 54	2 82	1 13	1 40	3 95	3 80	3 11	88	100.2	1.45	62
7	2 64	2 84	1 14	1 43	3 77	4 86	3 12	95	107.3	1.42	63
8	2 77	2 77	1 17	1 51	4 92	4 77	3 15	114	115.7	1.27	70

As it may be seen the productivity in relation to the initial system was raised about 4 times, while the costs per product were decreased by 44%, and the level of the utilisation of invested capital was increased by 28%.

From graphs in Fig. 8 it may be seen that with the rise of the productivity the costs of the system are increasing, i. e. the need for gretaer initial

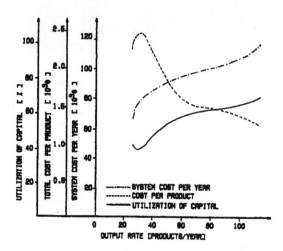

Fig. 8. System performances vs output rate

investments. However, due to the better utilisation of the equipment and the extension of the production the cost per product are cut down, which is bringing greater profit.

CONCLUSION

The proposed model offers significant assistance in the decision making process in the design of complex manufacturing systems. This model is applied in the elaboration of feasibility studies during the pre-investment phase of the design of industrial systems. Further research for the improvement of the presented method are in course, with the spesific orientation towards the linking of expert system for technology design (SAPT-EXPERT) and the development of the model for automatic technology-based layout generation.

REFERENCES

Colding, B. N. (1975). A cost model and performance index for a manufacturing systems. Annals of the CIRP, Vol. 24/1

Colding, B. N. (1980). Manufacturing performance criteria, optimization and the productivity mountain. 4th International Conference on Production Engineering, Tokyo.

Browne, J. and B. J. Davies (1984). The design and validation of a digital simulation model for job shop control decision making. Int. J. Prod. Res. 2, 335-357.

Milačić, R.V. (1981). The laboratory concept of modeling manufacturing cell. Annals of the CIRP. Vol. 30/1, 389-394

Milačić, R.V., Milojević, M. and P. Bojanić (1984) A contribution to the FMS design method based on utilization and economical efficiency approach. Proceeding of the 5th International Conference on Production Engineering. Tokyo, 788-794.

Milačić, R.V. (1987) Manufacturing systems design theory. Mechanical Engineering Faculty, Belgrade.

Milačić, R.V. and B. Babić (1988) A contribution to the simulation and optimization method for FMS. 20th International CIRP Seminar on Manufacturing Systems, Tbilisi.

Schmidt, B. (1977). GPSS-F-Einfeuhrung in die Simulatien Diskreter Systeme mit Hilfe eines FORTRAN - Programmpaketes. Springer-Verlag, Berlin

TASK EXECUTION SIMULATION OF ROBOT
APPLICATION PROGRAMS

G. Schreck, C. Willnow and Ch. Krause

*Fraunhofer Institute for Production Systems and Design Technology,
Department of Automation, Berlin, FRG*

Abstract: With the increasing use of industrial robots in manufacturing, off-line
programming becomes an economic factor. An important tool in an off-line
programming system is a simulator which allows to verify the correct task
execution. The principle aspects of such a simulation system are discussed. The
features required especially for the verification of assembly task execution are
outlined. The concepts of a realized system which allows the handling of time
variant kinematik linkages and the modelling of communication aspects is
described.

Keywords: Simulation, Industrial Robots, Assembling, Failure Detection,
Off-line Programming, Kinematic Modelling, Communication Modelling

INTRODUCTION

The use of industrial robots in manufacturing
systems requires computer aided planning tools
for the manufacturing system planning as well
as for robot programming. An example for such
tools is a simulation module within an off-line
programming system. It enables the user to test
and optimize off-line created application programs and can also be used for the planning and
verification of work cell layouts.

An important aspect of simulation systems is the
quality of their simulation results, i.e. to what
extend the simulation fits reality. This depends
on the computer internal modelling of the
behaviour of all work cell components. Further
the user interface plays an important role, i.e.
which knowledge and experience related to
manufacturing technology as well as system
operation is required by the user and what is
executed by the system automatically.

In the following these principle aspects of simulation systems will be discussed. The features
required especially for assembly task simulation
are outlined. A realized system mainly developed

as part of the ESPRIT-Project 623 which allows
the handling of time variant kinematic linkages
is described. Furthermore problems of the
sequential execution of parallel processes are
outlined.

THE PROBLEM OF PROGRAM
VERIFICATION IN TASK EXECUTION
SIMULATION

An important requirement for off-line programming systems is a test of the executability and
practicability of the created robot application
programs at the shop floor to a maximum extend
/1/. This practicability test contains testing of the
defined trajectories related to positions, orientations, velocities and accelerations. Furthermore end-effector commands and interactions
with peripherals have to be checked.

The following question arises: Has the application programmer to specify the effects he wants
to get as a result, or is it sufficient to input a robot
application program and the system shows the
effects of the program execution automatically?
In the first case an animation is executed while

in the second case an application program is checked and verified.

The general process of off-line programming is shown in Fig. 1. Out of a description of the production task which forms the information basis for the programmer, application programs are generated. These application programs can be tested by a simulator. The simulator itself is configurated to simulate the described production task and shows the effect the application program will have on the production environment. Also for this configuration of the simulator the task description is the main information source.

However, as both the application program and the simulation are derived from the same task description they will fit each other very well even if they are incorrect due to faults of the task description. So what errors can be found by using a simulation to verify an application program?

The problem which presents itself is that task descriptions - depending on their level of detail - contain ambiguities, lack completeness and consistency and are often not entirely formal. Therefore a formal basis for deriving application programs and their simulations is often not at hand. Consequently, the process of deriving application programs and their simulations has to be supported by humans who on the one hand are able to process ambiguous, incomplete and incorrect task descriptions but on the other hand lack accuracy. Even application programs and simulations which have been generated automatically do not feature the necessary level of completeness and therefore have to be checked and corrected.

AN APPROACH FOR TASK EXECUTION SIMULATION

One approach to the solution of the above mentioned problem lies in the realization that application programs as well as their simulations are formalizations of a task description. Both are of course incomplete, but when the application programs are run in a simulator, inconsistencies between these formalizations become evident. Consequently, application programs and simulation have to be seperated. This can be done in the following way:

The application program is the same program that will later be run in the real robot control. It influences the environment only by changing the output signals of the robot control but contains no information about the intended effects e.g. which part is gripped at which point of time during an assembly. The application program is a sequential algorithm and in this sense describes procedural aspects of the production task execution.

The introduced simulator program on the other hand describes declarative aspects of the production task. It defines which parts do exist, which of them can be gripped under which conditions. This approach results in a test of procedural aspects of the simulation task against its declarative aspects. Its advantage is obvious: It does not only visualise the intention of the programmer, it visualises the reaction of the robots environment to the statements in the robot application program.

SIMULATION MODELS

The simulator requires computer internal representations of all components of the work cell and their behavior according to predefined criteria. This repesentations are called simulation models and are structured in the following way /2/:

- control models which describe the motion behavior of components;
- kinematic models which contain the frame relations of the different links;
- shape models which describe the graphical representation of components.

Figure 2 shows the functional connection of these models for the example of a robot motion simulation. The robot application program is loaded into the control model of the IR. This control model interprets the application program and supplies the kinematic model with the joint values of the links. Within the kinematic model the frames describing position and orientation of each robot part are calculated. The connection of these frames to the relevant shape models enables the visualization of the robot motion on a graphic system.

Generally the above described procedure is also applicable for other components. To simulate the task execution of an entire work cell, models of all components are required.

TIME VARIANT KINEMATICS

A lot of production tasks e.g. assembly tasks need the handling of time variant kinematics. An example of gripping a part is shown in Fig. 3. The gripper is positioned above the part (state 1), approaches it (state 2) and then closes (state 3). At this stage, two different cases can occur: Either the part is linked to the gripper and both will move upwards together (state 4a) - which is what we would expect - or the part is not linked to the gripper and consequently will remain in position (state 4b). In reality the laws of physics decide which case occurs; in a simulation this decision is made by an algorithm.

This decision is often made by including some special statements to relink the simulated parts within the application program which controls the gripper's movements. However, in a simulation this may form an obstacle for error detection. If such a statement is executed, the part will be linked to the gripper regardless of the gripper's state or position.

In accordance with the proposed approach, another solution was chosen:

The simulation contains declarative statements which define the linkstate of parts and models physics in an abstract way /3/. A suitable formalisation of such statements is predicate logic. The predicates are dual and thus make it possible to decide whether a link is valid or not; their values depend on the state of the simulation. The level of detail can be chosen by the programmer, but it must be highly abstract in order to achieve a real-time behaviour of the simulation. For instance predicates like

$$distance(part, gripper) < \varepsilon \wedge closed(gripper)$$

(that means: if the distance between the part to be gripped, and the gripper itself is less than a declared distance, and the gripper is closed, then the part will be gripped)

can be sufficient.

The applied kinematic model is based on kinematic chains modelled by the method of Denavit-Hartenberg /4/ which provides rotational and translatory joints. It was extended by fixed joints and branches which results in kinematic tree structures. Such a tree describes the actual kinematic of a simulation state. The concept of defining several joints for a link and determining a currently valid one by predicates leads to a structure of a partially ordered graph which is the graph of possible trees. Fig. 4 shows such a graph (a) and its trees (b,c).

The links are denoted by blocks (A,B,C,D), the joints are denoted by arrows. The continuous lined arrows between links AB and AC denote time invariant joints, the dotted arrows between the links BD and CD denote time variant joints. The validity of the joints is determined by the predicates "p" and "¬p". The actual tree for the validity of p is shown in Fig. 4b, the tree for the validity of ¬p is shown in Fig. 4c.

The mapping of links to joints must be unique and competely determined for each point of time while running the simulation. This was achieved in the example by defining the second predicate as the negation of the first one. For more complex cases this must be guaranteed by the system. The following method was chosen:

To prevent ambiguity for each link priorities are assigned to the possible joints and their predicates. The predicates are evaluated with descending priority and the joint of the first true predicate is chosen.

To guarantee completeness a default joint and a tautology are added with lowest priority to the joints of each link. It places the link at an error position. Parts appearing at this error position indicate failure in task execution.

COMMUNICATION MODEL

The discussion of simulation systems generally deals with kinematics and controls, but the connection between them which is formed by the information exchange is not mentioned. The manager for the information exchange is called the comunication model. In the presented system it will play a central role, since it has a strong influence on system behaviour.

Controls and kinematic form a system of parallel processes. Because this system is simulated on a sequentially processing computer, the following problems of synchronization do occur:

a, b, and c are processes which compute the functions $y=a(x)$, $y=b(x)$ and $y=c(x)$. A, B, C and D are communication buffers of the processes, and the processes are linked by the buffers as shown in Fig. 5a.

If $y = v(\,t,\,x\,)$ is the function which provides the value y of buffer x at the point in time t, t_i is the point in time before and t_{i+1} the point in time after execution of the process, then the following statement is true for A with the execution sequence a; b; c

$$v(\,A,\,t_{i+1}\,) = a(\,v(\,B,\,t_i\,)\,),$$

but with the execution sequence c; b; a

$$v(\,A,\,t_{i+1}\,) = a(\,b(\,c(\,v(\,D,\,t_i\,)\,)\,)\,)$$

is true.

The execution sequences a;b;c and c;b;a evidently can lead to different results; the system doesn't operate in a determinate fashion.

Moreover, in one case more execution cycles are necessary for the information contained in buffer D to have an effect on buffer A than in the other. This difference in signal run time has a disadvantageous effect on processes with differentiating capabilities.

To overcome this effect, each buffer X is replaced by an input buffer X_i and an output buffer X_o (Fig. 5b) and the computation process is divided into two steps:

1. Each process computes the values of its input buffer and stores the result in the output buffer.

2. The values of the output buffers are copied to the corresponding input buffers. This copying process is done by series of processes designated as K.

Now the following statement for A is true with the execution sequence

$$a;b;c;k \quad v(\,A_o,\,t_{i+1}\,) = a(\,v(\,B_i,\,t_i\,)\,),$$

the same as with the execution sequence

$$c;b;a;k \quad v(\,A_o,\,t_{i+1}\,) = a(\,v(\,B_i,\,t_i\,)\,).$$

The system operates now in a determinate way, the run time of signals is definitely specified.

In order to keep the operations of the simulation system determinate, communication buffers are split up in input and output buffers and controlls and kinematic are executed in alternation with a copying process.

Moreover the in- and output buffer build the general interface between all components of the simulation controls, kinematic and graphic. The communication model makes available input and output buffers and lines of the necessary types of signals, e.g. boolean, real and frame. The interface of a simulation component consists of a set of input and output buffers. The connections between them are defined by lines.

The explicit formalization of communication aspects allows - with regards to the proposed approach - to verify the communication between controls in a work cell.

SIMULATION NETWORK

The data structure representing the controls, kinematic and communication model build a network called simulation network.

Figure 6 shows an example of a simulation network which includes a robot with three joints. Its kinematic consists of the socket, link1, link2 and link3. The robot is positioned on the floor and equipped with a gripper, consisting of two fingers. Its task is to grip a screw lying on a table.

Each joint of the robot is supplied with a joint angle entering the kinematic model through a named input channel (entering arrow). The joint angles are produced by a robot control denoted by a block. The joint angles leave the robot control through named output channels (leaving arrows). The input channels of the kinematic and the output channels of the control are connected by lines.

Beside the robot control is the gripper control. Its task is to transform the boolean signal from the robot control (close) to the real signals controlling the joints of the gripper's fingers. It models the motion behaviour of the gripper.

INTERNAL STRUCTURE OF THE SIMULATOR´S KERNEL

Figure 7 shows the modulare structure of the simulator´s kernel. The Communication Module provides several types of input and output buffers e.g. real, bool, integer, frame. It is the definite interface for the information interchange of the other kernel modules. The Process Module includes the modelling of controls and physical aspects. The Function Network Module provides fundamental operations like AND, OR, ADD, DISTANCE. It allows to define functions like the relinking predicates and other physical aspects. The Kinematic Module provides parts and joints of several types e.g. rotational, translatory, fixed. It contains the actual and possible kinematic trees. The Graphic System Module is the definite interface to the graphic monitor. It initializes the graphic system, loads the graphical description of the parts and transmits the actual position and orientation of the parts.

ENVIRONMENT OF THE SIMULATOR´S KERNEL

The external interfaces of the Simulator´s Kernel are shown in Fig. 8.

The Simulation Network (SN) is the complete set of data which are loaded into the Simulator´s Kernel to execute a simulation task. There are two ways to build the SN in the Simulator´s Kernel.

The Simulation Language Interpreter interprets text files written in a specific simulation language and builds the SN. An example of this language describing the network from Fig. 6 is given in Fig. 9. The DB-Description Loader is the equivalent interface to a Data Base System.

The State Saving & Loading allows to save a current simulation state and to reload it. Thus the simulation can be started not only from the beginning but also from a saved state e.g. for testing.

The Network Editor is an interactive information interface for the user.

The Consistency Check can detect inconsistencies like loops in the kinematic chains, ambiguity and incompleteness of the predicates determining the validity joints.

SUMMARY

Problems of the task execution simulation of robot application programs are discussed. A simulation concept testing the procedural aspects of the statements in the application program to the declarative aspects of the behavior of the robot environment is described. Especially the handling of time variant kinematics and the modelling of communication processes are presented. Realization aspects of the simulation system are outlined to show the flexible configuration to applications.

REFFERENCES

1.   Spur, et al:
     Planning and Programming of Robot
     Integrated Production Cells. ESPRIT
     Technical Conference, Brussels, Belgium,
     September 1987.

2.   Duelen, G.; Kirchhoff, O.; Bernhardt, R.;
     Schreck, G; Algorithmic Representation of
     Work Cells and Task Description for Off-line
     Programming. Robotics and Computer-
     Integrated Manufacturing, Vol3, No 2, pp.
     201-208 (1987)

3.   Sinowjew, A; Wessel, H.:
     Logische Sprachregeln. VEB Deutscher
     Verlag der Wissenschaften, Berlin (DDR),
     1975.

4.   Denavit, I; Hartenberg, R.S;
     A kinematic Notation for Lower Pair
     Mechanisms Based on Matrices, Journal of
     Applied Mechanics 77 (1955) pp. 215-221.

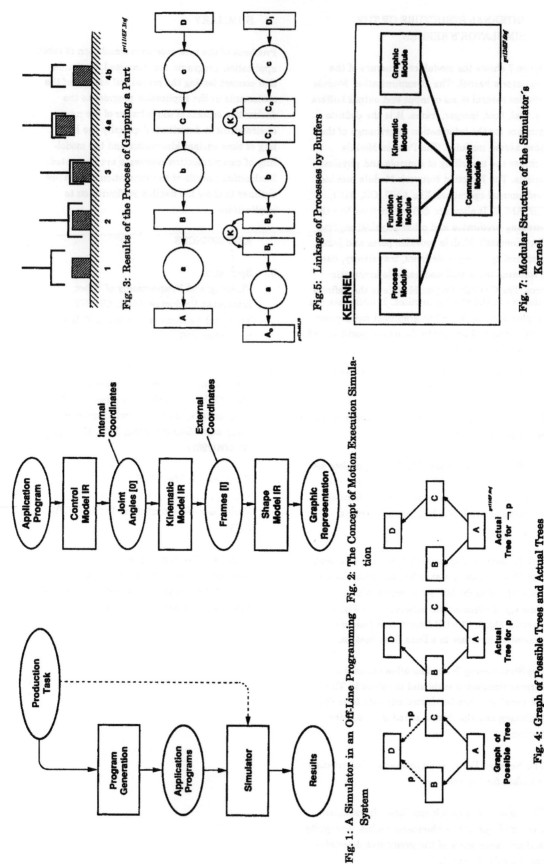

Fig. 1: A Simulator in an Off-Line Programming System

Fig. 2: The Concept of Motion Execution Simulation

Fig. 3: Results of the Process of Gripping a Part

Fig. 4: Graph of Possible Trees and Actual Trees

Fig. 5: Linkage of Processes by Buffers

Fig. 7: Modular Structure of the Simulator's Kernel

KERNEL

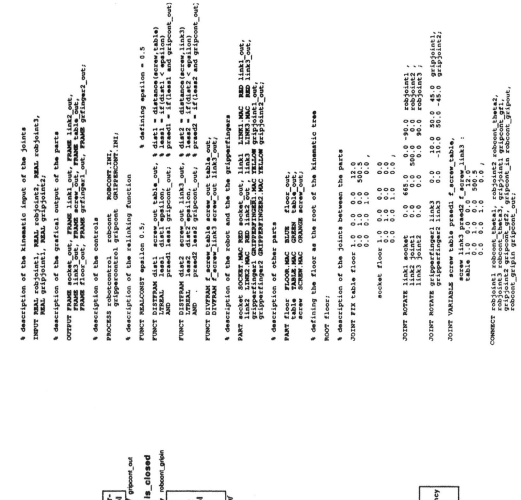

```
% description of the kinematic input of the joints

INPUT REAL robjoint1, REAL robjoint2, REAL robjoint3,
      REAL gripjoint1, REAL gripjoint2;

% description of the grafical output of the parts

OUTPUT FRAME socket_out, FRAME link1_out, FRAME link2_out,
       FRAME link3_out, FRAME screw_out, FRAME table_out,
       FRAME floor_out, FRAME grfinger1_out, FRAME grfinger2_out;

% description of the controls

PROCESS robotcontrol   robcont   ROBCONT.INI,
        grippercontrol gripcont  GRIPPERCONT.INI;

% description of the relinking function

FUNCT REALCONST epsilon 0.5;                  % defining epsilon = 0.5

FUNCT DISTFRAM dist1  screw_out table_out,    % dist1 = distance(screw,table)
      L7REAL   less1  dist1-epsilon,          % less1 = if(dist1 < epsilon)
      AND      praed1 less1 gripcont_out;     % praed1 = if(less1 and gripcont_out)

FUNCT DISTFRAM dist2  screw_out link3_out,    % dist2 = distance(screw,link3)
      L7REAL   less2  dist2-epsilon,          % less2 = if(dist2 < epsilon)
      AND .    praed2 less2 gripcont_out;     % praed2 = if(less2 and gripcont_out)

FUNCT DIVFRAM f_screw_table screw_out table_out,
      DIVFRAM f_screw_link3 screw_out link3_out;

% description of the robot and the grippfingers

PART socket SOCKET.MAC RED socket_out, link1 LINK1.MAC RED link1_out,
     link2 LINK2.MAC RED link2_out, link3 LINK3.MAC RED link3_out,
     gripperfinger1 GRIPPERFINGER1.MAC YELLOW gripjoint1_out,
     gripperfinger2 GRIPPERFINGER2.MAC YELLOW gripjoint2_out;

% description of other parts

PART floor FLOOR.MAC BLUE   floor_out,
     table TABLE.MAC GREEN  table_out,
     screw SCREW.MAC ORANGE screw_out;

% defining the floor as the root of the kinematic tree

ROOT floor;

% description of the joints between the parts

JOINT FIX table floor 1.0  0.0  0.0   12.0
                      0.0  1.0  0.0  500.5
                      0.0  0.0  1.0    0.0  ,

          socket floor 1.0  0.0  0.0   0.0
                       0.0  1.0  0.0   0.0
                       0.0  0.0  1.0   0.0  ;

JOINT ROTATE link1 socket  0.0  665.0   0.0  -90.0  robjoint1 ,
             link2 joint1  0.0    0.0 500.0    0.0  robjoint2 ,
             link3 joint2  0.0    0.0   0.0   90.0  robjoint3 ;

JOINT ROTATE gripperfinger1 link3  0.0   10.0  50.0   45.0  gripjoint1,
             gripperfinger2 link3  0.0  -10.0  50.0  -45.0  gripjoint2;

JOINT VARIABLE screw table praed1 f_screw_table,
               screw link3 praed2 f_screw_link3 :
               table 1.0  0.0  0.0   12.0
                     0.0  1.0  0.0  500.5
                     0.0  0.0  1.0    0.0  ;

CONNECT robjoint1 robcont_theta1, robjoint2 robcont_theta2,
        robjoint3 robcont_theta3, gripjoint1 gripcont_gf1,
        gripjoint2 gripcont_gf2, gripcont_in gripcont_gripout,
        robcont_gripin gripcont_out;

END;
```

Fig. 9: **Example of a Simulation Network Definition**

Fig. 6: **An Example of a Simulation Network**

Fig. 8: **Environment of the Simulator's Kernel**

COMPUTER-AIDED PROGRAMMING TOOL
FOR ROBOTICS

J. Warczynski

Institute of Control Engineering, Technical University of Poznan, Poznan, Poland

Abstract. There is a growing interest in the development of software tools
for off-line robot programming in the field of robotics aplication. The
use of such tools could simplify the time-and-labour-consuming task of
robot program development as compared with classical programming methods.
This could be particularly promising within a CAD — CAM allowing
integration robotics into the common planning and control system. The paper
describes such a software package aimed to assist off-line development and
debugging robot programs as well as model-based evaluation of robot
performance with respect to the movement dynamic phenomena. The system
operates in the framework of three main modes which are : modelling,
programming and simulation. Also the workcell layout and verification of
control strategies are possible.

Keywords. Robot; robot system modelling, off-line programming, simulation,
robot dynamics.

INTRODUCTION

The wide industrial application of robots
in the last decade has revealed serious
bottlenecks of true eccnomic use of them
when applied in increased variety of
tasks. The main difficulty is the lack of
appropriate software systems.

When robots are used for simple tasks with
operations like pick and place or spot
welding only, then the classical teach-in
programming with playback of the taught
positions could be sufficient. However, if
robots were to be applied to more extended
tasks, the labour-intensiv nature of such
programming method, connected with
relatively long production break which is
necessary for program testing and
debugging, would greatly limits such an
application because of obvious economic
reasons.

There have been developed some dozens of
large, general purpose CAD systems
concerning robotics. As a rule they
require powerful mainframes with sophisticated graphical workstations. Generally,
they seem to be based on a similar scheme
(Hocken, 1986). With their CAD graphical
facilities they can be used for modelling
the robot and workcell geometry. The
graphical part of a typical system is
coupled to other modules. Usually they
are destined for modelling robot
kinematics and for graphical determining
of the manipulator path within the
modelled workcell. An output generation
module converts the graphically developed
program into the robot control code which
can be then downloaded to a robot
controller of a specific manufacturer.

There is also another trend in conceiving
the problem today. It is based on
dedicated systems, devoted to

accomplishing some special tasks, which
are implemented on low-cost mini- or
microcomputers (Dombre et.al.,1986). The
outline of such an off-line programming
and simulation system for robot
structures in an open kinematic chain of
rigid links, with fifth-class pairs has
been described in this paper. In the
presented software package the emphasis
has been placed on modelling the robot
dynamics and model-based evaluation of the
robot system performance with respect to
the dynamic phenomena of manipulator
movement. This is achieved by emulating
the work of the robot systems.

SYSTEM ORGANIZATION

An outline of the system structure is
presented in Fig.1 which shows its
modular organization. There are five
program modules of different functional
destination in it:

- manipulator dynamics will be modelled
  using dynamic model generator and

- module for modelling of execution
  devices, sensors and control loops
  complements this aim,

- in order to prepare robot programs
  the programming module will be used
  and

- geometric models needed for program
  verification will be created with the
  geometric model editor,

- desired form of simulation results
  ensures the graphics presentation
  module

The modules cooperate with one another
and also with the database that proceeds

under control of a system managing program.

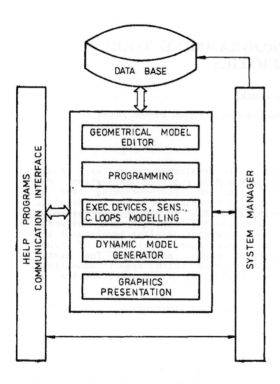

Fig.1. The system structure.

The above outlined structure is thought to be open so as to enable eventual extensions of the system by joining new program modules. The use of the database in the system supports, to certain extend, such possibilities. The application of the database enables also the storage of input data, system outputs or other information related, in some way, to the system. The memorized information can be retrieved when necessary and input, in the appropriate form, into a program module being in use.

Special endeavours were done to make handling the system as user-friendly as possible. A dialog manner in which a user interacts with the system serves this purpose. On request the user will be provided with comprehensive information about modes, functions and possibilities of the system, so as no knowledge of its work is required in advance. There are also some additional programs which could be used for preparing the input data or for processing the simulations outputs.

Generally the system operates within the framework of the following three main modes:

- modelling,
- programming and
- simulation.

MODELLING

In the modelling mode, robot system models of different nature will be created. This proceeds by

- automatic generation of manipulator kinematic and dynamic models,

- interactive creation of dynamic models of robot actuator, sensor and control systems as well as geometric models of a robot and a workcell.

Robot Kinematic and Dynamic Models

The evaluation of robot movement dynamic performance based on the simulation requires kinematic and dynamic models of the robot system.

The model of manipulator kinematics is necessary for coordinate transformations, i.e. for solving the inverse and direct manipulator kinematics. This model will be generated automatically after the input of Denavit – Hartenberg parameters for manipulator links.

The facilities for creation the robot dynamic models have been split over

- dynamic model generator and

- module for modelling of execution devices, sensors and control loops.

With the aid of the dynamic model generator the equations of the movement of the manipulator mechanical part will be generated automatically by applying a symbolic processing. This is accomplished according to the Lagrangian formulation with Q-Bejczy's matrices (Bejczy, 1974; Luh, 1983). The application of Q-matrices allows for significant simplifying of symbolic manipulations as the differentiation operation of homogeneous transformation matrices can be reduced to the symbolic matrix – matrix multiplication. The whole symbolic processing of the model generation process has been programmed with the use of a symbolic processor written in PROLOG. This program outputs symbolic expressions for the coefficients of the movement equations:

$$F_i = D_{ii}\, q_i + \sum_{j \neq i}^{N} D_{ij}\, q_j +$$

$$+ \sum_{j=1}^{N} \sum_{k=1}^{N} q_j\, C_{i,jk}\, q_k + G_i \tag{1}$$

where

q_i is the i-th joint coordinate,

F_i is the externally applied actuating joint force or torque,

D_{ii} is the coordinate-dependent effective inertia at the joint i,

D_{ij} is the coordinate-dependent coupling inertia between joints i and j

$C_{i,jk}$ are coordinate- and velocity-dependent centrifugal and Coriolis coefficients at the i-th joint,

G_i is the coordinate-dependent gravitational force at the joint i.

As soon as the symbolic expressions for the coefficients have been obtained, they are processed by a special, converting program, written in PROLOG, which transforms these expressions into a C-program computing their values for given values of the joint variables. By multiplying the computed coefficients respectively by instantaneous joint velocities and accelerations the manipulator inverse dynamic problem could be solved, i.e. the required forces or torques for the prescribed motion could be find. For simulation however, solving the forward dynamics is of the main importance as it allows for the evaluation of robot performance with respect to the modelled dynamic phenomena of its motion.

For emulation of the whole robot system the dynamics of the robot mechanical part as well as those of the actuator and control systems and their mutual interactions have to be taken into account. This should be done in the framework of a common complete dynamic model. In order to simplify modelling the actuator system we have assumed generally that contributions to the actuator torque or force yielded from feedforward control and negative feedback, arising due to the actuator velocity, are additive. Under this assumption the actuating forces or torques can be expressed in the following form:

$$F = B' u - H q \qquad (2)$$

where

u   is the control vector of actuator system,
B' is the control distribution matrix,
H   is the matrix of the viscous friction coefficients and constants related to the velocity-dependent reaction.

The mutual couplings between the dynamic models of the mechanical part and of the actuator system of the manipulator are shown in Fig.2.

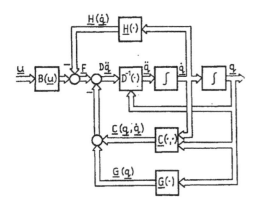

Fig.2 The complete manipulator dynamic model.

The modelling of an actuator proceeds by interactive setting the actuator state equations which are of the universal form:

$$x_i = A x_i + B u_i \qquad (3)$$

where

x_i is the state vector of the i-th actuator,
u_i is the control vector of the i-th actuator.

Besides of the model of the actuator system also the emulator of the robot system controller has to be created. Currently it can be accomplished by setting an appropriate emulating program. This solution is not convenient to the user. Another one is yet under development and this will be an object-oriented command language for unique modelling the actuators and controllers.

Geometric Models

It is obvious that in order to verify the programmed robot movements it is necessary to provide the system with the geometric model of the robot. It will be achieved by using the geometric model editor aimed at creation such a model within 2.5D computer graphics, i.e .as a wire-frame model displayed on a CRT terminal. This model editing process proceeds inter-actively and can be divided into two stages.

In the first stage the geometric primitives like cube, tetrahedron and other polyhedra are used for constructing the separate manipulator links and objects in a workcell. The dimensions of the primitives are parameterized which allows for changing them according to the user's desire so as to built the appropriate models. There are also ready-for-use models of parts which are of common use in robotics. Each primitive is defined in an interactive manner with a menu technique, which means that the operator selects the primitive type only and than enters the values of its geometric para-meters. Afterwards, the primitive will be located in the scene which can be done with the help of the appropriate transformation function. The primitives built into environment elements or separate manipulator links are rigidly attached together and, if necessary, will also be transformed together.

In the second editing stage a model of the whole manipulator will be combined from the links obtained in the first stage. This is different from modelling elements belonging to the workcell since it concerns relationships between geometric objects which are not constant. These relationships will be determined by establishing joint movement variables for the connected pairs of links, proceeding from the base link to the end-effector. According to this sequence a special list will be created which will be responsible for an appropriate transformation order of manipulator links during the animation process.

PROGRAMMING

For preparing robot programs a user is provided with the system programming module which enables textual programming in a high-level programming language or within the teach-in method. It is also possible to combine the both methods which sometimes simplifies the programming task. The module consists of the following programs (Fig.3.):

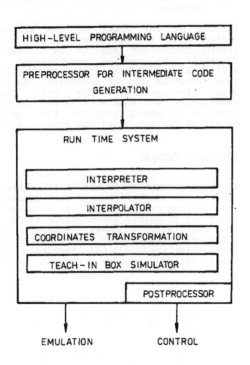

Fig.3. The programming module.

- program editor for preparing a robot program in a high-level programming language,
- programming language compiler,
- preprocessor,
- interpreter,
- interpolator,
- coordinate transformation program,
- teach-in box simulator.

The preprocessor placed on the input side of the run-time system transforms the compiler output code into an intermediate one. We have used here the idea of IRDATA (Rembold, Blume, Frommherz, 1985). Such an approach is aimed at allowing the use of different programming languages which make the system more universal. The execution of the program instructions included in the intermediate code will be controlled or initiated by the interpreter. More important tasks of the interpreter are trajectory interpolation and coordinates transformation. The coordinates transformation comprises computing the manipulator joint variables given the world coordinates of the tool center point. These world coordinates of each trajectory point are determined by the interpolator with the use of splines and the trajectory knots specified in the user's program.

SIMULATION

In the framework of the system simulation mode programs developed within the programming module will be executed using models of the robot system instead of the

real hardware. The data flow of the simulation process is shown in Fig.4.

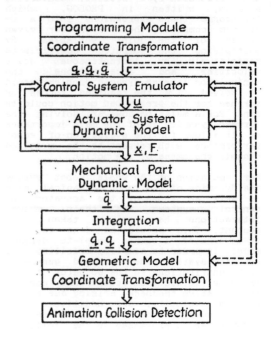

Fig.4. The data flow of the simulation.

On the basis of the simulation outputs like joint coordinates, state vectors or control commands it will be possible to evaluate the performance of each modelled part of the robot system. Moreover, the animated geometric model of the manipulator, placed within the workcell model, can assist in verifying the programmed robot actions and especially in detecting collisions with environment elements.

There are two possible levels of the simulation mode. The first one concerns the geometric and kinematic models only, without reference to the dynamic models. In the second level also the models of the actuator system, controllers and manipulator mechanical part are referred to.

The proper presentation of the system outputs assures the graphics presentation module.

In order to demonstrate some capabilities of the presented system let us consider several simulation outputs relating evaluation of dynamic performance for an example robot manipulator shown in Fig.5.

The input data referring to the kinematics and dynamics of this manipulator are defined in Table 1.

The results of the simulation (Fig.6 - 7) refer to a positioning task. The required generalized forces for accomplishing this task are shown in Fig.6. Fig. 7. shows the contributions to this forces caused respectively by the gravitation, effective inertias, Coriolis and centrifugal forces. All graphics show the normalized values.

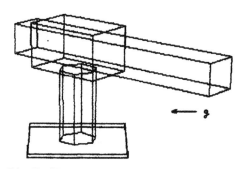

Fig.5. The example robot manipulator

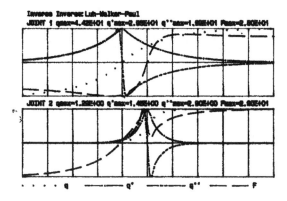

Fig.6. Simulation results for the positionig task

CONCLUSION

The robot simulation and off-line programming system has been developed to assist programming robots, workcell layout and especially model-based dynamic performance evaluation of the robot system work.

The main goals of the system could be pointed out as follows:

- Increase of the rate of robot operation time; this could be possible due to preparing a new program does not require the presence of a real robot.

- Decrease of programming costs.

- Increase of the safety factor during the program verification.

- Detecting errors in sequences of the programmed robot movement (for example collisions with environment elements) before a physical robot is set in motion.

- Evaluation of the robot dynamic performance.

- Verification of control methods and algorithms to be applied.

- Workcell layout and search for its best organization which could help in

choosing the optimal location of the robot within its environment.

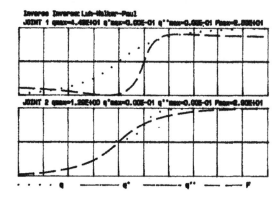

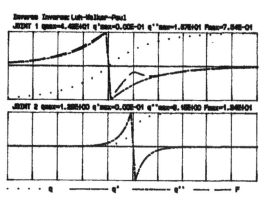

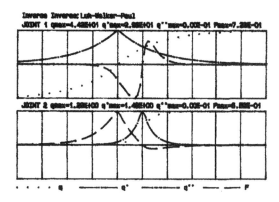

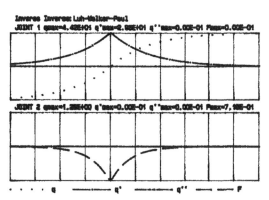

Fig.7. The simulation outputs

TABLE 1 Parameters of the Example Robot

Link	Mass	I_{xx}	I_{yy}	I_{zz}	I_a	cener of mass			D-H paramet.				fric.coeff.
	[kg]		[kgm^2]				[m]						[Ns/m;Nms]
						x	y	z	α	a	d	θ	
1	5.01	0.108	0.018	0.100	2.193	0.00	-0.105	0.00	-90.0	0.0	0.0	q_1	0.0825
2	4.25	2.510	2.510	0.006	0.782	0.00	0.00	-0.64	0.0	0.0	q_2	0.0	0.3098

REFERENCES

Bejczy,A.K. (1974). Robot arm dynamics and control. Tech. Memo.,Jet Propulsion Lab., Pasadena, pp. 33-669.

Dombre,E.,A.Fournier, C.Quaro and P.Borrel (1986).Trends in CAD / CAM Systems for Robotics. Proc. IEEE. Int.Conf. Robotics and Automation. San Francisco.

Hocken,R. and G.Morris (1983). An overview of off-line robot programming systems. Annals of the CARP.Vol.35, 2nd ed.

Luh,J.Y.S. (1983). Conventional Controller Design for Industrial Robots - A Tutorial.IEEE Trans. Sys. Man. Cbs. SMC-13, 328-315.

Rembold,U., C.Blume and B.J.Frommherz (1985). The proposed robot software interfaces SRL and IRDATA. Robotics and Computer - Int. Manuf.,2, 219-225.

Warczynski,J. (1987). Computer System for Simulation of Robot Kinematics and Dynamics. Tech. Report, Inst. of Control Eng., Tech. Univ. of Poznan, Poland. (In polish).

MODELLING INDUSTRIAL CHEMICAL PLANTS: SIMULATION VIA STELLA™

G. Jones and P. B. Taylor

*Department of Computing, Management Science, Mathematics and Statistics,
City of London Polytechnic, London, UK*

Abstract. A simple example is given which serves as a generic subsystem for models of industrial chemical plants. Some properties and operating policies are described and some of the difficulties of using an analytical approach noted. As an alternative, the authors consider the benefits and shortcomings of using a visual interactive modelling package, STELLA™.

Keywords. Chemical industry; modelling; simulation; simulation languages; stochastic systems.

INTRODUCTION

There is considerable interest within the chemical industry in evaluating the availability and performance of chemical plants. Many typical plants are very complex in nature. They tend to consist of processing stages (or units) together with buffer stocks (or tanks) with many or all of the units being subject to failure. A consequence of such failures is that units cannot always be operated independently of one another due to the presence of full or empty tanks which necessitate adjustments to flow-rates. Thus, we immediately see the need to be aware of an overall operating policy for the plant.

A complete description of a particular plant is seen therefore, to depend on many factors. These can include such aspects as the nature of the chemical processes involved, a detailed operating policy to be employed in respect of the plant, details of planned maintenance sessions and so on. Thus, it appears that any attempts at modelling individual plants are likely to falter if we seek to include all aspects of the particular plant involved. The resulting model would be so complex as to be intractable.

We therefore consider simplified models which aim to reflect the essential features to be found in typical plant configurations. We concentrate our attention on such aspects of plant behaviour as the failure and repair characteristics of units, capacities of tanks and maximum operating rates of units.

METHODS OF ANALYSIS

The two principal approaches which have been employed to date in the solution of the type of model under consideration are the analytic approach and simulation.

Hybrid approaches have also been advocated, see for example Cheng (1972) and Jones (1987), but these will not be discussed in this paper as we propose to concentrate on simulation.

The analytic approach proves to be very difficult for all but unrealistically simple models. Thus, even though several authors have approached the problem analytically, few examples appear in the literature.

The model to be considered in this paper consists of just two units together with a tank, arranged in series as shown in Fig. 1.

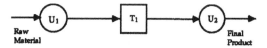

Fig.1 A simple unit-tank model

Clearly this is a very simple model and few industrial processes can be represented as simply. However it is easy to see that quite complex manufacturing plants can be represented realistically by using the model of Fig. 1 as a building block; networks of such elements, involving series, parallel and branching components, can be established and, given a suitable development environment, simulated.

Units

A unit may be subject to failure in which case it must be repaired or replaced. This means that at any given time t, a unit may be working, or capable of working, or it may have failed and be undergoing repair. Thus, we may think of a particular unit as being in one of two states, working or failed. We therefore refer to a unit which is working as being UP, and a failed unit is said to be DOWN.

We assume that a unit U_i has a maximum possible workrate, or rating, α_i. Then at some time t, its actual workrate $r_i(t)$ will satisfy the inequality $0 \leq r_i(t) \leq \alpha_i$. We note that $r_i(t)$ can be zero either due to U_i being down or because of congestion downline of U_i; for example, in Fig. 1 the tank T_1 may be full and unit U_2 may have failed so that we must set $r_1(t) = 0$. Similarly, we may need to set $r_i(t)$ less than its rating α_i in order to balance input and output in respect of a tank which is empty or one which is full, depending on the ratings of the units. Finally, we need to know something concerning the failure rate of a unit U_i and also the time to repair of a failed unit. Thus, we define

$$P(U_i \text{ fails in } (t, t + \delta t) | U_i \text{ up at } t)$$
$$= \mu_i \delta t + o(\delta t)$$

and, for a unit which is down at time t,

$$P(U_i \text{ has repair completed in}$$
$$(t, t + \delta t) | U_i \text{ down at } t)$$
$$= \lambda_i \delta t + o(\delta t).$$

Tanks

Each tank T_j has a finite capacity K_j. At any time t, the level of fluid in the tank, $S_j(t)$ say, will satisfy the inequality $0 \leq S_j(t) \leq K_j$.

Clearly, the tank may become empty or it may become full. When either of these states occur, it will be necessary to adjust the workrate of units connected to the tank.

The Problem

We have already introduced the concept of an overall operating policy for plants. Thus in order to compare competing policies we will adopt as our criterion that of expected throughput of material under the different operating policies. One such policy could be to run all units at maximum possible rate at all times, which we will call the *full-on* policy. Clearly, we are interested in such quantities as the distribution of stock levels in the tanks and also the availability of units.

THE ANALYTIC APPROACH

The model shown in Fig. 1 has been considered by a number of authors since the early 1960s, starting with Finch (1961). They have approached the problem from differing standpoints and one aspect worthy of mention is that analyses have been carried out under differing operating policies. In particular, we mention the work of Fox and Zerbe (1973) and more recently Malathronas et al (1983). As Malathronas et al remark, Fox and Zerbe assume a *tank-full* policy. This means that the units are operated in such a way that the tank always remains full, even if it means shutting down U_2. Other papers

advocate a *do-nothing* policy; this really means adopting a policy of limited non-intervention. That is, both units are run until one fails or the tank becomes full or empty when some action must be taken. This latter policy clearly coincides with Cheng's full-on policy provided the units are operated at the maximum rate. Malathronas et al also point out that the do-nothing policy produces a larger expected output than the tank-full policy and is the more realistic of the two. It would appear that they are unaware of the result due to Cheng (1972) which establishes the full-on policy as being optimal for models of this type, which he has called line-tanks models.

The various analyses undertaken highlight the complexity of the analytic approach for this simple model. They all deal with the steady-state solution which tends to be obtained under simplifying assumptions such as equal ratings, equal failure rates or equal repair rates for the units.

Jones (1987) has discussed in detail the transient case and its complexity is such that there is little virtue in reproducing any detail here. Suffice it to say that the solution for stock level alone involves the solution of a set of four simultaneous partial differential equations of the type found in single-server queueing theory. Further, if any of the relationships between ratings are altered the equations to be solved will also alter. That is, each case must be considered on its merits.

We therefore proceed to consider a simulation approach but before doing so we remark on the various events which affect the operation of the system. They are the failure of and completion of repairs to units, together with the tank becoming full or empty and these are the features we must incorporate in our simulation model which will be operated under the full-on policy.

Finally, we indicate in Fig. 2 some typical behaviour patterns of stock level with the passage of time. As can be seen from the figure, it is possible for the stock level to have the values zero and K, that is for the tank to be empty or full respectively, for significant periods of time.

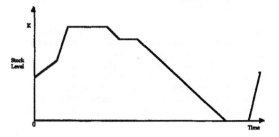

Fig. 2   Changes in stock level over time

STELLA™ AS A DEVELOPMENT TOOL

STELLA™ (Richmond *et al*, 1987) is a visual interactive simulation environment implemented on a Macintosh microcomputer. The interface uses a hydraulic metaphor with elements called *stocks*, *flows*, *converters* and *connectors* which can be assembled to form a structural diagram. Three tools are provided to assist the development: a *hand*, to select and move entities around the diagram; a *ghost* which clones stocks and converters and enables complex diagrams to be simplified topologically; and *dynamite*, which edits (by destruction) parts of the diagram. In the current version of STELLA™ (2.01) the usual Macintosh editing facilities of Cut, Copy and Paste can be used, resulting in a powerful development environment at diagram level.

The interface is shown in Fig. 3, together with an example of a very simple system. The similarity of the example to the conceptual model of Fig. 1 shows that, iconographically at least, STELLA™ promises to be a useful modelling device for systems such as those which have been described earlier in the paper.

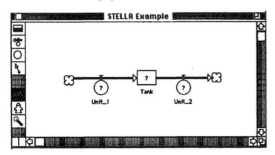

Fig. 3 The STELLA™ interface

Functional and logical relations between inputs and outputs are easily specified (beginning by 'double-clicking', in the Macintosh vernacular), the process being simplified by a variety of built-in selectable functions and by internal consistency checks which STELLA™ carries out interactively.

Debugging and testing of a completed (or partially completed) model are greatly facilitated by diagram animation and by using various windows, mainly for output, provided by STELLA™. The Equation window gives the model in equation form and permits easy checking and editing. The Graphical and Table windows are invaluable as a means of monitoring the workings of the model at run-time - the model can be run a number of times until a seemingly anomalous result is observed and it can then be interrupted and its logic scrutinised.

While STELLA™ has undeniable advantages in its visual and interactive operation, it is not without its limitations for this kind

of application; this perhaps results from its development as a tool primarily for system dynamics. It uses equal time increments and has no facilities for event- or activity-based simulation which would effect a considerable saving in computation time. Also, although it possesses the usual range of logical functions it has no logical variables nor, more importantly, logical control structures. Even so, its use as a development tool is unrivalled in the authors' experience.

THE STELLA™ MODEL

The complete model, as a structural diagram, is shown in Fig. 4 (the shadings on the Stock elements are part of the animation). It will be seen that the basic model, in the top part of the diagram is quite simple and 'readable', whereas the generation of breakdowns and their duration, shown in the lower part, seem more involved than the simple model would seem to merit. This is undoubtedly due to the lack of logical control structures.

A sample of the model's equations is shown below. It will be appreciated that considerable assistance is afforded to the modeller in their construction by way of prompts, syntax checks and variable selection.

```
Tank = Tank + DT * (r1 - r2)
INIT(Tank) = tank_cap/2
Throughput = Throughput + DT * (r2)
INIT(Throughput) = 0
Time_next_up_1 = Time_next_up_1
      + DT * (chg_up_time_1)
INIT(Time_next_up_1) = inter_bd_time_1
      + next_down_time_1
Time_of_next_bd_1 = Time_of_next_bd_1
      + DT * (chg_bd_time_1)
INIT(Time_of_next_bd_1) =
      inter_bd_time_1
alpha_1 = 1.2
alpha_2 = 1
chg_bd_time_1 =
      IF (Time_next_up_1 > TIME)
      THEN 0
      ELSE(TIME + inter_bd_time_1
      - Time_of_next_bd_1)/DT
chg_up_time_1 =
      IF (Time_next_up_1 > TIME)
      THEN 0
      ELSE (inter_bd_time_1
      + next_down_time_1)/DT
inter_bd_time_1 =
      -(1/mean_bd_rate_1)*LOGN(RANDOM)
mean_bd_rate_1 = 0.02
mean_down_time_1 = 1/0.03
next_down_time_1 =
      -(mean_down_time_1)*LOGN(RANDOM)
r1 = IF(Tank≥tank_cap-0.00001)
      THEN alpha_2*up_1*up_2
      ELSE alpha_1*up_1
r2 = IF (Tank<0.00001)
      THEN 0 ELSE alpha_2*up_2
tank_cap = 10
up_1 = IF(TIME>Time_of_next_bd_1)
      AND (TIME <Time_next_up_1)
      THEN 0 ELSE 1
```

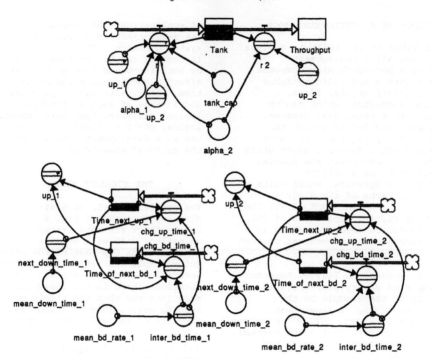

Fig 4 STELLA™ structural model

Finally, as an example of an extensive run Fig 5 shows the graph of tank level and states of the units against time.

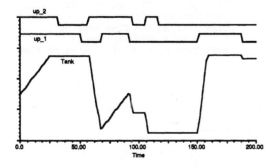

Fig. 5 Output of an extended run

CONCLUDING REMARKS

Experience with STELLA™ has shown its considerable usefulness as a development tool for simulation despite the fact that queueing models are not its particular forte. The ability to debug and verify a model on-line, gaining a feel for how the modelled system behaves, contributes greatly to the modeller's understanding. It is felt that it would be particularly fruitful to carry out participative model development together with an industrial client using the interactive environment.

While the powerful graphics interface is an obvious boon in development work, execution times can be disconcertingly long. It is certainly recommended that the Macintosh II version (making use of the 68020) is employed for all serious queueing applications.

REFERENCES

CHENG, R.C.H. (1972). Models for the simulation of industrial chemical plants. Ph.D. Thesis, University of Bath, Bath, UK

FINCH, P.D. (1961). Storage problems along a production line of continuous flow. *Annls. Univ. Scient. Bpest. Rolando Eötuös* **3-4**, 67-84.

FOX, R.J. and D.R. ZERBE (1974). Some practical system availability calculations. *AIIE Transactions*, **6**, 228-234

JONES, G. (1987). Analytic and Markov decision process models of industrial chemical plants. PhD Thesis, UWIST, University of Wales, Cardiff, UK

MALATHRONAS, J.P, J.D. PERKINS and R.L. SMITH (1983). The availability of a system of two unreliable machines connected by an intermediate storage tank. *I.I.E. Trans.*, **15**, 195-201.

RICHMOND, R.,S. PETERSON, and P. VESCUSO (1987) *An Academic User's Guide to STELLA™*. High Performance Systems, New Hampshire, USA

ACKNOWLEDGEMENT

The authors are glad to acknowledge the use of STELLA™ software and documentation from High Performance Systems of Lyme, New Hampshire, USA in the preparation of this paper.

Copyright © IFAC Information Control Problems in
Manufacturing Technology, Madrid, Spain 1989

KNOWLEDGE-BASED ORDER SPECIFIC NC DRILLING SYSTEM

L. F. Pau*, D. Paus** and T. Stokka**

*Technical University of Denmark, Lyngby, Denmark
**Nordic Intelligent Manufacturing, Stabekk, Norway

Abstract: The paper describes a knowledge based system which generates hole positions/types, as well as the corresponding NC macros, on custom order beams for trucks. This system exploits components data bases, CAD, and configuration procedures. It is implemented part in PROLOG, part in other languages, and uses a meta parser. The main benefits are in order customization, and reduced manufacturing times for such orders.

Keywords: Expert system, NC machines, Customized manufacturing, Automotive industry, Drilling, Planning.

1. INTRODUCTION

The background for the project below lies in the overall ambition of several major European car and truck manufacturers, to be able to customize their products to specific client requests, while minimizing costs and inventories. In this way, the same manufacturing facilities should be able to produce customized parts, obtained by revisions to basic CAD designs but checked for geometrical and assembly compatibility, (Pau, 1986).

This production philosophy is quite similar to the one presiding in microelectronics over ASIC's (Application specific integrated circuits), or printed circuit boards and cell libraries in particular. "Building block" CAD files are available from the library, but are assembled and simulated jointly to achieve customized circuit or board needs (Battelle, 1987).

To achieve the above ambitions, special emphasis must be put on the following techniques (Figure 1):

A components description database, with attributes for all variants parts

B specification procedures, with data structures encompassing all parts description databases

C descriptions of the uses of each component, and of its relations to other components

D order verification procedures, for parts configuration and checks against parts in inventory

E updates to CAD files

F updates to production schedules at the flow control level

G updates to NC macros

H distributed processing and data communication

The project carried out for one company, has been focussing on one specific aspect of the above concept, which is knowledge based order specific NC drilling in the main beams onboard of medium/heavy duty trucks. The geometrical layout of holes is today available, both in drawings, and in CAD files (under a Dassault Systems/IBM CATIA system). Custom specific hole layouts must be generated automatically to produce ready-to-use NC command macros. In addition, the designer must have explanations about how the customized layouts were verified, and about related production costs per beam.

Previously, the generation of NC macros was carried out manually; the operation was excluding customized new holes on the basis of his inspection of component descriptions. This process resulted in 3-10 erroneous holes being drilled per day; these errors had to be corrected by tedious welding, not to mention production delays and broken drills.

The knowledge based order specific NC drilling system (S), is primarily making use of the following information sources:

Information source	Implementation	Language
A Parts description database	IBM/MVS/ROSAM	PL/1
B Specification procedures	IBM/MVS/CAPRI	PL/1
C Fact base: use of each part and relation to other parts	MS/DOS	PROLOG II (see Giannesini, 1987)
E CAD files for holes	IBM/MVS/CATIA	FORTRAN
G NC macros	MS-DOS	NC command language
S Knowledge based order-specific NC drilling system	MS-DOS	PROLOG II + graphics commands

Table 1: Data and fact bases

This system (S) must be used interactively by the designer, and by the NC machine operator. It must give explanations to generated hole layouts, to generated macro commands, and to the use/justification of specific holes (in relation to parts assembly).

ARCHITECTURE OF THE KNOWLEDGE BASED SYSTEMS (S)

The architecture selected, is a hybrid knowledge based system, with interfaces to PL/1, FORTRAN, and the NC machine command language. The inference and explanation procedures are coded in PROLOG II, with interface procedures, running on a PC/AT compatible with RS-232 links to other mainframes.

Specific to (S) are:

i   a meta-parser, capable of parsing symbol strings of different types (both components descriptors, and hole geometries in the beam)

ii a knowledge representation, with fusion of all required information sources in fact or procedural form

iii a user interface, with specific user set-ups.

Knowledge and fact bases: The information sources, A,B,C,D,E,G lead to the following knowledge or fact bases, with each their specific data structures:

A1: Parts codes with variants and text descriptors of their uses; these are of the type nTmS where n is the parts number string, T the type, m the variant code, and S the text descriptor string.

C1: Description string of the use of each hole, and of the other parts mounted on the beam via this hole

E1: Nominal hole layout of a beam section, with coordinates, diameter, and type/profile of each hole

G1: Data structures for all NC drilling macros, written in meta-assembler syntax

S1: Lists of alternative hole types, for a given coordinate range on the beam; variants to the same due to differences in drilling tools and profiles for the NC machine

S2: Exclusion rules for eliminating tentative hole placements and/or type, in view of: placement procedures, of minimal distance to edge or other holes, etc.

S3: Cost evaluation procedures, in view of an optimized drilling path; the costs account for set-up time, drilling time, travel/alignment time, and tool changes

Knowledge representation: This is essentially in form of logic predicates (coded in PROLOG), operating on lists generated by the dedicated parsers, and with attributes attached to the list header. The parser describing hole geometry allows for portability of beam geometries into knowledge bases of type S 1/2/3.

User interface: It consists of drivers to parts description data base (A), to CAD system (E), and to NC machine controls (G). It is supplemented by a menu driven windows environment, with alternative set-ups for different user groups:

i   **Designers:** They get hole layouts encoded by hole types and origin (standard layout, customized variants); a mouse pointed at a hole generates an explanation for that placement and selection. The designer is also allowed to update knowledge bases, E, S through a simple editor. This editor is supplemented by a forward chaining inference procedure for the exploitation of all consequences of a given custom design. For each such design, a cost estimate is generated by S3.

ii **NC machine operators:** They get the sequence of NC macros in one window, and in another window a script over holes when drilled as a result of the drill sequence. The operator can edit this sequence, and shift some holes to adjust to tolerances etc. This edition is supplemented by a backward chaining inference to verify given hole locations or macros, and check the tool use sequence. A drilling sequence router and optimizer is implemented, which selects the hole sequence on the basis of hole attributes and drilling criteria (e.g. minimize number of tool changes); this planner uses exploration in attributed graphs.

PARSER	Meta Parser

This meta parser (see Table 2) is used to analyze and evaluate all strings in A1, C1, E1, and G1 for different selections of the terminal vocabularies.

Example

a)   For example, in the case of a four hole layout D1, this one could be described by the string:

$$\text{layout} = a(t1,d1) + (b\text{-}c)\,(t2,d2) + 3^*e(t3,d3)$$

where:

a,b,c,e: are given vectors in two dimensions, of lengths (x,y) projected according to a coordinate system on the beam

(ti,di):   type and diameter of i-th hole generated by the string to the left

+,- :   vector operations

3*:   scale factor

The parser then returns the structured decomposition of the layout, and the way in which these elements relate in terms of displacements of NC commands (+,-).

b)   In the case of a numerical calculation, the operator would enter a formula to be evaluated in string form:

cost =   HOLE(t,d)*5+RANGE(d)*256+ SETUP

where:

>HOLE(t,d): drilling cost per hole of type t and diameter d

>RANGE(d): number of tools, i.e. of hole diameters, in a given layout

>SETUP: set-up cost

The parser would then return in graph form, the way in which the functions operate on different data sets, and also the calculated value of the expression.

		Terminal vocabulary	
Application	Knowledge base	Elements in vocabulary	Separators
Parts codes	A1	Integers (n,m) Words (T) Strings of words (S)	Spaces
Hole uses	C1	Parts codes	Separator (·) and parenthesis ()
Hole layout	E1/E2	Vectors (x,y) giving transitions between hole positions with type and diameter attributes	Operations on vectors: +: addition -: substraction *: scaling (): parenthesis
Drilling (·) macros	G1/G2	Drilling macro numbers or names	Separator
Numerical calculations	S3	Numbers	Functions

Table 2: Knowledge bases and the corresponding vocabularies as used by the parser

CONCLUSION

Such a knowledge based order specific NC drilling system (S) is of strategic importance for the truck/car manufacturer for which the project is carried out, in view of the following new capabilities, made possible by the applications of artificial intelligence (Krakauer, 1987):

- bring to the NC machine shop floor integrated design and manufacturing data

- reduce the errors due to end product customization

- control costs due to customized layouts

- generate NC macro sequences

- increase the dialogue between designer and manufacturing engineers

- reduce bottle-necks around a truck component (here the main beam) of paramount importance to the total manufacturing cycle

- reduce handling of drawings for custom designs, and thus adjustment time

BIBLIOGRAPHY

1. Battelle Institute, Artificial intelligence in manufacturing, Geneva/Columbus (1987)

2. Giannesini, F., Kanoui, H., Pasero, R., Van Caneghem, M., PROLOG, Addison-Wesley Publ. (1987), or Inter Editions, Paris (1985)

3. Krakauer, J., (Ed.) Smart manufacturing with artificial intelligence, SME Publications, Dearborn, MI, (1987)

4. Pau, L.F., Artificial intelligence in economics and management, North Holland Publ., (1986)

Table 3:   **Meta-parser in PROLOG**

```
enter (p) → read (p) expression (e,<p,nil>)   draw-tree (e);

word (a,<a·x,x>) →;

expression (e,<x,y>) →              sum (e,<x,y>);

sum (e,<x,y>) →                     product (p,<x,z>)
                                    rest-of-sum (p,e,<z,y>);

product (p,<x,y>) →                 primary (f,<x,z>)
                                    rest-of-product (f,p,<z,y>);

primary (n,<x,y>) →                 word (n,<z,y>)    integer (n);

primary (e,<x,y>) →                 word ("(", <x,z>)
                                    expression (e,<z,t>)
                                    word (")", <t,y>);

rest-of-sum (p,e,<x,y>) →           word (0, <x,z>)
                                    op-add (0,0-a)
                                    product (p', <z,t>)
                                    rest-of-sum (<0-a,p,p'>,e,<t,y>);

rest-of-sum (e,e, <x,x>) →;
rest-of product (f,p,<x,y>) →       word (0,<x,z>)
                                    op-mul (0,0-m)
                                    primary (f, <z,t>)
                                    rest-of-product (<0-m,f,f'>),
                                    p, <t,y>);

rest-of-product (f,f,<x,x> →  ;

DEFINITIONS OF OPERATIONS:          Op-add, op-mul in relation to sym-
                                    bols +,-,*; or text: add, sub, mul

READING INPUT STRINGS for parsing

READ STRING ATTRIBUTES from external knowledge bases by drivers

INTEGRATE OTHER KNOWLEDGE BASES

EVALUATE EXPRESSION, AND OUTPUT RESULT including parsing
tree and colouring of beam layout
```

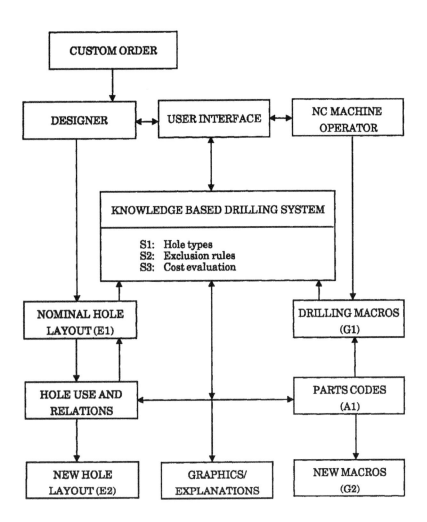

Figure 1: Knowledge based order specific NC drilling system (S).

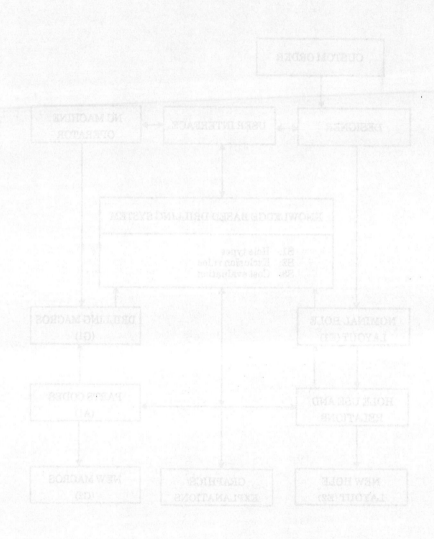

Figure 1. Knowledge-based order-specific NC drilling system (S).

A KNOWLEDGE BASED ENVIRONMENT FOR ARTIFICIAL INTELLIGENCE MODELING OF INDUSTRIAL PROBLEMS: PRELIMINARY CONCEPTS

J. Cuena and A. Salmerón

Facultad de Informática, Universidad Politécnica de Madrid, Madrid, Spain

Abstract. This paper presents a preliminary design of an environment based on Artificial Intelligence, developed to support applications of control and management in industrial installations. Firstly, problems of traditional knowledge representation in industrial installations are discussed. Afterwards the need of inference models for physical behaviour is detected and the two main areas of research are reviewed briefly: (1) qualitative simulation and (2) deep diagnosis. Specialized and mixed knowledge based systems is required. It is proposed a possible general structure for representation and reasoning. Finally, the main lines of the software environment to design and to implementation of this type of industrial applications are presented. This software environment is in course of development.

Keywords. Artificial Intelligence, Expert Systems, Knowledge Based Modeling, Simulation.

INTRODUCTION

The development of knowledge representation techniques has been supported by traditional simple domains applications (games, puzzles, blocks world...). During the sixties and the seventies an important progress has been made based on these domains (i.e. the traditional GPS and STRIPS approaches to problem solving and planning, the development of production systems frame based, semantic networks representations). During the last years some extensions have been developed of these traditional representation techniques (the SOAR system [Laird, 84] or the PRODIGY system [Minton et al., 88]) but as there is an increasing demand of more mature and more effective applications of Artificial Intelligence, in some fields, an important development effort should be made to cope with the real problems. In one of these fields, in the industrial applications, the traditional representation techiques (rules, frames or semantic networks) do not have, usually, power enough to deal with the physical engineering problems. New approaches have been developed in the knowledge representation field to meet this type of problems [Hayes, 79, 85] [De Kleer, Brown, 84], [Kuipers, 86].

Nowadays it is detected a need for a new generation of knowledge representation environments. These new approaches should be based on the integration of traditional reasoning paradigms, used for general aspects of problem solving, with other more specific paradigms. These paradigms will be applied to different types of objects which can be found in the professional practice of the different branches of engineering. This new type of knowledge representation environments should be, obviously, specialised to deal with each of the differents classes of problems.

In the next paragraphs will be discussed the difficulties of the traditional approach to knowledge representation in order to deal with industrial engineering problems. General features of new approaches: qualitative simulation and deep diagnostic methods; are also studied to meet the detected limitations. Finally, the general architecture and development of an environment based on knowledge is proposed. These aspects are now in course of research.

---

(1) Research supported under contract with FUE and IGC for the EUREKA FIABEX project.

THE PROBLEMS OF THE TRADITIONAL KNOWLEDGE ENGINEERING APPROACH

The knowledge engineering method, to develop an expert system, is based on the elicitation of human expert's knowledge. This method is made up of the following general steps:

- Identification of the conceptual frame of discernment (i.e. the conceptual language used to describe and to reason about the type of problems to be dealt by the system).

- Identification of the different reasoning lines used by the human expert to identify, explain and solve problems.

- Loop of:
  * Set of knowledge units deduced from the previous steps and represented in agreement with an available knowledge representation environment.

  * Simulation of reasoning with a set of training examples processed in the knowledge environment.

  * Evaluation by the experts of the solutions proposed and their explanations. As a result, modifications will be proposed and new knowledge units will be introduced.

When the problems to deal with are known deeply enough by human experts, this general method can lead to very accurate knowledge bases. These knowledge bases are able to reason about complex problems and they have an insight enough to be compared with the human expert performance.

However, the industrial installations are new, in most of the cases, or they are managed by human experts with a general or a basic professional knowledge. But this knowledge is not deep enough to describe, with simple knowledge representation techniques, the complexity of some industrial processes. An example, is the modeling of the evolution of the state variables, along time in a physical or chemical process, submitted to different external conditions. In these cases, human experts are able to give different general conditions in order to describe the complexity of the process, but they need to use a mathematical model of physical or chemical laws. In fact, the general approach used in engineering, is to represent the behaviour complex aspects as mathematical models. These models are used by simulating cases off-line, to improve the capability of these complex phenomena understanding.

If Artificial Intelligence were required to be used in assessment in almost real time, about the behaviour of an industrial installation, would be necessary to develop reasoning models that embody the knowledge about physical or chemical behaviours all together with other specific aspects used by human experts. This approach is described in the next paragraphs.

THE QUALITATIVE SIMULATION APPROACH

An aspect that always appears, when reasoning about the physical objects behaviour, is the modeling of transitions along time. That is: given an initial state and a prediction of the external actions, that may act on the object in a near future, deduce the possible future states of the object along time. This is a problem known by engineering professionals. They use, traditionally, a procedure of three steps based on quantitative numerical models. Finite differences or finite elements representation are used to define the quantitative model that describes the physical object behaviour. The procedures consist of the following steps:

(1) Definition of a sample of numeric data for simulation. With these data the quantitative model must cover enough the classes of possible external actions.

(2) Process of the defined sample by runs of the model.

(3) Analysis of the results by professional experts for the identification of the general features of behaviour.

In this process it is firstly produced a numerical specification step of the pattern of external actions. The actions are forecast by the professional expert. There is a last step of abstractions in qualitative terms, of the results given by the numerical simulation runs. This abstraction is justified because, although the model is able to work with very accurate values, the precision of the future environment must be defined as a qualitative type. The final reason is that the experts cannot believe the precise numerical results obtained by the model, and they propose a qualitative abstraction of these results.

This process has a disadvantage: the steps (1) and (3) are made by human experts, so to introduce the whole process on the computer it might be used two approaches:

- Modeling the steps (1) and (3) by computable processes. (1) Can be made by random number generation with prefixed distributions; (3) can be represented by a classification knowledge base. The knowledge base could be used to deduce different qualitative attributes from the numerical sample of the behaviour already obtained.

- Defining a qualitative quantity space in which it could be represented the aspects considered at steps (1) and (3). Afterwards, to define a qualitative formulation of the simulation process, in such way that it might be obtained straightly from the qualitative specification of (1) the set of possible qualitative statements about the future of (3).

The second approach is more efficient in the computer and more explanable than the first one. The Artificial Intelligence researchers have been conscious about the interest of this approach at an experimental level. There are, by the moment, three main paradigms:

a) Approach based on components [De Kleer, 84]. The theory of De Kleer and Brown assumes that a physical system consists of distinct parts, called components, connected together through conducts, the components process and transform the materials. And conducts only transport these materials.

The goal of this approach is to infer the behaviour of the composite device from models of its device components. For modeling the components' behaviour two concepts are used:

* Confluences, are qualitative differential equations that represent different competing tendencies between the physical system attributes and theirs time derivatives associated.

* Qualitative states, defined by the values of the system attributes.

The result of the qualitative simulation process is an envisioning graph that models the device behaviour. Each node in this graph is a qualitative state. Each link between nodes represent a possible transition between qualitative states. For the generation of the feasible transitions a constraint satisfaction method is applied.

b) Approach based on constraints [Kuipers 86, 88]. The theory of Kuipers is based on a qualitative view of the differential equations used on traditional physics. In this approach the behaviour of a system is modeled by continuous time-varying parameters.

The element of the model are predictions rules and constraints. Since the parameters are continuously differentiable, the mean value theorem restrict the way which the transitions between qualitatives value is produced. The evaluation along time of the state attributes can be derived by prediction rules. Constraint among attributes is used to filter the values predicted by the rules. There are several kinds of constraints: arithmetic, functional, derivative (first and second order), phase space view, ... The envisionment consists in a set of time-points representing the qualitatively distinct states of the system and a set of transitions between them. A qualitative state is defined through values of each attribute at each time point. There is a process called QSIM that builds an envisionment graph with a set of qualitative states and their transitions. QSIM work with a queue of qualitative states called ACTIVES. The ACTIVES queue, at the begining, is loaded with a set of initial qualitative states. The process QSIM consist mainly in a loop for every element in the QUEUE in the identification of the transitions consistent with the set of constraints of every state variable and the generation as next states by combination of possible values.

c) An approach based on processes [Forbus, 84]. The theory of Forbus provides a language to specify processes and to describe their effects. In this approach the changes in physical systems are always caused by processes. The situations of the physical world are modeled by: a set of objects, the attributes of the object and the objects relationship. Processes work over the objects changing the values of their attributes. As in Kuipers theory the attributes take values from a continuous range and are represented by quantity spaces. A process consist of:

* A enumeration of the objects affected by the process.
* A set of constraints and relations between objects attributes.
* A set of preconditions. A constraint can be applied in the circumstances specified by its preconditions asociated.
* A set of influences imposed by the process over the objects attributes.

DEEP DIAGNOSTIC

The diagnostic systems were one of the first applications of Artificial Intelligence. Many of them were applied in Medicine. These systems used a knowledge representation based on rules. A rule is a shallow relation cause-effect. These relations were given by an expert. Experts have a mental model about the problem universe and the rules are the shallow presentation of the results given by using this mental model.

When an industrial installation is new or recent, there are no human experts to provide the shallow rule based version of their knowledge so it is needed, to provide diagnostics, the introduction in some way of a representation of the deep mental model.

This deep knowledge has the advantage of being portable between industrial installations with common elements. The shallow knowledge is never portable because of the incidence of specific features in an installation.

The behaviour model can be obtained by abstraction (learning) of a representative set of envisionment trees generated by qualitative simulation. With this model of behaviour the process of diagnosis can be structured in three steps:

- Prediction of behaviour with an explanatory structure associated.
- Detection of symptoms by the differences between predicted and observed values.
- Search for explanation of the symptoms based on the explanatory structure.

[Reiter, 87] and [De Kleer, 87] have proposed diagnostics models based on these ideas. Reiter made an analysis based on a representation in first order logic with explanatory structure a proof tree. De Kleer uses an equational model and uses as explanatory structure an assumption truth maintenance system (ATMS).

AN APPROACH TO A KNOWLEDGE REPRESENTATION ENVIRONMENT FOR INDUSTRIAL MODELING

The paragraphs above showed that:

* It is necessary to work with state transitions models of physical components integrated with other knowledge units of a more traditional type, in way to model the general understanding of the system structure.

* It has also been shown that there is, in the deep diagnostic models, a need of using a logical behaviour model. With this behaviour model the diagnostic inference engine will be able to deduce the possible new states of the installation, just in case of absence of problems. The symptoms are detected by comparison between the states observed in the installation and the states that were predicted. The diagnostic process explores the behaviour model to infer possible failure hypothesis.

* The knowledge representation applied to physical modeling is settled as constraints between attributes. That means that in installations, with a significant number of elements, the problem of constraint satisfaction may become very cumbersome. Thus, it is very important to find decomposition criteria of the general model and to apply the satisfaction process to differents models of the installation. These models must be defined at differents levels of representation.

The next paragraphs introduce the specifications for an environment, taking account of these concepts and, also, introduce the basic modelling elements and its interactions.

Environment specifications

The environment must combine: the traditional paradigms of knowledge representation and other paradigms more specific to deal with knowledge elements in the industrial field.

a) For the simple or general elements (i.e. making predictions or evaluations) the environment must be able of providing traditional solutions. It is well known the efectiveness of these solutions. These solutions belong to the way of working of traditional expert systems of first generation. So the environment must admit as representation element the tuple: <object, attribute, value>. It will be included reasoning mechanism based on rules: An adapted version of the Rete algorithm [Forgy 82] is implemented for forward reasoning.

For backward reasoning the methods MYCIN, PROSPECTOR and DEMPSTER-SHAFER are used.

b) In complex applications, the environment must supply more advanced knowledge representation elements. These elements model industrial components and installations. The installations are considered as a components set. These elements are used to build models of the indutrial installation at different levels of detail. Components can model differen industrials processes as qualitative Forbus processes. The installation, considered as a component set,

is able to infer its own behaviour. In order to make possible this inference constraints satisfaction algorithm has been implemented, that is a version of the Waltz algorithm.

Finally the environment is oriented to be used in the profesional domains. So it is obliged to run in general purpose computer. It has been selected the Unix System V operating system and the C programing language.

General description of the representation

In the modeling of a physical object, in both natural and industrial systems, different elements of representation are needed to construct a model in a structured way. Most problems in industrial systems modelling are produced by the bidirectional influences between the descriptive atributes of the system. A part of these bidirectional influences are feedbacks. A way to solve these problems, specially in industries, is to use two modeling elements: the components and the installation. The component has knowledge about its individual behaviour. The installation has knowledge about the physical relation among components, and about the mutual constraints produced by the components behaviour. The inference at component level deduces active processes and its modeling constraints. The inference at the installation level produces the attribute value set for the whole installation by constraint satisfaction.

Two frames are defined: the component frame and the installation frame. The component frame has a set of input attributes, output atributes and internal atributes. The component, also, has a set of processes. A process is a qualitative model based on rules or constraints. Each process model the different operating modes for states transitions. Every component belong to one instalation. The Figure 1 shows the structure of the component.

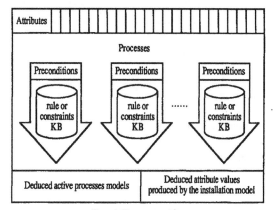

Attributes

Figure 1: The component frame

The component, as a modeling element, has a knowledge base for the selection of the active process at any moment. Once a component has selected one of its processes, this process is delivered to the installation. The selection will made in the following way: the process whose preconditions are true in the initial state, provided by the installation, are selected. The knowledge base of every process deduces an instance of the constraints and rules model of the process that is consistent with the initial state. The knowledge base of a component can not make references to an attribute of other different component. The relations between attributes, that belong to different components, are stored at installation level. This kind relations represent connections between components.

The installation frame knowledge base represents (1) the topology, by the graph of connections between components, (2) the behavior of the connection elements. For example a constraint can express: "the pressure at output O_3 of component C_3 must be greater than the pressure at input I_1 of component C_1".

Figure 2 shows the structure of the installation. The qualitative simulation is made in two steps: (1) constraint satisfaction at installation level that produces the set of possible changes at the different attribute values, (2) state transition made at component level based on local models using the possible changes received from the installation level.

With this procedure it can be obtained for every component the possible future transitions from the initial state. To produce a second level of transitions an approximate time assignment is required for every component that allows the deduction of a set of possible simultaneous future states. Every element of this set can be used as initial state in a new transition reasoning step. This aspect is now in course of analysis.

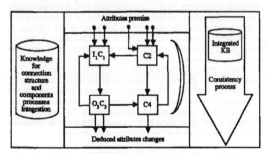

Figure 2: The installation frame

The result of several steps of qualitative simulations over the installation is a tree of transitions between state that represent a partial view of the whole graph of envisioning. Several models of the installation, at diferent levels of detail, are necessary to produce a hierarchical qualitative simulation. The approach is top-down. Firstly the high level model is explored. A second qualitative simulation, at one more level of detail, will be necessary if differences appear between the behaviour predicted and the real installation behaviour, or some dangerous transition are detected. This process continues until a satisfactory level, of explanations about the troubles, had been obtained.

Finally, the construction of the differents models of the industrial installation will be supported by a library of industrial components. This library is constituted by a set of classes of components.

A installation can be modeled at different detail levels that are used if necessary, i.e.: a maximum aggregate model can be used to detect possible failures in a macrocomponent which leads to the detailed simulation of this macrocomponent, etc. This process of hierarchical detail simulation can be continued until a prefixed limit or a previous sufficient level is attained (Fig.3). A detail level is sufficient if it is possible to generate information, about symptoms, detailed enough for the diagnostic model. A symptom is a discordance between predicted an observed data. A diagnostic model produces hypothesis of failure components that can explain the symptoms. The definitive structure of the diagnostic environment is now in course of development.

CONCLUSIONS

An environment for hierarchical modeling has been defined for industrial installations. This structure is based on two model elements: components and installation. The component has knowledge about its individual behaviour. The installation has kowledge about the physical connection of the components. Several models of the industrial installation, at different levels of detail, are necessary to produce a hierarchical qualitative simulation, to be used in failure explanations.

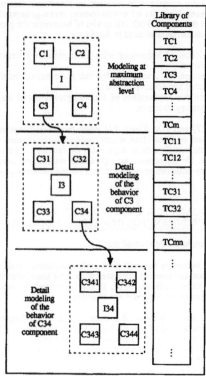

Figure 3: Hierarchy of Models

REFERENCES

Cuena J. (1988): "The Qualitative Modeling of Axis-Based Flow Systems" European Conference on Artificial Intelligence. Munich (ECAI-88). Pitman.

Davis R. (1984): "Diagnosic Reasoning Based on Structure and Behaviour". Artificial Intelligence 24, 347-410 (1984).

De Kleer J., Brown J.S. (1984) : "A Qualitative Physics Based on Confluences". Artificial Intelligence 24, 7-83 (1984).

De Kleer J., Brown J.S. (1984): "Qualitative Reasoning about Phisical Systems". Artificial Intelligence (1984).

De Kleer J., Williams B.C. (1987): "Diagnosing Multiple Faults" Artificial Intelligence 32 (1987).

Forbus K.D. (1984): "Qualitative Process Theory" Artificial Intelligence 24: 85-108. 1984.

Forgy D.L. (1982): "Rete: A Fast Algorithm for the Many Pattern/Many Object Pattern Match Problem". Artificial Intelligence 19 (1982).

Genesereth M.R. (1984): "The Use of Design Descriptions in Automated Diagnosis". Artificial Intelligence 24, 411-436.

Hayes P.J. (1979): "The Naive Physics Manifesto" in "Expert Systems in the Electronic Age" D. Michie ed. Edinburgh University Press, 1979.

Hayes P.J. (1985): "The Second Naive Physics Manifesto" in "Formal Theories of Commonsense World". Hobbs and Moore eds. Ablex, 1985.

Kuipers B. (1986): "Qualitative Simulation" Artificial Intelligence 29, (1986).

Kuipers B. (1986): "Qualitative Simulation" Journal of Artificial Intelligence 29, 1986.

Kuipers B.J., Lee W.W. (1988): "Non-Intersection of Frajectories in Qualitative Phase Space: A Global Constraint for Qualitative Simulation" AAAI-88. The Seventh National Conference on Artificial Intelligence. 1988.

Laird, J.E. (1984): "The SOAR2 User's Manual" Computer Science Department, Carnegie-Mellon University, 1984.

Minton S. et al (1988): "PRODIGY 1.0: The Manual and Tutorial" (Tech. Rep.) Computer Science Deparment. Carnegie Mellon University, 1988.

Quinlan J.R. (1979): "Discovering Rules by Induction from Large Collections of Examples" in "Expert Systems in the Microelectronic Age" D. Michie ed. Edinburgh University Press, 1979.

Reiter R. (1987): "A Theory of Diagnostic from First Principles" Artificial Intelligence 32, 57-95. 1987.

Winston P.H. (1984): "Artificial Intelligence" 2nd edition. Addison Wesley.

KNOWLEDGE BASED SYSTEMS FOR
ENGINEERING DESIGN AND MANUFACTURE

P. J. Rayment and D. T. Pham

School of Electrical, Electronic and Systems Engineering, University of Wales,
Cardiff, UK

ABSTRACT

In this paper, the technology of knowledge-based systems is first reviewed. The two
key elements of these systems, the knowledge base and the inference engine, are
examined in detail. The knowledge base is seen as being a collection of facts and
rules expressing experiential knowledge about a problem domain, a hierarchy of frames
storing information on stereotyped objects or generic concepts and specific instances
of these, or a set of nodes representing these objects and concepts linked into a
network by arcs symbolising relations between them. The inference engine is found to
operate by forward chaining from available data, backward chaining towards the goal,
or adopting a mixture of these strategies.

Following the review, recent applications of knowledge based systems in engineering
design and manufacture are described. These cover a wide range of activities
including equipment and process selection, process planning, production modelling,
machine fault diagnosis, real time monitoring and control, and design of machine
components and systems.

1. INTRODUCTION

Intelligent knowledge-based systems (IKBS),
knowledge-based expert systems, or simply expert
systems, are computer programs embodying human
expertise for solving difficult tasks normally
requiring human specialists.
Early expert systems achieved remarkable success
at handling practical problems of significant
scientific and social implications, mainly in
chemistry, geology, mathematics and medicine.
This has created widespread interest among workers
in other areas in applying expert systems to their
particular domains. The relatively recent
availability of low-cost mini and microcomputers
has facilitated a rapid expansion of expert
systems development activities in these areas.

In this paper, the technology underlying knowledge
engineering is reviewed to show what expert
systems are, how they are structured, and how
their key elements are designed and operated. The
review serves to introduce the reader to the
subject. More detailed information can be found
elsewhere (see for example[1-12]). Following the
review, some of the recent applications of expert
systems in engineering design and manufacture are
described. It will be seen that engineering
expert systems are addressing a wide range of
practical tasks, from routine selection and
diagnosis through to creative synthesis and
design.

2. THE STRUCTURE OF EXPERT SYSTEMS

An expert system generally consists of the
following components:

o A knowledge base containing knowledge (facts,
information, rules of judgement) about a problem
domain.

o An inference mechanism (also known as inference
engine, control structure, or reasoning mechanism)
for manipulating the stored   knowledge to
produce solutions to problems.
o A user interface (or explanation module) to
handle communication with the user in natural
language.
o A knowledge acquisition module to assist with
the development of the knowledge base.

The knowledge base and inference mechanism
constitute the kernel or core of the system and
thus are the essential parts of it. They will be
examined in more detail in later sections. A user
interface of varying degrees of sophistication is
generally present in most systems, whereas the
knowledge acquisition facility may not always be
available. Note also that the knowledge base and
inference mechanism are interacting but separate
entities. This sharp distinction between domain
knowledge and problem-solving methodology is
fundamental to the nature of an expert system. It
has two important implications[13]. First, it
means that a knowledge base concerning a new task
can be substituted for the existing knowledge
base, producing a different system as a result.
Second, it facilitates the multiple use of the
same knowledge by the expert system in diverse
activities such as searching for solutions,
providing explanations, and acquiring new
knowledge. This is because, with the
problem-solving methodology implemented
separately, the knowledge base can be constructed
so as to store only what the expert system should
know rather than what it should do[13]. This in
turn enables the problem domain knowledge to be
expressed in a pure form excluding all references
as to how or where it will be used.

3. THE KNOWLEDGE BASE

There is little doubt that the power of an expert
system resides in the knowledge that it possesses.

As stated earlier, this knowledge is embodied in
the system's knowledge base. The latter could be
likened to a conventional database storing
information on a particular subject matter.
However, knowledge is more than a mere collection
of data, the term 'knowledge' carrying with it the
concepts of 'structure' and 'understanding'.
How structure and understanding are achieved to
give an expert system its aforementioned power is
a direct result of how knowledge is represented.
In this section, the main techniques of knowledge
representation will be reviewed. Since knowledge
can be represented only if it exists, the
extraction of knowledge about a problem domain in
the first instance is an important step in the
building of a knowledge base. This will be
briefly addressed in the present section.

Knowledge Representation
There are several ways of representing
knowledge[1]. The three most popular are rules,
frames and semantic nets[7]. Rule-based
representation is a surface representation,
whereas schemes using frames and semantic nets are
deep representations.

Rule-based representation[14-15]. In a rule-based
system, knowledge is represented in terms of facts
pertinent to a problem area and rules for
manipulating the facts. Many systems also
incorporate information about when or how to apply
the rules (that is, meta-knowledge, or knowledge
about knowledge).

Facts are asserted in statements which explicitly
classify objects or specify the relationships
between them, such as "A PUMA 560 is an Industrial
Robot", "An Industrial Robot is a Robot", "A Robot
has an End Effector", "A Robot is a Machine", and
"A Machine is an Artefact".

Rules are modular 'chunks' of knowledge of the
form 'IF Antecedent THEN Consequent', or 'IF
Situation THEN Action' meaning 'If the situation
described in the antecedent part of the rule is
true, then produce the action specified in the
consequent part', hence the names 'IF-THEN Rules',
'Situation-Action Rules' or 'Production Rules'.
Examples of rules are:

(a)   IF "the task is drilling" THEN "use a
      point-to-point robot"
(b)   IF "several robots can technically
      perform the required task equally well"
      THEN "select the cheapest robot"
(c)   IF "a task needs a high level of sensory
      perception" AND "product batch sizes are
      very small" THEN "there is a strong
      possibility <0.9> that the task is not
      suitable for robotisation".

Rule (c) illustrates two points. First, the
consequent part of a rule does not have to specify
an action. Instead it can be an assertion,
hypothesis, or conclusion which will be added to
the knowledge base if the antecedent part is
satisfied. Second, certainty factors can be used
in a rule (or a fact) to indicate the degree of
confidence attached to it. This enables the
expert system to deal with information which is
inexact or not completely reliable.

It has been mentioned that a system may also
possess meta-knowledge to guide its own reasoning.
In a rule-based system this is expressed as
meta-rules. An example of a metarule is:

IF "the task is electronic component
assembly" AND "the components are all
integrated circuits or passive devices of
standard dimensions" THEN "skip rules for
assessing technical feasibility and
invoke those for determining cost
effectiveness".

When diverse types of knowledge have to be
handled, the rules (including meta-rules) are
sometimes grouped into specialised independent
sets, each corresponding to one type of knowledge.
These so-called knowledge sources all operate on a
common central database, the blackboard, and
communicate their results to one another via this
blackboard.

The transparency of rule-based representation has
made it the choice representation scheme for many
expert system development projects, especially in
situations where the domain expertise is founded
on empirical observation of past associations.
However, rule-based representation tends to be
shallow, in other words, unable to describe
adequately the fundamental principles in a problem
area.

Frame and semantic net-based representations. In
contrast, representation schemes using frames or
semantic nets allow a deeper insight into
underlying concepts and causal relationships and
facilitate the implementation of deeper-level
reasoning such as abstraction and analogy. A
frame [16] (or, its near equivalents, concept,
schema, and unit) is a record-like data structure,
a form for encoding information on a sterotyped
situation, a class of objects, a general concept
or a specific instance of any of these. For
example, one frame might represent a particular
type of machine(a robot), another, a whole range
of machines(computers, robots, machine tools
etc.), and yet another, the more general class of
artefacts(bridges, machines, buildings etc.)(See
Figure 1).

Associated with each frame is a set of attributes,
the descriptions or values of which are contained
in slots. For instance, the frame for the
'Industrial Robot' sub-class of robots might have
slots entitled 'manufacturer', 'degrees of
freedom', 'payload and speed' (See Figure 2). In
addition to attribute values, slots can also store
other information such as constraints on the
nature and range of these values ('value class',
cardinality min', 'cardinality max') and procedures
(arbitrary pieces of computer code) which are
executed when the values are changed. Frames are
usually organised into a hierarchy, as already
seen in Figure 1, with those at the upper levels
representing more generic classes and those at the
lower levels, specialisations of these classes.
The strength of frame-based systems derives from
this hierarchical structure which enables frames
to inherit attributes from other frames located
above them in the hierarchy.

Knowledge representation schemes using semantic
nets are similar to those based on frames. A
semantic net is a network of nodes linked together
by arcs. Nodes stand for general concepts (or
types), specific objects (or tokens), general
events (or prototype events), or specific events.
Arcs describe relations between nodes. A simple
semantic net expressing the concept of a machine
is illustrated in Figure 3. There, arcs represent
'is-a' and 'has-part' relations which establish
the same kind of property inheritance hierarchy as
in a frame-based system.

Knowledge Acquisition.

An expert system can acquire knowledge by being told facts, rules, concepts, or relations, by including such information from examples, or by learning from observation and discovery[19].

The first form of knowledge acquisition is the commonest one. At the most basic level, rules and facts have to be written by the expert system designer (or 'knowledge engineer') in an AI language such as LISP[20], a programming language for processing lists of symbols, or PROLOG[21], a programming system built upon First-Order Predicate Logic. In recent years, development tools [22-25] have become available that allow information to be entered into the knowledge base in more natural, English-like sentences. These development tools include expert system shells which are expert systems without a knowledge base and knowledge engineering environments or hybrid tools combining different knowledge representation formalisms and reasoning mechanisms. In addition to facilitating the knowledge acquisition process, some advanced tools can also check the incoming knowledge for possible inconsistencies and redundancies.

There are a few induction programs that help an expert system extract rules from case examples presented to it by a human expert[26-28]. The examples are selected to contain some underlying pattern and the rules are explicit generalisations of that pattern. For instance, the examples might be the sequence of numbers, 1, 2, 3, 5, 7, 11, 13 and a correctly induced rule might state that the sequence concerns the ordered set of positive prime numbers. Although systems using induced knowledge can perform better than those employing knowledge which has been conventionally acquired[29], they have the disadvantage of being less transparent, in other words, less able to explain their behaviour[19].

Finally, discovery systems which autonomously acquire new knowledge and increase their knowledge base by applying guided rules for adding information to a minimal initial knowledge base are very rare. This is perhaps because, whereas the first two forms of knowledge acquisition discussed above are first-order logic activities, discovery tends to be a second-order logic process[19] and practical computing techniques for handling the latter kind of logic are still under development.

4. THE INFERENCE MECHANISM

Although, as mentioned previously, the knowledge base is the most important component in an expert system, the latter will not be useful unless it has a good inference mechanism to enable it to apply the stored knowledge.

Different inference mechanisms are possible, depending on the type of knowledge representation adopted. In a rule-based system, the inference mechanism, also called rule interpreter, examines facts and executes rules contained in the knowledge base according to set logical inference and control procedures. Some of the commonly encountered inference procedures (or inference rules) are: modus ponendo ponens (or modus ponens), modus tollendo tollens (or modus tollens), and resolution[30,31] (See Figure 4.). The application of inference rules to a knowledge base enables goals (or conclusions) to be proved or disproved, or new facts and rules to be created. For instance, using the modus ponens rule, it is possible to infer the new fact "a robot has a drive system" from the fact "a robot is a machine" and the rule

"If X is a machine, Then X has a drive system".

By referring to the modus tollens rule, the conclusion that an industrial robot is an aircraft can be disproved from the fact that "An industrial robot has a control system, a drive system, a mechanical structure and an end-effector", and the rule "If X is an aircraft, Then X has a fuselage, a tailplane and wings". Finally, the resolution rule allows the new rule "If X is a home robot, Then X has an end-effector" to be obtained from the rules "If X is a home robot, Then X is a robot" and "If X is a robot, Then X has an end effector".

Reasoning by the exercising of inference rules can proceed in different ways according to different control procedures. One strategy is to start with a set of facts or given data and to look for rules in the knowledge base the 'If' portion of which matches the data. When such rules are found, one of them is selected based upon an appropriate conflict resolution criterion and executed or 'fired'. This generates new facts and data in the knowledge base which in turn causes other rules to fire. The reasoning operation stops when no more new rules can fire. This kind of reasoning is known as 'forward chaining' or data-driven inferencing'. It is illustrated in Figure 5.

An alternative approach is to begin with the goal to be proved and try to establish the facts needed to prove it by examining the rules with the desired goal as the 'Then' portion. If such facts are not available in the knowledge base, they are set up as sub-goals. The process continues until all the required facts are found, in which case the original goal is proved, or the situation is reached when one of the subgoals cannot be satisfied, in which case the original goal is disproved. This method of reasoning is called 'backward chaining' or 'goal directed-inferencing'. An example of backward chaining is given in Figure 6. Note that with backward chaining, the reasoning strategy is focused: facts irrelevant to the desired goal, such as "A PUMA 560 has a teach pendant" are not created. Backward chaining is therefore favoured when the number of rules is large and forward chaining could lead to a 'combinatorial explosion'. In practice, forward and backward chaining are sometimes integrated and an iterative convergence process is used to join these opposite lines of reasoning together at some point to yield a problem solution[32]. This corresponds closely with the strategy, adopted by systems engineers, of using both the 'top down' and 'botton up' approaches in designing complex systems. The terms 'top down' and 'bottom up' refer to the direction of the search for a solution: top-down search begins from goal states and proceeds to start states and bottom-up search takes place in the opposite sense. Top-down search is thus implemented in backward chaining and bottom-up search, in forward chaining.

Regardless of the search direction, from any given goal state or start state there are usually several alternative paths leading to different possible solutions. These paths constitute the branches of a 'search tree' rooted at the goal state or start state in question. The search tree may be explored 'depth first' or breadth first'. A depth-first search plunges from the root deep into the tree, considering a sequence of successors to a state until the path (line of reasoning) is exhausted. The search then proceeds to the next branch of the tree, exploring it in depth. With breadth-first search, all possible alternatives at the root are generated, then the alternatives at the next level are produced, and so on. The search is thus conducted in breadth across the tree.

In contrast with rule interpreters used in rule-based systems, which are general-purpose deduction mechanisms, the inference engine in a frame/net-based system has a much more limited scope, being part of the system's knowledge assertion and retrieval facilities. Because of this limited scope and of the fact that it utilises inferences founded on the structural properties of the representation scheme and 'wired in' to the latter, the inference engine in a frame/net-based system can operate much faster than rule interpreters[17].

The most powerful automatic inference methods available to a frame/net inference engine apply the inheritance characteristic of frame and net structures mentioned previously. For instance, since the "Industrial Robot" frame in Figure 1, has a sub-class link to the "Robot" frame and the latter has a sub-class link to the "Machine" frame, the engine will automatically 'retrieve' the belief that "Industrial Robot" is a sub-class of "Machine" and as a consequence inherits attributes from the latter. Other inference methods, readily implemented in the frame formalism, use constraints such as specifications of the nature (value-class) and range (cardinality) of attribute values to determine whether a given item could be a value of a given slot. For example, when a value is being added to the 'Manufacturer' slot in the FS-2 frame (Figure 2.), the value is automatically rejected if it is not the name of a company. Reasoning by propagation of constraints has been used to limit search and refine plans in some advanced expert systems[33,34].

5. KNOWLEDGE-BASED SYSTEMS FOR ENGINEERING DESIGN AND MANUFACTURE

This section focuses on some of the recent knowledge-based expert systems developed for applications in engineering design and manufacture. As with expert systems in other fields, they can be grouped into three broad classes depending on the nature of the problems that they solve.

At one end of the spectrum are 'derivation' (or 'classification') problems which include selection of machine elements, tools, equipment and processes, fault diagnosis and real-time machine control and monitoring. In such problems, the problem conditions are stated as parts of the description of the solution[35]. All the possible outcomes are known and the knowledge base is used to complete the solution. The problem-solving process could be viewed as the identification of a path leading from given conditions to desired outcomes.

At the other end of the spectrum are 'formation' or 'synthesis' problems which cover all design activities (for example, component design, tool design, system design) where the solution space is completely open (infinite) and problem conditions are specified as properties that a solution must possess as a whole[35]. Usually, there are several alternative outcomes, none of which is known a priori.

Between the two extremes, are problems such as process planning, production scheduling, and system configuring, which have features belonging to both the derivation and formation types. That is, the solution to these problems generally consist of combining elements drawn from a finite set. Although the outcome of the solution process thus implemented might not have been contained in advance in the knowledge base, all of its elements are.

Expert Systems for Derivation Problems

There are several prototype expert systems designed to aid the selection of machine elements, tooling, equipment, and processes. For example, BETSY is an expert system described by Hasle et al[36] for bearing selection. The system was developed with a hybrid tool written in LISP which supports both rule-based and frame-based programming. Rules are used to represent heuristic knowledge, and frames, to store quantitative information concerning the various types of bearings and their applications. The system employs forward chaining through sets of rules to generate a preliminary list of possible bearing types and then processes the list to remove those bearings that violate design constraints. Backward chaining is used to arrive at advice on application details such as the appropriate ISO fits and internal clearances for the selected bearing.

GEAREX, reported by Bose and Krishnamoorthy[37], is a system for selecting standard 'off-the-shelf' gearboxes for general industrial applications. Implemented in C under the Unix operating system, GEAREX is also both rule-based and a frame-based program. A feature of GEAREX is its effective use of a relational database to store static knowledge (such as the technical specifications found in commercial gearbox catalogues).

Tooling selection is addressed in the expert systems by Pham and Yeo[38] and Mahabala et al[39]. The former system guides a robot user in selecting grippers to handle different workpieces. The system also provides advice on where to grip a workpiece. When interacting with an expert user the system is capable of modifying its behaviour to learn from his experience. Although the system is coded in PROLOG, it is basically procedure-orientated. However, it does resort to the inherent backward chaining facility of PROLOG in explaining its line of reasoning.

The expert system by Mahabala et al is a rule-based program which recommends appropriate drills for drilling operations. The system can select from some 200 standard and special-purpose drills. It contains both knowledge rules and meta-rules. The latter are used to make problem solving (a backward chaining process) more efficient by reducing or structuring the search space so that unnecessary searches are avoided. The meta-rules can be divided into two types, those for specifying mutually exclusive classes and those for setting search priorities. The specification of mutually exclusive classes (for instance, the classes of twist drills, spade drills, etc.) enables search space reduction in that as soon as the hypothesis about one class is true, the rest can be disregarded. The application of rules for setting search priorities, such as 'examine drills with a straight shank before those with a Morse taper if the drill diameter is less than 17.5mm', clearly leads to higher efficiency since preferred or more likely solution paths are tried first.

KBSES is a knowledge-based system developed by Kusiak and Heragu[40] for selecting machine tools and materials handling equipment. The system is based on rules for capturing procedural knowledge and frames for holding information on parts and machines. In addition, it also has a set of optimisation models and algorithms. Using rules, it selects the model appropriate for the task in hand. Then it generates the data required by the model and goes through the selected optimisation procedures to yield the list of equipment best suited to that task.

For process selection, there are the examples of
the systems by Fukuda[41] and by Hon and
Ismail[42]. The first system, WELDA, provides
advice on choosing welding methods for the
fabrication of prototype structures. It also
recommends the correct material for the structures
and the preparation techniques (cutting and
bending) associated with the chosen method of
welding. WELDA is implemented in OPS83, a
production-rule language featuring forward
chaining and procedural control. Numerical
computing is carried out using a Fortran
subroutine which is linked to the main OPS83
program using the call-by-reference facility of
OPS83[41].

The second system is a rule-based program which
selects hole-making and finishing processes
(drilling, punching, boring, reaming, etc.) to
suit the technological features of the machined
components (material, geometry, tolerances, etc.).
The system is written in PROLOG. The inference
method is backward chaining.

In contrast with the 'selection' expert systems
discussed so far, which all operate off-line, the
signal-interpreting system described by Huang et
al[43] is a real-time program. It is used by an
arc-welding robot to interpret weld seam images
captured by a CCD camera and recognise meaningful
geometrical features corresponding to the various
surfaces meeting at the weld joint. The system is
rule-based and is written in PASCAL. Its
inferencing technique is forward chaining through
rules in a depth-first mode.

Another real-time system is that proposed by
Villa[44] for monitoring tool wear and planning
tool maintenance in automated machining. The
system consists of a module for estimating the
tool decay level based on observation of the
surface profile of the machined part and a module
for selecting tool maintenance decisions from a
database of maintenance strategies. The selection
is driven by data produced by a
tool-decay-dynamics predictor. The latter has a
stochastic model of the tool wear rate, the
parameters of which are identified from the afore
mentioned tool decay level estimates. Tool decay
level recognition, essentially a diagnosis
problem, is carried out by means of rules for
matching decay levels with wear-out data.

Diagnostic expert systems differ from condition
monitoring systems such as that outlined, in that
they are for fault finding as opposed to fault
prevention. In other words, they are invoked
after a failure has occurred rather than before.
Examples of diagnostic systems include those by
Zheng et al[45], Puetz and Eichhorn[46] and Snoeys
and Dekeyser[47].

The expert system by Zheng et al is for diagnosing
faults in automobile engines. It adopts a
hierarchical and modular approach to describing
the structure, operation and possible faults of an
engine. Consequently, a complex diagnosis task
can be divided into simpler components for which
solutions are readily known. The system is
written in LISP and possesses a hybrid
architecture featuring rules, meta-rules, frames,
and a blackboard. It makes use of both deep and
shallow knowledge in its reasoning. The latter
could be described as a combination of forward and
backward chaining. Forward chaining is employed
initially to obtain hypotheses about the location
and cause of a fault. Backward chaining is then
used to verify the hypotheses. This procedure is
repeated until the fault has been identified to
the desired level of detail.

The interactive expert systems by Puetz and
Eichhorn are dedicated to the diagnosis of faults
in CNC machine tools. The prototype system was
developed in a LISP environment. However, the
target system which is integrated with the machine
tool controller is written in C. The systems are
both rule based, although the target system also
used a frame-like technique
(object-attribute-value) to represent the
structure of the machine tool and the properties
of the relationship between its components.
Conflict resolution is implemented by means of
heuristics (cost of answering a question,
frequency of faults, available evidence to support
a given conclusion, etc.).

The system developed by Snoeys and Dekeyser is for
diagnosing faults in a wire-EDM machine. The
system is rule-based and performs both forward and
backward chaining. The latter inference method is
employed to provide clear explanations of
conclusions and to predict the consequences of an
operator's actions. Thus the system is also an
effective aid for training inexperienced operators
in the use of the machine.

Finally, an example of a real-time expert machine
control system is that by the same two authors
Snoeys and Dekeyser[47] for controlling their wire
EDM machine. The system uses heuristics as well as
theoretically based rules for adjusting the
machining parameters (reference voltage and pulse
frequency), avoiding unstable machining and wire
rupture, and implementing its adaptive control
constraint strategy. Both the diagnosis and
control systems are written in PROLOG.

It should be noted that in the discussion of
real-time for this last case, and indeed for the
two previous real-time cases above[43,44], one is
not necessarily referring to fast computing[48].
A more accurate definition for a real-time system
is one that states that there must be a strict
time limit in which the system must have produced
a response to an environmental change[49]. A
decision must be available at or before the time
that it is needed[50], and this must be the case
irrespective of the algorithm employed, and
whether or not it be knowledge-based'. Thus,
real-time is essentially a relative concept.

Expert Systems for Formation Problems

As mentioned earlier, formation problems are those
involving design. Genuine design expert systems
are still rare and most of the so-called design
systems are in fact only selection (derivation)
programs for solving the large class of problems
in between the derivation-formation spectrum.

An example of a system approaching a true design
system is the Mechanically Intelligent Designer
(MIND) developed by Lembeck and Velinsky[51] for
wire rope design. The system is heavily biased
towards design analysis. It incorporates both
rules (including meta-rules) and analytical
procedures and uses the former to control the
latter. A design task is normally divided into
subtasks each of which is solved using a small set
of pertinent rules. This division avoids the need
to have all encompassing rule sets and speeds up
the design process. The system is written in C
and performs forward as well as backward chaining.

Another instance of a mechanical design system is
CDDES by Wang et al[52]. This is a program for
roller chain drive design. CDDES has a blackboard
architecture with a number of specialised modules
communicating among themselves via a common
memory. The modules perform various functions,
both analytical and

symbolic, such as preliminary design, structural design, strength evaluation, and redesign. The system is written in LISP and encapsulates design knowledge in the form of rules, frames and mathematical procedures.

Holdex, developed by Lim and Knight[53], is an expert system primarily for designing fixturing devices. The design of a fixture to hold a workpiece involves four steps: transforming the geometrical parameters associated with the workpiece into non-geometrical entities which are more easily handled in design rules (for example transforming a tolerance specification of 0.005mm on the diameter of a hole into a requirement that the latter can be a dowel hole), abstraction of other non-geometrical entities (e.g. batch size, initial condition of the workpiece material, existing machined features), deducing the geometrical parameters of the fixture from the non-geometrical information obtained in the previous two steps, and finally decomposing the deduced geometrical parameters of the fixtures into a 2D geometric drawing for subsequent production and assembly. Holdex is written in LISP and PROLOG and is integrated with a geometrical modelling CAD package. The system is rule-based and employs a combination of forward and backward chaining.

The expert systems outlined so far in this section are all geared to narrow problem areas. In contrast, Dominic, by Howe et al[54], is conceived as a domain-independent program suitable for any design task which can be formulated as an optimisation problem. Dominic requires to be given an initial design and a description of the relationship between the design goals and the design variables. It proceeds to evaluating that design and identifying its weaknesses, then assesses the effect of changing a selected variable on the overall performance of the design. It implements that change if the effect is positive and continues the evaluation and redesign cycle until the specified goals are met. Dominic has been tried on two completely different design tasks, namely the design of V-belt drive systems and extruded aluminium heat sinks. It has in both cases exhibited a performance which compares favourably with that of a human expert and a domain - specific expert system. Dominic is implemented in LISP and resorts to purpose-written FORTRAN analysis routines for solving particular design problems.

Finally, a flavour of future intelligent design systems can be captured from the EDISON program by Dyer et al[55]. This is a program for creating novel mechanical devices, using knowledge of qualitative physical relationships, symbolic reasoning, and discovery rules. The program has two modes of operation: brainstorming and problem solving. Brainstorming involves starting with an existing object and creating new objects through processes of mutation, generalisation, and drawing analogies. During problem solving, a specified goal such as improving a given design is achieved by planning and satisfying individual sub-goals. EDISON is designed using RHAPSODY, an AI program development environment incorporating a variety of software tools, including a logic programming language, a frame representation language and message-passing semantics.

Other Expert Systems

These include systems for process planning, system configuring, and production scheduling,

which as previously mentioned are problems involving a degree of synthesis as well as derivation.

Several process planning expert systems have been developed[56]. Some recent examples are: CIMS, a package by Iwata and Sugimura[57] which integrates CAD and computer-aided process planning; CUTTECH, a system by Barkocy and Zdeblick[58] for selecting cutting tools, planning operation sequences, and determining cutting speeds and feed rates for producing a given part; and EXCAP-Y, a system by Davies et al[59] for creating the machining sequence for turned components. Generally, all of the systems developed so far can be grouped into two categories, variant and generative. In a variant system, the process plan for a new part is obtained by adapting an existing plan for a similar part that has been previously machined. Recognition of part similarity is by matching part codes or logic descriptions[60]. A generative system does not employ ready-made plans but devises a special plan for each part to be machined, using knowledge of its processing requirements, available operations, and process planning principles. A feature recognition facility is provided in some advanced systems[61] to extract processing requirements directly from the CAD models of the part and blank. Most process planning systems are rule based and adopt either the forward or backward inferencing technique. For a more detailed review of process planning systems, see[62] or [63].

An instance of sophisticated expert programs for configuring machine systems, programs that are often misclassified as design expert systems at the formation end of the derivation-formation spectrum, is that by Akagi et al[64]. The program is aimed at configuring marine power plants, a task involving the selection of a variety of equipment (main engine, diesel generator, turbo generator, shaft generator, auxiliary boiler, exhaust gas economiser) and building them into a complex system. Coded in LISP, the program is a hybrid system combining AI, CAD as well as design optimisation techniques. The program incorporates design rules, a frame-type database (for storing information on the different types of machines), and mathematical optimisation procedures.

Modern Computer Integrated Manufacturing (CIM) environments provide a particularly rich application area for knowledge-based systems; Kusiak[65] gives an excellent set of overviews of the implications of applying AI techniques to such environments. A useful starting point in the consideration of the computer integration tasks involved is to first view a manufacturing system as consisting of a hierarchy of inter-linked production planning and control levels. Using previous work, Rogers[66] describes a five-level hierarchical model, based primarily on the concepts of planning and control time horizons; these comprise from highest to lowest levels: Facility (Factory), Shop, Cell, Workstation and Equipment. The time horizons vary from milliseconds at the equipment level to months at the top level.

Opinions vary as to where the Flexible Manufacturing System (FMS) fits into this overall scheme. Some[67] regard the FMS as being in essence the same as a CIM system while others [68,69] think of them as falling within the lower end (production control regions) of the hierarchical model.

Currently, knowledge-based system techniques are being applied to all levels of the CIM hierarchy.

However, in the brief reviews to be presented here, discussion will be restricted to some interesting work being carried out mainly at the middle levels, as represented by the shop, cell and workstation zones. This represents a time horizon from seconds to days. Possibly the most important characteristic of this region is scheduling for manufacturing control; scheduling is a common theme found in all the brief review discussions below.

The functions of scheduling basically bridge the gap between production planning and shop-floor control operations; as such, schedules serve as a guide for production and for establishing the resource requirements to maintain the effectiveness of the manufacturing enterprise. Newman[70] provides a very useful and well written introduction to the subject of scheduling as applied to CIM systems, and also broadly discusses the knowledge-based techniques that may be employed to implement them. In particular, he highlights the importance of real-time dynamic scheduling; these run during process execution, and are required in modern FMS and "Just-in-time" manufacturing processes in order for them to operate efficiently.

As contributions to ESPRIT project 932 "Knowledge-Based Realtime Supervision in CIM", Meyer et al[71,72], have been developing knowledge-based techniques to the supervision of an intelligent workshop controller. The controller architecture is developed into several sub-modules for solving such tasks as interpretation, diagnosis, planning and plan execution, which in turn control such functions as production planning, quality control and preventative maintenance. An experimental prototype workcell controller has been realised for a surface-mounted-device workcell of an audio factory, the software being developed on a KEE environment running on a Symbolics LISP machine. A flexible planning model to optimise machine scheduling (a dynamic scheduling procedure) has been utilised, the heuristics to solve the optimisation problem being partly contained in rules and partly made explicit in approximative algorithms.

For comparison, the workcell controller is also being developed on a special research blackboard building tool[73], more pre-structured for this type of problem. Run time efficiency is reported as being rather weak for both cases.

Dynamic scheduling forms a fundamental real-time control level for an experimental rule-based FMS operation control strategy being developed by Ranky[68,74]. In the representational model of a CIM structure presented in this case, FMS cells and workstations, from a control point of view, are designed as nodes in a distributed computer network (FMS islands) rather than in the hierarchical structure outlined previously. A major goal of the project is to test the applicability of this system for a FMS capable of executing viable routes for parts and other control strategies in real-time, as generated by process planning staff off-line. The production rule bases are built up with special IF-THEN rules written in PROLOG or Pascal-like formats. During operation, the dynamic scheduling program can access the rule-base, choose from the offered alternative routes and make real-time control decisions based on fast secondary optimisation routines. During simulation tests, cell scheduler speeds of between 2 - 8 seconds have been claimed.

A final example of an intelligent production scheduling system is that by Bruno et al[75].

This is a package for arranging the sequence of operations in an FMS. The forward chaining rule-based package written in OPS5 can take into account various types of real life constraints (production constraint, resource constraint and capacity constraint). Its efficiency is enhanced by the provision of queuing network analysis (implemented in FORTRAN 77) which enables first evaluation of the performance of the FMS, the production of which is being scheduled.

CONCLUSION

The field of knowledge engineering has been reviewed to highlight the nature of expert systems and their internal organisation. The review has focused on the two key elements of these intelligent programs, the knowledge base and the inference mechanism.

It has been found that expert systems can essentially be divided into two types, rule-based and frame-/net-based systems. The names assigned to these systems reflect the knowledge representation formalisms they use. Rule-based systems adopt a shallow representation of knowledge in terms of associative IF-THEN rules. Frame-/net-based systems employ a deep representation scheme with a hierarchical structure for holding information regarding concepts or objects and relations between these. The choice of knowledge representation techniques influences the design of the inference mechanism. Rule-based systems use logical inference methods with forward and backward chaining of the rules to search for solutions. Frame-/net-based systems carry out their deductive reasoning as part of their knowledge assertion and acquisition activities.

Rule-based systems have so far been the most widely used. However, frame-based systems and, in particular, hybrid systems employing both frames and rules are also becoming more common. For example, of all the new engineering expert systems surveyed in this paper, almost half are either hybrid or frame-based.

Finally, as can be seen in the paper, expert systems have moved well into engineering from their original application domains of chemistry, geology, mathematics and medicine. In engineering design and manufacture, they are now being used or contemplated for such a diverse range of tasks as machine selection, equipment condition monitoring, fault diagnosis, real-time control, component design, tool design, process planning, production control, and so on. Thus, although knowledge engineering technology still has to mature - better means of representing, acquiring, and using knowledge have yet to evolve - it is true to state that its potential as a powerful tool for information processing and decision making in engineering design and manufacture is now widely recognised.

ACKNOWLEDGEMENT

The authors gratefully acknowledge the help and encouragement given to them by the late A.Saeed in the preparation of this paper.

REFERENCES

1. A.Barr.and E.A.Feigenbaum (eds.),The Handbook of Artificial Intelligence, 1 & 2, Kaufmann, Los Altos, CA, 1981.

2. D.Mitchie (ed.),Introductory Readings in Expert Systems, Gordon and Breach, New York, 1982.

3. F.Hayes-Roth, D.A.Waterman, and D.B.Lenat (eds.), Building Expert Systems, Addison-Wesley, Reading, MA, 1983.

4. S.M.Weiss and C.A.Kulikowski, A Practical Guide to Designing Expert Systems, Chapman and Hall, London, 1983.

5. R.Forsyth (ed.), Expert Systems: Principle and Case Studies, Chapman and Hall, London, 1984.

6. J.L.Alty and M.J.Coombs, Expert Systems Concepts and Examples, NCC, Manchester, 1984.

7. D.A.Watermann, A Guide to Expert Systems, Addison-Wesley, Reading, MA, 1985.

8. P.Harmon and D.King Expert Systems: AI in Business, Wiley, New York, 1985.

9. W.J.Black, Intelligent Knowledge-Based Systems: An Introduction, Van Nostrand-Reinhold, Wokingham, 1986.

10. R.H.Michaelson, D.Michie and A.Boulanger, 'The technology of expert systems', BYTE, 10 (4), April 1985, pp.303-312, 1985.

11. E.A.Feigenbaum, Knowledge Engineering for the 1980s, Computer Science Department, Stanford University, Stanford, CA, 1982.

12. B.G.Buchanan, 'Expert systems', Journal of Automated Reasoning, 1 (1), pp.28-35, 1985.

13. R.Davis, 'Knowledge-based systems', Science 231(4841), 28 February, pp.957-963, 1986.

14. F.Hayes-Roth, 'Rule-based systems', ACM Communications, 28 (9), September, pp.921-932, 1985.

15. M.D.Rychener, 'Expert systems for engineering design', Expert Systems, 2 (1), January, pp.30-44, 1985.

16. M.Minsky, 'A framework for representing knowledge', in The Psychology of Computer Vision, P.H.Winston (ed.), McGraw Hill, New York, 1975.

17. R.Fikes and T.Kehler, 'The role of frame-based representation in reasoning', ACM Communications, 28 (9), September, pp.904-920, 1985.

18. N.J.Nilsson, Principles of Artificial Intelligence, Tioga, Palo Alto, California, 1980.

19. A.Walker, 'Knowledge systems: principles and practice', IBM Journal of Research and Development, 30 (1) January, pp.2-13, 1986.

20. P.H.Winston and B.K.P.Horn, LISP, Addison-Wesley, Reading, MA, 1984.

21. W.F.Clocksin and C.S.Mellish, Programming in Prolog, 2nd Ed., Springer-Verlag, Berlin, 1984.

22. M.H.Richer, 'An evaluation of expert system development tools', Expert Systems, 3 (3), July, pp.166-183, 1986.

23. J.Van Koppen, 'A survey of expert system development tools', Proceedings 2nd International Expert Systems Conference, London, September, pp.157-173, 1986.

24. P.Jackson, Introduction to Expert Systems, Addison-Wesley, Reading, MA, 1986.

25. L.Ortolano and C.D.Perman, 'Software for expert systems development', ASCE Journal of Computing in Civil Engineering, 1 (4), October, pp.225-240, 1987.

26. J.R.Quinlan, 'Discerning rules by induction from large collections of examples', in 'Expert Systems for the Microelectronic Age' D.Michie (ed.), Edinburgh University Press, Edinburgh, pp.168-201, 1979.

27. R.S.Michalski, 'A theory and methodology of inductive learning', in 'Machine Learning, an Artificial Intelligence Approach', R.J.Michalski, J.G.Carbonell and T.M.Mitchell (eds.), Kaufmann, Los Altos,CA, pp.83-134, 1983.

28. A.D.Shapiro, Structured Induction in Expert Systems, Turing Institute Press, Glasgow, Addison-Wesley, Wokingham, 1987.

29. R.S.Michalski and R.L.Chilauski, 'Learning by being told and learning from examples: an experimental comparison of the two methods of knowledge acquisition in the context of developing an expert system for soybean disease diagnosis', Journal of Policy Analysis and Information Systems,4 (2), June, pp.125-161.

30. W.F.Hodges, Logic, Penguin, Harmondsworth, 1978.

31. M.R.Genesereth and M.L.Ginsberg, 'Logic programming', ACM Communications, 28 (9), September, pp.933-941, 1985.

32. W.B.Gevarter, Artificial Intelligence, Expert Systems, Computer Vision and Natural Language Processing, Noyes Publications, New Jersey, 1984.

33. M.J.Stefik, 'Planning with constraints', PhD dissertation, Heuristic Programming Project, Computer Science Department, Stanford University, Stanford, CA, 1980.

34. D.Sriram,'Knowledge-based approaches for structural design', PhD dissertation, Department of Civil Engineering, Carnegie-Mellon University, Pittsburg, Pennsylvania, 1986.

35. D.Sriram and R.Joobbani (eds.), Special issue 'AI in engineering', SIGART Newsletter, April (92), pp.38-127, 1985.

36. G.Hasle, G.Stokke and M.Kloster, 'BETSY - an expert system for the selection of bearings', in 'Knowledge Based Expert Systems for Engineering: Classification, Education Control', D.Sriram and R.Adey (eds.), Computational Mechanics Publication (Southampton, K and Boston, Mass., USA), pp.221-237, 1987.

37. G.Bose and C.S.Krishnamoorthy, 'GEAREX - a Unix based gearbox selection expert system', in 'Knowledge Based Expert Systems for Engineering: Classification, Education and Control', ibid, pp.345-358, 1987.

38. D.T.Pham and S.H.Yeo, 'An adaptive knowledge based system for selecting robot grippers', in 'Expert Systems in Engineering', D.T.Pham (ed.), IFS Publications (Bedford, UK) and Springer-Verlag (Berlin, West Germany), pp.408-424, 1988.

39. H.N.Mahabala, G.Raviprakash, M.Prakash and B.Ramanadh, 'Expert system for selection of drills: a case study for use of meta-knowledge and consitency checks', in 'Knowledge Based Expert Systems for Engineering Classification, Education and Control', D.Sriram and R.Adey (eds.), Computational Mechanics Publications (Southampton, UK and Boston, Mass, USA), pp.371-386, 1987.

40. A.Kusiak and S.S.Heragu, 'KBSES: a knowledge-based system for equipment selection', The International Journal of Advanced Manufacturing Technology, 3 (3), Special Issue on Knowledge-Based Systems, D.T.Pham (ed.), pp.97-109, 1988.

41. A.Fukuda, 'Development of WELDA: an advisor expert system for welding', in 'Knowledge Based Expert Systems for Engineering:Classification, Education and Control', D.Sriram and R.Adey (eds.), (eds.), Computational Mechanics Publication (Southampton, UK and Boston, Mass, USA), 285-296, 1987.

42. K.K.B.Hon and H.Ismail, 'A knowledge-based system for the selection of hole-making processes', in 'Expert Systems in Engineering', D.T.Pham (ed.), IFS Publications (Bedford, U.K.) and Springer-Verlag (Berlin, West Germany), pp.444-451, 1988.

43. N.Huang, R.J.Beattie and P.G.Davey, 'A rule-based system for interpreting weld seam images', The International Journal of Advanced Manufacturing Technology, 3 (3), Special Issue on Knowledge-Based Systems, D.T.Pham (ed.), pp.111-121, 1988.

44. A.Villa, 'An expert system for tool monitoring and maintenance in automated machining process', in 'Artificial Intelligence Implications for CIM', Kusiak (ed.), IFS Publications (Bedford, UK) and Springer-Verlag (Berlin, West Germany), pp.516-527, 1988.

45. X.J.Zheng, S.Z.Yang, A.F.Zhou and H.M.Shi, 'A knowledged-based diagnosis system for automobile engines', The International Journal of Advanced Manufacturing Technology, 3 (3), Special Issue on Knowledged-Based Systems, D.T.Pham (ed.), pp.159-169, 1988.

46. R.D.Puetz and R.Eichhorn, 'Expert systems for fault diagnosis on CNC machines', in 'Knowledge Based Expert Systems for Engineering:Classification, Education and Control', D.Sriram and R.Adey (eds.), Computational Mechanics Publication (Southampton, UK and Boston, Mass, USA), pp.387-396, 1987.

47. R.Snoeys and W.Dekeyser, 'Knowledge-based system for wire EDM', The International Journal of Advanced Manufacturing Technology 3 (3), Special Issue on Knowledge-Based Systems, D.T.Pham (ed.), pp.83-96, 1988.

48. J.A.Stankovic, 'Misconceptions about real-time computing', IEEE Computer, pp.10-19, October 1988.

49. C.A.O'Reilly and A.S.Cromarty, 'Fast is not real-time: designing effective real-time AI systems', SPIE Vol. 548, pp.249-257, 1985.

50. M.Lattimer-Wright, 'An expert system for real-time control', IEEE Software, pp.16-24, March 1986.

51. M.F.Lembeck and S.A.Velinsky, 'A prototype expert system for wire rope based mechanical design', Proceedings 7th World Congress on the Theory of Machines and Mechanisms, IFToMM, Sevilla, Spain, September 1987, pp.1029-1032, 1987.

52. Q.Wang, J.Zhou and J.Yu, 'A chain-drive design expert system and CAD system', in 'Expert Systems in Engineering', D.T.Pham (ed.), IFS Publications (Bedford, UK) and Springer-Verlag (Berlin, West Germany), pp.315-323, 1988.

53. B.S.Lim and J.A.G.Knight, 'Holdex - holding device expert system', Proceedings 1st International Conference on Applications of Artificial Intelligence in Engineering Problems, D.Sriram and R.Adey (eds.) Computational Mechanics Publication (Southampton UK) and Springer(Berlin, West Germany), pp.483-501, 1986.

54. A.Howe, P.Cohen and J.Dixon, 'Dominic: domain independent program for mechanical engineering design', Proceedings 1st International Conference on Applications of Artificial Intelligence in Engineering Problems, ibid, pp.289-299, 1986.

55. M.G.Dyer, M.Flowers and J.Hodges, 'Edison: an engineering design invention system operating naively', Proceedings 1st International Conference on Application of Artificial Intelligence in Engineering Problems, ibid, pp..327-341, 1986.

56. C.R.Liu, T.C.Chang and R.Komanduri (eds.),Computer-Aided/Intelligent Process Planning, ASME (New York), 1985.

57. K.Iwata and N.Sugimura, 'An integrated CAD/CAPP system with "know-hows" on machining accuracies of parts', ASME Trans, Journal of Engineering for Industry, 109 (2), May, pp.128-133, 1987.

58. B.E.Barkocy and W.J.Zdeblick, 'A knowledge-based system for machining operation planning', Proceedings AUTOFACT 6, SME (Dearborn, Michigan), Tech Paper MS84-716, 1984.

59. B.J.Davies, I.L.Darbyshire and A.J.Wright, 'The integration of process planning with CAD CAM, including the use of expert systems', Proceedings International Conference on Computer Aided Production Engineering, IMechE, Edinburgh, April, pp.35-40, 1986.

60. P.C.Y.Sheu and R.L.Kashyap, 'Object-based knowledge bases in automatic manufacturing environments', The International Journal of Advanced Manufacturing Technology, 3 (3), Special Issue on Knowledge-Based Systems, D.T.Pham (ed)., pp.39-52, 1988.

61. A.H.Van't Erve and H.J.J.Kals, 'XPLANE, a knowledge-based driven process planning expert system', Proceedings International Conference on Computer Aided Production Engineering, IMechE, Edinburgh, April, pp.41-46, 1986.

62. S.S.Heragu and A.Kusiak, 'Analysis of expert systems in manufacturing design', IEE Trans on Systems, Man and Cybernetics, SMC-17 (6), November-December, pp.898-912, 1987.

63. S.R.T.Kumara, S.Joshi,R.L.Kashyap, C.L.Moodie and T.C.Chang, 'Expert systems in industrial engineering', in 'Expert Systems in Engineering', D.T.Pham (ed.), IFS Publications (Bedford, UK), and Springer-Verlag (Berlin,, West Germany), pp.387-407, 1988.

64. S.Akagi, T.Tanaka and K.Kubonishi, 'An expert CAD system for the design of marine power plants using artificial intelligence', JSME International Journal Series 3, 31 (1), March, pp.149-156, 1988.

65. A.Kusiak, (ed.), Artificial Intelligence: Implications for CIM, IFS(Bedford, UK), 1988.

66. P.Rogers and D.J.Williams, 'Knowledge-based control in manufacturing automation', International Journal of CIM, 1 (1), Jan-March, pp.21-30, 1988.

67. A.Kusiak, 'Artificial Intelligence and CIM Systems', in 'Artificial Intelligence: Implications for CIM', A.Kusiak (ed.), IFS(Bedford, UK), pp.3-44, 1988.

68. P.G.Ranky, 'A real-time, rule-based FMS operation control strategy in CIM environment - Part 1', International Journal of CIM, 1(1), Jan-March, pp.55-71, 1988

69. R.Gross and K.Ravi Kumar, 'Intelligent control systems', in 'Artificial Intelligence: Implications for CIM', A.Kusiak (ed.). IFS(Bedford, UK), pp.347-358. 1988.

70. P.A.Newman, 'Scheduling in CIM systems', ibid, pp.359-402.

71. W.Meyer, 'Knowledge-based real-time supervision in CIM: The Workcell Controller'. ESPRIT'86: Results and Achievements; pp.33-52, CEC 1987.

72. W.Meyer, R.Isenberg and M.Huebner, 'Knowledge-based factory supervision: The CIM shell', International Journal of CIM, 1 (1), Jan-March, pp.31-43, 1988.

73. R.Isenberg, 'Comparison of BBI and KEE for Building a Production Planning Expert System', 3rd Int. Expert Systems Conference,London, pp.407-421, June 1987.

74. P.G.Ranky, 'A real-time, rule-based FMS operation control strategy in CIM environment - Part 2', International Journal of CIM, 1 (3), July-Sept. pp.185-196, 1988.

75. G.Bruno, A.Elia and P.Laface, 'A rule-based system to schedule production', in 'Expert Systems in Engineering', D.T.Pham (ed.), IFS(Bedford, UK), pp.452-465, 1988.

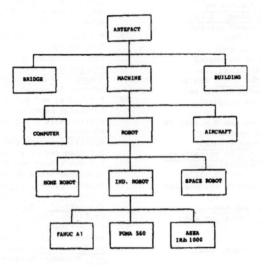

Figure 1.   Frame representation of part of the hierarchy of artefacts

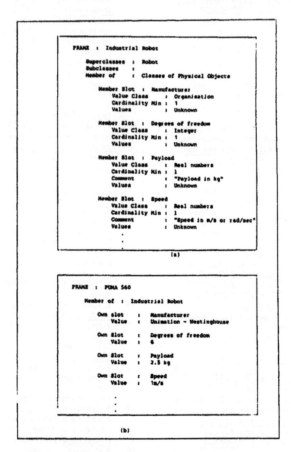

Figure 2.   Frames representing the 'Industrial Robot' subclass of robots
and object Puma-560 in that subclass.

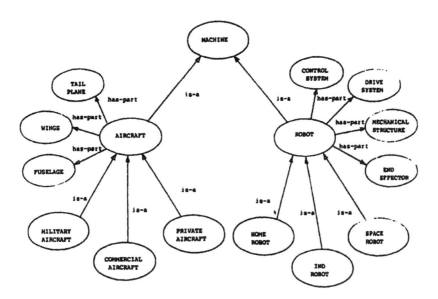

Figure 3. A simple net partially representing the concept of a machine.

```
1) Modus ponendo ponens :

        A ⟹ B   ,   A ⊢ B

        (If A is TRUE        Then        B is TRUE
            A is TRUE      Therefore      B is TRUE


2) Modus tollendo tollens :

        A ⟹ B   ,   B̄ ⊢ Ā

        (If A is TRUE        Then        B is TRUE
            B is NOT TRUE  Therefore  A is NOT TRUE


3) Resolution

        (A ⟹ B ; B ⟹ C) ⟹ (A ⟹ C)
```

Figure 4. Three basic inference rules.

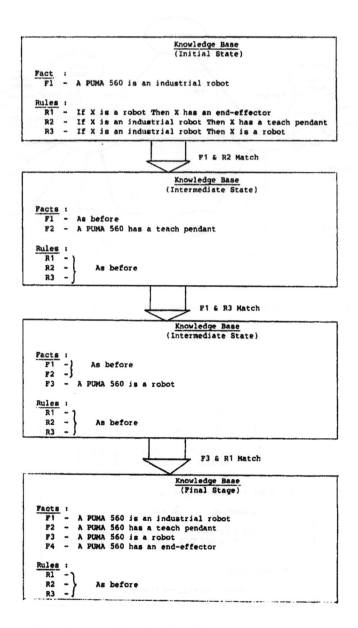

Figure 5. An example of forward chaining.

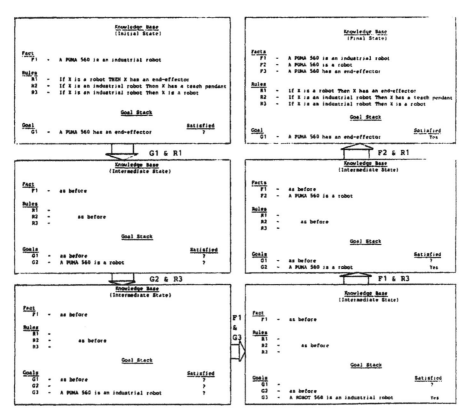

Figure 6. An example of backward chaining (the goal stack stores the goals to be satisfied).

AUTOMATIC FINE-MOTION PLANNING BASED ON POSITION/FORCE STATES

R. Suárez and L. Basáñez

Institut de Cibernètica (CSIC-UPC), Barcelona, Spain

Abstract – This paper outlines a new approach to fine motion planning in presence of uncertainty. The concept of position/force space is introduced, and the uncertainty sources briefly reviewed. With these elements a new model of the task is proposed using position/force states, and based on them, the procedure to obtain the plan and to execute it, is described. The method has no theoretical constraints on the number or type of the degrees of freedom, and can be used with different position/force control types.

Keywords – Robots; assembling; automation; fine motion; planning.

INTRODUCTION

The Problem

Assembly tasks with robots requires performing fine motions when the pieces are close up or in physical contact. One of the main problems in dealing with fine motion planning and execution is the uncertainty inherent in real world object dimensions and positions, in sensory information, and in robot positioning. This uncertainty seriously affects the performance of some types of assembly tasks using robots, unless a properly planned strategy is used. Decreasing the amount of uncertainty largely increases the cost of the system, and anyway technical limitations always remain.

This paper outlines a new approach to fine motion planning in presence of uncertainty. A new model of the task based on position/force states is proposed, and the procedure to obtain and to execute the plan is described. The method has no theoretical constraints on the number or type of the degrees of freedom (*dof.*), or on the position/force control type.

The proposed method forms part of an automatic cell programming and monitoring system for assembly tasks presently under development at the Institute of Cybernetics of Barcelona (Basáñez and others, 1988a, 1988b).

Previous Work

Significant contributions dealing with the problem of automatic fine motion planning have been reported.

Mason (1981) provides a way to determine natural and artificial constraints in order to use hybrid control. Dufay (1984) proposes an automatic planning method, dealing with uncertainty, in which, after multiple task executions during a training phase, an induction phase supplies the general plan. Lee (1985) suggests a technique based on two-dimensional cut diagrams to generate the compliance vector for hybrid control. Turk (1985) proposes a fine motion planning algorithm based on geometric states (regions) and assuming damping control.

*This work was partially supported by Fundación Ramón Areces under the
project SEPETER

Lozano Perez (1984) proposes a formal approach to the synthesis of compliant motion strategies. It is based on the concept of *pre-image* obtained from the task geometric information, goal positions and commanded velocities. Erdmann (1984) suggests a method for planning motions in presence of uncertainty, based on the concepts of *pre-image* and *back-projection* and, on the same line, Buckley (1987) presents an interactive system to build a compliant motion strategy, and a planner capable of dealing with simple problems.

TASK POSITION AND FORCE MODEL

Configuration Space

Describing the position of a rigid object requires the specification of all its degrees of freedom, both translations and rotations. This can be done by mean of a set of independent parameters, called its *configuration*. The number n of independent parameters is equal to the number of degrees of freedom the object has. The n-dimensional space defined by these parameters is called *Configuration space (C-space)* (Lozano Perez, 1983).

Therefore, the position of a rigid object in the real world is represented by a point in the C-space. This means that the problem of manipulating a rigid body in the real world can be translated into the problem of manipulating a point in the C-space. When there are obstacles in the environment, the object is only free to move in some ranges of its degrees of freedom; thus, only a subspace of the C-space represents valid configurations in which there is no collision with the obstacles. This subspace is called *free space*. The boundary between the free space and the subspace of invalid configurations is represented by hypersurfaces called *C-surfaces*.

The main advantage of transforming the problem of moving a real object among real obstacles into the problem of moving a point among transformed obstacles is that motion constraints appear explicitly, and it is easier to deal with them. However, computing the exact C-space for a high dimensional problem may be a hard work, specially if rotations are involved.

Position/Force Space

The $2n$-dimensional *Position/Force Space (PF-space)* will be defined as an extension of the n-dimensional C-space, by attaching to each point of the C-space the n-dimensional static reaction force that appear when the object becomes in contact with the obstacles.

In the case of rigid object and rigid obstacles, it is clear that there can only be finite non-zero forces in configurations corresponding to C-surfaces, while zero force will be attached to free space configurations and arbitrarily large force will be associated to invalid configurations. In the case of elastic object and obstacles, the valid points of PF-space will depend on the elastic properties, but the above definition also applies. In this paper, we only deal with the rigid case.

Using an appropriate reference system and in static situation, reaction forces will be normal to C-surfaces (Erdmann, 1984), thus force parameter values will depend on C-surfaces.

The set of all points in PF-space whose projections into C-space belong to a C-surface will be called *Contact Subspace (CS)*, and the set of points of PF-space whose projections belong to free space in C-space will be called *PF-free space*.

UNCERTAINTY

Uncertainty Sources

The sources of uncertainty can be categorized into three major groups:

a) *Geometric tolerances in the object dimensions.* All industrial manufactured mechanical parts have tolerances in their dimensions. Three main tolerance specifications can be mentioned: size tolerance, form tolerance and relative features position tolerance (Requicha, 1983).

b) *Inaccuracy in object position.* Relative position between the object to be inserted and the place where it must be inserted could be generated by three causes: inaccuracy in the absolute location of objects in the environment, inaccuracy in the object position in the robot gripper, and inaccuracy in the robot positioning (Brooks, 1982; Day, 1988).

c) *Inaccuracy in forces measurement.* Inaccuracy in forces measurement is due to the limited resolution of force sensors and also to their physical location in the system, which sometimes implies complex transformations to obtain the resultant (e.g. sensors in robot joints).

Forces measurements are often used to decrease position uncertainty. Nevertheless, the need of estimating friction coefficients to reduce orientation uncertainty seriously limits this method.

Groups a) and b) affect position parameters, while group c) affects force parameters.

Uncertainty Model

The uncertainty in the rigid object position relative to the environment is equivalent to the uncertainty in the position parameters of C-space and PF-space. In the same way, the uncertainty in the determination of reaction forces in the real world is equivalent to the uncertainty in the force parameters of PF-space. Therefore, we will consider and model uncertainty directly in the PF-space.

All the uncertainty sources mentioned above must be taken into account together to obtain the *worst case uncertainty values* in both position and force parameters. With these values it is possible to construct uncertainty envelopes giving rise to uncertainty regions in PF-space. This means that the actual static location of a certain point in the PF-space can be any other inside the uncertainty region attached to that point. The form of these uncertainty regions depends on the type of parameters chosen to specify position and reaction forces in PF-space, and also on the desired model of uncertainty. This uncertainty model is based on those described by Requicha (1983) and Benhabib (1987).

Once the uncertainty in the PF-space has been defined, is easy to obtain the subspace of PF-space containing possible contact points in presence of uncertainty. This subspace will be called *Uncertain Contact Subspace (UCS)*, and can be obtained by making the union of the uncertainty regions associated to each point of CS. The "expansion" of CS to UCS implies the diminution of the PF-free space and of the subspace of invalid configurations.

POSITION/FORCE STATES

Position/Force States Definition

Moving through the diminished PF-free space has no risk of collision despite uncertainty, so that movements in this subset can be considered and treated as *gross motion*. On the contrary, moving through UCS, it is not possible to know with precision where a collision will occur and then, purely position control is not adequate in this subspace. Movements and actions tending to solve the task in UCS is what we consider as *fine motion*.

In our approach, UCS is partitioned into subspaces called *Position/Force States (PF-states)*. PF-states must satisfy two properties to avoid ambiguous situations:

    a) The union of all PF-states is equal to UCS.

    b) Any two PF-states are disjoint.

There are no more restrictions in PF-states selection, but some criteria are necessary in order to do an appropriate partition of UCS into a useful set of states. Assuming rigid objects, the criteria to partition UCS could be:

    a) PF-states projections from PF-space over C-space, called *states position projections (SP)* must be disjoints or completely equivalents.

    b) Different PF-states with equal SP must have different ranges of force directions.

    c) Each PF-states will be associated with some geometric features of the C-space (vertices, edges and faces).

Contiguous PF-states Graph

PF-states will be represented in a *contiguity graph (CGraph)*, in which the nodes are PF-states and the links connect the contiguous ones. Two PF-states with a frontier dimension less than $(2n-1)$ are not considered as contiguous. This CGraph is a useful representation tool for the planning work.

STATE TRANSITION OPERATORS

Definition

A *State Transition Operator (T)* is a command for the robot control system in order to produce the transition from a PF-state to another contiguous one. The form of T depends on the type of the robot control system and on the PF-states definition.

A robot position/force control system can accept both position and force commands, depending on the control scheme adopted and the desired behaviour (Suárez, 1988). PF-states approach can be used to determine possible natural and artificial constraints in task geometry for *hybrid control* (Mason, 1981; Raibert, 1981), as well as the direction of movement for *stiffness* (Salisbury, 1980) or *damping control* (Whitney, 1977).

In this paper, a robot control system working in damping control mode is assumed, therefore T must be a commanded velocity.

Determination of T

Determining T is equivalent to obtain the motion direction parameters and, after that, fix the module of the commanded velocity. Because more than one T can exist between two contiguous states they will be grouped into *State Transition Operators Sets (TS)*. Under this assumption, the rules for automatically obtaining TS directions for the two degrees of freedom problem are given in Appendix A. Directions will be represented by unitary vectors δ, and those permitting sliding on an edge are considered as valid ones to change the state.

FINE-MOTION PLANNING AND EXECUTION

Figure 1 shows a flow chart with the steps of the planning and execution phases. On-line decision work is intended to be reduce to a minimum, in order to allow major operation velocity.

After constructing the PF-states and knowing the procedure to obtaining the associated operators, the next steps must be followed.

PF-states Sequence

A sequence of contiguous PF-states, linking initial and goal PF-states, will be established using any search strategy in CGraph. The initial PF-state can be determined by sensory information, and the goal PF-state can be easily obtained from the final desired conditions. Different criteria can guide the search through CGraph (e.g. minimum PF-states number in the sequence, minimum PF-states number with non-zero force,...).

Operators Sequence

Once a PF-states sequence has been selected, the set of operators (TS) to pass from one state to the following in the sequence must be determined. It is possible to select only a subset of all feasibles state transition operators according to different criteria (e.g. higher directions range or minimum number of possible successor states).

Filtered Operators Sequence

Two consecutive TS may have a partially coincident range of directions, so they can be intersected and then replaced by the

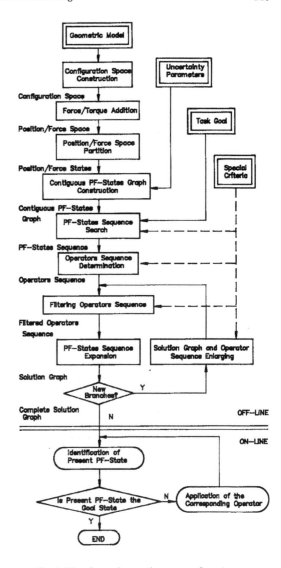

Fig. 1. Planning and executing process flow chart.

intersection result. This operation will be done beginning with the first TS and ending when the last one is reached or when the intersection becomes null. When the procedure ends with the last TS, the intersection set of directions is the goal TS that can solve the task directly. On the contrary, if a null intersection results, the procedure must be reinitialized taking as the first TS the last one considered. This means that to solve the task, a command change must be done when the corresponding PF-state of the sequence is reached. Others criteria can be simultaneously taken into account, e.g. if some states must be specially avoided, the operators that consider them as possible successor states can be discarded.

Branch PF-states Sequence Expansion

After filtering the operators sequence, the resulting TS may allow for transition to others PF-states besides those of the desired sequence. These PF-states may occur during the task execution so that they must be considered in the plan. This is done taking them as initial states and executing the planning procedure again (PF-states sequence, TS sequence, filtered TS sequence, and branch PF-states sequence expansion). When no more branches appear, the expansion has finished with the result of a directed subgraph of CGraph called *complete solution graph (SOLGraph)*.

SOLGraph may have closed loops of two differents types: *pseudo closed loops* and *repetitive closed loops*. The first ones are those in which the union of their associated TS sequence does not have directions with opposite components (more than 90° between them); this means that although the loops in the SOL-Graph actually exist, during plan execution they will be automatically solved by transition operators, or even they will not appear. The second type of loops are those in which the union of their associated TS sequence have directions with opposite components; these loops really give rise to vicious circles in the plan execution and must be monitored during on-line work. If they actually appear, another plan must be executed, taking as initial PF-state one of those in the loop and following a different strategy. If possible, repetitive closed loops must be avoided in the plan.

Task Execution

SOLGraph has the necessary information to guide the execution: PF-states that may appear and TS to go through them. The plan execution consists of identifying the present PF-state from sensory information and applying an operator T from the proper TS until a new PF-state is detected, repeating this process until the goal PF-state is reached. When a repetitive closed loop is expected, the actual PF-state sequence must be monitored in order to detect and avoid vicious circles.

If an unexpected PF-state appears (e.g by accident or any external action) the task must be re-planned from this PF-state. Special error recovering strategies can be formulated to avoid re-planning all the sequence.

SIMPLE CASE EXAMPLE

Suppose the simple task of positioning a square block in a corner (Fig. 2), considering only two translational degrees of freedom. Taking point A of the block as reference, the resulting C-space is shown in Fig. 3. The PF-space is 4-dimensional, so it is illustrated in Fig. 4 as two projections of dimension 3, representing forces by module M and phase Φ. Supposing the worst uncertainty values as U_x, U_y, U_M and U_Φ, the PF-space in presence of uncertainty is shown in Fig. 5. The partition of UCS of PF-space into PF-states is described in Fig. 6, and the resulting CGraph is shown in Fig. 7.

Suppose that the initial PF-state, reached by gross motion, is $a0$. The final desired state is $abAB$, thus, a sequence of PF-states that solve the task is: $a0$, $ab0$ and $abAB$.

The operators to follow that PF-state sequence, selected with the criterium of minimun target PF-states number, and the initial filtered result are shown in Fig. 8. Branching PF-states sequence with this resulted TS gives the directed graph in Fig. 9. Repeating the planning procedure from states aA, abA, and abB (search a sequence of states and operators, filtering, and branching) the remaining operators on Fig. 10 are found. The SOLGraph of Fig. 11 is obtained with the attached TS of Fig. 12. So, the task will be solved despite uncertainty using any T from TS showed in Fig. 12. The reduction of the initial filtered TS in Fig. 8 to that in Fig. 12, is the result of intersecting it with the operators to pass from aA to abA, and from abA and abB to $abAB$ (Fig. 10).

CONCLUSIONS

A new automatic fine-motion planner, specially oriented to insertion tasks in presence of uncertainty, has been proposed. The plan, including uncertainty both in position and force parameters, is elaborated from a special model of the task based on position/force states and a set of operators to change from one state to another. The output is a set of commands for the robot control system that ensures the success in the assembly execution. The approach has no constraints in the number or type of the degrees of freedom, and can be used with different position/force control types.

REFERENCES

Basáñez L., R. Kelley, M. Moed, and C. Torras (1988a). A least-commitment approach to intelligent robotic assembly. *Proc. of the 1988 IEEE Int. Conf. on Robotics and Automation*, Pennsylvania, USA, 1318-1319.

Basáñez, L., C. Torras, A. Sanfeliu, and J. Ilari (1988b). Automatic cell programming and monitoring through the cooperative interplay of operation specialists. *Proc. of the Second Int. Symposium on Robotics and Manufacturing*, New Mexico, USA, 1067-1074.

Benhabib, B., R. Fenton, and A. Goldenberg (1987). Computer-aided joint error analysis of robots. *IEEE Journal of Robotics and Automation*, RA-3 (4), 317-322.

Brooks, R.A. (1982). Symbolic error analysis and robot planning. *The Int. Journal of Robotics Research*, 1 (4), 29-68.

Buckley, S.J. (1987). Planning and teaching compliant motion strategies. *MIT Art. Intel. Lab. report AI-TR-936 (Ph.D Thesis)*.

Day, C. (1988). Robot accuracy issues and methods of improvement. *Robotics Today*, 1 (1), 1-9.

Dufay, B., and J. Latombe (1984). An approach to automatic robot programming based on inductive learning. In M. Brady and P. Paul (Ed.), *Robotics Research: The First Int. Symposium*. The MIT Press. pp. 97-115.

Erdmann, M. (1984). On motion planning with uncertainty. *MIT Art. Intel. Lab. report AI-TR-810 (Master Thesis)*

Lee, C.S.G., and D. Huang (1985). A geometric approach to deriving position/force trajectory in fine motion. *Proc. of the 1985 IEEE Int. Conf. on Robotics and Automation*, St. Louis, USA, 691-697,

Lozano Perez, T. (1983). Spatial planning: a configuration space approach. *IEEE Trans. on Computers*, C-32 (2), 108-120.

Lozano Perez, T., M. Mason, and R. Taylor (1984). Automatic synthesis of fine-motion strategies for robots. *The Int. Journal of Robotics Research*, 3 (1), 3-24.

Mason, M. T. (1981). Compliance and force control for computer controlled manipulators. *IEEE Trans. on Systems, Man, and Cybernetics*, SMC-11 (6), 418-432.

Raibert, M.H., and J.J. Craig (1981). Hybrid position/force control of manipulators. *Trans. of ASME Journal of Dyn. Syst., Meas., and Control*, 102, 126-133.

Requicha, A.A. (1983). Toward a theory of geometric tolerancing. *The Int. Journal of Robotics Research*, 4 (2), 45-60.

Salisbury, J.K. (1980). Active stiffness control of a manipulator in Cartesian coordinates. *Proc. of the 19th IEEE Conf. on Decision and Control*, 95-100.

Suárez, R. (1988). Control de fuerza en robótica. *Automática e Instrumentación*, *185*, 221-235.

Turk, M.A. (1985). A fine-motion planning algorithm. *Intelligent Robots and Computer Vision - SPIE*, *579*.

Whitney, D. (1977). Force feedback control of manipulator fine motions. *Trans. of ASME Journal of Dyn. Syst., Meas., and Control*, (June 1977), 91-97.

APPENDIX A:
OPERATORS DETERMINATION

In this Appendix, the rules for automatically obtaining TS directions for the two degrees of freedom problem are given. The following nomenclature will be used:

E : PF-state with attached force.

$\overline{E}$: PF-state with null attached force.

SP : state position projection of E or $\overline{E}$.

n : normal vector to the frontier between the SP of the present PF-state and any contiguous SP, external to the former.

Δd : set of unitary vectors with the directions of the attached forces of E.

d : unitary vector belonging to Δd.

δ : unitary vector representing the operator direction.

$(\cdot)$: dot product.

$(\times)$: cross product.

and the following subindices:

p : corresponding to the present PF-state.

s : corresponding to any non-present PF-state with $SP_s = SP_p$.

c : corresponding to any PF-state with SP_c contiguous to SP_p.

Two different cases are initially possible, depending on the present PF-state reaction force:

Case A. Present PF-state $\overline{E}_p$

1. δ may produce the transition from $\overline{E}_p$ to $\overline{E}_c$ if it satisfies $(\delta \cdot n) \geq 0$. This condition also applies to transitions to PF-free space.
2. δ may produce the transition from $\overline{E}_p$ to E_s if it satisfies $(\delta \cdot d_s) \leq 0$.

Case B. Present PF-state E_p.

Two cases must then be contemplated:

Case B1. E_p is associated only with edges.

1. δ may produce the transition from E_p to $\overline{E}_s$ (contact lost) if it satisfies $(\delta \cdot d_p) \geq 0$.
2. δ may produce the transition from E_p to E_c with $\Delta d_c \cup \Delta d_p \neq \phi$, if it satisfies $(\delta \cdot d_p) \leq 0$ and $(\delta \cdot d_c) \geq 0$.
3. δ may produce jamming in E_p if it satisfies $(-\delta) \in \Delta d_p$.

Case B2. E_p is associated with a vertex. A vertex generates three non-null force states, two (E_p and E_s) are associated with the normal directions to the edges that intersect in the vertex, and the third state (E_s') is associated with the range of directions between those of the others two.

If E_p is associated with a convex vertex the following transitions are possible:

1. δ may produce the transition from E_p to $\overline{E}_s$ (contact lost) if it satisfies $(\delta \cdot d_p) \geq 0$ or $(\delta \cdot d_s) \geq 0$.
2. Idem B1-2
3. Idem B1-3
4. δ may produce the transition from E_p to E_s if it satisfies $(\delta \cdot l) \geq 0$ and $(\delta \cdot d_s) \leq 0$, where l is a unitary vector that satisfy $(d_p \cdot l) = 0$ and $(d_s \cdot l) \leq 0$.

In the convex vertex case, E_s' is an unstable state, thus no operator to reach it is considered.

If E_p is associated with a concave vertex the following transitions are possible:

1. Idem B1-1
2. Idem B1-2
3. Idem B1-3
4. δ may produce the transition from E_p to E_s if it satisfies $(\delta \cdot k) \geq 0$ and $(\delta \cdot d_s) \leq 0$, where k is a unitary vector that satisfy $(d_p \cdot k) \geq 0$ and $(d_s \cdot k) = 0$.
5. δ may produce the transition from E_p to E_s' if it satisfies $(-\delta) \in \Delta d_s$ or $(-\delta) \in \Delta d_p$ or $((-d_p \times \delta) \cdot (-d_s \times \delta)) \leq 0$.

Table 1 summarizes the operator conditions to change PF-state.

TABLE 1 Operator conditions for the transition between PF-states in the 2 dof. case

From \ To			E_p	$\overline{E}_s$	E_s	E_s'	$\overline{E}_c$	E_c
$\overline{E}_p$			n.a.	n.a.	$(\delta \cdot d_s) \leq 0$	n.a.	$(\delta \cdot n) \geq 0$	n.a.
E_p	Edges			n.a.	n.a.	n.a.	n.a.	$(\delta \cdot d_p) \leq 0$ and $(\delta \cdot d_c) \geq 0$ with $\Delta d_c \cup \Delta d_p \neq \phi$
	Vertex	Concave	$(-\delta) \in \Delta d_p$	$(\delta \cdot d_p) \geq 0$	$(\delta \cdot k) \geq 0$ and $(\delta \cdot d_s) \leq 0$; with $(d_p \cdot k) \geq 0$ and $(d_s \cdot k) = 0$	$(-\delta) \in \Delta d_s$ or $(-\delta) \in \Delta d_p$ or $((-d_p \times \delta) \cdot (-d_s \times \delta)) \leq 0$		
		Convex		$(\delta \cdot d_p) \geq 0$ or $(\delta \cdot d_s) \geq 0$;	$(\delta \cdot l) \geq 0$ and $(\delta \cdot d_s) \leq 0$; with $(d_p \cdot l) = 0$ and $(d_s \cdot l) \leq 0$	n.a.		

n.a. : non applicable

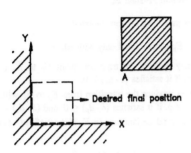

Fig. 2. Real description of the task.

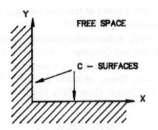

Fig. 3. Configuration space (C-space).

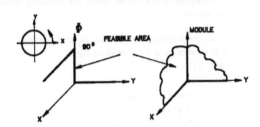

Fig. 4. Position/Force space (PF-space).

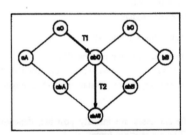

Fig. 5. PF-space with uncertainty.

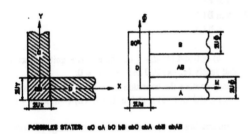

Fig. 6. Position/Force states description (PF-states).

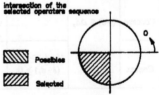

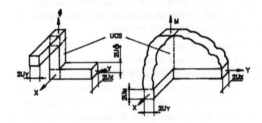

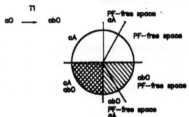

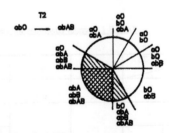

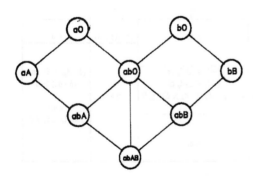

Fig. 7. Contiguous PF-states Graph (CGraph).

Fig. 8. PF-states sequence and corresponding operators, and operators filtering result.

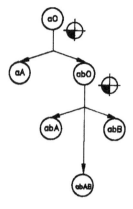

Fig. 9. Branched initial PF-states sequence.

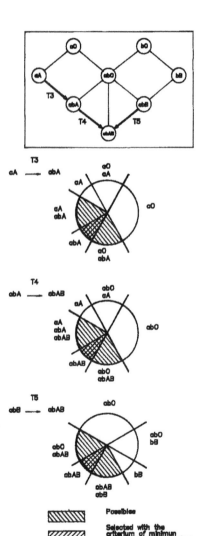

Fig. 10. Operators to pass from states *abA* and *abB* to *abAB*,
and from *abB* to *abAB*.

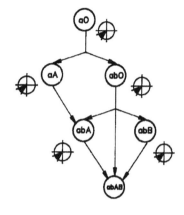

Fig. 11. Complete Solution Graph (SOLGraph).

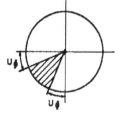

Fig. 12. Directions of the solution operators set.

Fig. 9. Backtracked AND/OR search sequence.

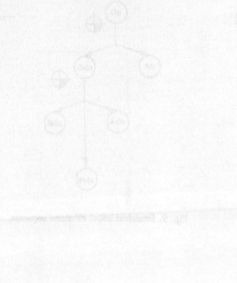

Fig. 11. Complete solution graph (SG1 loop).

Fig. 12. Directions of the solution-operator set.

Fig. 10. Operators in free state space and that drive to state and force drive to state.

INTELLIGENT CONTROL FOR DESIGN-BASED AUTOMATED ASSEMBLY

R. Shoureshi*, M. Momot* and J. Feddma**

*School of Mechanical Engineering, Purdue University, West Lafayette, USA
**School of Electrical Engineering, Purdue University, West Lafayette, USA

Abstract. Automated assembly is the most difficult task in the development of a manufacturing automation system. A large percentage of the direct labor content of a typical product is in the assembly operations. In order to develop an automated assembly cell, an integrated scheme from design to assembly is presented. This scheme represents a design-based automated assembly where the designer's intent is integrated into the execution of the assembly. An assembly planner utilizes the designer output to generate an optimal assembly plan based on the resources constraints. A task planner converts the high-level assembly plans into a sequence of robotic manipulations. An intelligent control scheme executes the assembly operations based on the task planner's results and the sensory feedback. Details of the intelligent controller and communication network is presented in this paper, and some of the implementation results are discussed.

Keywords. A.I., CAM, Computer Control, Intelligent Machines, Manufacturing Process, Sensors.

INTRODUCTION

Most manufacturing products require some form of assembly, and that is indicated by the fact that a large percentage of the direct labor content of a typical product is in the assembly operations. Assembly is the most demanding part of manufacturing and the hardest to automate. Automated assembly requires dexterity, flexibility, and intelligence. These elements are part of the human body from infancy, that is, a child can easily put together Legos and assemble various objects. However, attempting to automate assembly as humans do is not a feasible task for common factory use. Instead, the effort should go into examining assembly from the design stage, and trying to alter the product and the environment to make automated assembly a feasible task. After all, the factory-of-the-future is a man-made environment and can be altered to suit the needs of technology.

Assembly automation requires integration of several technologies including robotics, sensing, actuation, control, computational geometry and artificial intelligence. The idea is to reduce assembly turn around time, shorten the set up time, and to be able to adapt to radical production variations, and incorporate changes in products. In order to achieve these goals, there has to be a large trend toward more and more machine intelligence.

DESIGN-BASED ASSEMBLY

One of the major objectives for the factory of the future is reduction in the total manufacturing cycle from design through assembly and delivery by an order of magnitude. This would result in better control and lower costs. Reduction in total manufacturing cycle will manifest many problems which are not now regarded as significant and will emerge as serious bottlenecks, because they are currently concealed by long lead times. In order to eliminate these problems, an integrated approach from the design stage to manufacturing has to be taken. The total manufacturing cycle includes other stages such as a customer initiating an order all the way to shipping of the finished product. However, in this paper we will focus only on the design and manufacturing portion of the cycle.

An integrated approach to design-based assembly is proposed in Fig. 1. As shown, the designer initiates this operation by developing a design for different parts. The results is a part and assembly model database. The critical issues involved in this stage are: how the designer should create parts and assemblies, how to capture design intent and facilitate design for assembly feedback.

The designer develops part designs and a method for assembling these parts. Then the next step, i.e., assembly planning will be initiated. At this level, methodologies to close the gap between the CAD model (product design) and the task-level plans will be developed. The objective is an automatic generation of optimal task-level assembly plans based on the CAD models of the parts. The optimization issues involve determination of minimum product development cycle while satisfying precedence constraints. Other limitations include available resources for performing the required assembly, such as tools, fixtures, robots, etc.

The results of the assembly planner will be transferred to a task planner which converts high-level assembly functions into a sequence of robotic manipulations. The task planner takes into account kinematics constraints and the path constraints generated by objects in the assembly workcell. The task planner output would be a set of desired motions that will be communicated to the motion controller. The motion controller (Local Planner) receives the motion plans and uses sensor feedbacks to execute assembly operations. Figure 2 shows the input/output relationships for the local planner. It consists of an intelligent controller (Motivator) and a machine controller. Due to the required communication between the local planner and other modules, an information network is required. This network is described in the following section.

COMMUNICATION

Communication code resides on the local planner (motivator) in most part, and partially distributed over the other modules. To establish communication between the corresponding receiver and transmitter module a unified code is developed. Whenever each routine wishes to send a message, it determines the correct code and gives the code and any information that goes along with the code to the subroutine. The communication subroutine then sends the message and the information to the specified recipient(s). The code consists of a sequence of numbers identifying the transmitter, the intended receiver and the type of message. The message could be as simple as an acknowledgment that a job has been completed. Table 1 shows a sample of these codes.

A Star Network is used to establish communication between different modules. This network is centered around a blackboard, which stores information of general value to all routines, such as the actual position of the manipulator or the status of a module, namely, which job has been completed. This information can be accessed by any module at any time and is updated periodically by an assigned module. The blackboard adds to the flexibility of the system and adds the capability of adding more modules to the system.

INTELLIGENT CONTROL

Machine intelligence is a requirement for a design-based assembly. An intelligent controller has to perform four tasks: 1) control information flow between different modules and data bases, 2) make decisions in response to unexpected events, 3) command and synchronize different assembly machines, robots, and active fixtures to perform an assembly task, 4) provide a friendly human interface. These tasks present new challenges in the general area of intelligent control systems. Integration usually brings complexity in software and hardware, and that may cause the manufacturing system to be vulnerable to failures. Therefore, control for automated assembly has to be fault tolerant, namely, it should be intelligent enough to allow the system to function successfully even when portions are inoperative. For example, metal cutting systems are inherently subject to many breakdowns and disturbances. The unavoidable heavy forces place a strain on mechanical components, loose chips and cutting fluids get into mechanisms and electronics, friction and heat produce wear, in short, the environment is extremely hostile to even very well-constructed equipment. Therefore, fault tolerance is a required feature of the control system. This paper focuses on the control strategies required for the proposed design-based assembly.

The control scheme consists of both a supervisor (motivator) and a low level (machine) controller. The machine controller receives a desired position from the task planner and computes the actuation inputs. In the present study the current required to move the robot to the desired position is the control output to each joint. As shown by Shoureshi, et al. (1987) and Momot (1987), this control scheme is capable of compensating for uncertainties by using the maximum and minimum values of the parameters of the model to add an uncertainty compensation factor (UCF) to a proportional scheme. Figure 3 shows the closed loop block diagram of this controller. For a large state error, which is defined by the operator, the total UCF is applied so that the control emulates a bang-bang controller. For a small state error, a gain of less than the UCF is added to the input. This method has been successfully applied to other cases and manipu-

lators , see Shoureshi, et al. (1987). The control scheme is flexible in that it is designed around the norm of the system and incorporates information about the load and nonlinear terms such as friction. It is also fast since only the information of the maximum and minimum values of the uncertainties are used. Speed is the most important consideration in this system, since it must run at 35.7 Hz (a hardware requirement by the vision system).

In the high level control of an assembly workcell, it is necessary to coordinate many activities and sensory inputs with what is happening externally. For example, vision provides information on the location of parts and their motions, a low level control subroutine provides information on the position and condition of the robot and another subroutine generates the trajectory necessary to move from point to point. In order to provide some intelligence to the operation and assure that everything runs smoothly, the task manager (Motivator) was developed. The motivator provides global control. It introduces intelligence to the system where it is required. In this case, the motivator has been partitioned into three subprograms: the user interface, the task manager, and the error recovery routines.

The task manager is based on "STRIPS". It contains four main rules (or procedures): Find (object), Goto (object/position), Get (object), and Place-on (object1, object2). The absence of an Unstack rule is due to the fact that no disassembly is required in the present operation. However, it could easily be added. There are also three types of states: position, gripper, and object. Under each state type, there are two or more states. Position can either be at home or at object(s). Gripper is either open or closed. Object can either be not-found, found, in gripper, or on object(s). A rule can be executed when the precondition states of that rule form a subset of the current states of the world. As more objects are added to the world, more states and state types can be added.

To get from the initial state (as determined by the motivator) to the final state (given by the operator), a decision tree (based on rule execution) is searched using a depth-first procedure. A branch of the tree is searched until one of the following occurs: the goal is found, the depth bound is met, or the current state matches a previous state. In the last two cases, a different branch is then searched. If the goal is found, the rules necessary to obtain the goal are executed. The actual execution of these rules is orchestrated by the motivator, too. Rules are allowed to be executed concurrently to allow for a faster overall operation. For example, while the local controller moves the robot toward an object, the vision can be looking for another object.

Error recovery also provides some intelligence to the operation. The error recovery program makes periodic checks on the system. If too much current is input to the robot the motivator senses that via an operation time-out and informs the other modules and the operator. If the vision senses a dropped object (possibly due to a drop in the air pressure), it takes corrective action by trying to retrieve the object or getting another one. Another sequential failure alerts the operator.

The operation of the task manager is based on a Prolog program called STRIPS. STRIPS uses the state of the system to determine the next operation to be performed, similar to low level control. To do this, rules are composed of three lists. These lists are defined in Prolog, as a prerequisite list, an add list, and a delete list.

The rules are fired depending on the prerequisite list. If the state of the system meets all of the requirements in the prerequisite list for a rule, that rule is fired. When a rule is fired, the state of the system changes according to the add list (states are added to the present list of states) and delete list (states are deleted). For example, the robot should not attempt to assemble part A onto part B until it has located part B, and has part A in its gripper. These are the preconditions. The rule that might be fired could be called "place part A onto part B". After this rule fires, the state of the gripper is "open" and "empty" and the state of the robot is "at-part B". Therefore, the delete list would contain something like "gripper closed" and "part A-in-hand". The add list would contain "gripper open" and "empty".

In this manner, the task manager can step through the whole routine without actually performing any of the operations. This is necessary, because it is not assured that the rules will always lead directly to the desired goal. By doing a depth-first search with a limit on the depth, many possible procedures can be evaluated until the one that leads to the desired goal is obtained. This is also helpful in error recovery. By comparing the desired state with the actual state, the task manager can detect errors, similar to low level control. Once an error is detected, the task manager can re-evaluate the situation from the actual state of the system. Of course, the task manager is only as good as the number of states that it can detect. If a fuse blows and there is no way to sense this, there is no way for the task manager to know what went wrong. Even if it could determine this error, however, it could only overcome this difficulty by alerting the operator.

Unfortunately Prolog does not interface well with the C routines required to set-up and maintain the sockets (the means of communication between the different devices). Therefore, the general idea of STRIPS was translated into a C program. The main difficulty here is the fact that Prolog is optimized for list processing and C is not. In order to overcome this difficulty, the states were defined as places in a bit field. For example, "gripper open" is 2, "at part B" is 4 and "at part A" is 8. This limits the number of states that can be defined because each state is either ON or OFF and the bit field is not large (32 spots). However, once this is done, the number crunching capabilities of C makes the program as fast as the same routine in Prolog (both determine the rules required to obtain the goal in about 0.56 seconds). Instead of lists for the prerequisites, addenda, and deletions bit fields are used and exclusive-or'ed then and'ed with the present state of the system. If the result is zero, the precondition has been met. The rule is then put in an array (with an integer representation) for later recall.

This is by no means an optimal solution to the problem, since Prolog possesses many nice features that C doesn't. The Prolog program is only 50 lines long, whereas the C program (just the part for the task management, not the communications) is over 4 times longer. Prolog offers more flexibility, is concise and easier to read and understand. However, for I/O operations and real-time control the C routines are more appropriate.

HARDWARE FOR IMPLEMENTATION AND RESULTS

Figure 4 shows the configuration and some of the hardware used to implement the vision-based intelligent control scheme. Three Sun workstations are used for three of the modules: the motivator, the vision module, and the task planner. All three processes are carried out concurrently, with the motivator making sure that all processes flow smoothly. The modules are connected with primitive socket routines based on INET sockets, allowing interprocess communication across the 10 Mbytes/sec client ethernet. The controller translates desired positions into motor currents and outputs to a PUMA 560 manipulator.

The PUMA 560 is a 6 degree of freedom robot, all joints being rotational joints. The gripper and last link were modified as shown in Fig. 5. This modification permits mounting the local reference frame camera to the robot. Two cameras are used to provide world information and local features. The camera is mounted to joint 6 so that it is always directed at the gripper. The extension of link 6 also permits any fine movements of joints 4 and 5 to be translated into roughly linear motions. This is especially important when performing an assembly. During this operation, the orientation joints (joints 4 and 5) must account for any gross perturbations of the positional joints (joints 1, 2 and 3) and variations in the speed of the conveyor belt.

Two cameras are used to give object positions in a world and local coordinate systems. This information can be extracted by the vision module in approximately 33 msec and is passed to the task planner, which completes the loop. Figure 6 shows the locations of the two cameras and the conveyor belt carrying the carburetor.

Based on the described hardware the vision-based intelligent controller was implemented using the PUMA to perform the desired assembly operation, namely, positioning of a flexible gasket on the moving carburetor. The objective is to track the gasket using vision feedback until the robot's gripper is positioned accurately enough to pick up the gasket. The gasket's motion is constrained to three degrees of freedom: x and y position, and x-y plane orientation, θ. There is no previous knowledge of the motion of the part's feeder within this plane. The robot's motion is constrained to four degrees of freedom: the three mentioned previously plus the z position. The z position is added to show how the distance between two feature points in an image may be used to control this dimension if the desired distance is known.

Before the task begins, a desired image of the gasket (see Fig. 7) is taught via teach-by-showing techniques. A menu-driven program prompts the use for the desired image processing steps. The user is required to select the features to be used for object recognition. In this case, the circles in the gasket are used as key features. For simplicity and speed, feature extraction is performed on a threshold image. The white objects in the image are traced, and seven Fourier descriptor coefficients (FDCs) are computed for each object from its crack code. This technique is described by Mitchell (1984). Circles are identified as having higher order FDCs near zero. The center of each circle and its radius are computed from the crack code's starting point and the zeroth other FDC. These locations and radii are stored for latter use in the recognition stage.

The task playback mode consists of two stages: object recognition and high speed vision feedback. The role of the object recognition stage is to identify all features in the image, in this case circles. An exhaustive matching of circle radii and distances between pairs is performed. A probability measure is used to rank the matches between actual and desired circles. Although time consuming (approximately 5 seconds), this method could locate the gasket even if only half of the circles are visible.

The methods of feature extraction also change during the vision feedback stage as compared to the recognition stage. While crack codes and FDCs are used in the recognition stage, simple scan lines are used in the vision feedback stage. To locate the two circles in real-time, eight scan lines originating at the estimated origin of the circle are used to find the circle's edge. The center of the circle could be determined if three or more points on the edge are found. Using this method, we are able to obtain a vision cycle time of approximately 60 to 70 milliseconds. If the object moves too fast, this method fails. In which case, an area of interest window twice the size of the circle is searched using crack code methods. If this search fails, the window is double until the circle is found. If more than one circle is found, the circle radius and the distance to the other circle are checked for correct identification.

Figure 8 shows the desired and actual (x,y) position of the robot end-effector as it tracked the gasket around a quarter turn of the parts feeder. As expected, the actual position lags the desired position. The amount of lag depended on the gasket's velocity and the feature extraction time. In this figure, the angular velocity of the parts feeder was approximately 0.1 rad/s. When this velocity increased, the tracking accuracy decreased. At approximately 1 rad/s, the vision module could no longer follow the gasket. Feature prediction was not used in these experimental results in order to show the characteristics of the feature-based trajectory generator. In the future, we hope to increase the current tracking speed by using feature prediction to estimate the motion of the part.

Figure 9 shows the resulting orientation of the end-effector versus time. The slight change in orientation in the beginning was caused by an initial misalignment. The robot's motion was fairly smooth while the gasket was moving, and as seen in the latter half of the graph, the robot settled in on the steady state value with very small overshoot. Similar results were obtained for the other three features.

CONCULSIONS

Results of an integrated study on design-based for automated assembly were presented. It is shown that integration of design, assembly planning, task planning, and motion control provides faster response, and more flexibility in terms of system uncertainties compared to typical manufacturing systems. For automatic assembly operations, information about assembly machine and its environment are very crucial. Vision system is one of the most appropriate tools for providing such information. This study shows that by establishing a proper communication network and integration of sensory information into the operation of a real-time intelligent controller, automatic assembly in a dynamic and unstructured environment can be achieved.

ACKNOWLEDGMENT

This research is supported by the Purdue NSF-Engineering Research Center on Intelligent Manufacturing Systems.

REFERENCES

1. Mitchell, O.R., Grogan, T.A. (1984) "Global and Par-tial Shape Discrimination for Computer Vision," Optical Engineering Vol. 23, No. 5, pp. 484-491.

2. Momot, M.E. (1987) "Implementation of a Bounded Uncertainty-Based Robot Controller," MSME Thesis, Purdue University.

3. Shoureshi, R., Momot, M., Roesler, M.D. (1987) "Robust Control for Manipulators with Uncertain Dynamics," Proceedings of the 10th IFAC World Congress on Automatic Control, Vol. 8, pp. 358-363.

4. Shoureshi, R., Roesler, M.D., Corless, J. (1987) "Control of Industrial Manipulators with Bounded Uncertainties," ASME Journal of Dynamic Systems Measurement and Control, Vol. 109, pp. 53-59.

Table 1 Sample of Communication Codes for Different Modules

Transmitter	Receiver	Code	Description
Blackboard	Vision	13 B x,y,z (float)	Object Orientation
Motivator	Network	20 A	Request to Shutdown Network
Motivator	Vision	23 A	Update World Coordinate
Trajectory	Vision	42 A	Acknowledgment
Vision	Motivator	32 C	Failed Operation
Trajectory	Controller	45 F	Exit-Stop Process
Controller	Trajectory	54 B	Exit-Error

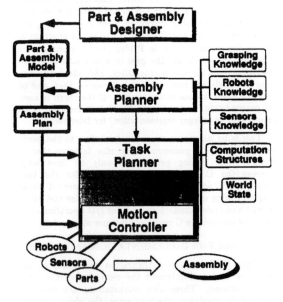

Fig. 1. Design-Based Automated Assembly System Architecture

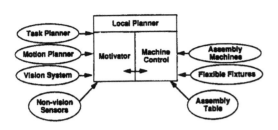

Fig. 2. Operational Block Diagram of Local Planner

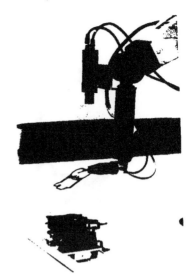

Fig. 5. Gripper and Modification of Last Link.

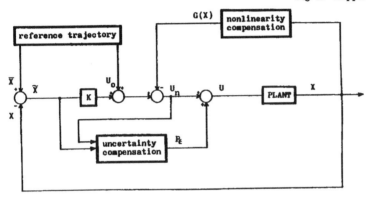

Fig. 3. Closed Loop Block Diagram of Uncertainty Controller

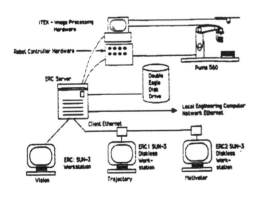

Fig. 4. Configuration and Hardware Used for Implementation.

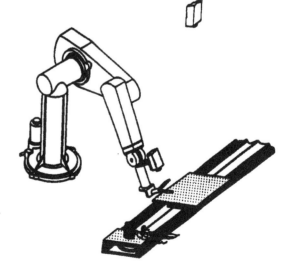

Fig. 6. Location of Two Cameras.

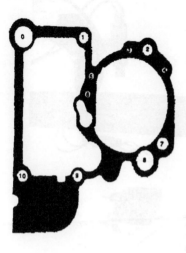

Fig. 7. Image of Gasket

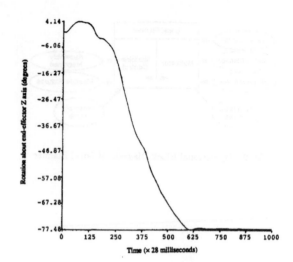

Fig. 9. Orientation of End-Effector

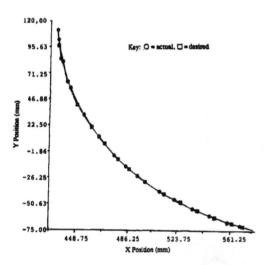

Fig. 8. Desired and Actual Position of End-Effector

A COMPUTER INTEGRATED MANUFACTURING SYSTEM FOR SHEET METAL FORMING

M. Tisza* and T. Kassay**

*Department of Mechanical Engineering, Technical University of Heavy Industry,
Miskolc, Hungary
**Department of Metal Forming, Machine Building Factory Digep, Diósgyór,
Hungary

Abstract. In recent years, the increasing market demand for new products in shorter
periods has dramatically influenced production requirements. All around the world, lot
sizes are decreasing and the varieties of components are increasing. Since production
requirements are permanently changing, flexibility is emphasized in manufacturing processes. Besides these requirements, reliability, quality and high-productivity should
be provided, as well. Flexible Manufacturing Systems (FMS) and Computer Integrated
Manufacturing (CIM) can respond to these requirements. In this paper, the development
and realization of a Computer Integrated Manufacturing System for sheet metal forming
will be analysed, as a case study. The general concept and the main objectives of the
realization of the system will be given. Particularly, the main differences of sheet
metal forming systems in comparison with FMS in machining operations will be emphasized. Finally, the main advantages of the Sheet Metal Forming Flexible Manufacturing
System will be summarized.

Keywords. CAD/CAM; flexible manufacturing; warehouse automation; sheet metal forming.

INTRODUCTION

The development and wide-spreading application of
NC/CNC forming machines can be regarded as the
first step towards Flexible Manufacturing Systems
(FMS) (Leslie, 1973). The second important step
was the automatization of transport, loading and
unloading processes, assuring programmed storage
and material handling capabilities (Kean, 1978).
The third - and from the viewpoint of Flexible
Manufacturing Systems decisive - step was the introduction of low-cost, high-speed computers which
provided the opportunity to convert machine control units into CNC systems with enhanced program-
-storage and communication capabilities (Merchant,
1983). Finally, the combination of all these forementioned developments into integrated systems can
be regarded as the final step in establishing
Flexible Manufacturing Systems (Hartley, 1983).

The previously described tendencies have significantly influenced the Hungarian Sheet Metal Industry, as well. The first numerically controlled
sheet metal forming machines were imported from
leading West-European firms (RASKIN, TRUMPF,
PULLMAX, WIEDEMANN, etc.) at the end of the 60's.
The development of home-made NC/CNC plate-shears
and press brakes at the beginning of 70's can be
regarded as another important step in the automatization tendencies of Hungarian Sheet Metal Industry (Erdősi, 1982).

From the middle of the seventies the research interest turned towards the development of Flexible
Manufacturing Cells and Systems in the Hungarian
Sheet Metal Industry, as well. (It should be
noted, that in metal-cutting industry significant
results were achieved: a number of numerically
controlled manufacturing centres and flexible
manufacturing cells have been exported even into
industrially developed countries.)

The first Flexible Manufacturing System for Sheet
Metal Forming was installed at the largest Hungarian Electronic Enterprise (VIDEOTON) by the
financial support of the National Committee for
Technical Development (Rabb, 1984). This system is
mostly based on machines and devices imported from
leading West-European firms, but the computer system and software controlling the production system
are home-made.

In the meantime, the results achieved in the domestic forming-machine industry (e.g. the development of sheet-metal manufacturing centre LMC-250
with laser contour-cutting, the enhanced capacity
and reliability of home-made numerical control
systems), as well as the favourable changes both
in the hardware and software conditions in the
Hungarian Computer Industry and the significant
results achieved in Computer Aided Design and
Manufacturing made it possible to set as an aim
the realization of a complex Sheet Metal Forming
Flexible Manufacturing System based exclusively on
home-made machines and devices.

This project has been realized at one of the
largest Hungarian Machine Building Factories
(DIGÉP) by the financial support of Ministry of
Industry in cooperation with scientific and
research institutes.

MAIN OBJECTIVES OF REALIZATION OF FLEXIBLE MANUFACTURING SYSTEM

The main objectives of this project can be summarized as follows:

(1) The system should serve as a reference one of
domestic sheet metal forming NC/CNC machines,
manufacturing centres, transport and storage
devices, computer hardware and software facilities.

(2) The system as a whole and different part of it
as Flexible Manufacturing Cells should be
suitable for marketing.

(3) Following from its reference character, it
    should also serve for training and education
    of specialists in Computer Aided Design and
    Manufacturing, in Computer Aided Process
    Planning, Computer Integrated Manufacturing,
    etc.

(4) It should provide valuable information and
    experimental facilities for further develop-
    ment of FMS both in sheet metal forming and in
    other branches of industry, as well.

(5) Besides the reference character of the system,
    it should satisfy every demand of sheet metal
    forming production of small and medium size
    series emerging at the factory where the
    system is implemented.

GENERAL CONCEPT OF THE SYSTEM

Among the main objectives of the system outlined
in the previous section, it was also stated that
the system - besides its reference character -
should meet every demand of sheet metal forming
production emerging at the factory where the
system is implemented. The sheet metal forming
production of the factory may be characterized by
the large number of components with a relatively
small sizes of series. It follows from the main
profile of production since the factory DIGÉP
where the system installed is involved first of
all in manufacturing machines for metal forming
industry. Mechanical and hydraulic presses, plate-
-shears, press brakes and various types of wire-
and cable-drawing machines are world-wide known
products of DIGÉP.

The number of sheet components to be produced is
over 2200 a year. The production series of various
components changes in a wide interval depending on
the series-size of the basic product (the formerly
listed forming machines) and on the incidence rate
of given component in each basic product. Lot
sizes from hundred to some thousands can be
considered to be characteristic.

The Sheet Metal Forming Manufacturing System is
capable of working sheets with the thickness of
0.5-10 mm. The most characteristic size is
1500-3000 mm with the thickness of 1-6 mm. The
total production volume is about 800-1000
tons/year.

Following from the facts mentioned above, this
Sheet Metal Forming Manufacturing System is designed for small- and medium-size series. (Sheet
metal parts produced in large series or mass
production will be manufactured in the future too,
in sheet metal forming workshops in batch-produc-
tion requiring conventional technological and
production control methods.)

Elaborating the working principle, it seems to be
practical to apply the same ideas unambiguously
accepted in metal-cutting (machining) systems,
namely: the most reliable operation can be assured
by the highest possible degree of automation in
the total production verticum, minimizing the necessity of human intervention and by this means
excluding the subjective errors as much as
possible.

In reality, following from the special nature of
sheet metal forming, a number of different circumstances should be taken into consideration,
which result in a fundamentally new concept. The
main reasons of applying new concept can be summarized as follows (Tisza, 1986):

- The geometry of sheet metal components differs

basically from that of the components produced
by machining processes. This difference reveals
first of all in the significant shape changes
occurring during forming operations (like bending, deep-drawing, etc.). Due to this fact, the
originally two dimensional sheet parts often become three dimensional requiring significantly
more place for storage and another kind of
transport and material handling.

- Following from the characteristics of small- and
medium-size series and from the special nature
of auxiliary technologies (e.g. straightening,
grinding, welding, etc.), it is obvious that
depending on the components to be produced some
manual operation may be inevitable which can not
be neglected economically due to the small
batches.

- Though the type of components can be considered
to be significantly large, the type of raw materials is relatively small. Due to this fact,
it is reasonable to carry out the cutting
(shearing) of sheet panels both in time and
place in a concentrated manner using computerized optimization methods to minimize both the
material waste and the necessary raw material
stock.

- At the same time, the following should also be
taken into consideration: in Hungary both the
prices of raw materials and wages are significantly lower in international comparison, so the
cost of production is lower even in conventional manufacturing than that of for machined
parts. Therefore, a compromise is necessary
between automatization and rentability not to
give up economical operation for self-contained
automation.

Taking the forementioned conditions into consideration, the main concept of the Integrated
Sheet Metal Manufacturing System can be summarized as follows:

(1) On the ground of threefold requirements of
    rentability-reliability- and automatization,
    the computer integrated production control in-
    volves
    - some preparatory processes of production,
    - the technological design and process
      planning,
    - the elaboration of NC control programs and
      the control of manufacturing machines,
    - the material handling (transport and storage
      processes), and
    - the computer aided quality control.
    Some sorts of preparatory and finishing opera-
    tions (like straightening, grinding, etc.) are
    independent of the computerized production
    control, though they are located as integrated
    parts of the total system in separated prepa-
    ratory cells.

(2) Complete automation is applied only there,
    where it is indispensible from the viewpoint
    of continuous operation of the system (e.g.
    the material transportation between the ware-
    house and manufacturing cells, loading and
    unloading processes, control of manufacturing
    machines, etc.).

(3) The elimination of subjective errors - as far
    as possible - is provided by the computerized
    production control. This is solved by the so-
    -called production condition test. It means
    that any command to carry out any operation
    can be given out only in that case if all the
    necessary conditions for production (e.g. raw
    materials, tools, machines, control programs,
    etc.) are available.

GENERAL STRUCTURE OF SHEET METAL MANUFACTURING SYSTEM

The principal scheme of computerized production control and the simplified process of material flow in the sheet metal manufacturing system can be seen in Fig. 1. The sheet metal manufacturing system is in direct connection with the central computer system of the factory through the production control system. The short-range production program is based on the long-range production plan of the factory elaborated on the ground of production orders, development targets and industrial strategy.

Thus, the short-range production program containing the production tasks for a given period can be regarded as the basic input of the manufacturing system. It includes the types of components (with their characteristic properties) to be produced within the given period, the date required or other priorities inforceable in production. These are the necessary input information required and the production control program determines all the other parameters (e.g. sorting the parts by their date-wanted code into individual groups of components of the same material type and thickness, determination of the necessary volume of sheets and the optimum material utilization by a nesting process, the required turret configuration, etc.).

The main tasks of the computer control system can be summarized as follows:

- Starting out of the basic input information arising from the short-range production program (such as part number, geometric and material characteristic of components to be produced, the types and sizes of raw materials available, the date wanted or other priorities and limitations - like grain orientation, etc.) to carry out the technological and process planning including the determination of the formerly listed optimum turret configuration and toolpath, etc.

- Elaboration of numerical control programs for manufacturing machines and other numerically controlled devices and postprocessing them (fitting to the control system of manufacturing machines).

- Determination of production control program involving both manufacturing and material handling processes and to carry out the computer aided manufacturing and quality control tasks.

- To assure documentation in structured data-files at any level of production. The system should also be capable for automatic updating of data-files.

- The computer control system is directly linked to the main-frame computer of the factory and in this way to other CAD/CAM facilities available within the firm to carry out different computer aided design tasks (eg. geometric design of sheet components, determination of optimum material utilization, etc.).

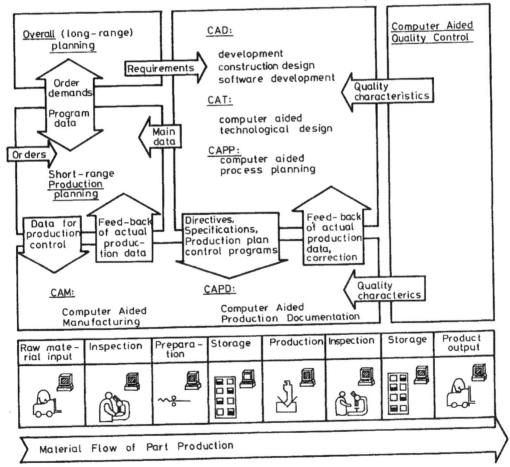

Fig. 1. The principal scheme of production control and the simplified process of material flow in the sheet metal manufacturing system

The raw material (sheet panels, blanks, etc.) can be regarded as the physical input to the system, as it is shown in the bottom part of Fig. 1. The principal scheme of material flow can also be seen which includes the primary quality inspection, preparatory operations, manufacturing and storage processes from the raw material input until the output of the final products.

The general scheme of elaborated sheet metal manufacturing system can be seen in Fig. 2. Considering the structure of Computer Integrated Sheet Metal Manufacturing System, a modular approach can be observed. According to Fig. 2., the following main modules can be distinguished:

(1) Manufacturing modules involving three manufacturing cells (a shearing cell, a complex sheet metal manufacturing cell, and a bending cell).

(2) Material-handling module including also various submodules for storage, transport, loading and unloading tasks.

(3) The computer control system for carrying out the abovementioned production planning and control tasks.

(4) The so-called connected subsystems which may be further subdivided into two main-groups

　　a) internal linkages which belong to the computer production control and carry out indispensable supplementory tasks for the system,

　　b) external linkages which are independent of the computer production control but their work is also essential for the continuous operation of the total system.

In the following only the manufacturing modules will be described in detail where actually the forming operations occur.

Manufacturing modules

The installed metal forming machines and the applied technological structure are capable of carrying out all the usual metal forming operations even in case of components of complicated geometry. Obviously, in an actual production process only certain part of the system is involved in the manufacturing process depending on the complexity of the part to be produced.

As it has already been mentioned, the manufacturing modules may basically be divided into three main groups:

　　- the so-called shearing cell,
　　- the complex sheet metal manufacturing cell, and
　　- the bending cell.

The main function of the shearing cell to cut the panels into prescribed sizes necessary for further processing in the system. Sheet to be processed are handled completely automatically including the transportation from and to the automated warehouse by AGVs, as well as the loading, feeding, unloading and sorting of sheared parts.

The central manufacturing unit in this cell is a CNC plate-shear, type DTO 10 (produced by DIGÉP). Sheets to be cut are processed from intermediate stacks brought from the warehouse by AGVs according to the manufacturing program. (As it has already been mentioned shearing is carried out on groups of components of the same material type and

thickness minimizing the scrap and necessary stock, alike. This is done by the shear nesting program called SCOPT which is able to handle a large number of specifications and restrictions (like the required number of parts to be processed, stock materials available, grain orientation, etc.). Material utilization is optimized using the method of linear programming applying heuiristic principles, as well. If the prescribed material utilization cannot be obtained the operator can manually intervene to avoid high scrap-rate percentage. If the target is met, the shear patterns are processed into CNC control programs.)

The completely automated operation of shearing cell is assured by the auxiliary devices controlled also by the computer production control. Blank sheets are automatically positioned by a CNC controlled sheet loader type DFM-350 LB. It picks up the sheet with suction cups using magnetic separator to ensure that only one sheet is lifted. After loading the sheet, it is fed by a programmable cut-to-length feeding device type DLA-10. Sheared parts can be removed either by a staging conveyor or special lifting device (type DLK-10) possessing many pneumatically operated cups to handle a wide variety of components. The sheared parts are transferred to the automatic sorting system (type DLR-10) which selects and automatically assigns the sort bin locations.

Components possessing complicated blanked and punched contours are processed in the complex sheet-metal manufacturing cell where an LMC-250 manufacturing centre with a mechanical turret configuration for 20 tools and laser-contouring facilities is installed. (Besides punching, nibbling and laser-contouring, this manufacturing centre is capable of milling, drilling and tapping, as well.)

The production control program determines the necessary turret configuration, the optimum arrangement of parts on the sheet for minimum material waste and the optimum tool path.

This manufacturing cell is also supplied by all the necessary auxiliary devices for automatic loading, feeding, unloading described formerly.

The third manufacturing cell is based on a DEC 100E type, CNC controlled press brake for bending operations. This machine is supplied with all the necessary facilities making it very comfortable for use in FMS systems. (Continuously adjustable bending force facility according to the requirements of technological processes, three axis numerically controlled back-stop system, automatic depth setting control possibility for producing accurate bending angles, quick and reliable hydraulic tool change system, etc.).

This manufacturing cell is also supplied by all the necessary auxiliary devices for automatic loading, feeding, unloading described formerly.

The computer production control provides possibility for computer aided quality control either at the end of the manufacturing processes or at each forming operation depending on the requirements prescribed for the components.

Material handling (transport and storage) system

The flexibility and reliability of Sheet Metal Manufacturing System is strongly dependent on the applied transport and storage devices.

In this system an ASR warehouse (high-shelved

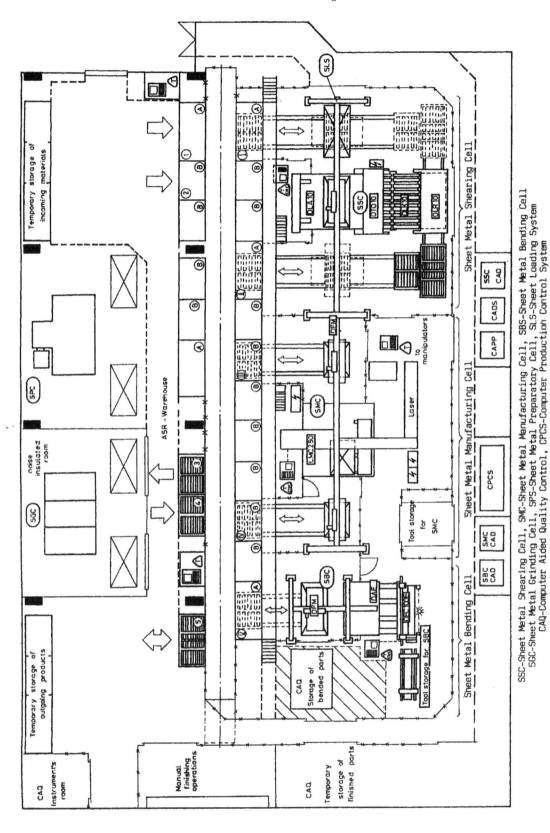

Fig. 2. General scheme of Sheet Metal Forming Flexible Manufacturing System

SSC-Sheet Metal Shearing Cell, SMC-Sheet Metal Manufacturing Cell, SBS-Sheet Metal Bending Cell
SGC-Sheet Metal Grinding Cell, SPS-Sheet Metal Preparatory Cell, SLS-Sheet Loading System
CAQ-Computer Aided Quality Control, CPCS-Computer Production Control System

type) is used to store the sheets and components. It is capable of automatic storage and retrieval of sheets and blank materials. From it, AGVs can deliver sheets to loading devices and they can take them to the next stages or back to the warehouse according to the production program. In each manufacturing cell, as it was analysed formerly, different types of loader, unloader, feeder and sorting devices are also installed.

A special attention should be devoted to the dimensioning of the capacity of ASR warehouse for sheet metal forming flexible manufacturing systems due to the particularities of sheet forming processes. As it has already been mentioned, for intermediate storage of sheet parts and components a high-shelved type ASR warehouse is used with appropriate pallets and boxes suitable both for storage and transportation. Following from the special nature of sheet forming processes, the following main requirements should be taken into consideration:

- each pallet or box can contain a certain part or the total amount of a component's series;
- due to the different sizes of various components the pallets (boxes) used can also be different in size, but it is desirable that at least one of their sizes should be equal to the width of the shelves of the ASR warehouse;
- after some forming operations (like bending, deep-drawing, etc.) significant shape and dimensional changes can occur. Due to this fact, the originally two dimensional sheet parts often can become three dimensional requiring significantly more place for storage.

The elaborated theoretical method for dimensioning the ASR warehouse involves the following main steps:

(1) The optimum arrangement of a given component on a pallet or in a box.
(2) The arrangement of each component on given pallets (or in given boxes).
(3) Determination of necessary type and number of storage media.
(4) Simulation, modification and selection of optimum solution.

For the determination of optimum solution several target functions can be set up, eg:

- the maximum value of volume utilization or
- the minimum number of storage media, etc.

In the derivation of the solution, various limitations can also be handled, eg:

- the maximum blank sizes,
- the number of various components put on a given type of pallet,
- the necessary number of pallets and boxes for a given type of blanks,
- the maximum number of pallets which can be applied, etc.

Applying the beforementioned target functions and limitations a mathematical method has been elaborated for theoretical determination of ASR warehouse capacity and for that of the necessary number and types of storage media. The mathematical description of the elaborated method can be found in detail in a formerly published paper (Rabb, 1984).

CONCLUSIONS

Among the main objectives of establishing this Flexible Sheet Metal Manufacturing System, it was emphasized that this system should serve as a reference one of domestic sheet metal forming NC/CNC machines, manufacturing centres, transport and storage devices, computer hardware and software facilities and it should be suitable for marketing as a whole or different modules of it both inland and in foreign countries.

The main advantages of the realized sheet metal forming flexible manufacturing system may be summarized as follows:

- First of all the flexibility should be emphasized. New products can be introduced within significantly shorter period: an extremely quick respond to market requirements can be achieved.

- A significant decrease in direct and indirect production costs may be realized by the following:

  * the radical reduction of manufacturing times,
  * the significant increase in productivity,
  * the material savings due to improved material utilization,
  * the significant reduction in inventory costs, etc.

- Increased reliability, better product quality and higher productivity eliminating the subjective errors can also be achieved.

Taking all these benefits into consideration, it may be stated that Flexible Manufacturing Systems in sheet metal forming can also respond to the increased market demands emerging in recent years.

REFERENCES

Crestin, J. P. - Waters, J. F. (1986). Software for discrete manufacturing, North Holland, Amsterdam.
Erdosi, J.- Fülep, I. (1982). Computer Technique in Sheet Metal Forming, Gépgyártástechnológia, 10. 458-464. (in Hungarian)
Hartley, J. (1983). FMS at Work, North Holland Publ. Co, New York.
Kean, C. G. (1978). NC in Sheet Metal Working, Part 1. Sheet Metal Industry, 8., 934-939.
Kusiak, A. (1986). Modelling and Design of Flexible Manufacturing Systems, Elsevier, Amsterdam.
Leslie, W. H. P. (1973). Numerical Control users' handbook, McGraw Hill, London.
Merchant, E. M. (1983). Current Status of, and Potential for, Automation in the Metalworking Manufacturing Industry, Annals of CIRP, 32., p. 519.
Rabb, L. - Sveda, B. (1984). Development of Flexible Manufacturing System for Sheet Metal Working at the VIDEOTON Electronic Enterprise, Gépgyártástechnológia, 9. 416-421.
Schmidt, V. (1978). Anwendung der NC-Technik bei Blechbearbeitungsmaschinen, Zeitschrift für ind. Fertigungstechnik, 68., 571-574.
Tisza, M.- Romvári, P.- Rácz, P. (1986). A Complete CAD/CAM Package for Sheet Metal Forming, 26th MTDR Conference, Manchester, 31-39.

Copyright © IFAC Information Control Problems in
Manufacturing Technology, Madrid, Spain 1989

FACCS: THE FLEXIBLE ASSEMBLY CELL CONTROL SYSTEM

P. Valckenaers and H. Van Brussel

Department of Mechanical Engineering, Katholieke Universiteit, Leuven, Belgium

Abstract. At the K.U.Leuven a flexible assembly cell is being developed, comprising
part feeding equipment, a transport system and multiple robots. A set of networked
computers, programmable logic controllers and robot controllers is organised into a hierarchical control system. This paper discusses the control system design. It is
generic with respect to a large class of flexible production systems. The effort required to adapt the control system to a new class member is minimal.
The control system schedules the operations in an efficient and robust fashion. Failure recovery and resource management are handled by the control system assisted by di-
rectives in the product definitions.

Keywords. Automation; control system design; distributed control; hierarchical intelligent control; production control.

INTRODUCTION

The flexibility of the assembly hardware, underlying a control system, is continuously increasing
(Arnstrom, 1988; Van Brussel, 1986). This flexibility offers the control system numerous options
for each decision it has to make. It renders the
development of a control system that employs that
freedom to achieve high productivity, reliability
and programmability quite complex. Furthermore,
the users want to be able to program at a high
level while the system takes care of the details
automatically.

The control decisions that will determine how a
production system will behave are taken in several
stages. Fox and Kempf (1987) distinguish *process
planning*, *off-line scheduling* (production planning)
and finally *on-line scheduling* (dispatching). Process planning uses information tied to the product
and general knowledge about production technologies
and facilities. Off-line scheduling adds timeinvariant information to a specific production
facility. On-line scheduling completes this with
the observable time-variant aspects of the production system. The three stages go from a situation
with partial information and a lot of time for the
decisions towards a state with maximal information
but considerable time-pressure on the decision making process.

The control system FACCS, that is discussed in this
paper, is situated in the *on-line scheduling* stage.
The input programs for FACCS are produced by other
systems earlier in the decision making process.
Because FACCS will be time-pressed to make decisions, the program formats allow the off-line systems to reduce the computational load of the
on-line system. They support two types of information. The first one consists of constraints. The
corresponding formats promote the avoidance of
overconstraining since this would unnecessarily de-
prive the on-line control system from useful options. The second type consists of rankings.
Rankings are decisions made off-line based on
reasonable assumptions about the production system.
The on-line system will use them as advise. Unex-

pected events may disprove the assumptions. When
this happens it may be impossible or unwise to follow the advise. FACCS will detect this and take
the necessary decisions by itself.

The control system, discussed in this paper, is designed to be generic with respect to a large set of
production systems. For all members a few basic
assumptions must hold. First, the transport system
must provide random and independent routing of the
mobile resources between the stations where they
are needed. The transportation delays should be
neglectable. Finally, the operations scheduled by
the control system must have a bounded execution
time and always leave the partial products in a
transportable and storable state. The design and
implementation of those operations is considered to
be outside the control system.

WORKFLOW CONTROL

The assembly cell consists of a set of workstations. Each of those workstations is capable of
executing a number of operations. The same operation may be available at more than one station.

The workload of the assembly system resembles a
shopping list : an arbitrary mix of the available
products. Items are removed from the list when
they have been produced. Items are added by messages from the user. Each message adds a production order for a single product. The control system assigns a unique identifier to each production
order in the system. Each order has also a *production order state set*. This set is initially empty.

For each product there is a definition. The definition for a simple product is shown in Fig. 1. A
product definition consists of a set of production
steps. Each step specifies a *guard set*, an operation identifier, a list of resource variables and
an *enabler sequence*. The resource variables are
used to tell the control system whether identical
or different resources are used by the operations.
The operation identifier is used to start the execution of the step. The actual execution is done
by processes outside the control system.

Step Id	Guard Set	Operation Identifier	Resource Variables	Enabler Sequence
00	{}	LoadBase	aPallet	1, 10, 0
01	{10}	Repair_B	aPallet	1, 0
02	{1}	Insert_1	aPallet	2, 20, 0
03	{20}	Repair_1	aPallet	2, 0
04	{1}	Insert_2	aPallet	3, 30, 0
05	{30}	Repair_2	aPallet	3, 0
06	{1}	Insert_3	aPallet	4, 40, 0
07	{40}	Repair_3	aPallet	4, 0
08	{2,3,4}	PutLid	aPallet	5, 50, 0
09	{50}	Repair_L	aPallet	5, 0
10	{5}	Unload	aPallet	6, 0
11	{6}	Eop		
12	{0}	Abort		6

Fig. 1. A simple product definition.

The *guard set* is a set of numbers used to determine when its step has become eligible for execution. A production step is not eligible for execution until its guard set has become a (possibly improper) subset of the *production order state set*. In the example above exactly one production step is eligible or *enabled* at the start of a production order. Indeed, only the step 00 has an empty guard set. The operation *LoadBase* will put the base plate of the product on a pallet. The resource variable *aPallet* is then bound to that physical pallet.

The *enabler sequence* is an ordered sequence of numbers. When an operation is executed it will select elements of the enabler sequence. These will be added to the *production order state set*. The operation selects the elements by their position. In the example above the operations select the first element in case of success, the last element in case of a catastrophic failure. When there are three elements, the second element is added in case of repairable faults. In general, an operation may add zero or more elements during its execution.

Thus, normally *LoadBase* will add the element 1 to the state set. This will enable the steps 02, 04 and 06 that will insert components. When everything goes as expected, these three steps add the elements 2, 3 and 4 to the state set. The state set becomes {1,2,3,4}. Then production proceeds with *PutLid* followed by *Unload*. Finally *Eop* (end of production) is executed. This operation informs the user and the entire system that the order was completed. *Eop* does not perform any visible action. The *Abort* operation must be available on all systems. It will stop the production as soon as possible and recover the resources held by the order.

The workstations exchange the production order state sets directly without the assistance from the cell control level. The *enabling* of production steps is however not sufficient. Resources are needed for their execution. Through these the cell level controls the stations. This will be discussed in more detail in the following paragraph. When the necessary resources are available, the steps that are *enabled* become *ready* for execution.

Until now the workstation control has only obeyed the constraints imposed by the input programs from the off-line system and the observed system state. With multiple steps in the *ready* state the workstation must choose one for execution. It must take decisions. In normal circumstances the rankings from the off-line system can be used. For each production order there is an advised sequence of steps transmitted to each station.

Unexpected events (e.g. empty feeders) may force station control to deviate from the advise. The station will first try to select a production step

with an operation in the *current operations set*. The operations in such a set do not require a significant amount of non-productive actions (e.g. gripper exchange) when they are selected for execution directly after the current operation. When there are no such steps in the ready state, an arbitrary ready step is chosen.

When an operation is selected for execution it must update the current operations set. In our implementation this is done by mapping each operation onto a set constant or onto the current set. In the latter case the current set is not modified when the operation is selected. In the former case the set constant is assigned to the current set. This updating method can be customized without affecting the overall design.

RESOURCE FLOW CONTROL

Faccs recognizes three different resource categories. First there are the *local* or *static* resources belonging to a workstation. Each station manages its local resources by itself. The second class of resources consists of the *mobile* or *shared* resources. The *semaphores* constitute the last category. The latter two are controlled at the cell level. Deadlock and starvation are the major concerns of the resource control systems.

Local Resources

Workstation control may only request non-local resources for an enabled production step when the local resources are available. Examples of unavailable local resources are empty feeders and grippers that were removed from the exchange store. Workstation control will send a message to the execution control process (outside Faccs). This message contains the operation identifier, the order identifier, the product identifier and the step identifier. Execution control will send a positive reply when the resources are in working order for as far as can be observed. This reply will change the state of the production step from *enabled* to *requesting_semaphore* or to *requesting_mobile_resource_block* (Fig. 2).

When execution control detects that a local resource becomes unavailable, it sends an unsolicited message to workstation control that identifies all the affected steps. Those steps go back to the enabled state unless they where already selected for execution. Their requests for non-local resources are cancelled. The allocated non-local resources are released. If a step is already in the executing state it is no longer under workstation control. An executing step can however stop itself with an exit code that will put it back into the *enabled* state instead of the *done* state. The step must make sure that the mobile resources are in a transportable state so they can be released. This will for instance happen when a sensor in the gripper detects an empty feeder.

At the assembly cell in our department execution control uses only the operation identifier to check whether the local resources are available. In general, all the elements of the message are needed because they are also contained in the message that starts the execution of production step. Indeed, the product identifier and the step identifier enable the operations to retrieve any information tied to the product. The order identifier enables the exchange of information amongst operations from the same production order. The benefits of the use of these elements would be greatly reduced when the set of needed resources cannot depend on them.

Semaphores

Semaphores are only needed by operations that are available in more than one station. Their function is to make sure that steps that specify such an operation are executed once and only once. The allocation of the semaphores is done by cell control. The allocation scheme for the semaphores and the mobile resources is identical. It will be discussed in the paragraph on the latter one.

A semaphore will be requested by workstation control from cell control for each step that is enabled, with the local resources in working order and with the operation available at more than one station. There is exactly one semaphore for each order identifier and step identifier combination. A semaphore is a virtual resource. It only exists in software. Originally a semaphore has the state *not-done*. A step that receives the semaphore in this state goes to the *requesting_mobile_resource_block* state. Steps that do not need a semaphore go directly to this state. When a step receives the semaphore in the *already_done* state it goes directly to the done state. The semaphore goes from the *not-done* state to the *already_done* state when the step holding the semaphore sends a message to workstation control during execution. The semaphore will be released. All other stations receive the semaphore and discard their step.

Mobile Resources

Typical examples of mobile resources are the pallets on which the products are assembled. More exotic ones are pallets containing tools or components (Arnstrom 1988). The allocation of these resources is accomplished through a cooperation of the stations with cell control.

Station level. When a step reaches the *requesting_mobile_resource_block* state, workstation control will notify cell control that it has use for the mobile resources tied to that step. When the available buffer space for resource blocks exceeds a given threshold, the station will also notify cell control when the amount of work it can perform with the resources it currently holds drops below another threshold. Cell control will not allocate any resources until it receives such a message. The threshold values should be high enough to avoid unnecessary idling of the station while it waits for the arrival of the resources. On the other hand, the values should be as low as possible because resources should not be queued at one station while another station is idling and could use these resources. When the resource block is allocated, the steps go to the *ready* state.

Cell control can also pre-empt resource blocks. This forces the corresponding steps out of the ready state. Notice that pre-emption of the semaphores or the detection that a local resource is no longer available will also force workstation control to release the mobile resources. The resources are put onto the transport system as soon as possible without disturbing the on-going operations. When other resources block the station exit the pre-empted resources will have to wait until these are released. Operations in the *executing* state are not effected by pre-emptions.

The workstation will also notify cell control when it is holding so few resource blocks that it can obey a pre-emption with a very short delay. With the *FIFO* resource buffers at the stations in our assembly system, this means that station control has just released its only resource block. Cell control may then allocate a resource block on the condition that only short operations can be selected for execution.

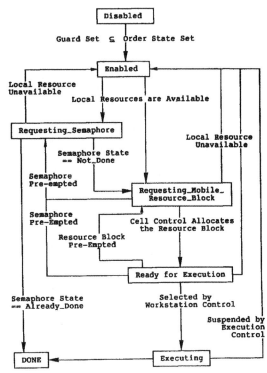

Fig. 2. Production Step State Transition Diagram.

Cell level. Cell control receives requests from the workstations for resource blocks. Such a block consists of all the resources needed by a step. The allocation will permit the step to execute to completion without the need for additional resources. Therefore, the *hold-and-wait* condition (Peterson and Silberschatz, 1987) will never occur. This implies that deadlock situations over mobile resources are prevented from happening.

In a different way the linear ordering (i.e. local, semaphore, mobile) imposed on the acquisition of resources of different types ensures that the *circular-wait* condition never occurs. This prevents deadlock over resources of different types. Deadlock prevention for semaphores is not needed since each step needs at most one semaphore. Local resources are allocated by execution control. Unless the station is capable of executing operations in parallel, all local resources are allocated to the currently executing step.

The block allocation scheme is known to have two main disadvantages : low resource utilization and starvation are both possible. The former is not a problem in FACCS. When the threshold values mentioned earlier are not to high, allocation of resources that remain unused for long periods will not occur. Starvation will be the main concern in the allocation scheme that is discussed underneath.

Normally cell control will simply follow off-line advise when messages from the stations indicate that they are in need of additional resource allocations. This advise consists of a sequence of recommended allocations for each station. Each recommended allocation has a time-stamp attached to it. These time-stamps are only used to solve conflicts between stations. The oldest advised allocation, for which a request was received from a station, reserves all the resources it needs as soon as they become available. The remaining resources are assigned to the second oldest advise when it can use them and so on. When a request for an older advise is received it takes resources away from the younger ones. This way the most important scenario for starvation is eliminated : a request for several popular resources cannot be starved by requests for a few of those.

The dependence on the ability of the off-line system to produce good time-stamps is obviously a disadvantage. However, when a request for a station is not served because an older advise holds some of the required resources and the station signals that it is able to release the resources at short notice, the request will be granted under the condition that no lengthy operations may be started. When the older advise must be followed later on, the resources will be pre-empted. The two mechanisms together make for a scheme that is armed against starvation and inaccurate advise.

Events that are not anticipated by the off-line system may make it impossible to follow the advise. Production orders affected by such events are considered to have become *abnormal*. Advise for those orders is ignored. Requests for such orders are inserted in the advised sequence just in front of the oldest advised allocation for which no request was received at the arrival time of the abnormal request. It gets the same time-stamp. This solution avoids starvation of the abnormal order by the normal ones and vice versa. The insertion method for abnormal requests can of course be replaced by a more sophisticated one. This would not affect the overall design of the control system.

Orders become abnormal when one of their operations signals that it did not select the *normal* enabler elements during execution. The unavailability of local resources will also make an order abnormal when the advise wants to direct the semaphore to the affected station. This will create the opportunity to execute the operation at a station where the local resources are still available. Finally, a request with no matching advise will also cause an order to become abnormal.

BINDING OF RESOURCE VARIABLES

The resource block allocation scheme assigns resources to workstations for short periods. There is however another resource allocation problem. The first time a resource is used by a production order, it has to be bound to a resource variable. The operation definition specifies the resource type. When the variable is the special anonymous one (i.e. _), no special action is required. This anonymous variable is used for resources that are released in their initial state by the operations (e.g. tool pallets). When a normal identifier is used (e.g. *aPallet*), the mobile resource is bound to the production order as long as a step that stipulates this identifier might still be executed. The pallets with the partial products on them are typical examples. If those would be shared by several orders, disaster would occur.

The development of a single deadlock and starvation handling scheme - for the binding of the variables to the resources - that is suitable for most production systems, is unrealistic. First, there is a lot of interaction with the external world. This world can vary considerably. An exotic example is a printed circuit board assembly system where the board is transported through the system without a pallet underneath. The number of mobile resources shrinks and grows all the time. Secondly, the implementations of sophisticated schemes have a significant computational complexity (e.g. the banker's algorithm in Peterson and Silberschatz (1987)). In other words : they consume a lot of computer power. If your mobile resources are not scarce, the optimal scheme will be a very simple one. If they are scarce, the investment in computer power and customized software may be profitable. The sophisticated binding methods can use the product definitions and the operation definitions as sources for information. Together they describe all the possible sequences a production order may use.

CONCLUSION

The on-line control system FACCS, described above, is capable of handling the major exception types in production systems. The first one consists of the operations that do not produce their normal result. The second one is the unavailability of resources at unexpected times for unknown durations. The latter is handled automatically. The former exception type is managed with the aid of the information in the product definitions.

Secondly, the system utilises opportunities that were not foreseen by the off-line systems. Thus, productivity is made robust with respect to the unpredictable aspects of the production system.

The constraints input for the control system consists of a production system description, product definitions and the workload. The user is not burdened with details.

The off-line systems have the opportunity to exert strict control on the behaviour of the production system through the advise. Strict control requires a good model to predict the behaviour of the production system. An off-line scheduler, implemented in Prolog, is under development at our department. It will generate the advise. This scheduler will not be entirely off-line. It will receive feedback information about the state of the production system and use that to synchronize its internal simulation model.

Finally, the control system need not to be changed to handle change-overs and maintenance. The former consists simply of removing some resources and adding others. Maintenance simply means that some resources are unavailable for a certain period. Notice also that *virtual* operations, products and resources can be used for monitoring functions, planned maintenance and special control functions.

REFERENCES

Arnstrom, A., P. Grondhal, and G. Sohlenius (1988). Advantages of sub-batch principle in flexible automatic assembly as used in the IVF-KTH concept Mark II. In *Annals of the CIRP 1988*, Vol. 37/1/1988. Hallwag Verlag, Bern. pp. 9-12.

Van Brussel H., D. De Winter, P. Valckenaers, and H. Claus (1986). Introducing flexibility in assembly systems. In H. Van Brussel (Ed.), *Proceedings of the 16th International Symposium on Industrial Robots*. IFS (Publications) Ltd, Bedford. pp. 557-567.

Fox, B. R., and K. G. Kempf (1987), Reasoning about Opportunistic Schedules. In Y. C. Ho (Ed.), *1987 IEEE International Conference on Robotics and Automation*, Vol. 3. Computer Society Press of the IEEE, Washingthon, D.C. pp. 1876-1882.

Peterson, J. L., and A. Silberschatz (1987). In M. A. Harrison (Ed.), *Operating System Concepts*, 2nd Ed. Addison-Wesley, London. Chap. 8, pp 257-286.

APPENDIX 1

A GRAPHIC REPRESENTATION OF PRODUCT DEFINITIONS

A graphic equivalent for the sample product definition (Figure 2 in the paper) is shown in Fig. 3. It is similar to a precedence graph in the sense that operations (i.e. the nodes of the graph) are enabled for execution when all the incoming edges have been *activated*. It is dissimilar to a normal precedence graph in that the execution of an operation does not activate all outgoing edges automatically. The operations will activate zero or more outgoing edges *at run time*. The edges are selected by their position (left to right). An alternative would be to label the edges and have the operations select them by the label. We call this a *dynamic precedence graph*. Finally, when two edges join, the activation of a single incoming edge activates the outgoing edge and, when an edge splits in two or more edges, the activation of the incoming edge activates all the outgoing edges. Figure 4 shows the edges that are activated for a production order where one operation (Put_Lid) needed repair.

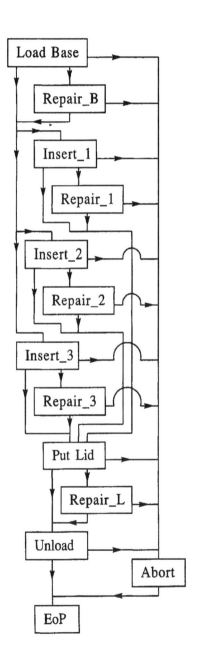

Fig. 3. A sample product graph.

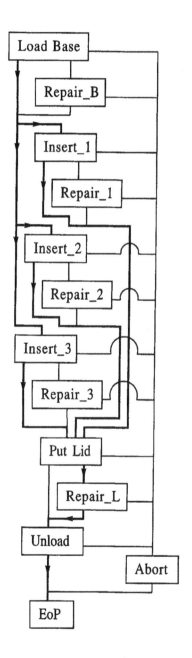

Fig. 4. An execution trace example.

THE SCHEDULER'S INFORMATION SYSTEM: WHAT IS GOING ON? INSIGHTS FOR AUTOMATED ENVIRONMENTS

K. N. McKay, J. A. Buzacott and F. R. Safayeni

Department of Management Sciences, University of Waterloo, Ontario, Canada

Abstract. The scheduling and planning tasks are key facets in traditional and modern manufacturing information systems. As the manufacturing systems move towards automation, the topology and operating characteristics of the scheduling task must change and address a hybrid situation of man-machine interaction. Some of the tasks currently performed by humans will not be needed at all, others will be added, some will change, and some will be performed by automated decision support systems. This paper briefly describes the information system used by schedulers in traditional environments and details the transformation requirements as the task is more automated. The paper concentrates on the types of information used in the scheduling process and how they can be addressed via decision support concepts. The information structures described are being embodied in an advanced scheduling platform (WATPASS-II) currently under development by the Waterloo Management of Integrated Manufacturing Systems (WATMIMS) Research Group.

Keywords. Manufacturing processes; Man-machine systems; Management systems; Inventory control.

INTRODUCTION

The scheduling and planning task is a vital component of the manufacturing environment and is one of the components often considered for integration in the drive towards factory automation. This task is part of the larger information system required for manufacturing and can be described by its topology, processes, and corresponding data streams.

There are three significant manufacturing scenarios that impact the scheduling and planning component of the information system:

- partial or limited automation where human operators perform the majority of information gathering, dispersion, operations, material handling, etc.

- integrated cad/cam systems -- from design to the computer controlled machines and material handling systems on the factory floor -- with human skill complementing the process due to internal or external variances inherent in the environment, market, supply chain, or process

- 'lights out' factories that can run predictably and reliably with minimal human intervention

Each of these scenarios imply information system requirements, design issues, compromises, and control problems. The first scenario represents the majority of companies today, and everyone agrees that things should be changed and can be improved. The third scenario offers the best potential for automating the planning and scheduling functions -- unfortunately, it describes very few factories and has been difficult to achieve. The third scenario is also not financially feasible for small companies and also depends on a very stable situation found usually in high volume repetitive manufacturing; thus ruling out one-of-a-kind job shops or prototyping environments. This leaves the second scenario to consider. It is the

most common situation in automated factories and is the goal of many others. As a hybrid system, the second scenario has components and issues from both of the other scenarios and is the one concentrated upon in this paper.

In the hybrid system, the topology, processes, and inputs/outputs may not be the same as in a non-automated situation. For example, there may be shop floor tracking systems, finite capacity scheduling tools, or rough cut capacity systems for order entry. The hybrid structure may have some functions removed and others added, while others may be moved to different points in the topology. The inputs and outputs may also undergo a transformation in terms of content and quantity (i.e., the information may not be subjectively filtered by the sender or receiver).

How does one determine the transformation and what is required in the hybrid structure? It is necessary to understand each information system and then combine the two. For example, high quality process information may be automatically collected and consolidated without human intervention, but cause and effect relationships may be difficult to ascertain and still require human interpretation. All of the information system components (topology, processes, and data stream) are closely linked and changing one will affect the others. While there are many examples of concepts for automated systems using traditional or AI techniques, there are few (e.g., Becker, 1985; Rickel, 1988) that analyse the informal structure existing in the real world. The focus of the authors' research has been on the informal structure presently in the real world and has provided numerous insights into the situation (McKay, 1987). For example, McKay and co-workers (1989) describe a decisional structure oriented towards the hybrid situation that deals with the topology and process issues -- what can be automated and what cannot. The topics addressed in this paper are complementary to McKay and co-workers (1989) and pertain to the third component

327

of information systems -- the data streams.

The following section briefly reviews the overall
scheduler's information system. Subsequent sections delve into the type of information used in
the scheduler's information system. While not all
of the information used by schedulers can be captured and manipulated by a computer, there are
some generic concepts that can be imbedded in a
scheduling decision support system which would
greatly assist the scheduler. These concepts are
currently being implemented in a scheduling decision support system called WATPASS-II.

PROBLEM FORMULATION

McKay (1987) showed that in a traditional setting,
the scheduler uses a vast array of information
sources. Over twenty verbal and written sources
were identified -- the majority being verbal. The
written sources included the typical manufacturing
information: routings, historical logs, standards,
and so forth, as well as non-manufacturing sources
such as newspapers. The verbal sources provided
real-time status of the system's health, external
influences, and a variety of inter-relationships.
The verbal paths also carried the critical information used in predictive and reactive scheduling
decisions. Furthermore, the majority of the
information sources were output destinations for
status distribution or decision results. Additional input sources included the scheudler's
other senses: auditory, sight, smell, and touch.
As the scheduler moves through the shop, data
collection is constantly taking place.

Why does a scheduler or planner use all of these
sources? To actually schedule and optimize production in the strict sense? Of course not! The
schedulers spend very little of their time scheduling and preparing an official plan according to
methods associated with the Job Shop Scheduling
Problem as defined and studied extensively in
academic circles (Conway and co-workers, 1967;
French, 1982; Graves, 1981; Panwalkar and co-
workers, 1977). The schedulers are problem
solvers who create alternative routings and processes on the fly. The schedulers attempt to
produce a reasonable loading on the shop, not
optimized sequences, since there are so many possible things that can happen between the time work
is released on the factory floor and it is completed. They use their experience and skill, not
mathematical algorithms, to make the feasible and
reasonable decisions. To assist them in their
decision making, the schedulers need to know what
is happening at any moment, so that when a problem
occurs, they have a reasonable view of what is
happening in the shop and what the options are (if
any) that exist.

The information on each path can be precise data
or rough estimates. For example, in a shop with
skilled tradesmen, the routing does not describe
every operation in detail; it assumes the worker
will read the drawing and make the appropriate
decisions. In prototype or one of a kind shops,
the details and predictive accuracy are very vague.
Some paths can also carry information that is
qualitative in nature -- a worker is tired, it is
too hot, things are just not working right.

Contrast the above with the ideal automated factory
situation represented by the third scenario described in the introduction. First, all data is
not quantifiable. Second, all processes are not
internally free from perturbations and disruptions.
Third, the shop is not isolated from disturbances
from the outside world. Fourth, all processes can
not be predicted in terms of processing require-

ments. Fifth, not all goals and objectives are
quantified and visible. Hopefully, the second
scenario is somewhere between the two extremes and
is not as dynamic as the traditional environment.
However, until the shop is ideal, it is the authors'
opinion that the partially automated shop will have
periodic examples of everything that can be found
in the traditional information system. When these
events occur and the information must be obtained
and processed, the automated scheduling and planning tools should provide the scheduler with concepts and functions not possible in the traditional
manual world of scheduling.

SOLUTION FRAMEWORK

In any scheduling system there are hundreds of data
elements that must be maintained and manipulated.
The typical information systems include information
about:

- calendars
- shifts
- resource capabilities
- route sheets
- client information
- financial data
- work in progress
- work pending
- work assignments
- assembly or bill of material structure

These major data elements can contain several hundred minor or individual fields. This data can be
readily supplied in most CIM environments. The
accuracy and reliability of the data may not be
absolute, but something can be normally provided.
However, there are other types of data that the
scheduler may need when called upon to create an
alternate solution or make a scheduling decision.
In a CIM situation, it is unlikely that the
scheduler will be on the shop floor as much as in
the past and must rely on the computer system to
collect and organize some of the key information.
It is also possible for the computerized scheduling
system to utilize portions of this information in
scheduling decisions.

A major difficulty to overcome is the variety faced
by the information systems. For example, each
factory will present different challenges as the
type and meaning of data will be diverse. Not all
of the variety can be dealt with automatically,
but there are three concepts that have been developed by the authors that will potentially address
part of this diversity -- ladles, historical cause
and effect tracking, and a structured knowledge
base driven by a specialized inference engine.

Ladles

In the manufacturing setting, the scheduler needs
to keep track of a large variety of information
related to the shop status and schedule. The types
of information are very diverse and are often
linked to inventory, operations, resources, etc.
for control purposes. For example, schedulers were
observed to keep track of the number of operations
a machine performed per week -- it had been determined that the number would be a constraint due to
maintenance problems. This type of information
usage provides the first need: dynamic information
that is used to control work assignment at the
tactical level.

Information generated during the scheduling process is also used in the strategic decision making
process. For example, in one factory, the plant
manager had to report to superiors on gross

tonnage shipped (past, present, and future) instead of the dollar value or number of units. Order acceptance, market forecasting, and facility planning was based on this measurement. In this same example, the scheduler did not use tonnage at the tactical level; he had to deal with the individual items and specific customer orders. Hence, the second need is for flexible data collection and reporting. This must be tailorable to each situation, as each shop and each management level have different desires.

Information collected from the floor and during planning may act as a stimulus in the scheduling and planning cycle. For example, material consumption during the manufacturing cycle implies the creation of suitable work orders or purchase orders to ensure blockage or starvation does not occur in the factory during production. Consequently, there is a need for special information processing in addition to constraint control and information gathering. This third need corresponds to the dynamic creation of work orders representing net demands for materials and subcomponents.

Special information structures called ladles satisfy the above needs. The ladles are named entities accessible to the scheduler on operations, machine setups, scheduling constraints, and reports. In a sense, ladles are software structures similar to ladles, cups, bowls, and buckets in the physical world. It is possible to put things into a ladle and pour them out later. There are different strategies for filling, emptying, and monitoring the contents of ladles. The numbers and semantic usage of ladles is controlled by the scheduler. There are three types of ladles:

- Scalar Ladles are single entities which can be incremented or decremented. For example, a scalar ladle can be incremented everytime a certain part is operated upon by a specific machine for tracking purposes, or a scalar can be used for accumulating special costing information such as hydro usage.

- Accumulating Vector Ladles are multiple entities with time fences of effect -- the vector elements are never reset. The scheduler sets the time fence to be used for each element in the vector and the system automatically increments or decrements from the appropriate element. For example, the work in process inventory level per week can be monitored using a vector ladle.

- Refillable Vector Ladles are multiple entities with time fences of effect -- the vector elements are reset to zero or a specified level at the start of each fence boundary. These vectors can be used to control the number of occurrences of a special event. For example, the number of furnaces that can be simultaneously turned on can be controlled with this type of ladle.

Increments and decrements of any ladle can be fixed or variable. Variable values are based on elapsed time, processing time, setup time, teardown time, quantity input, or output quantity. Ladles can be global or local relative to work or resource entities. Furthermore, costs can be attributed to quantities being placed or removed from ladles, or whenever the ladle is reset automatically.

The three types of ladles have in turn, three major uses: information, flow control, and order control.

Information ladles are available in any scheduling mode of WATPASS-II (Operations Research, Artificial Intelligence, or Simulation). Any of the three ladle types can be used for information purposes and they can be summed, combined, or mathematically manipulated. The information ladles can be accessed by the report generator or screen formatter.

Flow control ladles are only used in the Artificial Intelligence and Simulation modes of WATPASS-II. It is possible to delay the start of any operation until a ladle has reached a certain level (i.e., equivalence relationships: gt, eq, ne, ge, lt, le). The flow control ladles can be used to emulate constraints across jobs, special assembly relationships and the like. It is possible to use any of the three ladle types and the system will automatically access the appropriate vector element based on the simulated clock.

Order control ladles are also only used in the Artificial Intelligence and Simulation modes. These ladles automatically create work orders or purchase orders to satisfy demand occuring in the 'next' fenced boundary. The generated orders are initially backwards loaded from the point of demand using fixed leadtimes or dynamic times based on operations using finite resources. If the order cannot be filled, the system attempts to satisfy as much of the demand as possible and pushes out the remainder. The order control ladles must be vectors and they have three additional qualifiers associated with them -- the routing sheet to use, minimum and maximum batch sizes to make, and the maximum allowed threshold to delay any order if the minimum is not satisfied. This special class of ladles can provide an interface to MRP systems for the purpose of generating net demands in an MRP bucket based on specific/pegged work order demands.

Historical Cause and Effect Tracking

Humans have been collecting variance and production data for quite a while. The information is used in efficiency, quality, and process analysis. Automated systems can also collect a significant quantity of data over time. While the humans can collect, generate, and process both quantitative and qualitative information which is continuous or discrete, the computer systems have been limited to a subset of the discrete quantitative data. The existence of a partial database implies that the computer system may have difficulty determining cause and effect relationships. For example, the predicted time for a specific job may be twenty hours and ultimately take thirty hours. The reason for the variance may be due to many reasons, some related to the actual job (e.g., estimates are wrong, material is hard to work with), or not related to the job at all (e.g., the operator was not feeling well). It would be wrong for the system to automatically scale the standard hours for the next similar job based on this singular sample. If the human is integrated into the system as an active feedback element, the computer system can filter and track the cause and effect elements not possible before.

When a significant variance occurs between the planned and actual results, the human can be prompted to provide some insight into the problem. Some standard categories of adjectives are provided that relate to time, geographical/department, environment, cultural, and personnel. However, other categories are completely at the mercy of the scheduler -- numbers and contents. This allows the scheduler to develop and use localized terminology and groupings of events.

Keeping historical data is useful if there is something to compare it against. At various times, the scheduler can attach adjectives to any of the elements in the scheduling system -- work, resources, shifts, etc. and have these matched by historical combinations. The scheduler would only update those areas which are expected to cause exceptions and perturbations. The system will then notify the scheduler of past variances which were significant and permit the scheduler to adjust the planned time based on historical results. For example, in one foundry it was noted that at specific times of the summer, high humidity was very common. The high humidity in combination with a certain process (type of sand, alloy, and casting shape) affected yield by 100%. This particular combination may occur once every two years, but when it does, several weeks of production are dramatically affected.

Concepts such as this utilize the complementary skills of the human and computer -- the computer keeping track of large amounts of data and the human providing purposeful decision making.

Knowledge Bases and Inferences

Knowledge based scheduling can be characterized by three typical approaches -- constraint-based and other search algorithms programmed in an AI environment (e.g., Fox, 1983; Ow, 1986; Steffen and Greene, 1986; Elleby and co-workers, 1989). IF-THEN rule based scheduling where all rules exist together and the scope of the rule is controlled from within the rules themselves (e.g., Randhawa and McDowell, 1988; Karni and Hayeems, 1989), and hybrid systems which combine analytical techniques with some rule based structures or heuristics (e.g., Villa, 1989). All of these approaches are complex and strive for optimality. Believing that feasibility is the first objective, the authors propose a simple composite system combining both procedural and IF-THEN logic that specifically does not optimize but allows the scheduler to specify the critical constraints and relationships that matter. The procedural logic would maintain and control all of the housekeeping tasks (i.e., data structure linkages, etc.) and the IF-THEN logic would exist as a structured rule base organized around the flow of work through the shop.

The rule-base is composed of the traditional rules -- IF condition THEN action -- where the condition has access via a grammar to all database structures used for scheduling purposes and the action can manipulate the database in addition to making scheduling decisions. The rules are associated to either work or resources. Whenever a resource is about to perform an action, a selected set of rules will be evaluated. That is, whenever work arrives at a machine, a batch is started, part is finished, and so forth, a set of rules linked to that specific resource will be fired. Similar structures exist for work as it is started, moved, worked upon, and finally completed.

The condition side of the rule can be formed of conjunctions and disjunctions to represent multiple conditions before the action is fired. The rule grammar is designed specifically to match the manufacturing hierarchies and substructures -- contracts, jobs, work orders, operations, departments, work areas, machines, etc. The grammar can access any field read/write that is normally available to the user through the data entry and scheduling system (e.g., quantities, priorities, ladles). In addition, the grammar can access read/only the system status fields such as total hours currently loaded into a department. The grammar has meta names for the current operation,

current machine, and so forth to permit generic rule subroutine libraries. It is possible for a condition to automatically search through a number of data elements -- e.g., does the work order have a specific type of operation.

The action side also supports multiple results and the full grammar. In addition, the action side can perform any of the functions available to the user through the data entry or scheduling. For example, a rule could move an operation to a specific location, shift another by two days. Other specialized actions are provided, an example of which is to reset the scheduling algorithm back to a certain point in time.

The inference engine will support interactive debugging and tracing techniques to ensure that the scheduler or algorithm designer will have an intuitive confidence level regarding the validity and quality of the generated schedule. This is very important as our experience has indicated that unless the scheduler can intuitively understand why a certain operation was placed on the schedule at a specific time and why not somewhere else, the scheduler will not trust the system and will not use it. This observation actually led to the removal of several sophisticated scheduling algorithms in WATPASS-I.

The structured knowledge base and inference engine serves four main purposes. First, the user can collect and use the various quantitative rules of thumb as they become apparent. Second, it helps the scheduler to organize the rules and maintain them. Third, it is more efficient in all senses to look at only those rules which are relevant. Fourth, it permits researchers to develop a wide variety of algorithms and test them without the need of programming in low-level languages such as 'C', FORTRAN, PASCAL, or MODULA-2.

CONCLUSION

It has been proposed in this paper that realistic automated systems will have to support human decision makers whenever the unexpected occurs. This implies a number of flexible structures that will permit the scheduler to collect information and control the scheduling support system. Three major information system concepts have been developed based on insights gained from studying real world scheduling behavior, and experience with the Waterloo Planning and Scheduling System (WATPASS-I). These structures -- ladles, historical cause and effect tracking, and a structured knowledge base -- have been proposed as techniques that will provide the necessary flexibility to support the scheduler in an automated setting. Development is underway on WATPASS-II which will provide a research vehicle for field evaluation of these concepts.

REFERENCES

Becker, R.A. (1985). All factories are not the same. Interfaces, 15-3, 85-93.

Conway, R.W., W.L. Maxwell, and L.W. Miller (1967). Theory of Scheduling. Addison-Wesley, Reading, Massachusetts.

Elleby, P., H.E. Fargher, and T.R. Addis (1989). A constraint based scheduling system for VLSI wafer fabrication. In J. Browne (Ed.), Knowledge Based Production Management Systems. North-Holland, Amsterdam. pp. 107-114.

Fox, M.S. (1983). Constraint-Directed Search: A Case Study of Job Shop Scheduling. PhD diss., Computer Science Department, Carnegie-Mellon University.

French, S. (1982). Sequencing and Scheduling: An Introduction to the Mathematics of the Job-Shop. Ellis-Horwood, Chichester.

Karni, R. and I. Hayeems (1989). An intelligent system for parallel machine scheduling. In J. Browne (Ed.), Knowledge Based Production Management Systems. North-Holland, Amsterdam. pp. 207-222.

McKay, K.N. (1987). Conceptual Framework for Job Shop Scheduling. MASc Dissertation, Department of Management Sciences, University of Waterloo.

McKay, K.N., J.A. Buzacott, and F.R. Safayeni (1989). The scheduler's desk -- can it be automated? To appear: IFAC Workshop on DSAMS, Genova, Italy, September 1989.

Ow, P.S. (1986). Experiments in knowledge-based scheduling. Working paper No. 50-85-86, Carnegie-Mellon University.

Panwalkar, S.S. and W. Iskandar (1977). A survey of scheduling rules. Operations Research, 25-1, 45-61.

Randhawa, S.U., and E.D. McDowell (1988). An investigation of the applicability of expert systems to job shop scheduling. Submitted to International Journal of Man-Machine Studies.

Rickel, J. (1988). Issues in the design of scheduling systems. In M. Oliff (Ed.), Expert Systems and Intelligent Manufacturing. Elsevier Science. pp. 70-89.

Steffen, M.S., and T.J. Greene (1986). A prototype system for scheduling parallel processors using artificial intelligence methods. IEE 1986 Spring Conference.

Villa, A. (1989). Hybrid knowledge-based/analytical approach to production management systems design. In J. Browne (Ed.), Knowledge Based Production Management Systems. North-Holland, Amsterdam. pp. 133-152.

OPTIMIZATION TECHNIQUES APPLIED TO JOB-SHOP SCHEDULING

H. Bera and M. J. Gannon

Department of Mechanical Engineering, South Bank Polytechnic, London, UK

ABSTRACT

This paper presents a method of minimising the total elapsed time (or makespan) of n jobs going through m different processes. The order of processes for the jobs are not necessarily identical. The method not only minimises the total elapsed time but also reduces the idle times and waiting times of jobs in-process. This condition is very important in some chemical and foundry processes where the temperature differences between two processes is critical.
Keywords. Job-Shop, schedule, waiting time, idle time, elapse time.

1.0 Introduction

For medium to large size industrial types of scheduling problems it is not possible to optimise because of the large number of feasible schedules that exist. This is of the order (n!)m. Scheduling problems can be easily formulated by typical operational research techniques such as dynamic programming, integer programming but determining the optimal solution requires considerable computational resources. Therefore in practice it is necessary to employ simulation techniques to investigate the possibility of an optimal schedule.

Unfortunately one of the limitations of employing discrete-event simulation for investigating various scheduling strategies is that there is no "in-built" optimization procedures. Any attempt to locate an optimum solution to the problem, relies on searching the relevant factor space. This search often degenerates into a "trial-and-error" process, often due to financial and time constraints imposed by management or project controllers. It is therefore necessary to consider an alternative method, one which allows determination of optimal schedules in a cost and time effective manner.

Optimising required performance measure(s) within a schedule is critical to many of the process industries. For instance it may be of importance to maintain a constant temperature between two processes which in scheduling terminology implies that there must be "Zero waiting time in-process" for a job. Naturally management have an interest in optimization as long as it results in increased production which in turn results in increased profits. In this paper, A job-shop problem is used to illustrate an optimization methodology, which not only optimises the elapsed time but also reduces the idle time and waiting time in process of the schedule.

2.0 Classification Of Scheduling Problems

Suppose, that there are n jobs (J=1,2,...n) that has to be processed through p processes (q=1,2...p). The general scheduling problem dictates that the jobs should be sequenced through the processes in such a manner that some required measure of performance is determined

(Bera, 1984, 1985,1986,1988a,1988b).

The three types of distinct scheduling problems are as follows:
1. Job-Shop : The processing sequences of jobs is not identical.
2. Flow-shop: The processing sequences of jobs is identical.
3. Open-shop: The processing sequence of jobs has no relevance.

Such schedules are necessary for the overall production planning of:
  -processes.
  -manpower.
  -ordering parts, tools, supplies, etc.
  -cost budgeting.

3.0 Initial Feasible solution

Clearly, as previously mentioned in section (1.0) there is a requirement for the application of optimization techniques to scheduling. Typical performance measures of interest to the process industries are as follows:
- Total elapsed time.
- Total idle time.
- Total idle time with zero idling in-process.
The interested reader is referred to Bera (1984, 1986) for a more detailed description of the optimization methodology.
An example is now presented to illustrate the determination of the optimal makespan and reduction of the idle and waiting time in-process of an initial feasible schedule.

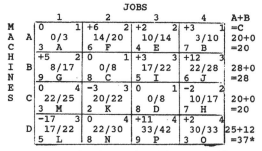

		JOBS 1	2	3	4	A+B
M	A	0 ⎸ 1 / 0/3 / 3 A	+6 ⎸ 2 / 14/20 / 6 F	+2 ⎸ 2 / 10/14 / 4 E	+3 ⎸ 1 / 3/10 / 7 B	=C / 20+0 / =20
A C H I N E S	B	+5 ⎸ 2 / 8/17 / 9 G	0 ⎸ 1 / 0/8 / 8 C	+3 ⎸ 3 / 17/22 / 5 I	+12 ⎸ 3 / 22/28 / 6 J	=28 / 28+0 / =28
	C	0 ⎸ 4 / 22/25 / 3 M	-3 ⎸ 3 / 20/22 / 2 K	0 ⎸ 1 / 0/8 / 8 D	-2 ⎸ 2 / 10/17 / 7 H	=20 / 20+0 / =20
	D	-17 ⎸ 3 / 17/22 / 5 L	0 ⎸ 4 / 22/30 / 8 N	+11 ⎸ 4 / 33/42 / 9 P	+2 ⎸ 4 / 30/33 / 3 O	=37* / 25+12

TABLE 1: Processing Times, starting, finishing time, Idle or waiting times for an initial Feasible solution.

Key:

Y		Z
	S/F	
X		

X = Processing time.
Y = Waiting time if+ve,
    Idle time if=-ve.
S = Starting time.
F = Finishing time.
Z = Processing sequence number
    (or operation number)

A = Minimum time required on each machine.
B = Earliest possible starting time on each machine.
C = Earliest possible completion time on each machine.

The solution shown in TABLE 1 is obtained by calculating the starting/Finishing time of each operation on each machine. The starting/finishing times are calculated in ascending order of operation numbers of all jobs i.e. the starting/finishing time of all 1st operations of jobs are calculated before the calculation of those for the 2nd operations of all jobs. If more than one job require a particular machine for a given operation, jobs are selected in ascending order of processing times for the operations concerned or the earliest possible time (when the processing times are the same) Bera (1984).

The above table (TABLE 1) shows that the total elapsed time is 42 units and this is the completion time of the last job on machine D. This 42 units of elapsed time can only be minimised by reducing the idle time on this machine. However this machine has 17 units (-17) of idle time before the beginning of the first job on this machine (job 1) and it is required to investigate whether it is possible to reduce this idle time.

4.0 Optimality Test

*Earliest completion time on all machines=37 units, refer TABLE 1 and this is the completion time of last job on machine D. However there are 4 jobs which can be selected as the last job on machine A and the outcome of selecting one of these jobs is as follows:-
Let, the last job on machine A be:

Job	C	D	E
1	20 + 17	=	37
2**	20 + 10	=	30**
3	20 + 14	=	34
4	20 + 16	=	36

MINIMUM=30 units

Clearly, the minimum value is 30 units for Job 2. Where, C=Sum of processing times on machine A.
D=Sum of processing times of remaining operations.
E=Completion time of each job.

The above calculation shows that all jobs may be completed in 30 units of time if job 2 is selected as the last job on machine A. But it has been previously seen that the machine need at least 37 units of time to complete all jobs. Therefore the earliest possible completion time is not less than 37 units.

Reduction of Idle Time Of Machine D.

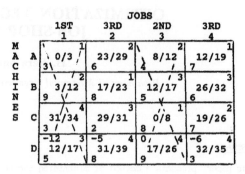

TABLE 2: showing starting/finishing times when the starting time of first job (job 1) is 12 and the starting time of the 2nd job (job 3) is 17 on machine D.

The figures in column (1) of TABLE 2 shows that the idle time of machine D has been reduced from 17 units to 12 units, there is a gain of 5 units of idle time of machine D. Now it is required to select the remaining jobs so that the idle time in-process on machine D is less than 5 units. It could be seen from TABLE 2 that if job 3 is selected as the second job then the idle time in-process would be zero units.
the 2nd job = 3
Now the third job can be either job 2 or job 4.
Let, the third job=2.
Idle time in-process=5 units (refer column(2)).
Not optimal
Let, the third job=4.
Idle time in-process on machine D=7 units (refer column (4)). Not Optimal.

6.0 Alternative starting time of 2nd job

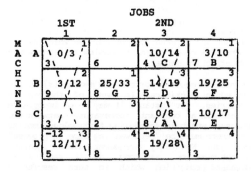

TABLE 3. Showing completion times when the second job starts at 19 and the third job starts at 28 on machine D.

The order of calculation in TABLE 3 is shown by letters (A) - (G). The calculation at stage (G) has been terminated because it shows that the total elapsed time >>42 (Not Optimal) because completion time of Job 2 is 33+ (6+2+8)=49. (See column 2)

conference   of   the   Irish   Manufacturing
committee, Belfast.

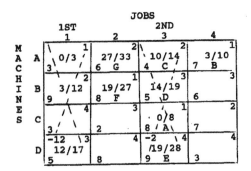

TABLE 4. Showing completion time when the second
and the third jobs are 3 and 2 respectively.

The order of calculations in TABLE 4 is shown by
the letters (A) - (G). The calculation at stage
(G) has been terminated because it shows that the
total elapsed time is >42.   (33+2+8 = 43 See
Column 2).

It can be seen from the figure in TABLE 1 that no
job can start on machine D before 12 units.

It also shows that no other job can be selected
as the second job with idle time in-process less
than 5 units. Therefore the optimal solution for
elapsed time is 42 units.

7.0 Conclusion

It has been seen in many cases that the initial
feasible solution is the optimal solution. If it
is not, the optimal solution can be found by the
method shown in this paper.   It is also found
that the difference between the optimal and the
initial feasible solution is negligibly small [2]
which means companies do not bother about the
optimality test only to reduce a small amount of
elapsed   time.   The proportion of optimal
solutions to the near optimal solutions is at
present under investigation.

References:

1.   Bera,H.(1984). Cost prediction Modelling of
     batch Production in a Manufacturing System
     for sheet metal fabrication. PhD Thesis,
     South Bank Polytechnic.
2.   Bera, H.(1985). Job-Shop Scheduling. 3rd
     joint International Conference in Mechanical
     Engineering, Cairo.
3.   Gannon M.J.,Bera, H.,Rahnegat H., Dobbs P.
     J.,(1986) Simulation for large multi-batch
     Production control. 2nd National Conference
     on Production Research, Napier College,
     Edinburgh.
4.   Bera, H.(1986). Optimal Sum of idle time of
     production facilities. FOCOMP, Krakow,
     Poland.
5.   Bera, H.(1988). An unorthodox approach to
     Job-Scheduling. 3rd International Conference
     on CAD/CAM, Automation, Michigan U.S.A.
6.   Bera, H. & Gill, R.(1988). A computer Aided
     Technique   for   optimal Scheduling. CIRP
     Seminar, Singapore.
7.   Bera, H.(1988). Optimal Waiting time of jobs
          with zero waiting in process. AMSE
          Conference   on   Modelling   &
          Simulation, Istanbul, Turkey.
8.   Bera, H.(1985). Optimal Sequence for waiting
     times of jobs in process. Proc of 2nd

PRODUCTION CHANGE MULTICRITERIA OPTIMIZATION WITH DUE-DATE CONSTRAINTS

A. Gonthier* and J. B. Cavaille**

*BSN Emballage, Villeurbanne, France
**CERT-DERA, Toulouse, France

Abstract
The control of a manufacturing line presenting both continuous and discrete features has been
reduced to a scheduling problem. We propose a decision aid environment using Little algorithm
to solve it. The application to an industrial case is presented.

Key words
Manufacturing Processes, Multi Objective Optimization, Operations Research, Set-up Costs,
Decision-aid tool

INTRODUCTION

A continuous process is feeding a discontinuous manufacturing line.

The problem appears to be complex due to the large number of criteria to be considered.

A complex modelling could only lead us to develop a very complicated tool, difficult to run in a actual industrial context. We have chosen instead a rather easy to run algorithm in a decision-aid approach.

Step by step, the scheduler is reacting the solution that he prefers. Later, the tool will be integrated in an AI environment which would provide some aid to the operator during his best scheduling search.

STATEMENT OF THE PROBLEM

In a manufacturing environment, the problem may in fact be reduced to scheduling n jobs / 1 operation on single machines with sequence dependent multicriteria set-up costs, making allowance for due-date constraints.

Manufacturing context

In industrial on-line processes, production changes involve costly discontinuities. The question is to optimize the scheduling of those changes, fitting in tricky sales forecasts and thus strict delivery dates and concurrently minimizing the change-over costs of these job changes. Criteria relevant to their evaluations are numerous and sometimes opposed. For instance process flow stability can be opposed to job change technical operations or machine speed synchronization.

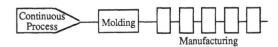

Decentralized planning

The whole planning system is assumed to be properly built with an upper centralized module ensuring the global cohesion of the whole, giving for instance a long term planning framework with aggregated data, preserving the adequate short term autonomy which is necessary for our system.

The latter will make sense if used as a short term optimization tool on a rather restricted horizon for local scheduling, which could be for example set up in a decentralized manufacturing plant.

The local scheduling system

We present here the entire local scheduling system as proposed in previous works (Gonthier 88a). The following 3 blocks : Data Analysis, Local Knowledge, Expert Supervisor are of no interest here. We just focus on the Optimization module, the structure of which we can detail as follows.

Informations arising from the data Analysis block, allow the Supervisor to choose among the available algorithms and to send the good parameters taking account of Local Knowledge.

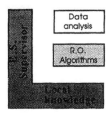

METHODOLOGY

Multicriteria TSP modelling

As presented, the pure sequencing problem - without delay considerations - including sequence dependent costs can be modelled with the help of the classical Traveling Salesman Problem formulation.

Indeed, the best path through the cost matrix built with all the possible previous-and-next couples of patterns models, will give the optimal sequence.

Thus a representative criterion has to be found that is a valid "distance" between every different couple of manufactured models. At this point there appears the undoubtedly industrial feature of the problem which prevents from selection of only one representative and objective criterion.

The solution is then to set a multicriteria interactive TSP. The local decision maker will have to adjust the weights given to the different criteria in order to satisfy his own and present view of a "good change".

Available global tools : TCTSP

We now take into account the whole problem mixing sequencing and delay considerations. This problem belongs to the Time Constrained Traveling Salesman Problem. The only difference with the standard formulation being that the cost is in this case a weighted sum of different basic costs.

We know that the introduction of time constraints substantially complicates the TSP resolution, even for a rather low number of "towns".

Solutions are given by dynamic programming methods (Christofides , Psarfatis) or by modelling the precedence constraints using graphic techniques (Nemhauser 62, Lenstra 77).

The resolution is globally carried out in a rather complex way. Since the due-date respect is considered as a hard constraint, no overtime is allowed which could however result in considerable winnings in change-over costs.

Available partial tools

* The simple TSP we previously mentionned has been paid a lot of attention to in Operational Research. Known as being NP-complete, it is often solved with different heuristics or Branch and Bounds methods still limited to a few tens of jobs.

* Appart from this objective of minimizing change-over costs, it is also possible to only minimize the maximum tardiness of each job, which seems one of the most realistic criteria. For the other criteria (mean tardiness,...) the problem is also NP-complete.

Selected approach

Rather than implementing a global sophisticated algorithm, tricky to use, and giving only one strict solution, we prefer the "decision-aid tool" approach, with a fast, simple and easy to run algorithm allowing the operator, by means of different weights and penalties, to converge towards a solution satisfying his time varying and multiple objectives.

Among the two former partial tools, we decided to privilege the first one according to the essential feature that is optimal sequencing.

The other choice would have favoured delay satisfaction, and approached methods such as groups of switchable jobs (Roubellat, Thomas).

Moreover we can expect for our whole planning system enough local autonomy to get rid of these time constraints at least in the short term.

Implementation

The research of an optimal sequence is similar to the single TSP where, as previously presented, towns are here jobs and the traveller is the on-line single machine. Change-over cost from job i to job j evaluated by the k the criterion is $C_{k,ij}$.

The rather low number of jobs to be scheduled (<30) allows the use of an optimal algorithm.

The algorithm Little

The assignement of the initial job on the process is assured by the addition of a dummy job in the matrix.

The weights proposed by the decision maker are A_k.

The resultant matrix is then built with $\sum_{ij} A_k \times C_{k,ij}$

Due-date penalties

In the case where due-dates (T_i) pose constraints on the problem, some sequences become unacceptable.

The adopted method consists in penalizing the incompatible sequences - those which tend to violate the due-date constraints - by modifying the initial cost matrix with an appropriate weight.

Three penalization rules have been implemented and tested :

Rule 1

$$C_{k}{'}{,}ij = C_{k,ij} + P \times O(T_i-T_j) \times M$$

where $O(T_i-T_j) = 1$ if $T_i>T_j$
$\qquad\qquad\quad = 0$ if $T_i \leq T_j$

Rule 2

$$C_{k}{'}{,}ij = C_{k,ij} + P \times Max(0,T_i-T_j) \times M$$

Rule 3
$$C_{k}{'}{,}ij = C_{k,ij} + P \times Max(0,(T_i+D_i) - (T_j+D_j)) \times M$$

where T_i : is the due-date for beginning manufacturing of job i
D_i : is the size or duration of job i
P : is a weight to be adapted by the decision-maker
M is the matrix arithmetic average

RESULTS

Data used

They arise directly from an industrial example.

• Work-shop level modelling
The set-up cost is represented by 7 criteria which can be set in 3 different classes :
- the variations of material flow in the continuous part of the process, C1
- the variations of the molding machine cadence at continuous/discrete transition, C2
- the variations of product shapes which involve changes on the manufacturing stations.

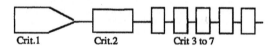

criteria 1 and 2 are calculated directly from flow and cadence measurements.
criteria 3 to 7 lead to a matrix filled by 1 or 0, depending on whether or not a change in the shape parameter is involved.

• Production to be scheduled
Data issued from the upper planning level give the size or duration in days for each job and the last possible date for starting production.

Different configurations have been tested.

Multicriteria decision-aid approach

In this section, we take no account of the due-date constraint.

• We first present the variations of two criteria Ci and Cj when the weights Ai and Aj are linked by : $A_i = 1 - A_j = \alpha$
α varies from 0 to 1.

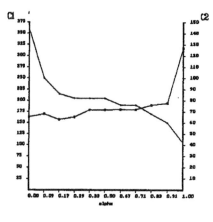

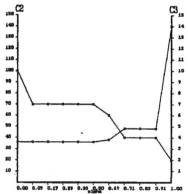

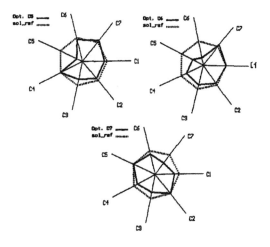

The 3 diagrams that follow give good compromises as could be found by the decision maker employing appropriate weight balancing. The dot on each scale represents the optimal result expectable for each criterion.

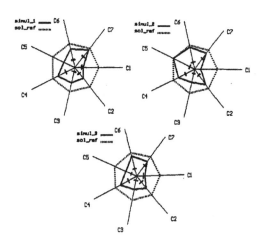

These pictures illustrated the research of an adequate compromise between two criteria.

• We represent all the criteria and plot the solutions on a 7-branchs diagram.

On each diagram, dotted lines depict the initial "natural" schedule given by the upper level. The first 7 diagrams show the schedule found by minimizing only one criterion.

Results appear to be interesting, give a good representation of solution space to the operator in order to guide his research. We may also notice that there is no real benefit in searching the best improvement of C1, because of the important loss on the other criteria.

Due dates penalization

Criteria relevant to the problem of delay constraints have been chosen as
- the number of late jobs N
- their cumulated tardiness T

To test our penalization approach and compare the different rules, we kept two different examples previously mentionned, and tested various growing weights P

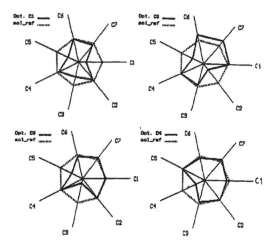

With a good one-criterion optimization

Rule 1

C1	C2	C3	C4	C5	C6	C7	N	T	P
14970	134	7	12	2	13	11	10	479	0
12098	156	7	13	4	13	12	11	507	1
12098	156	7	13	4	13	12	11	507	2
14594	150	8	12	4	12	12	8	513	4
14594	150	8	12	4	12	12	8	513	6
14594	150	8	12	4	12	12	8	513	10
14594	150	8	12	4	12	12	8	513	20
36555	146	9	11	4	15	13	5	305	50
36553	126	11	12	4	15	12	16	43	99

Rule 2

C1	C2	C3	C4	C5	C6	C7	N	T	P
14970	134	7	12	2	13	11	10	479	0
46183	130	9	11	4	12	11	7	37	1
46183	130	9	11	4	12	11	7	37	2
62143	130	11	11	4	12	11	5	15	4

Rule 3

C1	C2	C3	C4	C5	C6	C7	N	T	P
14970	134	7	12	2	13	11	10	479	0
31709	156	10	12	4	12	15	12	155	1

with a "good" compromise

Rule 1

C1	C2	C3	C4	C5	C6	C7	N	T	P
17226	90	6	10	2	11	9	11	522	0
15082	146	8	12	4	12	11	7	443	1
37491	142	9	11	4	12	10	9	160	2
37491	142	9	11	4	12	10	9	160	4
37491	142	9	11	4	12	10	9	160	6
36553	126	11	12	4	15	12	16	43	10
36553	126	11	12	4	15	12	16	43	20
36553	126	11	12	4	15	12	16	43	50
36553	126	11	12	4	15	12	16	43	99

Rule 2

C1	C2	C3	C4	C5	C6	C7	N	T	P
17226	90	6	10	2	11	9	11	522	0
28993	126	9	12	4	15	11	16	44	1
36553	126	11	12	4	15	12	16	43	2
36553	126	11	12	4	15	12	16	43	4

Rule 3

C1	C2	C3	C4	C5	C6	C7	N	T	P
17226	90	6	10	2	11	9	11	522	0
10695	146	10	12	4	12	12	6	23	1
17877	130	11	12	4	14	12	14	52	2

As a first conclusion, we may say that the rule n° 1 seems quite adapted to the objective of minimizing delays.

For a better validation of the efficiency of the two other rules, we should investigate further tests with more detailed values of P.

CONCLUSION

This method has been implemented on a PC micro computer which gives the visualisation of each criterion and its evolution during the compromise search of the decision maker.

The main contribution may be summed up as :
- multicriteria optimization which would formalize the difficult compromise between several criteria issued from different sectors of the manufacturing site and all partly relevant to the problem of job change modelling.

- an original means to mix both pure sequencing and delay constraints, with an iterative and simple tool instead of a sophisticated algorithm.

- above all, a real decision-aid tool for manufacturing to solve this quite complex problem of scheduling with multiple objectives and due-date constraints.

REFERENCES

Baker, E.K (1974). Introduction to sequencing and scheduling. Wiley 74

Baker E.K. (1982). An exact Algorithm for the Time Constrained Traveling Salesman Problem. Technical Notes - July 82

Driscoll, W.C., Emmons, H. (1976). Scheduling production on one machine with changeover costs. AIIE Transactions - Vol 9 - n° 4

A. Gonthier, A, Binder Z. (1988a). Methodological Approach of a Hierarchical Scheduling Problem as set in Glass Industry. AFCET Conference on Manufacturing Systems Grenoble 1988

Gonthier, A., Rochefort, J.F., Sarthou, S., Cavaillé J.B.(1988b). Optimisation des coûts de changement de production avec prise en compte des délais au moyen d'un algorithme simple de Recherche Opérationnelle. Rapport de stage Sup Aéro

Little J.D.C, Murty, K.G, Sweeney, D.W., Karel C. (1963).- An algorithm for the Traveling Salesman Problem. Operations Research - Vol 11.

Picard, J.C., M.Queyranne M. The Time-Dependent Travelling Salesman Problem and Application to the Tardiness Problem in one Machine Scheduling. Dept de Génie Industriel - Ecole Polytechnique - U de Montréal

Roubellat, F., Thomas V. (1987). Une méthode et un logiciel pour l'Ordonnancement en temps réel d'ateliers. 2° Conférence Internationale INRIA - Système de Production - Paris.

White,C.H., Wilson R.C. (1977). Sequence dependent set-up times and job sequencing. Int. Journal Prod. Research. - Vol 15 - n° 2

CELLULAR FACILITY DESIGN MADE EASY

A. J. Wells

CSIRO Division of Manufacturing Technology, Woodville, Australia

Abstract: A suite of computer programs have been developed to assist in the complex
task of designing a cellular manufacturing plant. Use of the system can yield a basic
multi-cell facility design in a matter of hours, and requires only standard manufacturing
information as input.
Manufacturing and industrial engineers have used the interactive design system to help
plan the restructuring of a medium-sized repetitive manufacturing plant into twelve
work cells. Simulation is used to determine the most appropriate operating strategies
for the cells, as well as the manning levels required for a range of production volumes.
For all of the cells that have been implemented to date, the manufacturing performance
is extremely favourable, as indicated by production statistics. As one component of a
major effort to update the working conditions and operating philosophies of the plant,
the work cells have enjoyed excellent support from the shop floor personnel, management, and trade unions.

Keywords: Cellular manufacturing; plant layout; computer-aided design; modelling;
scheduling

INTRODUCTION

This paper describes work carried out by CSIRO
in collaboration with General Motors Holdens
Automotive Limited, Elizabeth, South Australia.
The project was to restructure the operations of the
Small Parts Fabrication Plant, which is essentially
a moderately large job shop. The "products" of
this enterprise are simple assemblies comprising
from one up to twenty internally-made or purchased components. Some 600 components and
assemblies are manufactured on basically a two-
week cycle. Approximately 200 machines are
involved, covering nearly 100 distinct operations
such as pressing, welding, heat treating, deburring,
drilling and surface coating.

Group Technology strategy supported by a cellular
manufacturing arrangement was deemed to be the
most appropriate approach. The strategy is by no
means new, but application of the strategy is a
non-trivial task, particularly when the problem
possesses the indicated dimensions. This difficulty
has contributed to the slow rate at which the principles have been consciously adopted by Western
industrial enterprises (Wemmerlov and Hyer,
1987).

Methods developed previously for tackling the
fundamental task of grouping parts and processes
failed to cope satisfactorily with all aspects of the
problem. A novel method of distilling the essential
elements from the mass of data was required,
together with a design method that could fully
utilise knowledge of shop-floor constraints. New

software tools have been developed within this
project to assist in this analysis of manufacturing
data and the interactive design of a multi-cell
facility. The tools and their successful application
are described in the following sections.

THE GROUPING PROBLEM

How can an appropriate grouping of products and
processes be determined?

One method, Part Classification and Coding,
encodes significant properties of parts - geometry
and material type - and then groups parts on the
basis of the compound codes. Similarity of codes
implies similarity of process requirements, thus
limiting the range of machines required to handle
such a group of parts.

An alternative method, called Process Flow
Analysis (PFA), concentrates on the existing
manufacturing process chains. It identifies parts
that require similar sequences or sets of operations.

The first method requires the analysis of design
engineering information and the construction of a
suitable coding scheme. If associated with an
engineering database, the impact of this approach
on the Product Design and Engineering function
can be enormous through support for product
rationalisation actions, and by providing discip-
lined access to existing design data.

The second method, assumes a reasonable standard
of existing Process Planning, and does little to

guide or help the product design engineer. It can accommodate subassemblies, whereas Classification and Coding is usually applied to single components. This PFA approach concentrates primarily on the Manufacturing facility, and provides substantial benefits for less risk and effort than the multi-department alternative.

The second method was selected for this project for reasons of both the limitation of scope and the accommodation of subassemblies. Furthermore, computerised process routing information and bill-of-materials were available from existing systems.

The determination of appropriate groupings is non-trivial. Most methods published for the PFA approach depend primarily on a part-machine incidence matrix, ignoring much of the available information such as machine loading, machine multiplicity and assembly relationships (King and Nakornchai, 1982). We have adopted a constraint based allocation method for the generation of part-machine groups that ultimately develop into manufacturing cells.

PROCESS FLOW ANALYSIS (PFA) PRELIMINARIES

The goals of Cellular Manufacturing pertinent to this project required that the plant should:

- Process individual products (and their components) entirely within one cell;

- Manufacture any component in one cell only;

- Require each purchased component in one cell only.

Total achievement of these goals implies no inter-cell material movement, consistent processing for all components, and cell-level accountability for every product. Events in any cell do not directly affect the remaining cells.

Unfortunately, the system linkages implied by these goals would generally lead to one big "cell" comprising all the existing equipment - no improvement at all! Alternatively, several obvious cells that can be dedicated to specific products may be all that can be carved off easily.

So, even at this early stage, compromise is required. The first constraint relaxation for this project was to ignore the third goal completely. The manufactured quality of purchased components is not affected by the layout of the using facility. The distribution task for purchased material *is* affected, but was not of major consequence within this project .

Secondly, some operations do not lend themselves to dispersion amongst a number of cells. A heat treatment facility is an example of such an operation. There may exist, therefore, a number of "service" facilities which are "fixed" in location and require that a part leaves the confines of a cell. We have taken the view that, in general, the part may then proceed for further processing in any cell, since the tight local control over that part is already interrupted.

Thirdly, process plans often specify a particular machine for an operation on a part when several other machines would be equally suitable. Since the design of cells involves the process requirements rather than the stated machine requirements, there may be a need to equate several machines to one that is representative of their process capability.

Finally, some operations specified in the process plans may not represent significant capital investment. They could be supported at many additional locations for little cost. Other operations may be widely distributable (for little cost) if there are a large number of machines available to support them. Neither of these categories of operation influences the initial stages of cell development. Temporarily, eliminating them as irrelevant to the analysis can simplify the task considerably. These processes are reinstated as required after the basic design has taken shape.

Hence, before detailed analysis begins, some of these strategic decisions need to be made. They simplify the analysis and, in fact, are probably necessary for a solution at all. They also tailor the solution to conform to some major physical constraints.

PFA TECHNIQUE

The "heavy" analysis starts at this point! Part routing information is used to fully develop every sequence of processes that is to be supported. This requires some careful tracking across part number changes that are inherent in assembly operations. The associations of common components and common assemblies, required by the Cellular Manufacturing goals, are retained.

Now for the first pragmatic simplification. The existence of "fixed" operations within the process chains permits us to break the chains into segments.

Secondly, previously nominated processes are replaced by their representative equivalents, or eliminated from the process chain segments if they are temporarily irrelevant.

The third simplification involves the elimination of repeated operations and order within the sequences. We restrict our attention to providing a process *capability* within a set. Routing considerations within a set are thus deferred at this stage.

Where unbroken process segments remain associated through assembly considerations, amalgamated process sets are formed.

Having reduced the process planning information in this manner, duplicate sets are eliminated, with compounding of the total load from the individual instances. All process sets which are a subset of another set are marked.

The remaining unmarked process sets represent *a basis which will support all of the initial process chains*. To support in full the original processing requirements, these basic sets may require to be linked via "fixed" operations and may require the reintroduction of operations previously classed as irrelevant and set aside.

The basic sets, as manufacturing cells in their own right or in combination, support both of the remaining goals except when "fixed" operations are involved. The only material traffic external to cells will be between cells and "fixed" operations.

PFA RESULTS

Typical results from a 600 part, 100 process example contain between thirty and fifty basic process sets. Without reintroducing "as-required" operations, the sets range in size from two to ten operations per set.

The variation seen in the number of basic sets is a direct consequence of the preliminary decisions regarding fixed, equivalent and irrelevant operations.

Part number "portfolios" are associated with the different process sets as they are developed, and may eventually be supported by one or more basic sets. Those supported by only one basic set represent a *mandatory workload* on that set. The remainder form a *distributable workload* which may be employed for either static or dynamic balancing. In this particular case, from thirty to fifty percent of the load was specifically assigned, with the remainder being distributable over two or more basic process sets.

This partial allocation to part-process groups is quite automatic, requiring no user input and acknowledging no constraints on machine incidence, process loading, or operation group size.

CELL BUILDING

Individually, the basic process sets, as developed by the Process Flow Analysis, are unlikely to represent satisfactory manufacturing cells for a number of reasons. Generally comprising relatively few processes, they offer a restricted range of capabilities; they are not "flexible" or broad enough in their capabilities. The process loads associated with their part "portfolios" may represent poor machine utilisation. To implement separately all of the basic sets is unrealistic in machine requirements.

The cell-building task is carried out using a constraint-based allocation technique. The basic sets are allocated to cells such that, perhaps, the demanded incidence of an operation is constrained to the availability of machines, and the individual machine loadings are reasonable.

Factors to consider during the development of cells are:

1. Possible capital outlay on new plant
2. Machine loading
3. Cell size
4. Material flow within cell
5. Material flow outside cells
6. Critical operations
7. Machine distribution across several cells
8. Multiple use of processes

These factors influence the operating performance of the cells; each cell's demand on supervisory skill, its robustness in the face of machine or tool or stock supply failures, and the full system's ability to accommodate new parts or alterations in product mix.

A computer-aided manufacturing cell design system has been developed to run on a SUN[†] graphics workstation using the windowing environment provided by the SunView package. This system has the capability of rapidly providing information to support design decisions by industrial and manufacturing engineers as the interactive design session proceeds.

LOAD ALLOCATION

Having allocated the processes to cells in such a way that all required processing can be achieved, as desired, within the multi-cell system, it is necessary to consider process loading. Part process segments and additional machines (not processes) are allocated to cells to obtain an acceptable balance of load, both within the individual cells, and across the total plant.

Process segments with only one possible location are immediately allocated, together with machines to provide sufficient capacity. The remaining process segments and machines are then allocated, in an iterative fashion, until a satisfactory distribution is obtained. At present, the process segments are allocated automatically by a balancing algorithm, but the placement of additional machines is left to the design engineer.

If a satisfactory solution cannot be found at this stage, modification of the design at the preceding cell-build phase may be necessary. As an example, the number of machines required to support a particular process in several cells, as required by the uniquely located process segments, may be within the availability of such machines. However, the "best" optionally allocated load distribution could require additional machines to provide

---

† SUN is a registered trademark of Sun Microsystems, Inc.

sufficient capacity in one or more cells, and this may take the requirement for such machines beyond the availability. Reallocation of some of the basic process sets provides a means of reducing the number of cells that require this particular process, and this in turn may lead to an acceptable total machine requirement. Naturally, an alternative solution is simply to add to the machine availability, and this path might be attractive in comparison with the consequences of restructuring the cells.

In our experience, starting from appropriately formatted and correct manufacturing data, the time required for Process Flow Analysis and preliminary multi-cell design for this plant is less than eight hours. This makes the examination of several alternative strategies a very attractive and realisable proposition.

SIMULATION

The final design tool is a simulator that enables the designer to examine the operating characteristics of each cell. A range of operating strategies and scheduling rules can be applied to determine the number of operators required, machine and operator utilisations, work-in-progress inventory levels, and maximum capacity. The results of these simulations can also indicate the necessity for further modification of the design to achieve desired operating goals.

IMPLEMENTATION EXPERIENCE

Pilot Cells
A number of cells have now been developed manually within the Plant, primarily for two reasons. Firstly, six cells have been developed to produce very restricted product families. In some of these, operations are generally well loaded by the total volume of the product family, and the sequences of operations in each cell are consistent enough across the product set that flow-line operation is appropriate. Advantages are low in-process inventory, a supply ability approaching Just-In-Time, improved quality control and high operator efficiency. For the others, the process sets are sufficiently unique to the product that allocation to a cell is justified, despite relatively low utilisation levels.

These are examples of easily identified segments of the total production that are separable and viable in isolation.

Secondly, two general purpose cells, comprising two or three closely coupled operations, were also implemented as trial general purpose systems. Simple as they were, control and efficiency improvements were sufficient to justify pursuit of the approach for a major portion of the operation. Further support came from simulation of more complex cell designs, compiled with the help of early stages of PFA development. Simulation results indicated that Work-In-Progress inventory savings would be of the order of 50%, with a

similar improvement in lead times.

Later in the PFA development, a formal pilot cell was defined, comprising seven machines and handling some 50 parts for 23 assemblies. The cell was supplied with basic components by early production operations which were considered "fixed" as far as a pilot study was concerned. It was thus not as "deep" a slice of the total operation as it might have been.

Implementation of this cell included a thorough overhaul of all equipment, as well as engineering improvements to machines and tools for ease of service, operation and set-up. Press hydraulics were relocated to allow floor level access to valves and filters. Press shut heights were made equal and bolsters were fitted with rollers, T-slots and die-locating pins. Resistance welders were fitted with PLC's to allow preset weld cycle parameters to be selected by production operators, and to monitor the weld process for GO / NO-GO indication.

Pilot Cell Operation
The cell operated successfully for 18 months, during which time it not only performed its essential manufacturing function, but served as a vehicle for promoting the concept within the organisation, and for exploring the many operational issues. Within this period, two-shift operation was commenced, with the introduction of additional parts, and cell management was transferred from the specially established project team to normal Production Control.

In addition to direct production operations (no processes were automatic), the three operators per shift handled all inspection and quality assessment, material handling, packaging, labelling and machine set-up activities. Set-up engineering improvements and reduced inter-process material handling effort offset this expanded work-load.

Much attention was paid to working closely with industrial unions to secure agreement on the expanded roles of shop-floor personnel, and on other work practice changes contributing to productivity, quality and flexibility.

Scheduling
Simulation of the operation of this cell quickly concentrated on the job scheduling aspects. A shop-floor PC-based interactive scheduler is now in use for cell level production control. This system uses standard process data, combined with user entered production volumes, machine availability constraints, operator availability and job priorities. Forward planning for a workload of twenty shifts takes just a few minutes.

In the majority of cases, where the machine complement permits, all operations that are required to produce an end item are set up concurrently. The total process is then proved out prior to commencing full production on this transient "flow-line". Such an approach has very favourable influences on Work-in-Progress(WIP) inventory, manufacturing lead time, scrap generation and quality. It does, however, require more sophisticated scheduling to minimise the impact on machine utilisation.

Benefits

People who are familiar with the literature on this subject are well aware of the large range of possible benefits. This particular implementation supports the claims one hundred percent! Some of the results arising from the revised operating methods are:

- Work-in-Progress within the cell is limited to jobs actually being processed (with operations overlapping).

- Operators handle all Quality Audit, material handling, packaging, and set-up activities. Set up engineering improvements and reduced inter-process material handling effort have offset this.

- Material handling associated with operations now within the cell has been reduced by 70%.

- Scrap levels reduced by 90%

- Set-up engineering improvements are reflected in process consistency.

- Traditionally "difficult" jobs are set up and processed with ease.

- Responsibility for components and assembly eliminates assembly problems.

The constructive attitudes of operators is demonstrated by:

- Acceptance of responsibility that has led to increased pride in always obtaining good product.

- Early detection of requirements for maintenance service and constructive assistance to maintenance engineers (i.e. improved maintenance efficiency).

- Reduction of the requirement for supervision to almost nil after only four weeks' operation.

Many of these benefits could have been achieved to some degree without the physical rearrangement of equipment. It is the opinion of personnel closely related to the project that little would have been achieved without the concentration and focus developed by the cell.

As in the ESPRIT philosophy of Human-Centred CIM, this reorganisation seeks to improve overall operations by capitalising on the human operator's natural aptitude for achieving complex goals within a somewhat unpredictable environment. This is facilitated by first providing adequate resources within a manageable domain, then placing the responsibility for achieving realistic goals squarely on the operators.

COMPLETE FACILITY DESIGN

The design system has been applied to develop a multi-cell arrangement of the total facility. In addition to six existing cells already devoted to very specific part groups, six further cells will complete the restructuring. The new cells each possess approximately fifteen machines, and require from seven to twelve operators, depending on production volumes. Implementation is proceeding on a progressive basis, to be completed by 1990.

CONCLUSION

A technique has been developed for efficiently analysing existing job shop process information to identify characteristic process combinations. An interactive procedure has been developed which uses these combinations as building blocks to design and simulate a number of manufacturing cells that, in combination, can support current and future manufacturing requirements.

The benefits claimed for Cellular Manufacturing have been demonstrated within a pilot installation, and all to a greater degree than had been initially contemplated. The operating improvements have been obtained not through high investment in new equipment, but through a change in methods coupled with a moderate expenditure on process engineering. In addition, though, what amounts to productivity benefits are occurring in indirect areas - maintenance, tooling, production control and material handling external to the cell. Supervisory demands are also minimal, paving the way for a more effective utilisation of management skills within the facility. The approach is an essential part of a broad restructuring exercise in the Plant. It has the constructive support of unions, workforce and management, and has proved itself to be a commercially viable proposition for the subject plant's operations for the foreseeable future.

References.

KING, J.R. and NAKORNCHAI, V. (1982). Machine-component group formation in group technology: review and extension. International Journal of Production Research, Vol. 20, #2.

WEMMERLOV, U and HYER, N. (1987). Research issues in cellular manufacturing. International Journal of Production Research, Vol. 25, #3, pp.413-431.

Acknowledgements.

Acknowledgement is made of the significant contributions to this project by members of the Integrated Manufacture Program (Adelaide), T. D. Seabrook, D. H. Jarvis, R. R. Lamacraft and A. Tharumarajah. Willing contributions by General Motors Holdens personnel have been equally important to the success of this project, and are similarly acknowledged.

PROTOTYPING OF F.M.S. FROM THE DESIGN OF A PREGRAPH BASED ON SOME EXTENDED PETRI-NETS

E. Castelain, J. P. Bourey and J. C. Gentina

*Institut Industriel du Nord, Laboratoire d'Automatique et Informatique Industrielle,
Villeneuve d'Ascq, France*

Abstract. The present paper is concerned with the design of control monitoring systems of Flexible
Manufacturing Systems (FMS) and particularly focuses on the first steps of a computer aided design which aims
to obtain both a formal description of the functional role of the production unit and some results on its dynamical
behaviour.

INTRODUCTION

Many research teams have selected high level or hybrid tools for modelling of FMS (/BAL 87//MAR 87//SAH 87//VIL 88/). The L.A.I.I. research group has decided to use an hybrid model composed of the three following parts to represent a flexible production cells :

- The Control Part (CP) describe by a Structured Adaptive Coloured Petri Net (SAC-PN) which ensures the proper scheduling and coordination of the elementary tasks.

- The Decisional Level (DL) composed of a declarative knowledge-based system intended to solve all the conflicts and indeterminisms bound to the flexible architecture of the Control Part with inference techniques, algorithms and high level decision scheduling.

- A descriptive model of the process, called the Operative Part (OP), to represent both effectors and sensors and all the material devices involved in the the production process (i.e. machines, robots, parts, tools, ...). This model is based on an object-oriented approach.

The Control System (C.S.) is composed of both the Control Part and the Decisional Level models.

The highly sophisticated models of the Control System can only be the results of progressive, modular top-down and computer-aided procedures. This is the reason why we have developed a whole computer-aided design methodology in the context of a larger plan called C.A.S.P.A.I.M. ("Conception Assistée de Systèmes de Production Automatisés pour l'Industrie Manufacturière").

In this paper, we will essentially illustrate the main idea recently developed in C.A.S.P.A.I.M. : according to some extension, the first model called Pregraph should be really efficient to prototype the production and the control structures.

In order to set such a problem in the context of the C.A.S.P.A.I.M. project, we will quickly describe its global frame in the first section. More details are given in /BOU 88a//BOU 88b//KAP 88/ /CRA 89/.

Then we will detail the step concerned with the elaboration of the Pregraph.

At last the extentions of this first model will be presented in order to prototype the F.M.S.

THE C.A.S.P.A.I.M. PLAN

The global frame and design of the control of FMS is composed of five essential phases (fig. 1) :

- The elaboration of the Pregraph. The aims of this step is to provide, from a formalized description of the elementary operations of each production sequence, an intermediate Coloured Petri-Net (called Pregraph) intended to be a consistent, complete and concise support for the further structured development of the Control Part. This step is detailed in section II.

- The Predimensionment Phase. As the Pregraph also recapitulates the functional role and the different routings of all the parts, a possible use of this tool according to several extensions of the model, can be expected in a prototyping phase of the material production architecture and in a preliminary evaluation of dynamic behaviour. This point will be discussed in section III.

- Elaboration of the three detailed models . The detailed models of the Operative - Control -, and Decisionals Parts are generated according to more precise specifications interactively supplied by the user.

The detailed model of the control part is obtained by a method which is similar to a compilation process. This development step translate the Pregraph structure into **Structured** Petri-net modules linked together. This class of PN which is more restrictive but more systemetic than "ordinary" PN was introduced in order :
(i) to give the properties, such as modularity, maintenability, ..., of the structured programming languages to the PN model
(ii) to facilitate the translation of the model of the control part into a code which is implementable on the control devices (Industrial Programmable Controller for example)
(iii) to show the elementary processes and links between them.

The main idea of the Structured development is to create a Structured PN module for each place of the pregraph and an elementary structured PN process to model the transfer of parts in the production system (Fig 2)

This step is more precisely described in previous publications /BOU 86/ /BOU 87/ /BOU 88b/ /CAS 87/ /KAP 87/.

- Detailed simulation /CAS 87/. The aim of this step is to provide qualitative and quantitative results on the detailed dynamic behaviour of the system.

- Implementation /CRA 89/. The problems are tackled which are bound to the implementation of the Control System.More precisely, constraints of implementation are taken into account both for the C.P. level by the choice of an implementation language (Grafcet by example) and for the D.L. .Hardware constraints such as distribution of the Control, choice of the Local Area Network, ..., should be also considered. The models generated and used for the description of the Control Part and the Decisional Level were proving well adapted to solve these problems in the framework of distributed Control System.

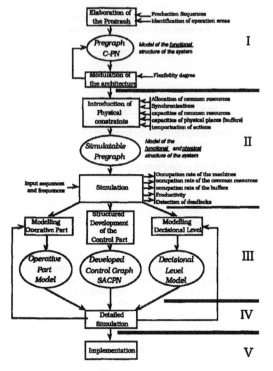

Fig. 1.

In order to see how the prototyping is performed in C.A.S.P.A.I.M. The first and second phases will be investigated now.

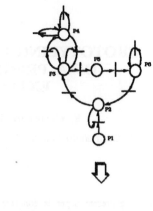

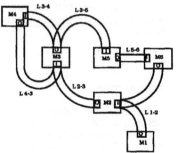

Mi = structured, coloured, adaptive Petri-net module representing the control process associated with th place Pi of the pregraph
Li-j = structured, coloured, adaptive Petri-net module corresponding to a control process coordinated by the control modules Mi and Mj

Fig. 2.

ELABORATION OF THE PREGRAPH

The first point consists in building a description of the functional[1] purpose of the production unit from operative rules describing the elementary steps of production operative sequences. Production rules are used as description tools, according to the following pattern :

$$premises \rightarrow results$$

where premises and results are composed of predicates. Some of these predicates must necessary have the following type :

$$presence (part , place) or presence (object , place)$$

which is for the moment the only kind of predicate that the design system needs to elaborate a basic version of the control graph of the production unit.

Four types of elementary rules are proposed to build a description of production operative sequences. The user may build variant and hybrid forms, or more complex rules.

However, the presented rules fulfil the requirements of various descriptions. As far as possible, the user is advised to confine the decomposition of the operative processes to these forms. An extra piece of information is sometimes added to the rules in order to remove any ambiguity in the further treatment, when building the final control system (for example, we must set apart machining from repositioning which are often described with the same rule frame). In order to give an accurate representation of the functional role of the production unit, we use this extra information to distinguish :

i) transformational actions (machinings, physical transformations) also called functional actions
ii) positional actions (transfers, positionings)
iii) informational actions (measures)

The four types of rules are the following :

Type 1 rule

presence (part, place 1) → presence (part, place 2)
type of action = positionnal
use = transfer, positioning-transfer
examples = conveyor, robot arm, Automated Guided Vehicle

Type 2 rule

presence (part1, place) → presence (part2, place)
type of action = transformational, positional, informational
use= physical transformations
examples = machinings, assembly, thermic treatments,
positionings, measures

Type 3 rule

presence (part 1, place 1) AND presence (part 2, place 2)
→ presence (part 3, place 3)
with eventually place1 = place 3
type of action = transformational, positional
use = agglomeration
examples = assembly positioning, palletization

Type 4 rule

presence (part 1, place 1) → presence (part 2, place 2)
AND presence (part 3, place 3)
with eventually place1 = place 3
type of action = transformational
examples = machinings with reprocessing of shavings,
depalletization,...

We give below two examples of variant or hybrid rules :

- Transfer and palletization (type 3, Fig 4a)

presence(part, place 1) AND presence (fixture, place 2)
→presence (part-fixture, place 2)
type of action = positional, use = assembly.

- Loosening from a machining platen - transfer - palletization
(hybrid rule, Fig 4b)

presence(part-platen, place 1) AND presence (fixture, place 2)
→ presence (platen, place 1) AND
presence (part-fixture, place 2)
type = positional, use = disassembly-assembly.

Thus, the user is required to select information of first importance among the requirements of the manufacturing file, and then to build a first functionnal synopsis of the operative sequences of the production unit. The other specifications of the manufacturing file will also be detailed in a progressive and structured way, at the right point to occur, in the further steps of design. This method avoids the mixing of the different kind of requirements.

Of course, the whole work of description of the operative sequences is completely aided by a software which is called "Completeness and Consistency Analyser". Its functioning is based on two modes : recapitulation and adjustment.

The software aids the reasoning of the conceptor while building step by step the production rules that describe the operative sequences of the production unit. It avoids omissions, description deficiencies about rules and initial facts (i.e. starting points of an operative sequence, for example presence (raw-part, input-buffer)) effectively used, final facts (i.e. final points of an operative sequence, for example presence (manufactured-part, output-buffer)) and the link between a possible progressively determined set of initial and final facts.

This software uses an hybrid reasoning made of mutually cooperating inferences of both forward and backward chaining. It is intended to clarify all the logical chaining and reasoning deficiencies, (for example omission of rules, initial or final facts, useless rules or facts, ...) to make up a consistent and complete description. Thus, an aided questionning of the user is performed until a rigorous and satisfying definition of the operative sequences is achieved. The task is over when all the production flows are defined, according to the user's wishes and to the conditions of completeness and consistency.

Then a translation and an interactive aggregation of the production rules are performed from the settled description to a coloured Petri-Net which we call the Pregraph

The respective Coloured Petri-Nets translations which correspond to the four elementary production rules are given in fig 3.

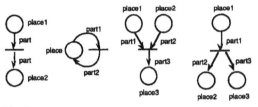

Fig. 3.

The coloured Petri-net corresponding to the palettization and hybrid rules are given fig 4.

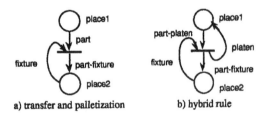

a) transfer and palletization   b) hybrid rule

Fig. 4.

A third step consists in modulating the previous description according to required flexibility level, to the specifications or to the user's wishes. This interactive modulation does not alter the consistency and the completeness of the description. At present, the modulation essentially focuses on :

- multiplication of operating places (i.e. machines, assembly stations, ..)
- multiplication of blocks, that is to say sets of associated places like lathe with palletization station
- addition of intermediate places such waiting buffers
- effective permutation of operations.

The arcs between the places are interactively duplicated, erased or modified. The user is called upon as little as possible and the risk of forgetting a detail of description is considerably reduced.

Thus the pregraph is a Petri-Net aggregated description of the functional role of the operative sequences including a maximum flexibility- and parallelism- level. The obtained model gives only the structural skeleton of both the production and the control. It could only be used to specify the scheduling of each part. In order to describe the paralell control and dynamic behaviour of the system, several developments should be performed (step 3). The Pregraph was proving a complete, consistent and concise support for the further steps of the design. In the following section, we will present some extentions of the Pregraph model for prototyping by means of a simulation.

PROTOTYPING OF FMS

Introduction : Why to use the Pregraph ?

In the original frame of the CASPAIM project, evaluation of the dynamical behaviour of the conceived Control System was completely made in the simulation phase (phase IV on figure 1). This simulation is carried out by a simulator which is able to interpret the hybrid model of the system : the SAC-PN of the CP, the production rules of the DL an the OP model. It enables anomalies of modelling to be detected and the different choices specially in the DL definition to be validated. It was also used to get the dimension of variable elements of the Operative Part (including machines, transportation and handling systems, buffers and storage areas) with respect to a modification - simulation loop. With this method, every time a redimensionment is performed, the SAC-PN of the CP has to be created again and the production rules of the DL have to be modified. Furthermore, due to a very detailed and therefore very large model of the production system, this simulation takes very much execution time. For these reasons we wanted to introduce a first level of dimensionment in order to avoid to develop too early the global model of the system. Hence, we decided to dimensionate the operative system as soon as possible in the design frame of CASPAIM by introducing analytical and simulation methods /CAV 87/ at the level of the Pregraph (phase II on figure 1).

Study of the "steady state"

First, let us say that flexible systems often work in transitory conditions because of the small or medium size of the possible production series. The steady state functioning corresponds to a large production of the same ratios of manufactured parts. The first idea is that the performances of any transitory state are linked to the performances of this steady state. In fact, this steady state is the asymptotic limit (figure 5) which can never be reached because of the "perturbations" which occur in the real behaviour. Breakdowns, delays, variations of the flow of parts entering the system (with regard to the steady state flow) imply that the production "response" will never reach the steady state.

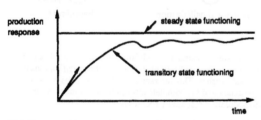

Fig. 5.

The first work then consists in correctly dimensioning the steady state functioning from an adjustement of the production architecture. The next step will then consist in a correct parameterization of the system in order to be as close as possible of this steady state (figure 6).

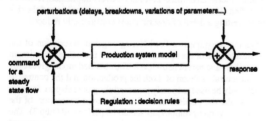

Fig. 6.

The study of the steady state is simply based on a computation of maximum flows through a net with the hypothesis that the flow is conservative (steady state hypothesis). This computation can easily be solved by Linear Programming methods. As the Pregraph recapitulates the different routings of all the parts and all the actions of their production sequences, this computation can be made on the Pregraph. It just needs to define the input ratios , the duration of actions and the way that physical resources are distributed (several transfers, for example, are performed by a same robot).

At this phase, bottlenecks can be put in a prominent position since they determines the maximum possible flow. So, if this flow does not satisfy the production purposes, the bottlenecks have to be resorbed : a second machine is added to increase the production capacity, a supplementary robot is added if the bottleneck is due to transport means.

Let us notice that the various storage areas such as the input and output buffers of the machine, the intermediate storage areas, have no effect upon the calculated flow because of the steady state hypothesis. This will be seen in the following step.

How to approach the steady state functioning

This second step consists in providing good properties to the real behaviour of the system (which is in fact a succession of transitory states) in order to approach as close as possible the asymptotic steady state functioning. This can be achieved owing to two different ways :

First, the size of the storage areas (the queues of the system) can be optimized to absorb variations with respect to the steady state. The size of these queues has to be precisely defined. If a queue is too small, either no-input events may occur on downstream machines (ie. no part to machine), or full-output events may occur on up-stream machines (ie. a part cannot leave a machine because its destination queue is full). On the contrary, if queues are too large, saturation phenomena may occur. In all cases, the performances of the system are degraded.

Besides this static optimization, good dynamic properties can be provided owing to a regulation loop made by means of decision rule (figure 6). The purpose of these rules is to solve undeterminisms and conflicts, to realize a good production planning in order to be not far from the steady state. For example, such a rule could be : " It is better to give priority to a part which is nearer the end of its production sequence because it avoids saturation phenomena ". Let us notice that these rules are included in the Decisional level when the 3 detailed models are developped (phase III on figure 1).

Why to simulate the Pregraph ?

Statistic analitycal methods (such as the analytical solvers proposed in CAN-Q or QNAP 2 /CAV 87/) based on queue net theory can be used to undertake the optimization of FMS. In these methods, approximations are made to make the model able to be solved, and generally, the refinement due to the limits of the analytical computation is made by simulation. We have not yet introduced this kind of analytical methods in CASPAIM and we directly use simulation. Rather than to use a specific simulation language like QNAP 2, RESQ or SLAM 2, we have prefered to develop a specific simulator based an extended model derived from the Pregraph and to perform the optimization with a simulation/modification loop. The reasons of this choice are the following :

- specific simulation languages cannot take into consideration synchronism and competitive phenomena as easily as a PN model

- these languages requires a very specific kind of modelling which cannot be easily integrated in CASPAIM plan. Let us recall CASPAIM tries to garantee the adequation between the specifications and the implementation and so the intermediate models must be quite homogenous.

The model of the simulatable pregraph

Let us recall that the original Pregraph is only a functional description model of the production sequences which is used to automatically develop the SAC-PN of the CP (phase III on figure 1). As it is just descriptive, it is impossible to directly perform a qualitative or quantitative analysis of the original Pregraph with simulation means. So the Pregraph has to be extented to be simulatable in the sense of parallel behaviour analysis. We are going to present these extensions made on the Coloured PN of the Pregraph to obtain the non-coloured, Stochastic, Capacity and Temporized PN of the simulatable pregraph. Each extension is introduced to model some specific aspect of either the OP, or the CP, or the DL and so to obtain a quite autonomous model.

Places and transitions with restricted capacities

To each place of the net, a capacity is assigned to indicate the maximum number of tokens this place may contain. This is the classical notion of Capacity PN which is introduced here to indicate that a buffer or a storage area of the OP can only contains a limited number of parts.

Each transition of the pregraph represents a tranfert or a transformation. We have introduced the notion of capacity associated to a transition to model the limited capacity of transportation or transformation means. So, the capacity of a transition indicates the maximum number of simultaneous firing of this transition.

Macro places

In the original Pregraph, each place represents one production unit (machine, stock,...) and can contain a same part in several states of its production sequence (figure 7 a).

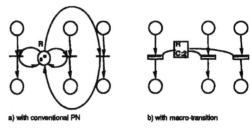

a) with conventional PN   b) with macro-transition

Fig. 8.

Temporization of the arcs

Delays on places or on transitions are usually used to model the dynamic behaviour of the O.P.. As the simulatable pregraph can be assimilated to an aggregated view of the C.P. graph, we needed a more complex way to temporize the actions. This temporization must be compatible with the capacity mechanism and with the control process which will be generated by structured development. So we choose to temporize each arc of a transition as described on fig. 9.

Let us notice that a place is occupied even when it contains a non available token and a transition is "occupied" until there are no longer an unvialable tokens in the connected places. This temporization mechanism is specially adapted when there are several arcs forward and backward the transition (assembly, disassembly, ...).

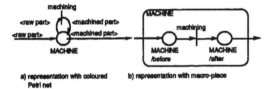

a) representation with coloured   b) representation with macro-place
Petri net

Fig. 7.

To be simulated, such places had to be developed. In the simulatable pregraph, each place corresponds to a special state of a part in a particular production unit (figure 7 b). So the notion of macro place (with a capacity) has to be introduced to model a physical position. The capacity of a macro place indicates the maximum number of parts that can be at the same time in a same physical place. In the example of figure 7, the capacity of the macro place "machine" is 1 to indicate only one part can be on the machining area. The sum of the occupations of the places included in the macro place must never exceed its capacity.

Macro transitions

For the representation of common resources, like transportation systems, which are used for different transfers, mutual exclusions for the concerned transitions have to be realized. Using conventional PN, this could be done by the structure shown in figure 8 a. We abbreviate this structure with the notion of macro transition (figure 8 b).

The capacity of a macro transition is the number of tokens contained by the critical resource (R). It indicates the maximum number of transitions assigned to the macro transition that can be simultaneously fired.

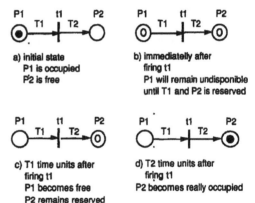

a) initial state
P1 is occupied
P2 is free

b) immediately after
firing t1
P1 will remain undisponible
until T1 and P2 is reserved

c) T1 time units after
firing t1
P1 becomes free
P2 remains reserved

d) T2 time units after
firing t1
P2 becomes really occupied

Fig. 9.

Stochastic decisions

The structured developed graph of the C.P. is deterministic. All the indeterminisms are solved either by the coloration or by the D.L.. As the simulatable pregraph is uncoloured and as D.L. is not defined at this point of design, all the indeterminisms have been solved owing to stochastic laws based on steady state calculation results.

Simulation.

Owing to the previous extension, the extended pregraph becomes a possible model of evaluation by simulation. This model is almost autonomous since only the part flow entering the system is missed. The input flow is emulated by a stochastic function which satisfy the production ratios and the flow of real parts from the precedent production units.

Then this complete model is imulated by an adaptation of the PN simulator which is used to simulate the developped model (phase

IV on figure 1). This simulation consists in an interpretation of the graph both by the firing of its transitions (evolution by activity) and by delayed events memorized in a bill-book (evolution by events).

Simulation gives two kinds of results.

At first, it gives the following statistical information :

- the occupation rate of production resources (machines, robot, ...)

- the occupation rate of intermediate buffer areas and the average number of parts laying there

- the average production time of a part,...

This statistical information can be then analyzed in order to modify the model and to optimize a set of chosen criteria owing to a modification/simulation loop.

Secondly, the experience showed that this simulation method is also an excellent mean to detect and analyse deadlocks situations. This is due to the large number of permutations of the system state generated by stochactic decisions. For each kind of deadlocks, the simulation can give the precise state which has lead to such a situation and statistical information about the frequency of apparition. This information is very useful to define the rules of the D.L. the purpose of which is to prevent these deadlocks.

CONCLUSION

The Pregraph is then an essential tool in the CASPAIM design frame.

At first, it provides a solid basis (consistent, complete and concise) from which the detailed and structured model of the Control Part can be systematically designed.

Secondly, the simulation of the pregraph is a very useful step in the CASPAIM project. It allows to get :

- quantitative information about the behaviour of the system

- a fast prototyping of the operative system

- qualitative information on the potential deadlocks

Then, the definition of the Decisional Level rules will be easier.

The prospective developments in the C.A.S.P.A.I.M. project are to integrate in this simulation model a minimal Decisional Level.

The purpose of this minimal DL is firstly to avoid deadlocks and secondly to optimize at this step, the strategies which must be developped to regulate the behaviour of the system (figure 6). This minimal DL will be the skeleton of the developped DL in the same way that the Pregraph is the skeleton of the developped SAC-PN of the Control system.

REFERENCES

/BAL 87/
Balbo, G., G. Chiola, G. Franceschinis and G. Molinar-Roet (1987). Generalized stochastic Petri nets for the performance evaluation of FMS. IEEE International conference on robotics and automation, RALEIGH (USA), pp. 1013-1018.

/MAR 87/
Martinez, J., P. Muro and M. Silva (1987). Modelling, validation and software implementation of production systems using high level Petri-nets. IEEE International conference on robotics and automation, RALEIGH (USA), pp. 1180-1185.

/SAH 87/
Sahraoui, A., H. Atabakhche, M. Courvoisier and R.Valette (1987). Joining Petri-Nets and knowledge based systems for monitoring purposes. IIEE International conference on robotics and automation, RALEIGH (USA), pp. 1160-1165.

/VIL 88/
Villarroel, J.L., J. Martinez and M. Silva (1988). GRAMAN : a graphic system for manufacturing system design. IMACS SMS'88, International Symposium on System modelling and Simulation, CETRARO (Italie), pp. 79-84.

/BOU 88a/
Bourey, J.P., E. Castelain, J.C. Gentina and M. Kapusta (1988a). C.A.S.P.A.I.M. : a computer aided design of the control of F.M.S.. 12th IMACS World Congress'88, PARIS, Vol. 3, pp. 517-521.

/BOU 88b/
Bourey, J.P. (1988b). Structuration de la partie procédurale du système de commande de cellules de production flexibles dans l'industrie manufacturière. Thèse de Doctorat d'Université de USTLFA, LILLE (France).

/KAP 88/
Kapusta, M. (1988). Génération assistée d'un graphe fonctionnel destiné à l'élaboration structurée du modèle de la partie commande pour les cellules de production flexibles dans l'industrie manufacturière. Thèse de Doctorat d'Université de USTLFA, LILLE (France).

/CRA 89/
Craye, E. (1989). De la modélisation à l'implantation automatisée de la commande hiérarchisée de cellules de de production flexibles dans l'industrie manufacturière. Thèse de Doctorat d'Université de USTLFA, LILLE (France).

/BOU 86/
Bourey, J.P., D. Corbeel, E. Craye and J.C. Gentina (1986). Adaptive and coloured structured Petri nets for description, analysis and synthesis of hierarchical control end reliability of flexible cells in manufacturing systems. 1rst European Workshop on fault diagnosics, reliability and related knowledge based approaches, RHODES, Vol. 1, pp. 281-295, D.Reidel Pub. Comp., 1987.

/BOU 87/
Bourey, J.P. and J.C. Gentina (1987). Computer aided design for structuration and representation of control of flexible manufacturing systems. 8th European Workshop on Application and Theory of Petri-Nets, ZARAGOSSA (Espagne), Proc. pp. 117-135.

/CAS 87/
Castelain, E. (1987). Modélisation et simulation interactive de cellules de production flexibles dans l'industrie manufac- turière. Thèse de Doctorat d'Université de USTLFA, LILLE (France).

/KAP 87/
Kapusta, M. and J.C. Gentina (1987). Introduction to a first step of the aided design of control system of flexible manufacturing cells. IEEE Montech'87 / Compint'87, MONTREAL, pp. 258- 262.

/CAV 87/
Cavaille, J.B. and J.M. Proth (1987). Pratique de la simulation en production discontinue. Conférences SIPRODIS, EC2 Edition, Colloques et Conseil, Novotique.

AN ULTRASONIC PHASED-ARRAY-SENSOR
FOR ROBOT ENVIRONMENT MODELLING
AND FAST DETECTION OF COLLISION
POSSIBILITY

L. Vietze and I. Hartmann

*Technical University Berlin, Fachgebiet Regelungstechnik und Systemdynanik, Berlin,
FRG*

__Abstract__ : An ultrasonic "8x8 phased–array–sensor" operating with a frequency of 40 kHz is
used for modeling the environment of the mobile robot. Sharp formed beams can be sent to
arbitrary directions. To determine the axial distances of the unknown obstacles the returned
signal is correlated with the so called "normalized echo" of a plane plate. Even with sidelobes
less than –10 dB of the steerable mainlobe the reflections from sidelobes could be bigger than
reflections from the mainlobe. As shown it is still possible, though, to determine the polar–
coordinates of the obstacle. Next a fast collision avoidance strategie is presented. The basic
idea to speed up processing is to concentrate on dangerous obstacles by classifying four different
"danger–classes". The axial speed components can be measured by the doppler shifted signals
and added as an additional information to the "danger– classes".

__Keywords__ : Ultrasonic transducers; phased–array; modeling; robots; automation.

INTRODUCTION

The demand for greater flexibility in industrial production
calls for more intelligent robots which are able to alter the
off–line predetermined proceedings. The modeling of the
robot environment is needed to obtain the information
about collision detection, navigation and the quality
control. To do all these tasks, a fast robot environment
modelling system and a decision system is necesessary.
The use of an ultrasonic phased–array–sensor for
environment modeling and fast collision detection will be
described. At first the ultrasonic phased–array–sensor and
the environment modeling will be explained, the collision
detection will be shown later on.

PHASED–ARRAY–SENSOR

The basic idea of the phased–array–sensor, well known
from modern radar technology, is to drive an array of
transmitters with different phase shifted signals on the
same frequency. By this way sharp formed beams can be
sent to arbitrary directions without any mechanically
moving parts. To obtain a steerable beam in the range of ±
60° with small sidelobes the distance between the
transmitters has to be smaller than half the wavelength
which could be shown theoretically or by simulation
(Käs 1981). Increasing the number of transmitters of the
phased–array– sensor improves the beamforming and decreases the sidelobes. The quality of the steerable
beamforming phased–array–sensor could be estimated by
comparing the width of the main beam angle and the
power and angle of the sidelobes for different layouts.

Simulations showed a practical solution for an 8x8
transmitter array with a distance of half the wavelength.
The transmitters of the constructed phased–array–sensor
have a center frequency of 40 kHz (the wavelength in air is
about 8.6 mm), the distance between the transmitters is
4 mm. The designed logical–hardware is controlled by a
TS 800–transputersystem. As shown in Fig. 1 the designed
hardware receives data from the transputer before sending
the ultrasonic pulse.

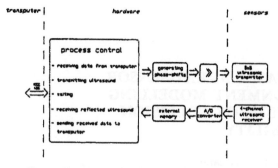

Fig. 1 : Structure of the electronical
phased–array–sensor hardware

By this way

- the sending direction of the pulsed beam,
- the length of the pulse,
- the waiting time before receiving,
- the length of the receiving time,
- the sample rate of the A/D–Converter
  (up to 660 kHz) and
- the amplifying factor for the returned signal

will be adjusted. The pulsed beam can be steered with a resolution of 256 steps so that the smallest steerable angle is about half a degree. Fig. 2 presents a measured radiaton–characteristic for a steered beam of the 40kHz 8x8 ultrasonic phased–array–transmitter.

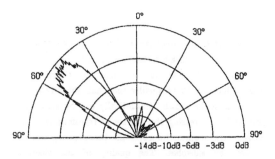

Fig. 2 : Radiation characteristic of 40kHz
8x8 transmitter

The returned signal is amplified, converted into the digital form and stored in an external RAM memory. After receiving the whole signal the received data is sent to the transputersystem and a new measurement begins. If the approximate distance to an obstacle is known and large enough, the time of flight of the new measurement could be used for sending the received data from the last measurement to the transputer.

After having filtered the signal by an adjustable digital

bandpassfilter the signal is filtered by a direction–filter. Instead of using a phased–array–receiver only four receivers in a distance of ten wavelenghts to each other are used. This way the hardware costs have been reduced. By adding the four signals of the four receivers with a definite time shift (not only a phase shift) it can be shown that the signal to noise ratio is getting better instead of using only one receiver.

3–D ENVIRONMENT–MODULATION

The ultrasonic–sensor sends a short ultrasonic pulse to a selected direction. In case of any obstacle an echo will be returned and digitalized. After having filtered the signal the time of flight and the frequency–shift due to the Doppler–effect will be determined. With a known velocity of sound the radial distance of the measured obstacle is calculated. As it is well known, the velocity of sound depends on the temperature of the air and the atmospheric pressure. Instead of measuring both of them or using another ultrasonic–sensor for velocity determination the same phased–array–sensor is used to measure the speed of sound. By using an angle of 60^0 the time of flight to a definite objekt fixed on the mobile robot is measured. To obtain a good accuracy of this measurement a cross–correlation with the known undisturbed echo will be done. This kind of time of flight measurement will be done every few seconds to have the right value of the velocity of sound constantly.

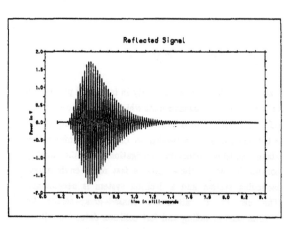

Fig. 3 : Reflection on a large even plate
(normalized echo $e_0(t)$)

The reflection of a 20 amplitude long ultrasonic burst on a large smooth and plane surface is shown in Fig. 3 and is called the normalized echo $e_0(t)$. As known, diffuse reflections on smooth surfaces like in industrial environment can be neglected. In this way the received signal could be understood as an ensemble of echoes from even reflecting

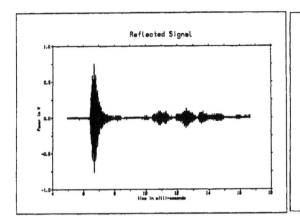

Fig. 4 : Reflected signal $e_0(t)$

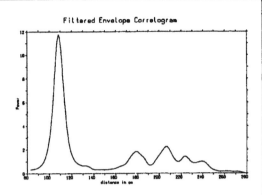

Fig. 6 : Envelope of the cross–correlation
$e(t)$ with $e_0(t)$

plates. The received echo $e(t)$ can be written approximately as a superposition of p normalized echos $e_0(t)$ with different time of flights t_i and power coefficents a_i to

$$e(t) = \sum_{t=0}^{p} a_i \cdot e_0(t{-}t_i) \qquad (1)$$

The power coefficent a_i could be calculated with the reflection coefficent a_{ri} of the reflecting obstacle and the radiation characteristic into the obstacle direction a_{ci} by

$$a_i = a_{ri} \cdot a_{ci} \qquad (2)$$

Note that the reflection coefficent a_{ri} does not change while scanning stationary obstacles with a stationary sensor. To measure the axial distance of the different objects even in noisy industrial environment cross–correlation with the normalized echo $e_0(t)$ will be done (Fig. 4). At the maximums of the correlation factor the radial distance to the objects can be measured if the differences of the axial distances are large compared with the wavelength (Fig. 5; Fig. 6).

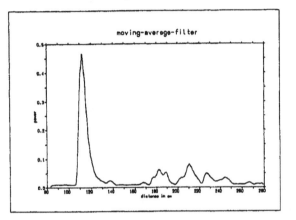

Fig. 7 : Moving–average–filter $\bar{e}(t)$

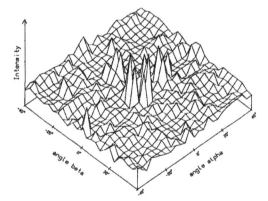

Fig. 8 : Moving–average–filter concerning
all sending angles for the closest object

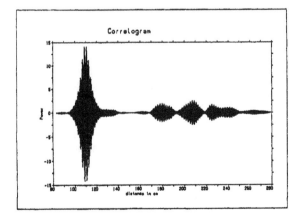

Fig. 5 : Cross–correlation $e(t)$ with $e_0(t)$

In case the reflection coefficient a_{ri} of the obstacle i is big even sidelobes less than −10dB of the mainlobe could produce bigger reflections than the mainlobe. To obtain the accurate direction of the obstacle i the maximums of the correlation–curve could be plotted on the two sending angles. The peak on this plot would give the information about the direction of the obstacle. Instead of calculating the cross–correlation for all the sending directions a moving–average filter

$$\tilde{e}_i = \frac{1}{2 \cdot n} \sum_{j=i-n}^{i+n} \left| e_j \right| \qquad (3)$$

is used. The direction of the obstacle is obtained by plotting the peaks of one time t_i on the two sending angles. By these means calculation time can be saved. For the closest obstacle (the first peak in Fig. 7) the obstacle direction can be seen in Fig. 8 .

FAST COLLISION POSSIBILITY DETECTION

With the knowledge of the current velocity of sound the 3–dimensional position of an obstacle is determined by the two sending angles α and β and the time of flight (Auer 1986), (Löschberger, Magori 1987). The knowledge about the actual speed components of the objects is the next important information for modelling the robot environment.

To calculate the future trajectory of the own robot the starting place and the speed and acceleration components for the next future have to be known. The 3–D–coordinates of the robot $\underline{x}_r(t)$ for any future time t can be calculated by the equations

$$\underline{\dot{x}}_r(t) = \underline{A}\ \underline{x}_r(t) \qquad (4.1)$$

$$\underline{y}_r(t) = \underline{C}\ \underline{x}_r(t) \qquad (4.2)$$

with $\underline{x}_r \in \mathbb{R}^9$. With the speed components v_{ix}, v_{iy} and v_{iz} and the assumption of constant accelerations a_{ix}, a_{iy} and a_{iz} over the time interval $\left[t_i - \Delta t_i\ ,\ t_i \right]$ the 3–D–coordinates can be calculated by

$$x_r(t_i) = x_r(t_i - \Delta t_i) + v_{ix} \cdot (t_i - \Delta t_i) + \frac{1}{2} \cdot a_{ix} \cdot (\Delta t_i)^2 \qquad (5.1)$$

$$y_r(t_i) = y_r(t_i - \Delta t_i) + v_{iy} \cdot (t_i - \Delta t_i) + \frac{1}{2} \cdot a_{iy} \cdot (\Delta t_i)^2 \qquad (5.2)$$

$$z_r(t_i) = z_r(t_i - \Delta t_i) + v_{iz} \cdot (t_i - \Delta t_i) + \frac{1}{2} \cdot a_{iz} \cdot (\Delta t_i)^2 \qquad . \qquad (5.3)$$

The moving or standing obstacles with the number j at the time t_i have the coordinates $x_j(t_i)$, $y_j(t_i)$ and $z_j(t_i)$. A collision would occur if the distance $r_j(t_i)$ were be smaller than the so called security distance. The distance $r_j(t_i)$

can be calculated by the equation

$$r_j^2(t_i) = (x_r(t_i) - x_j(t_i))^2 + (y_r(t_i) - y_j(t_i))^2 + (z_r(t_i) - z_j(t_i))^2 \qquad (6)$$

To solve this equation the exact speed components of all the obstacles have to be known. In practice a lot of standing and moving obstacles like machines, boxes, other robots or humans are detected. To reduce the calculation and measuring costs four danger– classes of collision will be defined:

class 1: moving or standing obstacles
that would collide with the
own robot,

class 2: moving obstacles which could
cross the trajectory of the own
robot to an unknown time,

class 3: moving and standing obstacles
which do not collide with the
own robot.

The possibility of collision increases the lower the danger–class is. By these means the measuring and calculation times can be used better for the "dangerous" obstacles. If an object is of class 1 the trajectory of the own robot has to be changed, solutions for planning a collision free trajectory can be found in (Gerke 1985). Obstacles of class 2 are moving with unknown or not completely known speed components. If the speed components of the obstacle are known better, the obstacle can be classified to class 1 or to class 3 . This model of danger–classes will fit to any sensorsystem, but the use of an ultrasonic sensor speeds up the classification by using the additional information about the axial speed components.

The mobile robot moves along his path with the speed

$$\vec{v} = \begin{pmatrix} v_x \\ v_y \\ v_z \end{pmatrix} \qquad (7)$$

The axial speed $\vec{u}_a$ of the obstacle moving with the speed $\vec{u}$ could be determined by measuring the doppler shift of the returned signal (Seto 1971).

The doppler shift Δf of the obstacle is given by the sending frequency f_s and the sending angles α, β :

α : projection of the ultrasonic beam in
the x,y–plane,

β : direction of the ultrasonic beam
vertical to the x,y–plane,

to

$$\Delta f = f_s \cdot \left[\frac{c - v_a(\alpha,\beta)}{c + u_a(\alpha,\beta)} \cdot \frac{c + u_a(\alpha,\beta)}{c - v_a(\alpha,\beta)} - 1 \right] \qquad (8)$$

The speed components $v(\alpha,\beta)$ and $u(\alpha,\beta)$ into the beam (axial) direction can be described by

$$v_a(\alpha,\beta) = v_x \cdot \cos\alpha \cdot \cos\beta + v_y \cdot \sin\alpha \cdot \cos\beta + v_z \cdot \sin\beta \qquad (9)$$

$$u_a(\alpha,\beta) = u_x \cdot \cos\alpha \cdot \cos\beta + u_y \cdot \sin\alpha \cdot \cos\beta + u_z \cdot \sin\beta \quad . \qquad (10)$$

The Eq. (8) results in

$$\Delta f = 2 \cdot f_s \cdot \left[\frac{u_a - v_a}{c + v_a - u_a - (u_a \cdot v_a)/c} \right] \quad . \qquad (11)$$

In a first approximation the speed of the own robot and the moving obstacle is low in comparision with the velocity of sound. The axial speed towards the sensor can than be expressed by

$$u_a = \frac{\Delta f}{2 \cdot f_s} \cdot c - v_a \qquad (12)$$

Of course there are three possibilities of speed of the obstacle:

class A: the axial speed of the obstacle
is positive,
class B: the axial speed of the obstacle
is negative,
class C: no move in axial direction.

It is interesting to have a look at Fig. 9 to the different possibilities of moving obstacles. Box 1 has no axial speed toward the sensor. Of course, nothing can be said about the lateral speed component of this obstacle. So, after the first measurement the most unfavorable case is assumed, namely that the obstacle (box 1) will move in the tangential direction. If this tangential direction is not crossing the trajectory of the own robot in the next time the obstacle can be put to class 3C. If the tangential direction were crossing the trajectory of the own robot, the obstacle would be classified as 2C.

The robot 2 in the Fig. 9 has a speed measured negative in the axial direction. The earliest possibility of collision would result from a high tangential speed towards the trajectory. If the tangential line does not cross the

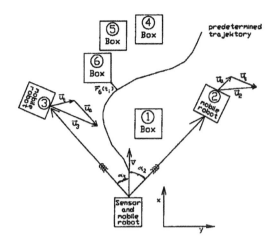

Fig. 9 : Collision detection in practice

trajectory in the next future, it can be put to class 3B. If the tangential line is crossing, the obstacle is to be put to class 2B.

The remaining possibility as seen in robot 3 is that there is an axial speed toward the sensor. In that case the obstruction has to be classified 2A anyway.

As long as an obstacle is graded class 2A, 2B or 2C it has to be watched over for clear classification to classes 1 or 3. As second step the transversal motion can be predicted with the Kalman-filter if the measurment equation

$$\underline{y}_j(t_i) = \underline{C} \cdot \underline{x}_j(t_i) + n(t_i) \qquad (13)$$

will be used. But already the previous informations about the axial speed components help to speed up the classification.

CONCLUSION

The ultrasonic phased-array-sensor presented is able to model the industrial environment and to detect the possibility of collision. Further investigations will be carried out to improve accuracy, to make the sensor more noise resistent and to speed up calculation time.

REFERENCES

Auer,B.(1986). Bildkonstruktion kleiner
Körper durch digitale Verarbeitung
reflektierter Ultraschallsignale, TU Berlin 1986.

Gerke, W.(1985). Die dynamische Pro—
        grammierung zur Planung kollisions—
        freier Bahnen der Industrieroboter.
        Robotersysteme Bd.1 43—52
        Springer Verlag Berlin
Käs, G.(1981). Radartechnik.
        Expert Verlag Grafenau/Württ.
Löschberger, J., Magori,V.(1987).
        Ultrasonic Robotic Sensor with
        lateral Resolution.
        IEEE Ultrasonic Symposium 1987
        14.10—16.10.1987 Denver, Co., USA
Seto, W.W.(1971). Theory and Problems
        of Acoustics. McGraw—Hill New York

MODULAR SIMULATION MODEL OF ROBOT CELLS

H. J. Warnecke*, K. H. Kuk**, H. S. Cho***

*Fraunhofer-Institut fuer Produktionstechnik und Automatisierung, Stuttgart, FRG
**Robotics Laboratory, Korea Institute of Machinery & Metals, Kyungnam, Korea
***Department of Production Engineering, Korea Advanced Institute of Science
& Technology, Seoul, Korea

Abstract. In this paper, a model for the simulation of the material flow not
only inside a robot cell with flexible handling sequence but also between
robot cells is presented. A method for the connection of special simulation
programs for the production subsystems has been developed.
For the systematic development of simulation programs a logic models is
employed, which links a real system and a simulation model.

Keywords. Simulation Models ; Programming ; Robot Cells ; Entity/Relation-
ship-Model; Petri-Net.

INTRODUCTION

A recent trend of production automation is a
building of robot cells. Robots in these cells are
used for handling and transport tasks. Although
robots have been used for several years, robots are
mainly used for long-term repetition of relatively
simple tasks. The current manufacturing environment, small batch manufacturing, has caused the
robot cell to increase the flexibility.

The feature of the flexible robot cells is the
variable sequence of robot tasks. In order to minimize the investment risks for such robot cells
and to install these cells step by step, a simulation model is required, which can model the
relations not only in a flexible robot cell but
also between the cells.
In this paper, a simulation model with such function is developed for robot user and planners of
robot application. The feature of the developed
model is that the model can be run on a simple
microcomputer so that it has a good application
chance. Besides, the simulation model can be
easily modificable for the control of robot cells.

PRESENT SITUATION

So far as we know, a special simulation model for a
flexible robot cell has been as yet not reported,
because such robot cells with a variable task
sequence are recently applied. There are also few
simulation models for a robot cell with fixed task
sequence. The reasons for that are : Firstly, two
model groups, one group for the graphic simulation
of robot motion and another group for the study of
material flow in a production system, were independently developed. Secondly, the time of robot application is shorter in comparison with other elements of production system.

The restrictions of the analyzed simulation models
for the study of material flow of a production system are :
- high cost for development and maintenance of simulation models,
- difficulty of connection with other FMS software,
- high cost for modification of the simulation models for the purpose of using the models as control software.

REQUIREMENTS FOR SIMULATION SYSTEM

The following requirements for a model development
are determined from the analysis of the developed
simulation models.

Prior Requirements for a Model Development

Since an extending function of the simulation
models for the future use has been neglected in
contrast to task functions of the models, the cost
of development and maintenance of models is very
high. Following requirements are selected to improve an extending function of models.
● working out a method for the development of a
  relative large simulation system using a microcomputer
● making out a design standard for the effective
  realization of this method

Requirements for Modelling a Real System

In modelling a dynamic flow among robot cells the
requirements can be divided into as following.
● modelling the material flow
● modelling the decision making process for the
  material flow

Requirements for Design of a Simulation System

- design of the module and data structure
- possibility of the simulation system modification for the using this simulation system as a control software

WORKING OUT A DEVELOPMENT METHOD

A Method for the Development of a Simulation System

A simulation system for the interconnected robot cells can be developed using the following two methods :

A system elementoriented method
This method gives the first consideration to the elements of a production system.
In case of using this method a simulation program is developed for the production system. In this simulation system a production system is not divided into the production subsystems (workpiece manufacturing, transport, assembly and warehouse) and the elements of a production system do not belong to a specific production subsystem.
A production-subsystem oriented method
This gives the first consideration to the production subsystems. In case of using this method the independent simulation programs for each production subsystem are developed and combined using the interface module.

The first method is as yet frequently used and the second method presented in this paper(Fig. 1).

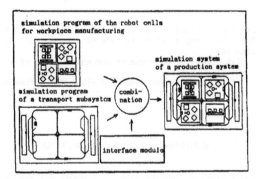

Fig. 1. Combination of the Independent Simulation Programs

The method based on the production subsystem has an advantage in using small micro computer, because the individual simulation models for each production subsystem can be combined in accordance with main memory capacities of computers.
Two methods are compared each other in Fig. 2 with a viewpoint of model development and its use.

Design Standard for the Combination of the Independent Simulation Programs

The following two standards are written out not for the purpose that the simulation programs for each production subsystem can be effectively developed and also that the independent simulation programs can be combined.

Standard for the design of each special simulation program :

	method evaluation item	method 1	method 2
model development	satisfaction of model accuracy for each production subsystem	-	+
	independant module development of a simulation model	0	+
model use	work scale of programming	+	-
	relational analysis among production subsystems	0	+
	relational analysis among elements of production system	+	0
	preparation cost of input data	0	+
	model size adaptation to the capacity of computer main memory	-	+
	usage of a simulation model as control software	0	+

method 1 : method based on system elements

method 2 : method based on production subsystems

+ : good, 0 : middle, - : bad

Fig. 2. Comparison of Two Different Methods

- classification of element attributes in a production system
- classification of relations among system elements
- description method of the relations among system elements
- description method of event flows in a production system
- classification of control rules for a production system
- dividing a simulation program into modules
- method and format of data input and output for a simulation program

The first five items are directly related to the system analysis for development of a simulation program.

Standards for a program control of each special simulation program :

- control method of time advancement
- management method of event
- method for deciding the duration of initial system state
- protocal method of simulation flow
- completion method of simulation flow

DESIGN OF THE SPECIAL PROGRAMS USING A LOGIC MODEL

For the effective development of the two special simulation programs, a program for the workpiece manufacturing subsystem and another program for the transport subsystem, a logic model between a real system and a simulation program(Fig. 3.) is employed.

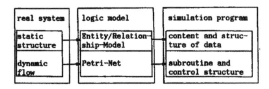

Fig. 3. Logic Model as a Linking Element

Data Structure

For deciding the content and structure of data in a simulation program a describing method is as yet not used. But modelling the relations between the elements of a production system has became as important as modelling the system elements themselves, because the production system is becoming more complex.

The Entity/Relationship-Model is a good supporting tool to systematically analyse and check the relations between the system elements.

An Entity/Relationship-Model for a transport subsystem, which links the robot cells, is depicted in Fig. 4.

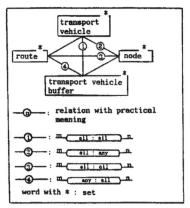

Fig. 4. Entity/Relationship-Model for a Transport Subsystem

Program Structure

The Petri-Net is used not only to improve the reliability of a simulation program, but also to support the communication between the program developers.

The Petri-Net facilitates the work to acquire the basic states, events and the interaction of the system elements through a transition in a real system. These informations can be used for the following tasks :

- basic states for modelling of event generation
- basic events for modelling of event execution
- interaction of the system elements through a transition for modelling the event selection

In order to obtain such informations from a Petri-Net of a real system the following conditions for the real system should be satisfied :

- The activity of the real system must include the possible relations between the system elements.
- The activity of the real system must include the possible events on the system boundary.

Fig. 5 shows a Petri-Net of a robot cell having a robot, two buffers of cell I/O(Input/Output) and a production machine.

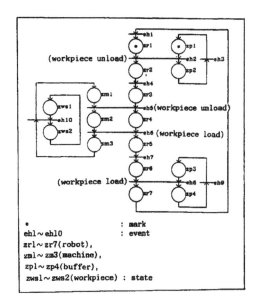

Fig. 5. Petri-Net of a Robot Cell for Manufacturing a Workpiece

CONTROL RULE AND PROGRAM FLOW

Control Rule

The handling sequence in a robot cell is decided by a heuristic method. For a special situation, in which two handling tasks, a task with two cell elements, to be tended by a robot and another task with three cell elements can be simultaneously selected, a priority rule for choosing a task is determined. In order to avoid a lockout in a transport subsystem, in which a transport in two transport directios is for each transport route possible, a control rule for a transport route selection is determined.

Program Flow

After the combination of the two independent simulation programs there are frequent interactions between two production subsystems, workpiece manufacturing and transport, during the program running. These interactions mean that a workpiece is moved from the workpiece manufacturing subsystem to the transport subsystem or vice versa(Fig. 6).

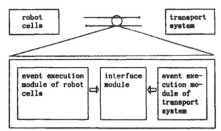

Fig. 6. Interaction between the Two Production Subsystems

APPLICATION

In this section, an example is given to illustrate how the developed simulation system is used to model a production system consisting of several robot cells. Fig. 7. shows an extension of the present production system by introducing a robot cell and a transport system between the robot cells.

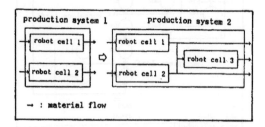

Fig. 7. Introducing a Robot Cell and a Transport System special machine tool in each cell.

The production system 1 has two robot cells designed by group technology. The simulation results of this system show a low utilizing rate of a

In order to increase the utilizing rate of this special machine, a supplementary robot cell should be planned. By introducing a supplementary cell, rearrangement of workpieces assigned to each cell is done. For this, a certain transport system like automated guided vehicle may be necessary. The new transport system 2 consists of three robot cells and a transport system.
Each robot cell in the production systems 1 and 2 has its own elements and structure(Fig. 8).

production system	cell element (i/j/k)	cell layout
system 1 cell 1	1/5/6	|—a—|—b—|—a—|—b—|—a—|—b—|—a—|—b—|—a—| P1 M1 R2 M2 P3 M3 P4 M4 P5 M5 P6 IR →
system 1 cell 2	1/4/5	|—a—|—b—|—a—|—b—|—a—|—b—|—a—| P1 M1 P2 M2 P3 M3 P4 M4 P5 IR
system 2 cell 1	1/4/5	|—a—|—b—|—a—|—b—|—a—|—b—|—a—| P1 M1 P2 M2 P3 M3 P4 M4 P5 IR
system 2 cell 2	1/3/4	|—a—|—b—|—a—|—b—|—a—| P1 M1 P2 M2 P3 M3 P4 IR
system 2 cell 3	1/1/2	M1 P1 IR P2

i/j/k : (i: number of robot, j: number of machine, k: number of buffer)

robot : transport velocity : 1.0m/s
        load and unload time at machine(= 5s)
        load and unload time at buffer (= 3s)

Fig. 8. Element and Layout of Robot Cells

The fifth machine in cell No. 1 and the fourth machine in cell No. 2 of production system 1 are identical special machine tools. In a case of production system 2, two special machine tools used in robot cells 1 and 2 are regrouped into one machine in robot cell 3. The first and last buffers in each robot cell is input and output buffers of a cell respectively.

The workpiece families according to the group technology consist of two types(Fig. 9). The transport net of the production system 2 is depicted in Fig. 10.

production system	workpiece type	sum of machinning stations	machinning time(s)				
			MA1	MA2	MA3	MA4	MA5
system 1 cell 1	1	5	103	122	105	137	[135]
	2	4	139	97	94	107	–
system 1 cell 2	3	5	98	115	142	–	–
	4	4	112	105	132	[127]	–
system 2 cell 1	1	5	103	122	105	137	–
	2	4	139	97	94	107	–
system 2 cell 2	3	5	98	115	142	–	–
	4	5	112	105	132	–	–
system 2 cell 3	1	1	[135]	–	–	–	–
	4	1	[127]	–	–	–	–

MAn: machine with number n
[___]: machinning on the special machine

Fig. 9. Machinning Order for Each Robot Cell

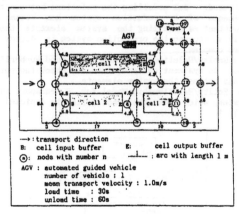

→ : transport direction
B: cell input buffer   E: cell output buffer
⊙ : node with number n   —⊥— : arc with length l m
AGV : automated guided vehicle
     number of vehicle : 1
     mean transport velocity : 1.0m/s
     load time : 30s
     unload time : 60s

Fig. 10. Transport Net in Production System 2

The simulation result showed the utilizing rate of the system elements (Fig. 11) and the mean of flow times of workpieces to go through the system.

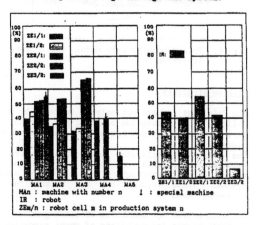

MAn : machine with number n   l : special machine
IR : robot
ZEm/n : robot cell m in production system n

Fig. 11. Utilization of the System Elements

CONCLUSION

This paper presents a new simulation model for modelling the interconnected robot cells with a variable handling sequence. This simulation model can be a supporting tool for the planning and control of such robot cells. However, the new method for the development of a simulation program and the logic model can be widely used. The subsystem-oriented development method has enlarged the traditional module concept from the module in a computer program to the module in a program system.

Further research needs are the program development for the other two production subsystems, assembly and warehouse, which can increase greatly the cases of the program combination in a simulation system, and the development of a simulation program generator using the logic model.

REFERENCES

Eversheim, W., Bette, B., Stolz, N.(1985). Einsatz von Handhabungsgeraeten in komplexen Mehrmaschinensystemen. VDI-Z Bd. 127 Nr. 14, 513-517.

Gossens(1987). : Die flexible Automatisierung des Produktions-Prozesses durch den Einsatz von Industrie-Robotern. Europa Seminar 1987., Wasserburg, 20. Mai.

Handke, G.(1987). Der Roboter als CIM-Komponente. Praktiker-Tagung "Prozess-integrierter Robotereinsatz", Muenchen, 26-27 Maerz.

H.S., Cho : H.J., Warnecke ; D.G., Gweon(1987). Robotic Assembly : a Synthesizing Overview, Robotica, Vol. 5, 153-165.

Kuk, K.H.(1988). Modulares Simulationsmodell fuer die Ablaeufe in verketteten Fertigungszellen mit Industrierobotern. Diss. Uni. Stuttgart.

Medeiros, D.J. Sadowski, R.P.(1983). Simulation of robotic manufacturing cells : a modular approach. SIMULATION, January, 3-12.

Milberg, J.(1985). Entwicklungstondenzen in der automatisierten Produktion. Technische Rundschau 37, 42-43.

o.V.(1984). Maschinenverkettung mit dem Roboter. Flexible Automation 3, 43-46.

Pritschow, G.(1985). Die flexible Fertigungszelle. Fertigungstechnisches kolloquium, Stuttgart, 10-12 Oktober.

Schellenberger, D., Scheibner, H., Bahmann, W., Moldenhauser, H.G.(1986). Moeglichkeiten und Trends bei der automatischen Werkstueck-handhabung mit Industrierobotern. wt-z. ind. Fertig. 76 Nr. 10, 585-589.

Schmidt-Streier, U.(1982). Methode zur rechnerunterstuetzten Einsatzplanug von programmierbaren Handhabungsgeraeten. Diss. Uni. Stuttgart.

Spur, G., Furgac, I., Deutschlaender, A., Browne, J., O' Gorman, P. Robot Planning System, Robotics & Computer-Integrated Manufacturing, Vol. 2, No. 2, 115-123.

Steinhilper, R.(1984). FFS-geeignete Teilfamilien und Fertigungsaufgaben : 10 Empfehlungen zu Planung und Realisicrung. Boeblingen, 11-13 Sep..

Warnecke, H.J., Abele, E., Walther, J.(1983). Programmable assembly cell for automotive parts und units. International Conference on Advanced Robotics, Tokyo, Japan, 29-37.

Warnecke, H.J., Schraft, R.D.(1980). Systematic, Computer-Aided Planning of Industrial Robot Application. Annals of the CIRP Vol.29, 339-343.

Warnecke, H.J., Kuk, K.H.(1986). SIZES, ein System zur Planung und Steuerung von Fertigungssystemen mit Industrierobotern. Angewandte Informatik 12, 511-523.

Zachau, H., Rebentrost, A.(1984). Simulationsuntersuchung einer flexiblen Fertigungszelle fuer die spanende Bearbeitung. Fertigungstechnik und Betrieb 34 Nr. 10, 615-619.

AUTOMATED SEWING WITH DIRECT DRIVE
MANIPULATOR

T. Arai*, T. Nakamura* and M. Sato**

Mechanical Engineering Laboratory, Tsukuba, Ibaraki, Japan
**Juki Co., Ltd., Chofu, Tokyo, Japan*

Abstract. A 3-D automated sewing system with a 6 DOF direct drive
manipulator is introduced to improve productivity and to promote automation in the apparel industries. A smooth trace technique is adopted by
improving the conventional position/force hybrid control method to attain
high sewing accuracy as well as to eliminate burdensome motion teaching.
The experimental results show the proposed system is excellent with
respect to sewing accuracy and applicability comparing with the conventional teaching play-back method.

Keywords. Automated sewing; direct drive manipulator; hybrid control;
compliance control; smooth trace;

INTRODUCTION

Production automation in the apparel industry is far behind comparing with other
manufacturing industries. The development
of the automated sewing system is required
to improve productivity and to promote
automation in the apparel industry. The
Ministry of International Trade and Industry started the Automated Sewing
Project in 1982. In order to realize the
system a flexible sewing technique for
various sizes and kinds of garments has to
be developed. A 3-D automated sewing system using a smooth trace motion with a 6
DOF direct-drive manipulator is introduced
for such a purpose. As opposed to the
conventional sewing machine into which a
piece of fabric is placed by human hands,
a manipulator moves a compact sewing
machine attached on its tip to sew fabric
fixed on a fabric fixture. Generally the
precision of a sewing line should be less
than ± 0.5mm in the apparel industries.
The precision is dependent on the relative
positioning accuracy between the controlled sewing machine and the fabric fixture. It is expectedly difficult for a
conventional teaching play-back method to
attain this sewing accuracy. And even if
it were possible, it is very hard to teach
the sewing line accurately. If a fixture
is-used as a guide for the sewing machine,
the manipulator can move it by applying
smooth trace motion with the hybrid
position/force control technique
(Raibert,1981). In the experiment, a
small force, produced when the sewing
machine traces the fixture, is detected by
a five axes force sensor attached on the
tip of the sewing machine, and by feeding
back this force information, smooth trace
motion is realized. In the following the
principle to realize the sewing task by
smooth contact motion and the experimental
results are discussed, and the conventional teaching play-back method and the

proposed method are compared in terms of
sewing accuracy.

3-D SEWING SYSTEM

Fig.1 shows the configuration of the
automated sewing system. It consists of a
direct drive manipulator with 6 DOF, a
compact sewing machine attached to the tip
of the manipulator, a flexible fabric fixture, a visual sensor, and a control subsystem implemented in several microcomputers.
The manipulator has been originally
developed for this system (Yano, 1987).
Each joint is actuated by a high performance torque motor directly, therefore it
is capable of controlling output force
and/or torque as well as attaining fine
positioning. The compact sewing machine
is designed by scaling down the conventional sewing machine head and its weight
is only 2 kg. The maximum sewing speed is
5cm/s. The flexible fabrics fixture is
now being developed. It is designed to
fix fabrics of various sizes flexibly and
to get the complex sewing curve stably.
The visual sensor is composed of two MOS
image sensors and they are used for locating the start point and several reference
points of the sewing curve.
In the control subsystem there are
several processes described as a block
diagram in Fig.1. The apparel data base
has a standard size and shape of a sewing
curve of each cloths. However, actually
the shape is slightly different from the
standard one on account of the shape of
the fixture. Moreover, the three dimensional position changes every time the
cloths are fixed. Usually, the sewing
curve is determined relative to the hem of
the cloth. Thus exact position and shape
of the hem should be determined before
sewing is taken place. The object recognition process locates several reference

points of fabrics fixed on the fixture in the task coordinates using a stereo scope method.   Based on these data the task scheduler decides the rough sewing path and orientation of the sewing needle for the motion controller.   The motion controller controls each joint of the manipulator to attain the exact sewing motion by using the idea of smooth tracing as described in the following and this paper mostly focuses on this part.

The motion control is performed by the computer system composed of seven commercially available personal computers.   The six computers are assigned to each joint as a sub-computer for servo control and other basic operations, and the seventh computer acts as a host   computer to manage the six sub-computers and to calculate kinematics and statics of the manipulator.   The computers communicate with each other using high speed communication memories (Kokaji, 1986).   Fig. 2 shows the block diagram of this computer system.

SMOOTH TRACE MOTION

In the proposed system a part of the fabric fixture, assumed to be parallel to the sewing line, serves as a guide for the sewing machine.   The manipulator is controlled to move the sewing machine at constant speed along the guide by touching smoothly using the contact force sensor

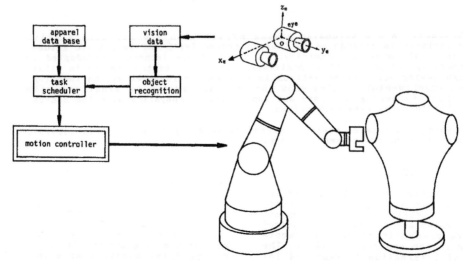

Fig.1 Configuration of the 3-D automated sewing system

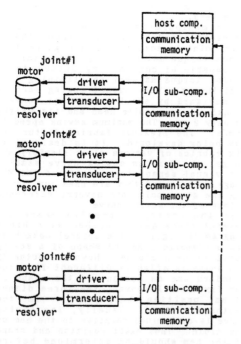

Fig.2 Block diagram of the computer system

output.   Cartesian coordinates are fixed on the tip of the sewing machine as shown in Fig.3.   If forces in x and z axes and moments around x, y and z axes can be controlled and speed in y axis can be maintained by using position and force sensors, smooth trace motion along the guide will be realized.   This is called a hybrid position/force control method. Generally hybrid control requires many computations, for example, position and force coordinate transformations, dynamics of the manipulator, joint servos and so on.   To reduce the computational burden and to increase the stability in the proposed system a simplified method is adopted, where an error of contact force is added to a servo error for trajectory control.   Let F^m_r be the nominal contact force.   The error of contact force ΔF^m is given as follows by using the output of the contact force sensor F^m.

$$\Delta F^m = F^m_r - F^m \tag{1}$$

The torque produced at each joint to cancel this error is solved by using Jacobian J defined in the manipulator kinematics.

$$\Delta T = J^T \Delta F^m \tag{2}$$

The torque required for the trajectory control is given as follows.

$$T_\theta = K_p(\theta_r - \theta) \tag{3}$$

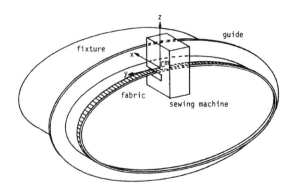

Fig.3 Sewing by means of smooth trace motion

where θ_r is the reference trajectory data roughly taught. T_d is provided to compensate for the manipulator dynamics and T_c to compensate for Coulomb friction. The required torque at each joint for smooth trace motion is solved as follows.

$$T = K_p(\theta_r - \theta) + K_f \Delta T \\ + T_d + T_c \mathrm{sgn}(\theta) \qquad (4)$$

where $\mathrm{sgn}(\theta)$ is $(\mathrm{sgn}(\theta_1), \ldots, \mathrm{sgn}(\theta_6))^T$.

Fig 4 shows the block diagram of the hybrid control system for smooth trace sewing motion. The calculation of servo for each joint is done in the sub-computer assigned to each joint and the calculations required for processing the contact force sensor output, coordinate transformations and Jacobian are done in the host-computer. The sampling rate is 0.8ms in joints 1-4, and 0.6ms in joints 5 and 6. The force feedback term described in equations (1) and (2) is calculated at 30ms intervals and the data is transferred to the sub-computer through the communication memory. The contact force of the sewing machine to the guide is detected by a newly developed force sensor attached on the tip of the sewing machine. Fig.5 shows this force sensor. It consists of five small cantilevers which has a strain gauge to sense small bending motions caused by contact force. When each element touches a guide, the force $\mathbf{f}^m$ and the moment $\mathbf{m}^m$ acting in the coordinates C^m fixed on the sewing machine are estimated as follows. As in Fig.6, let p_{ti} and z_{ti} be a position and a normal vectors of the i-th cantilever of the force sensor. Then, the contact force $\mathbf{F}^{mT}=(\mathbf{f}^{mT}, \mathbf{m}^{mT})$ is solved by using the output of i-th sensor element f_{ti}.

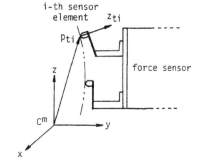

(a)

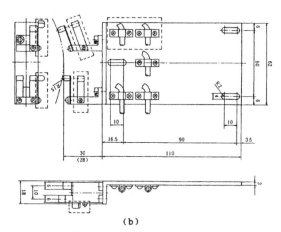

(b)

Fig.5 (a)View of the contact force sensor
(b)Configuration of the sensor

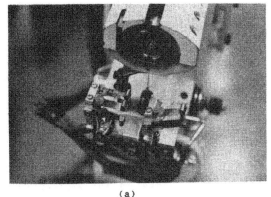

Fig.6 Force acting on the i-th sensor element

$$f^m = \sum f_{ti} z_{ti} \qquad (5)$$
$$m^m = \sum p_{ti} \times f_{ti} z_{ti} \qquad (6)$$

where $\times$ is a vector product. The equations are combined as follows.

$$\mathbf{F}^m = J_t f_t \qquad (7)$$

where

$$J_t = \begin{bmatrix} z_{t1} & \ldots & z_{t5} \\ p_{t1} \times z_{t1} & \ldots & p_{t5} \times z_{t5} \end{bmatrix} \qquad (8)$$

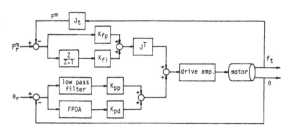

Fig.4 Block diagram of the hybrid control system

$$f_t = (f_{t1}, \ldots, f_{t5})^T \qquad (9)$$

J_t is a 6×5 constant valued matrix and it can be calculated in advance.

EXPERIMENTS

Two kinds of sewing experiments using teaching play-back and smooth trace methods are performed to demonstrate the feasibility and the effectiveness of the proposed system. In the trajectory control of the manipulator, joint servos are to function effectively to attain quick response and high accuracy. Fig.7 shows the block diagram of the joint servo system. It is based on the PD control method. Since the manipulator joints are directly driven by a high performance torque motor except joint 3, where a steel belt is used for transmission, each joint has a small amount of friction and therefore requires rate feedback in the joint servo to increase the damping factor and to maintain adequate joint stiffness. The following problems arise. First, a difference operation in calculating rate causes an emphasis of noise. Second, a high gain is required to increase stiffness, but its upper limit is dependent on the sampling rate. If the sampling rate is fast, it becomes higher but the resolution of the rate deteriorates. Taking these two points into considerations, the servo system has two kinds of filters. One is a usual low pass filter in the proportional gain, and the other is the four point difference algorithm(FPDA) in the derivative gain. The interval of the difference operation for the derivative is longer than the sampling time. The parameters in the servo are tuned by observing the output response of each joint in a trial and error manner. Table 1 shows the list of gain parameters for optimum tuning.

The sewing task for a circle line is done by trajectory control based on joint servos. A circle trajectory based on three manually taught points on the fixture is calculated and the joint data corresponding to that circle is stored. Fig.8 shows the results of the sewing and the sewing accuracy is around ± 1.0mm.

In the second kind of the experiment, the sewing task with smooth trace motion is carried out. As for the force control, an integral term is added to position one in equation (4). This causes a complete separation of position and force axes in the work coordinates statically. Furthermore, the dynamic compensation Td is excluded, since the manipulator is mechanically well balanced so that each joint is hardly effected by the gravitational burden (Yano, 1987), and since its moving speed is slow enough that non-linear term can be ignored. Five forces and moments in coordinates Cm are calculated by referring $f_z = -50.0$gf to realize a trace motion. This results in that $f_x = 34.4$ gf, $m_x = 0.133 \times 10^{-3}$kgfm, $m_y = 1.71 \times 10^{-3}$kgfm and $m_z = 0.611 \times 10^{-3}$kgfm. In the y axis the reference speed is 6.7mm/s for trajectory control. Fig.9 shows the output of forces and moments during the trace motion with the gains listed in Table 2. Fig.10 shows the results of the sewing. It is clear that the sewing accuracy is greatly improved and is less than 0.5mm.

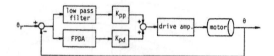

Fig.7 Block diagram of the joint servo system

Joint#	1	2	3	4	5	6
sampling rate(ms)	0.8				0.6	
difference interval(ms)	4.8		3.2	6.4	1.2	
filter cut off frequency(Hz)	100				none	
proportional gain	4.5	5.0	7.5	4.5	1.3	1.0
derivative gain	.818	.552	.688	.251	.061	.015

TABLE 1 Gain parameters in the joint servos

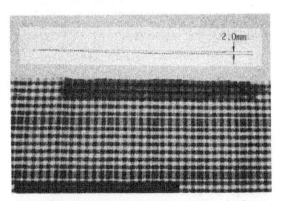

Fig.8 Results of sewing by means of play-back motion

	proportional	integral
f_x	50.	500.
f_z	50.	500.
m_x	500.	3355.
m_y	50.	500.
m_z	500.	3355.

TABLE 2 Gain parameters in the force control

CONCLUSIONS

A 3-D automated sewing system using a direct-drive manipulator and smooth trace motion is proposed to realize a sophisticated automation of sewing task. When the appropriate fixture is used, the proposed method is flexible enough to ensure the high sewing accuracy required in the apparel industries with no burdensome and time consuming teaching task. The remaining problems are the development of a flexible fabric fixture suitable to various kinds and sizes of garments and the development of a method for setting pieces of fabrics on the fixture automatically. Fig.11 is the view of the sewing motion.

ACKNOWLEDGMENT

The authors would like to express their appreciations to the office for National R & D Programs of the Agency of Industrial Science & Technology for their support.

REFERENCES

Kokaji, S.(1986). Collision-free Control of a Manipulator with a Controller Composed of Sixty-four Microprocessors, IEEE Control Systems, 6, 9-14.
Raibert, M.H. and J.J. Craig(1981). Hybrid Position/Force Control of Manipulators, Trans. ASME, Journal of DSMC, 102, 126-133.
Yano, T., T.Arai, T. Nakamura, E. Nakano, R. Hashimoto, I. Takeyama, S. Sugioka, and J. Takahashi(1987). Development of Direct-Drive Manipulator for Automated sewing System, Proc. 17th ISIR, 17.55-17.61.

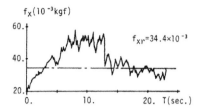
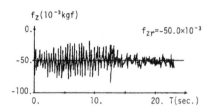
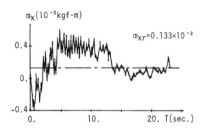
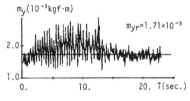
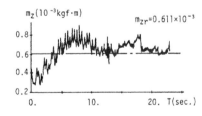

Fig.9 Responses of forces and moments in the work coordinates in the trace motion

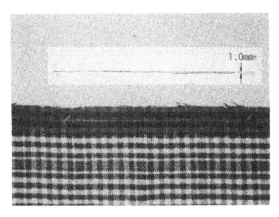

Fig.10 Results of sewing by means of smooth trace motion

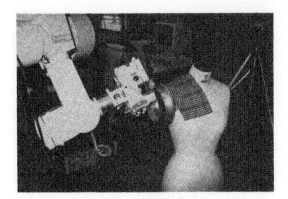

Fig.11 View of the developed sewing system

A MODULAR APPROACH TO SENSOR
INTEGRATION IN ROBOTIC ASSEMBLY

J. J. Rowland and H. R. Nicholls

*Department of Computer Science, University College of Wales, Aberystwyth, Wales,
UK*

Abstract. We describe the use of functional abstraction for the design of modular sensor integration
systems that provide task supervisors with the sensor information required to perform robotic
assembly tasks or subtasks. We offer a versatile approach to high-performance sensor integration.
This is consistent with the low-cost robotic assembly cells that will be needed to make automated
assembly attractive to a wider sector of manufacturing industry.

Keywords. Industrial robots; sensors; flexible manufacturing; computer interfacing; data handling;
microprocessors; software engineering; real time computer systems.

INTRODUCTION

Many workers have used intelligent sensing in robotic
assembly. We are developing methods for the design and
organisation of versatile sensing subsystems. Our work is
based on a concept of *virtual sensors* and is proceeding in
the context of the research into assembly workcell
supervisors here at Aberystwyth. The term *virtual sensors*
is also used by other workers to describe abstraction in
sensing, for example Milovanovic (1987) and Ghani (1988).

An assembly robot uses multiple sensors mainly to confirm
expectations or to detect and quantify positioning errors
and component tolerances. Only a subset of the potentially
available sensory information is useful at any stage in a
task, although different subsets are relevant at different
stages. In many other applications involving multiple
sensors, such as a mobile robot, the demands on the sensor
system are greater because most of the available
information is potentially relevant most of the time (e.g.
Brooks, 1987).

The real-time constraints of sensory processing make it
inappropriate to include it within an assembly task
supervisor. The use of an intelligent sensor system is
desirable in all but the simplest of sensing strategies. It
forms a virtual machine that lies between the physical
sensing devices and the supervisor and provides sensor
knowledge in the form required by the supervisor. It thus
removes from the supervisor the low-level processing and
stringent real-time constraints associated with processing
the outputs of the physical sensing devices themselves.

Different physical sensing devices may be used in different
ways at different stages of a task. They may be used
individually, to give *low-level* information, or they may be
used in combination to produce sensing functions at a
higher level of interpretation. In this case the sensory

information is combined by computational modules, each
of which provides an abstract but specific sensor function.
Sensory information can thus be provided at a level
appropriate to the current requirements of the supervisor.

The design, structure and architecture of the sensor
integration system is considerably simplified by this
approach. There is also an increase in the system's ability
to perform all *relevant* sensory data processing within the
real-time constraints imposed by the task. Application of
functional abstraction in this way results in a design that
is highly modular, giving benefits of increased resilience
and reliability, as well as ease of maintenance and
reconfiguration for different sensors and workcells. It is a
methodical and modular approach to sensor integration.

Other application areas that also have well defined sensor
integration requirements include monitoring of process
plant and aircraft engines. Whilst this paper concentrates
on robotic assembly because it is a major interest of the
authors and is a good case study, we intend our approach
to be sufficiently generic that it could be applied to areas
such as these.

BACKGROUND

Sensing in Assembly

It is reasonable to assume that a robotic assembly cell is
part of an integrated manufacturing process, rather than
operating in isolation. Although this is not universal at
present it is essential if robotic (flexible) assembly is to
become widely established in manufacturing industry.
Major requirements for the use of flexible assembly cells
are that, to some extent at least, components are designed
for assembly by robot and are fed with relatively accurate
position and orientation, either palletised or by feeder

(Redford & Lo, 1986). Parts whose unique orientation is not achievable should be fed so as to minimise the number of possible orientations and hence simplify orientation sensing.

The major role of sensors in this environment is to compare reality with expectations and to evaluate discrepancies. Essentially this means detecting presence or absence of correct parts and measuring to allow for component tolerances and positioning errors during relevant stages of handling and assembly. Sensing to inspect product quality may sometimes be required, but again involves comparison with expectations: visual inspection is often used, for which techniques are well established (Batchelor, Hill & Hodgson, 1985).

Sensor Interfacing

Even in the relatively predictable environment of an assembly cell, numerous problems have to be overcome in establishing a sensing strategy. It is likely that the cell will contain diverse sensor types. There may binary sensors, such as microswitches or proximity sensors, and analogue sensors that measure parameters such as force and displacement (perhaps in a gripper). Parameters such as end-effector coordinates, probably derived from the manipulator control system, and time may constitute further sensory inputs. Tactile pads or fingers, as well conventional vision systems, may be used for inspection and determining part orientation.

All these sensors will differ in electrical characteristics and physical interfacing requirements. The format and meaning of the information they provide will differ, leading to widely differing computational demands: a proximity sensor provides simply a *presence/absence* indication, while a vision system may output a two-dimensional array of grey levels. Each has very different computational requirements.

Sensor Interpretation and Integration

Each stage in handling the sensory information requires processing, either electrical or computational. At the lowest level, the incoming electrical signals must be transformed so that they can be manipulated computationally. For example, the output of a proximity sensor may be transformed so as to become a Boolean, or logical, value.

The low-level sensory data will be subject to interpretation in the context, for example, of the current workcell state and its history. A higher level of processing may ascribe further meaning to such data, based on the properties of the sensor, its position in the workcell, and on knowledge gained from other sensors. For example, a robot may deposit an object near to a proximity sensor. This sensor can be used to confirm that the deposit occurred. Further, the combination of arm position data with data from sensors on the gripper can yield the orientation of the deposited object. Thus, in our example, ambiguity in orientation can be resolved through sensor combinations that provide an *orientation sensor* function.

Combining outputs of sensors to give "new" information (*sensor integration*) is required in many fields. As well as those already mentioned, it arises in the military context where plotting a potential target, and its identification as friend or foe, relies on correct and sufficient integration of possibly conflicting information from a number of different radar-based sensors.

Sensor integration involves progressive refining of sensory and context information and suggests the possibility of a highly structured and generic approach to the computation involved. However, the design of a generic approach is not easy and workers in robotic assembly have most often used application-specific, rather than generic, sensor integration methods. In the comparative absence of generic workcells this *application-specific* approach has few apparent disadvantages. However, as technology develops, sensing systems that are modular, maintainable and extensible will be required to improve reliability, reduce system integration time and reduce cost.

Henderson and Shilcrat (1984) describe a concept of *logical sensors* which can form a basis for the design of generic sensor systems. The original emphasis was resilience through reconfiguration but later work is concerned with integrating sensory information (Henderson, Hansen and Bhanu, 1985). This logical sensor concept has also been applied to a gripper (Luo & Henderson, 1986).

We are developing a unified approach to the design and organisation of sensory systems that is based on functional abstraction. This gives a framework for unifying sensory data handling. It can also help determine which sensing devices should be fitted in a cell, through functional decomposition of the high-level knowledge requirements of the task supervisor. The resulting structure of functional modules, that connects the supervisor interface with physical sensors, can translate directly to an appropriate software structure and machine architecture for sensor integration.

TASK SUPERVISORS AND SENSING

A task supervisor is concerned with the high level aspects of task progress. Rather than deal with capture and processing of low-level sensory data, it is most appropriate for the supervisor to be presented with sensory data in a suitably 'high-level', or *abstract* form. These high-level functions may indicate task progress, including evidence of assembly errors, and may provide component information such as orientation or the effect of machining tolerances.

The sensors that a supervisor accesses directly should therefore be virtual, rather than physical, so that low-level computation and the accompanying stringent real-time contraints are removed from, and hidden from, the supervisor. This allows a given supervisor to operate with different types and configurations of physical sensors, provided that the appropriate low-level sensing functionality is available to allow the abstractions to be made.

There will be cases where sensory data must interact directly with actuator control systems (rather than through the supervisor). Such cases will be principally in *reflex* actions (such as emergency stop) and in low-level closed-loop control. A benefit of functionally modular sensing systems is that a design framework may be established that supports such interaction, and also allows actuation to be incorporated into sensing functions. An example of a sensing function that incorporates actuation is described by Beni, Hackwood & Rin (1983).

Hardy, Barnes & Lee (1987) present a classification of purposes for which sensor data are used in robot control. Their main categories are:-

- monitoring progress
- collection of *deferred data*, i.e. data unavailable at the time of cell programming
- guarded actions
- closed-loop control.

While this classification is well suited to treatment of sensor data by supervisory systems (Hardy's intended purpose), a modified classification that reflects the real-time nature of the various classes is perhaps more appropriate when considering the design of the sensory processing systems themselves. A classification on this basis might be:-

- closed-loop control and *reflex* actions
- alarm, where damage is imminent (also applicable to guarded actions)
- warning, where there is a danger of incorrect task completion
- monitoring progress, on request from the supervisor (as Hardy)
- collection of deferred data, again on request from the supervisor (as Hardy).

All these constitute real-time operations but the critical time within which sensor data of each class must be available will vary: broadly, the real-time criticality reduces from top to bottom in this list. For example, while the lower level aspects of the task, such as arm and gripper control, are under the *overall* direction of the supervisor, the sensing functions immediately relevant to the low-level aspects of control are likely to have the most stringent real-time constraints. This is because they are directly within the feedback loops of dynamic systems.

A further basis for classification that is implied in the above list is the source of the *event* that provokes update of the *sensed* task-world. This event may be, simply, a supervisor request; alternatively, it may be an occurrence in the real world, be it a transition in a binary sensor, the crossing of a threshold in an analog sensor, some complex combination of the outputs of several cooperating sensors, or simply the passage of (real) time.

There are, therefore, three main functional considerations in designing an intelligent sensory system:-

- the high-level sensing functions determined by **task** requirements, and the low-level sensing requirements that these imply
- the real-time constraints that task requirements place upon each sensory function
- the *event* that provokes the availability of sensory information from each function.

Other considerations are those that simply constitute good practice in designing hardware/software systems.

DESIGN USING FUNCTIONAL MODULARITY

Let us consider a simple task: an arm-mounted gripper is to be moved quickly until it is close to a part that is to be grasped. It then moves into position and grasps the part, checks that only one part is held and inspects to check that the part has been correctly manufactured. We will consider the role of two sensors in this operation: a proximity sensor and a tactile array (Mott, Lee & Nicholls, 1985; Nicholls & Lee, 1989), both of which are mounted on the gripper. In performing this task the supervisor needs to abstract information from the raw data provided by these sensors:-

1. Initially the supervisor needs to know when the gripper achieves close proximity to the part, and then when contact is achieved.

2. It needs to determine that only one part is held.

3. For inspection of the object it needs to know whether, on the basis of predetermined criteria, the part has been correctly manufactured.

In Fig. 1 these functions are shown, similarly numbered, as being available to the supervisor. Fig. 1 also shows that they are derived from further abstractions of the raw sensor data. The functions may contain expectations such as, in our example, the criteria that determine *correct manufacture* and criteria that distinguish *rough* and *smooth*.

The functions available to the supervisor are outputs of a network of computational elements whose purpose is to transform the information obtained from physical sensing devices into the form required by the supervisor. Hence, the supervisor receives the transformed information as if the information is being provided directly by (abstract) sensors with the appropriate capability. Information from a physical sensor may, on its way to the supervisor, undergo transformation by a number of nodes in this network of computational elements.

We thus have a data-flow network, where nodes may obtain information from, and feed information to, other nodes so that integration as well as transformation of sensor information can be achieved. We disallow cyclic paths in the network, to prevent deadlock at the dataflow level, and so the network is a directed acyclic graph.

As in Fig. 1, each node in the graph is a computational element that has a single, identifiable, function. Each node is termed a *virtual sensor* and may take as its inputs

information from other virtual sensors; this information is
then integrated to provide "new" information. The
processing encapsulated in a virtual sensor may be
algorithmic or knowledge-based.

All virtual sensors operate concurrently; they process and
provide information. These concurrent functional modules
form a real-time system that may be implemented either on
individual embedded processors, or as concurrent processes
on one or more processors. Data flow between the virtual
sensors must take place only along protected data paths so
as to ensure data integrity. In this way the system becomes
truly modular, and provides the maximum of versatility in
re-use of the modules in different sensing configurations.
Such an approach also enhances the reliability of complex
systems such as these. There are several real-time design
methods that can be used to design and represent the
detailed structure of virtual sensor systems.

A fundamental requirement in sensing design is the
provision of appropriate physical sensing capability to
meet both the low-level and high-level needs of the task.
Information provided by the physical sensors must be such
as to allow synthesis of the high-level functions required by
the supervisor. Our approach to functional abstraction
also gives a basis for physical sensor selection: functional
decomposition of the high-level functions needed by the
supervisor will help reveal the physical sensing functions
that must be made available in the cell.

PROPERTIES OF VIRTUAL SENSORS

Virtual sensors are computational entities that are
supplied with information by, or can request information
from, other sensors. Each virtual sensor is autonomous and
simply reacts to information it receives, according to its
specified function. This is in many ways analogous to the
way in which a simple physical sensor operates. However,
as we saw above, physical sensors vary greatly in the way
in which they interface to other parts of a system; an
important feature of virtual sensors is that they exhibit a
defined interface that allows them to communicate with
other virtual sensors. The processing necessary to perform
a virtual sensor function is therefore encapsulated within a
computational entity that has a uniform but versatile
interface specification.

Some of the benefits that virtual sensors can offer include:-

- the ability to integrate information from multiple
  sensors

- the ability to define a uniform information interface
  for different sensors

- the ability to treat *time* as a sensor input

- the ability to incorporate expectations and to
  respond according to whether or not the expectations
  are met

- the ability to respond to *sequences* of sensor events
  or conditions

- the ability to select different sensing functions during
  a task

- the ability to provide functions that reveal trends in
  workcell parameters, over successive assemblies, and
  over combinations of sensing devices.

A SENSORY GRIPPER AS A TEST-BED

As a test-bed for the virtual sensor concept we have
designed and implemented a controller for a sensory
gripper. We have interfaced the controller to an electrically
operated gripper known as the Modular Tactile Gripping
System (MTGS). The gripper itself was designed and built
at the *Institute for Production Automation* in Stuttgart
(Warnecke & Haaf, 1981). However, since the design of our
controller is based on the modular concept of *virtual
devices*, it is relatively straightforward to adapt the
controller to operate with other electric grippers.

The controller is based on an embedded 68008
microprocessor. Interfaces to the gripper itself measure the
motor current, monitor an encoder on the motor shaft, and
have access to proximity sensors. However, the interface
between the controller and the task supervisor provides the
supervisor with other sensory functions. These include
gripping force and *jaw separation* that are derived from the
low-level sensed parameters of current and encoder output.
Further sensors can be integrated into the controller.

In addition to providing high-level sensing functions, the
controller gives the supervisor the ability to control the
gripper through high-level commands. The supervisor can
specify desired jaw separation and gripping force. The
feedback necessary to achieve the desired conditions is
implemented within the controller at a low-level so as to
meet the real-time requirements of mechanical control.
Further details of the controller are described in Nicholls,
Rowland & Sharp (1989).

This controller also allows us to investigate *active* virtual
sensing (defined earlier). As a demonstration we have
implemented a virtual sensor function that combines the
force and jaw-separation sensing with control of actuation.
This allows measurement, in a single virtual sensor
function, of the compressibility of a spring held in the
gripper, and could therefore perform component inspection
(of springs) during an assembly operation. The active
virtual sensor requests a number of different
jaw-separations and makes a force measurement at each;
some example data, and the straight-line fit from which
the spring constant is determined, are shown in Fig. 2.

This approach to control also suggests a related concept of
virtual actuators that is outside the scope of this paper.

CONCLUSIONS

In our concept of *virtual sensors* we have the basis of a
modular sensing strategy that allows considerable
versatility of configuration, and relative ease of design.
The concept allows the implementation of sensing systems
from reusable components. The complexity of the sensing
system may therefore be optimised taking into account the
requirements of task supervision and real-time
performance.

We have referred to a system we have implemented that serves as a test-bed. It provides us with a basis for work aimed at further developing the generic aspects of virtual sensors and specifying a semi-formal method for designing and implementing virtual sensor systems for industrial applications.

ACKNOWLEDGEMENTS

Some support for this work has been received under a research programme funded by SERC (ACME) grant no. GR/D 37852.

We would like to thank Kevin Sharp for his work on the gripper controller and the active sensing demonstration.

REFERENCES

Batchelor, B.G., D.A. Hill and D.C. Hodgson (1985). *Automated Visual Inspection*. IFS Publ., Bedford, U.K.

Beni, G., S. Hackwood and L. Rin (1983). Dynamic sensing for robots – an analysis and implementation, In *Proc. 3rd. Intl. Conf. on Robot Vision & Sensory Controls*. pp. 249–255. IFS Publ., Bedford, U.K.

Brooks, R.A (1987). A hardware retargetable distributed layered architecture for mobile robot control. In *Proc. IEEE Conference on Robotics and Automation*. pp. 106–110.

Ghani, N., (1988). Sensor integration in Esprit. In *SYROCO '88 (Symposium on Robot Control)*. IFS Publ., Bedford, U.K.

Hardy, N.W., D.P. Barnes and M.H. Lee (1987). Declarative sensor knowledge in a robot monitoring system. In in U. Rembold and K. Hormann (Eds.) *NATO ASI Series, 29, Languages for Sensor-Based Control in Robotics*, Springer Verlag, Berlin, Heidelberg, pp. 169–187.

Henderson, T.C. and E. Shilcrat (1984). Logical sensor systems. *J. Robotic Systems*, 1(2), 169–193.

Luo, R.C. and T.C. Henderson (1986). A servo-controlled robot gripper with multiple sensors and its logical specification. *J. Robotic Systems*, 3(4), 409–420.

Milovanovic, R. (1987). Towards sensor-based general purpose robot programming language. *Robotica* 5, 309–316.

Mott, D.H., M.H. Lee and H.R. Nicholls (1985). An experimental very high resolution tactile sensor array. In A. Pugh (Ed.), *Robot Sensors Volume 2: Tactile and Non-Vision*. IFS Publ, Bedford, U.K.

Nicholls, H.R. and M.H. Lee (1989). A survey of robot tactile sensing technology. *Int. J. Robotics Research*, 8(3). In press.

Nicholls, H.R., J.J. Rowland and K.A.I. Sharp (1989). Virtual devices and intelligent gripper control in robotics. *Robotica*. In press.

Redford, A., and E. Lo (1986). *Robots in Assembly*. Open University Press, U.K.

Warnecke, H.-J. and D. Haaf (1981). Components for programmable assembly – grippers, sensors, conveying systems for an industrial robot engaged together with fixed automation. In *Proc. 2nd Int. Conf. on Assembly Automation, Brighton, U.K.*, pp. 225–233.

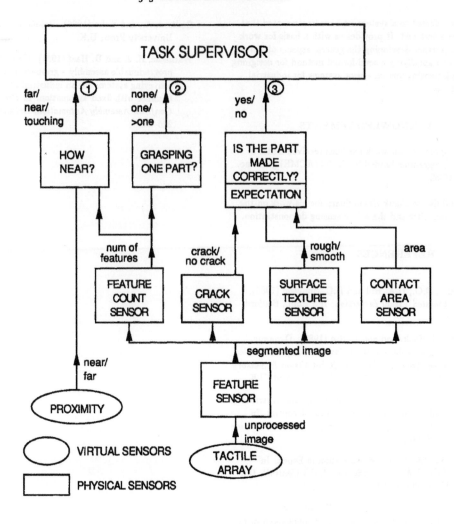

Fig. 1. Virtual sensing – an example

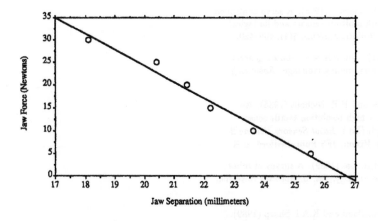

Fig. 2. Data from the active spring sensor.

FORCE INTERACTION AND CONTROL OF TWO-ARM MANIPULATORS

J. Z. Sasiadek* and M. B. Zaremba**

Department of Mechanical and Aeronautical Engineering, Carleton University, Ottawa, Canada
**Dept. d'informatique, Université du Québec á Hull, Québec, Canada*

Abstract. This paper presents a model of force interaction between the two-arm robot and a load handled by the robot. The model is applied to dual force-position control of a two-arm robot. Since position control and force control are closely interrelated, the design and implementation of such a hybrid (force + position) control system is a comlex task. When two-arm manipulators are used to handle objects, a force control system has to be employed in order to protect the body from slippage out of grasp. The force control works simultaneously with a position control.
A multi-level adaptive control system based on a reference model is presented in this paper. A simulation study has been performed to evaluate the performance of the proposed control scheme.

Keywords. Robots, manipulation, force control, Model reference control.

INTRODUCTION

The rapid growth in robots applications poses new problems and challanges. In particular, handling large and unwieldy objects can be very difficult for a single-arm robot. In that case a two-arm robot or two robots working simultaneously offer better performance and are sometimes the only acceptable solution. Space applications are the most visible example of the necessity of employing dual-arm manipulators and two or more robots working simultaneously on one task.

The control problems in dual-arm manipulators have been described in several papers. One of the most important questions addressed is related to simultaneous control of force and robot's position. Adaptive control has been proposed in decentralized (Choi and Bien, 1988) and hierarchical scheme (Sasiadek and Srinivasan, 1987, 1988). These papers dealt mostly with position control, although simplified force control has been considered as well (Sasiadek and Srinivasan, 1988). Others were concerned with force control only. Contact force control was modeled in (Laroussi and others, 1988), but little consideration was given to the optimization of contact forces. Contact stability problems were described in (Nakamura, 1988). Special, task-oriented approach was given in (Li and Sastry, 1988). Various aspects of dual-arm control were discussed in (Alford and Belyen, 1984, Zheng and Luh, 1985, Tarn and others, 1986). Nonlinear optimization for dual-arm manipulators was presented in (Carignan and Akin, 1988).

This paper presents a general model of force interaction between a load and the end-effectors of two robot arms. This force model is subsequently incorporated into a multi-level control system. The system controls simultaneously forces and positions of the robots. The position control is based on Model Reference Adaptive Control (MRAC) which has been described in detail in (Sasiadek and Srinivasan, 1988).

If we classify tasks assigned to two-arm manipulators into three groups:
1. Tasks being done by the two arms together, where each arm has an equal contribution,
2. Tasks being done by the two arms together, where one arm serves as the leading arm and the second performes an auxilliary role,
3. Tasks being done independently by each arm, the control system proposed in (Sasiadek and Srinivasan, 1988) could fall into category no. 2. The system which is presented in this paper could be classified into group no. 1.

PROBLEM STATEMENT

Handling large and irregular objects using two-arm manipulators requires complex control systems. These systems have to incorporate two separate but interrelated loops:
1. position control,
2. force control. Both loops have to generate appropriate torques for each joint driven by a motor.

Dynamics of a manipulator, used in position control, is described by:

$$\tau'(t) = H(q) \cdot \ddot{q} + c(q, \dot{q}) \tag{1}$$

where:

τ' - torque generated by position control,
H - actual general inertia matrix of the manipulator,
c - nonlinear coupling vector,

377

q - position vector,
q̇ - velocity vector,
q̈ - acceleration vector.

Torques generated by force control are calculated from the following formula:

$$\tau''(t) = J \cdot f \qquad (2)$$

where:

τ'' - torque related to force control,
J - Jacobian of the manipulator,
f - force exerted by robot on the body.

Torques $\tau''(t)$ generated by the force control system are responsible for holding, lifting and moving objects. Force f in equation (2) could be either measured or calculated from the force model. The latter case is more general and it allows to operate robots without the feedback from force sensors. This is especially important in space applications or when robots are working in a hostile environment. The purpose of this paper is to formulate a mathematical model of force interaction, subsequently used in robot control sysytem.

CONTROL SCHEME

The proposed control scheme is shown in Fig. 1. The input values to the control system are desired trajectory of the load $(x_m(t), y_m(t))$ and its orientation $(\alpha(t))$. It has been assumed that kinematics of the robot and the mass of the load being handled is known.

As previously mentioned, the contribution of robot A is equal to the contribution of robot B. However, it has to be understood that forces exerted by two robots may not be equal.

The desired trajectory of the load is an input to the force model. It also allows us to calculate the desired robot trajectory, and, through the inverse kinematics, gives the joint angles. The joint angles are subsequently used to calculate the torques required to position the robots. They are also used to calculate the Jacobian and and resulting torques in force control. The torques for position control and force control are added to provide the total required torque .

FORCE MODEL

The arms hold the load as shown in Fig. 2.

The motion of the object is defined by the trajectory of its center of gravity M_c from point P1 to point P2 in the robot R_A coordinate system

$$x_m = f(t)$$
$$y_m = g(t) \qquad (3)$$
$$\alpha = h(t) \; ; \quad t \in [t_{P1}, t_{P2}]$$

Force F acting on the object of mass m consists of two elements:
F_a - force due to linear acceleration of the object, and gravity force G.

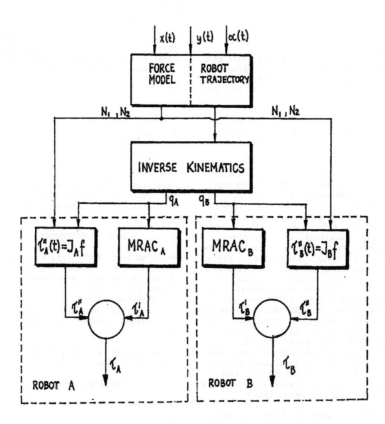

Fig. 1. Scheme of the control system.

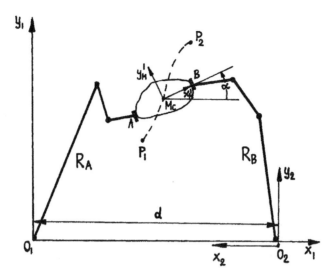

Fig. 2. Dual-arm manipulator.

$$F = F_a + G \qquad (4)$$

where:
$$G = m \cdot g \quad ; \qquad (5)$$

g - gravity acceleration

Force F_a can be calculated using equation (3) as:

$$F_a = F_{ax} + F_{ay} \qquad (6)$$

where:

$$F_{ax} = m \cdot \ddot{x}_m \qquad (7)$$

$$F_{ay} = m \cdot \ddot{y}_m \qquad (8)$$

Force $F[F_x, F_y]$ calculated as above is defined in the world coordinate system Oxy. The same force expressed in the load coordinated system is

$$F_\alpha = \begin{bmatrix} F_{x\alpha} \\ F_{y\alpha} \end{bmatrix} = \begin{bmatrix} \cos\alpha & \sin\alpha \\ -\sin\alpha & \cos\alpha \end{bmatrix} \begin{bmatrix} F_{ax} \\ F_{ay} \end{bmatrix} \qquad (9)$$

or:

$$F_\alpha = R(\alpha) \cdot F_a \qquad (10)$$

Figure 3 shows the forces acting on the load. $N_{1\alpha}$ and $N_{2\alpha}$ are normal forces applied to the load by the robots R_A and R_B. $T_{1\alpha}$ and $T_{2\alpha}$ are tangential forces due to friction μ.

The dynamic equations of linear and angular motion of the load are:

$$\sum_{i=1}^{2} N_{i\alpha} = F_{x\alpha} \qquad (11)$$

$$\sum_{i=1}^{2} T_{i\alpha} = F_{y\alpha} \qquad (12)$$

$$\sum_{i=1}^{2} (T_{i\alpha} \cdot l_i + N_{i\alpha} \cdot k_i) = I \cdot \ddot{\alpha} \qquad (13)$$

where:
I - moment of inertia calculated about $M_c(x,y)$

Thus, we have 3 equations (11),(12),(13) and 4 variables: $N_{1\alpha}$, $N_{2\alpha}$, $T_{1\alpha}$ and $T_{2\alpha}$. The fourth required equation is obtained from the relation between tangential and normal forces.

In order to minimize the forces, it is assumed that at any particular time one of the robots is responsible for exerting the force that accelerates the load along x' axis, whereas the other one applies a force only sufficient not to let the load slip out of the end effector.

It cannot be immediately determined, which robot is the leading one. The sign of $F_{x\alpha}$ has been used as an indicator for preliminary selection of the leading robot. Thus, the fourth required equation is chosen as follows:

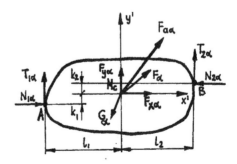

Fig. 3. Forces acting on the load.

If $F_{x\alpha} \geq 0$, then

$$N_{2\alpha} = \frac{T_{2\alpha}}{\mu} \qquad (14a)$$

If $F_{x\alpha} < 0$, then

$$N_{1\alpha} = \frac{T_{1\alpha}}{\mu} \qquad (14b)$$

Finally, equations (11), (12), (13), and (14a) or (14b) serve to obtain $N_{1\alpha}$ and $N_{2\alpha}$.

However, the results have to be verified in order to avoid slippage of the load at the end effector of the second arm. This can occur in situations when particular mass distribution of the load affects the grasping forces to more extent than the horizontal acceleration does.

The verification procedure of the results for the first case, when equations (11), (12), (13), and (14a) are used, is as follows:

If $N_{1\alpha} > \frac{T_1}{\mu}$, then the results are correct.

If $N_{1\alpha} < \frac{T_{1\alpha}}{\mu}$, then the final results are calculated

by solving the set of equations (11), (12), (13),

and (14b). Physically, it means that forces $N_{1\alpha}$

and $N_{2\alpha}$ have to be increased.

Similar operation should be performed in the case when $F_{x\alpha} < 0$.

Once forces $N_{1\alpha}$ and $N_{2\alpha}$ applied by the robots to the load are calculated in a load coordinate system, they are transformed to the coordinate systems related to the base of each robot. The transformation to the coordinate system of the first robot (equivalent to the world coordinate system) is

$$N_1 = R^{-1}(\alpha) \cdot N_{1\alpha} \qquad (15)$$

For the second robot:

$$N_2 = R^{-1}(\alpha + 180°) \cdot N_{2\alpha} \qquad (16)$$

Forces N_1 and N_2 obtained from equations (15) and (16) are used in force control of the robot arms.

The coordinates of points A and B required for position control algorithms can be immediately obtained from the following transformations:

$$A = R^{-1}(\alpha) \cdot \begin{bmatrix} -1_1 \\ -k_1 \end{bmatrix} + \begin{bmatrix} x_m \\ y_m \end{bmatrix} \qquad (17)$$

$$B = \begin{bmatrix} -1 & 0 \\ 0 & 1 \end{bmatrix} \left\{ R^{-1}(\alpha) \begin{bmatrix} 1_2 \\ k_2 \end{bmatrix} + \begin{bmatrix} x_m - d \\ y_1 \end{bmatrix} \right\} \qquad (18)$$

For robots with 3 or more degrees of freedom, the orientation of the end effector should be in accordance with the orientation of normal forces N_1 and N_2 given by equations (15) and (16).

CONCLUSIONS

The proposed control system has been simulated on an Apollo workstation. Performance of the system was investigated for various load trajectories, and load masses. It has been proven that the scheme works properly and can be implemented in practice. However, it has to be noted that some combinations of position and torques cause vibrations unacceptable from practical point of view.

REFERENCES

Alford, C., and Belyen, S. (1984). Coordinated Control of Two Arms. *Proc. IEEE Int. Conf. on Robotics and Automation*, Atlanta, Georgia.

Carignan, C.R., and Akin, D.L. (1988). Cooperative Control of Two Arms in the Transport of an Inertial Load in Zero Gravity. *IEEE J. Robotics and Automation*, 4/4..

Choi, Y.K., and Bien, Z. (1988). Decentralized Adaptive Control Scheme for Control of Multi-Arm-Type Robot. *Int. J. Control*, 48/4.

Laroussi, K., and others (1988). Coordination of Two Planar Robots in Lifting. *IEEE J. Robotics and Automation*, 4/1.

Li Zexiang, and Sastry, S.S. (1988). Task-Oriented Optimal Grasping by Multi-fingered Robot Hands. *IEEE J. Robotics and Automation*, 4/1.

Nakamura, Y. (1988). Contact Stability Measure and Optimal Finger Force Control of Multi-Fingered Robot Hands. *Proc. USA-Japan Symp. on Flexible Automation*, Minneapolis, MI.

Sasiadek, J.Z. and Srinivasan, R. (1987). Adaptive Control of Dual Arms. Proc. IEEE Int. Conf. Systems, Man and Cybernetics. Alexandria, VI.

Sasiadek, J.Z. and Srinivasan, R. (1988). Hierarchical Adaptive Control of Dual Arm Manipulators. *Proc. USA-Japan Symp. on Flexible Automation*, Minneapolis, MI.

Tarn, T., and others (1986). Coordinated Control of Two Robot Arms. *Proc. IEEE Int. Conf. on Robotics and Automation*, San Francisco, CA.

Zheng, Y.F., and Luh, J.Y.S. (1985). Control of Two Coordinated Robots in Motion. *Proc. 24th IEEE Int. Conf. on Decision and Control*, Fort Lauderdale, Florida.

Copyright © IFAC Information Control Problems in
Manufacturing Technology, Madrid, Spain 1989

A NEW ENERGY CONTROL SYSTEM USING REAL TIME EXPERT SYSTEMS

S. Nakamura, M. Yoshino, T. Yamada, T. Goto and Y. Hanji

Keihin Works, NKK Corporation, Kawasaki, Japan

Abstract. NKK has developed a new energy control system for Keihin Works using real time expert systems. The most noteworthy method in recent years of saving energy from integrated iron and steel works is to minimize the total energy cost. There are many plants and control factors in the energy processes, so it is difficult to organize such energy systems according to their complexity. NKK has solved the problems of organization by using a local area network and real time expert systems. The new energy control system contains the energy operation decision and the LD-gas distribution control, and it is now on-line running satisfactorily.

Keywords. Steel industry; Energy control; Energy planning; LD-gas distribution control; Local area network; Expert systems; Real time computer systems; Software development.

INTRODUCTION

There are many varied production processes in Keihin Works that are being automated and being systematized to a very high level. In such a situation, the development rate of software in Keihin Works needs to be increased, in order to promote greater efficiency. Computers, therefore, will play a much greater role in process development. On the other hand, the new technology that has been paid much attention in past years is an expert system that is a type of computer system using AI (artificial intelligence), and its use is widespread. The energy operation of iron works, as the theme of this paper, is the main factor in the production cost, and therefore, more rationalization is needed and the detailed management of its complicated processes is long overdue. In reality, however, there is no method available which does not demand the operator's ability in the optimum energy operation in choosing various types of energy data. Expert systems have been adopted for the energy process control in Keihin Works that demand such advanced human judgment.

APPLYING REAL TIME EXPERT SYSTEMS

The iron and steel industries are largely-energy-consuming industries, and saving energy is therefore of great concern. Especially varied in present times is the necessity to minimize the total energy cost of iron works, and so many energy planning systems, based upon hourly or daily unit data collected by the energy relation management are developed by steel companies across the world. However, differences between the actual result and the plan of the energy relation management often occur, and since the plan of the equipment production frequently changes, a timely amendment to the energy operation decision system is needed. The problems are summarized as the following.

(1) Difficulties occur in connecting directly from one position to all the energy relation management points for correcting all the data real timely in a unified way.

(2) Energy relation planning is too complex for usual programmes constructed by planning logic.

(3) Extreme fluctuation in short cycles causes LD-gas production to be unmanageable either by daily or hourly unit.

The former energy control system of Keihin Works had to be replaced to solve these problems. The new system has the following characteristics;

(1) A connected energy management by using an exclusive optical LAN (local area network).

(2) A decision of the optimum energy operation by using a real time expert system.

(3) An automatic control of the LD-gas distribution by using a real time expert system.

The basic concepts of the replacement are shown in Table 1. Expert systems, a type of new technology of computer systems, have advantages in that are able to create an expression from the operator's intelligence and experience, and in the simplicity in maintaining the software for the logic alteration.
A real time expert system that adopts the property of real time (that is a subject of the expert system development in former times) has been applied. The summary of the new system is mentioned in the following.

THE ENERGY FLOW OF KEIHIN WORKS

Keihin Works is an integrated iron and steel works, and naturally its energy consumption is supplied by directly purchased energy and by-product energy. The outline of its energy flow is shown in Fig.1.

The Gas Balance

The operating plants which produce gas are the blast furnaces, the coke ovens and the LD converters, and each of them produces by-product gas (B-gas, C-gas and LD-gas). The consumer plants are the heat furnaces of Keihin Works, and each of them consumes mono-by-product gas or mixed-by-product gas (M-gas). The surplus by-product gas is brought to the steam power station and it generates electricity. There are also gas holders, acting as buffers for each by-product gas to absorb their fluctuation.

The Steam Balance

The major plants which produce steam are the CDQ (coke dry quench) equipments and the exhaust gas boilers of the heat furnaces. The consumer plants include the vacuum degassing equipment of the steel making process. The surplus steam by-product is brought to the steam power station, and the electricity shortage is made up there.

The Electricity Balance

The major plants which generate electricity are the TRT (top pressure recovery turbine of the blast furnace) equipments, and as mentioned above, the steam power station also generates electricity by the by-product gas and steam. The consumer plants include the hot strip mill and the cold strip mill. The surplus electricity is supplied to the electric company. The shortage is made up by purchasing from the electric company or by increasing electric generation with purchased oil.

For controlling this energy flow real timely in order to manage all the production and consumption quantity of by-product gas, steam and electricity, and in order to obtain the solution to minimize the total energy cost, the data is needed to be gathered.

THE CONCENTRATED ENERGY CONTROL SYSTEM

There are many points existing for measuring or controlling the energy quantity or condition from energy sources, such as by-product gas, steam and electricity. In fact, the total of measuring points counts up to about 16,000. Therefore, it is difficult to connect directly all the points individually to the energy center (the unified energy managing function of Keihin Works). With the new system, however this problem is solved by introducing the exclusive high-speed (10 Mbps) optical LAN. There are about 20 DDC stations located in each small area of energy management on the LAN, and the CRT terminals in the control room of the energy center can indicate and set the data, supervise and control the loops and operate the energy equipments. The energy relation management is turned unified by this system.

THE ENERGY OPERATION DECISION SYSTEM

The energy operation decision system creates a solution that can minimize the total energy cost based on all the energy relation management data from Keihin Works. The definition of the total energy cost is shown in Fig. 3.
Ideally, energy balance, the total energy production (Ep) is always equal to the total energy consumption (Ec). The condition with no surplus and no shortage is desirable, and therefore, it becomes the purpose of the energy relation plan since a surplus becomes a supply of electricity to the electric company, by sacrificing by-product gas and steam, and a shortage demands the purchasing of oil or the purchasing of electricity from an electric company. In fact, the energy loss in surplus and shortage appear like

Fig. 3., because the equipment production plan alters. In minimizing these energy losses, the problems faced in constructing a system that tenders the best solution reflected by the alteration of the equipment production plan, include the complexity of the logic, and the difficulty of software maintenance of the reconstruction dependent on the alteration of the energy circumstances. The new system solved these problems by using a real time expert system. The outline of the algorithm is shown in Fig. 4.

The Gas Balance Calculation

The production and the consumption of each by-product gas can be estimated, based on the equipment production plan and the actual result of the production and the consumption. In the minor loop of the M-gas mixer, WI (Wobble index), which is a calorie index, is constant. The gas holder's level must be higher than the low limit, and lower than the high limit. According to these, the by-product gas quantity to the power station can be calculated.

$$G_i^P - G_i^C = G_i^M + G_i^E + G_i^H + G_i^S \qquad (1)$$

$$G_M^S = \sum_{i=1}^{3} G_i^M \qquad (2)$$

$$WI = \frac{\sum_{i=1}^{3} H_i G_i^M}{\sqrt{\sum_{i=1}^{3} R_i G_i^M}} = \text{const.} \qquad (3)$$

$$L_i^L \leqq L_i^O + \int_0^t G_i^H dt \leqq L_i^H \qquad (4)$$

$$
\begin{aligned}
i &= 1 : \text{B-gas,} \\
&\quad 2 : \text{C-gas,} \\
&\quad 3 : \text{LD-gas,}
\end{aligned}
$$

where, G_i^P : each gas production (total),

G_i^C : each gas consumption (total),

G_M^C : M-gas consumption (total),

G_i^M : each gas quantity to the M-gas mixer,

G_i^E : each gas quantity to the power station,

G_i^H : each gas quantity to each gas holder,

G_i^S : each gas quantity to be sacrificed,

H_i : each gas calorie,

R_i : each gas density,

L_i^O : initial level of each gas holder,

L_i^H : high level limit of each gas holder,

L_i^L : low level limit of each gas holder.

The Steam Balance Calculation

The production and the consumption of by-product steam can be estimated based on the equipment production plan and the actual result of the production and the consumption. According to these figures, the quantity of steam to or from the power station can be calculated.

$$S^P - S^C = S^E + S^S \qquad (5)$$

where, S^P : steam production (total),
S^C : steam consumption (total),

S^E : steam quantity to or from the power station,

S^S : steam quantity to be sacrificed.

The Electricity Balance Calculation

The production and the consumption of electricity can be estimated based on the equipment production plan and the actual result of the production and the consumption. According to these figure, the result of the gas balance calculation and the steam balance calculation, the surplus or the shortage of electricity can be calculated.

$$E^C = E^P + \sum_{i=1}^{3} k_i G_i^E + k_S S^E + k_O O + E^X - E^Y \quad (6)$$

where, E^P : electricity production (total),
E^C : electricity consumption (total),
E^X : purchased electricity from the electric co.,
E^Y : supplied electricity to the electric co.,
O : oil quantity to the power station,
k_i, k_S, k_O : coefficients.

Minimization of the Total Energy Cost

Theoretically, the total energy cost is calculated like;

$$C_X E^X + C_Y E^Y + C_O O + \sum_{i=1}^{3} C_i G_i^S + C_S S^S \quad (7)$$

where, C_X, C_Y, C_O, C_i, C_S : coefficients.

To minimize the surplus or the shortage of electricity, in consideration of the unit price of oil and the contents of the power rate contract with the electric company. Accordingly, the most reasonable method can be chosen from the followings;

(1) Methods for minimizing the electricity surplus

Decrease the consumption of oil.
Decrease the consumption of by-product gas at the power station (re-calculate the gas balance).
Supply electricity to the electric company.

(2) Methods for minimizing the electricity shortage

Increase the consumption of by-product gas at the power station (re-calculate the gas balance).
Increase the consumption of oil.
Purchase electricity from the electric company.

The energy operation can be decided from the above algorithm. This system calculates over a period of 3 hours at 4 minutes intervals, and in 4 minutes unit, because of the energy balance's momentary fluctuation. Fig. 5 shows a result of the calculation of the energy operation.

THE LD-GAS DISTRIBUTION CONTROL SYSTEM

Aforementioned, the energy operation decision system is a real time amendment of the energy relation plan, but there remains a necessity to manage LD-gas production more closely because of its minute by minute fluctuation. Dealing with these sudden changes puts many demands upon the operator's time, so automation would seem to be desirable. The production of LD-gas is intermitted, owing to the blowing time of the LD converters, and its flow rate is very high. Accordingly, the operators must supervise the condition of the LD converters most of the time, and adjust the LD-gas distribution (to the M-gas mixer or the power station) for defending the LD-gas holder's level depends upon his intuition and experience in this field.

This system includes an expert system that can construct a logic based on the operator's know-how as it is, and it has the property of real time. Fig. 6 shows the construction of the system, and its algorithm is mentioned in the following;

Estimation of the Blowing Time

The blowing times of the LD converters can be estimated based on the blowing plan, and the estimated times can be amended based on the actual result. So, if the actual result time of the blowing was later than estimated in the plan, the amended plan might be postponed.

$$t_B' = t_B + \Delta t \quad (8)$$

where, t_B : planned blowing start time,
t_B' : amended blowing start time,
Δt : delay time.

The Decision Concerning the Total Consumption of LD-gas

The total consumption of LD-gas can be decided and derived from the above results, from each by-product gas flow to the power station and each gas holder's level. In this system, a fuzzy function is applied for this calculation.

$$Q' = f(Q, L, t_1, t_2) \quad (9)$$

where, f : fuzzy function,
Q : total LD-gas consumption,
Q' : altered total LD-gas consumption,
L : LD-gas holder level,
t_1 : time length to next blowing position (estimated)
t_2 : time length to next blowing position after t_1. (estimated)

The Decision Concerning the LD-gas Distribution

The LD-gas distribution can be derived from the above results, from each by-product gas flow to the power station and each gas holder's level.

$$Q^M + Q^P = Q \quad (10)$$

where, Q^M : LD-gas flow to the M-gas mixer,
Q^P : LD-gas flow to the power station.

This control system calculates every 30 seconds to identify sudden changes in the process. The LD-gas flow to the M-gas mixer is controlled automatically based on the calculation result because there is frequent alteration. Fig. 7 shows an example of the result of this system.

EFFECTS OF THE NEW SYSTEM

The applicable results of real time energy control expert systems are mentioned in the following;

The Energy Operation Decision System

Labor saving; The number of personnel for energy center has been saved by half.

The LD-gas Distribution Control System

Automation; The total percentage of automatic control application is about 70%.

This system is almost full automated, but will not be successful when the estimation of the blowing time is upset. The blowing operation might be out of plan if an unexpected accident has happened or a special operation has been done, such as "reblows". This system therefore needs some logic that considers these probabilities to raise percentage in near future.

CONCLUSION

The new energy control system of Keihin Works includes a developed real time expert systems, and has come to be widely used in practice. To control real time complex processes such as those in Keihin Works, high-grade systematizers and high-grade software management are needed, but it was considered that too much software might be dangerous. The software management of an expert system is said to be comparatively simple, and so far, expert systems have been used for development of batch systems of human support. An example in point is the accident analyzing system. The success of this system showed that an expert system is sufficiently capable of being applied in real time control, and in the future, it can be expected that expert systems will play a big role in new system developments.

REFERENCES

Baggley, Glenn W. and Smyser, Harry E. (1982). An Organization Approach to Energy Management. Iron and Steel Engineer, October 1982, 48–52.

Tsunozaki, Y., Takekoshi, A., Hashimoto, K., Aoki, T., Wakimoto, K. and Sakurai, M. (1987). An Expert System for Blast Furnace Control at Fukuyama Works. NKK Technical Report, 119, 1–8.

Momoeda, K., Aoki, S. and Ishitobi, J. (1988). Expert System-Based Diagnostic System for Thermal Power Plant. Toshiba Review, 43, 9, 710–714.

Takekoshi, A., Aoki, T., Takahashi, I., Tomishima, H. and Yoshino, M. (1989). Application of Knowledge Engineering for Iron and Steel Works. NKK Technical Report, 125, 2–9.

Yada, M. (1987). Expert Systems. Artificial Intelligence Comprehensive Review, 139–183.

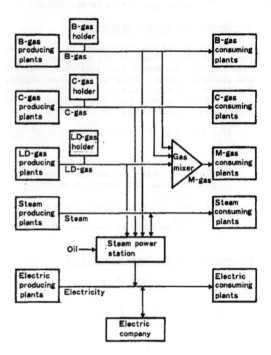

Fig.1. The energy flow of Keihin Works.

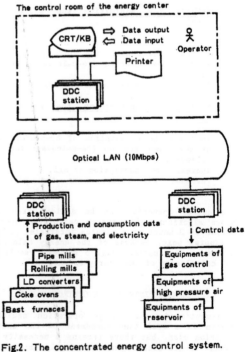

Fig.2. The concentrated energy control system.

TABLE 1. Basic Concepts of Replacing the Energy Control System

Item	Before Replacement	After Replacement
1.Data collecting	Data collectors/ Controllers of each energy control points	Optical LAN and DDC system
2.Judgment (Calculation)	Operator	Real time expert system
3.Control	Controllers of each energy control points	Optical LAN and DDC system

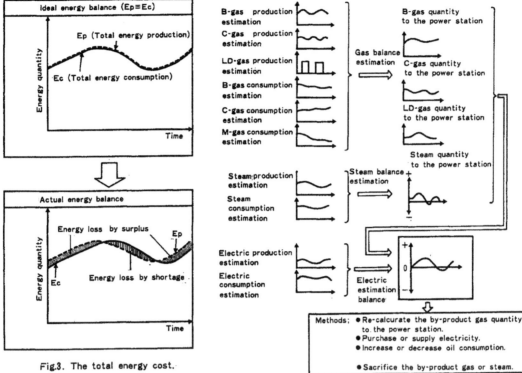

Fig.3. The total energy cost.

Fig.4. The energy operation decision system.

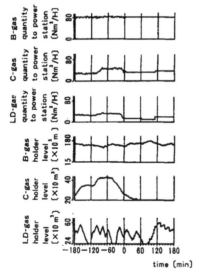

Fig.5. A result of the energy operation decision system.

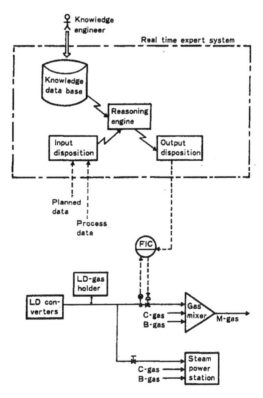

Fig.6. The LD-gas distribution control system.

386   S. Nakamura *et al.*

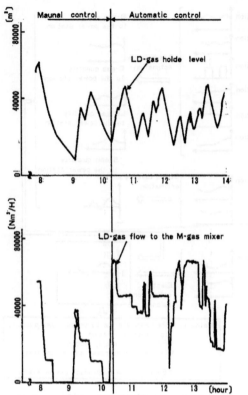

Fig.7. A result of the LD-gas distribution control.

PRODUCTION SCHEDULING USING AI TECHNIQUES

B. Rodríguez-Somoza, R. Galán and E. A. Puente

Departamento Automática, E.T.S.Ing. Industriales, Madrid, Spain

Abstract Within the CIM environment, one of the main areas of investigation at now is the scheduling system. Usually, the scheduling problem was solved in an inneficient way due to the complexity of the problem. Recently, Artificial Intelligence is begining to be apllied to this kind of problems and seems to be a good solution. This paper describes a new methodology based on Artificial Intelligence, and applied to FMS. The main goal in the development of this method, was to achieve minimun machine idle time, with a small CPU time that allow the apllication of the method to real world systems. Examples of different size are given, from seven jobs on five machines to nine jobs on nine machines. The CPU time per task are in all the examples under 1 s.

Keywords Artificial Intelligence; Flexible Manufacturing; Heuristic programing;

INTRODUCTION.

Within CIM environment, one of the main goals is to get flexibility, which means be able to produce several different goods in the same production unit at the same time, without loosing efficience.

To get such goal the hardware needed can be clasified as follows:

Hardware
- Not physically related with Production (Computer, Networks, etc)
- Physically related in Production (Machines, AGV, robots etc.)

A device, can include both classes of hardware. For instance, a robot can include an arm, and an interface with a network or computer.

About the software implied, there is a great number of clasification criterias. Also, there is diferences between the theorical functions that must be included, and the assignement of funtions to the parts of a implemented sofware system. Taking into account the functionalities that must be included, the sofware can be classified as follows:

Software

Real time
- Interfacing with Hardware
- Interfacing with other software

Non Real Time
- Design Software (CAD)
- Engineering Software (CAM)
- Quality control Software
- Production Management software

The Real time software that interfaces with other software, is named usually the scheduling and control software.

The scheduler is the upper level module of the real time software, and must decide which tasks and when will take place. The main goal of the scheduler is to get the maximum work from the manufacturing system, taking into account the production policies established at a higher level and off-line by the management system. Usually the scheduler is composed by off-line and on-line packages, but this separation is only due to real time restrictions, and not to other considerations.

The scheduling problem is mathematically a NP-hard one, which means that no analitical solution can be obtained for it. The simplest method to solve the problem, is to search in a tree, but such a tree will be a very large one, and then for normal problems this method is unapllyable.

A factory is not a deterministic environment, several kinds of disturbances (breakdowns of machines, tools), and unknown factors makes imposible to use a off-line schedule without modifing it during the production process.

The most usual solution, solves some part of the problem (parts loading, Routing, etc.) off-line, and uses very simple dispatching policies on-line (LPT, SPT, EST, EFT, etc.), this methods gives a solution far from the optimum one, but simplifies the problem (Brandolese,1983; Buckley,1988; Mohamed,1984; Morutsu,1983; Singh,1984). Those dispatching rules, has a great influence in the scheduling results, and so a lot of investigation was do about they (Edghill,1985; Graves,1981; Panwalker,1977).

About theoretical solutions, that cannot be apllied on-line but helps to get an idea about the optimum schedule, some solutions was obtained for flow-shops of a limited size but no for job-shops. So it can be deducted that a job-shop is quite more complex than a flow-shop.

Also a number of interactive systems was developed that allows a human operator to define and test scheduling procedures (Nakamura, 1988; Spur,1985).

Another class of solutions begin to appear now, those methods try to aplly in several ways AI techniques. AI is specially well suited to the problem, since it can use approximate reasoning to obtain good solutions for very complex problems. Various approaches can be actually considered, constrait directed search(Smith,1986), Expert Systems (Sauve,1987), mixed heuristic functions-rules (Chryssolouris,1988), and Predicate Logic (Bullers,1980). All of this method until today was off-line methods.

This work develops a scheduling system for job-shops, that operates in real-time without the need of off-line scheduling, and using AI techniques.

The optimization goal, is the minimum idle time in the machines. For small systems (for instance four machines and four jobs), usually a optimun solution (losses = 0) is obtained.

At this time there is no posibility to take into account WIP, buffer size, or due dates. The method doesn't accept variable paths for each job. Those thinks, will be considered in the future. Also, due to the kind of factory for what the method is prepared, the setup times is considered as negligible.

The method use rules to select the tasks that will be considered as alternatives in each step. Also, a set of rules is used to create a priority order between those tasks. When the tasks are selected and ordered, a heuristic function evaluates each one, and a master module chosse one of it.

Three kinds of search can be used. The first, suposes that when a task cannot begin at a time less than the beginning of the previous. The second one, make the same hypothesis when chosing a task, but don't modifies the minumum times when the task is scheduled. The third one forgets all about real-time.

ALGORITHM.

The general flowgraph for the scheduling problem, is the following.

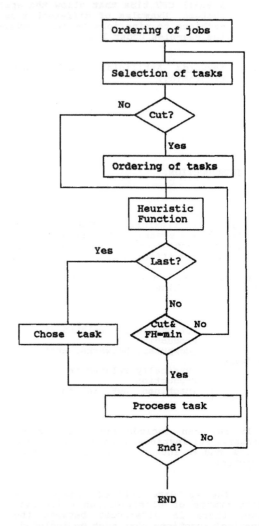

At the begining of the development, there was observed a great dependence of the solution with the order of the jobs, so a previous ordering method was developed that helps to find good solutions. At present the quality of the solution is only weak coupled with this order, but the ordering method can help to get solutions in less time, and have some interest about behavoir of the problem. The method is not used usually because the goal of the work is to develop an only real-time method, and previous ordering must take place off-line.

The method run in real-time, so when a machine finish his work, the computer look out for new tasks. First, it select a set of tasks from the ones that can be processed at the time or later.

When apllying the heuristic function, there is a minumum value that can be obtained, so the algorithm takes the first task that give this value. Due to the fact that several task can give the minumum value of the heuristic function, and that one of these task can be better than other, it is desirable to order the set of tasks in order of preference.

The heuristic function need a few time of CPU, so it must be used as limitted as possible. If only one heuristic criteria is used, and the minumum value of it is know in each step, it is possible to take the first task that give this value, and don't lose a better choice.

There is important to note that losses is not the same that idle time. The idle time can be necessary. For example, if there are only one job with two tasks, 5 sec at machine 1 and 3 sec at machine 2 in this order, there will be 2 sec of idle time each time a piece is procesed.

1.- Previous ordering of jobs.

This module was used only to search into the behavoir of the problem, but usually it is'nt used because it imposses a constrait that cannot be usually present into real systems. The solutions obtained with previous ordering is usually better than the ones obtained without it.

To make this ordering, a tree is build. Each node of the tree represent a set of jobs including from 1 to the total number, and going from one node to another, implies add a new job to the set. The order in what the jobs are added, has an important meaning.

The method look out for a path throught the tree in what the total idle time and the idle time increases from the first to the final node (the full set of jobs). To simplify searching, the objetive function is to minimize the sum of idle time of each node. The search begins with the two job sets. The idle time can be considered with two quantities, the absolute value or the percent value.

2.-Selection of tasks.

Several selection criterias was implemented during development. Those criterias use the concept of first interval in several implementations. The notion of first interval tries to reflect the fact that not all of the possible tasks can be considered as good alternatives, and then the system can save the time to evaluate it.

A valid first interval, is a condition that allows to build a set of tasks, and complies that a task not included in the set is not a valid alternative to those included. Such a condition is not very difficult to design, the problem is to find a good one.

A task is a not valid alternative, if when evaluated with the heuristic function, the value obtained is worse than the value of any valid alternative. To analize a first interval criteria, common sense is used and not the heuristic function.

As an example, one of the criterias is explained.

Real-time first interval (RTFI) definition.

This first interval criteria, is designed to be used when real time is needed. The main hypothesis is that the starting time of the task that is scheduled, will be automatically the real time of the system, and so the task that is scheduled next must start at this time or later.

Let us see a example of which is the RTFI in one particular case.

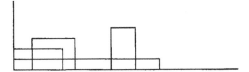

In the previous figure, each task is represented by a rectangle. The height is a integer that gives the number of the job that need this task. The X-axis represent time. There are four tasks in the figure. In this particular case, it has no sense to take into account task 4 when chosing a task. If we chosse task 4 it is clear that tasks 2 or 3 can be processed prior with no interaction.

A more formal definition of the RTFI is the following: " The real-time first interval for a particular situation in a factory, gives the set of tasks that must be considered when chossing the next task. The RTFI includes all the tasks so that it beginning time is less or equal the first end time of all the tasks possible".

The explanation is valid for the tasks of one machine, but can be generalized to several machines, applying it as if the tasks pertains all to the same machine.

3- Ordering of tasks.

As we state previously, there is desirable to order the tasks when a method with cut is used.

Task ordering can be used for two purposes, expend less time while evaluating alternatives (find a good one as soon as posible) or help the heuristic function to get a good schedule.
In this work the second one is considered.

The ordering methods, was obtained from usual dispatching rules. When the first rule cannot chosse in a great number of situations, it is necesary to add another criteria. Such is the case of EST. The EST criteria, fail to chosse when the are a number of tasks that can be started inmediatly, when there are no idle time, given that this is a usual situation, it is desirable to add a secondary criteria.

The ordering criterias implemented at now are the following:

- .- Earliest Starting Time & Default order. The tasks that begins at a earlier time, was considered first. When two tasks has the same begining time, the one that was included previously in the list of tasks, was considered first.
- .- Earliest Starting Time & Minumum proportion produced. Same as EST, but when two tasks begin at the same time, the one that was produced in less proportion of the demand, is chossed.
- .- Earliest Starting Time & Largest Processing Time. Same as EST, but when two tasks begin at the same time, the one that has greater processing time was selected.
- .- Largest Processing Time.
- .- Earliest Starting Time & Shortest Processing Time.
- .- Shortest Processing Time.
- .- Largest Remaining Processing Time. The task for which the processing time of the remaining tasks in the job was greater, is selected.
- .- Largest Remaining idle time. The remaining idle time is the time that the machines must wait to end the tasks that follows the task considered in the job, if the job is the only that will be produced in the system. The time of the machines is measured begining at the finish time of the task been evaluated

4.- Heuristic función.

This heuristic function estimates the minumum finishing time for each machine.

Previous to begin scheduling, a estimation is obtained that will be used as a objetive, and to measures the final losses.

The minumum time of each machine considered alone, is estimated first. With the minumum time for each machine separately, and apllying the optimum criteria of minumum total time expended by the system, another finishing times for the machines is obtained.

The function calculates the theorical optimum (not sure feasible) for each situation. The first step calculates the optimum for each machine separately and a second one for the complet system and taking into account the optimization criteria. The optimum criteria is to minimize the sum of finishing times. What measures the convenience to execute a task, is the difference between the estimation for finishing times prior an following the execution.

With the heuristic function implemented, the idle time is not the real idle time of each machine, but the difference between the minumum idle time and the real idle time.

A very important comment is that the heuristic function used, tend to give no idle time to the machines with more work, as the examples show. So the span for the production system is usually optimun or very near.

It must be noted that the goal calculated is ever better or equal than the real optimum (the feasible one). Even if it can be possible to calculate the finishing time with more precission, due to time restrictions, at now a less exactly approach is used. This must be taked into account in the examples, because the loosses are not real but only with reference to the calculated optimum. The calculated optimum is more far from the real one as the size of the example increases.

About the CPU Time for each step, it increases linearly with the demand of pieces.

RESULTS.

The program was developed in a HP 520 with HP-UX Operating System and using C language. And on AT compatible computer with DOS 3.3 and using Turbo C of Borland.

From the the examples used to test the program, four of them with different complexities are included. The CPU times was for the 286 machine.

Due to the different number of options that the program allows, only the results for one set of options is given. There are not one method that gives the best solution for all the examples, so the one that gives the best media is considered.

The option set considered is:

Main function method	:Real Time
Previous ordering	:Not used
Selection	:RTFI
Ordering	:LRIT

CASE 1:

Total CPU time	55 s
Tasks	442
Average time per task	.12 s

Job1(20 units)	M1(6)	M3(4)		
Job2(34 units)	M2(9)	M4(10)	M3(5)	
Job3(20 units)	M1(8)	M2(5)	M3(3)	M4(8)
Job4(20 units)	M2(3)	M1(7)		
Job5(15 units)	M1(14)	M2(8)	M3(9)	M4(11)
Job6(20 units)	M1(4)	M2(5)	M3(10)	M4(7)
Job7(10 units)	M1(6)	M2(9)	M3(10)	M4(3)
Job8(5 units)	M1(8)	M2(10)	M3(6)	M4(7)

Machine	minT	RminT	FinalT	Loss	%
1	740	740	740	0	0.00
2	796	796	796	0	0.00
3	775	799	807	8	0.99
4	870	879	883	4	0.45
TOTAL		3214	3225	12	0.37

CASE 2:

Total CPU time	75 s
Tasks	360
Average time per task	0.21 s

Job1(10 units)	M1(5)	M3(10)		
Job2(10 units)	M1(6)	M2(10)	M4(10)	M3(11)
Job3(10 units)	M2(5)	M3(10)	M4(9)	
Job4(10 units)	M1(3)	M2(4)	M3(9)	M4(2)
Job5(10 units)	M1(4)	M2(12)	M3(6)	
Job6(10 units)	M1(9)	M2(10)	M3(6)	M4(3)
Job7(10 units)	M2(9)	M1(3)	M3(10)	M4(1)
Job8(10 units)	M1(10)	M2(9)	M3(8)	M4(7)
Job9(10 units)	M1(2)	M2(3)	M3(4)	M4(5)
Job10(10 units)	M1(5)	M2(2)	M3(7)	M4(10)

Machine	minT	RminT	Ftime	Loss	%
1	470	470	470	0	0.00
2	670	670	670	0	0.00
3	810	815	823	8	0.97
4	470	559	566	7	1.24
TOTAL		2514	2529	15	0.59

CASE 3:

Total CPU time	40 s
Tasks	290
Average time per task	0.14 s

Job1(10 units)	M1(5)	M2(9)	M3(10)	M4(10)
	M5(8)			
Job2(10 units)	M1(7)	M2(9)	M3(3)	M4(10)
	M5(12)			
Job3(10 units)	M2(10)	M1(9)	M3(11)	M4(12)
	M5(13)			
Job4(10 units)	M1(7)	M2(9)	M3(5)	M4(6)
	M5(3)			
Job5(10 units)	M2(9)	M5(10)	M4(7)	M1(5)
Job6(10 units)	M1(5)	M3(7)	M5(9)	
Job7(10 units)	M4(10)	M3(8)		

Machine	minT	RminT	Ftime	Loss	%
1	380	380	380	0	0.00
2	460	460	460	0	0.00
3	440	463	465	2	0.43
4	550	550	551	1	0.18
5	550	559	570	11	1.93
TOTAL		2412	2426	14	0.58

CASE 4:

Total CPU time	490 s
Tasks	600
Average time per task	0.81 s

J1(10 units)	M1(5)	M2(9)	M3(10)	M4(10)
	M5(8)	M6(10)		
J2(10 units)	M1(7)	M7(9)	M3(3)	M4(10)
	M2(5)	M5(7)	M8(3)	M9(14)
	M6(1)			
J3(10 units)	M2(10)	M1(9)	M3(11)	M9(7)
	M5(13)			
J4(10 units)	M1(7)	M2(9)	M3(5)	M4(6)
	M5(3)	M9(10)	M8(6)	M7(14)
J5(10 units)	M2(9)	M5(10)	M4(7)	M1(5)
	M8(10)	M6(5)		
J6(10 units)	M1(5)	M3(7)	M5(9)	
J7(10 units)	M4(10)	M3(8)	M2(7)	M1(13)
	M8(6)	M7(7)	M5(10)	
J8(10 units)	M5(10)	M2(15)	M9(7)	M8(10)
	M7(12)	M6(12)	M4(6)	M3(9)
	M1(1)			
J9(10 units)	M9(10)	M8(11)	M7(9)	M5(14)

Machine	minT	RminT	Ftime	Loss	%
1	525	724	736	12	1.63
2	715	715	715	0	0.00
3	575	735	751	16	2.13
4	520	520	598	78	13.04
5	1030	1030	1035	5	0.48
6	280	480	555	75	13.51
7	720	727	727	0	0.43
8	620	630	715	85	11.89
9	615	742	758	16	2.11
TOTAL		6303	6590	287	4.36

BIBLIOGRAPHY

Bensana,E.,G.Bel, and D.Dubois (1988). OPAL: A multi-Knowledge-based system for industrial job-shop scheduling. Int. J. Prod. Res.Vol 26, N 5

Buckley,J.,A.Chan,U.Graefe,J.Neelamkavil, ,M.Serrer,V.Thomson (1988). An integrated production planning and scheduling system for manufacturing plants.Robotics & Computer integrated manufacturing.Vol 4, N 3/4,1988

Brandolese,A.,M.Garetti (1983). FMS Control systems: design criteria and performance analysis. Proc of 2nd Int. Conf. on Flexible Manufacturing Systems. London,October

Bullers,W.I.,S.Y.Nof, and A.B.Whinston (1980). Artificial Intelligence in Manufacturing Planning and Control. AIIE Trans. Vol 12, N 4.Dec

Chryssolouris,G.,K.Wright,J.Pierce,W.Cobb (1988). Manufacturing systems: dispatch rules versus intelligent control. Robotics & Computer integrated manufacturing.Vol 4, N 3/4.1988

Dato,M.A.,F.Ciaffi, and P.Cigna (1983). A generalised job scheduling for FMS. Proc of 2nd Int. Conf. on Flexible Manufacturing Systems. Londres.October

Edghill,J.S., and C.Cresswell (1985). FMS control strategy - a survey of the determining characteristics. Proc. of 4th Int. Conf. on Flexible Manufacturing Systems. Stockholm.

Graves, S.C. (1981). A Review of Production Scheduling. Operations Research Quarterly.Vol 29, N 4.July-August.

Mohamed,N.S.,M.F.Abdin,and A.S.El Sabbagh (1984). Production Scheduling of Job-Shop Type Flexible Manufacturing. Robotics & Factories of the future, International Conference, Charlotte, North Carolina U.S.A. Dec.

Morutsu,Y., and F.Oba (1983). A Production Scheduling System for Flexible Manufacturing Systems. 1st International IFIP Conference on Computer Applications in Production and Engineering, Amsterdam.April.

Nakamura,N., and G.Salvendy (1988). An experimental study of human decision-making in computer-based scheduling of flexible manufacturing systems. Int. J. Prod. Res.Vol 26, N 4.

Panwalker,S.S., and W.Iskander (1977). A survey of Scheduling Rules. Operations Research.Vol 25, N 1.

Puente,E.A.,R.Aracil,C.Balaguer, and J.M. Sebastian (1986). Control and management structure of the FMS DISAM/2. IPAC'86.

Ranky, P.G. (1985). Computer Integrated Manufacturing. Prentice Hall International.

Rembold,U.,C.Blume, and R.Dillman (1985). Computer Integrated Manufacturing and Systems Technology.Ed Dekker.

Sauve,B., and A.Collinot (1987). An Expert System for Scheduling in a Flexible Manufacturing System. Robotics & Computer Integrated Manufacturing. Vol 3, N 2.

Shaw,M. (1988). Knowledge-based scheduling in flexible manufacturing systems: An integration of pattern-directed inference and heuristic search. Int. J. Prod. Res.Vol 26, N 5.

Singh,V. (1984). Robotics and Scheduling in Flexible Manufacturing System. Robotics and Factories of the future, International Conference, Chartlotte, North Carolina U.S.A. Dec.

Slomp,J.,G.J.C.Gaalman, and W.M.Nawijn (1988). Quasi on-line scheduling procedures for flexible manufacturing systems. Int. J. Prod. Res.Vol 26, N 5.

Smith,S.F.,M.S.Fox, and P.S.Ow (1986). Constructing and Maintaining Detailed Production Plants, Investigations into the Development of Knowledge-Based Factory Scheduling Systems. IA Magazine.FALL.

Spur, G., and K.Mertins (1985). Strategy based interactive production control for flexible automated systems. Proc. of 4th Int. Conf. on Flexible Manufacturing Systems. Stockholm.

Weck,M., and G.Kiratli (1987). Applicability of expert systems to flexible manufacturing. Robotics & Computer Integrated Manufacturing. Vol 3, N 1.

PRODUCTION SCHEDULING AND SHOP-FLOOR CONTROL USING A RELATIONAL DATA BASE MANAGEMENT SYSTEM

B. Benhabib, A. Cupillari, M. Charania and R. G. Fenton

Department of Mechanical Engineering, University of Toronto, Toronto, Canada

Abstract

The software discussed in this paper has been developed utilizing a relational data base management system. It consists of thirteen application programs, which provide a tool of organizing information for efficient production scheduling and shop-floor control. The programs are designed to cover all manufacturing operations of a job from a proposal stage to the final testing stage.

Shop orders and dispatch lists are created using the software for effective and prioritized shop-floor management. Shop status and job status reports are generated based on feedback information received through time card entries. These status reports are viewed by foremen and management in detailed or summary forms.

INTRODUCTION

The production planning related information processing activities in a manufacturing environment include: process planning, production scheduling, requirements planning, and capacity planning, (Groover, 1987). Production control, on the other hand, is mainly concerned with monitoring and controlling the physical operations on the shop-floor to implement production plans, (Groover, 1987). The flow of information between planning and control should of be closed-loop type for improved manufacturing efficiency.

During the last decade extensive research and development took place in the area of computerized production planning and control systems, (Arthor, 1986; Jurgen, 1983; Knox, 1983). Recent technological advances in computer aided design and manufacturing (CAD/CAM) systems, when combined with more traditional production philosophies such as Group Technology, provide today's manufacturers with effective tools to link their design and planning departments to the shop-floor.

The main functions of a computerized production planning and control system in a job-shop manufacturing environment are as follows, (Knox, 1983):

Production scheduling. The processing of customer orders.

Priority planning. Scheduling the sequence of orders on the shop-floor.

Shop-floor control. Monitoring the progress of job orders on the shop-floor and reporting the status of each order to management so that effective control can be exercised.

The backbone of the above production planning and control functions is the engineering data base, containing the data required to produce the components and assemble the final product. It includes: product designs, material specifications, bill of materials, process plans, etc. The special computer software that would effectively manipulate this data is commonly referred to as a data base management system (DBMS), (Tsichritzis, 1982).

The customized production scheduling and shop-floor control software developed and described in this paper

uses a relational DBMS. This management software tool represents the first phase of the information flow automation process in a medium size company where it was implemented. The main product line of this company is injection molds for thin walled plastic containers.

Each mold is composed of approximately forty different (unique) "manufactured parts", twenty different "standard parts" which are common to all molds, and forty different "purchased parts". The manufactured and the standard parts are produced in house. The standard parts are considered as manufactured parts during their production and as purchased parts when they are used for compiling the bill of materials of the molds. The company's intention is to automate the processing of the following information: bill of materials (BOM), detailed shop-floor instructions, and accounting. In response to this goal a three-year program was undertaken, of which the first completed phase is described in this paper.

This paper consists of four main sections. Section 2 provides an insight to relational DBMSs and specifically discusses the Empress DBMS used for the software development. Section 3 details the entity relationship diagram structured for efficient data manipulation for the specific company, and describes the application programs developed and used for this purpose. Section 4 summarizes the development and presents the conclusions.

RELATIONAL DATA BASE MANAGEMENT SYSTEMS

A data base management system (DBMS) is a software tool for developing application programs requiring access to shared information. It provides various functions via a high level transactions specification language (3GL and 4GL). Users can generate transactions in this language and the DBMS assumes the responsibility for managing physical data access and access rights, recovery from systems failures, etc., (Tsichritzis, 1982).

The first step in designing a data base scheme is the creation of an Entity Relationship Diagram (ERD) (Goldstein, 1985). An entity is a "thing" that can be distinctively identified. Some entities relevant to manufacturing data are: jobs, employees, and customers. A relationship is an association among entities. A relationship relevant to manufacturing data, is: the relationship between the entities Shop_Orders and Employees, specified as (employees) "work-on" (shop orders). In a typical ERD, entities are identified by rectangles, and relationships are identified by diamond shapes placed between entities. The ERD can also explicitly represent an "existence dependency constraint", which is identified by a double rectangle. For example, a constraint can be defined as, "the existence of the Time_Card entity depends on the existence of the associated Employee entity".

In a relational DBMS model there is only one basic construct, the "relation", which is used to represent both entities and relationships, (Goldstein, 1985; Vernadat, 1983). A relationship can be thought of as a two dimensional table. A column in this table corresponds to each attribute of the entity or relationship type, and a row (tuple) or record for each occurrence of it.
The Empress/32 relational DBMS was used in the development of the production scheduling and shop-floor control software, (Rhodnius Inc., 1987). This particular DBMS includes: a version of the structured query language (Empress/32 SQL); a powerful report writer (M-Writer); a form-style screen interface; a 4GL application builder (M-Builder); 3GL applications (MX and MR) ; and, standard interfaces to high level languages, such as, C, Fortran, and Cobol. The software configuration is illustrated in Fig. 1.

The application builder (M-Builder) is a high level operating environment for developing and running applications. It provides tools for screen painting, windowing, and function key definition, which make it particularly suitable for the development of screen intensive applications driven by menus and function keys. Applications are programmed in a command language that provides commands to access data base tables; commands to control windowing and screen display; and, commands to invoke the operating system and the Empress SQL.

The fully integrated report generator (M-Writer) can be employed to produce custom formatted reports using records from the data base tables, standard files, and other application programs. Values can be displayed in a variety of ways, using default or picture formats. Arithmetic operations can be applied to the data (grouped or sub-grouped), where sorting or sub-totalling can be carried out for any level of grouping.

The MX and MR applications are languages allowing to modify a record at a time. For example summing up the attributes of a record or changing the value of on attribute.

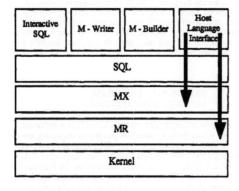

Fig. 1. Software configuration

SOFTWARE DEVELOPMENT

Entity Relationship Diagram

The entity relationship diagram (ERD) given in Fig. 2 was developed after determining the necessary manufacturing information flow. The ERD illustrates the relationships between the most important entities used in the software.

The Job_Desc entity is used to group all attributes concerning the description of a job and associate them with the customers. Since each job may belong to only one customer, but each customer may have more than one job order, this relationship has a 1:N cardinality.

The Job_Dates entity relates completion dates to jobs. Since each job would have at most one completion date, this relationship has 1:1 cardinality.

The Job_Parts entity is used to prepare BOM for a job, which may consist of manufactured, standard, and purchased parts. Since each manufactured part may belong to only one job, but a job may have more than one manufactured part, this relationship has a 1:N cardinality. Standard and purchased parts, on the other hand, are not identified by a specific job number, thus, this relationship has an N:M cardinality. Since each job would have only one BOM, and vice versa, the relationship between Jobs_Desc and Job_Parts has a 1:1 cardinality.

Shop orders are created for manufactured parts. These may consist of more than one operation (tuple). Thus, "make mfg parts" and "make standard parts" relationships between Shop_Orders and Job_Parts have a 1:N cardinality. Since several employees can work on a single shop order operation, the corresponding relationship has an N:M cardinality. Similarly, a single operation can be completed in more than one shift (i.e., more than one time card), thus, the corresponding relationship has an N:M cardinality.

Process plans provide a detailed breakdown (sub-operations) for every shop order operation, thus, yielding a relationship which has a 1:N cardinality.

All information concerning machine groups, machine tools, and attachments are stored in Tools_Desc. The relationship between Tools_Desc and Shop_Orders has a 1:N cardinality, while the relationship between Tools_Desc and Process_Plans has an M:N cardinality.

Application Programs

The flow diagram given in Figure 3 illustrates the necessary interfaces between individual modules - application programs - that manipulate the manufacturing data. The thirteen modules can be categorized into the following six sub-classes for clarity:

A. Entries. Modules used to enter basic data that would enable the information flow to commence:
- *Employee entry*
- *Customer/Supplier entry*
- *Machine tool and attachments entry*

B. Production scheduling. Modules used to initiate production scheduling:
- *Job entry*
- *Parts coding and entry*
- *Bill of materials entry*

C. Production in an open-loop. Modules used to schedule production in an open-loop mode:
- *Shop orders entry*
- *Process plans entry*

D. Shop floor. Modules used to collect data from the shop-floor, interpret, and prioritize scheduling, thus, closing the loop of information flow:

- *Shop-floor status (and dispatch list generation)*
- *Time card entry*

E. Management. Modules used to provide management with an overview of the production status:
- *Job/Production status*

F. Information storage. Modules used to store information on completed jobs:
- *Shop orders history*
- *Time cards history*

A.1. Employee Entry. This application is used to insert and/or update employee data (number, name, etc.). This program can also be used to sort and display (or print) employee records based on the selected attribute - numerical sorting according to employee number, and alphabetical sorting according to employee (last) name.

A.2. Customer/Supplier Entry. This application is used to insert and/or update customer/supplier data (number, name, etc.). This program can also be used to sort and display (or print) customer/supplier records based on the selected attribute - numerical sorting according to customer/supplier number, and alphabetical sorting according to customer/ supplier name.

A.3. Tool Entry. This application is used to insert and/or update machine tool and attachments data. The insertion process is hierarchical, every machine tool record inserted must belong to a "group" (cell), and each group must be assigned a unique group number. Similarly, attachments are assigned identification numbers and a machine tool number which they belong to. This program can also be used to sort and display (or print) data according to groups, or provide complete numerically sorted lists of all selected machine tools and attachments.

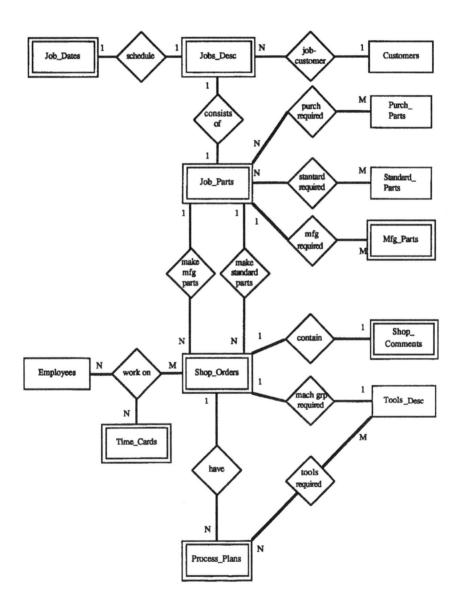

Fig. 2. Entity Relationship Diagram

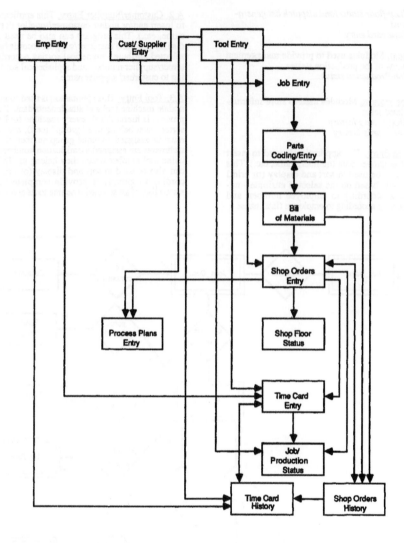

Fig. 3. Information Flow Diagram for Application Program Interface.

B.1. Job Entry. This application is used to insert and/or update proposal and job data. Every job is assigned a unique job number, beside other relevant information on the customer. Once a job is entered, its expected completion date also has to be entered. If an update is made to the completion date of a specific job, all shop order dates belonging to this job are also automatically updated. Dates are issued based on calendar year working days (Monday to Friday).

Manufacturing priorities in terms of Rush (R), Medium (M), and Late (L) are assigned to operations through this application program as well. The late priority (L) is automatically assigned to all operations which have not commenced by their planned completion dates.

Priorities can be assigned to all the shop orders of a particular job, to a particular shop order, or to a particular operation of a particular shop order. It should be noted that, when a completion date of a job is changed, all its priorities are set to null and will have to be reset if necessary (except the "L" priorities, which are set automatically).

In addition to issuing priorities, this program releases shop orders automatically on their specified release dates. It can also be used to sort and display (or print) job data according to job number, job type, customer number, and customer name.

B.2. Parts Coding and Entry. This application is used to code parts and assign a part number, through a user friendly menu system.

Every *manufactured part* is first coded based on a customized Group Technology (GT) classification/coding (C/C) scheme developed for that purpose (Benhabib and others, 1989). Secondly, the manufactured part is also assigned a unique engineering part number. Thirdly, the part is assigned a job number.

Every *standard part* is first coded based on a customized Group Technology (GT) classification/coding (C/C) scheme developed for that purpose. Secondly the standard part is also assigned a unique engineering part number. Thirdly, the part is assigned a temporary job number which will facilitate the retrieval of the part in the compilation of the BOM.

Every *purchased part* is first assigned a unique engineering part number. Secondly, the part is assigned a temporary job number which will facilitate the retrieval of the part in the compilation of BOM.

Part data can be inserted and/or updated only through this program. A part's data record includes attributes such as: job number, engineering part number (for shop-floor and inventory control), GT number (for design and process planning), description, size, etc. This program can also be used to numerically sort and display part data based on the specified attributes.

Once a GT code is determined for a manufactured part or standard part and the record insertion process is initiated, this application program automatically runs a search-routine which determines the closest GT number in the data base of existing parts. This GT number is permanently stored in the record of the current part. It represents the most similar part in the data base selected on the basis of seven GT attributes (geometry / material / manufacturing process / etc.), (Benhabib and others, 1989).

B.3. Bill of Materials. This application program is used to create BOMs for already existing jobs (entered through the Job Entry program). Since during part entry manufactured parts have already been assigned job numbers, once a job number is selected all corresponding parts are automatically selected into the BOM. Standard and purchased parts can be selected by the temporary job number assigned to them in the part entry application program. Also, since both application programs share a common data base, any changes made herein to part records are reflected into the Parts Coding/Entry application program. For example, when a manufactured part record is deleted from the BOM that record is also removed from the part records table.

Once the BOM of a part is created and inserted, it can be displayed or printed using this program.

C.1. Shop Orders Entry. This application program is used to create shop orders for manufactured parts. The shop order for a specific part includes standard attributes describing that part, as well as a routing plan for its production. Each operation entry includes: machine tool used, general manufacturing description, estimated manufacturing time per part, priority settings, etc.

This program is also used to sort and display (or print) shop orders according to job number, part number, and/or release date. Shop orders for a completed job can be transferred to the history table, if desired, using the Shop Order History application program (described in (F.1)).

C.2. Process Plans Entry. This application program is used to create process plans for manufactured parts. Process plans provide a mean of specifying the details of how to carry out a machining operation listed in the shop order. Thus, the process plan can be considered a detailed version of the shop order, where operations are further detailed as sub-operations specifying machining attributes such as depth of cut, feed rate, cutting speed, tools and fixtures required.

The GT code of the part, for which the process plan is prepared, can be used in this program to search the data base for the most similar existing process plan of a previously manufactured part. It should be noted that, the process plans of completed jobs should be accessible in order to conduct an efficient and useful "search & modify" scheme.

D.1. Shop Floor Status. This application program is used to display all incomplete shop order operations. The data can be sorted and displayed (or printed) according to selected attributes, such as, job number, machine number, specific date, etc. When a shop order status is displayed, the sum of all operation times up-to-date is also given, and, the estimated remaining manufacturing time to complete that part is indicated.

This application program can also be used to generate priority ordered *dispatch lists* for all machine tools (or machine tool groups), for a particular period of time. These lists also indicate the sum of manufacturing times for each priority group. Job summary lists can also be generated in the same fashion.

A *shop loading schedule (SLS)*, summarizing weekly loading of all machines on the shop-floor can also be generated using this program. The SLS can be generated for a specific job, or for all the jobs in-progress. The first and last weeks for the SLS can be either specified or default values can be accepted.

D.2. Time Card Entry. This application program is used to enter time cards completed by the employees at the end of each shift. (Work is in progress to completely automate this function for real-time data entry from the shop-floor). Once this data is entered, all data base tables are updated automatically.

This application program can also be used to sort and display (or print) detailed or summary job lists (identifying employees who worked on them), detailed or summary employee lists (identifying machines that the employees worked on) according to attributes, such as, employee number, job number, part number, date, etc. The summary versions of these lists are for accounting purposes. For all completed jobs, corresponding time card records can be transferred to the history tables, if desired, using the Time Card History application program (described in (F.2)).

E.1. Job/Production Status. This application program is used to display (or print) proposal and job summary lists for records inserted through the Job Entry program. A detailed or summary list of completed shop operations can also be displayed using this program, showing estimated versus actual "production times" and "machines used". Additional part summary or job summary lists, which display the percentage "completeness" of the work, can also be generated.

F.1. Shop Order History. This application program is used to display shop orders of completed jobs which have been transferred into the history tables. The selection process is one-by-one as in the case of the shop order entry application program. A more comprehensive list of the shop orders can also be generated using this program, sorted according to selection attributes such as: job number, release date, etc. Shop orders can be reactivated, when necessary, by transferring them back into the main data base tables.

F.2. Time Card History. This application program is used to display time card records of completed jobs which have been transferred into the history tables. The selection process is one-by-one as in the case of time card entry application program.

SUMMARY

A production scheduling and shop-floor control software developed using the relational data base management system Empress was presented in this paper. The application package consists of 13 main application programs which process data from the tables of the data base. The programs are designed to trace all manufacturing operations of a job from its initial proposal stage to its final completion stage.

Prior to inserting the first job into the Job Entry application program, the following three programs have to be employed to insert the necessary engineering data: the Employee Entry; the Customer/ Supplier Entry; and, the Tool Entry application programs. Once sufficient data has been entered through these programs, a job and its corresponding completion date can be entered. The Parts Coding/Entry program can then be used to insert manufactured, standard, and purchased parts into the data base. These parts are then inserted into the BOM using the BOM application program.

The Shop Orders Entry application program is used to generate manufacturing shop orders, to plan machine loadings, and to determine job completion dates. The Process Plans Entry is used to provide more detailed information concerning tools and fixtures, and machining conditions for each operation on the manufacturing of a part specified in the shop order. To plan day-to-day shop floor activities and generate dispatch lists, the Shop Floor Status program is utilized.

The Time Card Entry program is used to close the loop of information flow between the planning department and the shop-floor. The Job/Production Status program provides authorized management with vital status information on the shop activities. The software also provides the possibility of storing completed job information on "history" tables, which can be re-activated whenever necessary.

The software developed and discussed in this paper provides an opportunity to introduce automated production scheduling and shop-floor control in medium size manufacturing company.

REFERENCES

Arthor, W.E. (1986). Shop floor control - the first step to CIM. CASA-SME. Technical Paper.. MS86. 227.

Benhabib, B., R.G. Fenton, K. Jagadisan, T. Kolovos, and I.B. Turksen (1989). Development of a group technology based classification and coding system for injection mold parts. 8th Canadian CAD/CAM and Robotics Conf.. Toronto, In Print.

Goldstein, R.C. (1985). Database Technology and Management. John Wiley & Sons, New York.

Groover, M.P. (1987). Automation. Production Systems. and Computer Integrated Manufacturing. Prentice Hall Inc., New Jersey.

Jurgen, R.K. (1983). Computers and Manufacturing Productivity. IEEE Press Inc., New York.

Knox, C.S. (1983). Organizing Data for CIM Applications. Marcel Dekker Inc., New York.

Rhodnius Inc. (1987). EMPRESS/32. Reference User's Manual, Documentation Version 2.4 for Unix, Vol. 1,2, and 3.

Tsichritzis, D.C. and F.H. Lochovsky (1982). Data Models. Prentice Hall Inc., New Jersey.

Vernadat, F. (1983). Manufacturing Databases, National Research Council Canada. Division of Electrical Engineering. Technical paper.

Acknowledgements

The authors would like to thank Tradesco Mold Ltd., Rexdale, Ontario,Canada, for their confidence and financial support which allowed the successful completion of this research project. Special thanks are extended to N. Travaglini, H. Goul, and P.F. Rycroft.

A PRODUCT INFORMATION SYSTEM TO IMPROVE THE YIELD OF A MANUFACTURING PROCESS

G. M. Geary*, H. Mehdi*, J. W. Chamberlayne**

*School of Engineering and Applied Science, Durham University, Durham, UK
**Phillips Components, Belmont Industrial Estate, Durham, UK

Abstract. This paper describes a pilot project for shopfloor data collection
currently underway at the Durham factory of Philips Components Ltd. (UK), where colour
TV tubes are manufactured. The project is managed under a joint project with Durham
University and aims to replace existing manual methods of quality control data with
more efficient automated methods. The data will be collected with industrially robust
intelligent terminals networked together via a low cost local area network.

The system will analyse the data collected and provide continuous screen displays on
production performance and an extensive on-line query facility for production
supervisors.

An analysis of the benefits of the system has shown that a saving of over £0.5m per
annum should be made by its introduction.

Keywords. Quality control; CAM; industrial control; production control; industrial
yield improvement; management systems.

INTRODUCTION

Philips Components Ltd., Durham, have been
manufacturing television tubes since the seventies
at their UK site in Durham. With a factory
capacity of 1.5 million tubes per annum the
company's turnover is over £70 million per year.
Recently the company invested in a new production
line to produce high quality 14" colour monitor
tubes for PC terminals (the only manufacturing
facility of such in Europe).

A primary concern to the company is yield
improvement. A significant number of automation
projects rest on achieving high enough yield
levels for good return on investment. At present
the factory depends on inspectors and analysts to
block the flow of defect products to the later
stages of production. These inspectors and
analysts provide the primary source of information
on the quality checks performed. This information
is the key component to reducing reject levels and
thus improving yields.

DEVELOPMENT OF SHOPFLOOR INFORMATION SYSTEMS

Shopfloor information systems at Philips in Durham
have been slowly evolving for a number of years
now. However, except for one or two automated
processes, the data is collected manually over a
24 hour period. At the end of that period the
data is collected and entered into a microcomputer
(VAX 3500) where it is analysed and various
reports are generated. An inherent disadvantage
with this method is that it relies on an extensive
paperwork system which is prone to errors and
duplication of information. Also the long time
between recording the information and getting the
feedback reports means that the production
supervisors are not reacting to problems as
quickly as they would like.

Recently with the introduction of the project with
Durham University, shopfloor information systems
started developing more rapidly. The information
systems introduced so far, however, have been
based as single user PC systems to improve data
collection in various parts of the factory and
have contributed significantly to the efficiency
of data handling and analysis.

The company's information systems strategy is now
moving towards the establishment of a factory-wide
unique product identification system which can
track the flow of goods through the various stages
of production, route products to the appropriate
manufacturing processes, collect quality
information and generate management reports among
many other things.

In achieving this long-term strategy a pilot
project looking at some aspects of this strategy
will provide an opportunity to assess the future
implications and highlight areas requiring special
consideration. The SCRIPT system described in
this paper will play a major role in this.

SCRIPT SYSTEM OVERVIEW

SCRIPT is an acronym for Screen Information
Processing and Transactions. The system is to be
implemented in the flowcoating and
lacquer/aluminising areas of the Philips
television tube assembly factory. The flowcoating
process involves the application of green, blue
and red phosphor suspensions alternately to the
inside surface of a colour television screen.
After flowing evenly in suspension over the screen
surface and dried, the phosphor is developed in a
photographic process to harden off the required
stripes of phosphor and the excess is washed away.
The same process is repeated for each of the three
phosphor colours. The lacquer/aluminising process

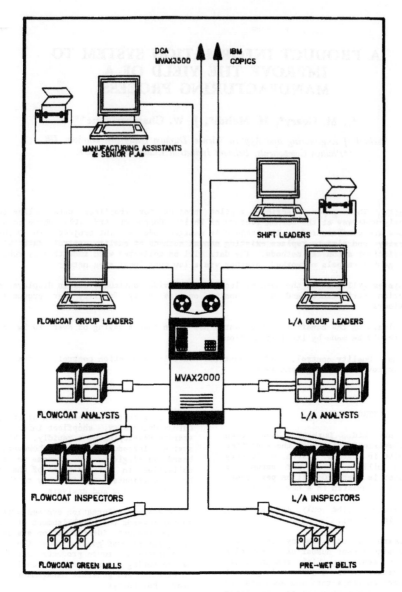

Fig. 1 Overview of the SCRIPT System

immediately follows the flowcoating process and applies a thin layer of aluminium on the inside of the screen over a uniform lacquer layer. The screens are inspected for faults at the end of the flowcoating and lacquer aluminising processes.

The SCRIPT system will replace existing manual data recording methods. The system collects information related to screen reject analysis and throughputs. The information will then be processed to provide various departmental personnel with instantaneous information on the performance of the production process in those two areas. A facility to generate reports is also being provided.

Figure 1 shows an overview of the various components of the system. At the bottom level automatic counters are used to count the number of screens that have started the process in the flowcoat and lacquer/aluminise areas. On the next level inputs from the flowcoat and lacquer/aluminise inspectors are taken into the system. The inspectors will sign on the system so that the inspection results are recorded against their personal identification number. The reject screens from inspection are regarded as suspect rejects until they are presented to the analysts who would confirm the inspection results or over-rule it with a different decision. In the two areas there are a total of six inspection and five analyst stations.

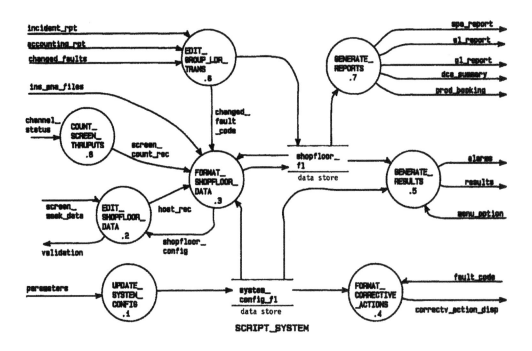

Fig. 2 SCRIPT Data Flow Diagram

The inputs from the inspectors and analysts are fed to the host system where the data is processed and analysed. The results are updated continuously and presented to the group leaders in both areas and more senior management personnel at the top levels.

A link is provided to the factory's DCA (Data Collection and Analysis) department to transfer summary reports. A similar link to COPICS* is used to upload production figures for the control of orders and production scheduling.

Figure 2 shows the top level data flow diagram for the SCRIPT system. The Edit Shopfloor Data process deals with the inputs from the inspectors and analysts which include signing on to the system, entering screen type, inspection or analysis result (this can be either 'good product' or one of a series of fault codes) and the location of department and line in which the screen was processed. The process generates validation outputs in the form of error messages or descriptions of entries made. The collected data is transmitted continuously to the host system where it is processed initially by the Format Shopfloor Data process. This is then stored in the shopfloor file data store.

Count Screen Thruputs deals with the inputs from the automatic counters in both the flowcoat and lacquer/aluminise areas. The Update System Configuration process allows the system manager to edit the fault codes or screen types accepted by the system or to generate the production calendar and update the collective actions database in the system configuration file.

Edit Group Leader Transactions allows the group leaders to generate shiftly accounting sheets and incident reports and to edit fault codes entered by the analysts.

Finally the Generate Results and Generate Reports processes produce the continuous on-line production monitoring displays, alarms, graphs and various daily, weekly and monthly reports.

SYSTEM HARDWARE ARCHITECTURE

Figure 3 shows the hardware configuration for the SCRIPT system. The system is configured around two networks, the LOCAN network which links all the shopfloor PCs together and the existing ETHERNET network on which the MVAX2000 resides. The terminals and the shopfloor data collection terminals are linked to the MVAX2000 via a terminal server. These are described below

Shopfloor data collection terminals

These terminals support the digital inputs from the Philips programmable logic controllers to the system and the inputs from the inspectors and analysts. The hardware will be based on the STE industry bus standard. The terminals used run under one 8088 processor chip and are networked together via the Philips LOCAN network to the host computer. LOCAN is a low cost local area network providing interconnection of terminals. The cabling is based on a seven twisted pair flat cable. The terminals are connected to the network via LOCAN adaptors which can support up to two terminals each. Network control is handled by a single LOCAN master box connected at any point along the network.

---

* COPICS is an MRP system developed by IBM and used by Philips factories.

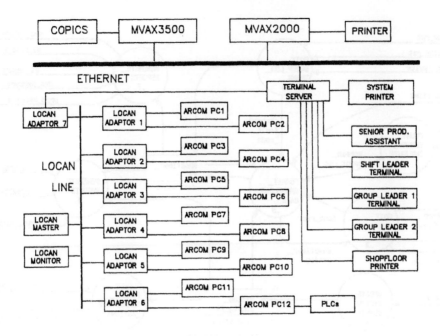

Fig. 3   SCRIPT Hardware Configuration

Supervisor Terminals

These terminals support the group and shift leaders and the senior production assistants. Each will have a terminal linked to the MVAX2000 and can be used to generate the various graphical and monitoring displays and reports mentioned earlier.

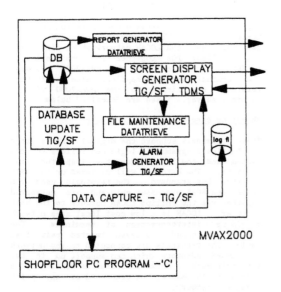

Fig. 4   SCRIPT Software Architecture

SYSTEM SOFTWARE ARCHITECTURE

Figure 4 shows the SCRIPT software architecture. The shopfloor PC program is a generic subsystem written in 'C'. Which modules are in use at any time depends on the location of the shopfloor terminal and the entries required from each inspector or analyst.

On the MVAX2000 a data capture module coded in TIG/SF (a transaction generator written by the Philips CAM group) will handle all data received and transmitted via the LOCAN network. This module will also send all transactions to a log file. Several other modules also run in parallel. These include the database update module which formats the database files and monitors alarm levels; the alarm generator module is activited when necessary to display alarms to the group leaders; the screen display generator module accesses the data from the database and acts as the basic user interface. File maintenance and report generation modules are written in Datatrieve.

CONCLUSIONS

The expected benefits from SCRIPT can be looked at in the light of the current losses experienced by the factory due to the performance of the flowcoating and lacquer/aluminising areas. These losses total about £5 million per annum. The factory estimates that the benefits gained from the SCRIPT system will enable an improvement in direct yield amounting to an annual saving of £0.5 million a year. These savings are attributed to the following:

(a) an improvement in the overall response time to production problem solving by highlighting and analysing these problems at an earlier stage;

(b) on-line analysis of data resulting in the generation of alarms to group leaders;

(c) more accurate accounting of products into each of the areas within the system boundaries;

(d) better and instantaneous presentation of performance results to senior production personnel which would concentrate their efforts on fighting rejects rather than analysing data;

(e) results on how operators are performing in terms of fault identification would highlight where and when training of operators is required;

(f) more reliable data gathering methods resulting in less errors and less duplication of information flow.

In addition to the above SCRIPT will be providing a wealth of in-house experience in automatic shopfloor data collection systems and computer networks on the shopfloor. This would greatly aid in the overall factory strategy in moving towards a factory wide automatic shopfloor data collection system.

It is worth noting that initial trials on the shopfloor has produced very positive feedback from the operators and resulted in useful changes being made to the system.

COMBINING TIME-OUT AND NON-DETERMINISM IN REAL-TIME PROGRAMMING

J. A. Cerrada, M. Collado, R. Morales and J. J. Moreno

Departamento de Lenguajes y Sistems Informáticos, Facultad de Informática, Madrid, Spain

Abstract. Non-deterministic programming by using the CSP scheme is at present widely used in real-time programming. The classical time-out mechanism is also used frequently in order to program timed operations. Modern programming languages like Ada or Occam include CSP constructs as a foundation, combined with a certain form of the time-out mechanism. This paper study how to combine both mechanisms, and the relative adequacy of several combining forms with respect to programming needs in practice. Combining forms are established in a systematic way, and practical examples are analyzed, determining which combining form is better adapted to each case.

Keywords. Non-determinism; time-out; CSP; Real-Time

INTRODUCTION

Real-time programming appears as an extension of concurrent programming by establishing time limits for the execution of certain tasks. The classical time-out mechanism remains one of the most useful to detect the validity of real-time operations.

A real-time program must adapt to its external environment. Its responses should follow the random behaviour of the interacting system and should be given in a short time with respect to the evolution speed of its environment, otherwise the response will not remain valid. A real-time programming system must provide support to fulfill this goal.

Historically, operating systems have been the first and most important real-time programs. Within an operating system there are time-out mechanisms in order to limit the duration of input/output operations. This mechanism ensures that communication with the external environment finishes within a given interval. The communication is assumed correct if it is achieved before the time-out limit. At the end of this limit the communication is assumed wrong and a correcting action can be carried out.

In order to adapt the behaviour of a real-time program to the random nature of the real world, non-deterministic programming structures have been proposed. The CSP scheme proposed in (Hoare, 1978) is a widely accepted one. However, it does not take into account real-time restrictions, and must be extended in order to fulfill the needs of practical software development.

Programming languages like Ada (U.S. DoD, 1983) or Occam (INMOS, 1987) include CSP structures as a foundation adding a certain form of a time-out mechanism. However, the programming structures provided by these or other languages do not seem sufficient to satisfy the needs in this field. Studies about formal use of time as in (Lamport, 1978) and (Lee, 1987) can be easily found, but it is not the same for practical or methodological aspects.

PROBLEM DEFINITION

The time-out policy is widely used to detect hardware errors, communication failures, computer overhead, etc., and so it appears as an important mechanism in process control systems. Every programming language intended to be used in this kind of application should provide facilities for using this mechanism. Regretably, in most cases the use of a time-out operation is not defined at the programming language level, and relies on calls to the operating system kernel.

The time-out mechanism is frequently bound to input/output operations establishing a limit for its duration. If the operation does not finish within the specified interval, it is assumed that something is wrong, and either it must be retried or an alternate error correction action should be undertaken.

A typical example in FORTRAN, adopted from the RTE operating system running on HP-1000 minicomputers, and using system executive calls, is as follows:

```
C   Set time-out
        CALL EXEC(Control, Timeout, Period, Units)
C   Receive data
        CALL EXEC(Read, Channel, Data, MaxLength)
C   Test for success
        CALL ABREG(Status, Length)
        IF ( Length.LE.0 ) GOTO error-handling
```

The time-out concept does not need to be related only to error detection. Time-out or similar mechanisms can also be used to schedule operations at specific time.

Non-determinism is implemented in programming languages like Ada or Occam by means of a SELECT statement (ALT in Occam), that is a form of a "guarded command".

An Ada SELECT statement includes several alternative actions or branches, controlled by the so called "guards", that consist of boolean conditions and/or ACCEPT clauses. The execution of a SELECT statement is acomplished by executing exactly one alternative. If several guards are fulfilled at a given time, the alternative is selected nondeterministically among them. A SELECT structure looks like this:

```
SELECT
    WHEN  Condition1 ->
      ACCEPT  Entry1( Parameters1 )  DO
              Action1s;     -- synchronous action part
      END Entry1;
      Action1a;             -- asynchronous action part
  OR WHEN  Condition2 ->
      ACCEPT  Entry2( Parameters2 )  DO
              Action2s;     -- synchronous action part
      END Entry2;
      Action2a;             -- asynchronous action part
  OR WHEN  ConditionT ->
      DELAY  Timeout-Limit;
      ActionT;              -- timed out action
END SELECT;
```

Ada and Occam allow the combination of a guarded command with a time-out detection. This is accomplished by using DELAY clauses in guards. This construction implies that the action bound to a DELAY clause is selected if and only if no other guard is fulfilled prior to the given time limit.

When the time-out policy is used for communication error detection, it is desirable to maintain both input-output operation and time-out detection logically and textually as close as possible. The SELECT structure of Ada is somewhat unclear because of the separation of ACCEPT and DELAY clauses.

Non-determinism and time-out can be combined in several forms, that can be implemented as different program structures and are studied in the following sections. Their adequacy is analyzed by estimating their usefulness in practical cases and illustrated with examples.

COMBINING FORMS

Keeping the CSP and the Ada SELECT as the basic models, alternatives for time-out/non-determinism combining forms can be established by considering the following aspects:

1 - the place in which time-outs are specified (globally or in every branch).

2 - the place where the action to execute when the time-out is reached is specified (globally or in every branch).

3 - the strategy used to decide which time-out is reached among a set of them (minimum or maximum).

This yields the following table:

Time Out Limit	Time Out Action	Strategy	Name
Global	Global	----	SELECT
Local	Global	Minimum	SELECTMIN
Local	Local	Minimum	SELECTFIRST
Local	Global	Maximum	SELECTMAX
Local	Local	Maximum	SELECTLAST

The remaining symmetric component on the table (global time-out limit and local time-out actions) has no clear meaning and is not considered here.

For every proposal in the table we give a syntax proposal (combining an Ada style with a CSP scheme) and a brief comment about its behaviour.

SELECT

This structure behaves like the Ada SELECT. A global time-out is used. When the time-out is reached without satisfying any reception, the global time-out action is executed.

If a Guard is defined as

```
Boolean Condition                          or
Receive( channel, message )                or
Boolean Condition AND Receive( channel, message )
```

we could write

```
SELECT
    WHEN  << Guard >> DO
          << Action >>
          .
          .
          .
    WHEN  << Guard >> DO
          << Action >>
    WHEN  Delay( timer ) DO
          << Time-Out Action >>
    OTHERWISE << Default Action >>
END SELECT
```

In this structure, the time-out limit and the time-out action are both specified globally. The intuitive meaning of the time-out action is an alternative to the execution of any normal action.

SELECTMIN and SELECTFIRST

Defining a General Guard as:

```
Boolean Condition                          or
Boolean Condition AND << Timed Reception >>
```

where a Timed Reception has the form

```
Receive( channel, message )                or
Delay( timeout_value )                     or
Receive( channel, message )
    WITH Delay( timeout_value )
```

the following structures can be established:

```
SELECTMIN
    WHEN  << General Guard >> DO
          << Action >>
          . . .
    WHEN  << General Guard >> DO
          << Action >>
    WHEN  TimeOut DO
          << Time-Out Handler >>
    OTHERWISE << Default Action >>
END SELECTMIN
```

The SELECTMIN structure allows the specification of a time-out in every branch by using a Delay command. The time-out action is global and is executed when the minimum of all the local time-outs is reached.

Notice that delays without receptions are allowed in guards.

The intuitive meaning of the SELECTMIN time-out limits is the detection of individual failures of every branch. The global time-out action appears as a common recovery alternative in any case.

The SELECTFIRST structure is similar to SELECTMIN, but has local time-out actions. The minimum time-out among all the time-outs specified in guards with true boolean conditions is selected. When it is reached, the time-out action bound to its guard is executed.

```
SELECTFIRST
    WHEN << General Guard >> DO
        IF TimeOut
            THEN << Time-Out Handler >>
            ELSE << Action >>
        END IF
    . . .
    WHEN << General Guard >> DO
        IF TimeOut
            THEN << Time-Out Handler >>
            ELSE << Action >>
        END IF
    OTHERWISE << Default Action >>
END SELECTFIRST
```

The time-out mechanism is used here for detecting individual failures in every branch, with independent recovery actions for each one. The first detected error is handled inmediatly.

The SELECTFIRST policy has been adopted in (Cerrada, 1988) and (Collado, 1989).

SELECTMAX and SELECTLAST

The SELECTMAX structure waits for receptions in every guard until the maximum time of all the time-outs is reached. In this case a global time-out action is executed.

If we define General Guard as before, this variant can be expressed as:

```
SELECTMAX
    WHEN << General Guard >> DO
        << Action >>
        .
        .
        .
    WHEN << General Guard >> DO
        << Action >>
    WHEN TimeOut DO
        << Time-Out Handler >>
    OTHERWISE << Default Action >>
END SELECTMAX
```

The time-out limits appear as valid response times of branch guards. The global time-out recovery action is only undertaken after waiting for the slowest valid reception.

We call this structure a non-restricted version, in opposition to another version of this structure called restricted. In the restricted version receptions after the time specified in a guard are not allowed.

Notice that in the non-restricted version a time-out in a guard does not impose a time limit for this reception. This limit is the same for all the guards and they are calculated over all the time-outs specified.

The SELECTLAST structure is quite similar to the SELECTMAX structure but it executes the individual time-out action corresponding to the maximum time-out specified (the last one reached).

```
SELECTLAST
    WHEN << General Guard >> DO
        IF TimeOut
            THEN << Time-Out Handler >>
            ELSE << Action >>
        END IF
    . . .
```

```
    WHEN << General Guard >> DO
        IF TimeOut
            THEN << Time-Out Handler >>
            ELSE << Action >>
        END IF
    OTHERWISE << Default Action >>
END SELECTLAST
```

The meaning of the SELECTLAST structure is in some ways strange. It can be explained as if every faulty action is not recovered, except the last one. Both rectricted and non-restricted versions can also be defined for this structure.

In every structure described here the selection between two guards in the same condition (i.e. two simultaneous receptions, two simultaneous time-outs reached, etc.) is made in a non-deterministic way.

In the sequel we shall refer to the non-restricted versions unless we make the opposite explicit. Moreover, we shall discuss the restricted version later. A syntax for the restricted version could be the same as the non restricted one. Only the name of the structure is changed to:

```
SELECTMAX-RESTRICTED      SELECTLAST-RESTRICTED
```

respectively.

EXAMPLES

A set of examples has been collected and analyzed. Their core structures have been identified and programmed by using the proposed combining forms.

The examples have been collected in order to cover a wide range of cases, including previously published ones as well as real applications. Some of them are shown in this section.

Data Logger

The following example summarizes the handling of several data acquisition subsystems (DAS) from a control room in a real project. A chronological record is to be maintained, including all available information at regular intervals.

Every subsystem monitors a set of sensors, this sending changes asynchronously to the central control system. Some of these changes imply alarm conditions and must be recorded individually. In addition, some indirect measurements are computed periodically at the central system and the whole set of values (measured and computed) is recorded.

The SELECT approach matches this policy, because of the global nature of the periodic time limit, which is not related to a specific kind of signal.

```
LOOP
    SELECT
        WHEN Receive(DAS1_alarm_channel, message) DO
            << record alarm >>
        WHEN Receive(DAS1_data_channel, message) DO
            << store data >>
            .
            .
        WHEN Receive(DASn_alarm_channel, message) DO
            << record alarm >>
        WHEN Receive(DASn_data_channel, message) DO
            << store data >>
        WHEN Delay(end_of_period) DO
            << compute indirect measures >>
            << record measures >>
    END SELECT
END LOOP;
```

Traffic Lights Control

The next example, taken from (Cerrada, 1986), simulates the control of a number of crossroads along a one way street, as shown in Figure 1.

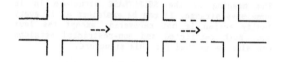

Figure 1. Traffic schema.

Traffic lights at every cross must be synchronized in order to obtain a "green wave" along the street. If synchronization fails, each cross must operate by itself. The traffic ligth cycle is shown in Figure 2. Time intervals are set from the central station. That includes the duration of the green light and the limits for the red one. The duration of the yellow state is constant. Synchronization of cycles is achieved by tuning the duration of the red state, inside the specified limits.

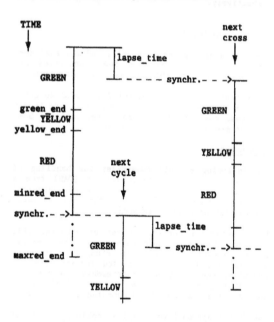

Figure 2. Traffic lights cycle.

Every crossroad is controlled by an independent task. Another task is devoted to coordination from the supervisor console. Each crossroad task receives synchronization messages (null) from the previous one and sends new signals to the next one. The coordinating task sends messages to crossroads in order to set the operation mode to automatic (normal) or manual (disconnected, flashing), and the values of the red and green intervals.

The behaviour of a SELECTFIRST structure is the most suitable in this case. Most guards include waits for a given time, and other guards include receptions with or without time-out. Every received message or time limit reached implies the execution of a different action.

```
TYPE lights_state IS (green, yellow, red,
                        conditional_red, flashing);
    order IS RECORD
                automatic: boolean;
                green_time, red_min,
                red_max, lapse_time: time
            END RECORD;
  . . .

VAR state: lights_state;
    green_start, green_end, yellow_end,
    min_red_end, max_red_end, synchr_time: time;
    send_next: boolean;
  . . .

start_cycle;
WITH  actualorder  DO
  LOOP
    SELECTFIRST
      WHEN (state = green) AND
          Delay(green_end) DO
        state := yellow;
        yellow_end := green_end + yellow_time;
        << set yellow >>

      WHEN (state = yellow) AND
          Delay(yellow_end) DO
        state := red;
        min_red_end := yellow_end + redmin;
        << set red >>

      WHEN (state = red) AND
          Delay(min_red_end) DO
        state := conditional_red;
        max_red_end := yellow_end + red_max;

      WHEN (state = conditional_red) AND
          Receive( previous_cross, null) WITH
          Delay(max_red_end) DO
        state := green;
        IF TimeOut THEN
          green_start := max_red_end;
        ELSE
          << get actual_time >>
          green_start := actual_time;
        END IF
        green_end := green_start + green_time;
        synchr_time := green_start + lapse_time;
        send_next := true;
        << set green >>

      WHEN send_next AND Delay(synchr_time)  DO
        send_next := false;
        Send( next_cross, null )

      WHEN Receive(orderport, actualorder) DO
        IF  automatic  THEN
          start_cycle
        ELSE
          state := flashing;
          << set flashing >>
        END IF
    END SELECTFIRST;
  END LOOP;
END WITH;
```

Please note that this program fragment combines Ada and Pascal notations, and also that time values are absolute values and not relative time-out durations.

Heterogenous servers

The next example illustrates how to program different kinds of service requests to server agents of different capabilities. After requesting a service, the client process must wait until some server acknowledges the request, signalling that it is ready to carry out the service. If there is no response within a reasonable time, the client can decide that the service cannot be accomplished, and do some alternative action.

This scheme can be found, for example, in LAN systems with several printers: laser printers, line printers, impact matrix printers, etc. Only some of them are adequate for mass printing (laser and line printers), and also only some of them can have graphic capabilities (laser and matrix printers). In addition, their response times are very different.

The intuitive solution is to wait for the longest allowed response time and establish a global alternate action in the case of time-out. According to this, the SELECTMAX structure appears as a good program scheme.

```
<< broadcast the service request >>;
SELECTMAX
    WHEN << server_1 could do the service >> AND
         Receive(server_1, acknowledge)
         WITH Delay(timer_1) DO
            << set server_1 task >>
    ...
    WHEN << server_n could do the service >> AND
         Receive(server_n, acknowledge)
         WITH Delay(timer_n) DO
            << set server_n task >>
    WHEN TimeOut DO
            << alternate action, no service
               available now >>
    OTHERWISE
            << alternate action, service impossible >>
END SELECT;
```

DISCUSSION

The first attempt to introduce time-out in a non-deterministic programming structure leads to two possible models:

1 - The waiting time is global (i.e. this time is specified by the process that waits).

2 - The waiting time is local (i.e. it is specified by the different behaviour of the receptions).

In the first model the time-out could be fixed independently of the nature of the receptions. In the second one a time-out needs to be fixed for every reception.

Problems of the first class appear very often in real time programming. They can be solved by using the SELECT structure described before.

The second class of problems also appears frequently. A subdivision can be made depending on when the time-out action must be executed. The simplest alternatives are to execute it after the minimum delay or after the maximum one.

Looking at the examples considered here (and others found in practice) it can be deduced the need of both models. However, it can be noticed that the class of examples that use the minimum strategy is wider than the class that uses the maximum strategy.

The minimum delay strategy appears in a natural way when the time-outs are determined by the nature of the reception channels, and the first one that reaches its limit signals the need to execute the time-out action. The SELECTFIRST structure seems to be more useful than the SELECTMIN structure for many of these cases.

The maximum delay strategy is meaningful when it is necessary to wait until all the channels notice that the communication is not possible to execute the time-out action. In this case the SELECTMAX structure seems more adequate than the SELECTLAST for this kind of problems.

Finally we could discuss the restricted versions of the maximum structures. Notice that we have not presented here any example of their use. There does not seem to be a general model for a large class of real time problems. However they can be useful to solve certain problems where the specifications impose some "laws" to the receptions. For instance, we could describe a system of administrative requests where for fairness reasons and depending on the nature of the communication channels, the maximum time for receptions of requests is specific for every channel that follows this kind of rules.

REFERENCES

Cerrada, J.A., M. Collado (1988). Implementation of a CSP-based extension of Pascal. Microprocessing and Microprogramming, 22, 233-242.

Cerrada, J.A., M. Collado (1986). A programming language for distributed systems. SOCOCO'86: 4th. IFAC/IFIP Symposium on Software for Computer Control. Graz, Austria, May 20-23.

Collado, M., R. Morales, J.J. Moreno (1989). Conversation modules: a methodological approach to the design of real time applications. Euromicro 89, Cologne, Sep. 4-8.

Hoare, C.A.R. (1978). Communicating sequential processes. Comm. ACM, 21, 666-677.

INMOS Limited (1987). Occam-2 programming manual. Prentice Hall.

Lamport, L. (1978). Time, clocks and the ordering of events in a distributed system. Comm. ACM, 21, 558-565.

Lee, I., S.B. Davidson (1987). Adding time to synchronous process communications. IEEE Trans. on Comp., C-36, 941-948.

U.S. DoD (1983). Reference manual for the Ada programming language. ANSI/MIL-STD 1851 A.

ROBOT CONTROL IN MANUFACTURING: COMBINING REFERENCE INFORMATION WITH ONLINE SENSOR CORRECTION

E. Freund and Ch. Bühler

Institut für Roboterforschung, Universität Dortmund, Dortmund, FRG

Abstract. A control concept is presented which takes into consideration the nonlinear system dynamics of robots as well as the available reference information about the path. Various levels of reference information can be introduced in the concept e.g. base reference and online sensor correction as well as partial information about the path. If no path information is available in advance a zero reference is used. In this case the concept leads to the base equations of the nonlinear decoupling and control method.

Keywords. Robots; flexible manufacturing; nonlinear control systems; control system design; sensors.

INTRODUCTION

The position control of industrial robots has been subject of intensive work in the past. Much effort has been spent to achieve a fast, accurate and aperiodic behaviour of the position and orientation of the tool center point.

Due to their multiple link construction the robots can be understood as a multiple control problem with nonlinear dynamics and nonlinear couplings, which depend on the mechanical structure of the system. Therefore the classical theory of linear, time invariant control systems is as an approximation only of limited use. Nowadays in industrial robot control linear single axis controllers are widely spread and dynamic couplings are neglected. So the design of this kind of controllers can be done based on very simple linear models of the system without much effort for mathematical modelling.

For specific applications linearization with respect to reference trajectories are carried out, which permit a linear controller design. The algorithms obtained can easily be computed in realtime, because the major part can be examined offline in the linearization procedure. To carry out the linearization, however, an adequate mathematical model of the robot system is needed. The concept can be employed only based on an apriori known reference path, as couplings arising from possible deviations from the reference are not considered in the control law. Therefore the online inclusion of additional sensor information, which e.g. in case of collision danger in multi robot systems may cause considerable evasive actions leaving the reference (Freund, Hoyer, 1988), is not possible.

Applying the nonlinear decoupling and control method (Freund, 1977, 1982) the axes of the robot can be decoupled in the whole working space leading to a timeinvariant linear dynamic behaviour for each axis with arbitrary pole placement. This control law works on the measured position and velocities of all axes and without a path reference. Therefore this control method allows the inclusion of online path modification e.g. by means of external sensors.

In addition to these concepts a number of adaptive and robust approaches should be mentioned, which either use very simple linear models (Koivo, Guo, 1983; Moya, Seraji, 1987) or are based on the complete nonlinear system description (Furuta, Kosuge, 1984; Slotine, Li, 1987).

In this paper a concept is proposed, which makes use of all apriori available information about the model of the system as well as the intended movements. The mathematical description of the robot has to be carried out once with a sufficient accuracy, while the reference information is introduced depending on the given application in various forms.

NONLINEAR DECOUPLING AND CONTROL WITH CHANGING REFERENCE PATH

Overall Structure

The principle of nonlinear decoupling and control (Freund, 1982) has successfully been applied to position control of robots in the past (Klein, 1986). By means of a compensation of nonlinear dynamic couplings an input to output decoupling of the dynamics is achieved, providing a uniform system behaviour in the whole working area. Of course this can only be achieved if no boundary conditions of the state variables or the actuators are hurt and the model of the system is of sufficient accuracy. By the introduction of additional feedback an arbitrary pole placement for each decoupled subsystem is possible, which allows to choose the dynamics for all axes within the bounds of the available actuator torque. For a fast and asymptotic movement of this arm, multiple poles on the real axis are chosen, usually.

This kind of controller design equals the method of pole placement for linear systems which guarantees an asymptotical transition of the output without a steady-state error for an input change. This input mode is equivalent to PTP-mode in robotics, while the CP-mode can not be interpreted as a sequence of such input switchings, because the time between two changes is too small for a stationary system response. Therefore aperiodic behaviour can not be guaranteed in general. On the other hand some control concepts are based on a dynamic input reference and use the linearization method. While the first method leads to a unique dynamic behaviour of the system in the whole working space the latter is restricted to the reference and not useful elsewhere.

The concept presented here tries to provide high flexibility to online changes as generated by additional sensors involving as much reference information as possible at the same time. This is explained for a typical robot working sequence:

— movement to bench for tool exchange

— tool exchange

— movement to sensed workpiece location

— processing of the workpiece

The location of the exchange bench is fixed in the working area, i.e. apriori known. So the first task is a typical case for PTP—mode. The following tool exchange is a determined procedure with no need for adaption.

The movement to the workpiece location depends on the current sensor information and can be carried out in PTP— or CP—mode. Position and orientation of the workpiece are introduced to the controller online from external sensors. On the other hand type and dimensions of the workpiece as well as the processing rules are known in advance. So, for this part of the working sequence the geometrical path corresponding to the workpiece dimensions and the velocity profile of the TCP corresponding to the processing rule are given as apriori knowledge.

A typical example is shown in fig. 1. Two cars are shown in an overlied draft which were sequentially transferred to a robot work cell. Because of the positioning properties of the transport vehicles the cars are located in different positions with respect to the robot. So the position and orientation of the cars vary, whereas the geometry of the cars ist not changed.

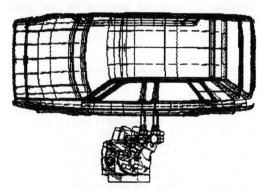

Fig. 1.   Overlied draft of two cars in a robot work cell

Now the given apriori knowledge should be included in the overall system, where the case of full reference information is covered as well as the case of no reference. In the case of no reference information the controller confines himself again to the base equations of the nonlinear decoupling and control method.

To provide the integration of sensors a general structure shown in fig. 2 is proposed. Here apriori information about the path is introduced by means of a reference as well as online changes by means of sensor feedback. The nonlinear control law has to be modified for this concept to a nonlinear error controller with separated reference part. The resulting controller is still nonlinear, because he should work well even in the case of considerable deviations from the reference path, which is not possible for a linearized controller. If a zero reference is introduced

the error controller again leads to the original equations of the nonlinear decoupling and control law.

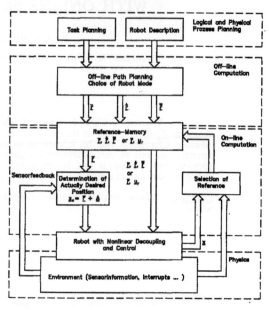

Fig. 2.   Overall structure

The Principle of Nonlinear Decoupling and Control

The mathematical description of an industrial robot can be derived e.g. with the help of the Newton—Euler algorithms or the method of Lagrange (Silver, 1982). The equations of motion for a robot with m rigid links are given in the form:

$$\underline{M}(t) = \underline{T}(\underline{\vartheta})\,\underline{\ddot{\vartheta}} + \underline{h}(\underline{\vartheta}, \underline{\dot{\vartheta}}) + \underline{g}(\underline{\vartheta}). \tag{1}$$

where the dimensions of $\underline{M}(t)$, $\underline{\vartheta}(t)$, $\underline{\dot{\vartheta}}(t)$, $\underline{\ddot{\vartheta}}(t)$, $\underline{h}(\underline{\vartheta})$ and $\underline{g}(t)$ are [m, 1] and $\underline{T}(t)$ [m, m], respectively. $\underline{M}(t)$ describes the generalized forces (forces, torques) and $\underline{\vartheta}(t)$ the generalized variable of motion (positions, angles), with $\underline{\dot{\vartheta}}(t)$, $\underline{\ddot{\vartheta}}(t)$ as the derivation of $\underline{\vartheta}(t)$ with respect to time. $\underline{h}(t)$ represents the Coriolis— and Centrifugal—terms and $\underline{g}(\underline{\vartheta})$ the terms of gravity.

Using the following correspondences:

$$
\begin{aligned}
\underline{u}(t) &= \underline{M}(t) \\[4pt]
x_{(2i-1)}(t) &= \vartheta_i(t) \\[8pt]
x_{2i}(t) &= \dot{\vartheta}_i(t) \\[4pt]
y_i(t) &= \vartheta_i(t)
\end{aligned}
\qquad i = 1, \ldots, m
\tag{2}
$$

one obtains the nonlinear state space description of the system

$$\dot{\underline{x}}(t) = \underline{a}(\underline{x}) + \underline{B}(\underline{x}) \cdot \underline{u}(t) \, , \, \underline{y}(t) = \underline{c}(\underline{x}) \qquad (3)$$

where the vector $\underline{a}(\underline{x})$ represents the system dynamics and the output vector $\underline{C}(\underline{x})$ and the input matrix $\underline{B}(\underline{x})$ directly result from the equation of motion (1).

Choosing the nonlinear control law

$$\underline{u}(t) = \underline{F}(\underline{x}) + \underline{G}(\underline{x}) \cdot \underline{w}(t) \qquad (4)$$

with

$$\underline{F}(\underline{x}) = -\underline{D}^{*-1}(\underline{x}) \, [\underline{C}^{*}(\underline{x}) + \underline{M}^{*}(\underline{x})] \, , \qquad (5)$$
$$\underline{G}(\underline{x}) = \underline{D}^{*-1}(\underline{x}) \cdot \underline{\Lambda}$$

the system becomes input to output decoupled ($\underline{w} \rightarrow \underline{y}$), i.e. in the closed loop system the i–th input $w_i(t)$ influences solely the i–th output $y_i(t)$. The resulting dynamics can be described by a linear differential equation for each input output set y_i, where the poles can be placed arbitrarily (Freund, 1977).

The decoupling is done by the term $\underline{D}^{*-1}(\underline{x}) \, \underline{C}^{*}(\underline{x})$ of the feedback vector $\underline{F}(\underline{x})$, while the dynamics are determined by $\underline{D}^{*-1}(\underline{x}) \, \underline{M}^{*}(\underline{x})$. The feedback matrix $\underline{G}(\underline{x})$ introduces the new input vector $\underline{w}(t)$ and an input gain matrix $\underline{\Lambda}$ keeping the decoupling conditions.

A comparison of the control law (5) with the equation of motion (1) yields:

$$\underline{D}^{*-1}(\underline{x}) = \underline{T}(\underline{\textit{y}})$$
$$\underline{C}^{*}(\underline{x}) = \underline{T}^{-1}(\underline{\textit{y}}) \cdot (\underline{h}(\underline{\textit{y}},\dot{\underline{\textit{y}}}) + \underline{g}(\underline{\textit{y}})) \, . \qquad (6)$$

The matrix $\underline{M}^{*}(\underline{x})$, which determines the dynamics, is given for desired linear dynamics by

$$M_i^{*} = \sum_{k=0}^{d_{i-1}} \alpha_{k,i} \, x_{(2i-1+k)} \, , \, i = 1, ..., m \, ,$$

where d_i represents the order of the differential equation of the i–th decoupled input to output subsystem. Regarding the rigid body model of a robot holds $d_i = 2$ ($i = 1, .., m$) and each axis can be described by a second order differential equation of the form

$$\ddot{y}_i = \alpha_{1i}\dot{y}_i + \alpha_{0i} \, y_i + \lambda_i w_i \quad (i = 1, ..., m) \, . \qquad (7)$$

Controller Design for Path Tracking with Sensor Correction

In this chapter a controller is developed based on the nonlinear decoupling and control method, which provides the flexibility of this concept, but which on the other hand

is able to make use of apriori information about the movement.

In principle each movement can be separatet in three information categories:

1. Position of a base point of the path and orientation of a cartesian coordinate system in this point

2. Geometrical path, e.g. given in the cartesian coordinate system fixed in the base point

3. Desired velocity–profile along the path, e.g. as a function of the natural path parameter

These information should be introduced in the control concept in various levels, depending on the information apriori available.

In a first step a control law for a given reference is designed and in a second step the controller for an online modified path is presented.

Nonlinear Decoupling and Control for the Base Reference Path

A desired trajectory $\underline{r}_B(t)$ in joint coordinates is given, which is called base reference, where $r_{Bi}(t)$ describes the desired value for axis i ($i = 1, ..., m$). The derivation of $\underline{r}_B(t)$ with respect to time yields the velocity of the base reference $\dot{\underline{r}}_B(t)$, which makes up the state of the base reference $\underline{\varrho}_B(t)$

$$\underline{\varrho}_B^{T}(t) = \left[r_{B1}(t), \, \dot{r}_{B1}(t) \, ... \, r_{Bm}(t), \, \dot{r}_{Bm}(t) \right] \qquad (8)$$

Introducing this in the equation of motion (1) one obtains

$$\ddot{\underline{r}}_B(t) \qquad = \underline{co}(\underline{\varrho}_B(t)) + \underline{IN}(\underline{r}_B(t)) \cdot \underline{u}_B(t)$$
$$\underline{co}(\underline{\varrho}_B(t)) \qquad = -\underline{T}^{-1}(\underline{r}_B) \cdot (-\underline{h}(\underline{\varrho}_B) + \underline{g}(\underline{r}_B)), \qquad (9)$$
$$\underline{IN}(\underline{r}_B) \qquad = \underline{T}^{-1}(\underline{r}_B).$$

The [m, 1] vector $\underline{co}(\underline{\varrho}_B)$ represents the nonlinear couplings of the system, the [m, m] matrix contains the input couplings and $\underline{u}_B(t)$ describes the corresponding reference input. In analogy with eq. (3) the corresponding state space representation is given by

$$\dot{\underline{\varrho}}_B(t) = \underline{a}(\underline{\varrho}_B) + \underline{B}(\underline{\varrho}_B) \cdot \underline{u}_B \qquad (10)$$

$\underline{u}_B$ can be obtained, solving the inverse problem if $\underline{r}_B(t)$ is known

$$\underline{u}_B(t) = \underline{IN}^{-1} \cdot (\ddot{\underline{r}}_B(t) - \underline{co}(\underline{\varrho}_B)) \, . \qquad (11)$$

The difference between actual the state vector $\underline{x}(t)$ and base reference $\underline{\varrho}_B$ results to

$$\underline{\Delta x}(t) = \underline{x}(t) - \underline{\varrho}_B(t),$$

$$\underline{\Delta \dot{x}}(t) = \underline{\dot{x}}(t) - \underline{\dot{\varrho}}_B(t) . \tag{12}$$

For reasons of clarity of presentation the dependency on time of all variables is suppressed except in special cases.

Introducing eq. (10) into eq. (3) using $\underline{\Delta x}$ of eq. (12) and replacing $\underline{u}$ by $\underline{u} = \underline{u}_B + \underline{u}_d$, where $\underline{u}_d$ is the difference between input $\underline{u}$ and reference input $\underline{u}_B$ one gets:

$$\underline{\dot{x}} = \underline{a}(\underline{\varrho}_B + \underline{\Delta x}) + \underline{B}(\underline{\varrho}_B + \underline{\Delta x})\underline{u}_B + \underline{B}(\underline{\varrho}_B + \underline{\Delta x})\underline{u}_d \tag{13}$$

$$\underline{\Delta \dot{x}} = \underline{a}(\underline{\varrho}_B + \underline{\Delta x}) - \underline{a}(\underline{\varrho}_B)$$

$$+ \left[\underline{B}(\underline{\varrho}_B + \underline{\Delta x}) - \underline{B}(\underline{\varrho}_B) \right] \underline{u}_B \tag{14}$$

$$+ \underline{B}(\underline{\varrho}_B + \underline{\Delta x}) \cdot \underline{u}_d .$$

Combining all terms depending on the reference leads to

$$\underline{\Delta \dot{x}} = \underline{\bar{a}}(\underline{\Delta x}, (\underline{\varrho}_B, \underline{u}_B)) + \underline{\bar{B}}(\underline{\Delta x}, (\underline{\varrho}_B)) \cdot \underline{u}_d \tag{15}$$

where the dependency on $\underline{\varrho}_B$ and $\underline{u}_B$ is known.

Another representation of eq. 14, which clarifies the relation to the state space representation (3) is

$$(\underline{\varrho}_B + \underline{\Delta x})' = \underline{a}(\underline{x}) + \underline{B}(\underline{x}) \cdot (\underline{u}_d + \underline{u}_B) . \tag{16}$$

Based on eq. (15) a decoupling law for $\underline{\Delta x}$ can be derived in the same manner as in eqs. (4)–(7)

$$\underline{u}_d = -\underline{\bar{D}}^{*-1}(\underline{\Delta x}, (\underline{\varrho}_B)) \cdot \left\{ \underline{\bar{\mathbb{C}}}^{*}(\underline{\Delta x}, (\underline{\varrho}_B, \underline{u}_B)) + \right.$$

$$\left. [\underline{\alpha}] \underline{\Delta x} + \underline{\Lambda} \cdot \underline{d}(t) \right\} . \tag{17}$$

Here $\underline{\alpha}$ is a matrix of dimension $[m, 2m]$ which contains the coefficients $\alpha_{0,i}$, $\alpha_{1,i}$ for all axes, $\underline{\Lambda}$ is an diagonal matrix of dimension $[m, m]$ representing input gains and $\underline{d}$ represents the new inputs. Then the resulting decoupled closed–loop system for each axis is given by

$$\begin{bmatrix} \Delta x_{2i-1} \\ \Delta x_{2i} \end{bmatrix}^{\cdot} = \underline{\Delta \dot{x}}_i = \begin{bmatrix} 0 & 1 \\ -\alpha_{0,i} & -\alpha_{1,i} \end{bmatrix} \underline{\Delta x}_i + \lambda_i \begin{bmatrix} 0 \\ 1 \end{bmatrix} d_i(t)$$

$$i = 1, ..., m . \tag{18}$$

Choosing the new inputs $d_i \equiv 0$ $(i = 1, ..., m)$ eq. (18) becomes a homogeneous, linear differential equation. With a set of eigenvalues on the left side of the complex plane $\underline{\Delta x}_i$ tends to zero asymptotically. By an appropriate setting of the dynamics the error $\underline{\Delta x}$ nearly becomes zero in a finite time intervall.

Looking again at eq. (13) under this assumption, one obtains

$$\underline{\dot{x}} = \lim_{\underline{\Delta x} \to \underline{0}} \left\{ \underline{a}(\underline{\varrho}_B + \underline{\Delta x}) + \underline{B}(\underline{\varrho}_B + \underline{\Delta x})\underline{u}_B + \underline{B}(\underline{\varrho}_B + \underline{\Delta x}) \underline{u}_d \right\} \tag{19}$$

and with eq. (14) and eq. (17) one gets

$$\underline{\dot{x}} = \underline{a}(\underline{\varrho}_B) + \underline{B}(\underline{\varrho}_B) \cdot \underline{u}_B = \underline{\dot{\varrho}}_B . \tag{20}$$

In eq. (20) it can be seen, that $\underline{x}$ is identical to the state of the base reference $\underline{\varrho}_B$, i.e. the robot system tracks along the reference trajectory. If the initial error $\underline{\Delta x}(0)$ is not too big this identity is held very close. In the case of considerable big initial errors the behaviour is less ideal, but unlike with the linearized controller the system here is still drawn back to the reference.

Looking at eq. (15) without reference, i.e. $\underline{\varrho}_B \equiv \underline{0}$ and $\underline{d} = \underline{r}_B(t)$, it follows $\underline{u}_B \equiv \underline{0}$ from eq. (11) and $\underline{\Delta x} = \underline{x}$ from eq. (12).

In this case eq. (14) yields in the form

$$\underline{\dot{x}} = \underline{\Delta \dot{x}} = \underline{a}(\underline{\Delta x}) + \underline{B}(\underline{\Delta x}) \cdot \underline{u}_d \tag{21}$$

and the corresponding $\underline{u}_d$ from eq. (17)

$$\underline{u}_d = -\underline{D}^{*-1}(\underline{\Delta x}) \cdot \left\{ \underline{\mathbb{C}}^{*}(\underline{\Delta x}) + [\underline{\alpha}]\underline{\Delta x} + \underline{\Lambda} \cdot \underline{d}(t) \right\} \tag{22}$$

with

$$\underline{D}^{*-1} = \underline{I}\underline{N}^{-1}(\underline{\Delta x}) , \qquad \underline{\mathbb{C}}^{*} = \underline{\mathbb{c}}\underline{o}(\underline{\Delta x}) .$$

Eq. (21), (22) are equal to the original equations of the nonlinear decoupling and control. Therefore the derived controller results in the known control law of Freund (1982) for the case of no reference.

Nonlinear Decoupling and Control with Base Reference and Online Path Correction by Means of Sensor Information

Based on the control law (17) the problem can be investigated by a consideration of eq. (18), which describes the closed loop behaviour for one subsystem.

The online sensor correction of the path $\underline{r}_B$ is expressed by means of a new reference $\underline{r}_A$. If, for example, the sensor information is given as difference values $(\Delta x, \Delta y, ...)$ in a cartesian system, a $\underline{\Delta}(t)$ in robot joint coordinates can be obtained by the transformation of Δx, Δy ... . $\underline{\Delta}(t)$ can also be expressed as the difference between base reference and new reference:

$$\underline{\Delta}(t) = \underline{r}_A(t) - \underline{r}_B(t) . \tag{23}$$

Accordingly one obtains the error between the states of the references

$$\underline{\Delta \varrho}(t) = \underline{\varrho}_A(t) - \underline{\varrho}_B(t) , \tag{24}$$

where $\underline{\varrho}_B$ is derived from $\underline{r}_A$ in the same manner as $\underline{\varrho}_B$ from $\underline{r}_B$ (cf. eq. (8)).

The error $\underline{\delta x}(t)$ between the actual state vector and state of the new reference is given by

$$\underline{\delta x}(t) = \underline{x}(t) - \underline{\varrho}_A(t) . \tag{25}$$

In this section a law for $d_i(t)$ $(i = 1, ..., m)$ of eq. (18) is derived, which leads the error $\underline{\delta x}(t)$ to zero. Replacing $\underline{\Delta x}$ in eq. (18) by $\underline{\delta x}$ and $\underline{\Delta \varrho}$ (cf. eq. (12), (25)) one obtains:

$$\begin{bmatrix} \Delta x_{2i-1} \\ \Delta x_{2i} \end{bmatrix}^{\cdot} = \begin{bmatrix} 0 & 1 \\ -\alpha_{0,i} & -\alpha_{1,i} \end{bmatrix} \begin{bmatrix} \delta x_{2i-1} + \Delta\rho_{2i-1} \\ \delta x_{2i} + \Delta\rho_{2i} \end{bmatrix}$$

$$+ \begin{bmatrix} 0 \\ \lambda_i \end{bmatrix} d_i(t) \qquad (26)$$

for $i = 1, ..., m$.

With the identities

$$\Delta\rho_{2i} = \Delta\dot\rho_{2i-1} \text{ and } \Delta\dot x_{2i} - \Delta\dot\rho_{2i} = \delta x_{2i}$$

it follows

$$\begin{bmatrix} \delta x_{2i-1} \\ \delta x_{2i} \end{bmatrix}^{\cdot} = \begin{bmatrix} 0 & 1 \\ -\alpha_{0,1} & -\alpha_{1,i} \end{bmatrix} \begin{bmatrix} \delta x_{2i-1} \\ \delta x_{2i} \end{bmatrix}$$

$$+ \begin{bmatrix} 0 \\ -\alpha_{0,i}\Delta\rho_{2i-1} - \alpha_{1,i}\Delta\rho_{2i} - \Delta\rho_{2i} \end{bmatrix}$$

$$+ \begin{bmatrix} 0 \\ \lambda_i \end{bmatrix} d_i(t). \qquad (27)$$

Eq. (27) describes the error between the state of the new reference and the actual state of the system explicitely.

With the choice

$$d_i(t) = \frac{1}{\lambda_i} (\alpha_{0,i}\Delta\rho_{2i-1} + \alpha_{1,i}\Delta\rho_{2i} + \Delta\dot\rho_{2i}) \qquad (28)$$

for $i = 1, ..., m$, eq. (27) becomes a linear homogeneous differential equation, where the solution tends to zero for stable eigenvalues. Therefore for sufficient large time intervalls δx becomes zero as demanded. Eq. (28) can also be expressed in terms of the sensor information and its derivation:

$$d_i(t) = \frac{1}{\lambda_i} \left\{ \alpha_{0,i}\Delta_i(t) + \alpha_{1,i} \cdot \dot\Delta_i(t) + \ddot\Delta_i(t) \right\} \qquad (29)$$

Eq. (29) shows that the derivation of Δ_i is required. For the derivations cannot be measured they have to be approximated by difference quotients. This is of special importance because even a static deviation of the path in world coordinates, produces dynamic forms of Δ_i caused by the examined transformation from sensor to robot coordinates. The derivations can only be omitted if the deviation is static in the robot coordinates, which is a very specific case.

The complete set of the control law for apriori reference and online sensor correction can be summarized as follows

$$\underline{u}(t) = \underline{u}_B(t) + \underline{u}_d(t),$$

with

$$\underline{u}_B(t) = \underline{IN}^{-1}(\underline{r}_B) \cdot ((\underline{\ddot r}_B) - \underline{CO}(\underline{\varrho}_B))$$

$$\underline{u}_D(t) = -\underline{D}^{*-1}(\Delta x, (\underline{\varrho}_B)) \cdot \left\{ \underline{C}^*(\Delta x, (\underline{\varrho}_B), \underline{u}_B)) + [\underline{\alpha}]\Delta x + \underline{\Lambda}\, \underline{d}(t) \right\}$$

with

$$d_i(t) = \frac{1}{\lambda_i}(\alpha_{0,i}\, \Delta\rho_{2i-1} + \alpha_{1,i}\, \Delta\rho_{2i} + \Delta\dot\rho_{2i})$$

$$i = 1, ..., m.$$

In these equations all parts, which depend solely on $\underline{\varrho}_B$ can be examined offline, which is symbolized by additional brackets.

SIMULATION RESULTS

The presented algorithms have been investigated for different cases of apriori reference. The following simulations refer to a robot with a cylindrical working space especially regarding the linear and prismatic joint moving the robot in the x–y plane.

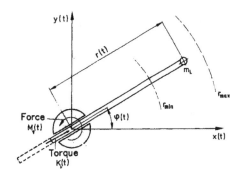

Fig. 3.   x–y view of the robot

As assumed, a very close path and velocity tracking was achieved in the case of full reference based control without online changes and without hurting physical constraints by the desired reference and velocity profile.

Performing a swinging movement of the robot arm different deviations from the apriori given reference were investigated. Figure 4 demonstrates the performance of a static turn of the path, but no change of the geometry of the path and of the velocity profile. In fig. 5 the same base reference is modified by online sensor correction, where the shape of the path is altered. In both cases the position error keeps very small. In the second case the deviation of the path causes some changes in the velocity profile, because the distance covered in the actual movement differs from the distance covered in the reference path.

The investigation of a straight line as reference path leads to similar results. Without online sensor correction by use of the reference based controller only nearly ideal path tracking was found.

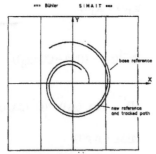

Fig. 4.   Path tracking for static shift of the reference by means of online sensor information

Shifting the new reference in comparison to the apriori given reference 0.2 m along the positive x–axis, the resulting velocity profile was close to the desired one.

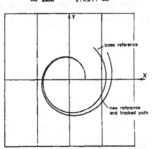

Fig. 5.   Path tracking for distortion of the reference by means of online sensor correction

Caused by realistic effects (Bühler, 1987), which were not introduced to the control law, some deviations arose.

Figure 6 presents the base reference, the shifted new reference and the actual movement of the robot. A path error caused by disturbances arises especially in the middle part.

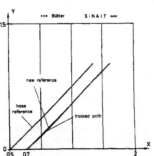

Fig. 6.   Study of straight line reference shifted in cartesian space

In fig. 7 the velocity profile of the TCP in comparison to the desired profile is presented.

In fig. 8 the influences of the different parts of the control law are outlined for the translational joint with respect to time. For this axis the reference input dominates the initial and finishing phase, while the error controller works in the non accelerated phase. The decoupling part is of low importance, because of the decoupling part which is included in the reference part.

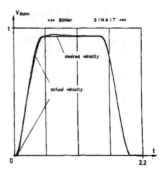

Fig. 7.   Velocity of tool center point

This is no more valid for the prismatic joint, as it can be seen in fig. 9. Here the error controller and the online decoupling part dominates in all phases of the movement.

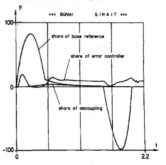

Fig. 8.   Partition of control output for the rotational movement

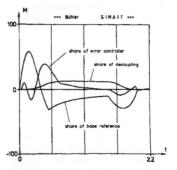

Fig. 9.   Partition of control output for the rotational movement

As a result of the investigation the path tracking can be improved by introduction of apriori knowledge about the path. Even in the case of online sensor correction causing considerable large deviation good results can be obtained.

CONCLUSION

A control concept is presented, which takes into consideration all the apriori available information about the path and which is able to react on online sensor corrections. As a base controller the nonlinear decoupling and control method is used, while in a second control layer online sensor correction is introduced.

Even in the case of considerable deviations of the actual desired paths a good performance is obtained. Switching between different references, including a zero reference for no apriori information, different levels of apriori information can be introduced.

ACKNOWLEDGEMENT

This research was supported by a grant of the Minister für Wissenschaft und Forschung des Landes Nordrhein–Westfalen, West Germany.

REFERENCES

Bühler, Ch. (1987). Strukturumschaltung und Empfind-lichkeitsanalyse bei der nichtlinearen Regelung für Industrieroboter. Abschlußbericht über das gleichnamige Forschungsvorhaben des Landes NRW.

Freund, E. (1977). A nonlinear control concept for computer controlled manipulators. Proc. 4th IFAC Symp. on Multivariable Techn. Systems, Fredericton, Kanada.

Freund, E. (1982). Fast nonlinear control with arbitrary pole placement for industrial robots and manipulators. Int. J. of Robotics Research, Vol 1, No. 1.

Freund, E., and H. Hoyer (1988). Real Time Pathfinding in Multirobot Systems Including Obstacles Avoidance. Int. J. of Robotics Research, Vol. 7, No. 1.

Furuta, K., and K. Kosuge (1984).Robust control of a robot manipulator with nonlinearity. Robotica, Vol. 2.

Klein, H.–J. (1986). Praktische Anwendung und Ergeb-nisse einer nichtlinearen Regelung für Industrieroboter. Robotersysteme Band 2, Heft 4.

Koivo, A.J. and T.–H. Guo (1983). Adaptive Controller for Robotic Manipulators. IEEE Trans. on Automatic Control, Vol. 18, N 2.

Moya, M. and H. Seraji (1987). Robot Control Systems: A Survey. Int. J. Robotics, North Holland, No. 3.

Silver, W.M. (1982). On the Equivalence of Lagrangian and Newton–Euler Dynamics for Manipulators. Int. J. of Robotics Research, Vol. 1, No. 2.

Slotine, J.–J.E. and W. Li (1987). On the Adaptive Control of Robot Manipulator. Int. J. of Robotics Research, Vol. 6, No. 3.

DATA STRUCTURES FOR COMPUTER-AIDED
ASSEMBLY PLANNING: A SURVEY

A. Delchambre

CRIF/WTCM, Industrial Automation Department, Brussels, Belgium

Abstract. This paper presents different existing data structures
used for the automatic generation of assembly plans.
It describes the information required for generating these plans :

- the geometric description of the final assembly and of the
components,

- the relational information between the components.

After this survey, a new approach is presented and discussed.

Keywords. Data structures; assembling; robots; CAD; assembly
sequences.

INTRODUCTION

The automatic generation and choice of
assembly plans are greatly interesting :
it provides actual flexibility in a
robotized assembly cell.
This work deals with the integration of
the design and the production of
assembled products. The final goal
consists in the direct generation of the
assembly plans from collected information
about the final assembly (which is
defined in a CAD database).
Such integrated solutions exist already
for machining processes but this approach
is quite different : a machined part is
totally defined by its shape. It is not
the case for a set of assembled
components.
The first and perhaps one of the most
important steps of an assembly planner
consists in the product analysis. The
modelization of the product has to
contain all the necessary information.
Then, the planner determines the feasible
operational sequences.

ASSEMBLY MODELIZATION :
STATE OF THE ART

This modelization has to describe the
geometry of the final assembly and the
connexions between the components.

Geometric description

It defines the geometry of each component
and its relative position in the final
assembly. The components are frequently
described as polyhedra. This means, in
principle, no strong restriction, since
bodies outside of this class can be
approximated.
This approximation is generally used in
the CAD systems.

Relational information

There exist different descriptions of the
connexions between the components. They
can be classified in four groups :

- the functional graphs which are easily
generated from the geometric analysis of
the final assembly,

- the models used by several high level
programming systems for computer
controlled mechanical assembly,

- classifications of parts and assembly
techniques which deal with the "Group
Technology".

Functional graphs. This formalism was
proposed by BOURJAULT in [1].
It represents the final assembly as a
graph whose nodes are the components and
whose links correspond to the physical
liaisons between the components.
Figure 1a shows a draft of an oil pump
and Fig. 1b its corresponding functional
graph.

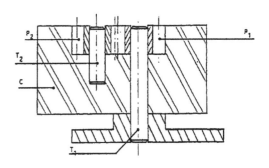

Fig. 1a Oil pump.

Figures 2a and 2b give an example of this structure.

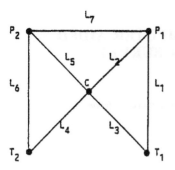

Fig. 1b   Functional graph.

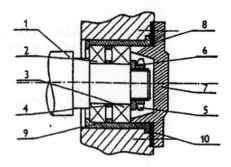

Fig. 2a   Gear assembly.

It is easy to generate automatically this formalism : a link is created between two components when they have at least one physical contact.

KROGH and SANDERSON , in [5], propose a similar model. They decompose the assembly into subtasks, each one involving only two subassemblies. The subtasks are equivalent to the Bourjault's functional links but each link does not correspond to a subtask. It is the reason why it is more difficult to generate automatically the Krogh and Sanderson's formalism. Nevertheless, this last representation is easily handled by a planner because there are no closed loops in the functional graph.

The major drawback of these two formalisms lies in the lack of information about the type of contact between two components.

VAGIN proposes, in [11], a first tentative to solve this problem. The links, in the functional graph, are classified by their geometrical characteristic which is the form of the part surface coming into contact with another surface in the assembly. The relations of connections are divided into three classes :

- the shaft and bush connection where cylindrical surfaces are used (R1),
- the contact between two plane surfaces (R2),
- a screwing fixation between two parts (R4).

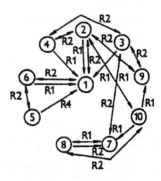

Fig. 2b   Connection graph.

JENTSCH, in [4], describes a similar representation. He distinguishes three types of contact, which are relevant to the applications restricted to translational dismounting only.

We see that each author employs his own representation but it depends much of the application each one has to treat. No general solutions are proposed for different examples.

World Models for computer controlled assembly. High level programming systems for computer controlled mechanical assembly, such as AUTOPASS ([7],[9]), modelize the real world thanks to a database called the world model. This world model includes the representation of both geometric information (such as the shape of an object and its location in the assembly world) and physical information (such as stability of objects, support

relationships and attachment
relationships between objects).
All this information is contained in a
graph-structure where each node
represents a volumetric entity and where
the links are directed and labelled to
indicate the relationships between the
entities.
An object node may represent three types
of entities :

- a three dimensional part : its shape is
  fixed but its position and orientation
  may change dynamically,
- a sub-part : the parts are decomposed
  in such basic polyhedra,
- an aggregate : it is an assembly of
  parts.

There are three kinds of relationships :

- part-of : represents the logical
  containment of one object in another.
  This relation introduces a hierarchy in
  the model and induces a tree structured
  world model,
- constraint : represents physical
  constraints of one object on another
  (rotational or translational),
- attachment : there are three types of
  it :

  - rigid : no relative motion can occur
    between the two related objects,
  - non-rigid : the objects cannot be
    separated by a large distance but a
    relative motion is possible between
    them,
  - conditional : objects are fitted
    together only by gravity but are not
    strictly attached.

Figure 3 shows an example of a world
model data structure.

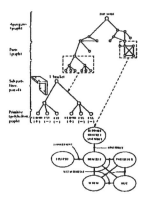

Fig. 3   World model data structure.

There is another object oriented language
for assembly : RAPT ([8]).It uses three
spatial relationships : against, fits and
coplanar.

All these representations contain the
complete information concerning the
geometric and spatial aspects but it does
not specify the assembly techniques. This

last information may be necessary if the
assembly planner wants to choose among
the different feasible sequences.

Group Technology. A lot of different sets
of components constitutes one of the
highest obstacles for assembly
classification.
Nevertheless, some attempts have been
made in order to classify the assembly
techniques.

KONDOLEON (in [12]) has obtained some
basic information on assembly tasks by
disassembling and reassembling a number
of typical products : a bicycle brake, a
refrigerator compressor, an electric
jigsaw, a small electric motor and an
electric toaster-oven.
The tasks, shown on Fig. 4, are
identified.

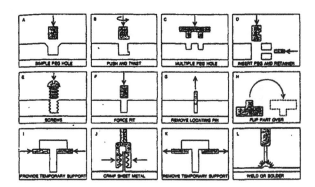

Fig. 4   Basic tasks identified by
Kondoleon.

In this classification, there are some
generic operations (screwing), but other
tasks are specific to particular parts
(insert single peg or insert peg and
retainer).
Some actions only differ in the object
shapes (insert simple peg or multiple
peg). Kondoleon does not distinguish the
really assembly operations (screwing)
from the positioning operations (flip
part over).
For all these reasons, this
classification is difficult to use for
the representation of the assembly.

SEKIGUCHI (in [10]) divides the connexion
relations into two main
groups :

- Fit relations which implied a pair of
  external and internal cylindrical
  surfaces.

They are decomposed in :

- pressure fit (Pr),
- push fit (Pu),
- screw fit (Sc),
- taper fit (Ta),
- position fit (Po),
- movable fit (Mo),
- gear coupling (Ge),
- ring fit (Ri),
- key fit (Ke).

- Contact relations between two side faces.
They are decomposed in :

- clamp contact (Cl),
- taper contact (Ta),
- plane contact (Pl),
- gear meshing (Ge),
- gap plane (Ga).

They are classified in the preceding groups using a downward degree of difficulty. It is determined by the combination of the freedom of motion and the required force to change the relative position of parts in assembly and/or disassembly.

We find again in this representation the two basic geometric relations between the parts : the fit-relations and the against-relations.
A connective matrix is built for each assembly direction (X, Y and Z) and for each type of connexion (contact and fit). Figure 5a shows a draft of a spindle head unit and Fig. 5b the corresponding connective matrices.

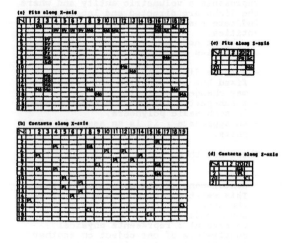

Fig. 5b   Connective matrices.

This classification contains perhaps the most information.

CRIF/WTCM APPROACH

The database of the CRIF/WTCM describes the geometry of the final assembly and the liaisons between the components.

Geometric description

In order to reduce the computation time, we modelized each part by some judiciously disposed parallelipipeds (3D) or rectangles (2D), whose edges are parallel to the coordinates axes. The dimensions and the positions of the parallelipipeds are entirely described by the coordinates of their two extreme vertices. At present, we suppose that the assembly directions are parallel to the coordinates axes and that each part of the assembly can be assembled in one straight motion.

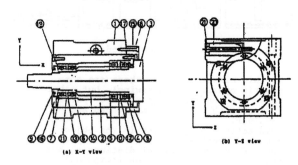

Fig. 5a   Spindle head unit.

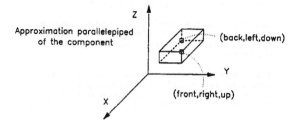

Figures 6.a and 6.b show a part of an electrical signalling relay and its modelization.

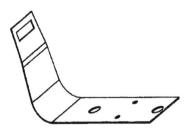

Fig. 6.a Real return spring.

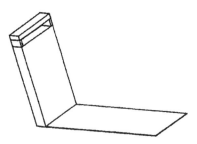

Fig. 6.b Modelized return spring.

We represent a component by the following fact (in Prolog):

```
part(name of the component,
     label of the parallelipiped,
     coordinates of point1,
     coordinates of point2).
```

For example, the return spring :

```
part(rs,para1,point1(16.0,27.0,11.0),
     point2(36.0,92.0,11.25)).
part(rs,para2,point1(16.0,92.0,11.0),
     point2(36.0,97.5,47.0)).
part(rs,para3,point1(16.0,92.0,50.0),
     point2(17.0,97.5,47.0)).
part(rs,para4,point1(35.0,92.0,50.0),
     point2(36.0,97.5,47.0)).
```

Relational information

The relational database consists in a liaison graph.
The links are labelled, in our approach, by the qualifiers :

- putting on (p) : the related objects are not strictly attached together, they are only in contact by some of their sides. This kind of link is characterized by more than one degree of freedom and is unstable in all the directions,

- insertion (i) : the related objects are fitted together. This liaison has only one translational degree of freedom and is stable in one direction,

- screwing (sn) : the parts are fixed together by several screws. No relative motion is possible between the two objects and the liaison is stable in all directions. n is the number of the screw,

- clipping (ci) : the parts are clipped together. The clipping consists in a non rigid attachment : the components cannot be separated by a large distance but relative small motion is possible between them. This liaison is non-reversible : there are always one clipping and one clipped parts. For example, in the relay, the return spring is the clipping part and the contacts support is clipped. We have to assemble first the clipping part.

Figure 7.a shows the assembly graph of the electrical signalling relay, presented at Fig. 7.b.

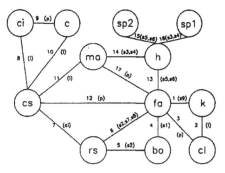

Fig. 7.a Liaison graph of the relay.

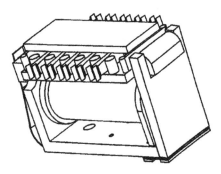

Fig. 7.b View of the relay.

This liaison graph is implemented in
Prolog by the following
facts :

```
link(1,[fa,k],s([s9])).
link(2,[cl,k],i).
    .
    .
    .
link(13,[h,fa],s([s5,s6])).
    .
    .
    .
link(17,[ma,fa],p).
```

CONCLUSIONS

The integration of the geometric and the
relational information constitutes the
most important characteristic of a
database for an automatic assembly
planner.
The geometric information describes the
final assembly and each component of it.
The relational information defines the
topology and the types of assembly
techniques used in the product.
With all this information, the system
must be able to generate the precedence
constraints between the components during
the assembly.

FURTHER WORK

In a very near future, we will :

- extract the geometric data from a CAD
sytem,

- automatically elaborate the functional
graph.

REFERENCES

[1] Bourjault, A. (1980). Contribution à
une approche méthodologique de
l'assemblage automatisé : élaboration
automatique des séquences opératoires.
Thèse d'Etat, Université de Besançon
Franche-Comté.

[2] Delchambre, A., D.Coupez, P.Gaspart
(1989). Knowledge based process planning
in robotized assembly. Proc. of SPIE VII
Conference on Applications of Artificial
Intelligence, The International Society
for Optical Engineering, Washington, pp.
768-776.

[3] Hornberger, J., B.Frommherz (1988).
Automatic generation of precedence
graphs. Proceedings of 18th International
Symposium on Industrial Robots, IFS
Publications, UK, pp. 453-466.

[4] Jentsch, W., F.Kaden (1983).
Automatic generation of assembly
sequences. Proceedings of the third
International Conference on AI and
Information Control Systems of Robots,
North-Holland, Amsterdam, pp. 197-200.

[5] Krogh, B.H., A.C.Sanderson (1985).
Modeling and control of assembly tasks
and systems. CMU-RI-TR-86-1, Flexible
Assembly Laboratory, The Robotics
Institute, Carnegie-Mellon University,
Pittsburgh, Pennsylvania.

[6] Lee, K., D.C.Gossard (1985). A
hierarchical data structure for
representing assemblies : part I. CAD,
17(1), 15-19.

[7] Liebermann, L.I., M.A.Lavin,
M.A.Wesley, T.Lozano-Perez, D.D.Grossman
(1980). A geometric modeling system for
automated mechanical assembly. IBM
Journal of Research and Development,
24(1), 64-74.

[8] Popplestone, R.J., A.P.Ambler,
I.M.Bellos (1980). An interpreter for a
language for describing assemblies.
Artificial Intelligence, 14, 79-107.

[9] Rocheleau, D.N., K.Lee (1987). System
for interactive assembly modelling. CAD,
19(2), 65-78.

[10] Sekiguchi, H., K.Kojima, K.Inoue
(1983). Study on automatic determination
of assembly sequence. Annals of the CIRP,
32(1), 371-374.

[11] Vagin, N.V., E.E.Klinov,
T.F.Lebedeva (1984). On a formal approach
to modelling mechanical assembly
arrangment in CAD/CAM. Computers and AI,
3(3), 273-278.

[12] Withney, D.E., T.L.De Fazio,
R.E.Gustavson, S.C.Graves, C.A.Holmes,
C.J.Klein (1986). Computer-aided design
of flexible assembly systems. R-1947, The
Charles Stark Draper Laboratory,
Cambridge, Massachusetts.

[13] Withney, D.E., T.L.De Fazio (1987).
Simplified generation of all mechanical
assembly sequences. IEEE Journal of
Robotics and Automation, 3(6), 640-658

LOGICAL STRUCTURE OF TOOLING SYSTEM DESIGN — FUNDAMENTALS OF TOOLING SELECTION EXPERT SYSTEM

V. R. Milacic and G. D. Putnik

Mechanical Engineering Faculty, University of Beograd, Beograd, Yugoslavia

Abstract. The paper presents a new approach to the tooling system design as the "classical" methods do not produce satisfactory solutions. This particularly applies to the tool selection problem in CAPP systems, as well as to the problem of the planning and control of tooling systems in FMS, that is to the overall CIM factory concept. The tooling system design is considered as a hierarchical system with four levels. Two of them are described in the paper, that is tooling system technological recognition and the tooling system composition logic level.

Keywords. Manufacturing processes, Flexible Manufacturing, Expert systems, Knowledge engineering.

INTRODUCTION

Tooling system design for manufacturing is a part of the manufacturing process design system. In the manufacturing process planning the relations between the following set of entitites are of particular interest. Set of workpieces (WP), set of type forms (TF), set of tools (cutting modules) (T), set of tool holders (TH), set of fixtures (FIX), set of machine tools (MT).

The above sets represent the structure mutually interlinked by relations. The structure of the above sets may be represented by a graph (automaton), Fig. 1, (Milačić, 1982).

The tooling system design is primarily tools-surface relation problem, that is the design of a tooling system in fact is the definition of a part of the superautomata given in Fig. 1. The part of the superautomata relevant for tooling system design is presented in Figure 2.

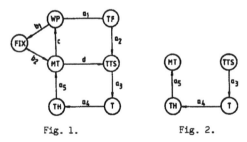

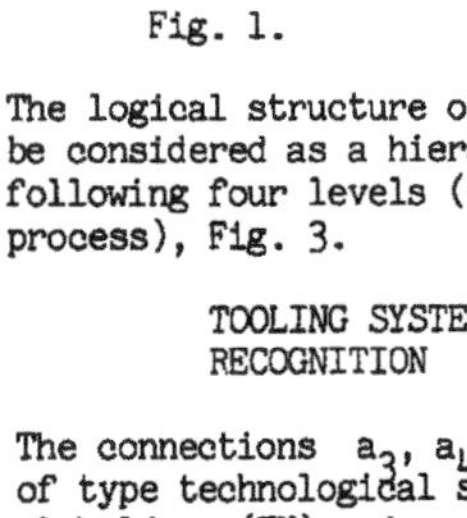

Fig. 1. 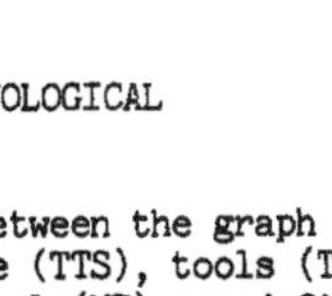   Fig. 2.

The logical structure of tooling system design may be considered as a hierarchical structure with the following four levels (similar to the manufacturing process), Fig. 3.

TOOLING SYSTEM TECHNOLOGICAL RECOGNITION

The connections a_3, a_4, a_5 between the graph nodes of type technological sequence (TTS), tools (T) tool holders (TH) and machine tool (MT) are defined by way of:

$$a_i = \{\text{GEOMETRY, DIMENSIONS, CHARACTERISTICS}\},$$

$$(i = 3, 4, 5)$$

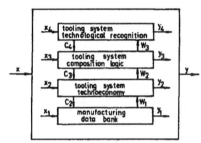

Fig. 3.

The geometry includes geometrical forms and their interconnections (topology), while dimensions involve all linear and angle measures (metrics). The characteristics are related to the tool material, that is tool parts, tool life, cutting regimes, statics, kinematics, dynamics, etc.

Technological recognition of geometry, dimensions and characteristics is performed in the operation plane through:

$$a_i^o = \{\text{COUPLING, COMPOSING}\}, \qquad (i = 3, 4, 5)$$

The cascade (connecting) operator (Mesarović, 1975) that is cascade composition is of special interest to us.

Cascade composition

There are two subsystems (Fig. 4a):

$$S_1 \subset X_1 \times Y_1$$
$$S_2 \subset X_2 \times Y_2$$

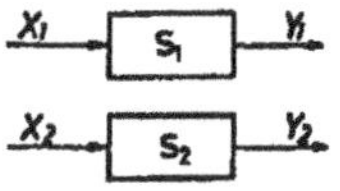

Fig. 4a.

Subsystems S_1 and S_2 are composed into the system S_3:

$$S_1 \circ S_2 = S_3$$

if:

$$Y_1 = Y_1^{\ast} \times Z_{x1}$$
$$X_2 = X_2^{\ast} \times Z_{y2}$$

where:

$$Z_{x1} = Z_{y2} = Z \quad \text{so that}$$
$$S_3 \subset (X_1 \times X_2^{\ast}) \times (Y_1^{\ast} \times Y_2)$$

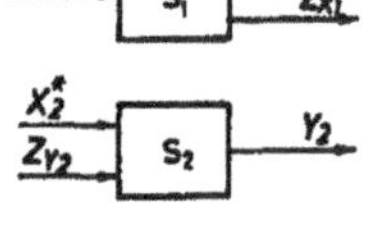

Fig. 4b.

and

$$((X_1,X_2),(Y_1,Y_2)) \in S_3 \leftrightarrow (\exists Z)(X_1,(Y_1,Z)) \in S_1 \wedge$$
$$\wedge ((X_2Z),Y_2) \in S_2)$$

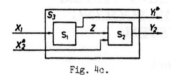

Fig. 4c.

To recognize the tool or tooling system for a machine tool means to determine whether the cascade composing operator may be applied to the machine tool system S_1 and the tools - tooling system S_2, that is to define (find, retrieve) all tools-tooling systems which may be cascadingly composed with the machine tool.

If no standard tools, cutting module, or a modular tooling system can satisfy the conditions for cascade composition, the cascade operator may be successively applied on several subsystems, that is on the following subsystems:

1. machine tool
2. tool holder
3. extension
4. cutting module/standard tool
5. machining pass (Fig. 5).

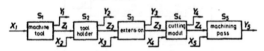

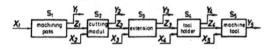

Fig. 5.

The recognition function F defined in this way has true values 1 or 0 (true or false):

$$F(S_3) = \begin{cases} 1, & (\exists z : z \text{ couples } S_1 \text{ and } S_2 \text{ into } S_3) \\ 0, & (\not\exists z : z \text{ couples } S_1 \text{ and } S_2 \text{ into } S_3) \end{cases}$$

The recognition function F may be also termed as coupling or as the coupling function. We may also state that it has, in the above case, two measures (values), that is it may be said that the coupling measure of the system S_1 and S_2 is 1 or 0.

The coupling function whose measure may have only two extreme values 1 and 0 enables the recognition process at the level of nominal elementary geometrical forms. However, in case of the technological level with additional characteristics, such as tolerance, machined surface quality and the like, the coupling function defined as above is insufficient.

In this way one tooling system may be represented by a graph as in Fig. 6a and 6b where the connection of individual elements is defined by the coupling function. The graph in Fig. 6b represents the network structure also. For the tooling system defined in the above way the knowledge base is defined with the tooling system recognition rules in the following form:

⟨RULE⟩ ::= IF ⟨ANTECEDENT⟩ THEN ⟨DESCEDENT⟩
HAS-HYPOTHESIS-VALUE ⟨Z_i⟩

TOOLING SYSTEM COMPOSITION LOGIC

The designing of tools and tooling systems is a composition process. First the type technological sequence [1] (latent technological transformation

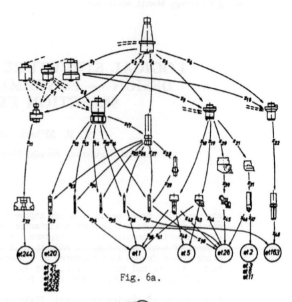

Fig. 6a.

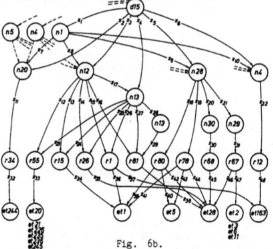

Fig. 6b.

set T_i by which type form TF_{i+1} is translated into type form TF_i) is decomposed to the elementary transformations ET_i. [2]. The elementary transformations are defined by tool-workpiece kinematic, i. e. the tool entity (Fig. 7).

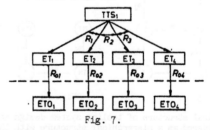

Fig. 7.

Definition A1: Each elementary transformation ET_i (technological passage TP_i) has a corresponding elementary tool entity, or elementary set of tools, or elementary tooling system, marked as ETO_i.

The connection between elementary transformation ET_i and the elementary tool ETO_i is established

---
[1] Type technological sequence represents the sequence of technological operations by which type form is achieved, as in example in Fig. 11.

[2] Elementary transformation corresponds to the concept of technological operations or machining pass.

through the production rule j for the criterion k (R_j^k):

$$(ET_i \ R_j^k \ ETO_i) , \qquad (i,j,k = 1,2,3,...)$$

If the abstract model is considered then the elementary tool consists of a sequence of tooling system elements. When coupled they represent elementary tool, for example TOOL HOLDER - EXTENSION - CUTTING MODULE , Fig. 8[3]. This means that elementary tool in the model is defined as the path between two graph nodes with determined constraints such as that the first node must belong to the tool holder set for a machine tool in question, while the last node must belong to the set of cutting modules. The constraints may also relate to other different criteria (for instance dimensions and the weight of the cutting module, or to the entire sequence).

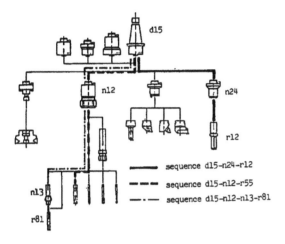

sequence d15-n24-r12

sequence d15-n12-r55

sequence d15-n12-n13-r81

Fig. 8.

The composing process is further applied to elementary sets, tooling systems ETO_i.

Definition A2: Each type technological sequence TTS has a corresponding type tool entity, or the type tooling system, marked as $\overline{TTO}_i$.

The above definition may be represented as in Figure 9a and 9b, which is in fact a condensed representation of Fig. 9a.

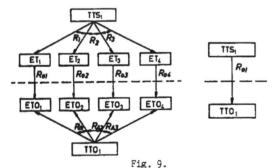

Fig. 9.

The composition of type tools is executed with the execution rule j for the criterion k (R_{aj}^k):

$$(ETO_i \ R_{Aj}^k \ ETO_{i+1}), \qquad (i,j,k = 1,2,3,...), \text{ or}$$
$$(ET_i \ R_j^k ETO_i) \ R_{Aj}^k \ (ET_{i+1} \ R_j^k \ ETO_{i+1})$$

The composing process is further applied to type

sets, tooling systems.

Definition A3: Each group of type technological sequences $GTTS_i(TO_i)$ has a corresponding group type tool entity, marked as $GTTO_i$.

The composing process is further applied.

Definition A4: Each complete set of technological operations KTO_i has a corresponding complete set of group tools marked as CTO_i.

Definition A5: To each technological process, marked as MPP (technological process is made of several complete sets of technological operations, that is, the technological process is the set of all technological transformations) corresponds the tooling system entity, marked as STO.

We shall retain here the term tooling system as the most general one, as it embraces all previously defined concepts, i.e.:

$$ETO \subseteq TTO \subseteq GTTO \subseteq CTO \subseteq STO$$

On the basis of technological recognition and technological process logic on one, i.e. technological process decomposition and technological tool recognition and tooling system composition logic on the other hand, a unique graph is obtained, as illustrated in Fig. 10. The production rules define the composition proces, i.e. the passing from one to the other level. The entire tooling system design graph may be understood as the decomposition and composition process. With the desomposition of the technological process to the level of elementary transformations, that is machining passes, the conditions are created for the definition of elementary tools, type tools and through to the tooling systems.

AN EXAMPLE OF THE TOOLING COMPOSITION FOR A TYPE FORM AND A HYPOTHETICAL PART

For type form tf15 the type technological sequence was defined, Fig. 11.

For these elementary transformations the following elementary tools may be composed (Fig. 12).

The variable block contains only the structural blocks in which no coupling measure between two successive blocks is false. The variable block follows:

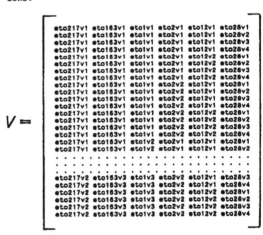

In figures 13, 14 and 15 three type tooling systems for type form tf15 are represented, that is for type technological sequence tts_{tf15}.

Further some criteria for the grouping of group ty-

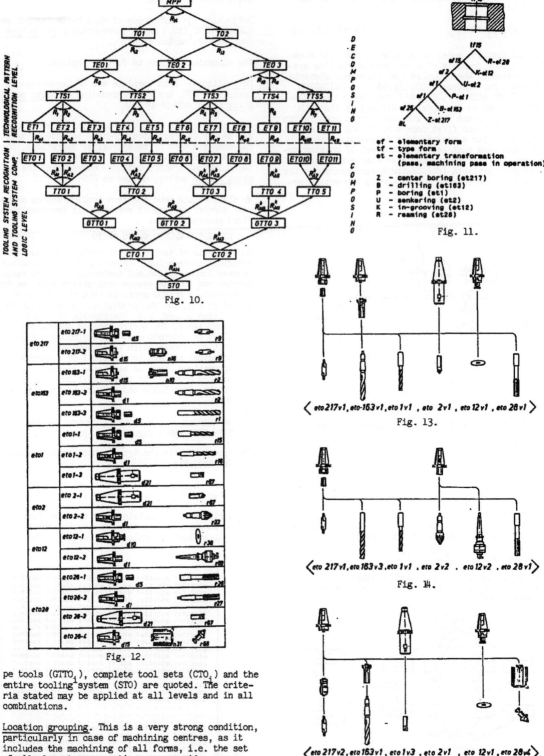

Fig. 10.

Fig. 11.

ef – elementary form
tf – type form
et – elementary transformation
(pass, machining pass in operation)

Z – center boring (et217)
B – drilling (et163)
P – boring (et1)
U – senkering (et2)
K – in-grooving (et12)
R – reaming (et28)

Fig. 12.

⟨ eto 217v1, eto-163v1, eto 1v1 , eto 2v1 , eto 12v1 , eto 28v1 ⟩
Fig. 13.

⟨ eto 217v1, eto 163v3, eto 1v1 , eto 2v2 , eto 12v2 , eto 28v1 ⟩
Fig. 14.

⟨ eto 217v2, eto163v1, eto 1v3 , eto 2v1 , eto 12v1 , eto 28v4 ⟩
Fig. 15.

pe tools ($GTTO_i$), complete tool sets (CTO_i) and the entire tooling system (STO) are quoted. The criteria stated may be applied at all levels and in all combinations.

Location grouping. This is a very strong condition, particularly in case of machining centres, as it includes the machining of all forms, i.e. the set of all elementary operations on one side, or per one workpiece, or according to one workpiece classification group, as it decreases manipulation times. It should be particularly stressed in large-scale production.

Similar form grouping. This condition is strong in view of tool planning and control, but it is weaker from the aspect of the production as it requires the transport of working conditions from one mach-

ining centre to another in a larger extent than it is in principle necessary, or a frequent exchange of tool magazines, which also represents a disadvantage.

Operation type grouping. This condition is weak for part production, but very strong for tool producers from the aspect of the planning of their production programs.

The application of the above mentioned criteria is illustrated at the example of a hypotechnical part (Fig. 16).

Location grouping (Fig. 17).

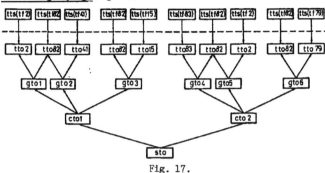

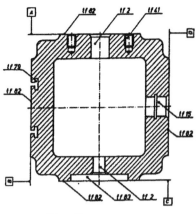

Fig. 16.

Fig. 17.

If the following type tools[3] are adopted

```
tto2n2(r9,r1,r15,d5,r67,d20).
tto2n2(r9,d5,r2,d1,r79,d15,r67,d20,r68,d7).

tto82n2(r31,d10).
tto82n2(r35,d11).

tto41n2(r9,r1,r15,d5,r67,d20).
tto41n2(r9,d5,r2,d1,r79,d15,r68,d7,r67,d20).

tto15n2(r9,r1,r15,d5,r67,d20,r38,d10).
tto15n2(r9,d5,r2,d1,r79,d15,r68,d7,r67,d20,
        r38,d10).

tto83n2(r35,d11,r55,d5).
tto83n2(r35,d11,r56,d1).

tto79n2(r35,d11,r55,r49,d5).
tto79n2(r35,d11,r56,d1,r49,d5).
```

. the following higher level tool sets (selected solutions are given for tool assortments and final tooling, systems).

```
gtto1(r9,r1,r15,d5,r67,d20,r31,d10).
gtto1(r9,d1,r15,d5,r67,d20,r35,d11).
gtto1(r9,d5,r2,d1,r79,d15,r67,d20,r68,d7,
      r31,d10).
gtto1(r9,d5,r2,d1,r79,d15,r67,d20,r68,d7,
      r35,d11).

gtto2(r31,d10,r9,r1,r15,d5,r67,d20).
gtto2(r31,d10,r9,d5,r2,d1,r79,d15,r68,d7,
      r67,d20).
gtto2(r35,d11,r9,r1,r15,d5,r67,d20).
gtto2(r35,d11,r9,d5,r2,d1,r79,d15,r68,d7,
      r67,d20).

gtto3(r31,r9,r1,r15,d5,r67,d20,r38,d10).
gtto3(r31,r9,d5,r2,d1,r79,d15,r68,d7,r67,
      d20,r38,d10).
gtto3(r35,d11,r9,r1,r15,d5,r67,d20,r38,
      d10).
gtto3(r35,d11,r9,d5,r2,d1,r79,d15,r68,d7,
      r67,d20,r38,d10).

gtto4(r35,d11,r55,d5,r31,d10).
gtto4(r55,d5,r35,d11).
gtto4(r35,d11,r56,d1,r31,d10).
gtto4(r56,d1,r35,d11).

gtto5(r31,d10,r9,r1,r15,d5,r67,d20).
gtto5(r31,d10,r9,d5,r2,d1,r79,d15,r67,
      d20,r68,d7).
gtto5(r35,d11,r9,r1,r15,d5,r67,d20).
gtto5(r35,d11,r9,d5,r2,d1,r79,d15,r67,
      d20,r68,d7).
```

```
gtto6(r31,d10,r35,d11,r55,r49,d5).
gtto6(r31,d10,r35,d11,r56,d1,r49,d5).
gtto6(r35,d11,r55,r49,d5).
gtto6(r35,d11,r56,d1,r49,d5).
```

```
cto1(r31,r9,r1,r15,d5,r67,d20,r38,d10).
cto1(r1,r15,r31,r9,d5,r2,d1,r79,d15,r68,
     d7,r67,d20,r38,d10).
cto1(r31,r35,d11,r9,r1,r15,d5,r67,d20,
     r38,d10).
cto1(r31,r1,r15,r35,d11,r9,d5,r2,d1,r79,
     d15,r68,d7,r67,d20,r38,d10).

cto2(r9,r1,r15,r676,d20,r31,d10,r35,d11,
     r55,r49,d5).
cto2(r55,r9,r1,r15,r67,d20,r31,d10,r35,d11,
     r56,d1,r49,d5).
cto2(r55,r9,r2,r79,d15,r67,d20,r60,d7,r31,
     d10,r35,d11,r56,d1,r49,d5).
cto2(r9,r1,r15,r67,d20,r35,d11,r55,r49,d5).
```

```
sto1(r38,r9,r1,r15,r67,d20,r31,d10,r35,d11,
     r55,r49,d5).
sto1(r38,r55,r9,r1,r15,r67,d20,r31,d10,r35,
     d11,r56,d1,r49,d5).
ato1(r1,r15,r38,r55,r9,r2,r79,d15,r67,d20,
     r68,d7,r31,d10,r35,d11,r56,d1,
     r49,d5).
sto1(r38,d10,r9,r1,r15,r67,d20,r35,d11,r55,
     r49,d5).
```

Similar form grouping (Fig. 18).

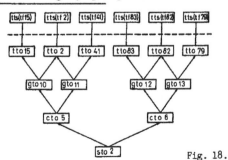

Fig. 18.

---

[3] Set of tools are represented by the list of tool elements, i.e. with the sets of tool holders, extensions and cutting modules. The names of tool elements are defined by the first character defining the type of tool element and by the number of the element itself (the number in the tool element catalogue). The character used for marking tool holder is "d", the character for extension is "n", while the character for cutting module is "r". For example r9 is designing the cutting module under the catalogue number 9.

If the same type tools are applied as in proceding
criteria the following higher level sets of tools
are obtained:

```
gtto10(r38,d10,r9,r1,r15,d5,r67,d20).
gtto10(r1,r15,r38,d10,r9,d5,r2,d1,r79,d15,
       r67,d20,r68,d7).
gtto10(r2,d1,r79,d15,r68,d7,r38,d10,r9,r1,
       r15,d5,r67,d20).
gtto10(r38,d10,r9,d5,r2,d1,r79,d15,r67,d20,
       r68,d7).

gtto11(r9,r1,r15,d5,r67,d20).
gtto11(r1,r15,r9,d5,r2,d1,r79,d15,r68,d7,
       r67,d20).
gtto11(r2,d1,r79,d15,r68,d7,r9,r1,r15,d5,
       r67,d20).
gtto11(r9,d5,r2,d1,r79,d15,r68,d7,r67,d20).

gtto12(r35,d11,r55,d5,r31,d10).
gtto12(r55,d5,r35,d11).
gtto12(r35,d11,r56,d1,r31,d10).
gtto12(r56,d1,r35,d11).

gtto13(r31,d10,r35,d11,r55,r49,d5).
gtto13(r31,d10,r35,d11,r56,d1,r49,d5).
gtto13(r35,d11,r55,r49,d5).
gtto13(r35,d11,r56,d1,r49,d5).

  cto5(r38,d10,r9,r1,r15,d5,r67,d20).
  cto5(r38,d10,r1,r15,r9,d5,r2,d1,r79,d15,r68,
       d7,r67,d20).
  cto5(r38,d10,r2,d1,r79,d15,r68,d7,r9,d1,r15,
       d5,r67,d20).
  cto5(r38,d10,r9,d5,r2,d1,r79,d15,r68,d7,r67,
       d20).

  cto6(r31,d10,r35,d11,r55,r49,d5).
  cto6(r55,r31,d10,r35,d11,r56,d1,r49,d5).
  cto6(r35,d11,r56,d1,r49,d5).

        sto2(r38,r9,r1,r15,r67,d20,r31,d10,r35,d11,
             r55,r49,d5).
        sto2(r38,r9,r1,r15,r67,d20,r55,r31,d10,r35,
             d11,r56,d1,r49,d5).
        sto2(r38,d10,r9,r1,r15,r67,d20,r35,d11,r56,
             d1,r49,d5).
        sto2(r38,r1,r15,r9,r2,r79,d15,r68,d7,r67,d20,
             r55,r31,d10,r35,d11,r56,d1,r49,d5).
```

THE COMPOSITION OF TOOLING SYSTEMS OVER
THE SELECTED GROUP OF REAL   PARTS (AN
EXAMPLE FROM INDUSTRY)

In order to test the selected concept the group of
8 prismatic parts machined on the machining center.
The complexity of parts, that is their geometrical
and technologial structure is given by different
number of type forms and elementary transformations
(technological operations) composing part i.e. with
the part is manufactured (naturally, the same type
form or transformation is repeated several times on
the same part with different characteristic geometrical values). This is shown in Table 1.

TABLE 1

Part No.	No. of type forms for part (TF)	No. of elementary transformations for part (ET)
1	6	10
2	3	7
3	4	6
4	5	9
5	4	6
6	7	10
7	10	10
8	7	9

The composition of the tooling system over elementary tools and type tools was tested. The influence
of three rules for the composition of elementary
tools was also tested, i.e. ETOMIN (elementary tool
has minimum number of elements - 2), ETON3 (elementary tool has 3 elements) and ETOMAX (elementary
tool has maximum number of elements). The results
of the composition including the minimum and the
maximum number of tool elements, that is their assortment within the tooling system separately for
each part is presented in Tables 2 and 3.

The results show the justifiability of the linking

TABLE 2

Part number	The composition of STO over ETO					
	The number of tool elements in the tooling system					
	ETOMIN rule		ETON3 rule		ETOMAX rule	
	min	max	min	max	min	max
1	11	18	12	19	18	22
2	10	14	11	15	14	19
3	10	12	10	13	14	17
4	10	16	11	18	16	22
5	5	10	6	13	13	17
6	10	17	11	19	16	21
7	12	18	14	19	19	23
8	11	15	13	16	17	22

TABLE 3

Part number	The composition of STO over TTO					
	The number of tool elements in the tooling system					
	ETOMIN rule		ETON3 rule		ETOMAX rule	
	min	max	min	max	min	max
1	11	19	14	21	20	23
2	9	15	11	16	16	19
3	9	17	12	18	19	20
4	16	22	15	23	21	26
5	8	17	9	17	15	18
6	14	23	17	23	22	28
7	13	20	13	19	17	22
8	12	20	13	19	17	22

of the group of tools with the groups of geometrical forms and the groups of technological operations at different levels of complexity. The differences between the solutions with the minimum and
the maximum number of elements in the tooling system, and the influence of different rules (criteria) on the tooling system composition point out to
the need for the introduction of an expert system
for tooling system design the basis of which may be
the concept presented here.

CONCLUSION

The paper describes a new approach to the tooling
system design. New concepts are introduced such as
"elementary tools", "type tools","group tools", etc.
by which the more complex structure of the tool concept are defined distant to the "classical" methods where under the term "tools" "elementary tools" are understood. Such extension of the tool concept enables more simple linking of tools with simple and complex technological tasks further enabling increased technoeconomic effects in CAPP, FMS
and CIM systems during tool selection, planning and
control phases. In the tooling system composition
examples for a hypothetical part, the obtaining of
several variants of tooling systems is presented,
subject to the applied tooling system composition
criteria. This means that one of the essential elements of tooling system design is the definition of
composing criteria. A defined logical structure of
tooling system design represent the basis for its
formalization, one of the principal conditions for
the building of tooling system design ES. The set
of rules for technological tooling system recognition, as well as the set of rules for the composition of complex tool structures represents the
knowledge base of the tooling system design ES.

REFERENCES

Mesarović,D., Takahara,Y. (1975). "General System
    Theory: Mathematical Foundations", Academic
    Press, New York
Milačić, V. (1982). "Computer-Based Information of
    Manufacturing Engineering Activities", Int. J.
    Prod. Res., Vol. 20, No. 3.
Milačić, V. (1987). "Manufacturing System Design The-
    ory-Production Systems III", Beograd University,
    Mech. Eng. Faculty, Beograd, (in Serbo-Croatian)
Putnik, G. (1988). "A Contribution to the ES Building
    for Machining Center Tool Selection", M.Sc.the-
    sis, Beograd University, Mechanical Engineering
    Faculty, Beograd (in Serbo-Croatian)

AUTOMATIC GENERATION OF MANUFACTURING CONTROL INSTRUCTIONS — AN EXPERT SYSTEMS APPROACH

J. Neelamkavil and U. Graefe

National Research Council Canada, Ottawa, Canada

Abstract. In a CIM environment, the requests for parts manufacture are usually given through a shop schedule and a supply of necessary NC programs for NC machine tools, programs for material handlers (robots, for example), and instructions, verbal or written, to operators of manually operated machines. Very few aids exist to conveniently orchestrate the parts production in such a shop, particularly, when production runs are very short and the orchestration changes fairly quickly.

At the National Research Council of Canada, we have embarked on a project which, when completed, will have the capability of specifying the required orchestration of production activities by means of manufacturing control instructions (ie, a manufacturing control language) generated automatically by an expert system software module. The design of the controller and the control language is done in such a manner that it breaks down the manufacturing task into a hierarchy of operations, operational steps (such as setup, transform, teardown), suboperations and machine primitives. The structure of this language also allows for the easy inclusion of exception handling procedures such as NC program requested tool changes by a robot or tool breakage, for example. This paper will provide an overview of the manufacturing shop controller and a brief description of the expert system that is being developed for the automatic generation of the control instructions.

Keywords. Cell controller; database; expert systems; flexible manufacturing; hierarchical systems; knowledge base; process plan.

INTRODUCTION

Requests for the manufacture of parts in a manufacturing shop are usually given through a production schedule and a supply of necessary NC programs for NC machine tools, programs for material handlers such as robots, and instructions, verbal or written, to operators of manually operated machines. The actual orchestration of the production of parts is usually left to shop foremen and the operators of individual machine tools and material handlers, who employ their own experience and visual clues to attempt to do the right thing at the right time. Coordination with out-of-sight areas of the shop is difficult. The gathering and reporting of up-to-date shop status and performance data either does not exist, or is inaccurate, incomplete and badly out of date. In some "automated" shops local orchestration of small groups of machine tools and material handlers is performed by programmable logic controllers. Such installations often lack integration and often do not provide the flexibility for rapid change-over for handling very short production runs.

NEED FOR MANUFACTURING CONTROL LANGUAGE

In order to see where the need for a manufacturing orchestration language arises, consider the answers to the three questions that may be asked about a request for the manufacturing of a set of parts. The questions are:

WHEN do things have to be done?
WHAT has to be done? and ...
HOW is it done?

The answer to the "When?" is given by the production schedule, which specifies at what time each of the operations in a job will commence.

The answer to the "HOW?" question is provided by NC programs, material handler programs and textual instructions to human operators.

The answer to the "WHAT?" is the missing link which is supposed to give instructions on the type of items to be produced, the sequence of operations to be performed, the necessary machining and handling resources needed and the identification of corresponding NC programs or textual instructions required by these resources. This paper suggests answering this question by a set of instructions in the form of a language contained in a "Bill of Operations" as proposed in a paper presented at the MAPL conference (Graefe, 1988).

HIERARCHY OF MANUFACTURING TASK

At any given time a manufacturing shop may be processing several jobs simultaneously, as dictated by a given manufacturing schedule. Each job in turn may consist of several operations which define the manufacturing activities on the various machines or manual workstations that a part has to pass through during its manufacturing cycle. Each operation can consist of several steps such as: setup, part transformation and teardown. A step of an operation can be further broken down into a number of suboperations. For example, a part transformation may be broken down into the three suboperations of part loading, part machining and part unloading. A suboperation

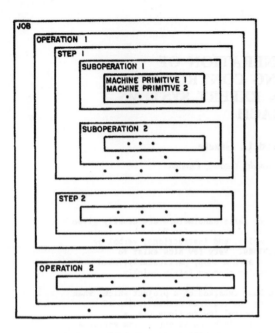

Fig. 1.   Structure of bill of operations

in turn can be broken down into a number of
machine primitive commands such as 'send an NC
program to a machine tool',or 'close a clamp or
gripper', or 'start the execution of an NC program
or robot program' etc. The logical breakdown of
the shop activities into jobs, operations etc.
leads naturally to a hierarchical structure of the
shop orchestration language, consisting of the
five layers:

1.   Job
2.   Operation
3.   Operational Step
4.   Suboperation
5.   Machine Primitive

For each job for the production of a number of
units of a particular part there will thus be a
description of "WHAT" has to be done, written in
the orchestration language, and contained in what
might be called a "Bill of Operations" for that
particular part. The Bill of Operations (see Fig.
1) will contain one "Job" statement, followed
(although not contiguously) by a statement for
each operation in the job. Each "Operation"
statement is then followed by its "Step"
statements, which in turn will be followed by
corresponding "Suboperation" statements and those
in turn by sets of "Machine Primitive" statements
relating to them. The following provides some
more detail about the statements at the various
levels in the hierarchy:

Job

There will be one "Job" statement for every job to
be processed in a shop. Every job has its own
Bill of Operation. The "Job" statement lists a
number of prerequisites which will have to be
fulfilled before the job can be released for
production on the production floor.

Operation

Within each job there will be one "Operation"
statement per operation which specifies an
operation number and the resource required for the

specified operation, as well as a number of
prerequisites necessary before the operation can
be started. Some of these conditions could, for
instance, ensure that the required resource has
completed its previous operation and is now idle;
that all necessary resource control programs (such
as NC programs, setup instructions etc.) are
available; that all necessary tooling or fixturing
is at the resource, in transit or at least in the
tool crib or ready in the fixture preparation
area; or that parts to be processed are at hand or
are not expected to be unduly delayed.

Step

For every operation there may be a number of
"Step" statements which refer to setup, transform
or teardown steps, as required by the operation.
They provide a step number and the resource
identification, as well as the type of step to be
performed, that is: Setup, Transform or Teardown.
Additional information given with the step
statement depends on the step type.

Setup. Each "Setup" statement refers to a unique
setup identification. This information is used at
the next lower layer for the purpose of skipping
setup completely, if the resource is already set
up for this particular operation from a previous
job's operation. It is assumed that the setup
step is performed at most once per operation.

Transform. The "Transform" step covers the actual
transformation of a part during the current
operation, including (as will be seen under
suboperations) loading, machining and unloading.
This statement requires one or more additional
conditions which must be satisfied before the
actual processing on discrete parts can commence.
One such condition could be, for instance, that
the resource has to be idle and the correct setup
has been completed. It is assumed that the
controller executing this instruction will do so
as many times as there are parts to be processed
for this operation.

Teardown. This statement may require additional
conditions to be met before teardown is to begin,
or conditions under which teardown is to be
skipped. Teardown will be performed at most once
per operation.

Suboperation

"Suboperations" provide the detail for the steps
of an operation. There are many possible types of
suboperations, each identified by a type such as:
Load, Machine, Unload, Transfer and Branch.
Additional parameters required depend on the type.
A set of suboperations in sequence describe the
actions to be performed by the parent step
instruction.

Load. The "load" instruction specifies which type
of part is to be loaded, and where it can be
obtained from. Hence one has to specify the id.
of the item to be loaded, and the location where
the item is to be loaded from.

Machine. The "machine" instruction is responsible
for the physical changes to be made to the part
during the current operation. It has to specify
the id. of the part which is to be machined.

Unload. The instruction for "unloading" a part has
to specify the part type id. Unloading implies
executing machine primitive instructions for
removing a part from a resource and placing it
into a standard storage location as given to the
material handler via its resource control program.

Transfer. This instruction is similar to the unload instruction except that the part is being transferred from the current resource to the resource which will perform the next operation. The part type id. has to be specified.

Branch. A "branch" instruction is used when a choice of suboperations exist, with the choice depending on some given conditions. For example, after finishing a machining suboperation the part is to be either unloaded into storage or transferred onto the next resource, depending on the status of the resource for the next operation. Thus the "branch" instruction has to provide the condition to be tested and an instruction address to which control is to be transferred, should the given condition be satisfied.

Machine Primitive

A set of "Machine Primitives" in sequence describe the actions to be performed by the parent suboperation instruction. The many possible types of machine primitives that can be specified include: Reserve, Send, Start, Resume, Open, Close, Release and Goto. Additional parameters required depend on the type.

Reserve. The "reserve" instruction reserves a particular resource for performing the parent suboperation. "Reserved" resources can not be used for other suboperations until they have been "released" from the current suboperation. For shared resources, such as a robot for material handling, the "reserve" instruction serves as a means of queuing up requests for service by competing suboperations.

Send. The "send" instruction requests the transmission of a specified resource control program, such as an NC program or text of setup instructions for example, to a specified resource.

Start. The "start" instruction commands a specified resource to start executing the currently loaded resource control program from the beginning, as given by the previous send instruction.

Resume. After the "start" instruction has been issued, the resource control program is assumed to be executed until the end or a programmed halt is encountered. The latter allows some other external action to be executed at this point in time. At the end of this external action one wants to continue executing the interrupted resource control program. This is achieved through the "resume" instruction. To illustrate this concept, assume that a robot has been given the instruction to start executing a program that picks up a part and moves it into the jaws of a clamp on a machining centre. At the point when the part is held in the open jaws of the clamp, the robot program execution has to be halted, while the machining centre is instructed by the "close" instruction (explained below) to close the clamp. After this action the robot is given the "resume" instruction to continue executing its program, opening its gripper and then removing its arm from the machining centre.

Open. The "open" instruction is used to command a resource to open some clamping device or robot gripper. This is usually employed when coordinating the interaction of two resources, such as a robot and a machine tool, for instance.

Close. The "close" instruction supplies the complement to the "open" instruction.

Release. The "release" instruction is used when the services of a particular resource are no longer required. For a shared resource such as a robot for instance, this would mean that the robot is now ready to be reserved for the next device in the waiting queue for this robot.

Goto. The "goto" instruction is used to transfer control to a specific area in the Bill of Operations. For example, code for the unload and transfer instructions may physically follow one another in a program. Since they are mutually exclusive options, a "goto" instruction at the end of the first option would facilitate skipping the second option.

FORMAT OF BILL OF OPERATIONS

There are three types of Bills of Operations. They describe how:

- Parts are to be made.
- Tools are to be changed.
- Exceptions are to be handled. This type is broken down into two groups:
    - Standard exception handling.
    - Exception handling for a specific job.

The format for all of these Bills of Operations will be similar, they only differ in the requirements for the required number of levels in the control hierarchy and optional fields. Bills of Operations should fulfill the following requirements:

a) A Bill of Operation should have a hierarchical structure, which:

- Breaks up a job into operations.
- Breaks up an operation into steps such as setup, part transformation and teardown.
- Breaks up a step into sub-operations.
- Breaks up a sub-operation into machine primitives.

Not all Bills of Operations require all levels of this hierarchy. For the manufacture of parts on a manually operated machine for instance, one would not require the sub-operation and machine primitive levels. For tool changing, one would not require the operation and step levels.

b) There should be a header for every operation level of a Bill of Operations (where such a level exists), which contains a list of all resources, resource control programs and resource mounted fixtures required for the operation, as well as a list of all required Bills of Operation for job specific exception handling procedures.

c) Each line item in a Bill of Operations shall have a unique label, an operator, one or more operands, optional conditions to be satisfied, and an identification number for a Bill of Operations to be used if exception handling should become necessary. Optional conditions to be satisfied are to include checks on the status of specified resources and the availability of parts or materiel in storage areas or on resources. Where no conditions are specified, default conditions shall be checked internally before execution of a line item commences. For example, the availability of a resource will have to be assured before commencing a new operation on that resource.

d) The label should indicate operation, step, sub-operation and machine primitive numbers. Where levels in the hierarchy are missing, appropriate zeroes are to be embedded.

e) The sets of operators to be used depend on the hierarchy level:

- At the step level operators should provide for the execution of setup, transform and teardown.

- At the sub-operations level operators should provide capabilities for loading, machining, processing, unloading and transferring, as well as conditional branching, in order to select appropriate actions depending on current shop status.

- At the machine primitive level operators should provide capabilities for reserving or releasing resources, sending of resource control programs, starting or resuming the execution of such programs and for the opening or closing of clamps or grippers, as well as for the unconditional branching to other specific line items in the Bill of Operations.

f) The number of operands required in a line item depends on the operator used. Operands should provide information on the resources required, setup identifications, resource program numbers, resource device numbers (such as clamp numbers), storage area numbers, fixture and part numbers as well as label numbers for conditional or unconditional branches.

METHOD OF GENERATING A BILL OF OPERATIONS

A Bill of Operations can be generated manually, once the language syntax has been defined. However, the number of instructions required is usually large, the work is very tedious and repetitive, and hence error prone. Machine generation of such a Bill of Operations is thus highly desirable. The input to such a generating program would consist of a process plan, as well as a database for resource capabilities and parameters, and a shop knowledge database. The process plan would provide the sequence of the operations involved, the resources which will perform the operations, as well as identify all the required resource control programs and textual instructions. Resource parameters would supply the name of clamping devices, as well as the identification of storage locations for parts before or after machining takes place. The shop knowledge database would provide the identification of material handlers to be used for on/off loading of parts as well as for transporting parts between machining resources as a function of part weight and dimensions.

The generating program itself could be built in two ways:

In the first method one could define the most general operation possible as a template to be used. The instructions for each required operation are then generated by deleting all non-applicable instructions from a copy of this template and then providing the relevant parameters for the identification of resources, resource control programs, parts etc.

A second method described in more detail in this paper would build up the instructions for each operation using an expert systems approach to determine the required steps, suboperations and machine primitives. This method, chosen as the development tool at the National Research Council of Canada, seems to be the preferred one, since it may be difficult to develop a template which is a generalization of all possible manufacturing operations.

DATA FOR THE BILL OF OPERATIONS GENERATION (BOG)

In an automated manufacturing facility, there is no room for any kind of ambiguity. As a result, the resource parameters specified in the bill of operations and used during the manufacturing operation must be known to the men and/or machines. This implies that at the time of the BOG all necessary data/information must either be obvious to the men/machines or should be automatically derivable. A Process plan can usually supply most of the manufacturing information.

Process Plan

The process plan contains the description of manufacturing operations that are required to convert a workpiece from its raw material state to finished condition. It provides the sequence of operations involved, the resources which will perform the operations as well as identifies all resource control programs and textual instructions. The information contained in a process plan can be categorized into three types, namely: required, default and logically derivable.

Required information. An example of the required type of information is the machine tool (i.e. the primary resource) that will be used for making the part. Other information in this category include the resource control program (rcp) number used (e.g, a NC program number) and precedence relations (e.g, previous and next operations).

Default information. Some information can be supplied through default in the sense that it needs not be given in a process plan, because the data associated with this category is very 'obvious' to the operator and/or machine. An example of this category is the storage location/s where tools are kept. The important thing to note here is that, if the missing pieces of information are to be treated as default within a process plan, then they must be obvious to the operator or to the machine who/which actually performs the task.

Derivable information. Once again, the information of this kind need not be given in a process plan. At the execution time, the operator or the system can infer the data using some logic. An example (in the context of material handlers) may be: if the weight of the part is more than 100 lbs. then use an overhead crane, else use a fork lift truck; or, if the AGV is broken down, then use a fork lift truck, etc.

Some information may be embedded within other information, usually occurring when the mode of manufacturing operation is automatic. The tool numbers and the material handlers specified within a NC program are good examples. In such cases they can be treated as required information.

BOG EXPERT SYSTEM

Work has been under way for the development of an expert system for the bill of operations generation. The system is being developed using the ART expert system shell. The system input requirements are filled from the process plan database, resource database and the shop knowledge base. Each of these will be briefly described.

Process plan database

All data associated with process plans is stored

in the process plan database. The process plan data includes all 'required' type information specified at the time of creating the process plan. It may also contain, if given specifically by the process planner, any 'default' and 'derivable' type information.

TABLE 1   Bill of Operations

```
BEGIN  20 00 00 00 LATHE-2, SC IDLE, LAST JOB COMP.

SETUP  20 10 00 00 LATHE-2

BRANCH 20 10 10 00, 20 20 0 0, SC IDLE, SETUP COMP.

OPERATE   20 10 20 00   LATHE-2
SEND      20 10 20 10   RCP11 LATHE-2
START     20 10 20 20   LATHE-2 RCP11

LOAD     20 10 30 00 FIXTURE1, FIXTURE1 IN STORE S1
RESERVE   20 10 30 10   OPERATOR
SEND      20 10 30 20   RCP22 OPERATOR
START     20 10 30 30   OPERATOR RCP22
CLOSE     20 10 30 40   [C1 C2 C3]   LATHE-2
RESUME    20 10 30 50   OPERATOR RCP22
RELEASE   20 10 30 60   OPERATOR

OPERATE   20 10 40 00   LATHE-2
SEND      20 10 40 10   RCP12 LATHE-2
START     20 10 40 20   LATHE-2 RCP12

TRANSFORM 20 20 00 00 LATHE-2, SC IDLE, SETUP COMP.
....................................
....................................

TRANSFER-UNLOAD 20 20 90 00    LATHE-2

OPERATE   20 20 100 00  LATHE-2
SEND      20 20 100 10  RCP71 LATHE-2
START     20 20 100 20  LATHE-2 RCP71

BRANCH 20 20 110 00, 20 20 130 0 SC OF LATHE-5 BUSY

TRANSFER  20 20 120 00  P1
RESERVE   20 20 120 10  TMH11
SEND      20 20 120 20  TR11 TMH11
START     20 20 120 30  TMH11 TR11
OPEN      20 20 120 40  [C11 C12]    LATHE-2
RESUME    20 20 120 50  TMH11 TR11
CLOSE     20 20 120 60  [C21 C22]    LATHE-5
RESUME    20 20 120 70  TMH11 TR11
RELEASE   20 20 120 80  TMH11
GOTO      20 20 120 90  20 20 140 00

UNLOAD    20 20 130 00  P1
RESERVE   20 20 130 10  UMH11

....................................
....................................

TEARDOWN  20 30 00 00   LATHE-2

BRANCH    20 30 10 00   20 00 00 0, DO NOT TEARDOWN

OPERATE   20 30 20 00   LATHE-2
SEND      20 30 20 10   RCP91 LATHE-2
START     20 30 20 20   LATHE-2 RCP91

UNLOAD    20 30 30 00   FIX1
RESERVE   20 30 30 10   UMH1
SEND      20 30 30 20   UFR1 UMH1
START     20 30 30 30   UMH1 UFR1
OPEN      20 30 30 40   [C1 C2 C3]   LATHE-2
RESUME    20 30 30 50   UMH1 UFR1
RELEASE   20 30 30 80   UMH1

OPERATE   20 30 40 00   LATHE-2
SEND      20 30 40 10   RCP101 LATHE-2
START     20 30 40 20   LATHE-2 RCP101

OPERN-COMPLETE  20 00 00 00    LATHE-2
```

Resource Database

Generally, all 'default' information is kept here. This can consist of the capabilities, tolerances, etc. of various resources, storage locations, information about 'routine exception handling', clamps, fixtures, and so on.

Shop Knowledge-base

This provides the identification of the material handlers to be used for on/off loading of parts as well as for transporting parts between machining resources as a function of part weight and dimensions, or it provides the name of alternate material handlers if the preferred one is not available for some reasons. In general, the shop knowledge-base will include all 'derivable' information described in the previous section.

BOG EXPERT SYSTEM OUTPUT

Table 1 corresponds to the bill of operations generated by the ART expert system shell. In each row of Table 1, the machine primitives appears first, as these are treated as 'objects' within the expert system. The primitives are followed by the instruction numbers, resource names/numbers, status codes, etc. It should be noted that the timing and the necessity for the execution of each instruction (i.e, each row in Table 1) is decided by the controller. For example, if the controller knows that the 'setup' has been done already, it will utilize the 'branch' instruction to skip to the next appropriate operational instruction. Similarly, if the item is manufactured in multiple quantities, the controller will signal the information that no 'teardown' will be necessary between successive copies of the part. It should also be noted that the status codes (SC) associated with the bill of operations will be updated in real time, and hence the actual instruction that will be activated will depend very much upon the context.

It is evident from Table 1 that, to generate this table using manual methods will be tedious and error prone. Also, many machine primitives appear in groups and in repetitive (cyclic) form. An example of such cyclic activity in a machining center can be of the form 'open the door of the machine' before a robot can access to the machine (whether it is for the part loading, unloading, tool removal, tool loading, etc.). Accordingly, a system which generates the control statements automatically is being developed.

Table 2 corresponds to a draft Process Plan Format (without any data) which is used for the initial development. For convenience, the process plan format is structured into different hierarchies (levels I to IV). Table 3 provides the preliminary data contained within a process plan, shown in a 'schema representation' form that is understood by the ART expert system shell. Table 3 along with a few production rules of the type "If ... Then ..." (which are meaningful to the ART shell) are used for generating the control instructions given in Table 1. The rules are arranged in such a way that the structure that is outlined in the next section will be followed.

It should be mentioned that the values corresponding to such information as 'loadmh' (ie, the material handler for loading the fixture), etc. were unspecified in the process plan and hence are not given in Table 3. This 'missing' information has been inferred using the data from the resource database and the shop knowledge-base.

TABLE 2   Process Plan Format

Level I.   Part Number

Level II.   Operation Number
                Machine id.
                Operation description (eg. "assembly")

Level III. Action Task: actions needed by operation
                RCP #   RCP #   etc.
                Previous RCP#   Previous RCP#

Level IV.   Fixture data: fixtr. needed by operation
                Fixture #
                Loading RCP#
                Clamp #s
                Storage_from #
                Loading material handler#
                Unloading material handler#
                Unloading RCP
                Storage-to #

Level IV.   Parts data: parts needed by operation
                Incoming Part#
                Loading RCP#
                Clamp #s
                Storage_from #
                Loading material handler#
                Reposition Part#
                Outgoing Part#

Level IV.   Reposition data: repositioning the part
                Reposition Part#
                Reposition RCP#
                Reposition material handler#
                Clamp #s

Level IV.   Trans./Unload: transfer to next machine
                Outgoing part#
                Transfer RCP #
                Transfer material handler#
                Clamps#-now
                Next machine id.
                Clamps#-next-machine
                Unload RCP #
                Unload material handler#
                Storage-to #

TABLE 3   Process Plan in Schema Form

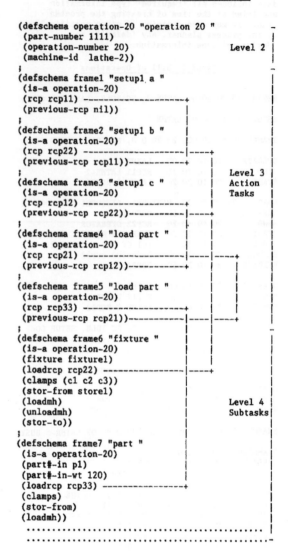

BOG EXPERT SYSTEM STRUCTURE

The system first checks the input data
requirements. If any of the 'required' type
information is missing it will ask the user
interactively for that data. If any other type of
data (default or derivable) are missing, the
system will automatically attempt to infer this
information. If for some reason, the system can
not infer the necessary information, it will
inform the user about the problem and prompt him
for this information.

The bill of operations are generated sequentially
by arranging all the machine primitives within the
context of the specific suboperation. The
procedure employed for the generation of the
control instructions (given in Table 1) is the
following:

Each operation (level II of Table 2) is made up of
a number of suboperations (level III). Each
suboperation is labeled with a unique rcp number
and is treated as an "action task". It also
contains information on the immediately preceding
action task rcp, labeled as the previous rcp#.
Each action task can also consist of a sub task
(level IV) which, if it exists, is labeled with
the same rcp number as the action task.

The rcp number and the previous rcp number
together provide the connecting link between the
two successive action tasks (as shown by dotted
lines in Table 3). If an action task has no
associated sub task, the system will generate
standard sequences of primitive control
instructions such as 'send an rcp program to a
resource', then 'start executing' it, for example.
If a subtask rcp number (example: rcp22 in Table
3), identical to the action task rcp number
exists, then all control instructions
corresponding to the subtask will be generated
(example: all fixture loading instructions for the
case rcp22 in Table 3) before generating the
standard sequences of primitive control
instructions. When all the action tasks (ie., the
rcp# and the previous rcp# chain) are executed,
the complete bill of operations appropriate to the
operation in question will have been generated.

REFERENCE

Graefe, P.W.U. (1988). A proposal for a discrete
    parts manufacturing control language. Symposium
    On Manufacturing Languages. Winnipeg, Canada.

Copyright © IFAC Information Control Problems in
Manufacturing Technology, Madrid, Spain 1989

A DECENTRALIZED ON-LINE SCHEDULING
STRATEGY FOR FMS

S. Engell

Fraunhofer Institute für Informations, Karlsruhe, FRG

Abstract: The paper deals with the scheduling problem in production
units which consist of several subsystems, e. g. workshops or flexible manufacturing cells. A two-layer architecture is proposed were
on the local level intelligent control units perform an optimization,
and the upper-level scheduler coordinates the subsystems.
The coordination problem is solved based on the mathematical description of the overall system in terms of the so-called minimax algebra.

Keywords: Flexible manufacturing. Decentralized control. Optimal
scheduling. On-line control of manufacturing systems. Distributed
control.

1. INTRODUCTION

This paper is concerned with the on-line
scheduling problem in medium or large
sized production units which consist of a
number of flexible work cells and workshops. The problem which is addressed is
optimal scheduling for just-in-time production, i. e. the finished parts or products are to be delivered on time but
production should start as late as possible to avoid stocks of half-finished
and finished products with the associated
capital cost.

This scheduling problem is of tremendous
complexity, in particular, if operations
can be scheduled on alternative machines
with possibly different effectivity. The
presently available techniques for production scheduling are not sufficiently
good (cf. [1]) and by and large are based
on simple priority rules. As the production environment is a very dynamical one,
because new orders are introduced, due
dates are changed, operation times vary,
problems with the delivery of external
parts occur, machines break down, workers
are absent etc., any optimization algorithm must be able to adjust the schedule
to the changing environment within reaction times of the order of magnitude of
seconds or minutes.

It is therefore a priori impossible to
perform an exact optimization even for a
small system of 6 machines with 50 operations to be scheduled within such reaction times, not to speak of an overall
system of up to 100 machines and hundreds
or thousands of operations associated
with hundred or more different jobs.

On the other hand, priortiy rules, as the
slack rule or the shortest imminent operation time rule (for a comparative study
see [2], [3]), have clear disadvantages
and should be replaced by more powerfull
algorithms especially in work cells with
multifunctional machines.

The introduction of the PC in production
environments has opened the road for new,
decentralized scheduling strategies. The
basic concept which is explored in this
paper is a decentralized two-layer architecture as shown in fig. 1. It consists
of "intelligent" subsystem controllers
which perform an optimization of the
schedules of small segments of the overall system, e. g. a workshop or a flexible production unit with 2 - 6 machines.
As the size of the local subproblems is
not too large, more sophisticated scheduling algorithms which minimize throughput time under the constraint to meet the
prescribed completion times can be applied, in contrast to the priority rules
which are used in centralized systems,
e. g. ABB's PRODUCAM system [9].

Clearly, these subsystems must be coordinated to achieve the desired just-in-time
production by the overall system. This is
the task of the upper-level scheduler,
and the main contribution of this paper
is to show that it can be done in a very
simple manner if a suitable mathematical
description of the problem is used.

This mathematical formalism is the so-
called minimax algebra on the extended
real numbers [4]. Its main benefit is the
fact that the controlled subsystems can
be described by a linear equation in minimax algebra notation. As such equations
can be inverted (in a special sense), coordination of the subsystems for JIT-
production can be achieved by simple algorithms.

Clearly, the resulting overall schedule
is still suboptimal, but the approach
here is a significant improvement over
the currently used priority rules.

The paper is organized in three main sections: Section 2 contains the mathematical description of FMS scheduling in terms
of the minimax algebra. Section 3 is concerned with the coordination of subsytems
of a large production process, and section
4 describes an effective local scheduling
algorithm.

2. MATHEMATICS OF FMS SCHEDULING

2.1 Definitions and basic results of minimax-algebra

In the sequel, we use the standard minimax algebra on the extended reals as discussed in [4], where $\dot{+}$ denotes the max operator and $\ddot{+}$ the min operator, $y \times z = y \dot{+} z = y + z$ for finite y, z, and adopt the conventions $y \dot{\times} \pm \infty = y \ddot{\times} \pm \infty = \pm \infty$ and $+\infty \dot{\times} -\infty = -\infty$, $+\infty \ddot{\times} -\infty = +\infty$. The matrix products $\underline{A} \times \underline{B}$, $\underline{A} \ddot{\times} \underline{B}$ are defined in correspondence to the definition in standard linear algebra with $+$ replaced by $\dot{+}$ resp. $\ddot{+}$ and $\cdot$ replaced by $\times$ resp. $\ddot{\times}$. Consequently, $\underline{E}$ resp. $\underline{\bar{E}}$ with

$$e_{ij} = \begin{cases} -\infty & i \neq j \\ 0 & i = j \end{cases}$$

$$\bar{e}_{ij} = \begin{cases} \infty & i \neq j \\ 0 & i = j \end{cases}$$

are the units of $\times$ resp. $\ddot{\times}$. Finally, the conjugate of a matrix is denoted by $\underline{\bar{A}}$ with

$$(\underline{\bar{A}})_{i,j} = -a_{ji}.$$

In the vector case, the $\dot{+}$ and $\ddot{+}$ operators are defined componentwise.

For our work, two basic results of minimax-algebra are needed:

Lemma 1 [4]

Let

$$\underline{x} = \underline{A} \times \underline{s}. \tag{1}$$

If

$$\underline{x} \leq \underline{x}_{max} \tag{2}$$

is prescribed, where the $\leq$ relation is defined as

$$\underline{a} \leq \underline{b} \iff a_i \leq b_i \text{ for all } i,$$

then all admissible solutions $\underline{s}$ satisfy

$$\underline{s} \leq \underline{\bar{A}} \ddot{\times} \underline{x}_{max} \tag{3}$$

and

$$\underline{A} \times \underline{s} \leq \underline{A} \times (\underline{\bar{A}} \ddot{\times} \underline{x}_{max}) \leq \underline{x}_{max}. \tag{4}$$

The vector

$$\underline{s}_{max} = \underline{\bar{A}} \ddot{\times} \underline{x}_{max} \tag{5}$$

is the maximal solution (componentwise) and produces a Chebychev-best-approximation to $\underline{x}_{max}$ as well as a best approximation in any norm of the form

$$||\underline{x}|| = \sum \sigma_i(x_i) \tag{6}$$

where the σ_i are monotonously increasing functions of x_i.

Lemma 2

Let

$$\underline{x} = \underline{a} \dot{+} \underline{b} \tag{7}$$

and $\underline{a}$ be given. Then if again (2) is prescribed,

$$\underline{b}_{max} = \underline{x}_{max} \ddot{+} \underline{a} \tag{8}$$

is the maximal vector which minimizes $\max_i(x_i - x_{max_i})$ as well as $||\underline{x} - \underline{x}_{max}||$ for any norm of the form (6).

Proof:

$x_i = \max(b_i, a_i) \geq b_i$, hence $\underline{b}_{max}$ which either produces $x_i = x_{max_i}$ or $x_i = a_i$ has the desired properties.

2.2 Minimax-algebra representation of FMS dynamics

The following discussion applies to an abstract FMS which may be a subsystem of a larger production system or the overall system itself.

We assume, that the inventory of the FMS under consideration consists of n running and N not yet started operations. K of these operations have exogenous starting conditions. The earliest possible times when these conditions are satisfied are the elements of the K-dimensional vector $\underline{s}$. L of these n+N operations are terminal operations, i. e. after this operation is performed, a part or product leaves the system resp. subsystem. For each of these terminal operations, a desired terminal time d_i is prescribed, the i-th component of the L-dimensional vector $\underline{d}$.

We assume that the FMS consists of M machines or workplaces, and that nominal operation times for each future operation on each of these machines resp. workplaces are known. For simplicity, we simply speak of machines in the sequel. These nominal operation times are gathered in the NxM-matrix $\underline{F}$. $(\underline{F})_{i,k} = f_{ik}$ is the nominal operation time for operation i on machine k. If operation i cannot be performed on machine k, $f_{ik} = +\infty$.

We assume, that a schedule for the N future operations is given, i. e. for each machine, an ordered sequence of operations has been determined such that all operations are contained once in one sequence.

Then P_i, the set of predecessors of operation i consists of

1) the last operation which has to be performed on the same machine before operation i
2) the operation which has to be performed on the same part before operation i
3) the last operations of all required parts in case of assembly operations.

The elements of the (NxN)-matrix $\underline{A}$ are defined as

$$a_{ij} = \begin{cases} -\infty & \text{if } j \neq P_i \\ f_{ik} & \text{if } j = P_i \end{cases} \tag{10}$$

where k is the machine on which operation i is scheduled.

The relation of the exogenous starting conditions to the operations is expressed by a matrix $\underline{B}$ of dimension NxK.

$$(\underline{B})_{i,j} = \begin{cases} f_{ik} & \text{if operation i has starting condition j} \\ -\infty & \text{if operation i has no exogenous starting condition.} \end{cases} \tag{11}$$

At most M operations can be running at a certain time. These may be terminal operations or operations which must be finished before one or more of the future operations can be started.

This is formulated mathematically via the NxM matrix $\underline{G}$:

$$(\underline{G})_{i,j} = \begin{cases} f_{ik} & \text{if operation i must wait for the termination of running on machine j} \\ -\infty & \text{otherwise.} \end{cases} \quad (12)$$

Let $\underline{x}_t$ be the vector of the predicted completion times of the future operations at time t, and $\underline{m}_t$ the vector of the predicted times when the running operations are finished resp. the machines are available again. If a machine is available but idle, the respective component of $\underline{m}$ is equal to the actual time t.

Then $\underline{x}_t$ is determined by

$$\underline{x}_t = \underline{A}_t \times \underline{x}_t + \underline{B}_t \times \underline{s}_t + \underline{G}_t \times \underline{m}_t. \quad (13)$$

The subscript t of all symbols in (13) indicates that the system is time-dependent and the information incorporated in (13) is the information available at time t.

In (13), the first term represents the interdependencies among the future operations, the second term the exogenous starting conditions, the third the availability of the machines and the dependencies on the currently running operations.

Each time when operations are terminated, the dimension and parameters of $\underline{A}_t$, $\underline{B}_t$, $\underline{F}_t$, $\underline{G}_t$ change. Furthermore, $\underline{s}_t$ and $\underline{m}_t$ may change at any time due to disturbances.

The completion times of the terminal operations are related to the vectors $\underline{x}_t$ and $\underline{m}_t$ via

$$\underline{y}_t = \underline{C}_t \times \underline{x}_t + \underline{D}_t \times \underline{m}_t \quad (14)$$

where

$$(\underline{C})_{i,j} = \begin{cases} 0 & \text{if operation j is the i-th terminal operation} \\ -\infty & \text{otherwise} \end{cases} \quad (15)$$

and

$$(\underline{D})_{i,j} = \begin{cases} 0 & \text{if the operation on machine j is the i-th terminal operation} \\ -\infty & \text{otherwise.} \end{cases} \quad (16)$$

Again, $\underline{C}_t$ and $\underline{D}_t$ change in a predetermined manner whenever an operation is terminated or new operations are scheduled.

The recursive equation (13) can be solved by at most N-fold iteration [4]

$$\underline{x}_t = \underline{R}_t \times \{ \underline{B}_t \times \underline{s}_t + \underline{G}_t \times \underline{m}_t \} \quad (17)$$

where the resolvent $\underline{R}_t$ is given by

$$\underline{R}_t = \underline{E} + \underline{A}_t + \underline{A}_t \times \underline{A}_t + \ldots + \underline{A}_t^N. \quad (18)$$

The calculation of $\underline{R}_t$ according to (18) however is computationally unefficient and can be avoided by use of search algorithms [8].

The completion times $\underline{y}_t$ then result from (17) by insertion of $\underline{x}_t$ into (14).

Using the inversion lemma of minimax algebra, we can calculate the latest possible completion times for the N future operations in the FMS such that the desired completion times $\underline{d}_t$ result under the given schedule:

$$\underline{x}_{t_{max}} = \bar{\underline{R}}_t \; \bar{\times} \; \bar{\underline{C}}_t \; \bar{\times} \; \underline{d}_t . \quad (19)$$

If

$$\underline{x}_{t_{max}} \geq \underline{B}_t \times \underline{s}_t + \underline{G}_t \times \underline{m}_t \quad (20a)$$

and

$$\underline{d}_t \geq \underline{D}_t \times \underline{m}_t, \quad (20b)$$

the due dates are met if the system operates according to the nominal conditions. Otherwise, delays are inevitable.

Assume now, that the vector $\underline{s}_t$ of external starting conditions can be split into two components

$$\underline{s}_t = \begin{bmatrix} \underline{s}_{1t} \\ \underline{s}_{2t} \end{bmatrix} \quad (21)$$

where $\underline{s}_{1t}$ is fixed and $\underline{s}_{2t}$ is free and should be maximized (componentwise) for JIT-production (start as late as possible). Then the completion times $\underline{y}_t$ are given by

$$\underline{y}_t = \underline{C}_t \times \underline{R}_t \times \underline{B}_t \times \begin{bmatrix} \underline{s}_{1t} \\ \underline{s}_{2t} \end{bmatrix} + \{\underline{C}_t \times \underline{G}_t + \underline{D}_t\} \times \underline{m}_t . \quad (22)$$

We partition the matrix $\underline{C}_t \times \underline{R}_t \times \underline{B}_t$ corresponding to the dimensions of $\underline{s}_{t1}$, $\underline{s}_{t2}$:

$$\underline{C}_t \times \underline{R}_t \times \underline{B}_t = [\underline{H}_{t1} \; \underline{H}_{t2}]. \quad (23)$$

Then from lemma 1 and lemma 2, the **vector of the latest free starting times which do not increase the lateness of the terminal operations** beyond the already inevitable lateness results as

$$\underline{s}_{t2_{max}} = \bar{\underline{H}}_{t2} \; \bar{\times} \; [\underline{d}_t + \underline{H}_{t1} \times \underline{s}_{t1} +$$
$$+ \{\underline{C}_t \times \underline{G}_t + \underline{D}_t\} \times \underline{m}_t]. \quad (24)$$

3. DECENTRALIZED SCHEDULING STRATEGY

3.1 Two-layer control strucure

The overall FMS control architecture is shown in fig. 1. Each subsystem (flexible cell or workshop) has a local controller, and these controllers are coordinated by the upper-lever scheduler.

The p local controllers periodically receive an inventory of jobs from the upper-level scheduler. Each job consists of

- the workplan (operation times on the machines, sequence in which the operations are to be performed)
- the external starting conditions (earliest possible starting times) for the operations of the job
- the due date for the completion of the last operation of the job
- a weighting factor or priority value.

For its inventory of jobs, the local controller determines an optimal schedule such that

a) the due dates are met resp. the delays are minimized
b) the completion times are minimized within this constraint.

After the local schedule is determined, the local controller returns a matrix $\underline{\underline{H}}_t^i$ and a vector $\underline{z}_t^i$ such that

$$\underline{y}_t^i = \underline{\underline{H}}_t^i \times \underline{s}_t^i + \underline{z}_t^i, \qquad (25)$$

where the components of $\underline{y}_t^i$ are the predicted completion times of the i-th subsystem and $\underline{s}_t^i$ contains the external starting conditions. $\underline{z}_t^i$ represents the effect of the currently running operations in the i-th subsystem.

$\underline{\underline{H}}_t^i$ and $\underline{z}_t^i$ result from the local schedules as described in the previous section:

$$\underline{\underline{H}}_t^i = \underline{\underline{C}}_t^i \times \underline{\underline{R}}_t^i \times \underline{\underline{B}}_t^i \qquad (26)$$

$$\underline{z}_t^i = \{\underline{\underline{C}}_t^i \times \underline{\underline{G}}_t^i + \underline{\underline{D}}_t^i\} \times \underline{m}_t^i. \qquad (27)$$

The task of the upper level control system is to coordinate the subsystems.

The fact that $\underline{y}_t^i$ in (25) is a linear function (in the minimax algebra) of $\underline{s}_t^i$ makes it very easy to coordinate the subsystems once their local schedules are known.

3.2 Coordination of the subsystems

3.2.1 Basic equations

Let

$$\underline{y}_t = [\underline{y}_t^{1T}, \underline{y}_t^{2T}, \, .. \, \underline{y}_t^{PT}]^T \qquad (28)$$

be the vector of all job completion times and form $\underline{z}_t$ from the local vectors $\underline{z}_t^i$ in the same manner.

Then $\underline{s}_t^i$ depends on $\underline{y}_t$ via another linear relation

$$\underline{s}_t^i = \underline{\underline{J}}_t^i \times \underline{y}_t + \underline{\underline{K}}_t^i \times \underline{s}_t \qquad (29)$$

where the vector $\underline{s}_t$ represents the external starting conditions for the overall system. The matrix $\underline{\underline{J}}_t^i$ represents the order in which the parts have to be processed by the subsystems and the transportation times between the subsystems.

Thus the overal system is governed by the recursive linear equation

$$\underline{y}_t = \begin{bmatrix} \underline{\underline{H}}_t^1 \times \underline{\underline{J}}_t^1 \\ \cdots \\ \underline{\underline{H}}_t^P \times \underline{\underline{J}}_t^P \end{bmatrix} \times \underline{y}_t + \begin{bmatrix} \underline{\underline{H}}_t^1 \times \underline{\underline{K}}_t^1 \\ \cdots \\ \underline{\underline{H}}_t^P \times \underline{\underline{K}}_t^P \end{bmatrix} \times \underline{s}_t + \underline{z}_t. \quad (30)$$

or

$$\underline{y}_t = \underline{\underline{\tilde{A}}}_t \times \underline{y}_t + \underline{\underline{\tilde{B}}}_t \times \underline{s}_t + \underline{z}_t.$$

The vector of the completion times of the parts which leave the overall system, $\underline{r}_t$, can be written as

$$\underline{r}_t = \underline{\underline{\tilde{C}}}_t \times \underline{y}_t + \underline{\underline{\tilde{D}}}_t \times \underline{z}_t. \qquad (31)$$

Denoting the resolvent of $\underline{\underline{\tilde{A}}}_t$ by $\underline{\underline{\tilde{R}}}_t$, we have

$$\underline{r}_t = \underline{\underline{\tilde{C}}}_t \times \underline{\underline{\tilde{R}}}_t \times \underline{\underline{\tilde{B}}}_t \times \underline{s}_t + \{\underline{\underline{\tilde{C}}}_t + \underline{\underline{\tilde{D}}}_t\} \times \underline{z}_t. \quad (32)$$

So if the local controllers compute the quantities $\underline{\underline{H}}_t^i$ and $\underline{z}_t^i$, and the dependencies among the local jobs including predicted transportation times are known, the completion times for the overall system can be easily computed as a function of the external starting conditions and the current "state" $\underline{z}_t$. Based on (32), those starting times which are still free can be maximized as described in section 2, equ. (21)-(24), for JIT-production.

Again, if

$$\underline{r}_t \leq \underline{d}_t \qquad (33)$$

is required, the latest possible completion times for the local jobs can be computed as

$$\underline{y}_{t\max} = \underline{\underline{\bar{R}}}_t \, \bar{\times} \, \underline{\underline{\bar{C}}}_t \, \bar{\times} \, \underline{d}_t. \qquad (34)$$

3.2.2 Introduction of new jobs

Let us assume that a new job is to be integrated into the overall schedule in addition to those jobs which are already incorporated in the local plans. We assume that the schedule for these jobs is satisfactory, i. e. the due dates are met.

The new job is first scheduled backwards from the desired completion time under the assumption that the difference of the earliest possible starting time and the completion time in each subsystem is equal to the nominal processing time plus some safety margin.

For all the other jobs, the local due dates are set equal to $\underline{y}_{t\max}$ calculated from (34) and the earliest possible starting times are taken as the vector which results from the insertion of $\underline{y}_{t\max}$ into (29).

Thus every subsystem can be assigned a complete enlarged inventory of jobs with earliest possible starting times and desired completion times. The local controllers return the data $\underline{\underline{H}}_t^i$, $\underline{z}_t^i$ from which the predicted completion times and the required externally determined starting conditions can be calculated as described above.

If the predicted completion times are unsatisfactory, the whole process can be iterated, i. e. the inventory of the local controllers is updated with due dates $\underline{y}_{t\max}$ from eq. (34) for the data $\underline{\underline{H}}_t^i$, $\underline{z}_t^i$ and the corresponding earliest possible starting times, and the decentralized optimization is restarted, eventually with different priorities.

For the outcome of this step, there are several possible cases:

a) The situation now is acceptable - fine.

b) The same jobs as before are delayed - the probality is small that more iterations will help. Look for technical alternatives (assigment of operations to other subsystems) or live with it.

c) The average delay is approximately the same but different jobs now are suffering - probably the system is overloaded and further optimization of the schedule can only distribute the delay among the jobs. Use priorities to do so and try to negociate the delivery dates unless you can enlarge the production capacity.

When the local schedules are accepted and the external starting conditions are fixed, the predicted vector y_t is calculated from (30) and the respective starting times result from (29). The local controllers receive these values as earliest possible starting times resp. completion dates, and the local schedules how to achieve these dates are known.

If delays seem invitable, the due dates for the subsystems should nonetheless not be later than those required for JIT production under ideal conditions, even if this seems unrealistic because of the external starting conditions. This will push the local controllers in the right direction when they adapt their schedules to new conditions.

3.2.3 Reaction to disturbances

One of the main reasons for decentralization is the possibility to localize the effect of disturbances. So if a disturbance occurs, the local controllers first try to minimize its effect on the local completion times. This is possible as long as the local controllers can command some free capacity in their subsystem. Whithin the framework here, the most adequate way to provide this ability to counteract against disturbances is to enlarge the nominal processing times by a factor which accounts for the probability of problems, e. g. by 20 %. If a disturbance occurs, the local controller affected by the disturbance tries to modify the schedule such that the prescribed completion times are achieved. If major delays are inevitable, the resulting earliest possible starting times are transmitted to the subsequent subsystems by the upper level scheduler, but their due dates are left unchanged. This process is continned until the local controllers of all subsystems which are affected by the disturbance have calculated a new schedule once. With this data, the new predicted delivery dates and the associated local starting times and due dates can be calculated and further steps can be taken as described in the previous section (iteration or technical modifications) if the resulting overall schedule is unsatisfactory.

Due to the fact that disturbances occur permanently, the transmission of the new predicted completion times should only be started

1) when a major disturbance (breakdown for a long period, problems with external parts, technical or quality related problems) occurs

2) periodically after the average duration of 2-4 operations to readjust the system to reality.

4. LOCAL SCHEDULING ALGORITHMS

Such a decentralized scheduling strategy must be based on adequate local optimiza-

tion strategies. The local algorithms must have the following properties

- dominant goal is the avoidance resp. minimization of delays
- within this predominant goal, the completion times are minimized
- on-line capability, i. e. a reasonable schedule is calculated very quickly and optimization takes place within some minutes at most; the best solution found so far is always available.

Intensive research on such algorithms has been performed at FhG-IITB in the last years.

The essential feature of the scheduling problem is that an exact optimization is impossible even for problems of moderate size like scheduling of 3 jobs with 4 operations each on 3 machines with fully overlapping functionality but different effectivity. This is due to the exponentially growing number of possible schedules and the extreme non monotonicity of the nonlinear search problem.

We have therefore proposed two-stage algorithms which satisfy the demands stated above [5, 6].

The low-level algorithms is a predictive version of the slack-rule [2, 3, 9]. The slack-rule simply is to asign the highest priority to the job where the lead-time i. e. the difference between the due date and the minimal required processing time, is minimal.

Among all possible priority rules, the slack rule is the one which is best suited for JIT-production because the due dates are directly taken into account [9].

The slack rule normally is used in order to decide which operation from a set of possible next operations should be performed on a certain machine. In an environment of multifunctional machines, also a decision has to be made where to perform the operation. Therefore the job with the highest priority is assigned to the next available slot on the machine on which the earliest completion time for the next operation results. By simulation into the future, a complete schedule is obtained.

The slack rule is a fast algorithm but it clearly is suboptimal because of the purely local nature of the decisions.

Therefore, a second optimization step is performed which uses information provided by the low-level algorithm.

Two possibilities have been investigated [5]:

a) to separate the problem in the time domain, i. e. to optimize within a set of "next" operations and to classify the initial schedules from the completion times which result if the subsequent operations are scheduled according to the slack rule [6]

b) to improve the schedule by recursive fitting of operations into idle periods of the machines (Moser [7], Peithner [8]).

So far, it seems that the latter algorithm is more effective than the separation in the time domain.

The cost function which we use to represent the hierarchy of goals defined above is

$$Q_i = \sum_j \alpha_j^i \cdot c \{y_j^i - d_j^i\} \qquad (35)$$

where d_j^i are the due dates, y_j^i are the predicted completion times, α_j^i is a parameter which reflects the importance of the job, and the function c is defined as

$$c(x) = \sum \begin{matrix} x & \text{for } x > 0 \\ 0.1x & \text{for } x < 0 \end{matrix} \qquad (36)$$

so that delays affect the cost functional much more than the early completion of jobs.

Example

As an example for the gains which are possible with improved local optimization algorithms over priority rules, we consider a system with 6 multifunctional machines with fully overlapping functionality which has to process 6 jobs with 10 operations each.

Figure 2 shows the Gantt-diagram which is produced by a predictive version of the slack rule. All due dates were assumed to be equal to 0, i. e. all jobs are late. The priority of all jobs is set to $\alpha_i = 1$. The resulting value of the cost functional (36) is Q = 1986.

We assume that machine 6 breaks down at t = 100 but is available again at t = 150. If scheduling is performed according to the slack rule (new schedules at t = 100 and at t = 150) the resulting sequence of operations is the one shown in fig. 3. Q increases to 2404.

If the slack rule is combined with the search algorithm which tries to fill the idling periods of the machines, the diagramm in fig. 4 results with Q = 1962 for the situation whe machine 6 is down from t = 100 to t = 150. The same algorithm gives Q = 1898 for the case without disturbance. Computing times for the improved algorithm are 2.13 sec at t = 0, 0.99 sec at t = 100 and 0.7 sec at t = 150 on an IBM PS2/80 with 20 MHz CPU and coprocessor.

5. CONCLUSIONS

We have presented a concept how improved optimization algorithms for the calculation of optimal schedules for small subsystems of a large production process can be coordinated in order to obtain good overall schedules. This coordination makes use of a representation of the decision-free subsystems in a minimax-algebra setting. The main benefit of this description is its linearity and invertibility (in the minimax sense). Due to this property, backwards and forwards calculations on the overall system level can be done without the necessity to consider the details of the local schedules.

Clearly, many questions concerning the details of such a strategy are still to be explored, e. g. how long the process should be iterated if the resulting overall schedule is unsatisfactory. Nonetheless, it is regarded as a very promising concept because

- there is no alternative to decentralization
- it is adapted to a distributed computer architecture and parallel computing is used
- the limitation to simple priority rules is overcome
- the coordination method is simple and elegant.

Acknowledgements

The author wishes to thank the reviewers of the draft paper whose critical remarks triggered a revision of the originally proposed concept. The work on the local optimization algorithms is supported by the Deutsche Forschungsgemeinschaft under grant DFG En 152/3. The programs for the simulation of the example were written by T. Kühn and K. Peithner under M. Moser's supervision, their dilligent work is greatly appreciated.

6. REFERENCES

[1] King, J.R.: Production Scheduling. In: A. Rolstadas (Ed.): Computer-Aided Production Management. Springer Verlag: Berlin, Heidelberg, 1988.
[2] Gere, W.S., jr.: Heuristics in job shop scheduling. Management Science 13, 167-191, 1966.
[3] Blackstone, J.H., jr.; Phillips, D. T.; Hogg, G.L.: A state-of-the-art survey of dispatching rules for manufacturing job shop operations. Int. J. Prod. Res. 20, 27-45, 1982.
[4] Cuninghame-Green, R.: Minimax-Algebra. Springer-Verlag: Berlin (Lecture Notes in Economics and Mathematical Systems Vol. 166), 1979.
[5] Engell, S.; Moser, M.: Regelung flexibler Fertigungssysteme (Control of flexible manufacturing systems). FhG-Berichte 2/89 (forthcoming).
[6] Engell, S.: Modelling and on-line scheduling of flexible manufacturing systems. Accepted for IFAC/IFORS/IMACS-Symposium "Large Scale Systems: Theory and Applications", Berlin, 1989.
[7] Moser, M.: Algorithmen zur Maschinenbelegungsplanung (Algorithms for FMS scheduling). Internal Report, FhG-IITB 1988.
[8] Peithner, K.: Algorithmen zur on-line Maschinenbelegungsplanung bei flexiblen Fertigungssystemen (Algorithms for on-line scheduling in flexible manufacturing systems). Studienarbeit, Fakultät für Informatik, Universität Karlsruhe, 1989.
[9] Beier, H.: Fertigungs- und Werkstattsteuerung der 4. Generation (Production and shop floor control of the fourth generation). Kommtech 88, paper 20.3, Essen, 7.-10.6.1988.

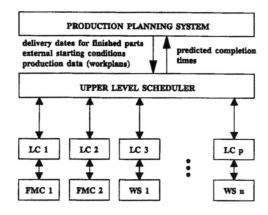

Fig. 1. Two-layer control architecture

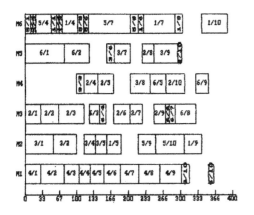

Fig.2.

Schedule obtained from the slack rule
without machine breakdown

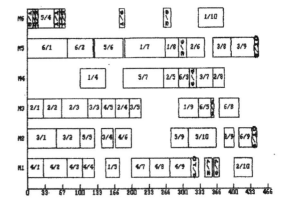

Fig. 3.

Schedule obtained from the slack rule
with breakdown of machine 6

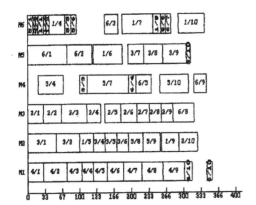

Fig. 4.

Schedule obtained from the optimization
algorithm whith breakdown of machine 6

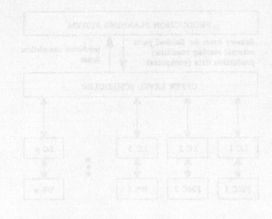

Fig. 1. Two-layer control architecture

Fig. 2.

Schedule obtained from the plant
without machine breakdown

Schedule obtained from the optimisation
with breakdown of machine 6

Fig. 3.

Schedule obtained from the optimisation
algorithm which breakdown of machine 6.

INTERACTIVE SCHEDULING FOR A HUMAN-OPERATED FLEXIBLE MACHINING CELL

L. Hatzikonstantis, M. Sahirad, M. Ristic and C. B. Besant

*Department of Mechanical Engineering, Imperial College of Science, Technology and
Medicine, London, UK*

Abstract. This paper describes an interactive scheduling system for a flexible,
human-operated machining cell. Attributing great importance to the operator's judgment,
especially under unfavourable conditions, emphasis has been placed upon the ability of the user
to influence the outcome of the system. Thus, preferences based on skill, past experience, or
locally available technological knowledge, that has not been incorporated into the system's
methods, can be assessed and implemented. Transparency in use and function, combined with
quick schedule generation should prove useful features, in order to facilitate decision making
regarding schedules, in the constantly changing manufacturing environment of small to medium
batch, high variety production. The structure and operation of the system are described and some
examples are presented.

Keywords: Flexible Manufacturing; Machining Cell; Production Scheduling; Computer Aided
Production Planning; Human Factors.

INTRODUCTION

Scheduling can be described as the allocation of resources
over time to perform a collection of tasks (Baker, 1974).
Within the manufacturing environment, scheduling plays a
key part, as it directly affects the efficient use of resources,
the ability of a manufacturing system to respond to changes
and the degree to which production targets are met. With
the proliferation of advanced manufacturing technology
(like CNC and automatic workhandling), the efficient use
of resources becomes even more important as this
technology represents great potential, but also high
investment. Furthermore, high variety and small to
medium batch production represents a high percentage of
industry around the world. This type of manufacturing is
dynamic in nature, therefore efficient planning and control
are of utmost priority. Scheduling can be seen as important
in order to achieve the purpose of economic production,
albeit not a stand-alone function, but rather a part of the
whole manufacturing design, planning and control strategy
of a company.

It is important to note that the attempt to achieve highly
automated production, through what is referred to as
Flexible Manufacturing Systems (FMS), has not always
yielded results as good as expected, in terms of flexibility
and reliability (Craven and Slatter, 1988). In contradiction
with trends towards completely unmanned cells, we
believe that the presence of skilled operator(s) is an
important factor contributing to the flexibility of a system,
especially in the highly dynamic environment of small to
medium batch, high variety production (Besant et al.,
1988). As only a small number of manufacturers can
afford large scale, highly automated systems, the way most
facilities can take advantage of advanced technology is to
implement such machinery in cells of varying sizes,
managed by a number of cell operators. Equipped with
appropriate decision making tools, the latter can provide an
important contribution to the overall flexibility of the
system.

A definition by Wemmerlow and Hyer (1987) describes a
cell as *'a collection of dissimilar machines or
manufacturing processes dedicated to the collection of
similar parts'.* Cellular manufacturing, as a way of
organising production, offers a number of advantages,
such as drastic reduction of totally unproductive
transportation times between operations and reduction of
machine setup times, by assigning jobs to cells according
to part family similarity. These advantages are particularly
important in the case of small to medium batch, high
variety production. The modularity of a cellular system
facilitates equipment modification and upgrading, but the
coordination of the whole manufacturing facility becomes
more complicated in an environment that may include cells,
independent machines in a functional layout and assembly
stations. Independence of the cells is one of the key factors
in such an environment. The targets of reduction of
transportation times and work-in-progress are served,
while the operation of the plant remains clear and easier to
control.

Within the cell, local decisions can be vital and one or a
number of skilled cell operators are probably best suited
to undertake a number of responsibilities, making use of
computerised decision support systems. Such
responsibilities may include local work scheduling and tool
management, process planning, generation and proving of
part programs and communication with workshop
coordination.

Particularly in the case of local work scheduling within the
cell, the presence of an experienced operator, aided by
concise and compliant software, can be an important asset.
Scheduling involves sequencing of operations on the
machines in the cell, in order to complete a number of
orders, achieving a number of production targets, within a
highly dynamic environment. This is an especially
demanding task, that could benefit from both computer
power and human judgment.

In the following paragraphs the development and operation
of a system built on the above guidelines are reported.

BACKGROUND

The flexible human-operated machining cell

Although the terms *flexible manufacturing system* (FMS) and *flexible machining cell* (FMC) have generally been associated with highly automated manufacturing, this is not the way in which they are used here. Indeed, human operators are expected to provide valuable contribution to the flexibility and responsiveness of a manufacturing system. Humans are seen in a complementary relation with technology and a manufacturing system should from the beginning be designed by simultaneously taking into account technological, as well as human factors. This does not mean that automation is discarded. On the contrary, it is seen as indispensable in order to achieve efficiency and free the operator from routine and repetitive tasks. However, the human operator should be in control of the manufacturing activity, making decisions on how to manage and plan the cell, especially in the case of unpredicted disruptions.

The cell will typically consist of a number of (preferably CNC) machines, auxiliary equipment like tools, fixtures, pallets, etc. and possibly a workhandling system, that will provide the ability for the cell to operate unattended for certain periods of time. The cell will be managed by one, or a number of operators. It will generally be a part of a larger manufacturing facility that may include other cells and probably machines in a functional layout.

Scheduling of a human-operated machining cell

A hierarchical planning and control factory structure is assumed, in which a high level workshop system assigns jobs to the cell. The cell is independent to the degree that it can then schedule locally the work, within the constraints of earliest start times and due dates assigned. The scheduling activity will typically be short term, in the order of one or two weeks. We assume that materials requirements have been dealt with at the higher planning level. At the cell level only machines are considered as schedulable resources. This means that it is up to the higher level workshop planner to feed the cell with tasks whose material requirements are satisfied. This is consistent with the idea of *cell independence*. However, tools, fixtures and other auxiliary resources, at a local level, should also be managed locally by the cell management system. Indeed, scheduling is only one of a group of functions, like process planning and tool management, that would constitute a complete computer aided production management system for the cell.

Objectives and Goals of the scheduling function

The efficient operation of any manufacturing system depends on many factors, scheduling being not the least important. The scheduling activity aims at satisfying a number of objectives/goals that can be:

- Multiple
- Subjective
- Complexly Interacting
- Conflicting
- Dynamic in Time

For example, the objectives of maximising machine utilisation and keeping work-in-progress low can be thought of as antagonistic, as low in-progress inventory might lead to machine down-times, thus reducing utilisation. On the other hand, objectives like reducing makespan and maximising utilisation can be thought as complementary, since by trying to achieve one, the other is also at least partially satisfied.

It is generally useful to distinguish among a number of discrete goal categories, the most important of which can be summarised as follows:

- On time completion of orders.
- Avoidance of in-progress inventories.
- Efficient utilisation of resources.
- Quality of manufactured products.

Any scheduling decision will have to take into account most or all of these objectives. Solutions based on a single criterion, although simplifying the problem, may easily fail to reflect a realistic manufacturing environment.

Interactive computer aided scheduling

Scheduling belongs to the category of *planning problems*, that are characterised by the ability of human experts to perform quite well in solving them, despite the combinatorial explosion that prevails. Apart from being able to distinguish good alternatives and rule out the most unpromising ones, a human expert has the ability to take into account multiple goals related to the scheduling function and base his decisions on realistic criteria. Furthermore, a human expert can recognise conflicts between goals and decide on how to resolve them. In other words, work out a satisfactory compromise among clashing requirements and especially when recovering from unexpected events. It is highly desirable to take advantage of such skills (Besant et al., 1988) and, in order to achieve this, human interactiveness is mandatory.

However, the human operator should not just provide skilled input to the computer system, but the latter should aim to relieve the human from hard computational work, in order to release creativity and enhance informed decision making capability.

SYSTEM OVERVIEW

A scheduling system for a Human-Operated Flexible Machining Cell based on the above guidelines will now be presented. This package has been developed on a Digital VAXstation II/GPX. It is part of a broader research project at Imperial College entitled "The Operation and Management of Flexible Human-Centred Turning Cells". A more general reference to all sides of this project, which includes integrating software functions such as Scheduling, Tool Management and Production Management in general, Process Planning, CAD/CAM, Quality Control and hardware control considerations, within a hierarchical factory structure, taking into account human factors in the design and development of the manufacturing environment, has been given by Besant et al. (1988).

Schedule generation by simulation

The idea behind the schedule generation system is to provide a tool that will help the operator make fast decisions, especially in case of unexpected events. Quick result generation, ease of use and transparent to the user operation are regarded as mandatory features, if the operator is to maintain control of scheduling, within the manufacturing environment of the cell.

Schedule generation is done by a discrete-event simulation based algorithm, that creates a *non-delay* schedule. This means that no machine is left idle, if there are operations waiting to be processed by it. *Priority rules* are used to load a certain operation to a machine, whenever a choice between alternatives has to be made. The latter happens, when a machine is "free" or "idle" and there is a number of operations, that have to be processed by this machine. In the case, when there is only one single operation awaiting processing by a free machine, this operation is loaded. This is not so, if a precedence constraint, imposed by the user, dictates that another operation must be loaded at that machine and it is specified that the machine must wait, even when there is another operation waiting. If this is the case, the generated schedule may no longer be non-delay.

A hierarchy of three priority rules is used. If the *main rule* cannot make a choice, because the value of the measure it uses for comparison is too close for more than one alternatives, the *first tie-breaker rule* is used and again, if needed, the *second tie-breaker rule*. To define "closeness" for a decision parameter a "similarity margin" is used. The use of tie-breaking rules ensures that decisions will not be based on marginal differences, while the "similarity margins" are user specified for added flexibility.

The package provides the user with the choice of some of the commonly used priority rules, for him to decide, according to case, what rules to use, so as to achieve the desired scheduling objectives. The following priority rules are provided:

- Most Operations Remaining (MOPNR)
- Most Work Remaining (MWKR)
- Least Work Remaining (LWKR)
- Earliest Due Date (EDD)
- Least Slack Time (LST)
- Least Slack per Operation (S/OPN)
- Least Critical Ratio (LCR)
- Shortest Processing Time (SPT)
- Longest Processing Time (LPT)
- Least Setup Time (LSUT)
- Least Work in Next Queue (WINQ)
- First Come First Served (FCFS)

The most suitable rules should be chosen according to specific case, in order to achieve the best results. For example, priority rules such as SPT and LWKR try to minimise Flow Time, which is of interest when jobs are required to be completed as soon as possible. MWKR tries to minimise Maximum Flow Time (or the makespan). The objective of meeting Due-Dates, which is approached by EDD or Slack Time rules, is usually very important. The least setup time rule (LSUT) selects low setup time jobs whenever a choice is needed. Thus, batches which enjoy common tooling and workholding are grouped together for machining. The LSUT rule is of particular importance in the scheduling of jobs on a turning machine/cell, as in turning operations the ratio of setup to processing times tends to be high.

Apart from simple priority rules, *combined priority rules* can also be used, trying to reach decisions, taking into account more than one decision parameters. For a turning cell, such a combined priority rule has been built, to choose jobs with low setup times, at the same time not letting jobs run late and if the two previous criteria are not conclusive, then select SPT jobs.

Schedule Modification by imposing Precedence Constraints

The idea behind this system is that a "good" schedule, according to various interrelated and subjective criteria, can be created, by a human scheduler, by modifying an already automatically generated "reasonably good" schedule. Heuristic generation through simple priority rules is known to give, in a short time, fairly good schedules, depending on the rules used. It is thus reasonable to quickly create a first schedule and then modify it to improve its performance, by applying experience, or local knowledge.

Modification is done by imposing precedence constraints. The following *types of precedence constraints* can be imposed:

FIRST : Schedule a particular operation of a job first on a certain machine.
LAST : Schedule a particular operation of a job last on a certain machine.
PRECEDENCE PAIRS : Schedule one job after another on a particular machine. The precedence can be *IMMEDIATE* or not. This means that, if after the first job, the second is not ready to start on the particular machine, the machine will wait, even if there are other jobs waiting. This can lead to a schedule that is not a non-delay one.

Another way of influencing the generated schedule is by defining *RUSH JOBS*. Up to three rush jobs can be defined, receiving priority whenever a choice is made, but not violating *precedence constraints,* which are considered higher in the constraint hierarchy.

The user can run the programme and create a schedule. He can then check parameters such as the schedule makespan, the average flowtime, the utilisation of machines, the number of tardy jobs, the average tardiness, etc. If the generated schedule is not satisfactory, he can modify it by imposing constraints, or using different priority rules. A new schedule is generated, which can again be modified, until an acceptable one is reached.

The user interface

Data information is easily accessible to the user through the system's front-end, which is menu-driven (fig.1). This kind of accessibility is useful, as decisions made by the operator have to rely on that data. Data regarding jobs, tooling requirements and machines are used to derive information about operations, like processing times and setup times. In addition, accessibility to the database is important when replanning is required. Thus, machines can be characterised as being out of order, and job routings can be changed. Even batch sizes and required tooling can be modified by the system, to recalculate related parameters.

The user interface has generally been designed to be friendly even to operators with no computing background. To ensure this, elaborate menu hierarchy has been avoided and there is only a single menu level. Thus, switching from one function to another is easy. Mouse activated soft buttons are used throughout to facilitate use of the package and reduce need for typing from the keyboard. Explanatory information is also provided in "help" menus. The result of scheduling is represented in the form of a Gantt chart. Process times are drawn as green (bright) and setup times as blue (dark) blocks. Statistical information, like makespan, tardy jobs etc. is presented in a separate text window (see fig. 1).

EXAMPLES AND CONCLUSIONS

As an example, a hypothetical turning cell consisting of a small and a bigger CNC Lathes and two CNC Milling machines is considered. Ten jobs, being machined by either a Lathe and a Mill or only a Lathe, are considered. They have different tooling, workholding, fixturing, due-dates, batch sizes, ready times, etc. Information regarding machines and jobs is presented in figure 2. Figures 3(a) and 3(b) show generated Gantt Charts. Rule similarity margins were set to 20% in all cases. In figure 3(a), Least Setup Time (LSUT) is the main rule, in order to reduce setup times, but there are tardy jobs. In figure 3(b), again LSUT is the main rule, but now there are no tardy jobs, by imposing the constraints: JOB-8 LAST on LATHE-1 ; JOB-6 FIRST on LATHE-2.

The process of interactively modifying schedules does not take more than a few minutes and is effective. Thus, the user can create a number of schedules, each performing well according to certain criteria, and choose the most appropriate compromise.

It is important to note that accuracy of information, like process times and setup times is essential for the scheduling system, as most scheduling decisions rely on such information. It is reasonable to assume that the most accurate data are available locally at the cell. Furthermore, in the case of unpredicted disruptions relevant information might only be available at the cell. Thus, it makes sense to plan cell activities locally.

Concluding, use of the system has shown that a fairly simple and fast schedule generator, based on simulation and simple priority rules, can give reasonable schedules, which can then be interactively modified by the operator to derive a more realistic and effective final schedule. Use of more elaborate computerised schedule generation techniques, performing (partial) optimisation, might improve the solution quality. However, in view of the difficulties in objective definition, the dynamic nature of the problem and the target of putting the operator in control, any automatic schedule generator should have to be combined with user interaction features, like the ones presented above. In that case, the improvements on the automatic schedule generation technique might have only a marginal overall effect on the whole system.

ACKNOWLEDGEMENTS

This research is part of the UK SERC (Science and Engineering Research Council) / ACME (Application of Computers to Manufacturing Engineering) research project entitled "Operation and Management of Human-Centred Turning Cells" at Imperial College (Ref. No. GR/E 13334).

REFERENCES

Baker, K. (1974). Introduction to sequencing and scheduling. J.Wiley and Sons.

Besant, C.B., M.Ristic, R.R.Slatter, M.Sahirad, I.Olama and L.Hatzikonstantis (1988). The Operation and Management of Flexible Human-Centred Turning Cells. *19th International Symposium on Automotive Technology and Automation (ISATA), Monte Carlo*, Vol. 1, 647-678.

Craven, F.W. and R.R.Slatter (1988). An overview of advanced manufacturing technology. *Applied Ergonomics*, **19**.1, 9-16.

Wemmerlow, U. and N.L.Hyer (1987). Research issues in cellular manufacturing. *International Journal of Production Research*, **25**, 413-431.

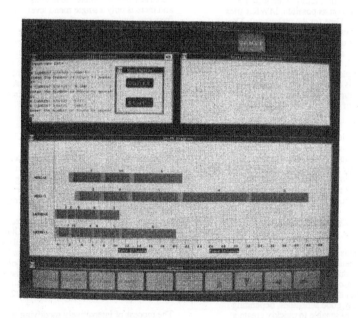

Fig.1 A photograph of the package's Front-End, where a generated Gantt Chart is displayed.

MACHINES:

NO.	NAME	TYPE	CAPACITY	OPERATIONAL
1	LATHE-1	HC/2-10	162 mm	YES
2	LATHE-2	HC/3	370 mm	YES
3	MILL-1	BP/2HP		YES
4	MILL-2	BP/2HP		YES

JOB INFORMATION:

NO.	DESCRIPTION	DUE-DATE	READY-TIME	BATCH-SIZE	ROUTING	RUSH
1	REAR. RING	21-07-88	-	60	2-4*	0
2	LEV. SCR.	17-07-88	-	30	1*	0
3	TRAN. PIN.	22-07-88	-	25	2-3*	0
4	TRAN. GR.	20-07-88	-	10	1-3*	0
5	SPINDLE	22-07-88	-	100	1-3*	0
6	FRO. RING	21-07-88	-	90	2-4*	0
7	CAM. SH.	19-07-88	-	30	1-4*	0
8	WR-AXIS	21-07-88	-	100	1*	0
9	SD-SCR.	20-07-88	-	15	2-3*	0
10	SHAFT-3	23-07-88	-	20	1-4*	0

Fig. 2 Machines of the Cell and Job Information for the Cell Case Study.

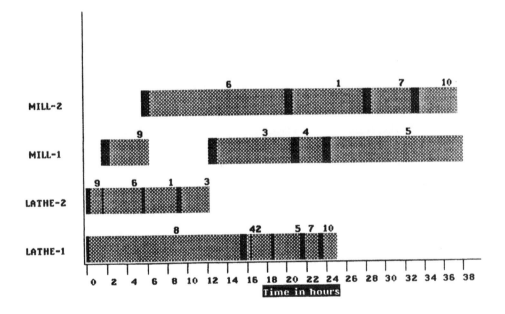

Fig. 3a Gantt Chart for the Case of 4 Machines and 10 Jobs Using the Following Rules.

Rules (Similarity Margin) :
1 : LSUT (20%), 2 : SPT (20%), 3 : LST
Makespan=38:00 hrs , Mean flow time= 24:00 hrs
Mean Tardiness = 9:35 hrs, Tardy Jobs = 2 & 7

Machine no.	Utilization	Setup/Processing Time
1	90.3%	10.8%
2	87.9%	13.8%
3	71.3%	12.3%
4	76.4%	11.6%

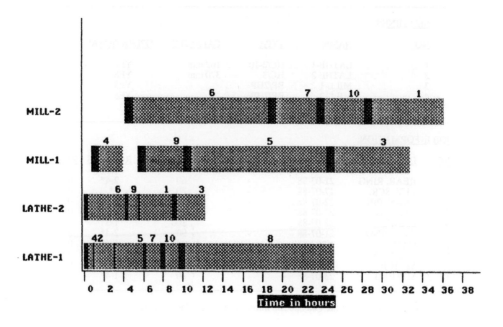

Fig. 3b Gantt Chart for the Case of 4 Machines and 10 Jobs Using the Following Rules.

<u>Rules (Similarity Margin)</u> :
1 : LSUT (20%), 2 : SPT (20%), 3 : LST
Precedence Constraints:
Machine 1:   Do   Job No. 8 Last
Machine 2:   Do   Job No. 6 First
Makespan=36:09 hrs , Mean flow time= 20:37 hrs
Mean Tardiness = 0.0 hrs, Tardy Jobs = None

Machine no.	Utilization	Setup/Processing Time
1	89.7%	11.4%
2	88.1%	13.5%
3	82.7%	12.3%
4	79.2%	11.6%

Copyright © IFAC Information Control Problems in
Manufacturing Technology, Madrid, Spain 1989

REACTIVE SCHEDULING OF A FLEXIBLE
MANUFACTURING CELL

J. M. Lázaro

The Robotics Institute, Carnegie Mellon University, Pittsburgh, USA

Abstract. Our main purposes with this work are to study the performance of the scheduling system
on a Flexible Machining Cell (FMC) cell and to gain insight for more general situations. Its
also our intention to identify the constraints which play a major role in the context of Flexible
Manufacturing Systems (FMSs). The scheduler described in this paper uses Constraint Satisfaction
Problem (CSP) techniques such as dependency directed backtracking, unary and binary classification
of constraints, for the task of scheduling the FMS. Special importance has been given to the
representation so that the constraints and algorithm used in this context can be easily applied
to other manufacturing cells. The possibility of changes while scheduling is also consider and
reactive techniques in the form of backtracking are used in order to avoid having to schedule
from scratch every time a change in either the state or the model comes up.

Keywords. Production control; Artificial intelligence; CAM; Industrial production systems;
Flexible manufacturing.

INTRODUCTION

Scheduling in a real plant is very complex and most of the schedulers
have to use predetermined rules of thumb to cope with it. Both flow-shop
and job-shop scheduling become NP-hard problems as soon as we only
have 3 successive operations or different machines. In order to get some
slight idea about its combinatorial complexity, let us consider a work
shop with M successive machines and O orders which require just one
operation and assume, for the sake of simplicity, that there is no time
interval between every two consecutive jobs. In this situation, the number
of possible schedules is limited by $(O!)^M$. For a work center with 10
machines which has to manufacture 10 elementary orders as those
described above, the number of possible sequences is over 10^{65}.

In addition to that, most real factories are known to be highly dynamic,
and some factors such as machine breakdowns, unexpected delays and the
arrival of new orders happen quite frequently. Therefore, plans make on
Tuesday may have already been invalidated by Thursday due to the
breakdown of a critical machine or the late arrival of any raw material.
For that reason, the system has to be flexible enough to take this new
information into account and patch its initial plan. A very important work
on this line in the context of a bigger system has been done with OPIS
(Ow ,Smith and Thiriez, 1988), where a set of heuristic procedures are the
knowledge sources which implement the reactive mechanisms in a
blackboard architecture. Another technique for reactive scheduling which
has recently been proposed can be found in the work of Elleby, Fargher
and Addis (1988). In this case, the capacity to react to changes is provided
by means of recording inconsistencies caused by *soft* constraints. *Soft*
constraints are those imposed during the process of Incremental
Constraint Satisfaction, and they are the only ones possible to retract. The
effect of retracting a *soft* constraint on the current schedule is to add all
those schedules which were declared inconsistent because of that
constraint.

In the following pages we describe a reactive incremental scheduler and
the domain on which it is going to be applied, a Flexible Manufacturing

System cell. In the following section we define the CSP paradigm and
show its close resemblance to scheduling, defined as the assignment of
time intervals to the resources required to machine a part. After that, we
describe the FMC, its constraints, and the representation of both of them.
In the section which deals with the scheduler, the system is defined and
the techniques used to cope with unexpected changes, mainly based on
the idea of i[dependency directed backtracking], are explained. Finally,
the last section includes some remarks and future extensions of the
current system.

DEFINITION

Given a set of variables $X_1,..., X_n$ with domains $D_1,..., D_n$, Constraint
Satisfaction Problems are those which deal with finding values for the
variables when they are subject to some constraints or relations among
those values. In this context, scheduling can be compared to a Constraint
Satisfaction Problem (CSP) where, instead of assigning a value to only
one variable at a time, we are finding values for several variables, which
are the resources and operations. For each operation some resources and
time intervals are found, and they are compatible with the domain defined
by the constraints. Scheduling is far more complex than the usual CSP
paradigm because each node is not a simple variable. On the contrary,
nodes have their own structure because decisions concerning several
variables are made each time the schedule is incremented. In this sense,
each node can be considered a CSP in itself.

The constraints which appear in the manufacturing environment are
identified, classified and broadly described by Fox (1987). As it is stated
there, some constraints are even contradictory among themselves. Cost is
a objective which must be minimized, but, under a wide range of
circumstances, meeting a due date may only be achieved with an increase
in the labor force or the overtime work and any of them is going to result
in a rise in the cost. Although the taxonomy is shown in Fox (1987)
includes most of the constraints, one aspect of them is not considered
there and it is the fact that, because the plant is a dynamic system, some
constraints are also highly dynamic. By this, we mean that these
constraints are known to be far more prompted to changes than others.
This problem causes the typical nonmonotonic situation, where the
addition of some information may result in an inconsistency with our

---

[1]Now in Labein, Cuesta de Olabeaga 16, 48013 Bilbao Vizcaya Spain

previous believes. At this point, there are two alternatives, rescheduling from scratch or trying to save and use as much information as possible from the previous plan and build our new schedule on that base. Our scheduler uses the second approach and responds to changes in the model of the cell or the assumptions under which a schedule was done by backtracking to a previous point in the search. In that process, after the detection and identification of a problem by the operator, the system backs up to a previous critical decision, and some intermediate assignments may be substituted by new ones, if necessary, depending on the type of problem which was encountered.

DESCRIPTION

Resources and Activities

There are four NC machines in the Flexible Manufacturing Cell. They are arranged by pairs on two workbeds, and each workbed has six bed positions, on which parts are set to be processed. Each bed position admits up to three parts, which can be processed by the same machine head. In each workbed, its bed positions can be divided into two groups of three, where each group is accessible to only one machine head.

There are usually eight or nine operations in the FMC for each part and they take times which go approximately from 2 to 7 hours. In general, operations require specific fixtures for machining a part. The common procedure for processing a part is the following. First of all, a compatible fixture is to be loaded onto the chosen free bed position. Then, the part is loaded onto that bed position and attached to it with the fixture. After that, the machine head will act upon that part. Once the machine head has finished, the part is ready to be unloaded and transferred to the next resource for another operation. Finally, the fixture can be unloaded. The fact that fixture changes are sequence dependent and the time it takes to load and unload them, which in some cases is much longer than the duration of some operations, are factors which describe the importance of a good sequencing of activities.

In order to represent that plant as well as the relationships in the model we use a semantic network with frames as nodes. These frames, which represent objects, activities and concepts such as machine head, operation and constraint, have slots which represent the characteristics or attributes of the entities associated to those frames. For instance, some attributes of a time interval are beginning, end, and duration. Attributes may have different values, depending on their domains, and they can be retrieved, modified, and so forth. As an example of a frame, the first operation of order-1, called operation-1-1, is represented as:

```
((operation-1-1
    IS-A: operation
    OPERATION-DURATION: 3.34
    RESOURCES-NEEDED: machine-head fixture workbed-position
    MACHINE-HEAD: any
    FIXTURE: f1-100-1 f1-100-2
    WORKBED-POSITION: any ))
```
Figure 1: The frame operation-1-1

In this initial stage, some simplifying assumptions have been made. The operations of loading and unloading the fixtures have not been considered yet, although the algorithm is independent of that assumption. We assign some definite duration to operations, although it would be more realistic to use an interval for that variable.

Constraints

At the beginning of the first section we reckoned the number of possible schedules for a simple model and obtained a huge figure, but this is nothing else than a mere number, because the constraints of the problem are going to drastically reduce the problem space and, therefore, the

search. The goal of any scheduler is to optimize some criteria while satisfying all the constraints, but in most of the real cases both tasks cannot be achieved simultaneously.

The main constraints which are relevant when making a plan can be classified, according to the type of domains involved, as:

- **Temporal**: Order due-dates and release times, demand windows or time bounds.
- **Relational**: Preferences among similar resources, precedences among operations.
- **Organizational**: Machine use, cost restrictions.
- **Physical**: Capacities, availabilities, incompatibilities.

But constraints can also be classified with respect to the number of variables which they affect, they can be unary constraints, binary, ternary and so forth. This kind of taxonomy has been used in the CSP literature for some time, Dechter and Pearl (1988) is an example, and our scheduler uses also this classification in the representation of those constraints attached to a resource or activity. We try to use a descriptive approach for the constraints so that the introduction or removal of some constraint can be done with high flexibility and no modification of the general algorithm, but some specific code is still needed in order to describe the way each constraint acts on the domain of its variables. Capacity, duration and demand window are all considered unary constraints, although, if we had had machines which processed parts with different sizes or geometries, capacity and duration would have also depended on the part. The fact that machines sharing a workbed cannot operate on certain positions at the same time is considered as a binary constraint between the time domains of both machine heads. In this FMS cell application, there are some additional constraints due to the fact that, for instance, machines sharing a workbed cannot operate on certain positions at the same time. Since this relationship constrains the scheduling of two variables, it is considered as a binary constraint. By using this approach, types of constraints involving more resources can be defined and very easily introduced into the problem solver.

On the other hand, to make things easier, there is only one process plan for each part type, and that simplifies enormously the operational precedence graph because that excludes the possibility of alternative routings for operations in every order. Because of that, relative precedence constraints have not been defined and are implicitly used. Anyway, defining routing for the orders seems more a task for a higher level, a production planner, than for the scheduler, which just has to assign times and reserve resources to operations.

SCHEDULER

We view production planning as an incremental CSP. The scheduler incrementally builds a plan taking into account the state of the already scheduled operations as well as the inconsistencies which might result from the following step. Since the instantiation order has a very important role in constraint satisfaction, the system focuses its attention on the most critical order, detected by means of a heuristic function. The value of that function depends on the due date, the time required for the remaining operations of that order and the priority of that order, which is based on the customer's demands and penalizations for late delivery. Its form is the following:

$$H_i = \frac{r_i}{s_i c_i}$$
$$P_i = I_i H_i$$

Where P_i is a coarse measure of the critical state of that order, I_i represents its importance as seen by the user from 0 to 1, r_i its remaining time, s_i its already scheduled time and c_i its normalized cost if not delivered on due date, either previously and cost is due to storage or with some tardiness, and cost is due to some penalty on late delivery.

The importance of such a utility function is related to the role it plays in guiding the search, as a main source of heuristics. The selection of a good function helps to obtain plans faster by preventing the system from getting to dead ends. As a result of that, the search is done more efficiently.

Once selected the order, the following operation, if still in the domain defined by a time horizon, is selected. The time horizon is introduced for two main reasons, to allow for unexpected changes, either in the state of the cell or in the number of orders to schedule, and to avoid narrowing down our problem space by taking too many hasty decisions. When dealing with a dynamic plant, where changes in the environment are likely to occur, it is not advisable to make scheduling decisions which affect operations and resources far ahead in time, because future changes may invalidate the plan and the system would have to schedule those operations again. Even more, as it is pointed by Elleby, Fargher and Addis (1988), the more decisions the scheduler makes, the smaller the number of remaining choices is. At any point, no more decisions must be taken than those which are strictly necessary. In an ideal situation, decisions should be made only when their lack would cause the factory to come to a halt, but in a real plant some knowledge about the following scheduling decisions is advisable to foresee situations which might result in dead ends for our search as well as for optimization purposes. This tradeoff between both tendencies can be handled by allowing some depth in the search by means of the time horizon. If the current order has been scheduled up that point, another order is selected. For that order, the following operation in its process plan is activated. Next, we estimate the demand window, concept used by Keng, Yun and Rossi (1988) or time bound, interval in which that operation should be performed. The demand window $[a_{im},b_{im}]$ for operation op_m of order or_i with n operations, based on the due date, d_d_i, the release date, r_d_i, or the end of the previous operation, $l\_e\_t_{im-1}$, the set of previous operations, S_{im-1}, and the duration of the unscheduled operations, d_{ij}, can be simply defined by

$$a_{im}=\begin{cases} r\_d_i & \text{if } S_{im-1}=\varnothing \\ l\_e\_t_{im-1} & \text{otherwise} \end{cases}$$

And if we call, for $j \neq n$,

$$D_i = \sum_{j=m+1}^{n} d_{ij}$$

then

$$b_{im}=\begin{cases} d\_d_i - D_i & \text{if } j \neq n \\ d\_d_i & \text{otherwise} \end{cases}$$

Once selected the interval, the scheduler focuses its attention on the different kinds of resources which are needed for that operation. That information is provided by elements as that seen in Figure 1. Constraints are sequentially evaluated to find intervals and specific resources which are still available.

As an example, let us consider the unary constraint which states that every operation should be done in its demand window and, for the sake of simplicity, let us assume that our current resource, r, has capacity one. For operation op_m, $DW_m=[a_m,b_m]$ is the calculated demand-window. If we call $S_{rn}=[u_{rn},v_{rn}]$ the scheduled intervals for that resource, with $n=1...$ p, and we assume that q of them are in the DW, then the intervals which will be tagged as available are those intervals $[c_{ri},d_{ri}]$ which are not selected yet and in the demand-window.

$$c_{ri}=\begin{cases} a_m & \text{if } i=1 \\ v_{ri} & \text{for } 1<i\leq q \end{cases}$$

$$d_{ri}=\begin{cases} u_{ri} & \text{for } 1\leq i<q \\ a_m & \text{if } i=q \end{cases}$$

If the orders have enough slack and no unexpected situation occurs, scheduling goes on until everything in the domain enclosed by the time horizon is scheduled. At the moment of selecting the best interval for the activity, we calculate its slack so that it may be helpful as a guide in backtracking. If due to some previous reservations a constraint cannot be satisfied and a conflict for an operation and a resource appears, the system comes to a temporary dead end and the scheduler tries to locally solve it by looking for an alternative resource in a breadth first search form. By using a declarative representation and this kind of algorithm, modifications of the model such as addition of new items or removal of existing ones can be done without any effect on the algorithm.

It may be the case that the state of the FMS is so contended that there are not any free intervals for the operation in the interval defined by the time bound. As a result of that, no solution can be found by merely extending its current state. The system can follow two approaches in this case. If contention is small, backtracking is the only process needed but, if that is not enough, relaxation of some constraint may also be required. The description of both processes is the following:

1. **Backtracking.** First, it tags as open that branch of the search tree, and makes an estimate of the cost of this branch by using the evaluation function. If it is found later that there are not any better paths in the search tree, that one will be considered again, as it is explained below. After that, it determines the type of the resource which caused the dead end and backs up. Operations in the demand window which have reserved that kind of resource are selected, and from them one is removed. The new operation takes its position, after checking for consistency with the applicable relationships. At this point, the system has two possible reactions, either to unschedule that operation and all of the following ones, and that means that it has to reschedule from that moment, or to unschedule that operation, accommodate the new one, and propagate changes where necessary. At this moment, only the first system is implemented.

   In Figure 2 we see a simple example which shows the differences between both types of backtracking. In situation (A) we see that the demand window for op-7 in that resource is already busy. In situation (B), op-4 and following operations are unscheduled, op-7 is inserted, and rescheduling of the removed operations is necessary. On the other hand, in case (C) operations are not removed, but only shifted. If there is enough slack, the shifting propagation should affect few operations and resources.

   If no solution is found under the previous circumstances, the system goes directly to the second procedure.

2. **Relaxing constraints.** When backtracking is not enough and no improvement is found by doing so, the only way to get some solution can be relaxing some constraint and the point to do it is that one with the best value for the evaluation function. Relaxing the due date solves the problem.

capacity

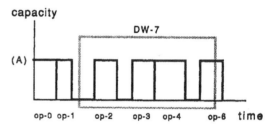

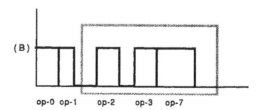

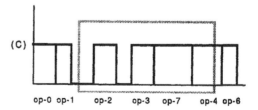

Figure 2: Backtracking due to op-7

Some constraints may happen to modify their values while the scheduler is still planning, but they can do it in different ways. More precisely, we could say that some constraints can change, while others can be changed. As it has been seen above, due dates may be changed by allowing relaxations when their previous values cannot be accomplished. The fact that an order cannot be delivered on the date it was due, does not mean that the schedule should be rejected as absolutely useless. Moreover, the penalty or cost from the customer might be very small. By allowing some changes we solve the problem. On the other hand, machine capacities or availabilities can change due to breakdowns or some unexpected delays may arise, invalidating the proposed schedule. It is obvious that this problem is usually far more serious than the former, since it affects plans which were already defined for many orders. Since both sorts of modifications are different, the system takes advantage of that and responds in a different form. The first case has already been described above. For the latter case, it seems that having to schedule from scratch every time a constraint changes is not the best way to deal with the situation. Some parts of the schedule developed so far might be still consistent with the new state of the plant. Therefore, after identifying the cause of the problem, the system performs a dependency directed backtracking so as to save as many results as possible from the old schedule. Let us see two examples:

- If there is a failure in the arrival of a raw material, the system unschedules all the operations which need that material and redefines as free the resources and intervals which were used by those activities.

- In the case of a machine breakdown, in principle, only operations which are scheduled on that resource after that moment should be unscheduled and, after that, rescheduled again on an alternative machine. If the slacks are enough, or as it has been suggested previously, the plan is not very far ahead in time from the breakdown moment, only a small number of modifications is expected to be necessary.

CONCLUSIONS

Some relevant features have not been represented in our model yet, although they will be included in the future. The most important of them is a resource-based module to perform detection and analysis of bottlenecks. This is expected to reduce backtracking because some dead ends may be foreseen and those branches of the search tree, avoided. A system with such a bottleneck detector has already been implemented and is described by Smith and Ow (1985).

Because the manufacturing environment is stochastic in nature, another important issue in a scheduling system is its capability to deal with stochastic events. With respect to that, knowledge gathered about the plant gives us some estimation about the frequency of certain relevant events, such as machine breakdowns, and this information can be used to define an optimal time horizon or depth in the search which, as it was explained above, will be our limit for making scheduling decisions. Some research in this area is also left for a posterior version of the system.

The system is still under development at the Robotics Institute of Carnegie Mellon University, with Knowledge Craft[1] running on top of Common Lisp.

ACKNOWLEDGMENTS

I do wish to thank Katia Sycara, Norman Sadeh, Mark Fox and the other members in the CORTES project for their valuable comments during the development of this work. The project to build the reactive scheduler was supported by a grant from the Departamento de Educacion, Universidades e Investigacion del Gobierno Vasco-Eusko Jaurlaritza and by The Robotics Institute of Carnegie Mellon University.

---

[1]Knowledge Craft is a trademark of Carnegie Group Inc.

REFERENCES

Dechter, R., and J. Pearl. Network-based heuristics for constraint-satisfaction problems. *Artificial Intelligence*,34,1,1-38.

Elleby, P.,H.E. Fargher, and T.R. Addis (1988). Reactive constraint-based job-shop scheduling. In M. D. Oliff (Ed.). *Expert Systems and Intelligent Manufacturing*, 1st ed., North Holland. pp 1-10.

Fox, M.S. (1987). *Constraint-Directed Search: A Case Study of Job-Shop Scheduling*. Pitman, London. pp 4-74.

Keng, N.P., D.Y.Y. Yun, and M. Rossi (1988). Interaction-sensitive planning system for job-shop scheduling. In M. D. Oliff (Ed.). *Expert Systems and Intelligent Manufacturing*, 1st ed., North Holland. pp 57-69.

Ow, P.S.,S.F. Smith, and A. Thiriez (1988). Reactive plan revision.*Proceedings AAAI88*, Vol. I,77-82.

Smith, S.F., and P.S. Ow (1985). The use of multiple problem decompositions in time constrained tasks. *Proceedings IJCAI85*, Vol. 2, 1013-15.

SIMULATING A FACTORY PRODUCTION PROCESS WITH AUTOMATED GUIDED VEHICLES

G. Petkovska, V. Nedeljkovic and M. Hovanec

Lola Institut, Beograd, Yugoslavia

ABSTRACT. Paper presents a simulation model of a workshop (factory) for the production of multi-layer printed circuit boards, which represents a flexible manufacturing system (FMS). Several automated guided vehicles perform the material handling.
The second version of GPSS-F simulation package was used for this simulation. The model enables solving of the following problems:
- determining the optimal capacity of the input, output and intermediate buffer stores of some departments in which certain segments of the production cycle (etching, lamination, drilling etc.) are carried out;
- determining the capacity of the machines, i.e. production equipment, within the departments where single fragments of production cycle are carried out;
- determining the optimal number of vehicles and their optimal trajectories for the transport of all resources needed to provide continuity of the production cycle.

Keywords. Computer simulation; GPSS-F; flexible manufacturing; materials handling; automatic guided vehicles

INTRODUCTION

Simulation as a "view into the working environment" is gaining an increasing significance in expensive production systems e.g. in flexible manufacturing systems. [1]
Workshop (factory) for the production of multi-layer printed circuit boards (Fig.1) consists of several production departments. Certain processes of both single-layer and complete multi-layer printed circuit boards production are carried out in different departments. This is defined by the technology of the printed boards production.
Following operations are needed for the production of single layers: material cutting, material preparing (clearing, grease removing), drilling of registration pinholes, applying photoresist, artwork image transfer, etching, etc.
The lamination of all prepared layers is followed by the final processing of the multi-layer printed circuit board, which consists of: drilling, plating, applying solder resist, control, etc.
Operations are performed in the departments which may have input, output and intermediate buffer stores. Printed boards are being delivered to the input buffers, where they remain until the department production equipment (machines) is released. After the processing is finished, the printed boards are being placed into the output or intermediate buffers, depending on the way of the

transfer to the next department: by using the automated guided vehicle or, when the departments are one next to another, by transferring the printed boards directly from the intermediate buffer of the first department to the input buffer of the next department. Thus, the printed boards remain in the buffers in case the needed resources are occupied: in the input buffer - when the department production equipment is occupied; in the intermediate buffer - when the input buffer of the department where the next operation is to be carried out, is occupied; in the output buffer - when either input buffer of the next (distant) department is occupied or until an empty vehicle could be provided. AGVs can transport a set of maximum twenty printed boards.
Within one department, more identical work stations (e.g. several identical drilling machines) could exist. In this way, several layers or complete printed boards could be simultaneously processed.
Several automated guided vehicles carry out the transport of sets of single layers and complete multi-layer printed boards. Host computer makes the decision which vehicle should carry out the transport. (If more than one AGV is used there is a need to separate them to avoid accidents.) Host computer assigns tasks to the computers controlling the vehicles. Criterion for decision making is the selection of the empty vehicle nearest to the place from where it should take over the set of printed boards (in case more

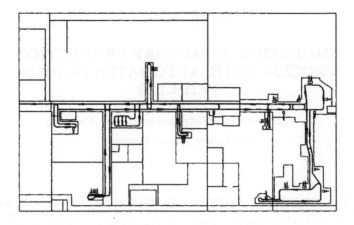

Fig.1. Layout showing linking through driverless
transport system

vehicles are not assigned tasks).
Trajectories are always chosen according
to the shortest distance criterion
(vehicles' trajectories allow more
different ways for passing from one halt
point to another). Positions of the input
and output buffers are defined as halt
points.
If there is another vehicle, without a
task, on the chosen trajectory of the
vehicle with task, then the vehicle
without a task is sent to a free halt
point.

PROGRAM PACKAGE DESCRIPTION

Fig.2 displays the structure of the
developed software package, which consists
of three programs. This package is
realized on the HP 9000/550 UNIX Version
5.61 in FORTRAN 77.

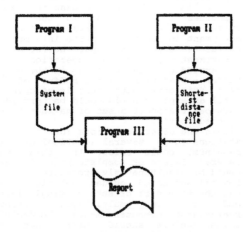

Fig.2. Structure of the developed
software package

PROGRAM I

Basic functions of this program are:
- initial definition of system
characteristics and processing routes of
the layers, i.e. complete printed boards;
- updating particular data;
- printout of basic system
characteristics.

Initial definition of system
characteristics includes:
- definition of the capacity of all
input, output and intermediate buffers;
capacities (number of identical work
stations within each department);
- definition of the transport system
(total number of transport vehicles,
average vehicle speed, initial vehicle
positions); etc.
Printed boards production process is built
into the so-called processing routes which
are given separately for each layer and
for the complete printed boards. This is
realized, first of all, by defining the
sequence of the departments where single
operations are performed, and processing
times within each department. All this
data is stored in the SYSTEM FILE and
represents the input data for the PROGRAM
III.
As already mentioned, PROGRAM I enables
simple and easy changing of both single
data and groups of data in the already
created SYSTEM FILE.
In a similar way, this program enables the
generation of the report about the entered
data. The report can be partial (e.g.
processing routes data for single layers)
or it can contain all entered data - about
the system and the boards.

PROGRAM II

Basic data used for the transport
simulation in FMS with several AGVs, i.e.
output of the PROGRAM II, are stored in
the SHORTEST DISTANCES FILE. This file
contains the following data:
- total number of halt points on the
vehicle's route
- number of segments defined in the

graph
    - lengths of all segments, ordered by
segment sequence number, as well as the
most important information:
        - lengths,
        - number of segments, and
        - sequences of the shortest distance
routes between each pair of halt points in
the given lay-out of the FMS, ordered by
the segments to be controlled in order to
avoid collision in the traffic of the FMS.
Lay-out of the workshop (factory) is
defined by the oriented graph, where nodes
represent the vehicles' halt points and/or
intersections (all points where the guide
paths diverge, merge or cross), and arcs
represent the trajectories which directly
connect the nodes (the aisles on which the
vehicles operate).

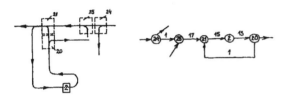

        [⁝]- intersection

        □ - halt point

Fig.3. a) Part of the lay-out of the
          printed boards factory
        b) Corresponding part of the graph

Characteristic of the arc is its length,
determined by the actual lay-out.
Intersections are collision zones - all
points on the vehicle's route where two or
more paths intersect in an arbitrary way -
defined by the technological requirements.
No more than one vehicle can occupy the
collision zone at any point in time.
Characteristic of the node is the "length"
of the intersection i.e. length on the
halt point seized for the vehicle (length
of the vehicle + estimated necessary
distance between vehicles). On this basis,
segments of the route, on which the
vehicle can be located, are now defined.
These segments must be reserved in order
to avoid collisions among two or more
vehicles; control is very easily achieved
thanks to the good features of the GPSS
simulation language. A segment represents
a node or an arc of the previously defined
graph. Hold-up time (the time required to
travel through a segment) is calculated
from the segment length and the speed of
the vehicle on that segment, i.e. the
duration of the task it has to carry out.
The shortest routes between halt points
are determined according to the algorithm
for determining k shortest routes [3].
This algorithm, beginning from a assumed
estimation vector, whose components are
lengths of k shortest routes which connect
observed node with the given original
node, successively reduces the estimation
by the double-search method. An optimal
estimation vector is generated as a result
of a finite number of iterations. In
addition, sequences of nodes for the k
shortest routes are determined.
According to this result, a sequence of
segments (nodes and arcs - alternately)

involved in the shortest route is
determined. This will enable precise
segment occupation (in time) and,
consequently, correct traffic control in
the given lay-out.

PROGRAM III

Main program is the model of the system,
which follows the flow-chart of the
printed boards motion through the workshop
independent motion and processing for
every layer of the printed board and,
afterwards, after lamination - the final
processing cycle for the complete multi-
layer printed circuit board.This is
achieved by describing each layer and
every complete printed board with the
independent transactions,which are
uniquely determined by GPSS-F assigned
system parameters, and private parameters.
If the transaction represents a printed
board layer, private parameters are:
    - current position in the workshop
(printed board layer - printed board is in
the input or output buffer, intermediate
buffer, on transport, in process);
    - total number of operations
(technological processing without
requirements for a transport vehicle is
considered as operation);
    - current operation number;
    - current elementary operation number
within one operation (operation can have
several elementary operations, each of
them is performed within a particular
department);
    - position in the buffer;
    - processing time (duration of the
elementary operation), etc.
In case the transaction represents a
complete printed board, it is described
with an additional parameter, which
indicates the number of layers composing
the printed board.
The vehicle is also described with the
independent transaction, whose private
parameters are:
    - halt point address where an empty
vehicle should come and take over the
printed boards prepared for the transport;
    - halt point address where the
printed boards should be transferred
(target point);
    - number of the transaction
corresponding to the transferred printed
boards set;
    - vehicle status (empty or full),
etc.
PROGRAM III determines which vehicle
should get the task, and then, using the
SHORTEST DISTANCES FILE data, the route of
the vehicle.
Work stations within one department are
represented as multifacilities. Number of
identical work stations within the
department is defined by multifacility
capacity.
Different statistical data are the output
of this program. Queues for input, output
and intermediate buffers are presented for
each department. In addition, queues for
the workshop input - output and transport
are presented. A queue is assigned to the
transport itself, as well. The queues are
defined with the following parameters:
    - current contents (number of
transactions in the queue)
    - maximum contents (maximum number of
transactions which the queue has

contained)
       - total number of transactions
arrived
          - total number of departures
          - number of zero-passes (passes
without waiting)
          - total waiting time of the departed
transactions
          - total waiting time of all
transactions
          - time of the last change
          - mean time in the queue
          - mean length of the queue.
By analyzing output data, it can be
determined which capacities in the
workshop are critical or insufficiently
utilized. By changing these critical data
in the SYSTEM FILE and by repeating the
simulation, the optimal capacities of all
resources (buffers, machines, number of
transport vehicles) can be quickly and
easily defined. These optimal capacities
provide maximal machines efficiency and
minimal waiting of the printed boards in
the buffers. (Considering the large number
of interdependencies, system optimization
is an iterative procedure, which is
repeated each time with a modified system
parameter while the other variables remain
constant.)

ANALYSIS OF SIMULATION RESULTS

Realized program package was used for a
concrete project of final lay-out
definition and resources optimization for
a multilayer printed boards factory
(Fig.1). In this chapter, several
illustrative simulations will be
presented. As mentioned, considering
multivariability of the system, all
simulations were performed by varying only
one parameter. Production of 100 three-
layer printed board sets with the defined
technological processing is considered.
Obtained results should be treated as the
best technical performance of the system,
because system failures were not taken
into consideration (in reality,
productivity is always lower).
In case of varying vehicles number in the
system, results are obtained as expected:
time required for the realization of the
plan is proportional to the number of the
vehicles (Fig.4). Choice of the number of
the vehicles used in the system is made
based on the analysis of the obtained
results, considering the vehicle's price
and its work- and maintenance cost.
Following result illustrates the influence
of the mutual departments position in case
that the whole transport is carried out
with 4 vehicles. Only two departments
exchanged their positions. Fig.5
represents time distribution hystograms
between finished products in both cases.
Mean values of these functions and their
dispersion point out the productivity of
the whole workshop. In the first case,
time for the plan realization was
45h25'34'', and after the exchange of the
departments position (case 2) this time
was 13.9 % longer, that is it was
51h45'34''.
In the following example number of
alternative machines within one
department, is varied. A department with
average throuput, in which all printed
board layers and complete printed boards
are processed, is selected. Number of

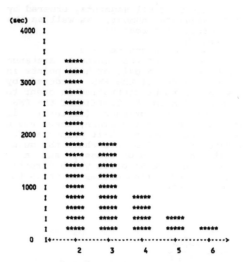

Fig.4. Dependence of transaction mean
     waiting time for the transport on
     the number of vehicles in the
     workshop

transport vehicles is 4, and the lay-out
of the departments corresponds to the more
favorable case of the above example. Fig.6
presents the comparative results table,
based on which the optimal number of
machines for the regarded department,
related to the defined technological
processing, can be determined.

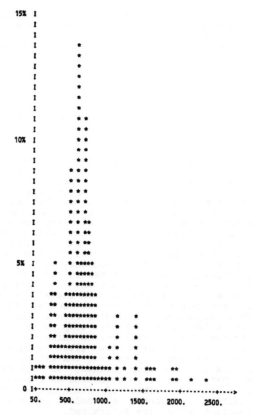

a) Case 1

REFERENCES:

[1] Bratley,P.,Fox,B.,Schrage,L., A
Guide to Simulation, Springer-Verlag, New
York Inc, 1987
[2] Schmidt,B., GPSS-FORTRAN. John
Wiley & Sons, 1980
[3] Phillips,D.,Garcia-Diaz,A.,
Fundamentals of Network Analysis,
Prentice-Hall,Inc.,Englewood Cliffs, 1981

b) Case 2

Fig.5. Time distribution function between finished products

No. of machines	3	6	9
Plan realization time (sec)	172524	163536	164488
average waiting time for the input buffer	2360.9	3.9	0
average waiting time for the output buffer	762.3	801.1	815

Fig.6. Comparative results

CONCLUSION

With the described model, factory
environments, in which technological
processings for particular elements (or
fits), as well as technological
processings for complete products are
defined, can be simulated.Such simulation
enables the optimization of required
capacities for particular technological
operations, of required buffer capacities,
number of engaged transport vehicles,
their trajectories, etc. Developed
software can be used in the designing
phase for defining a workshop lay-out,
i.e. the disposition of the complete
technological and transport equipment.

CONTRIBUTION TO COMPUTER-AIDED DESIGN OF FLEXIBLE ASSEMBLY SYSTEMS

J. M. Henrioud, A. Bourjault and D. Chappe

Laboratoire d'Automatique de Besançon, Institut de Productique, Besancon, France

Abstract . In this paper, the authors present a complete stategy to design a flexible assembly system
for any product. The basic idea is to divide the assembly actions in two classes; the dedicated ones
(specific of the product) and the positional ones (specific of the assembly process). The proposed
method includes an algorithm which generates all the valid combinations of the dedicated actions, in
the form of assembly trees ; the awkward ones are then rejected by the introduction of some set of
constraints. The remaining trees are then transformed by the assignment of equipments to its different
nodes, which introduces the positional actions. A comparison of the different solutions is then
achieved by the way of a simulation.

Keywords . Computer-aided design; flexible manufacturing; assembling; automation; simulation.

INTRODUCTION

The most important problem in manufacturing appears to day
to be an assembly problem ; the competition is very hard with
countries where manpower is cheap. The assembly tasks are
generally numerous and little automatised ; they contribute
considerably to the cost of a product (20 - 60 %). The set of
knowledge about machining is not here transferable to
assembling because assembly is specific.

Before the eighties, the automatisation of the assembly tasks
were most of the time the substitution of a man by a robot ;
that is an approach called "work-station" ; the economic
results were not always satisfying and today, it is preferable to
have a global vision, i.e. a system-like approach. In this way,
the most difficult is certainly to obtain and apply a
methodology which permits to design an efficient flexible
assembly system.

For several years we have been working on this problem,
more precisely, on the elaboration of a computer-aided workstation, to design flexible systems ; whose elements could
also be used to control the system on line.

The resulting method is divided in two pincipal stages
described in this paper: the determination of a set of assembly
plans, then, for each one, the determination of several
organisations; all the resulting assembly processes being
compared by means of a simulation. This approach was first
developped for single product assembly systems and is thus
presented in this paper; recently we have begun to extend it to
multi-products processes.

ASSEMBLY PLANS DETERMINATION

Up to this day, there are two efficient methods for the
exhaustive determination of the assembly plans. The first one
is known as the liaison-sequence method (Bourjault 1984, De
Fazio and Whitney 1986); we have already described it in
different papers. The second one, which could be named sub-
assemblies decomposition method, (Homen de Mello 1986
and Henrioud 1987), is the one used in this paper and is
succintly described here.

Product modelisation

Any industrial product resulting from the assembly of n
elementary components: $c_1,....,c_n$ can be considered as a set
of characters which have to be created in some convenient
order. These characters are listed hereafter in a classified way.
- the relational characters, specific of the assembly problem,
they are divided in two categories:
- the liaisons, there is a liaison between two elementary
components iff they have some common surface in the
assembled product.
- the joinings, each one adding cohesion to one or
several liaisons. The joinings are :
- welding
- screwing
- sticking
- deformation joining
- ...
- the complementary characters which are mainly:
- inspection and test
- labelling
- painting
- cleaning
- machining
- ......

Being given any product P, we list the set of all its characters
and we define a 5-uple:

$$< C,\Gamma,\Sigma,\Delta,f >$$

where
- C is the set of the elementary components

- Γ is the set of the (geometrical) liaisons, given as couples
$\{c_i, c_j\}$

- Σ is the set of the joinings

- Δ is the set of the complementary characters

- f is a function mapping $\Sigma \cup \Delta$ into $P(C) \times P(\Sigma) \times P(\Delta)$; it
defines, for each joining or each complementary
character, the set of elementary parts and characters
concerned.

The couple [C, Γ] defines a graph called the liaisons graph.

Any sub-assembly, or part, produced in the course of the assembly process is described by a 4-uple

$$< X, \gamma, \sigma, \delta >$$

with : $X \subseteq C$; $\gamma \subseteq G$; $\sigma \subseteq S$; $\delta \subseteq D$
and such that :
$[X, \gamma]$ being the subgraph of $[C, \Gamma]$ induced by X, it is connected and, if

$$f(\sigma \cup \delta) = (Y, \mu, \nu)$$

then : $Y \subseteq X$; $\mu \subseteq \sigma$; $\nu \subseteq \delta$

The function f has two purposes ; the first one is to define a necessary condition for the joinings and complementary characters to be set up, the second one being to define some precedence conditions between them. Thus if :

$$d \in D \quad \text{and} \quad f(d) = (X, \sigma, \delta)$$

every character in σ or in δ has to be set up before the character d.

Practically, this function f is such that for any joining j we have always :

$$f(j) = (X, \phi, \phi)$$

Any assembly process for a given product is a set of actions being realized in serie and/or in parallel. Among these actions we distinguish :

- the dedicated actions which create some characters of the product ; they are to be done whatever assembly process is involved
- the positional actions, which consist in moving the different parts between the equipments ; they depend upon the assembly process.

So, to evaluate the assembly plans we just consider the dedicated actions, postponing the introduction of the positional ones to the stage of process evaluation.

Figure 1 represents a quartz with its liaison graph ; it has:
three elementary components :
    -a crystal (a)
    -a base (b)
    -a cover (c).

two joinings :
    -welding of a and b : (w1)
    -welding of b and c : (w2)

and two complementary characters :
    -an electric control of the welding w1 : (el)
    -a labelling of the cover c : (la) ; this labelling assigns the quartz to a certain class of quality, depending upon the result of the electrical control, so we consider in the model that (la) is a labelling of the whole product.

The function f is thus defined by :

$$f(w1) = (\{a, b\}, \phi, \phi)$$
$$f(w2) = (\{b, c\}, \phi, \phi)$$
$$f(el) = (\{a, b\}, \{w1\}, \phi)$$
$$f(la) = (\{a, b, c\}, \{w1, w2\}, \{el\})$$

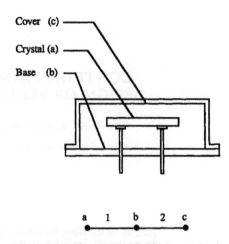

Fig . 1 Quartz given as exemple with its associated liaisons graph

Assembly trees

An assembly tree is a tree for which :
    - the root represents the product
    - the nodes represent the sub-assemblies or parts
    - the leaves represent the elementary parts

and such that each part associated to a node Ni can be produced from the parts associated to the successors of Ni ($k \in \{1, 2\}$) by the simultaneous execution of several dedicated actions.

In any assembly tree, each sub-tree including one node and its k successors defines an operation which is

    - geometrical if it creates some liaisons ($k \in \{1, 2\}$)
    - physical if it adds a joining ($k = 1$)
    - complementary if it adds a complementary characters ($k = 1$)

An assembly tree is the first description level of the assembly process, it includes it intrinsic part and doesn't involve the positional actions, which are to be added in a second stage, when the equipments and their assignment to the operations are chosen as it is shown in the second part of this paper.

Figure 2 shows the two assembly trees available for the quartz presented in Fig 1. The nodes are labeled by the character created by the associated operation.

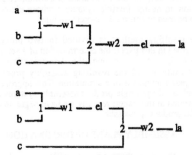

Fig . 2 Assembly trees available for the quartz presented above

The assembly constraints

The assembly constraints can be devided in two categories :

The operative constraints : these constraints define wether any given operation is feasible ; we distinguish :

- the geometrical constraints : when mating two parts, there must exist at least one collision-free trajectory to bring the two parts in contact. This constraint depends only upon the product and can be automatically evaluated

- the material constraints : the effective realization of any operation involves some equipments, there is at least one part to hold (two for the geometrical operations), which requires some suitable parts surfaces for the holding ; moreover, since every operation involves some spatial relation between two objects : each part with its holding equipment, or some external equipment for the non geometrical operations, there is again the necessity of a free-collision trajectory. But since the equipments are not defined at this stage, it belongs to an human expert to appreciate the feasability of the considered operation

- the stability constraint : each part produced in the course of the assembly process must be stable, i.e all its subparts have to keep exactly their relative positions. Stability can only be partly automatically evaluated since an unstable part can be temporarily hold by some outside device, which has to be designed.

The strategic constraints : We call strategic constraints those which deal with the operations sequencing ; they are mainly :

- operations grouping, because they require the same equipments or because they use the same assembly direction ;

- necessity that some sub-assembly appear in the assembly process (for inspection or storage purpose)

- delaying the introduction of some elementary parts (fragility, weight, volume)

- choice of a serial process (there is only adding of elementary parts) or inversely of a parallel one.

Assembly trees generation

In order to evaluate the assembly trees available for a given product we have elaborated a software LEGA based on a top-down algorithm. Starting from the product, we define all the valid operations having the product for result, which determines one or two parts ; then we reiterate for each of these parts till we obtain all the elementary components. An operation being valid when it agree with the set of the operative constraints.

The operative constraints are described by a database including a set of facts (the unfeasible operations) and a set of rules which formalize some properties of the operative constraints.

For each new operation produced by the generation tree process, the database is searched in order to find if this operation is feasible ; if the datas don't allow a positive or a negative answer then an human expert is questioned about this operation. His answer is used to complete the database.

At the beginning of the process the database contains no fact ; at the end it contains a complete description of the operative constraints for the considered product.

The software LEGA has been written in PROLOG which proved perfectly adapted to the problem discussed here.

Tree selection

The number of available assembly-trees is usually quite important (between 10^3 and 10^4 for products having about ten elementary components) so it is necessary to reject those which obviously correspond to awkward processes. To select these trees we introduce the strategic constraints ; they are to be chosen in some menu and the parameters are to be defined.

This step is certainly the most difficult of the method presented here ; there is still a lot of work to be done to improve its efficiency. Thus we are trying to associate a weight to each operation, according to its difficulty (presently they are weighted by O or 1) in order to compare the operative difficulty of the different trees. Such a weight can be evaluated by the mixing of several criterias, mainly :

- stability of the involved parts
- complexity of the trajectory
- morphology and size of the parts handling surfaces.

ASSEMBLY SYSTEM DESIGN AND SIMULATION

Each of the selected trees is transformed in a functional diagram using the symbols given in Fig 3. This step involves for each binary operation, i.e one mating two parts, the choice between the primary part (motionless) and the secondary part (moving). Some integrated rules based on the weight, the size and the shape of the two parts leads to a partial automatization of this step ; when no decision can be got from these rules, then the human expert is questioned. In each diagram all the necessary transports are introduced ; there is one for each elementary part and one for each of the secondary parts involved in the binary operations.

Then, equipments are assigned to the different operations ; there are usually several solutions for each diagram when several operations are liable to be realized by a same equipment ; moreover some buffers are introduced. This leads to a division of the assembly system into cells and work stations. A duration is roughly attributed to each operation, including the positional ones which are introduced at the same time as the equipments.

Since there are several trees selected at the precedent stage (a few tens) and several organizations for each tree, that means that there are some hundred solutions to compare. In order to perform this work we have developed a software, AISE.

The aim of AISE is to provide a quick simulation of each of the possible organizations defined at this stage. For each organization a Petri net is automatically deduced, which simulates its behaviour, it provides :

- the cycle time (i.e the out put period of the system)
- the occupation ratio of each equipment (working time / cycle time).
- the size evolution of the buffers.

CONCLUSION

This study is a contribution to computer-aided design of flexible assembly systems . Its main object is to encompass all the possible configurations; in order to face the size of this problem, we have split it up in several steps: deteminaton of all the available assembly plans, described by assembly trees, selection of the best ones, equipments assignment. Apart from the assembly trees determinaton, which is exhaustive, the whole method is essentially heuristic. At each stage, the human expert is presented with systematic choices, which forces him to examine solutions he wouldn't have thought of. Moreover each decision is registered, which is of great help for any discussion arising between the different engineers involved in the project. Under the condition that the good solutions have been selected at each stage, the final simulation allowing a quantitative comparison of the proposed configurations, the best one can be found.

The proposed method is available for products including about twelve elementary parts ; bigger products have to be split up in several sub-assemblies. It has been developped and tested in the field of mechanical industry, where the studied products were submitted to numerous operative constraints.

Our main objective is now the introduction of 3 D description of the products in our software, in order to provide it with some ability to detect automatically geometrical and stability constraints. We think that would help also to the operations evaluation and weighting, which would be very useful for the trees selection.

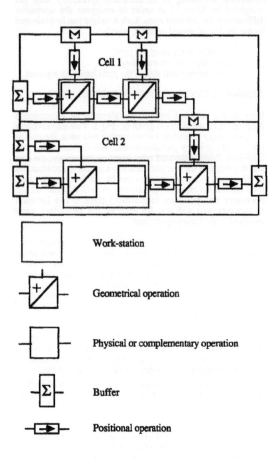

Work-station

Geometrical operation

Physical or complementary operation

Buffer

Positional operation

Fig . 3 Exemple of a functional diagram with symbols used

Homem de Mello, L.S., Sanderson A.C. (1986) AND/OR Graph Representation of Assembly Plans. Proceedings of AAAI-86, pp1113-1119. Morgan Kaufmann

Homem de Mello,L.S., Sanderson A.C. (1987) Task sequence Planning for Assembly IMACS World Congress on Scientific Computation Paris

Meunier M., Lhote F. (1987) Automatisation flexible de l'assemblage, Axes Robotiques

Lui, M.M. (1988) Generation and Evaluation of Mechanical Assembly Sequnces Using the Liaison-Sequence Method.Master's thesis Department of Mechanical Engineering, M.I.T.

REFERENCES

Bourjault, A. (1984) . Contribution à une approche méthodologique de l'Assemblage Automatisé : Elaboration Automatique des Séquences Opératoires Thèse d'Etat, Besançon

Bourjault ,A. (1987) Methodology of Assembly Automation: a new approach 2nd Int.Conf. on Robotics and Factories of the Future , San Diego

De Fazio T.L., Whitney D.E. (1987) Simplified Generation of all mechanical assembly sequences IEEE Robotics and Automation, Vol. RA.3, pp 640-658,

Henrioud J.M., Bourjault A., Chappe D.   (1987) Elaboration des gammes d'assemblage : Approche composants GAMI - Les Journées de la Productique

Henrioud J.M., Bourjault,A. (1987) Logic Programming Applied to Assembly Sequences Determination IASTED Robotics and Automation, Lugano

DEEP KNOWLEDGE FOR CONFIGURATION AND DIAGNOSIS IN A TECHNICAL ENVIRONMENT

K. Hinkelmann and D. Karagiannis

FAW–Ulm, Ulm, FRG

Abstract. Configuration of a production line and diagnosis of failures and inaccuracies are integrated by a common model of the machine and explicit modelling of causal relationships to assure the quality of the product and to prevent standstills. The organization of the knowledge base to be used for both problems and the interaction of deep and shallow knowledge are described. The system is applied on a synchronized production line.

Keywords. Artificial Intelligence; Expert systems; Failure detection; Configuration; Modeling; Quality control; Manufacturing processes

INTRODUCTION

Computer-Integrated Manufacturing offers a variety of examples to demonstrate how, starting from a deep model of the application area, an improvement in production may be achieved by configuration and diagnosis. The knowledge-based statement to be used on production-specific as well as on machine-specific level is an instrument for quality assurance and requires knowledge from the domain.

Production-specific tasks are

- the continuous quality assurance during production by permanent controlling as well as
- the prevention of rejections by preventive diagnosis.

Machine-specific tasks are

- the prevention of standstills by a knowledge-based configuration of machines,
- the reduction of non-productive time by a knowledge-based diagnosis and
- the acceleration of retooling phases by similarity analysis.

DEEP KNOWLEDGE FOR CONFIGURATION AND DIAGNOSIS

Configuration as well as diagnosis of a production installation are based on the comprehension of the production process, the knowledge about the processing stations and the quality requirements of the piece to be produced.

A knowledge-based system to be applied for both configuration and diagnosis tasks requires a design of the knowledge basis that is adequate for both problem areas to use knowledge about the configuration for the diagnosis and inversively to influence the reconfiguration after a failure diagnosis. If both components were supported by separate knowledge bases, it would be very difficult (if ever possible) to guarantee the consistency of knowledge.

Our deep modelling approach leans on an explicit representation of the production installation and the processing task.

Deep Knowledge

A distinction between deep and shallow models of expertise is made by several authors (Genesereth, 1982; Reiter, 1987; Klein, Finin, 1987). A definition denoting when a model M' is deeper than a model M is given in (Klein, Finin, 1987):

"Consider two models of expertise M and M'. We will say that M' is deeper than M if there exists some implicit knowledge in M which is explicitly represented or computed in M'."

In shallow models heuristic information plays a dominant role. Conclusions are drawn directly from observed facts that characterize a situation. While the design or structure of the real world domain is only weakly represented, the largest part of information is described in form of rules of thumb, rules of experience or statistical intuitions.

Deep models give a more detailed description of the real world domain, i. e. its design or structure together with an observation of the expected behaviour. They tend to be more robust than shallow models, handling problems not explicitly anticipated and exhibiting higher performance at the periphery of their knowledge (Klein, Finin, 1987).

The used formalism of knowledge representation has to enable the description of the statistical composition as well as the procedural and algorithmic behaviour. The pure predicate calculus is not adequate for this, as the knowledge of facts is represented completely unstructured as a lot of literals.

Modern formalisms offer structuring possibilities being directly interpreted by the inference mechanism. Examples are object-oriented programming languages (e. g. Smalltalk), Frames (Minsky, 1981), (e.g. KRL (Bobrow, Winograd, 1977)) or semantic nets

(example: KL-ONE (Brachman, Schmolze, 1985)) also offering the possibility of inheriting characteristics via specializing hierarchies. KRYPTON (Brachman, Fikes, Levesque, 1983) subdivides the knowledge representation into a T-box structuring the objects to be described, and into an A-box in which logic statements and rules about the objects defined in the taxonomy of the T-box are stored. Furthermore, the inclusion of functional formalisms offers the possibility to define algorithms already known and to use them if needed.

The use of special methods to describe parallel or sequential operations like Petri nets (Reisig, 1985), e.g. SAGE (Freedman, Malowany, 1988), or matrix-like representations of complex production processes in flexible manufacturing systems (compare (Shen, Chang, 1988)) are also possible. For the use of models in decision support see (Jarke, Radermacher, 1988).

The aim of modelling is to find a good balance between deep and shallow knowledge. While deep models are mostly used for diagnostic expert systems, we will show how they can be used for diagnostic as well as for configurating systems. The deep generic model of the physical device is the basis for both tasks, while heuristic and statistical rules lead the configuration and diagnostic process.

APPLICATION DOMAIN: A SYNCHRONIZED PRODUCTION PROCESS

As application area we regard a production line (PL) that reshapes a sheet steel blank to a hollow part. The complete forming operation is effected in various work stations (= production cell, PC) being connected by a transport system for production pieces (see Fig. 1). While the piece is passing through the production installation, there is no possibility to intervene in the processing, thus making a simulation of the installation necessary for verifying the course of manufacture.

Each of the production cells consists of two parts: a fixed lower part L and a movable upper part U; in each of the two parts a tool is installed. The pressure of the upper on the lower tool reshapes a production piece lying in between. The upper parts of all n levels can move only simultaneously; thus it is a

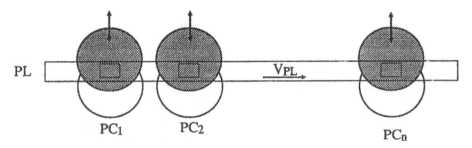

Fig. 1: Scheme of a production line

synchronized process with a clock frequency T_{PL}.

The clock frequency T_{PL} and the speed V_{PL} of the transport system are machine constraints and cannot be modified during the course of manufacture. The number n of active steps is also given. A production piece being processed at time t_1 at the first station will reach the second station at time $t_2 = t_1 + t$, whereby $t = 1/T$, leaving the installation at time $t_n = t_1 * (n-1)/T$.

For every piece to be produced it is fixed which production steps are to be effected in which order. In case that for the production of a piece less than n working cycles are necessary, the tools are removed from the active stations actually not needed. However, if more than n operating cycles are necessary, the production piece has to be treated intermediately before passing the next process with again not more than n working steps, as from the production-technical point of view it is not possible to reshape a material arbitrarily often without intermediate treatment.

The changeable factors on every production cell are

    a)   the pressure balance of the dies cushions in the lower part,

    b)   the vertical adjustment of the tool fixing device in the upper part as well as

    c)   the choice of tools.

While the type of tool is fixed for every level in relation to the piece to be produced, there are differences in the surface structure and small derivations in geometry, influencing the results of treatment.

This applies similarly for controlling the pressure for the reshaping. There is a total pressure given by the construction of the instal-

lation; however, this can be varied to certain limits for every step by adjusting die cushions and blocks.

The configuration process viewed in the following consists of choosing a concrete tool for every level as well as adjusting the pressure. If the concrete values of the produced piece differ from the required data, the responsible stations have to be identified (diagnosis process) and then newly configurated (configuration process).

MODEL DESIGN

Experience and mathematical model design are not sufficient to guarantee the quality of production, as they can only be used within a given frame. That's why here a modelling by the aid of deep knowledge on the application area is suggested, so that the system may also be used in situations beyond the foreseeable circumstances.

The Production Installation

We use an object-oriented approach based on frames to specify the production installation. Classes representing the pieces to be produced and the production line with all its relevant components are defined. Every piece which can be produced is described from a production-specific point of view giving its type, the demanded quality data with tolerances and a generic model of the production installation to produce that piece. This model, based on deep knowledge about the production process, describes of which components the installation consists, how these components are arranged and how they behave de-

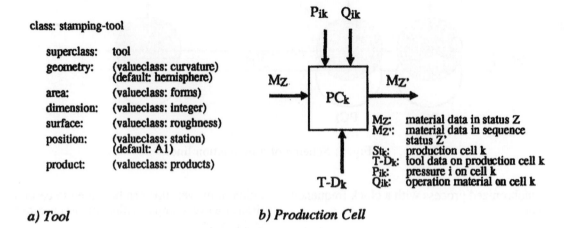

a) *Tool*   b) *Production Cell*

Fig. 2: Influencing parameters

pending on the relevant influencing quantities.

The functional elements of the installation are the tools. The geometric dimension of the bipartite, inversively formed tools is responsible for the transformation in the respective production step. There are several classes of tools for stamping, drilling, drawing etc. To every class there are several subclasses. This refinement is executed until there exists an own class of tools for every product at every production cell. For every class there are several instances varying slightly in regard of their properties like smoothness of their surface and curvature.

The next bigger unit is a production cell. The parameters influencing the processing on one level of the press are the pressing force, the tool data on this level, the material data and the means of production used (e. g. oil). Figure 2 shows a coarse model of the influence quantities of one single station of work.

In practice, production is effected on more than one level. For every product there exists a generic model of the production line giving the number of required production cell and their composition. For every production cell the intended class of tools is specified; the remaining paramenters are instantiated with default values. The ultimate instantiation, i.e. a selection of a concrete tool and an adjustment of the parameters, is done in the configuration process.

Starting from this model, restrictions may be portrayed via mathematical or physical formulas. Such formulas may be e.g. the description of the processing time $t_n = t_1 * (n-1)/T$ or of the total pressure as sum of the single pressure values of all production cells:

$$P = \sum_{i=1}^{n} P(S_i)$$

Furthermore, constraints are specified, that must be satisfied by the model. So the total pressure must not exceed the maximum pressure given by the machine.

Causality

Besides the knowledge concerning the production installation itself a representation of causalities is needed. These concern machine-specific aspects as well as causal relations of the production. So the pressure at a production cell influences the uniform thickness and roundness of the material. This causal knowledge does always refer to the model of the machine (Hudlicka, 1988).

Applying this knowledge to an instantiated model of the installation creates a model that considers the mutual dependencies between the levels. As you can see in Fig. 3, the reason for a defect arising on a certain level may be based on one or several of the levels lying

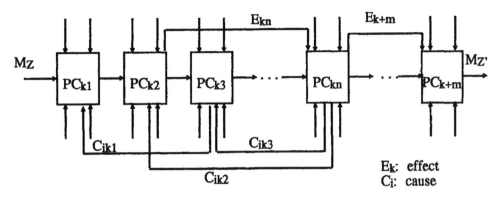

Fig. 3: Dependencies between production cells

MODEL-BASED CONFIGURATION AND DIAGNOSIS

Diagnosis and configuration are not independent, they are integrated by the underlying modell. While configuration instantiates the model and computes dependencies between the production cells, the diagnosis effects changes in the configuration of the installation.

Configuration

The configuration process concerns the selection of tools and the adjustment of the parameters at the respective production cells, depending on the material data of the production piece to be treated as well as - in the extreme case - on environmental influences such as room temperature and atmospheric humidity.

The configuration starts from a generic model of the production installation and the actual data about the wrought metal. The configuration process is thereby guided by heuristic and statistic knowledge and experience, being represented as shallow knowledge in form of rules. These rules generate from the generic model an instance that represents the actual configuration. In a second step the constraints are checked and causal knowledge is used to extend the model with a specification of dependencies between the production steps (see Fig. 3).

Diagnosis

After a production piece has once passed the installation with not more than n steps, it is measured and verified whether the quality data are within the requested tolerance. If this should not be the case, there is a defect in the configuration of the installation. The diagnosis problem now consists in fixing those components whose supposed abnormal functioning would explain the discrepancy between expected and actual behaviour.

Starting point of the diagnosis is the fully instantiated model of the production line with all its causal dependencies, as it is computed by the configuration process. Being a description of the normal behaviour of the installation, it is called a physiological model (like (Davis, 1984; Genesereth, 1984)) in contrary to a pathophysiological model describing the faulty behaviour (cp. (White, 1986)). A detected failure will now be traced back to identify the defective components. Due to the explicit use of causal relations the reconfiguration of the installation is guided.

The use of the system may become especially interesting when more than one run is necessary for the processing of a production piece. In case that after the first run the quality data of the piece are still within the admissible area, the processing afterwards may fail nevertheless. In this situation the aim is to identify in advance the future failure (which would mean a loss of production of several ten-thousand DM in the considered application) by preventive diagnosis and to prevent such failure by an adequate, fast new configuration.

If an unambiguous identification of the faulty components is not possible, the diagnosis does now use heuristic knowledge, experience from experts and knowledge from failure probabilities of single components to localize the defect.

CONCLUSION

The model-based approach offers the possibility to integrate configuration and diagnosis using a common knowledge source, thereby avoiding inconsistencies in the knowledge. The representation of causal relations as deep knowledge in combination with shallow knowledge permits the application of the expert system also in unforeseen circumstances.

The prototype of a knowledge-based system realizing this integrated process of diagnosis and configuration is being developped with the aid of the described methods at the FAW in Ulm in cooperation with industrial partners for a special production installation. As engineering environment for the model design the expert system shell KEE (Intelli-Corp, 1986) is used. The use of the special simulation tool SimKIT facilitates a pictorial simulation of the production process, basing on a model of the installation. This simulation offers a clear explanation of an eventual abnormal behaviour and additionally facilitates a refining of model design via an immediate feedback of design consequences.

Acknowldegement. We'd like to thank Christine Kimmel for her active contribution.

REFERENCES

Bobrow D. G., and T. Winograd (1977). An Overview of KRL, A Knowledge Representation Language. *Cognitive Science* 1 (1), 1977, 3-46.

Brachman R. J., D. Fikes, and H. Levesque (1983). KRYPTON: A Functional Approach to Knowledge Representation. *IEEE Computer* 16 (10), 1983, 67-73.

Brachman R. J., and J. Schmolze (1985). An Overview of the KL-ONE Knowledge Representation System. *Cognitive Science* 9 (2), 1985, 171-216.

Davis R. (1984). Diagnostic Reasoning Based on Structure and Behaviour. *Artificial Intelligence* 24, 1984, 347-410.

Freedman R., and A. Malowany (1988). SAGE: A Decision Support System for the Sequencing of Operations within a Robotic Workcell. *Decision Support Systems* 4, 1988.

Genesereth M. R. (1982). Diagnosis Using Hierarchical Design Models. *AAAI-82*.

Genesereth M. R. (1984). The Use of Design Descriptions in Automated Diagnosis. *Artificial Intelligence* 24, 1984, 411-436.

Hudlicka E. (1988). Construction and Use of a Causal Model for Diagnosis. *International Journal of Intelligent Systems* 3, 1988, 315-349.

Intellicorp Corp. (1986). KEE Software Development System User's Manual; KEE Version 3.0. Mountain View, CA, 1986.

Jarke M., and F.J. Radermacher (1988). The AI Potential of Model Management and its Central Role in Decision Support. *Decision Support Systems* 4, 1988.

Klein D., and T. Finin (1987). What's in a deep model? A Characterization of Knowledge Depth in Intelligent Safety Systems. *IJCAI-87*.

Minsky M. (1981). A Framework for Representing Knowledge. In: J. Haugeland (Ed.): *Mind Design*. Cambridge, MA: The MIT Press, 1981, 95-128.

White M. (1986). Automated Fault Diagnosis. In: E. Hollnagel et al. (eds): *Intelligent Decision Support in Process Environments*. Berlin/Heidelberg, 1986, 352-362.

Reisig W. (1985). *Petri Nets - an introduction*, Springer Verlag, 1985.

Reiter R. (1987). A Theory of Diagnosis from First Principles. *Artificial Intelligence* 32, 1987, 57-95.

Shen S., and Y.-L- Chang (1988). Schedule Generation in A Flexible Manufacturing System: A Knowledge-based Approach. *Decision Support Systems*, 4, 1988.

A KNOWLEDGE REPRESENTATION ENVIRONMENT FOR MANUFACTURING CONTROL SYSTEMS DESIGN AND PROTOTYPING

P. R. Muro, J. L. Villarroel and M. Silva

*Departamento Ingeniería Eléctrica e Informática, E.T.S.I.Industriales, Zaragoza,
Spain*

Abstract. This paper presents some key ideas of a project on a design environment for manufacturing systems control. It is based on a tool hierarchy which integrates different techniques. The kernel is a knowledge representation language, called KRON, which embeddes three methodologies: frame based representation techniques, object oriented programming and High Level Petri Nets (HLPN). MIKRON, a manufacturing oriented tool, is built on top of KRON. A single model built using MIKRON can be used at different levels of a control hierarchy (coordination, scheduling and planning). It offers the graphic and formal expresion capacity of HLPN, as a mean of systematizing, formalizing the interconnection, synchronization and causal relations, as well as the information flow produced by the activity execution. Facilitates the integration of AI techniques for solving scheduling and planning problems. In addition, this tool supports the model execution and simulation.

Keywords. Flexible manufacturing; Artificial intelligence; Petri nets; Modeling; Software tool.

INTRODUCTION

The design of control systems for manufacturing has shown to be a very complex task. A hierarchical approach has been a common solution to manage this problem (Dar-El, 1982). Four levels are generally identified on this hierarchy: local control, coordination, scheduling and planning. Several general purpose techniques (such as simulation languages, AI techniques, operation research models and algorithms, etc.) and application oriented methodologies (like Kanban and MRP Systems, group technology, etc.) have been used at different levels on this hierarchy. A good design system for the whole control hierarchy, should integrate several of these tools and methods in a single environment.

Several of the most promising integrated approaches to solve the above problems are based in Petri nets. The approach of Dileva & Giolito (Dileva, 1987) is based on an embedding of the net formalism into an Entity Relation data base schema. The approach of (Gentina, 1987; Bourey, 1988) leads to CASPAIM, a Lisp based tool that considers a channel-agency-like model building methodology. Bruno et al. (Bruno, 1986a; Bruno, 1986b) define PROT nets, a special kind of predicate/transition nets, and propose its application using object oriented programming, for building models of distributed systems, such as CIM systems.

The approach presented in this paper proposes a design environment for manufacturing systems control, based on a tool hierarchy, which integrates different techniques. The kernel of the system is based on a general purpose representation language, called KRON (Knowledge Representation Oriented Nets), which embeddes three methodologies: frame based representation techniques (Minsky, 1975), object oriented programming (Stefik, 1985) and high level Petri nets (HLPN) (Brauer, 1986). MIKRON (Manufacturing Intended KRON) is a manufacturing oriented tool, which is built on top of KRON. The tool hierarchy is expanded with a set of interfaces to provide a user friendly interaction (e.g. GRAMAN (Villarroel, 1988) provides a graphic interface at the coordination level).

The paper is structured as follows. The issues involved in KRON representation language are introduced in §2. In §3 the MIKRON basic characteristics are presented and ilustrated by

means of a simple example. The dynamic perspective is presented in §4 as a way to introduce additional features allowed by the model, like execution, simulation, information prediction and automatic data collection. Some conclusions are presented in §5.

KRON

The concept of 'frame' was first introduced by Minsky in 1975 as a way of stablishing stereotypical descriptions of situations and objects. A frame, is a data structure that includes declarative information, which is complemented by a rich control structure as procedural knowledge, in predefined internal relationships. The application of frame-based languages in the modelling of complex dynamic systems, has certain inconveniences due to the lack of a formalism to specify its dynamic behaviour (concerning both the states of the objects and the causal relationships between states and actions). Dynamic behaviour is mainly represented through procedural knowledge, which permits inconsistencies to arise such as the clasic frame-problem (or 'inter-frame-problem' as expresed in (Shoam, 1985)). These problems become important when the model reaches some complexity.

KRON (Knowledge Representation Oriented Nets) (Muro, 1989) is a frame based knowledge representation language intended for discrete event dynamic systems with parallel or concurrent behaviour. It belongs to the object oriented programming paradigm and integrates HLPN for the coordination of the dynamic aspects. The idea of adding the Petri net formalism for the modelling of information systems dynamics has been already pointed out in other domains such as data bases (Dileva, 1987).

Following the object oriented programming methodology, a model is composed by objects. An object can contain data, procedures, relations and metaknowledge. In addition to these features supported by other representation languages, KRON includes a set of semantic primitives (synchronization and net relations and specialized objects) implementing the high level Petri net formalism. The information related to a concept or physical entity, having a dynamic behaviour, is articulated around three classes of specialized objects:

- State-objects contain the state information and additional knowledge (physical description, relations, historical information, etc.). The state is saved in a set of state slots. To each state slot corresponds a single place of the HLPN.
- Marking-objects are mobile objects contained in the state slots. From the Petri net point of view they are the tokens evolving in the net.
- Action-objects define the possible activities of a kron-object that can produce a state change. They are equivalent to the transitions of a HLPN.

The Petri net pre and post incidence relations are supported in KRON by the denominates net-relations, which conect state slots and action-objects. Conditions for a action-object firing are defined by the pre net relations.

A firing mode is defined by each one of the possible set marking-objects verifying the causal conditions of an action-object. The set of firing modes of an action-object is equivalent to the color set associated with the transition in the underlying Petri net.

Following Petri net semantics, pre net-relations specify the marking-objects, that must be removed from the state slots when an action-object is fired with respect to a certain firing mode. The post net-relations, stablished from an action-object to the state slots, specify the marking-objects that must be saved in the state slots for a particular firing mode of the action-object.

KRON provides a simple way for interconecting objects by stablishing synchronization relations between their action-objects. The firing of two related action-objects is synchronized in a 'rendez vous'-like way. Both synchronized action-objects are handled as one single object. In this way, communications between different kron-objects, from a dynamic perspective, are supported by the same formalism that defines the internal dynamic of each object. Kron-objects can also be aggregated to form more complex kron-objects by multiple inheritance techniques.

KRON adds to the advantages of frame based knowledge representation and object oriented programming (such as inheritance, modularity, reusability, etc.) the ones apported by the formality of HLPN:

- A clear graphical representation of the dynamic processes executed by the objects.
- The modelling of important requirements in Change Theory (Shoam, 1985) like concurrent actions, inter- and intra-frame problems and suppressed causations.
- Eliminates the interpretation ambiguity fixing its semantics.
- Introduces a formal mechanism that facilitates consistency, precision and analysis.
- Supports the model execution and simulation.

MIKRON

A tool for manufacturing systems modelling must allow the description of: 1) production resources and relations between them; 2) operations to be performed on these resources and their inclusion in process plans, and 3) the set of elements moving through the production system (parts, tools, operators). Moreover, it must include concepts such as production planning and scheduling and allow the model simulation and verification.

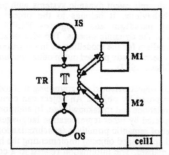

Fig. 1. Example of work-cell describing cell1 using GRAMAN sintax.

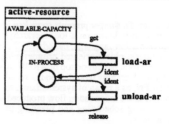

Fig. 2. Active-resource state-object and a representative action-object. They are writen down in schema-like syntax and underlying Petri net.

MIKRON appears as a specific modelling tool for manufacturing systems. It is implemented using KRON and offers the graphic and formal expresion capacity of HLPN, as a mean of systematizing and formalizing the interconnection, synchronization and causal relations, as well as the information flow produced by its activity execution. It covers tasks such as coordination, scheduling, planning, simulation, etc.. MIKRON appertains to the manufacturing systems representation methodology introduced in (Fox, 1983), which is extended and generalized to the project management domain in (Sathi, 1985).

The interest of considering resources and operations separately in manufacturing systems modelling has been shown in several works (Villarroel, 1988; Smith, 1986), in the context of manufacturing problem solving. Resources and operations are separately considered in MIKRON, making then possible the integration into one single model.

We will introduce some basic issues involved in the MIKRON modelling system semantics through a very simple example built using GRAMAN sintax. GRAMAN (Villarroel, 1988) is a graphic inteface used by MIKRON to provide an user friendly interaction. Consider Fig. 1, which schematically shows a work cell, named cell1, composed of two machines (M1,M2), a random-access input store (IS), another random-access output store (OS) and an immediate transport system (TR).

In MIKRON, a resource is composed by a state-object, that specifies possible states, and several action-objects modelling its activities (load, unload, processing start and end, etc.). Resource M1 shown in the example is modelled by M1 state-object (main object representing resource M1), and load-M1, start-op-M1, end-op-M1 and unload-M1 action-objects, which identify different resource activities (objects M1 and load-M1 are shown in Fig. 3). M1 is an instance of tool-machine (Fig. 4) that inherits the active-resource characteristics (shown in Fig. 2).

From all the acction-objects belonging to the resource, we will focus now in the ones modelling its inputs and outputs, called interface actions. In the case of M1, load-M1 models M1 input and unload-M1 models M1 output. These objects are important because the interconnection between objects are stablished through these interface actions. These may be objects containing a variable number of interface actions. This number depends on the particular connections of the object to other objects in its environment (the action-object is called in this case multiple interface action). The number of interface actions on transport resource TR depends on the manufacturing resources it is serving to. This fact does not change its behaviour rules. In the particular case of this example, TR holds three interface actions for input and another three for output (Fig. 5). The concept of multiple interface action allows the adaptation of the same object model to different environments.

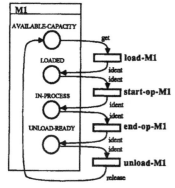

Fig. 3. M1 state-object and load-M1 action-object. Resource M1 is an instance of tool-machine.

A straightforward method to define complex resources aggregating simple objects is also provided by MIKRON using multiple inheritance mechanisms. Thi method can be used to define resources like machines with set-up, resources with shared operator or tool, machines with input or output self store, etc..

The structural description of a production facility is completed by stablishing interfaces between all the resources, making possible the flow of marking-objects through them. An example of resources connection is shown in Fig. 6 (M1 with transport resource TR). It is done by stablishing synchronization relations between interface actions of M1 and TR. This allows the substitution of load-M1 and output-TR-M1 action-objects, which are connected by a synchronization relation, by the new object, input-M1, that aggregates them (the same happens with unload-M1 and output-TR-M1 which are replaced by output-M1).

```
{tool-machine
    ; inherits features of active-resource
    IS-A active-resource
    LOADED    ; state slots
    UNLOADED-READY
    ; references to prototype actions-objects
    START-OP  &start-op-tm
    END-OP    &end-op-tm
}
```

Fig. 4. Tool-machine is-a related with active-resource.

Considering now an operation oriented perspective, a kron-object modelling a possible process plan P1 is shown in Fig. 7. State slots OP1, Op2 and OP3 indicate the possible states that a part can reach to carry out such a process plan (the set of parts trying to perform operation OP1 is identified by the marking-objects in OP1 slot). The possible causal relations between operations are determined by start-part-P1, end-op1-P1, end-

op2-P1, end-part-P1 action-objects and their net-relations. Each one of these action-objects represents the end of an operation and/or the beginning of the next one. In order to build a process plan, the modelling of configurations like sequences of operations, alternative processes, assembly/disassembly operations, quality control operations, or concurrency of operations are allowed.

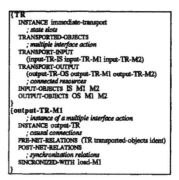

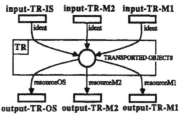

Fig. 5. TR is an instance of inmediate transport and it is connected with OS,M1 and M2 for output and with IS, M1 and M2 for input. Input-TR-IS, input-TR-M1 and input-TR-M2 are instances of a multiple interface action. Output-TR-OS, output-TR-M1 and output-TR-M2 are instances of another multiple interface action.

A coordination mechanism to synchronize the information flow, is needed in any resources/operations based description. This mechanism is built by means of partial synchronization relations. The partial synchronization relations are stablished between:

- the action-object modelling the end of operations on a resource and
- the action-objects in the process plans modelling the end of each operation to be performed on that resource.

The action-object of the process plans, and a firing mode of the resource action-objects (firing with respect to a particular operation), which are modelling the same fact (end of operation processing), are synchronized through one of these relationships. The partial synchronization relations provides a more complex concept than the synchronization relations introduced in §2. From a Petri net point of view, these relations imply the unfolding of the resource action-object in all possible firing modes. This partial synchronization is supported by special action-objects called conditional synchronization objects. They make possible to do this process transparent to the user.

DYNAMIC PERSPECTIVE

As we have seen, the interconnection process of kron-objects provides an incremental construction method of a model containing an underlying HLPN.

The underlying HLPN directly provides the model simulation/execution mechanism. This mechanism is implemented in MIKRON as the run/execution option. However since the model of a production system is not fully deterministic, there exists points in which decisions have to be taken (resource assignement to an operation, selection of a spare part, routing selection, etc). From the underlying Petri Net point of view, this decision point corresponds to a coupled conflict (lock-set concept extension to HLPNs (Nelson, 1983)). A coupled conflict can raise between several action-objects or firing modes. This suggest to create a specific class of object: the conflict. A conflict object groups the action-objects that are in a coupled conflict relation in the net. Its mission is to give a solution (decision) of the conflicts, by means of the associated control policy. This technique enables us to stablish a simple interface with a scheduler in the simulation/execution of the model (a similar solution is proposed in (Martínez, 1988)).

The use of MIKRON as a tool in the development of applications, requires historical data collection mechanisms, that allow the evaluation of relevant characteristics during design (modeling parts of the model, policies, algorithms, etc.). The user can collect information in some special state slots, called recording slots. The recording slots are connected to action objects only through post net-relations and thus do not alter the evolution of the net. These slots are useful during simulation and for the development of planification and scheduling application software.

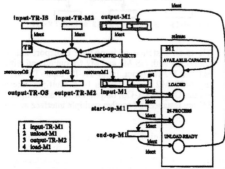

Fig. 6.   Connection of machine M1 and transport TR through their interface actions. The new interface actions are input-M1 and output-M1.

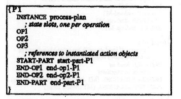

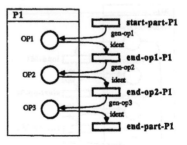

Fig. 7.   P1 process plan and underlying Petri net.

According to what we have established so far, the execution of a net follows the coupled conflict oriented approach. Whenever a conflict object receives an activation message, it in turn sends a message to each of its action-objects. These messages analize the possibility of firing the action-objects, that is, the set of their enabled firing modes. If, as a result of this analysis, any possibility of firing is detected, its conflict control policy is activated to return a solution: the involved action objects and their chosen firing modes. This result is communicated to the appropiate action objects in order to fire them.

It is frecuently useful to make a partial prediction of the model behavior up to a certain horizon (e.g. in planification and scheduling tasks). MIKRON includes a way to predict the evolution of the model called run-prediction. This prediction does not modify the actual state of the model. Instead it, generates possible future states in a certain horizon. It can be used as knowledge source for control policies that require an estimation of the evolution of the model (Martínez, 1988). Also, it can perform a partial evolution of the model: induce the evolution of one single operation, induce the evolution of process plans without constraints imposed by physical resources, etc.

CONCLUSIONS

Flexible manufacturing systems have complex dynamics. Although dynamic behaviour can be represented by usual knowledge representation languages, the information contained in these models can not be unambiguously interpreted without a large comprehension of the procedural knowledge.

In this paper the MIKRON system is presented. MIKRON is a tool that allows to model all the knowledge required for planning, scheduling and coordinating manufacturing systems. MIKRON is written in a general purpose knowledge representation language called KRON, that adds to the frame based knowledge representation and object oriented programming techniques a formalism based in HLPN to represent the system dynamics.

From a functional point of view, the techniques embedded in MIKRON allow to use the same model at several levels of the control hierarchy of the manufacturing system (Martínez, 1988): planning, scheduling and coordination.

MIKRON is being integrated in a software design environment for manufacturing systems. This environment allows the design, analysis and simulation of the model, as well as the production of code for control at different levels (Martínez, 1987; Colom, 1986; Villarroel, 1988).

We would like to highlight the following advantages of our approach:

- gives a clear graphical representation of the dynamic processes executed by the objects,
- provides a clear semantics of the dynamic model, removing then ambiguity from its interpretation,
- introduces a formal mechanism that facilitates consistency, precision and analysis,
- supports the model execution and simulation,
- facilitates the interconnection and synchronization of the objects dynamic activities,

- makes easier the model definition by specializing objects,
- encourages the reutilization of previously defined models,
- a single model can be used at different levels in the control hierarchy, and
- the model can be independently considered from the operation and resource points of view.

KRON and MIKRON are being implemented in a 1186 Xerox machine running LOOPS.

AKNOWLEDGEMENTS

We thank S.F. Smith from The Robotics Institute of Carnegie Mellon University for his helpful comments on previous versions of this work. This work has been realized under grants IT-2/86 of CONAI of Diputación General de Aragón and PA 86-0028 of the Comisión Interministerial de Ciencia y Tecnología. Pedro R. Muro and José L. Villarroel's research has been funded by Diputación General de Aragón.

REFERENCES

Bourey J.P. (1988). Structuration de la partie procedurale du syteme de commande de cellules de production flexibles dans l'industrie manufacturiere. These docteur d'Universite, Universite des Sciences et Techniques de Lille, Flanches Artois. Lille.

Bruno G., Marcheto G. (1986a). Process-translatable Petri nets for the rapid prototyping of proces control systems. IEEE Trans. on Software Engineering, Vol SE-12, No.2, pp.346-357. February.

Bruno G., Balsamo A. (1986b). Petri net-based object oriented programming modelling of distributed systems. Proc. of ACM Object-Oriented Programming Systems, Languajes and Applications Conference. Portland, Oregon (USA). pp.284-293.

Colom J.M., Martínez J., Silva M. (1986). Packages for Validating Discrete Production Systems Modeled with Petri Nets. Proc. of the 1st IMACS Symposium on Modelling and Simulation for Control of Lumped and Distributed Parameter Systems. Lille (France), June. pp. 529-536.

Dar-El E.M., Karni R. (1982). A review of production scheduling and its applications in on line hierarchical computer control systems". On-line Production Scheduling and Plant-Wide Control. KOMPASS & WILLIAMS eds.

Dileva A., Giolito P. (1987). High-Level Petri Nets for Production System Modelling. Proc. of the 8th European Workshop on Application and Theory of Petri Nets, Zaragoza (Spain). pp. 381-396.

Fox M.S. (1983). Constraint Directed Search: A Case Study of Job-shop Scheduling. Ph.D. dissertation, Dep. Computer Science, Carnegie-Mellon University, Pittsburg, PA (USA).

Gentina J.C, Corbeel D. (1987). Hierarchical control of Flexible Manufacturing Systems (FMS). Proc. of the IEEE International Conference on Robotics and Automation. Raleigh (USA), March. pp.1166-1173.

Hayes P.J. (1984). NaivePhysics 1 - Ontology for liquids. In Formal Theories of the Commonsense World (Hobbs, J.R. and Moore, R.C. eds), Ablex.

Martínez J., Muro P.R., Silva M. (1987). Modelling, Validation and Software Implementation of Production Systems Using High Level Petri Nets. Proc. of the IEEE International Conference on Robotics and Automation. Raleigh (USA), March. pp. 1180-1185.

Martínez J., Muro P.R., Silva M., Smith S.F., Villarroel J.L. (1988). Merging Artificial Intelligence Techniques and Petri Nets for Real Time Scheduling and Control of Productions Systems. Proc. of the 12th IMACS World Congress on Scientific Computation, Paris (France), July. Vol. 3, pp. 528-531.

Muro P.R., Villarroel J.L. (1989). KRON: Redes Orientadas a la Representación del Conocimiento. Internal Repport UZ-DIEI-89-IR5. Electrical Engineering and Computer Science Dep., University of Zaragoza, (Spain).

Nelson R., Haibt L., Sheridan P. (1983). Casting Petri-Nets into Programs. IEEE Trans. on Software Engineering, Vol SE-9, Nº 5, September, pp 590-602.

Sathi A., Fox M.S., Greenberg M. (1985). Representation of Activity Knowledge for Project Management. IEEE transactions on Pattern Analisys and Machine Itelligence, Vol. PAMI-7, Nº 5, September.

Smith S.F., Ow P.S., Le Pape C., McLaren B., Muscettola N. (1986). Integrating Multiple Scheduling Perspectives to Generate Detailed Production Plans. Proc. of the AI in Manufacturing Conference. SME, September.

Shoam Y. (1985). Ten Requirements for a Theory of Change. New Generation Computing 3. pp 467-477.

Villarroel J.L., Martínez J., Silva M. (1988). GRAMAN: A Graphic System for Manufacturing System Design. Proc. of The International Symposium on: System Modelling & Simulation. Cetraro (Italy), September. pp 79-84.

CIM IN CHOP- AND SPRAY-TECHNOLOGY FOR PRODUCTION IN GLASS-FIBER

M. Calandra* and L. Carrino**

*Italtriest S.p.A, Milano, Italy
**Universita'di Cassino, Facolta'di Ingegneria, Cassino, Italy

Abstract. Objective of the project was the development of an automatic plant for the production of workpieces in glass-fiber using "Chop and Spray" technology. This has been achieve by: spray guns trajectories optimization, process control, Water-Jet finish and CAD integration. The use of high technology systems and the move away from "hand-crafted" production criteria, brings significant benefits in terms of flexibility, quality and relative costs.

Keywords. "Chop and Spray" technology; industrial robot; process control; Water-Jet; CAD.

COMPOSITE MATERIALS: CHARACTERISTICS

Composite materials are composed of a matrix (thermoplastic or thermosetting resin) with a filling of reinforcing fibers (glass, kevlar, carbon and boron fiber).
The principal properties of these materials are:
- high mechanical resistance with respect to weight;
- the possibility of orienting the fibers, to adapt the structure of the part to the stresses to which it will be subjected;
- high resistance to corrosion and fatigue.

In early production techniques, a mix of resin and catalyst was applied by brush to a fiber fabric.
This technology has evolved over time into two different standard techniques:
- a pneumatic spraying system and chopper is used to lay a resin and fiber mat which is then compacted - "Chop and Spray" technique;
- the fiber, thermosetting resin and catalyst mix is prepared (prepreg) and then molded with the application of pressure and of temperatures which cause the resumption of the polymerization process (S.M.C., B.M.C. techniques).
This second solution is convenient for large-scale production of small to medium sized parts. The first, technique can be used, instead, to work on large molds without critical startup costs for a limited production.
The limitations which have prevented further development of this semi-manual form of the "chop and spray" method have been, on one hand, the undesirable working conditions involved and, on the other, the difficulty of obtaining the homogeneity, consistent quality and repetition of pieces required for an industrial product. This is a major drawback in the use of these materials for structural purposes.

Our objective was to overcome the limitations of the "Chop and Spray" system by employing an automatic facility where all process parameters - *temperature, pressure, load, trajectory and working speed* - would be rigorously controlled, to obtain a constant, quality product.

FEASIBILITY STUDY

The production cycle may be divided into 5 stages:
- laying of a gel-coat over the mold;
- laying of the glass resin (*chop and spray*);
- roller and brush rolling;
- U.V. polymerization;
- finishing.

Various factors were involved in the decision to automate:
- a large component of unskilled manual labor: of the 10 persons working on the entire cycle, only two were assigned to the spray stage;
- idle time calculated at about 10% of total working time and due to the nature and timing of the cycle itself;
- undesirable work environment and the physical labor entailed.
These factors and product quality standards led to the launching, in 1987, of plan for the total automation, in successive steps, of the entire plant.

DEVELOPMENT OF THE PROJECT

The plant is divided into three production cells which are entirely robot assisted, an automatic U.V. oven, an AGV system for moving the molds from cell to cell and from cell to storeroom.

It was decided to use general purpose robots rather than the hydraulic painting robots usually employed; this gave the robot system greater flexibility, together with increased programming possibilities.
The robots, suspended from a gantry at 180° (head down) move along a track which permits a longitudinal translation of approximately 2 meters, guaranteeing the possibility of working even with out-size molds.
The spray guns, for both the "Chop and Spray" and the gel coat systems, have not been changed. More pronounced has been the change in the rolling phase, where the conventional brushes and rollers have been replaced by a pneumatic device which applies to the part the pressure resulting from the combination of the

translation motion supplied by the robot and the *alternating motion* supplied by the pneumatic device itself.

Optimizing criteria were principally concerned with:
- the spray tool trajectories in relation with the desired *structural properties* of the workpiece: consistent thickness, good surface finish and porosity, and homogeneous mat distribution;
- geometric parameters relative to the distance between nozzle and mould, the diameter of the nozzle, the angle of the spray tool in relation to the mould;
- definition of the tools chosen to compact the layed mat with the robot.

The PLC chosen has a particularly broad hardware configuration so that it can govern several processes concerning different production stages simultaneously, as well as handle an extensive range of signals.
The AGV system - layout, timing, communication between control and vehicle, number of vehicles - is still at the experimental stage. The principal difficulties encontered are due to the restricted dimensions of the plant and the presence on the grounds of scrap which could interfere with the correct functioning of the vehicles.

THE PRODUCTION CYCLE

In giving the general description of how the cycle functions (*a*), we shall, for the sake of brevity, analyze only the chop and spray cell (*b*).

a) The vehicle picks up a virgin mold from storage, moves to the gel-coat cell, the door rolls up, the mold is deposited inside on a supporting platform which recognizes the work piece. If the piece is different from the preceding one, the robot loads from its *mass memory* the program for the part concerned. The vehicle withdraws, the door closes, the spray phase begins inside the cell. At the same time, another vehicle has brought a mold from the seasoning room to the chop and spray cell. When it has unloaded the work piece, the AGV is informed of a new mission: the pick up of a finished piece at the rolling cell.
If plant layout permits it, the same AGV which has brought the piece to the gel-coat cell may pick up the piece at the chop and spray cell and take it to the rolling cell.

b) Now let us look in detail at the chop and spray cell.
The PLC software constitutes the *link* between data coming from the process and the robot and control devices. The PLC reads the data entered, processes them, and acts on the regulating mechanism within tenths of a second.
However, the system on which the PLC operates is much slower to respond, since it is essentially composed of pneumatic or electropneumatic devices.
Regulation and activation are therefore based on the speed with which the plant can respond to a certain signal.
For example, the time required to regulate the flow rate of the resin pump, under normal operating conditions varies from 2 to 4 seconds, in proportion to the magnitude of the variation itself (Fig. 1).

The process logic has been divided into four parts which, although interacting, are substantially independent (Fig. 2):

- upstream controls;
- spray and regulating actions;

- controls during the cycle and exchanges with robot;
- washing.

The first point concerns the control of operating temperatures to obtain the viscosity, grade of polymerization and ease spraying.

The temperature is measured with a PT100 thermo-resistance in both the outer thermostatic container and in the resin tub itself (Fig. 3).

Let

T_{WA} = average water temperature,
T_R = resin temperature,

$$T_{WA} = \frac{T_{W1} + T_{W2}}{2}, \qquad (1)$$

at regime we must have:
$T_R \approx 25°C$ and $T_{WA} = 25°C$

$$T_R + T_{WA} <= 50°C. \qquad (2)$$

(2) is the law which allows the PLC to act on the valve regulating the flow of hot water to the thermostatic container and limiting water temperature to prevent physical or chemical modification of the resin. However T_R must always be over 20°C to have the minimum grade of viscosity for spraying.
Let us consider the mould positioned and the program concerned ready to start, and analyze the spray sequence.
The robot executes the programmed instructions moving smoothly over the entire surface of the mold. Its movements are constantly controlled by the PLC.
The quantity of resin and glass needed in every point of the part is regulated in two different ways in function of the properties and dimensions of the part concerned:
- by increasing or decreasing the speed of the robot's motion while maintaining the same flow;
- upon the request, sent to the PLC by the robot program, to change the flow rate and the consequent regulation in real time of the pressure on the resin pump and of the supply of compressed air to the glass chopper.
Obviously, in the event of variations in the flow caused by problems inside the spray system - a fall in pressure, an increase or decrease in the viscosity of the resin - the control provides for the regulation needed to restore the flow to the value set by the plant operator.
The pump regulation algorithm integrates data from an encoder connected mechanically to the pump piston by a pinion and rack.

Let
P= pump supply pressure
F_M= average measured resin flow rate
F_s= set resin flow rate
f(t)= instant flow rate (encoder measured).
To avoid high frequency oscillation f(t) is integrated over the pump cycle (4s):

$$\frac{1}{t2 - t1} \int_{t1}^{t2} f(t)dt = F_M \qquad (3)$$

which is the average flow rate of resin.
The supply pressure is controlled by the following law:
P= K*(F_s - F_M) where K is found experimentally to assure system stability.

This is what we mean by *upstream quality control*: the certainty that the necessary quantity of mat, or gel coat in the case of the gel coat cell, is deposited at every point of the workpiece.

The flow of the glass and the operating pressure are constantly controlled. Any mishaps (i.e. a break in the glass) cause an immediate halt and the transfer of the robot to a stand-by position at a distance from the mould.

The system then executes the washing sequence of the mixing chamber to avoid catalysis of the resin and catalyst mix inside the spray gun, which would create a serious maintenance problem. The problem must be corrected and the operator must give his consent before the robot can be positioned over the workpiece again and the spray stage resumed at the point in which it had been interrupted.

When this passage has been completed, the robot returns to its stand-by position and only the introduction of a new mould starts the cycle again.

INSERTION ON LINE AND START-UP PROBLEMS

The insertion of the individual cells in the production line was accomplished in steps, starting from the installation of the chop and spray unit, certainly the most complex for the quantity of parameters it controls. (The AGVs, as we have said, are still at an experimental stage.)

After a trial run of approximately three months, to gather a series of significant data in terms of process control and material's quality, the other two cells were installed almost simultaneously.

We encountered three different types of problems at the startup phase:
- integration of PLC and plant. It was not always easy to coordinate regulation and control speeds with process feedback. This meant that a large part of the trial period was spent in finding the proper timing. This stage has been particularly critical, since the response time of control is determinant for the positioning mode and the quality of the material produced;
- utilization of the robot. The problem here was to have the robot reproduce the operator's smooth movements, obviously optimizing the various positions assumed in the production of every piece. Study of the kinematics of the robot arms in positioning the mat were performed using CAD systems and graphic simulation to optimize the trajectories of the tools;
- the presence of the robots' brushless drives in a styrene rich atmosphere. The problem was eliminated at the origin by installing an aspirator system alongside the molds to draw off the gases as they are released.

Every effort was made to combine the company's extensive operating experience and the indications supplied by a theoretical model combining all the aspects of the process, including details that have been underestimated or not even identified during hand spraying (orientation and length of the fibers, positioning of the pistol with respect to the mold).

SYSTEM ARCHITECTURE

The system is composed of (Fig. 4):
- 3 ASEA IRB 2000 robots with 6 integrated axes and one external axis (track motion);
- 1 ASEA MasterPiece PLC (MP) in an extended configuration for all process control;
- 1 ASEA MasterView (MV), to visualize the process, with a MasterBus communication line connected with the PLC and slower lines to communicate with the Host Computer and with the AGV controls;
- 2 networking PC AT for control of the AGVs;
- 1 CNC card mounted on a PC AT to control a 3 axes X-Y board;
- 1 Flow System Water-Jet cutting system. Water-Jet nozzle is gantry mounted on a 3 axes X-Y board.

Particularly interesting is the MV's function as the highest level in the hierarchy of the whole system.

Through a series of synoptic tables and of lists, the process can be visualized and modifications made, altering parameters, timing and quantities in rapid interaction.

The alarms' list and the localization of failures identified by the PLC, are visualized in real time on the MV, permitting rapid and precise action to avoid particularly long and costly downtime.

We now analyze some of the more significant transducers present in the plant. By transducer, we mean all those elements acting as *physical interfaces* between the process and the PLC, that is, those elements which allow the PLC to acquire data concerning the process and act on plant regulators and controls:
- *tachometer*: glass flow transducer. Mounted on the glass chopper, it supplies a voltage proportional to the quantity of compressed air for the chopper. A simple proportion and an integration over a two second period, calculated by the PLC gives us the quantity of glass chopped.
- *glass flow sensor*: the sensor was obtained by exploiting the circular motion of the glass filament as it unwinds from the spool. If the filament breaks or is blocked, control is sent a digital signal.
- *encoder*: resin or gel coat flow transducer. The encoder, connected mechanically to the pump by a rack and pinion, is a three wire type: supply and feedback signal. The encoder's data, integrated over a sufficiently long period (about 4 seconds), gives the value of the resin flow.

SOFTWARE: CONTROL AND REGULATION STRUCTURE

The software which controls and regulates the plant is divided substantially into five parts:
- PLC: a program for each cell controlled and a number of sub-programs which can be called by the main program to act on the individual regulation of control logics;
- Robot: a program for each work piece, divided by cells;
- AGV: the AGV software was purchased together with the control and the vehicles. It consists essentially of a *data base* which organizes the routes followed in function of the series of data coming from both the AGVs' control stations and the MV. The exchange between Control and MV takes place via RS232, with extremely high priority;
- Supervisor: a graphic program for each stage of the production cycle and programs for communicating with the PLC, AGVs and Host Computer;
- CNC: a simple *Basic-like* program for each workpiece.

The Host Computer, interfaced with the MV through slow communication lines, receives production data at the end of the day.

In this way is possible to have constantly updated information regarding the production situation and can organize production, warehousing, and check production levels.

The plant operator must, in turn, organize the production cycle to optimize utilization of the equipment, and make sure that the cycle is correctly executed. He must also be ready to take the proper action in the event of process failure, performing the correct operations to disconnect the alerted cells, in order to avoid the consequences of a total halt.

The guidelines for this plant were to concentrate all controls and regulations in the PLC in order to facilitate software maintenance and operation in the event of process failure, which is immediately communicated by the supervision and monitoring system.

An additional reason for this orientation was the possibility of including operating programs for various molds in the robots' software. In this way one control software can handle production cycles of different workpieces.

In this context the robot becomes a *intelligent terminal* of the PLC, which can process even a complex series of instructions, but does not assume in the process a decision-making role, which is instead reserved to the controller.

WATER-JET FINISHING

After polymerization the workpieces are transferred to the finishing cell where the contour of the edge is cut by Water-Jet.

The Water-Jet cutting of the fiber glass products is an integral part of the project for automating the production cycle: the shaping of the contour of the piece requires, even at the CAD design stage, a thorough knowledge of the possible paths of the jet and of the optimal combinations of cutting parameters. Water-jet technology has been used to cut materials for about fifteen years, but only recently has this technique begun to be increasingly employed on an industrial scale.

Research on the subject has not to date resolved all the technological aspects of the problem, especially in the field of the processing of composite materials.

In this case the cutting parameters are: the water pressure, the velocity at which the workpiece advances, the diameter of the nozzle, the nature of the material and its geometry, the number and nature of the layers constituting the composite, the orientation of the fibers.

These factors determine process speed and the final quality of the cut.

There are at the moment no analytic models for the calculation of cutting parameters and nozzle wear, while the aspects determining cutting quality are also yet to be clarified. This situation has made it necessary to flank the automation project with another line of study dedicated to the water-jet cutting of glass-fiber, which has been inserted in the context of a broader program of research on the mechanical processing of composite materials conducted by the Milan Institute for Machine Tools of the National Research Council (Istituto di Macchine Utensili del CNR in Milano). The results of these studies shall be published in the near future.

The definition of optimal cutting parameters and the possibility of ensuring the desired cutting quality require a thorough knowledge of the meccanisms of interaction between jet and material. The mechanisms of the fracture of composite materials subjected to high impact forces for short periods of time, as is the case of cutting at high feed velocities, are very complex. The fractures encountered may be due to the effect of dynamic stress waves, static pressure and/or erosion. The reciprocal importance of fracture modalities is a function of both the properties of the material and cutting parameters.

The literature includes studies which treat the problem of the penetration of the jet as an almost static phenomenon, in which the material is subjected to hydrodynamic forces which can cause various types of damage to the material.

However, little has been done to date to determine the influence of the speed of advancement of the workpiece, which is a parmeter of great importance for industry. We may, however, affirm that there is limit value for feed velocity, below which the dominating mechanism of damage is the flow of the material in response to the hydrodynamic forces. Above that value, the mechanism of damage is similar to that of erosion.

The jet must last for a sufficiently long time (ratio between the diamter of the jet and feed velocity) in order to cut entirely through the material.

During this time the pressure exerted by the jet will vary between a minimum value, corresponding to the dynamic pressure, and a maximum value, sizably above the minimum, which corresponds to the impact pressure. This phenomenon reflects on the quality of the cut, which presents a cutting surface with a periodic morphology.

Delamination is also frequently found in the water-jet cutting of glass-fiber, but the defect can be reduced, in materials of equal thickness, by reducing feed velocity.

The effect on materials and cutting quality of the tranversely deflected part of the jet has also been thoroughly studied.

In conclusion, all the results regarding current research will be used to integrate basic plant data and the analytic models of the CAD/CAM area expressly dedicated to the contour shaping of the workpieces.

The plant concerned is composed of a water-jet system fed by a 60 HP Flow System pump and by an X-Y board with triaxial computerized numerical control, two of which axes (X and Y) are interpolating.

The specific production concerned here must have the possibility of cutting highly complex shapes. This is why the nozzle is gantry mounted on a 3-axis board.

The reaction thrust of the jet of water is relatively limited (F<50N). Nevertheless, the waves of pressure created by the action of the valves may cause vibrations in the structure of the board, negatively influencing the quality of the results. The nozzle, where the water pressure (300-400 MPa) is transformed into an extremely high speed jet (600-900m/s), presents a particularly important aspect of the problem. The material of the nozzle and its geometry appear to be critical elements in the wear of the nozzle and the profile of the jet.

FURTHER DEVELOPMENTS

At present the integration of the CAD system (CATIA, Dassault Systemes) with the automized cells is in the experimental phase. The objective is to increase the degree of vertical integration in order to obtain further benefits in terms of flexibility and control of process parameters.

Our first step has been to define the limits, degrees of freedom and kinematics of the system being concerned, using appropriate CAD supports for *simulation* and *animation* to create a synthetic image of the work environment.

On the basis of these definition we designed and subsequently evaluated the best trajectories for the robots, optimizing the path of the spray guns and of the rolling and finishing tools inside the cells.

Once the *model* of the operating environment has been created, the designer works directly at the computer: he first draws up a three-dimensional design of the workpiece and then simulates the paths of the tool to verify the compatibility of the piece with the environment - that is, if the machine can physically produce the piece - considering in particular, the quantity of scrap produced and the quality of the covering. During simulation a special function makes it possible to display the areas as they are covered.

While defining the geometry of the piece, the designer develops a *data base* containing all the characteristics of the product and process parameters: the fiber/matrix ratio, the thickness with respect to the structural properties of the piece, the tolerances allowed, the type and color of the surface coating (Gel Coat), the grade of finish, the cutting parameters for the Water-Jet system. This package of information, containing design and product characteristics, is post-processed with appropriate translating and interfacing programs before being sent on to the mass memory of the robots and of the CNC Water-Jet system.

The advantages are obvious: in the first place, the off-line programming of the individual pieces avoids long downtime for teaching work cycles to the robot on the shop floor, while the simulation stage practically eliminates trial runs.

The transfer of information from the data base to the PLC and to the CNC obviates the need for human intervention in setting process and cutting parameters.

CONCLUSION

Although it is still too early to assess overall results, all the data collected in these few trial months that the plant has been in operation do indicate the following benefits:
- production organization is more flexible;
- a large decrease in idle time during the production cycle;
- a considerable increase in productivity (10% in three months) at lower unit costs, due to increased Quality Control, with reduction in rejects and to optimization of available resources;
- the possibility of plannning maintenance through the processing of historical series relative to the functioning of the installation;
- improved working conditions.

REFERENCES

Crow, S.C. (1973). A theory of hydraulic rock cutting. Int. J. of Rock Mech. and Geomech.,10, 567-584.

Smohaupt, U.M. and Burns, D.J. (1974). Machining unreinforced polymers with high velocity water jets. Experimental Mechanics, 3, 152-157.

Hashish, M. and Reichman, J.M. (1980). Analysis of waterjet cutting. Proceedings of 5th ISJCT.

Koning, W. (1982). Elettroerosione e sistemi alternativi. Tecniche Nuove.

Di Ilio, A., Tagliaferri, V. and Crivelli V., I. (1984). Qualita' di lavorazione nel taglio dei materiali compositi con fascio laser e water-jet. Proceeding of Quality Control of Composite Materials, Ed. CMC, Napoli.

Koening, W. and others. (1985). Machining of fiber reinforced plastics. Annals of the CIRP, vol. 34/2. pp 537-548.

Takashi, K. (1986). Future of Water Jet cutting. Metalworking Engineering and Marketing, 3, 42-45.

Calandra, M. (1988). Robots, AGVs and PLC for production in fiberglass. Proceedings of the 19th ISATA, vol. 2. Allied Automation Limited, Croydon. pp. 209-219

Calandra, M. (1988). Robots and logic controller in glass-fibre applications. The Industrial Robot, 15, 201-202.

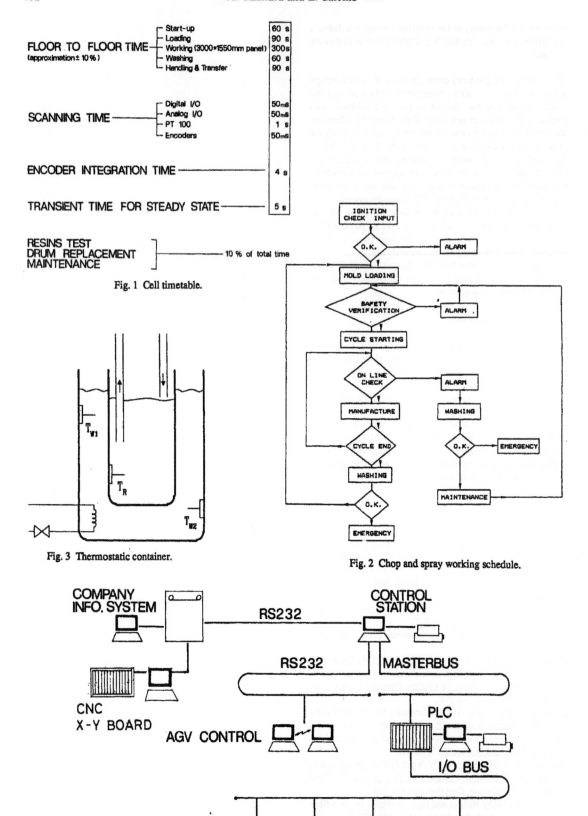

FLOOR TO FLOOR TIME — (approximation ± 10 %)
- Start-up — 60 s
- Loading — 90 s
- Working (3000×1550mm panel) — 300s
- Washing — 60 s
- Handling & Transfer — 90 s

SCANNING TIME
- Digital I/O — 50mS
- Analog I/O — 50mS
- PT 100 — 1 s
- Encoders — 50mS

ENCODER INTEGRATION TIME — 4 s

TRANSIENT TIME FOR STEADY STATE — 5 s

RESINS TEST
DRUM REPLACEMENT — 10 % of total time
MAINTENANCE

Fig. 1 Cell timetable.

Fig. 3 Thermostatic container.

Fig. 2 Chop and spray working schedule.

Fig. 4 System architecture.

MAXIMIZING PRODUCTIVITY FROM MACHINING CELLS

T. M. Hammond

Flexible Manufacturing Technology Limited, Brighton, UK

Abstract. With companies becoming increasingly more aware of the need to maintain profits in a competitive world market, FMT offer a solution which is available on a progressive step-by-step basis, which enables users to maximise the productivity for their machining cell.

This involves using proven FMT technology in simulating production orders and tooling requirements, linking with engineering design offices and other support facilities, to enable the machines to produce for periods either unmanned or with minimal manning, on a 24-hour basis.

The experience gained in approximately (40) cell installations is available to all our potential users, together with our proposal engineering services.

Keywords. CNC; Flexible manufacturing; Machine tools; Management systems; Production Control; Reliability; Simulation.

For most manufacturing companies the objective must be to focus the business and manufacturing strategy where they can achieve most benefits. This means they are interested in setting up manufacturing cells to produce high value added parts which are critical to the quality of end product.

With these companies, it is essential that they invest in cells of proven technology and install these, on a progressive basis as the equipment is technically absorbed in the company and achieves a payback.

Many large customers are now insisting that a qualified supplier must reduce prices by per cent per year, provide quality control charts and provide on-time supply and they will then sign 2-3 year contracts. Faced with these demands the smaller company is faced with two options - improve response time, quality, and Just in Time (JIT) delivery or succumb to competition and close down.

One problem they face is that the traditional methods of justifying capital investment in machine tools and manufacturing cells are too restrictive and take too narrow a view to identify all the major benefits of advanced manufacturing technology. Machining cells make the greatest impact outside the area of direct component machining costs, with the emphasis upon reduced handling and storage costs, easier assembly improved quality faster response to market demand and the two biggest factors of all - reduced inventory and work-in-progress.

For the UK based company Victor Products the choice they made was to move into machining cells and for their first they selected a FMT Two-Machine Cell as part of a £2 million modernisation plan.

The company produces mining accessories, service plugs and sockets and underground mains connectors. These existing products, machined mainly from brass and gun-metal, were produced on lathes, milling machines and conventional borers.

The FMT cell installed consists of two FMT Fleximatic machining centres linked by the FMT Roevatran which services 16 pallet stations. See Fig. 1.

The machines are FMT Fleximatic FM100 Horizontal Spindle CNC Machining Centres with a brief specification as follows:-

FMT Fleximatic FM100

Traverses	X Axis 750mm
	Y Axis 650mm
	Z Axis 500mm
Pallet Shuttle	Rotary with 2 pallets
Power	15 Kw
Speed Range	20 to 4500 rpm
Automatic Tool Changer	2 drums each of 40 tools with random selection.
Feedrate	1mm/min to 15 metres/min.
Features include	Deluge coolant Swarf Conveyor Stringent Guarding Broken Tool Detection Probe facility
Control System	Siemens Sinumerik 8M Part Programme Storage 128K
Transporter	The transporter linking the two Fleximatic FM100's is the FMT Roevatran rail guided vehicle.
Tool Setting	Tool Setting is by way of Frenco P40 machine. This is linked directly back to the host computer library of setting information.

Supervisory Computer — The host computer is a Hewlett Package HP1000 A600. The configuration installation gives a total of 2Mb main memory together with 39.5Mb of fixed disc storage. The functional system design for this supervisory computer was prepared by the Hoskyns Group to Victor Products' specification with the co-operation of FMT. The purpose of the computer system was to upgrade the control of the cell to carry out the following functions: direct numerical control; computer enhanced cell scheduling; computer controlled materials handling; tool management; cell management/status monitoring; production management/reporting.

The company committed itself step by step:-
- first one Machining Centre standing alone,
- then the second Machining Centre and the Transporter,
- finally and only after the system was working under the direction of the machine controller, the host computer and the full DNC linkage.

This Two Machine Cell reduced the value of work in progress from £225,000 with conventional machines to £125,000. Direct machining costs have been reduced by £75,000 and lead times halved from 12 weeks to 6 weeks for critical underground mains components. One important aspect of the Cell is the ability to operate 4.5 hours unmanned between shifts which is a critical aspect when cost justifying machining cells. The ability to simulate the operation of the cell and determine for the operator which longest cycle time fixtures should be loaded for the processing between shifts can mean for the Production Engineer the difference between success or failure of the cost justification. This vital factor together with the tangible and intangible benefits described has made simulation a vital factor in cost justification.

FMT have invested heavily in computer simulation and have developed generic simulation models which allow the cells and systems to be re-configured and re-costed quickly and easily.

The ability to vary batch sizes, number of different components, number of machines, cycle times, number of controlled vehicles and the number of operators and the utilisation of these at the load and unload station is vital if the overall business strategy is to be achieved. More important is the ability of the Chief Executive to obtain these alternatives quickly in order to create a dynamic planning process. As 24 hours or one week's production can be run in a matter of minutes the ability to create a dynamic planning process is greatly enhanced.

However, a machining cell is more than an number of machines linked by a component handling system. It is an all-embracing philosophy that impacts upon all aspects of a manufacturer's activities rather than machining cycle times in isolation.

That's why the first step to maximise a machining cell's productivity must start long before any hardware is even considered.

For many companies the installation of their embryo machining cell represents the first tangible evidence of their move towards Advanced Manufacturing Technology.

That's because they have followed an AMT rule of thumb that says, "Plan Down - Implement Up". To expand on this: Plan Down means review the company's product and manufacturing strategy for the next five years, in depth and in all aspects. Not just production engineering but also design; material resource planning; handling and storage; production planning and inventory; and last but not least, management data reporting to give better, more accurate control and faster response to an ever-changing market place.

Planning down starts with strategic data required by management to improve control of the company. It then works down through all aspects of the company activity - right down to the capacity and number of machines required on the shop floor and peripheral activities. Does this component stores really require heavy expenditure on automated handling equipment? Or will the dramatically reduced inventory promised by a machining cell negate this requirement? This could be a major indirect saving associated with a machining cell.

One of the most positive benefits of just the planning aspect of AMT is that inter-departmental barriers are broken down. When departments such as design, quality control, manufacture, finance and sales all work from a common data base, there is no room for isolationism.

Implement up often results in the machining cell being the most logical starting point for a company's first venture into AMT.

In theory a machining cell ought to be able to handle any component that fits in the physical capacity envelope of the machines.

In practice this amount of "elasticity" can put too great a demand on auxiliary aspects of the cell such as fixtures, tooling and programme store capacity.

Most cells start life dedicated to a family of components. At FMT for example we have worked together with customers on cells that produce such diverse components as valves; automobile engine parts; mining equipment; gearbox and transmission units; machining tools and electric motors.

The words "worked together" were carefully chosen. Maximum productivity for any machining cell can only be achieved with the closest co-operation between the machine tool manufacturer and the user over an extended period of time. This can be two to three years.

For most companies the leap from conventional machines, or even standard NC and CNC machines to a cell is just too big to be undertaken in one bold leap. It stretches both the technical and financial resources of the company. That's why FMT pioneered the "Step-by-Step" approach.

Today, proven modules of hardware and software can be progressively integrated into a cell. The pace of implementation is totally controlled by the user in both terms of number of machines and software-based manufacturing technology.

Implementation of hardware and software can be two parallel, but inter-related functions.

In general terms implementation of hardware means more capacity in the form of additional machine tools coupled with improvement component handling.

These additions usually start by simply duplicating the original machine to increase capacity. Then, specialist machines can be introduced. Cells are all about flexibility.

Two cases from FMT's experience spring to mind; automobile cylinder heads, and gate valves. The cylinder head design may change the inlet/outlet ports, but holes for the studs in the block are very unlikely to change.

On gate valves, bolt hole patterns are often governed by BS or ISO standards and are therefore unlikely to change. Both these cases can benefit from the productivity of multi-spindle drilling incorporated into the machining cell.

Similar arguments can be mustered for introducing the productivity of specialised machining stations such as gun-drilling, boring machines or high speed tapping units.

As a cell grows, productivity can be further improved by bringing additional functions under the control of the cell.
For example, washing followed by inspection insures that clean, correct components are available for assembly immediately after machining.

Turning to the progressive implementation of software, here the benefits can be enormous - but they are almost invariably indirect to the machining process. Faster design to production; better use of resources; dramatically reduced inventory are all quantifiable benefits, but again have little to do with the component cycle time on the machine.

Even the embryo cell must contain software capabilities such as tool monitoring, sister tool replacement, component identification, probing cycles etc. Together, these add up to its unmanned operation capability.

These software foundations can be built on. A DNC link with version control allows part-programmes to be stored away from the machines then transmitted to the machine in response to production schedules. The facility to transfer programmes back to the central computer system, after editing on the shop floor, provides a secure method of updating records.

As manufacturers move to Computer Aided Design (CAD) it makes sense to evaluate at the same time CAD/CAM (Computer Aided Manufacturing) link. This allows CAD data to be processed to produce part programmes. This, combined with the DNC facility provides a paperless environment from design to finished part.

Further software development can embrace a host computer which links the machining cell to other business systems. The ultimate goal is fully computer integrated manufacturing (CIM). Here, financial, commercial and manufacturing departments share a common data base. At Victor Products the HP1000 is linked directly to the company's MRP system and the cell driven dynamically by the overall company requirements.

If productivity is to be maximised, then it's useful to have some sort of yardstick against which to measure benefits that accrue.

This is where the NEDO report on "Advanced Manufacturing Technology", "The Impact of new technology on engineering batch production" is a useful, independent publication to consider.

It examines eight companies with turnovers of £3M - 20M and product mixes qualified as standard, customised or special.

The results are startling, inventory down by anything from a factor of 3 or 4. That means up to 75% reduction in dead money locked up in stock. Stock that could conceivably become worthless as advances in technology make it obsolete.

It is worth noting that industry in the UK has an estimated £16 billion locked up in inventory. Reduce that by an overall 25% (not unrealistic when viewed against examples of 75% already being achieved) and a massive £4 billion is available for investment - without going near the Banks for additional funds.

Compare that with an annual £2 billion investment in fixed assets expenditure in manufacturing industry and it is easy to see the potential for industrial revival.

It is probable that the figures for Spain are of a similar magnitude.

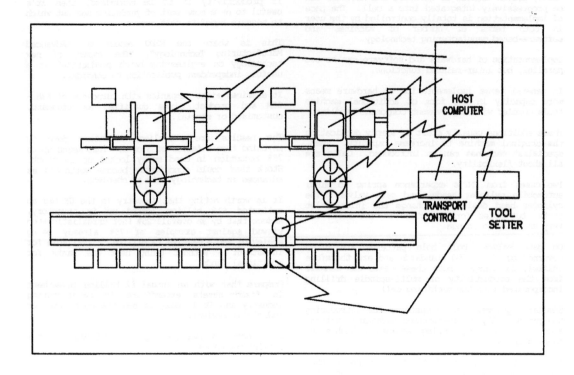

Figure 1: SCHEMATIC OF FLEXIBLE CELL

TRAJECTORY PLANNING METHOD FOR MOBILE ROBOTS

P. Campoy, G. F. Bobadilla, R. Aracil, M. A. Salichs and L. Moreno

Departamento de Automática, E.T.S.I.Industriales (DISAM), Madrid, Spain

Abstract. This paper accounts for the development of an original trajectory generation method for a two-dimensional mobile robot (i.e. capable of planar movement) designed to operate in industrial environments. In the last stage, a graph is searched to find an optimal solution. The nodes of this graph consist of three parameters, which determine absolute position and orientation of the robot.

A first issue is the generation of trajectories in collision-free space (union of consecutive nodes). Cubic-spline interpolation with optimization of the reference system with respect to a given cost function is used, where this cost or evaluation function has real physical meaning.

Very efficient and computationally inexpensive algorithms were developed to determine the optimal reference system and the cost function of a given piece of trajectory.

A second issue dealt with is that of graph-expanding. An on-line collision detection method is created, and subsequently, a method for generating intermediate nodes in the presence of a single obstacle.

Finally, an A-type algorithm is used to perform an on-line graph search, in order to determine the lowest cost trajectory in a fixed environment with obstacles.

The trajectories obtained offer an advantageous compromise between optimality and computational cost: while not sacrificing much of the first, possibility of on-line implementation is assured.

Keywords: Mobile robots, trajectory planning, obstacle avoidance, search graph, heuristics.

INTRODUCTION

The very spread method for finding collision-free trajectories are based on a point representation of the mobile robot. When working with such model of the mobile robot, there are basically two ways of finding the collision-free trajectory, each one operates in a different work space, which are :

1) Working in the two-dimensional physical space where the robot moves. In this case the robot is enclosed in a suitable circle and the obstacles are enlarged accordingly.

2) Working in the three-dimensional configuration space (C-space), whose coordinates are the planar position and the orientation, which determinate the state of the mobile robot. In this case obstacles are represented as three-dimensional constrains in such space.

Both operating ways are inefficient in the case of mobile robots thought to work in industrial indoor environment.

The first method implies to enlarge the obstacles' representation in such a distance that disappears most of the limited free space of an industrial environment where the mobile robot can go through (see Fig. 1). This way of working means using for a general case the solution which is apt for a robot with a circular cross-section, or which is made to maintain fixed orientation. Obviously, by doing so one wastes completely the manoeuvreing capability of the robot.

Fig. 1. Collision-free space in industrial environment

The second possibility of representing through a point is by working in the space of configuration (C-space) and has the following main backdraws in the case of study:

a) Increases of calculus complexity due to the augmentation of the dimensionality of the working space and to the fact of translating the obstacles into constrains in the C-space.

b) Forces to use fictitious cost functions in C-space which have not direct physical meaning in the movement of the mobile robot.

c) Difficults the control of the resulting trajectories due to the fact that every state parameter should follow a given time-function, independently of the others, without considering the actual mechanical relationships between them.

For these reasons, a new model of the robot is proposed, in which the robot is represented by means of an orientated segment.

MOBILE ROBOT MODELLING

Description of the proposed model

Due to the drawbacks of the methods mentioned above a new method of generating trajectories is presented in this paper, whose main advantage is the fact of generating trajectories which profit very well the free space in an indoor industrial environment and have as well a physical meaning, without increasing the complexity of calculus.

The proposed method is based on a segment representation of the mobile robot, which is adequate for most of the mechanical structures of industrial mobile robots, based on present AGVs (Automated Guided Vehicle) (Fig. 2). The chosen model is a segment with its origin at A (middle of the rear axis) and its "arrow" at B (middle of the front wheel). Note that this can be applied to any mobile robot whose plant section can be successfully enclosed in a "curved" rectangle as the one showed with little loss of extra space.

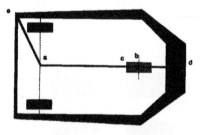

Fig. 2 Mechanical structure of an industrial mobile robot

Characteristics of the chosen model

The model presented here is specific for the studied case, where the bidimensional mobile robot has an structure similar to that in Fig. 2 and using the orientation manoeuvreing capability is required (e.g. as a transport unit in indoor environments). Accordingly this model doesn't intend to be a general solution in the trajectory generation of robots.

The advantages of this is model versus the point-model in C-space are the following:

- Good controllable feature of the resulting trajectories, where the robot orientation is not an independent variable, but it is always tangent to trajectories.

- Work-space is two-dimensional, while sacrificing only a small portion of collision-free space. This portion of "wasted area" is related to the difference between the apparent and real edge of the robot (see Fig. 2), and is only significant at the rear part, which is a negligible fact, if considering that the robot is allowed to move only forward, and with limitations in maximum swerving angles.

- Possibility of considering cost functions with physical meaning (time, length, weighted lengths), which is not feasible in C-space.

- Rich description of the interaction between the robot and its surroundings.

GENERATION OF THE TRAJECTORY OF UNION BETWEEN TWO NODES, (Trajectories in collision-free space)

Description of the method of interpolation

The method used is cubic-spline interpolation of the successive states of the robot, which are given by the position and orientation of the representing vector.

The position and orientation fixed at the "initial" and "final" states of the robot determine six constraints on the trajectory. In a given reference system, there is a unique cubic interpolant which satisfies these constraints. We use as an additional parameter the angle that the x-axis of the reference system forms to the x-axis of a given particular reference system, so that a single-parameter family of possible trajectories is offered at this stage.

Difficult cases (i.e. great changes of slope, or even orientations opposing the direction in which a natural trajectory would be traversed) can be solved admitting readjusting of unreasonable orientations, or simply by using a suitable intermediate state. This also applies for either very "small" or very "large" values of the parameter d/L (distance between points divided by the characteristic length of the robot).

The advantages of this method of interpolation are the following:

- Smoothness. Tangent is continuous in global trajectories, which are formed by juxtaposition of several interpolants. Also, with imposed slopes not very different from those of the line passing through the initial and final position, curvature along the trajectory is very small.

- "Global" solution. The trajectory is known in its whole and no incremental decisions are required.

- Very small computational cost: closed formula, with no iterative process.

- Last and very important is the fact of having very simple, explicit analytic expressions (for the position of a reference point) which completely determine the successive states of the robot. It must be noted that the simplicity of the algorithms presented below depends strongly on this fact.

The main drawback is that the generation of trajectories is performed statically, without any preview to the way in which the trajectories are going to be executed. There is no reference to the problem of control of trajectories and the controller will have to cope by itself with the dynamical aspects of the traversing.

A minor difficulty is that global trajectories do not have continuous second-derivative. So, there are not any discontinuities in the position of the robot, but the profiles of the swerving angles needed in the driving wheels are discontinuous. This feature does not represent any problem if trajectories are followed by a closed-loop controller.

In this method of interpolation, the cubic polynomial is somehow the generalization of the straight line used as trajectory between two points, which represent the initial and final states of a mobile robot in the fixed-orientation case.

Cost Function

For a robot with a structure similar to that sketched in Fig. 2, the cost function is defined as the length of the trajectory traversed by the front wheel, which is the only driving wheel.

Assuming a constant velocity on the driving wheel, the chosen cost function represents the time needed for traversing the given trajectory. This cost function generalizes the distance between the initial and final states of a mobil robot, when this is represented by a point.

The possibility of using efficiently this cost function is related to the method of interpolation, which allows for explicit analytic expressions for trajectories.

If (x_a, y_a) are the coordinates of the center of the front wheel, and $(x, f(x))$ the coordinates of the center of the rear axis (reference point), where $x \epsilon [x_0, x_1]$, the cost associated to a piece of trajectory is:

$$LONGD = \int_{x_0}^{x_1} \sqrt{x'_a(x)^2 + y'_a(x)^2}\ dx = \int_{x_0}^{x_1} \sqrt{1 + f'(x)^2 + L^2 \frac{f'(x)^2}{(1 + f'(x)^2)^2}}\ dx \quad [1]$$

A direct method of optimizing this function would be to use the Calculus of Variations to find the optimal function $y = f(x)$. Here only the one parameter cubic interpolants $y = f_\alpha(x)$ are used, where α is the angle which determines the reference system as explained below.

Exact computation of Cost Function: A Romberg numerical integration can be used to assure a maximum error bound. Integral expression of [1] is computed.

Approximate Computation of cost function: Let us define the "intrinsic reference system" for two given consecutive states of the robot as that whose x-axis passes through the initial and final point, so that the origin is at the initial point, the final point has positive abscissa and the orientation of the (x,y) system is positive (counter clock-wise x-y turn), see Fig. 3.

An α reference system is defined as one whose x-axis forms an angle α to the x-axis of the intrinsic reference system. Obviously, all α-reference systems produce the same trajectory as it is invariant under translations and does not depend on direction of traversing. However, two different angles α_1 and α_2 give rise to different cubic interpolants corresponding to the α_1 and α_2-systems.

Let $f\alpha(x)$ be the cubic-spline interpolant in an α-reference system.

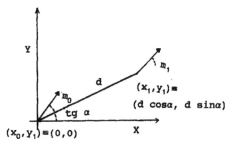

Fig. 3 Reference System

After algebraic manipulations in [1] (first order binomial Taylor series; averages value of integral with a positive weight function) the cost function can be approximated by:

$$LONGD_\alpha \approx d[1 + k_\alpha\ 5/48\ \cos^2\alpha(m_0^2 + m_1^2 + 2/5\ m_0 m_1)] + 2L^2/(d\cos\alpha)\ [1/\cos^2\alpha + 2/15\ (m_0^2 + m_1^2 + 1/2\ m_0 m_1)]^{-5/2} \quad [2]$$
$$(m_0^2 + m_1^2 + m_0 m_1)$$

where: α = angle which determines reference system.
  d = distance between points.
  $m_i = m_i \cdot tg\alpha$, (m_i is the slope in the α-system)
    $i = 0$: initial state
    $i = 1$: final state
  L = characteristics length of robot
      (length of model)
  K_α = constant depending of α and verifying $0 < K_\alpha < 1$, optimized empirically to adjust best to exact cost function.

The line followed here to estimate the validity of this approximate expression has not been to investigate the mathematical error bounds associated with the approximations (which are generally not reached by far), but to make a statistical study of errors, in order to determine the ranges of validity and associated errors. The imposed bound was an absolute value of relative error less than 2,5 %; and the average of relative errors obtained was 0,5 %; thus ranges of α and slope relative to intrinsic reference system were determined.

The approximate cost function using formula [2] involves computational cost which is 2/5 to 1/3 of exact computational cost (CPU: System Micro VAX under VMS), using Romberg integration with the same maximum error bound. Though not too large a difference, this is significant due to the great number of times that the cost function must be evaluated in the course of graph-search, as will be discuss below.

Out of the evaluated rang for formula [2], the Romberg algorithm is used.

Optimal Reference System

The problem is to find the value α_0 such that $f\alpha_0(x)$ attains minimum cost (LONGD(α) is minimum at α_0), given the initial and final nodes.

In practice, the problem is that given ϵ, find α such that $|LONGD(\alpha) - LONGD(\alpha_0)| < \epsilon$, instead of that $|\alpha - \alpha_0| < \epsilon$.

Exact Algorithm

A value of α as close as desired to α_0 can be computed, solving:

$$d/dt_\alpha \ LONGD(t_\alpha) = 0 \qquad\qquad [3]$$

(where $t_\alpha = \tan(\alpha)$ for simplicity)

The derivative of LONGD with respect to t_α can be computed through Leibuitz's formula of derivation of an integral with respect to a parameter, and it leads to a very cumbersome expression, (Derive [1] with respect to $\tan\alpha$, where f depends on α, and also x_1).

A secant algorithm, modified to avoid inflexion points, is used to determine the t_α which minimizes LONGD. (Obviously a Newton method would require further derivation to compute d^2/dt_α^2 LONGD and a much greater complexity).

However this exact algorithm is prohibitive in terms of computational cost, exceeding by far the simple computation of the exact cost function. Therefore it is used, only to test the heuristic method presented below.

Heuristic Algorithm

Let m_0, m_1 be the imposed slopes with respect to the intrinsic reference system. We distinguish between two cases:

a) $\text{sgn}(m_0) = \text{sgn}(m_1)$. The criterium used is intuitive: to minimize the area between trajectory and segment, which can be easily computed.

The value obtained is $\tan\alpha_0 = \dfrac{1}{2}\left(\dfrac{1}{m_0} + \dfrac{1}{m_1}\right)$

b) $\text{sgn}(m_0) = \text{sgn}(m_1)$. The previous criterium is not valid anymore, because the area can be made to approach zero as desired, while the cost function grows due to unlimited growth of second derivative in the initial and final points.

A simple closed formula has been developed, which interpolates optimal values computed exactly.

The heuristic algorithm has been tested using the exact algorithm, and the results are:

- Case a) is exact. The formula gives 0.00% error in α and LONGD for all cases. Reasonably, the conclusion is a theoretical one (minimum area implies minimum length of the trajectory of the front wheel).

- In case b) the formula has been optimized, so that for an extremely valid range of parameters (max $|m_1|$, d/L), the results are:

Average of absolute value of relative error: 0,02%
Maximum absolute value of relative error: 0,5%

The computational cost of this algorithm which reduces to the use of either one simple closed formula or another, is negligible with respect to any other computations involved.

The cubic interpolant finally obtained does not depend on any arbitrary reference system.

GRAPH EXPANSION METHOD. GENERATION OF INTERMEDIATE NODES.

In the presence of an obstacle, the interpolation described in section III might lead to a collision. Therefore, it is necessary to generate an intermediate node (position and orientation of robot) so that the consecutive union of the three nodes brings about two collision-free cubic interpolants, each one with its own reference system.

It is needed to design a collision-detection algorithm, and, based upon it, an algorithm to generate the intermediate node.

This algorithms are to be executed on-line, and it is here that the simplicity of analytic expressions of trajectories pays back.

Collision-detection algorithm

The inputs are of this algorithm are the trajectory of the reference point (point A in Fig. 1) which includes a cubic polynomial and limiting abscissa, and the obstacle modelled by a convex polygon. The output is a boolean variable which states if there is collision or not, of some of the successive positions of the segment (i.e. the model of the robot) with the polygon.

Basically, this involves numerical investigation of cross-intersection between a curve and a segment (where it is necessary to consider points of the curve where the derivative is equal to the slope of the segment). Additionally some quick ckeckings can be performed, which may serve to decide immediately that there is no intersection. If it is not so the more time-consuming routines are executed.

Let us define:

d_x = minimum distance from the center of the polygon to the cubic spline.
fmax = maximum radius of the polygon, which is computed off-line.
M = bound of difference in y-coordinate between the cubic spline (trajectory of the reference point) and the trajectory of the front wheel.

This quantities are easily computed. (For M, much better values than L can be found). Obviously, if the polygon does not intersect the cubic spline, and $d_x < r\text{max} + M$, there can be no intersection, saving the numeric checking of the section with the trajectory of the front wheel (which is more complex, as it is not a polynomial).

Also, if the polygon is small, and once it is seen that no edge of it intersects the two trajectories mentioned, checks are made that it is not enveloped between them.

It is possible to perform this numerical global checks due to the fact of having analytic expressions for the trajectories. However, the task is largely simplified as the trajectory of the reference point is a cubic polynomial.

Generation of intermediate nodes

In the presence of an obstacle with which the direct interpolation collides, a suitable intermediate state near the vertex is chosen, which represents an intermediate node in the search-graph. This intermediate state is chosen along the bisectrix of the correspondent angle of the polygon. Good initial estimates can be made as to the position of the free-collision-trajectory-generating intermediate node. The slope is a weighted average between the one at right angles with the bisectrix and the one that would appear naturally using clamped spline interpolation for the three points involved.

Though the algorithm is iterative, until no collision appears, two or three iterates at the most are normally needed.

Two versions of this algorithm were simulated: a) a simplified one, which used only one reference system, estimated as close to the optimum, for the two pieces of trajectory, and b) one slightly more complex, which optimized independently the angles of the reference systems of both implied splines. In forced cases of clear intersection with the polygon and large relative slopes, the advantage of the second algorithm is definite (see Fig. 4). As the computing time for optimum reference systems is negligible, the second version has been chosen.

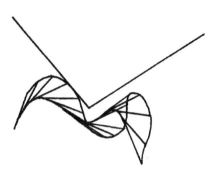

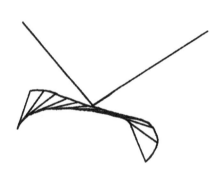

Fig. 4 Results' comparison using one (above) or two (below) optimal reference systems

RESULTS. GRAPH-SEARCH

All previous results are combined with the use of graph-search algorithms to find the optimal path.

While the graph is constructed on-line simultaneously to the graph search, the processing of the model of the surroundings, with obstacles appropriately enlarged, is performed off-line.

Evaluation function in the graph-search

The evaluation function is defined as the estimation of the cost of an optimal path conditioned to passing by a given node, reached throw its present path in the search tree.

It is computed as the addition of the cost needed to reach the present node (determined by all the previous nodes encountered, and calculated here as described in II.2) and an estimation of a minimal cost needed to reach the final node.

$$f' = g + h'$$

f': evaluation function
g : cost function to reach present node
h': estimation of remaining cost

The possible choices for h' are:

a) The same function as for g, as consequently with the trajectory generation method. It does not fulfill A* admissibility condition, and so the path found is not necessarily optimal.

Nevertheless, it has got great "heuristic power" (see ref. Nilsson 1985), and the search is quicker, as a limited number of nodes are expanded.

b) Distance between points. Obviously fulfills the admissibility condition, leading to optimal solutions, but the "heuristic power" is small. Inefficient graph search.

c) A weighted average between them, depending on the parameters p ("depth" of the node in the graph) and d_m/L, where d_m is a characteristic dimension of the surroundings. Obviously, for "deep" nodes, plausibly nearer the solution, the average must practically equal the function mentioned in a), as the constraint of slopes, makes the non-zero size of the robot be markedly felt. On the other hand, the distance between points is the major factor for a "shallow" node, as the size of the model may be negligible.

Results

The trajectories obtained are intuitively very close to the optimal solution one would sketch. Small computing times assure possibility of on-line implementation.

Figures 5 and 6 represent two exaples of the proposed algorithm, using the same initial and final position, but whit different orientation. The improvement of this algorithm versus a search algorithm where the cost function is based upon the distance between points are clearly seen.

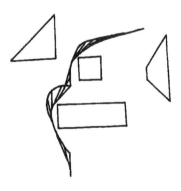

Fig.5 Example 1

Fig.6 Example 2

CONCLUSIONS

An original method has been presented in trajectory generation for mobile robots, using cubic interpolation.

The model chosen for the robot, as opposed to a point-model, has proved to be a complete success, and a good practical compromise solution for mobile robots designed to work in industrial environments, between computational cost and profiting robot maneuverability.

Efficient approximate algorithms have been presented that compute a chosen cost function and the reference system in which an optimal interpolant is obtained.

Also, an on-line collision detection algorithm has been outlined, and the method of generation of intermediate nodes (method of expansion of the graph) to which it applies. This algorithms take considerable advantage of the exact simple analytical description of trajectories mentioned above.

Finally, the solution of the global problem has been attained, combining this results with the classical graph-search algorithms. The graph is created and simultaneously explored for a collision-free trajectory, linking initial and final specified states (position and orientation) of the robot in an environment crowded with obstacles.

The trajectories obtained fulfill the two basic desired objectives: nearby optimal solution, computational economy assuring on-line implementation).

REFERENCES

Aracil, R.; P. Campoy y M.A. Salichs (61987) Robots Móviles. *Feria Robótica.* Zaragoza.

Brooks, Rodney A.; Tomás Lozano-Pérez. (1985) A Subdivision Algorithm in Configuration Space for Findpath with Rotation. *IEEE Transaction on Systems, Man, and Cybernetics,* Vol. SMC-15. No. 2.

Brooks, Rodney A. (1983) Solving the Find-Path Problem by Good Representation of Free Space. *IEEE Transactions on Systens, Man, and Cybernetics,* Vol. SMC-13, No. 3.

Canny, John. (1986) Collision Detection for Moving Polyhedra. *IEEE Transactions on pattern analysis and machine intelligence.* Vol, Pami-8. No. 2.1.

Chatila. (1986) Mobile robot navegation: space modelling and decisional processes. *Robotics Research 3.* MIT Press.

Chattergy, R. (1985) Some Heuristics for the Navigation of a Robot. *The International Journal of Robotics Research,* Vol. 4, No. 1, Spring.

Ihikawa, Yoshiaki; Norihiko Ozaki. (1985) A Heuristic Planner and a Executive for Mobile Robot Control. *IEEE Transactions on Systems, Man, and Cybernetics,* Vol. SMC-15, No. 4.

Iari i Valentí (1987) Study of new heuristics to compute collision-free paths of rigid bodies in a 2D universe. *Doctoral Thesis.*

Iyengar, S.S.; C.S. Jorgensen, S.V.N. Rao, C.R. Weisbin. (1985) Robot Navigation Algorithms Using Learned Spatial Graphs. *Research sponsored by U.s. Dept. of Energy Office of Basic Energy Sciences.*

Kambhampati, Subbarao; Larry S. Dacis. (1986) Multiresolution Path Planning for Mobile Robots. *IEEE Journal of Robotics and Automation,* Vol. RA-2. No.3.

Kant, Kamal; Steven W. Zucker. (1986) Toward Efficient Trajectory Planning: The Path-Velocity Decomposition. *The International Journal of Robotics Research,* Vol. 5, No. 3.

Keirsey, D.; J. Mitchell, B. Bullock, T. Nussmeier and D.Y. Tseng. (1985) Autonomus Vehicle Control Using AI Techniques. *IEEE.*

Kuan, Darwin T.; Rodney A. Brooks, James C. Zamiska, Mongobinda Das. (1984) Automatic Path Planning for a Mobile Robot Using a Mixed Representation of Free Space. *IEEE 1984.*

Lozano-Pérez, Tomás; Michael A. Wesley. Communications of the ACM. (1979) An Algorithm for Planning Collision-Free Paths Among Polyhedral Obstacles. *IEEE Transations on Computer* Vol. 22, No. 10.

Lozano-Perez, Tomás. (1983) Spatial Planning: A Configuration Space Approach. *IEEE Transactions on Computer,* Vol. C-32, No.2.

Nilsson, Nils J. (1987) *Principios de inteligencia artificial.* Ed. Diaz de Santos, S.A.

MOTION CONTROL OF AN AUTONOMOUS VEHICLE

K. Boukas

Ecole Polytechnique de Montréal, Dept. de génie mécanique, Montréal, Canada

ABSTRACT. The so called flexible manufacturing is at present becoming increasingly important in many fields of production. In manufacturing, the effort towards flexible automation include, in addition to the machine tool, more and more peripheral functions such as transport and handling. The use of autonomous vehicles for moving materials between work stations must be an important consideration in the overall design of a flexible automated factory. This paper describes the motion control problem of an autonomous vehicle. Simulation results characterizing the accuracy of the system are presented.

KEYWORDS. Control, Vehicle, Flexible Manufacturing.

1. INTRODUCTION

Robot carts are coming into increased use in automated factories and other such reasonably well structured environments. For the applications, the reader is refered to the Proceeding of the 2nd International Conference on Automated Guided Vehicle system held at Stuttgart (W. Germany) or the survey of Tsumura [1986]. Most of the automated carts in present use are not really autonomous (i.e, self guided) but rely on "trucks" imbedded in, or painted on, the factory or office floor to guide them from one station to another.

These non-autonomous carts present some inconveniences. For instance, if a cart during its navigation along a trajectory fails, part of the path will be not used since the cart stays on the same place. This will be traduced by stopping some work stations and reduces the productivity of the factory.

An alternative consists of using an autonomous cart which can navigate in the factory along any trajectory, and if a cart fails, it can be avoided by the other carts since their path can be changed. So this kind of cart is capable of navigating in structured environments such as offices and factories without external guide tracks. The on board control system uses odometry to provide the position and heading information needed to guide the cart along specified paths between stations.

In motion, the vehicle is a moving reference, and its position must be determined relative to fixed frame of reference somewhere in the environment. This fixed frame of reference must be provided for the vehicle, so that it can continuously refer to this frame using on board sensors (optical or ultrasonic imaging of the surrounding for example), and therefore know its precise position in the surroundings. Hence the vehicle can detect and correct for deviation in its path, in comparison to the command path stored on board.

There is significant interest in autonomous vehicles which are capable of intelligent motion (and action) without requiring either a guide to follow or teleoperator control. Potential applications of autonomous vehicles are many and include reconnaissance/exploratory vehicles for space (planatory exploration), undersea, land and air environments, remote repair and maintenance, material handling systems for the office and factory and even intelligent wheelchairs for the handicapped. Much of the current interest is centered around autonomous land vehicles for military applications in which the vehicle must operate in an unstructured (having few points of reference) and hostile

---

[1] This research has been supported by NSERC-Canada, Grants # OPG003 6444, A4952.

[2] K. Boukas is with the mechanical Engineering department, Ecole Polytechnique, GERAD, Montréal, Que., Canada.

environment. This is an extremely difficult problem which can be very much simplified by restricting the vehicle to operate in a structured, well characterized, environment such as a factory or office. This is a very strong restriction, yet a structured environment still presents challenging research and engineering problems with many potential applications.

Until now, different kinds of autonomous vehicles designed have been proposed and tested in different areas. Borenstein and Koren [1987] have studied the motion of a nursing robot. Nelson and Cox [1988] have studied the local path control system for an experiment cart "Blanche". This cart has also been studied by other authors (see Wilfong [1988] and Cox [1988]). European researchers, include the works of Rembold [1988], Julliere and al. [1983], and Campion and al. [1988]. Japanese contributions are summarised in the survey of Tsumura [1986]. In the field of manufacturing systems, the use of autonoumous vehicle is at its beginning, and it presents challenges.

During the autonomous vehicle operation, some parameters may vary as the mechanical load varies. These variations will be interpreted as a disturbance and it will introduce some error in the tracking trajectory problem, if the fixed controller settings used on the autonomous vehicle is not well designed. The disturbance can temporary or continuous. The temporary disturbance can be successfully corrected by a propotionnal controller. The continuous disturbance require an integral action.

The aim of this paper consists of applying this technique of control to an autonomous vehicle when tracking a given trajectory. The paper is organized as follow: In section 2, the autonomous vehicle model is established, and the control tracking problem is formulated. In section 3, the control algorithm is presented. In section 4, in order to demonstrate the efficiency of the algorithm, simulation results are presented.

2. MODEL

The autonomous vehicle described in this paper has the same design frequently used for computer controlled vehicles (see Borenstein and Koren [1987], Campion and al. [1988], Nelson and Cox [1988]). It consists of two drive wheels, each with its own controlled DC motor, and a free wheeling castors, which provides stability. The two DC motors with built-in reduction gears and incremental encoders, drive two wheels constituting the front axle of the vehicle. The resolution of the used encoders is such that one pulse represents 3 mm of tangential travel of the drive wheel. The motors are coupled to the wheel shafts through a gear ratio. In the rear, there is one free wheeling castor. A simplified model can be represented by the figure 2.1.

This kind of vehicle will be used to transport parts between work stations. The motion of the vehicle from one work station to another will determine the fixed points of the trajectory, which is generated by the scheduler. The trajectory generation problem is not considered in this paper, and the used trajectory is supposed given.

Our problem consists of driving the autonomous vehicle from a given location to another location along a given path. In order to represent the vehicle location relative to a fixed coordinate system, three values must be given: the x and y coordinates of the centerpoint G, and the angle θ between the vehicle's longitudinal axis and the X-axis. These parameters describe the state of the autonomous vehicle.

The control location of the autonomous vehicle is reduced to the control of the two DC motors. Let $\omega_l(t)$ be the speed of the left wheel at time t, and let $\omega_r(t)$ be the speed of the right wheel at time t. Let d represents the distance between the two wheels. Let r be the common diameter of the two drive wheels.

For a given angular left speed $\omega_l(t)$ and right speed $\omega_r(t)$, the linear displacement during the interval Δt of the left and the right wheels are given by the following equations (see Fig. 2.2):

$$\Delta L_l = (R + d)\Delta\alpha \qquad (2.1)$$

$$\Delta L_r = R\Delta\alpha \qquad (2.2)$$

where R is the radius of the curve described by the vehicle.

By combining these two equations we obtain the following the relationship between the variation of the orientation and the linear dislpacement of each wheel.

$$\Delta\alpha = \frac{(\Delta L_l - \Delta L_r)}{d} \qquad (2.3)$$

The linear displacement of the centerpoint G can be approximated by the following relationship.

$$\Delta L = \frac{(\Delta L_l + \Delta L_r)}{2} \qquad (2.4)$$

From Fig. 2.2, we can show by geometric manipulations, that $\Delta\alpha = \Delta\theta$. So the orientation equation can be writen as:

$$\Delta\theta = \Delta\alpha = \frac{\Delta L_l - \Delta L_r}{d} \qquad (2.5)$$

which gives the following relationship.

$$\frac{\Delta\theta}{\Delta t} = \frac{1}{d}[\frac{\Delta L_l}{\Delta t} - \frac{\Delta L_r}{\Delta t}] \qquad (2.6)$$

The projection of the equation (2.4) on X-axis and Y-axis of a fixed frame of reference, gives the following relationships:

$$\Delta x = \Delta L cos(\theta) \qquad (2.7)$$

$$\Delta y = \Delta L sin(\theta) \qquad (2.8)$$

The linear and angular displacements of the left and right wheel are linked by the following relationships:

$$\Delta L_l = r\omega_l(t)\Delta t \qquad (2.9)$$

$$\Delta L_r = r\omega_r(t)\Delta t \qquad (2.10)$$

If we let Δt converges to zero, and we substitute the longitudinal displacements in equations (2.6), (2.7), and (2.8) by their values given by relations (2.10) and (2.11), we obtain the following equations, which describe the dynamics of the vehicle (position and orientation).

$$\dot{\theta}(t) = \frac{r}{d}[\omega_l(t) - \omega_r(t)] \qquad (2.11)$$

$$\dot{x}(t) = \frac{r}{2}[\omega_l(t) + \omega_r(t)]cos(\theta(t)) \qquad (2.12)$$

$$\dot{y}(t) = \frac{r}{2}[\omega_l(t) + \omega_r(t)]sin(\theta(t)) \qquad (2.13)$$

This problem can not be solved by the well known classical control system theory because the dynamics of the vehicle is modeled by a nonlinear system. To control their mobile robot which has a similar difficulty to the one used here, Campion and al. [1988] have used the external linearization technique. Other technique was used in Julliere and al. [1983], Nelson and Cox [1988] and Borenstein and Koren [1987].

In our case, an alternative to avoid this difficulty (nonlinearity model), consists of driving the two DC actuators with the same speed, which permits one to simplify the model and the classical control theory can be used. This implies the separation of the control orientation and the control location. So to control the position, we have to force the speed of the two DC motors to be equal in amplitude and direction, and to control the orientation, we have to force the speed of the two DC motors to be equal in amplitude but with opposite direction. Thus the motion of the autonomous vehicle, are reduced to linear and angular rotation around the centerpoint G.

The block diagram of vehicle control is represented by Fig. 3.1. The motor is approximated as a first order lag. The equation in Laplace notation relating motor speed to armature voltage and load can be defined by the following equation (including the encoder):

$$F_i(s) = \frac{HK_1}{1 + \tau s} V_i(s) - \frac{HK_2}{1 + \tau s} T_i(s) \quad i = 1, 2 \qquad (2.14)$$

where $F_i(s)$ is the speed motor in pulses/V.s, $V_i(s)$ is the armature voltage, $T_i(s)$ is the load torque in N.m, H is the encoder gain in (pulses/rad), K_1 and K_2 are constants respectively in rad/V.s and V/N.m, and τ is the motor electromechanical time constant in s.

In the digital to analogue converter (DAC), the signal is held constant during the sampling interval T, and therefore its transfer function is that of a zero-order-hold (Z-O-H):

$$\frac{V_i(s)}{U_i(s)} = \frac{(1 - e^{-Ts})}{s} \qquad (2.15)$$

Substituting equation (2.15) into (2.14) yields

$$F_i(s) = \frac{HK_1(1 - e^{-Ts})}{s(\tau s + 1)} U_i(s) - \frac{HK_2}{\tau s + 1} T_i(s) \qquad (2.16)$$

Performing the z-transform on equation (2.16) yields

$$F_i(z) = \frac{b}{z - a} U_i(z) - \frac{b'}{z - a} T_i(z) \qquad (2.17)$$

where

$$b = HK_1(1 - a)$$
$$b' = \frac{HK_2}{\tau}$$
$$a = e^{-\frac{T}{\tau}}$$

In this paper, we try to solve the problem which consists of driving the autonomous vehicle from a given point to another given point along a given trajectory with precision. The initial point and the final point are together characterized by their location (x, y) relative to a fixed frame reference, and orientation θ which describes the angle between the vehicle's longitudinal axis and the X-axis.

We suppose that the given trajectory between the initial point $A(x_A, y_A, \theta_A)$ and the final point $B(x_B, y_B, \theta_B)$ is partitioned in $(n-1)$ portions, with n points, where the first point is the initial point and the nth point is the final point. So to drive the vehicle from the point $A_{k-1}(x_{k-1}, y_{k-1}, \theta_{k-1})$ to $A_k(x_k, y_k, \theta_k)$, $(k = 1, \ldots, n)$, where the position and the orientation are known, the following algorithm will be used.

Step 0. set $k = 2$;

Step 1. Calculate ϕ_k and l_k as

$$l_k = \sqrt{(x_k - x_{k-1})^2 + (y_k - y_{k-1})^2} \qquad (2.18)$$

$$\phi_k = arctg \frac{y_k - y_{k-1}}{x_k - x_{k-1}} \qquad (2.19)$$

Step 2. Send the reference control Φ_{k-1} to oriente the vehicle at point A_{k-1} as:

$$\Phi_{k-1} = \phi_k - \theta_{k-1} \qquad (2.20)$$

Step 3. Send the reference control position l_k to drive the vehicle to point A_k

Step 4. Test If $k < n$ set $k = k + 1$, $\theta_k = \Phi_{k-1}$ and go to step 1., else go to step 5.

Step 5. Adjust the final orientation to θ_B and Stop.

The control orientation or position will be presented in the next section.

3. CONTROL PROBLEM

The system to be controlled is represented by an equivalent model. The discrete model of the servomotor (left or right) can be obtained from the equation (2.17) and it is assumed to be of the form:

$$y(k) - ay(k - 1) = bv(k) - b'T(k) \qquad (3.1)$$

where

$y(k)$: is the servomotor speed at the kth sampling instant;

$v(k)$: is the control signal at the kth sampling instant;

and a, b are unknown parameters which remain to be estimated.

To identify the parameters a and b of the system, we have used the recursive least square technique. The parameters a and b were identified off-line.

In classic control theory, there is a vast well known array of design techniques for generating control strategies where the model is known. The design of the controller is generally based on a given set of specifications. These specifications are related to the transient response and the steady state response. In our case, the steady state error must be zero, and the maximum percent overshoot must be less than 5%. The time response must also be reduced.

During the vehicle motion, any temporary disturbance of the steady state velocities will be successfully corrected by a proportional controller. However, in order to correct a continuous disturbance, as might be caused by different friction forces in the bearings (e.g. due to an unsymetric load disturbance on the vehicle), an integral action is required as well. The PI controller is assumed to have a transfer function of the form:

$$C(z) = \frac{c_1 + c_2 z}{z - 1} \qquad (3.2)$$

where c_1 and c_2 are parameters to be determined.

The PI controller introduces a zero (in closed transfer function) that can be chosen to reduce the maximum percent overshoot of closed loop feedback. A feedforward compensation can be used to place the zero of the closed loop in the way to reduce the maximum percent overshoot.

The closed loop transfer funtion is given by:

$$F(z) = b \frac{(k_a + c_2)z + c_1 - k_a}{z^2 + (bc_2 - a - 1)z + a + bc_1} \qquad (3.3)$$

To determine the parameters c_1, c_2 and k_a, we have placed the poles and the zero of the system to a specified locations. These locations must satisfy the previous specifications.

For some well known considerations, it is often preferable to specify the desired characteristics in terms of continuous time features and to convert the root locations in the s domain into corresponding root locations in the z domain.

In the s-plane, the poles of the second order are given by:

$$s_{1,2} = -\zeta \omega_n \pm j \omega_n \sqrt{1 - \zeta^2} \qquad (3.4)$$

The associated poles in the z-plane are:

$$z_{1,2} = e^{-\zeta \omega_n T} e^{\pm j \omega_n T \sqrt{1 - \zeta^2}} \qquad (3.5)$$

The closed loop zero is supposed to be located at z_0.

The parameter c_1, c_2 and k_a associated to the given poles z_1 and z_2, and the zero z_0, are given by the following relationships:

$$c_1 = \frac{z_1 z_2 - a}{b} \tag{3.6}$$

$$c_2 = \frac{1 + a - (z_1 + z_2)}{b} \tag{3.7}$$

$$k_a = \frac{c_1 + c_2 z_0}{1 - z_0} \tag{3.8}$$

Note that there exists other techniques setting controller, which require more computations than the one used here. The reader can find more details in Goodwin and Sin [1984], Astrom and Wittenmark [1984] or Puthenpura and Macgregor [1987]. The block diagram of system control is represented by the Fig. 3.1.

4. SIMULATION

The autonomous vehicle controller has been simulated. The simulation program is written in C language and run on an IBM PC. To compare the efficiency of the proposed control algorithm, the autonomous vehicle was programmed to travel along a figure eight path. The same path has been used by Borenstein and Koren [1987] and Nelson and Cox [1988]. The path, as recorded by the control system is shown in Fig. 4.1.

The DC motors are supposed to have the same parameters in this simulation. These parameters are estimated and their values are $a = 0.82$ and $b = 0.2$.

The poles in the s-plane was chosen to be located at $s_{1,2} = -\zeta\omega_n \pm \zeta\omega_n\sqrt{1 - \zeta^2}$, with $\zeta = 0.707$ and $\omega_n = 10$

The zero z_0 was chosen by simulation and a value of z_0, which gives a good performance is $z_0 = -0.972$

The controller settings used provides a good transient response, and a nul steady stade error for a step input (in the deterministic case).

The simulation results based on the previous algorithm and the chosen control settings are illustrated by figures 4.2-4.8. These results compare favorably to the results of a similar experiment described in Borenstein and Koren [1987] and the references herein, but with different motion control algorithm.

5. CONCLUSION

The aim of this paper has been to drive an autonomous vehicle along a given trajectory. Under the assumption that the used model of the servomotor (left and right) is invariant, we have proposed a simple algorithm control, based on the classic control theory. The simulation results show the precision of the proposed approach in the deterministic case.

In reality, during the operation of the autonomous vehicle, some parameters may vary as the mechanical load varies. These variations will introduce errors in the model and consequently in the tracking trajectory. To avoid this kind of problem, the adaptive control can be employed to impose the system performance and help to optimize the control behavior over a wide range of parameters change (associated to disturbances). A similar approach was used by Boukas [1989].

6. REFERENCES

Goodwin, G. C., and Sin K. S., "Adaptive Filtering Prediction and Control", Englewood Cliffs, N. J.: Prentice Hall, 1984.

Astrom, K. J. and Wittenmark, B., "Computer Controlled Systems", Englewood Cliffs, N. J. Prentice Hall, 1984.

Borenstein, J. and Koren, Y., "Motion Control Analysis of a Mobile Robot", ASME, Journal of Dynamics Systems, Mesurements, and Control, vol. 109, june 1987, pp. 73-79.

Nelson, W. L. and I. J. Cox, "Local Path Control for An Autonomous Vehicle", Proc. 1988 IEEE, Int. Conf. on Robotics and automation, Philadelphia, PA.

Cox, I. J., 1988, "Blanche: An autonomous Robot Vehicle for Structured Environments", Proc. 1988 IEEE, Int. Conf. on Robotics and automation, Philadelphia, PA.

Wilfong, G. T., 1988, "Motion Planning for an Autonomous Vehicule", Proc. 1988, IEEE, Int. Conf. on Robotics and Automation, Philadelphia, PA.

Tsumura, T., 1986, "Survey of Automated Guided Vehicle in Japanese Factory", Proc. IEEE Int. Conf. on Robotics and Automation, San Francisco, CA, Vol. II, pp. 1329.

Rembold, U., "The Karlsruhe Autonomous Mobile Assembly Robot", Proc. IEEE Int. Conf. on Robotics and Automation, Philadelphia, PA.

Julliere, M., L. Marce, H. Perrichot, "A Guidance System For a Vehicle Which has to Follow a Memorized Path", Proc. of the 2nd Int. Conf. on Automated Guided Vehicle System, Stuttgart, W. Germany, 1983.

Campion, G., G. Bastin, D. Rollin, B. Raucent, "External Linearization Control for a Mobile Robot", IFAC, SYROCO 88 to appear.

Puthenpura, S. C. and J. F. Macgregor, "Pole-Zero Placement Controllers and Self tuning regulators With Better Tracking", IEE, Proc. Part D, Control Theory and Applications, Vol. 134, No. 1, January 1987, pp. 26-30.

Boukas, E. K., 1989, "Motion of an Autonomous vehicle: An Adaptive Approach", Submitted for publication.

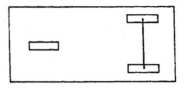

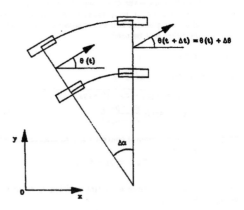

Fig. 2.1

Fig. 2.2

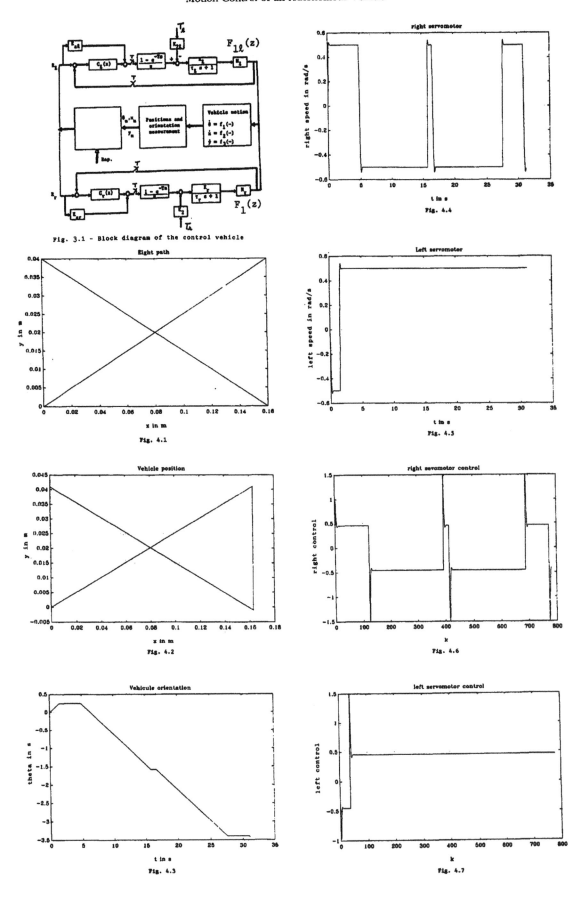

Fig. 3.1 - Block diagram of the control vehicle

Fig. 4.1 — Eight path

Fig. 4.2 — Vehicle position

Fig. 4.3 — Vehicule orientation

Fig. 4.4 — right servomotor

Fig. 4.5 — Left servomotor

Fig. 4.6 — right sevomotor control

Fig. 4.7 — left servomotor control

A RECURSIVE ALGORITHM FOR PATH PLANNING BETWEEN MONOTONE CHAINS

D. Lodares and M. Abellanas

Facultad de Informática, Universidad Politécnica, Madrid, Spain

Abstract. In this paper one gives a recursive algorithm to obtain the minimum length path connecting two points and avoiding collisions with obstacles which are monotone chains of rectilinear segments. The algorithm is shown to be $O(N^2)$ in time in the worst case and the results obtained after the implementation show that the expected case is $O(NlogN)$-time.

Key words: Obstacle Avoidance ; Path Planning ; Recursive Algorithms ; Robots ; Computational Methods ; Optimization ; Computational Geometry.

1. INTRODUCTION.

In this paper the following well known problem is consider: Given a robot and a set of obstacles, find a path, if any, to move the robot from a position A to a position B avoiding collisions with obstacles and with minimum length.

This kind of problems were treated by Wangdahl, Pollock and Woodward (1974) and by Lozano-Perez and Wesley (1979). The proposed solutions are based in the use of the so called visibility graph and, considering that the visibility graph with N nodes has, in general, $\Omega(N^2)$ edges, it results that $\Omega(N^2)$ is a lower bound for the computation time, being N the total amount of vertices of the obstacles.

In this paper it's given a solution to the following problem: Find the minimum length path between two points in the plane avoiding collisions with two chains which are monotone chains with respect to the same straigth line. (See Fig. 1).

The obtained method requires an $O(N^2)$ worst case time, and it's a recursive algorithm whose implementation is easy to perform.

The results obtained in practice with a random data generator shows that the expected case requires $O(NlogN)$-time.

Lee and Preparata (1984) also deal with some particular cases of the path planning problem in which the execution time can be improved using the fact that the total visibility graph is not required. These cases are: Find the minimum path between two interior points of a simple polygon, and find the minimum path between two points in the plane avoiding collisions with N disjoint and paralel line segments. In both cases they obtain $O(NlogN)$ time algorithms.

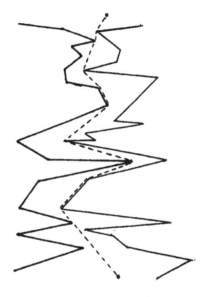

Fig. 1. Minimum path.

In a next paper (Lodares and Abellanas, 1989) we show that the worst case time can be improve for our problem to $O(N)$-time.

First of all we will enunciate precisely the problem. Secondarily we will give the algorithm and then we will prove the correctness and the worst case analysis.

2. STATEMENT OF THE PROBLEM.

In that paragraph we establish the problem and give the key properties that allow us to design our correct algorithm to solve it.

Problem P: Given two points in the plane A and B such that $y(A) > y(B)$ and given two disjoints chains of segments, monotones with respect to the y-axis

$$C_1 : v_1, ..., v_n \quad \text{and} \quad C_2 : u_1, ..., u_m \quad (n + m = N)$$

and such that $y(v_1) = y(A) = y(u_1)$, $y(v_n) = y(B) = y(u_m)$, $x(v_1) < x(A) < x(u_1)$ and $x(v_n) < x(B) < x(u_m)$, find a minimal length path between A and B avoiding collisions with the two chains.

We will define two new problems P_1 and P_2 and we will prove that the initial problem P is linear transformable to P_1 or P_2. Finally we will give recursive algorithms to solve such problems.

Problem P_1: It's the same as problem P but being $A = v_1$ and $B = v_n$.

Problem P_2: It's the same as problem P but being $A = v_1$ and $B = u_m$.

Proposition 1: The problem P is linear transformable to problems P_1 or P_2.

Proof: First we will prove that the input of problem P is linear transformable into an input for problem P_1 or P_2 and then, that the outputs of problems P_1 and P_2 can be transformed in constant time into an output for problem P. The second condition is obvious. For the first one we distinguish the following three cases:

a) If the line-segment $[A, B]$ does not cut C_1 and C_2, then we take the segment $[A, B]$ as chain C_1 and C_2 the same as given. It is clear that the solution to problem P_1 with these inputs give us the solution to problem P (that is the line segment $[A, B]$). The mentioned transformation requires a linear time to test the intersections between the segment $[A, B]$ and the chains C_1 and C_2.(see Lodares (1988)).

b) If the line segment $[A, B]$ cuts only one of the chains, for example C_1, let p and q be the first and the last intersection points. We consider the chain whose first segment is the line-segment $[A, p]$, whose last segment is the line-segment $[q, B]$ and that coincides with the given C_1 between p and q as chain C_1 for problem P_1. We consider chain C_2 as it was given. Now it's also clear that the solution for problem P_1 with these inputs give us the solution for problem P and that the required time for the transformation is dominated by the obtainment of p and q (this require a linear time; see Lodares again).

c) If the segment $[A, B]$ cuts both chains let p and q the first and last intersection points. If p and q belongs to the same chain we are in an analogous situation as in case b). If , for example, p belongs to C_1 and q belongs to C_2 we consider chain C_1 for problem P_2 the one whose first segment is $[A, p]$ and coincides with the given C_1 between p and v_n and chain C_2 the one which coincides with the given C_2 between u_1 and q and whose last segment is $[q, B]$. Once again the solution for problem P_2 with this input give us the solution for problem P and the transformation is dominated by the obtainment of the intersection points p and q as well.

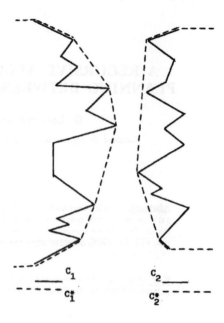

Fig. 2. The convex hulls.

Definition: Let C_1 and C_2 be two chains in the conditions of problem P. Let C_1' be the chain obtained by the joint to C_1 of the horizontal half-straigth lines with origin in the extreme points of C_1 towards the negative sense of x-axis, and let C_2' be the chain obtained by the joint to C_2 of the horizontal half-straigth lines with origin in the extreme points of C_2 towards the positive sense of the x-axis. We will call convex hulls of C_1 and C_2, and we will denote C_1^* and C_2^*, to the convex hull of C_1' and C_2' respectively. (See Fig. 2).

The next lemma is a key observation for the proof of the next proposition.

Lemma: Let C_1 and C_2 be two chains as in problem P_1 and a and b two consecutive vertices of C_1^*. One verifies that every point z of the path solution of the problem being $y(a) \geq y(z) \geq y(b)$ is on the left of segment $[a, b]$.

Proposition 2: The path solution to problem P_1 contains necessarily all the vertices of C_1^*.

Proof: It is clear that the path solution to problem P_1 contains all vertices of C_1^* that don't belong to C_2^*. Let b be a vertex of C_1^* belonging to C_2^* and let a and c the vertices of C_1 so that a, b, c are consecutive vertices in C_1^*; let p (resp. q) be the vertex of the path solution with minimum y-coordinate (resp. maximum) and such that $y(p) > y(b)$ (resp. $y(q) < y(b)$). By the preceding lemma point p is on the left of segment $[a, b]$ and q is on the left of segment $[b, c]$, therefore the minimum length path between p and q contains point b.

Definition: Let C_1 and C_2 be two chains in the conditions of problem P. For every two consecutive vertices a and b in C_1^* (resp. C_2^*) we will denote $A[(a, b)]$ the set of vertices v in C_2 (resp. C_1) such that $y(a) > y(v) > y(b)$ and $\det.(a, v, b) \geq 0$ (resp. $\det.(a, v, b) \leq 0$).

Proposition 3: For every two consecutive vertices a,b of C_1^*, the path solution to problem P_1 contains the vertex v of $A[(a, b)]$ of maximum distance to the segment $[a, b]$.

Proof: The result is clear when $[a, b]$ do not cut C_2. If $[a, b]$ cuts C_2, let p and q be the first and the last intersection points (p is different of q since there are at least two intersection points). Let C_3 be the chain whose vertices are $a, p, v_k, ..., v_{k+h}, q, b$ where v_i are the vertices of C_2 that are between p and q. Let C_4 be the chain whose vertices are $u, u_j, ..., u_{j+h}, w$ where u is the midpoint of the segment $[u_{j-1}, u_j]$ of C_1 such that $u_{j-1} = a$ and w is the midpoint of the segment $[u_{j+h}, u_{j+h+1}]$ of C_1 such that $u_{j+h+1} = b$. (See Fig. 3).
Now, we consider the problem P_1 where $C_1 = C_3$, $C_2 = C_4$, $B = a$ and $A = b$. As v is a vertex of C_3^*, and applying the proposition 2 to these inputs, we have that the path solution contains point v.

3. THE ALGORITHMS.

In the proof of proposition 1 we gave the procedure to transform problem P in one of the two problems P_1 or P_2. Therefore, to give algorithms to solve the last two problems it's only necessary.

Algorithm 1.
Input: The two chains C_1 and C_2 with n and m vertices respectively $(n + m) = N$ in the conditions of problem 1.

Step 1. Obtain the two lists L_1 and L_2 with the vertices of C_1^* and C_2^*.

Step 2. Merge the two lists L_1 and L_2 in a new list L_3.

Step 3. For every pair of consecutive vertices a and b in C_1^* consider the vertex v of L_3 being

$$y(a) > y(v) > y(b)$$

such that

$$det(a, v, b) \leq 0$$

and whose distance to the line-segment $[a, b]$ is maximum.

Step 4. Let L be the list obtained inserting all the vertices v obtained in step 3 in L_1.

Step 5. If $L = L_1$ this is the solution.

Step 6. If $L \neq L_1$ obtain the minimum path between every pair of consecutive vertices of L applying algorithm 1 or algorithm 2 (if both vertices belong to the same chain or not respectively).

Algorithm 2.
Input: The two chains C_1 and C_2 with n and m vertices respectively $(n + m) = N$ in the conditions of problem 2.

Step 1. Obtain C_1^* and C_2^*.

Step 2. If C_1^* do not intersect C_2^* then:

2a. Let u_1 and w_1 be the points in C_1 which belong to the common supporting lines to C_1^* and C_2^*, such that $y(w_1) > y(u_1)$ and let w_2 the vertex of C_2^* which belong to the supporting line that contain w_1.

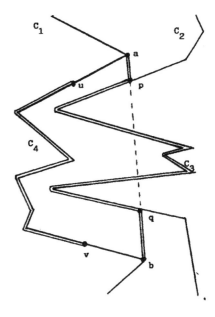

c_1 c_2

Fig. 3.

2b. The solution is given by chain C_1^* up to w_1 concatenated with segment $[w_1, w_2]$ and this one concatenated with chain C_2^* from w_2 to his end.

Step 3. If C_1^* and C_2^* do intersect then:

3a. Let p and q be the two points of intersection of their boundaries $(y(p) > y(q))$ and let z_1 be the last vertex in C_1^* with $y(z_1) > y(p)$ and let z_2 be the first vertex of C_2^* with $y(z_2) < y(q)$.

3b. For every pair (a, b) of vertices of C_1 which are consecutive in C_1^* obtain the vertex v belonging to $A[(a, b)]$ of maximum distance to the line-segment $[a, b]$. Do the same for C_2.

3c. Let $L : z_1, v_1, \cdots, v_k, z_2$ be a list where the v_i are the vertices obtained in 3b sorted by the y-direction.

3d. Apply algorithm 1 or algorithm 2 to every pair of consecutive vertices of L to obtain the minimum length path between these.

Step 4. The solution is obtained concatenating the part of the chain C_1^* from his origin to the point z_1 with the paths obtained in 3d and with the part of the chain C_2^* from z_2 to his end.

Analysis of the algorithms:

Both algorithms have a main loop which is procesed at most $O(N)$ times because in each of them is obtained at least one vertex of the solution which has at most $O(N)$ vertices.

As each step is linear in time, the total amount of time required by the algorithms in the worst case is $O(N^2)$.

One show (Fig. 4) one of the worst cases which never has been obtained in practice.

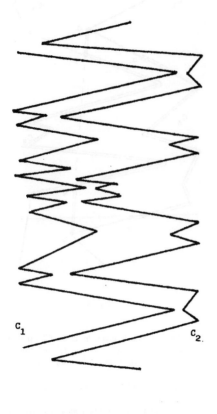

Fig. 4. The worst case.

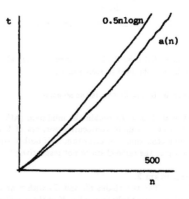

t(n)= time of data generation

d(n)= depth of recursion

Expected case analysis:

The previous algorithm has been implemented by J.Elices in C language completed with a random data generator. The generator give us two chains in the problem conditions whose vertices are uniformely distributed along the direction of monoticity.

The results obtained in practice with this generator are shown in Fig. 5.

CONCLUSION.

A recursive algorithm is obtained which is easy to perform and which runs in $O(N^2)$-time in the worst case. The practice after implementing shows that the expected time is $O(NlogN)$.

The problem P proposed in this paper appears in many situations in robotic path planning. One of the most usefull application appears in automatic cutting planning in textile industries.

The worst case time for this problem can be improved as we mentioned earlier to $O(N)$ time as we show (Lodares and Abellanas 1989) using a diferent non recursive technique.

a(n)= algorithm's time

Fig. 5. Results of practical analysis.

REFERENCES:

Lee,D.T. and F.P.Preparata(1984). Euclidean Shortest Paths in Presence of Rectilinear Barriers.Networks, Vol.14, 393-410.

Lodares,D.(1988). Un algoritmo óptimo para el cálculo de la intersección de cadenas monótonas.Actas de las XIII Jornadas Hispano-Lusas de Matemáticas. Valladolid, 1988.

Lodares,D.and M.Abellanas(1989). An optimal linear—time algorithm for path planning between monotone chains. Preprint, Dep. Mat. Apl. Facultad de Informática. Univ. Politécnica de Madrid.

Lozano—Perez,T.and M.A.Wesley(1979). An Algorithm for Planning Collision—free Paths Among Polihedral Obstacles.Comm. ACM 22 560-570.

Wangdalh,G.E.,S.M.Pollock,and J.B.Woodward (1974) Minimum trajector pipe routing. J. Ship Res. 18. 46-49.

INTERPOLATION ALGORITHMS FOR GENERAL NON-PARAMETRIC AND PARAMETRIC CURVES BY MATHEMATICAL PROGRAMMING

D. K. Kiritsis* and S. G. Papaioannou**

*Laboratoire de Conception Assistée par Ordinateur, Ecole Polytechnique Federale de
Lausanne, Lausanne, Switzerland
**Department of Mechanical Engineering, University of Patras, Patras, Greece

Abstract. Two new algorithms are proposed for generating implicitly
and parametrically defined curves. They differ from known algorithms
in the formulation and solution of the step selection problem. Unlike previous algorithms which rely on Boolean formulations, step selection is formulated as a simple integer programming problem, whose
solution can be obtained by inspection. This formulation is advantageous since it leads to eight-point interpolation algorithms, allows
the introduction of optimizing criteria and is easily extended to parametric curves. This last advantage makes the algorithms particularly useful for CAD applications.

Keywords. CAD; CAM; interpolation; mathematical programming; numerical control.

INTRODUCTION

This paper deals with the generation of non-parametric and parametric curves, using incremental steps
along fixed coordinate axes. This is important in
diverse areas as numerical control, digital plotters or graphic displays. Until now, interpolation
algorithms exist only for non-parametric curves.

Non-parametric curves can be generated by orthogonal (four-point) or non-orthogonal (eight-point)
interpolation algorithms.

Danielson [1] developed an orthogonal interpolation
algorithm using Boolean logic. This algorithm is
simple and general and selects among 4 possible
steps x_+, x_-, y_+, y_- using the criteria:

1) the step advances in the direction of interpolation, and 2) it points towards the curve.

Jordan et al [2], developed an eight-point algorithm for non-parametric curves which selects among
8 possible steps x_+, x_-, y_+, y_-, x_+y_+, x_+y_-, x_-y_-,
x_-y_+ using the criteria:

The step advances in the direction of interpolation, and it leads to the point with the minimum
distance from the curve.

To fulfil the second criterion, this algorithm
"looks ahead" one step and calculates the distances
at all possible steps that fulfil the first criterion. The points generated by this algorithm are
very close to the original curve since there is the
ability to step in any of the eight possible directions. The algorithm proposed here for generating
curves is simple and general and uses a simple mathematical programming method.

It is an eight-point interpolation algorithm, so,
smooth and well-defined curves are obtained,
slightly different from those obtained by the Jordan's et al algorithm.

We extended this algorithm to parametric curves and

obtained the same benefits. This means that there
exist now a simple and general algorithm for parametric curves useful in the area of CAD where most
commonly used curves are parametric.

EIGHT - POINT INTERPOLATION BY MATHEMATICAL PROGRAMMING

Let the curve be defined by an equation of the
form

$$f(x,y) = 0 \qquad (1)$$

and let P_i be the current position of the interpolation point defined by the position vector $\overline{P}_i$
(Fig. 1).

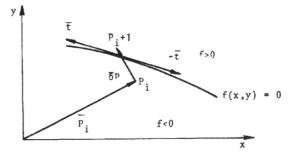

Fig. 1. Non-parametric interpolation

When the interpolating point advances to its next
position P_{i+1}, the vector $\overline{\delta p} = [\delta x, \delta y]$ describes
this advance.

Let

$$\overline{t} = [\frac{\partial f}{\partial y}, -\frac{\partial f}{\partial x}] \qquad (2)$$

be the tangent vector expressing the direction of interpolation. The algorithm is based on the following two conditions:

1. A step is a vector quantity $\overline{\delta p} = [\ \delta x,\ \delta y]$ so, the advance it produces in the direction of interpolation, is given by the inner product

$$\overline{t}.\overline{\delta p} = \frac{\partial f}{\partial y}\ \delta x - \frac{\partial f}{\partial x}\ \delta y \qquad (3)$$

This product must be positive, since the step must be consistent with the direction of interpolation.

Thus, the first condition is

$$\overline{t}.\overline{\delta p} = \frac{\partial f}{\partial y}\ \delta x - \frac{\partial f}{\partial x}\ \delta y > 0 \qquad (4)$$

2. The step must point towards the curve. We can use the value of $f(x_{pi},\ y_{pi})$ at any given point P_i as an indicator of proximity to the curve, and express the proximity indicating function $f(x,y)$ along the step $[\delta x,\ \delta y]$ as a Taylor series, assuming that, in the neighborhood of P_i, $f(x,y)$ does not possess extrema or saddle points.

Thus

$$f(x+\delta x, y+\delta y) - f(x,y) = \frac{\partial f}{\partial x}\ \delta x + \frac{\partial f}{\partial y}\ \delta y + \ldots \quad (5)$$

If all higher order terms are neglected we obtain:

$$f(x+\delta x,\ y+\delta y)-f(x,y) = \frac{\partial f}{\partial x}\ \delta x + \frac{\partial f}{\partial y}\ \delta y \qquad (6)$$

or

$$\delta f = \frac{\partial f}{\partial x}\ \delta x + \frac{\partial f}{\partial y}\ \delta y \qquad (7)$$

For a step to point towards the curve, it must drive the value of the function $f(x,y)$ towards 0. This occurs when $f(x,y)>0$ and $\delta f<0$ or when $f(x,y)<0$ and $\delta f>0$.

Thus, the second condition is

$$\begin{array}{llll} \text{or} & \text{if} \quad f > 0 & \text{then} & \delta f < 0 \\ & \text{if} \quad f < 0 & \text{then} & \delta f > 0 \end{array} \qquad (8)$$

A step is selected from the simultaneous fulfilment of the above two conditions (4) and (8).

Since these conditions are satisfied by more than one steps, we introduce a measure of optimization by requiring that $\overline{t}.\overline{\delta p}$ be maximized.

So, the next step is selected by solving the problem:

$$\begin{array}{ll} \text{maximize} & \overline{t}.\overline{\delta p} \\ \text{subject to} & f.\delta f < 0 \end{array} \qquad (9)$$

for interpolation in $\overline{t}$. If the direction of interpolation is $-\overline{t}$, $\overline{t}.\overline{\delta p}$ is minimized.

The optimal step gives the maximum advance along the curve and the distance of each position from the curve does not exceed the step size.

THE ALGORITHM

We may assume, without loss of generality, that the step size in each direction of movement is equal to 1. The step vector then is

$$\overline{\delta p} = [\ \delta x,\ \delta y\ ] \qquad (10)$$

where δx, δy take the discrete values -1,0,1.

The tangent vector is given by eq. (2) and the change δf in f by eq. (7).

Using the notation f_x for $\frac{\partial f}{\partial x}$ and f_y for $\frac{\partial f}{\partial y}$, the step selection problem (9) becomes:

$$\begin{array}{ll} \text{maximize} & f_y.\delta x - f_x.\delta y \\ \text{subject to} & \delta f = f_x.\delta x + f_y.\delta y \gtrless 0 \end{array} \qquad (11)$$

δx, δy take the discrete values -1,0,1

The inequality sign is taken to be > if at the current position $f<0$ and < otherwise. Observe that if the coefficients of one of the two variables in the objective function and the constraint are of the same sign, then the coefficients of the other will be of opposite sign.

Because of it, the optimal value of one of the two variables δx, δy is obtained directly depending on the sign of the inequality, and the value of the other by inspection. We give two examples to explain the method.

Example 1: Let the problem be:

$$\begin{array}{ll} \text{maximize} & f_y.\delta x - f_x.\delta y \\ \text{subject to} & f_x.\delta x + f_y.\delta y > 0 \end{array}$$

δx, δy take the discrete values -1,0,1

and f_x, $f_y > 0$

Since the coefficients of δx are positive and the sign of the inequality is >, the optimal value of δx must be 1. This is because any other value of δx will either violate the constraint or reduce the objective function or both. Then, since the coefficient of δy in the objective function is negative, δy must assume the smallest value which does not violate the constraint. If $|f_x| > |f_y|$, this value is -1, otherwise, it must be 0.

Example 2: Let the problem be:

$$\begin{array}{ll} \text{maximize} & f_x.\delta x - f_x.\delta y \\ \text{subject to} & f_x.\delta x + f_y.\delta y < 0 \end{array}$$

δx, δy take the discrete values -1,0,1

and $f_x > 0$, $f_y < 0$

In this case, we must have $\delta x = -1$, since the coefficients of δx in the objective function is negative and the sign of the inequality is <.

Any other value will again reduce the objective function and possibly violate the constraint. Then, if $|f_x| > |f_y|$ $\delta y = -1$, otherwise $\delta y = 0$.

The results of this analysis are summarized in Table 1, in which F,FX,FY,D are Boolean variables defined as follows:

$$\begin{array}{llll} F = 1 & \text{if and only if} & f>0 \\ FX = 1 & " & f_x>0 \\ FY = 1 & " & f_y>0 \\ D = 1 & " & |f_x|-|f_y|>0 \end{array}$$

In the case when the curve is traversed in the direction of $-\overline{t}$, the objective function must be minimized rather than maximized. Applying similar reasoning we obtain the results given in Table 2.

EXTENSION TO PARAMETRIC CURVES

Parametric curves are defined by

$$x = u(\phi)$$
$$y = v(\phi) \qquad (12)$$

We take the square of the distance function

$$f = d^2 = (x-u)^2 + (y-v)^2 \qquad (13)$$

as a proximity measure, since distance is a non-negative quantity.

For a fixed point p(x,y) the distance function f goes through a stationary point when the vector PN becomes normal to the curve (Fig. 2).

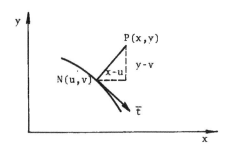

Fig. 2. Parametric interpolation.

Thus, the normality condition is

$$\frac{1}{2}\frac{df}{d\phi} = (x-u)u'+(y-v)v' = 0 \qquad (14)$$

At each step, ϕ is updated to satisfy condition (14).

The effect of a step $\overline{\delta p} = [\delta x, \delta y]$ on f is

$$\delta f = \frac{\partial f}{\partial x}\delta x + \frac{\partial f}{\partial y}\delta y \qquad (15)$$

Through the normality condition, ϕ is a function of x and y and we can compute the partial derivatives of (13) as follows:

$$\frac{\partial f}{\partial x} = 2(x-u)(1-u'\frac{\partial\phi}{\partial x})+2(y-v)(-v'\frac{\partial\phi}{\partial x})[(x-u)u'+$$
$$(y-v)v']$$

$$= 2(x-u)-2\frac{\partial\phi}{\partial x}[(x-u)u'+(y-v)v']$$

$$= 2(x-u)$$

Similarly

$$\frac{\partial f}{\partial y} = 2(y-v)$$

Thus,

$$\delta f = 2(x-u)\delta x+2(y-v)\delta y \qquad (16)$$

For interpolation in $\overline{t}$, the step selection problem is formulated as follows

maximize $\overline{t}.\overline{\delta p} = u'.\delta x+v'.\delta y$

subject to $(x-u).\delta x+(y-v).\delta y < 0 \qquad (17)$

δx, δy take the discrete values -1,0,1

Starting from the general condition (9), we see that the sign of the inequality is fixed in this case, since, by definition, f is always non-negative.

Due to the normality condition (14), the coefficients u', v', (x-u), (y-v) cannot be all of the same sign. Thus, the problem can be solved according to the same rules illustrated in example 2. The optimal solutions are listed in Table 3. The algorithm uses Table 3 to select the next step and then updates ϕ to maintain the normality condition. This is done by solving equation (14), using Newton's iterative formula, which in this case becomes

$$\phi_{i+1} = \phi_i + \frac{(x-u)u'+(y-v)v'}{(x-u)u''+(y-v)v''-u'^2-v'^2}$$

RESULTS

A circle and an ellipse generated by this algorithm are shown in Figure 3. The same curves generated by the Jordan's et al algorithm are shown in Figure 4 and by the Danielson's algorithm in Figure 5.

Although, this algorithm generates non-symmetric curves (in case the original curve is symmetric), it retains the advantages of an eight-point algorithm. It generates smoother curves and requires fewer steps than a four-point algorithm. The required number of the steps is generally equal or slightly more than in the Jordan's et al algorithm.

As we can see in the following examples (Fig. 3,4), this slight difference between the two algorithms arises at the points where the proximity function f(x,y) equals 0. Then, the algorithm gives the value 1 to the variable F, since then f>0.

If we separate the case f=0, there will be no difference in the number of steps between the two algorithms.

Nonetheless, we retain the case f>0 in order to use only binary variables, which facilitates a hardware implementation of our algorithm. Specifically, since the algorithm depends only on the signs of the values of the proximity function f(x,y) and its first derivatives and since a sign can be only positive or negative, we can consider these variables as binary variables which take only the value 1 or 0. Consequently the step selection decision as given in Table 1 for non-parametric curves can be implemented by the circuit of Fig. 6 while the step selection rules of Table 3 for parametric curves lead to the circuit of Fig. 7.

CONCLUSION

In this paper, two high-precision eight-point interpolation algorithms have been proposed applied for both non-parametric or non parametric continuous curves. The mathematical programming formulation has been used which has the advantage of simplicity and accuracy as well.

The potential of direct interpolation of general parametric curves makes these algorithms particularly useful for CAD/CAM and NC applications (e.g. drafting on plotters or "cutting" on NC machine tools of free form curves).

TABLE 1   Truth-table for non parametric interpolation (direction $\vec{t}$)

F	FX	FY	D	δx	δy	F	FX	FY	D	δx	δy
1	1	1	1	0	-1	0	1	1	1	1	-1
1	1	1	0	1	-1	0	1	1	0	1	0
1	1	0	1	-1	-1	0	1	0	1	0	-1
1	1	0	0	-1	0	0	1	0	0	-1	-1
1	0	1	1	1	1	0	0	1	1	0	1
1	0	1	0	1	0	0	0	1	0	1	1
1	0	0	1	0	1	0	0	0	1	-1	1
1	0	0	0	-1	1	0	0	0	0	-1	0

TABLE 2   Truth-table for non-parametric interpolation (direction $-\vec{t}$)

F	FX	FY	D	δx	δy	F	FX	FY	D	δx	δy
1	1	1	1	-1	1	0	1	1	1	0	1
1	1	1	0	-1	0	0	1	1	0	-1	1
1	1	0	1	0	1	0	1	0	1	1	1
1	1	0	0	1	1	0	1	0	0	1	0
1	0	1	1	0	-1	0	0	1	1	-1	-1
1	0	1	0	-1	-1	0	0	1	0	-1	0
1	0	0	1	1	-1	0	0	0	1	0	-1
1	0	0	0	1	0	0	0	0	0	1	-1

TABLE 3   Truth-table for parametric interpolation

x-u	y-v	D*	u'	v'	δx	δy
0	1	1	1	1	1	1
0	1	0	1	1	1	0
1	1	1	1	0	0	-1
1	1	0	1	0	1	-1
1	1	1	0	1	-1	1
1	1	0	0	1	-1	0
0	1	1	0	0	0	-1
0	1	0	0	0	-1	-1
1	0	1	1	1	0	1
1	0	0	1	1	1	1
0	0	1	1	0	1	-1
0	0	0	1	0	1	0
0	0	1	0	1	0	1
0	0	0	0	1	-1	1
1	0	1	0	0	-1	-1
1	0	0	0	0	-1	0

*D = ABS(x-u)-ABS(y-v)

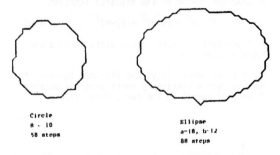

Fig. 3.   Curves generated by Math-Pro algorithms

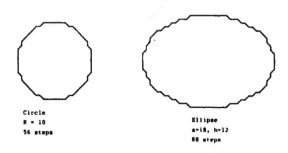

Fig. 4.   Curves generated by Jordan's et al algorithm

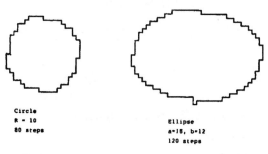

Fig. 5   Curves generated by Danielson's algorithm

REFERENCES

Danielson, P.E. (1970). Incremental Curve Generation. IEEE Trans. on Comp., vol. C-19, pp. 783-793.
Jordan, B.W., Lennon, W.J. and Holm, B.D. (1973). An Improved Algorithm for the Generation of Nonparametric Curves. IEEE Trans. on Comp., vol. C-22, pp. 1052-1060.
Papaioannou, S.G. (1979). Interpolation Algorithms for Numerical Control. Computers in Industry, vol. 1, pp. 22-40.

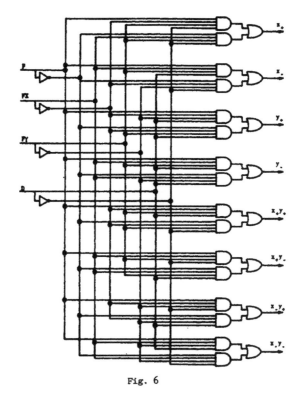

Fig. 6

Circuit for non-parametric curves

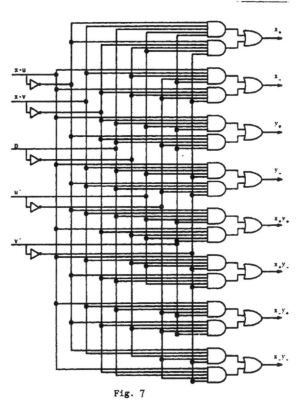

Fig. 7

Circuit for parametric curves

Fig. 6

Circuit for down-sample input

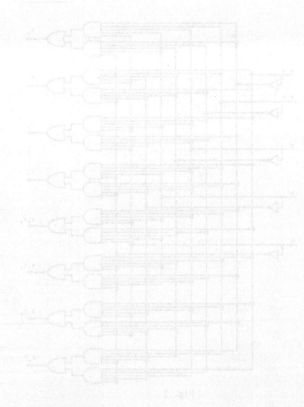

Fig. 7

Circuit for sample-rate output

RETROFITTING A CNC MACHINING CENTER WITH A MAGNETIC SPINDLE FOR TOOL PATH ERROR CONTROL

M. Anjanappa, D. K. Anand, J. A. Kirk, E. Zivi and
M. Woytowitz

Department of Mechanical Engineering, University of Maryland, College Park, USA

Abstract. The use of magnetic bearing spindle to machine thin rib components, such as microwave guides, provides the capability to minimize tool path errors along with the benefits of high speed machining. Tool path errors, defined as the distance-difference between the required and actual tool path, can be controlled using the unique features of magnetic bearing spindle. This paper deals with the control and interfacing issues involved in retrofitting a vertical machining center with an S2M-B25/500 magnetic spindle. The resulting facility, one of a kind in the United States, has been successfully tested by generating [i.e., cutting] a surface whose profile follows a control input signal.

Key words. CNC, Controllers, Cutting, Error compensation, Magnetic suspension

INTRODUCTION

Retrofitting a magnetic spindle to an existing CNC machining center provides several unique benefits unavailable in conventional machining centers (MTTF,1980). The current research project at the university of Maryland focusses on exploiting the following features of a magnetic bearing spindle in end milling operations;

. *The ability to translate and tilt the spindle rotor [thereby moving the tool attached to it] within the air gap, with appropriate user interface.*

. *Built-in three dimensional force and position sensors.*

. *High rotational speeds [30,000 revolutions per minute for S2M B25/500 spindle] of the spindle.*

Currently, the thrust of the overall project is to design and develop an on-line error minimization controller to increase the productivity of end milling of thin rib components. This controller, which is currently under development stage, when completed would use the cutting force information to translate and tilt the spindle rotor thereby compensating for tool path error.

To accomplish this, first of all, a CNC machine fitted with a magnetic spindle is needed. In addition, such a system must provide an user interface whereby the user can tap into the current status and be able to command the translation and tilt of the spindle rotor on-line in real time.

Currently, to the best of the authors knowledge, MFL Machine Tool Inc., is the only builder marketing a machining center fitted with a magnetic spindle. However, these machines do not provide the interface for translation and or tilting of the spindle rotor. Considering the situation, the decision was made to

retrofit an S2M-B25/500 magnetic spindle to an existing Matsuura MC500 CNC machining center at the University of Maryland. This paper presents the following research work not necessarily in the same order;

. *The retrofitting of Matsuura MC500 CNC machining center with the S2M B25/500 magnetic spindle.*

. *Interfacing and control of the spindle for safety and use in conjunction with the Yasnac controller of the machining center.*

. *Experimental validation of the interface to translate and tilt the spindle rotor.*

. *Development of the error minimization controller.*

Following is a brief discussion of the tool path error and the magnetic spindle.

TOOL PATH ERROR

The accuracy and surface finish of a machined part is a function of tool path error which is defined as the distance difference between the actual and the required tool path. Tool path error is classified, as shown in Fig.1, according to the source and the nature of the error as cutting force-independent and cutting force-dependent errors.

Static deterministic *position errors* are those repeatable errors which are a function of machine slide position. The cause for these errors are geometric inaccuracies of the slideways and the misalignment in the structural element assemblies. Dynamic deterministic *position errors* are those reproducible errors which are a function of the table feed rate. *Thermal Deformations* due to heat sources, both internal and external to machine tool system, are reproducible errors which causes a change in the required position of the tool relative to the workpiece. Thermal cycles of the spindle system, ambient temperature variation and friction are some

---

[1] University of Maryland-UMBC, Baltimore, MD 21228

example of the heat sources. *Weight deformations* are caused by changes in the weight of stationary objects which are firmly positioned on the machine tool table. These errors show up as reproducible static position errors and occur in addition to position errors.

	CUTTING FORCE-INDEPENDENT ERRORS			CUTTING FORCE-DEPENDENT ERRORS	
TYPE OF ERROR	POSITION ERROR	THERMAL DEFORM-ATION	WEIGHT DEFORM-ATION	CUTTING FORCE DEFORMATION	
NATURE OF ERROR	DETERMINISTIC			DETERMINISTIC	STOCHASTIC
FUNCTIONAL DEPENDENCY	POSITION FEEDRATE	TEMPERATURE	MASS	COMPLIANCE	DEPTH OF CUT PROCESS DYNAMICS

Fig. 1   Tool path error classification

The *cutting force deformations* are classified as (i) deterministic tool path errors due to compliance between the tool and the workpiece and (ii) stochastic tool path errors due to cutting process dynamics. The errors due to cutting force show up as position differences between the required and actual tool path relative to the workpiece and results in workpiece shapes which are not perfect. In summary, all the deterministic errors, both static and dynamic, are those errors that reoccur when an identical set of input parameters exist on a given machine tool structure. Stochastic errors, on the other hand, are defined as those errors which occur when a random input is presented to the machine tool. All these errors are discussed in detail by (Anand,1986; Anand,1987; Anjanappa,1987; Anjanappa,1988; Kirk,1987; Tlusty,1980).

MAGNETIC SPINDLE

The magnetic spindle currently available for use on machine tools was developed and built by Societe de Mecanique Magnetique (S2M) of France. Magnetic Bearings Inc. (MBI) is currently distributing the S2M spindle in the United States. At present there are three models of magnetic spindles available for milling purposes. These 3 models cover the maximum speed range between 30,000, and 60,000 rpm with a rated horsepower between 20 and 34 (Field,Harvey and Kahles,1982).

A magnetic spindle consist of a hollow shaft supported by contactless, active radial and thrust magnetic bearings, as is shown in Fig.2 (SKF,1981). In operation, the spindle shaft is magnetically suspended with no mechanical contact with the spindle housing. Position sensors placed around the shaft continuously monitor the displacement of shaft in three orthogonal directions. The sensor information is processed by a control unit and any variation in the position of the shaft are corrected by varying the current level in electro-magnetic coils, thereby forcing the spindle shaft to its original position. The magnetically floating spindle shaft can be rotated freely about its mass center even if the mass center deviates from the geometric axis.

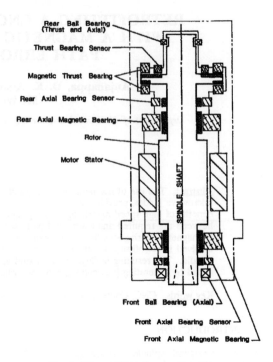

Fig. 2   Magnetic Bearing configuration

Conventional ball bearing (called touchdown bearing) are also provided on both ends of the spindle for supporting the shaft when the spindle is stopped and for serving as the touchdown bearings in case of a power failure.

There are numerous advantages in using magnetic spindles over conventional spindles and these are discussed in detail in (Anand,Kirk and Anjanappa,1986). Several investigators (Nimphius,1984; Raj Aggarwal,1984; Schultz,1984) have used magnetic spindles by retrofitting them on existing machine tools. Their primary focus was to use the magnetic bearing spindle to improve metal removal rate. In the present work it is suggested that the above two advantages of using a magnetically controlled spindle [to improve tool path errors] can take precedence over the advantage of high metal removal rate. Because of this magnetic bearing spindles will be useful for retrofitting on existing machine tools for tool path error minimization.

MECHANICAL INTERFACING

Mechanical retrofitting work required several modifications to the original machining center. It required disassembling the conventional spindle head from the machining center, cut off the existing spindle casting to leaving only the guide ways intact and then mount the new magnetic spindle in its place.

To maintain a weight balance between the original spindle head and the new spindle head all of the original components of the spindle head were cataloged and weighed. Before the spindle head could be removed, it was necessary to disconnect and remove the automatic tool changer and the spindle motor.

Moreover, all control lines to the original spindle drive system were reconfigured to allow machine operation once the spindle was removed.

After the spindle head was disassembled, the casting was cut to required size and the mounting surface was finish machined to accept a riser block. The riser block was necessary to configure the new spindle axis to its original position.the magnetic spindle. The riser block was designed to serve as an interface between the head and the new spindle. To preserve the rigidity of the head, some of the spindle housing was left intact, and the riser was designed to conform around the remaining portion of the housing.

The riser was designed to provide a rigid and accurate mounting surface for the new spindle. The major design considerations were rigidity and the preservation of the original tool position. The final design of the riser resembled an open box with the bottom side serving as the magnetic spindle's mounting surface and the open end enclosing the original spindle housing. To provide maximum rigidity it was decided to make the spacer from one piece of   stock. Availability and machinability dictated the use of 6061-T6 aluminum as the spacer material.

After the riser was completed, the spindle head was drilled and tapped and assembled with the riser. The spindle mounting surface was face milled flat relative to the slide ways of the head. To minimize the new spindle's alignment error, the spindle mounting bolt pattern was also aligned with the same slide ways.

The new head/riser assembly was then weighed. The careful design of the spacer produced a negligible five pound difference between the original spindle head and the modified magnetic spindle head. If it were to become necessary, it poses no problem since the counter weight could be appropriately adjusted.

The new spindle head was then assembled and all the lubrication lines were reestablished. Alignment tests were conducted using Hewlett Packard 5528 laser measurement system and the new riser block-magnetic bearing assembly was appropriately adjusted and pinned.

ELECTRICAL INTERFACING AND CONTROL

Implementation of the active magnetic bearing (AMB) error correction methodology involves the interaction and coordination of four independent controllers:

. *Existing CNC controller,*
. *Active magnetic bearing controller,*
. *Variable speed spindle drive,*
. *Online error minimization controller.*

While the existing Matsuura MC500 Yasnac CNC 3000G controller was retained, the latter three controllers were installed as part of the active magnetic bearing retrofit.

Derivation and implementation of the overall controller coordination scheme has accounted for the largest portion of the AMB retrofit. The functional requirements can be summarized as:

1. *Providing the operational control necessary to operate the CNC mill, Implementation of safety interlock measures,*

2. *Interfacing necessary for communication of real-time process monitoring and control data,*

3. *Coordination necessary to implement the error correction scheme.*

The nature and satisfaction of these requirements will be briefly reviewed in the following paragraphs.

The first objective, of the AMB retrofit process, was to provide the operational control necessary to support conventional milling functions. This effort involved the coordination of the CNC, AMB, and spindle drive controllers. This phase, which is now complete, allows for the operation of CNC, AMB, and spindle functions from the suitably modified CNC operator console. In addition, monitoring of AMB and spindle status, spindle speed, and spindle torque information has been provided on the operator console.

For CNC operation of a 25KW, 30,000 rpm active magnetic bearing spindle, safety is a critical concern. Although each of the CNC, AMB, and spindle drive controllers provide internal safety features, considerable effort has been invested into integration of various fault, interlock and error handling procedures. Levitation of the AMB system has been interlocked to satisfactory CNC operational status. Similarly, spindle rotation is interlocked to satisfactory AMB status. The S2M AMB controller provides a comprehensive set of monitoring functions including cooling, excessive spindle bearing load or displacement, and spindle drive faults, as reported by the spindle drive unit. AMB controller faults, in turn, invoke dynamic braking of spindle rotation and assert a table stop request signal. Table stop request has been implemented such that X, Y, and Z axis motion is frozen and a *SPINDLE ERROR* message displayed on the operator's display console

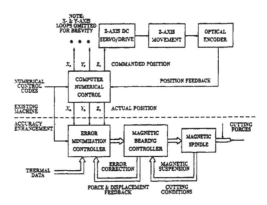

Fig. 3 Prototype multivariable control structure

CRT. Both the spindle fault and table stop request conditions must be reset by the operator, before operation is allowed to resume. Multiple spindle stop push bottoms, as well as the CNC emergency stop, are available to stop spindle rotation.

Central to this research, is the implementation of the error minimization controller. The basic elements of the accuracy enhancement retrofit are depicted below the dotted line of the simplified control system block diagram given in Fig.3. As indicated, various online process parameters are provided to the error minimization controller as inputs to the control algorithm. Based on these inputs, the error minimization control will use the predetermined error characterization to generate perturbational control signals to translate and tilt the AMB spindle to correct for dimensional errors. The theoretical development necessary to support the derivation of the robust, multivariable controller has been previously reported (Kirk and colleagues,1987). The input parameters currently under investigation include:

- *X, Y, Z axis position,*
- *X, Y, Z axis velocity commands,*
- *X, Y, Z axis servo lag,*
- *Spindle displacement and bearing forces,*
- *Thermal conditions.*

Extraction of real time X, Y, Z axis position, and potentially servo lag, data from the Yasnac 3000G CNC controller has particularly problematic. This information was not directly obtainable and very little documentation was available to facilitate its extraction. Since the desired information can be displayed on the CNC operator console CRT, various hardware and software debugging methods were applied to locate internal position data which could be extracted and provided to the error minimization controller. After several weeks of monitoring embedded microprocessor activity with a logic analyzer, a memory mapped, micron resolution, position interface was discovered within the CNC controller internals. After a complete disassembly of the embedded microprocessor read-only memory (ROM), several surgical ROM modifications were implemented to provide position at a predictable (8ms) update rate. A digital hardware position interface, modelled as a finite state machine, has been developed to capture, buffer and transmit the axis position data to the error minimization controller. The remaining error minimization controller inputs are comparatively straight forward analog signals. Due to the high levels of electromagnetic interference (EMI) and potential for ground loops, all digital signals are opto-isolated and analog signals differentially buffered.

Implementation of the error compensation methodology also requires mechanisms to coordinate and sequence CNC and error minimization controller operations. The present approach involves the down loading of the part program, in M & G codes, to both the CNC and error minimization controllers. In this scenario, the CNC and error minimization controller execute the same part program simultaneously. This allows the error minimization controller to cooperate with the existing CNC controller and completely defines the desired tool trajectory. Existing CNC support for optional M-codes M21 through M28 is currently being interfaced to the error minimization controller to provide sequencing and handshaking functions.

EXPERIMENTAL VALIDATION

In order to validate the error correction using the magnetic spindle a simple end milling operation was designed. In this approach a straight surface was machined [i.e. end milled] to half the length with no external control input. During the machining of the latter half portion, a sinusoidal bias voltage input was given to the spindle controller interface unit. The resulting machined surface had a straight surface for the half length and the remaining surface was sinusoidal. The frequency of the surface was same as the control input frequency as was predicted. The surfaces of both halves showed the same degree of surface roughness. This preliminary experiment proved that the spindle rotor [in turn the tool] can be both translated and tilted in a controlled manner at required frequencies.

FUTURE WORK

With the experimental facility fully operational, remaining efforts involve implementation of the actual error minimization methodology. These activities include:

- *Completion of CNC machine dynamics identification,*
- *Control system implementation and validation.*

Accurate identification of the AMB spindle and CNC machine dynamics is crucial to the development of the dynamical system models required for the control system design process. The spindle calibration and identification process is essentially complete. Identification of the CNC machine dynamics will proceed pending completion of the previously described real-time axis position interface.

Control system modelling is currently underway using a VAX based computer-aided control system design (CACSD) environment composed principally of the ACSL, MATLAB and MACSYMA software packages. This environment supports the robust, multivariable control design methodologies which will be applied, once the system identification phase is complete. Trial error minimization controller designs will also be evaluated, in this environment, using mixed continuous and discrete simulation, prior to microprocessor implementation.

The error minimization controller implementation will be performed using an IEEE 796 (Multibus) based computer system. This system utilizes an Intel 386/387 processor pair and executes the Intel RMX II operating system. High performance interface boards will be incorporated to ensure adequate real-time performance. In addition, the Datel ST-701 analog input board has been hardware and firmware modified to provide sufficient throughput. This error minimization computer system, currently being assembled, is similar to a system presently providing high speed system identification data acquisition services.

Evaluation of the error minimization capabilities of the magnetic spindle retrofitted Matsuura MC500 will be performed by comparing the dimensional accuracy of a sequence of sample parts. This sequence will use the MC500, with no error correction, as the baseline with which to evaluate various error map formulations and control system implementations. Dimensional accuracy will also be compared to a part milled on a conventional Matsuura MC510 vertical CNC mill.

ACKNOWLEDGEMENT

The research work reported in this paper represents a cooperative activity of the personnel from the University of Maryland, the National Institute of Standards and Technology [formerly the National Bureau of Standards], the Westinghouse Corporation, the David Taylor Naval Research Center and the Magnetic Bearing, Inc. This work has been supported by the National Science Foundation through grant NSF 8516218 the Engineering Research Center at the University of Maryland and ONR Program Element 61152N through the David Taylor Research Center.

REFERENCES

Anand, D.K., J.A. Kirk and M. Anjanappa(1986). Magnetic Bearing Spindles for Enhancing Tool Path Accuracy. *Advanced Manufacturing Processes*, Vol.1, No.1, pp.121-134.

Anand, D.K., J.A. Kirk, and M. Anjanappa(1987). Tool Path Error Control for End Milling of Microwave Guides. *Proceedings of the 7th World Congress on the Theory of Machines and Mechanisms*, Vol.3, pp.1499-1502.

Anjanappa, M., J.A. Kirk, and D.K. Anand(1987). Tool Path Error Control in Thin Rib Machining. *Proceedings of 15TH NAMRC*, pp. 485-492.

Anjanappa, M., D.K. Anand, J.A. Kirk, and S. Shyam(1988). Error Correction Methodologies and Control Strategies for Numerical Controlled Machining. *Control Methods for Manufacturing Processes*, DSC Vol. 9, pp. 41-49.

Field, M., S.M. Harvey, J.R. Kahles(1982). High Speed Machining Update, 1982: Production Experiences in the U.S.A.. *Metcut Research Associates Inc.*, Cincinnati, Ohio, USA.

Kirk, J.A., M. Anjanappa, D.K. Anand, and S. Shyam(1987). Accuracy Enhancement Methodologies in Thin Rib Machining. *Proceedings of the 14th NSF Conference on Manufacturing Research and Technology*, pp.9-14.

MTTF(1980). Technology of Machine Tools. *Machine Tool Task Force Report*, Vol.1-5.

Nimphius, J.J.(1984). A New Machine Tool Specially, Designed for Ultra High Speed Machining of Aluminum Alloys. *High Speed Machining*, pp. 321-328.

Raj Aggarwal, T.(1984). Research in Practical Aspects of High Speed Milling of Aluminum. *Cincinnati Milacron Technical Report*.

Schultz, H.(1984). High-Speed Milling of Aluminum Alloys. *High Speed Machining*, pp.241-244.

SKF (1981). Active Magnetic Bearing Spindle Systems for Machine Tools. *SKF Technology Services Report*.

Tlusty, J.(1980). Criteria for Static and Dynamic Stiffness of Structures, *Machine Tool Tssk Force Report*, Vol.3, Section 8.5.

ON A METHOD OF CUTTING ARBITRARY PLANE SHAPE BY USING A SMALL DRILL AND A PERSONAL COMPUTER

M. Itoh

Department of Mechanical Engineering, Ashikaga Institute of Technology, Tochigi, Japan

Abstract.   This paper deals with the brief description of an apparatus and its computer programs which are able to cut arbitrary plane shape at a job site in real time.   The apparatus is composed of a small training drill unit, a personal computer, a tablet (an electromagnetically scanning board) and a x-y plotter.   The values of the input signal to the personal computer are obtained by moving a contact-pen from a point to a point on the drawing of a work placed on the tablet.   The computer programs are written in BASIC.   Although the drill unit used in this study is a small training one, this system will be applicable to other machine tools including milling machines.

Keywords.   Personal computer;   drill;   man-machine interface;   interpolation;   tablet.

INTRODUCTION

Numerically controlled machine tools were first developed about thirty seven years ago. Since then, APT(Automatically Programmed Tools) and EXAPT(Extended subset-of APT) program languages have been developed and used by a number of manufactures and users of numerically controlled machine tools. For large numerically controlled machine tools, off-line programs are still used. However, in consonance with the progress of mini and micro-computers, small-sized numerically controlled machine tools with a built-in mini or micro-computer are now manufactured. In this paper, a trial apparatus which is able to cut arbitrary plane shape at a job site in real time is described. The apparatus is composed of a small training drill unit, a personal computer, a tablet and a x-y plotter.

OUTLINE OF THE APPARATUS

The block diagram of the apparatus is shown in Fig. 1. In the figure, the notations of D, C, T, and P indicate the drill unit, the personal computer, the tablet and the x-y plotter, respectively. The personal computer is com-

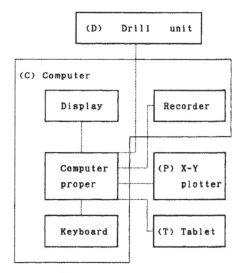

Fig. 1. Block diagram of the apparatus

posed of a display, a computer proper, keyboard and a recorder. The dotted lines in the figure show the electric cords to connect each of the instruments. It is desireable that C, T and P are set on the table of an operator. On the other hand, the drill unit D is usually set up at a place of some distance apart from the operator table. The manufactureres and the outlines of the instruments are shown in Table 1. The schematic diagram of the drill unit is shown in Fig. 2. In the figure, MX and MY denote the direct current motors to move the x and y axes screws of the table, respectively. The notation SD denotes the electromagnetic solenoid coil to move the drill downwards.

GX and GY are the gears to move the x and y axes screws of the table. DCX and DCY are the digital counters to detect the numbers of revolution of the screws of MX and MY. The displacement of x and y axes of the table are controlld by comparing the desired pulse numbers with the interrupted numbers of the photo sensors. The pulse numbers corresponding to the displacements of x or y displacement of the table are compared with those of the diggital computer.

PROGRAM OF THE PERSONAL COMPUTER

The computer program developed by the Pacific Co., Ltd., Japan was modified and used in this study. The flow chart of the program is shown in Fig. 5. In the figure, X(L,M,N) and Y(L,M,N) are the locations of the x and y axes of the table. These values of X(L,M,N), Y(L,M,N) correspond exactly to the values of the locations X,Y of the x-y plotter. The notations of L,M, N are parameters in the computer program. For example, if we should like to make a work like the shape of a leaf shown in thick line in Figure 6, first we draw the diagram of the work in thick line on a drawing paper. Then we draw a thin line which is outer than the thick line by

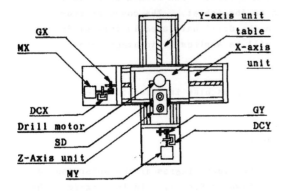

Fig. 2 Schematic diagram of the drill unit

Table 1 Manufactureres and outlines of each instrument

Instrument	Type	Manufacturer	Outline
Drill unit	PZ-AD1	Pacific Industrial Co., Ltd., Japan	Stroke : x-axis 170 mm y-axis 80 mm Feed rate : 100 pulse / mm

Personal computer
  Computer proper
  Display
  Keyboard
  Recorder
	
	PC-9801VX
	PC-KD871
	PC-9801V
	PC-PR201
		Nippon Electric Co., Ltd., Japan	
			CPU : 16 bit Microprocessor
			Size: 14 inch, colour display
			
			Speed:150char./sec,Pinhead:24

Tablet	DT-3203	Seiko Electronic Co., Ltd., Japan	Range of measurement :
			        x-axis 311 mm
			        y-amis 210 mm

X-Y Proyer	MP3200	GRAPHTEC Corp., Japan	Range of drawing : x-axis 404 mm, y-axis 285 mm

the amount of the radius of the fraise cutter.   The maximum value of L is the number of singular points of the drawing
shown in Fig. 6.   In the case of Fig. 6,
the value of L is equal to 12 including
the original point where L=1. In this case
the distance between the point L=2 and the
point L=12 is 74 mm, and the distance between the point L=1 and the point L=7 is
23.5 mm in length, respectively.   The
notation M means the number of arbitrary
data points from the L=1 point to the L=2
point in Fig. 6. In Fig. 6, we select the
value of M to be 5 including original point.

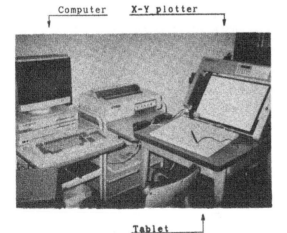

Fig. 3   Photo of the drill unit

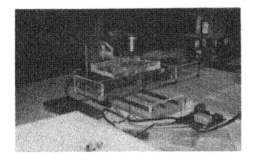

Fig. 4   Photo of the apparatus

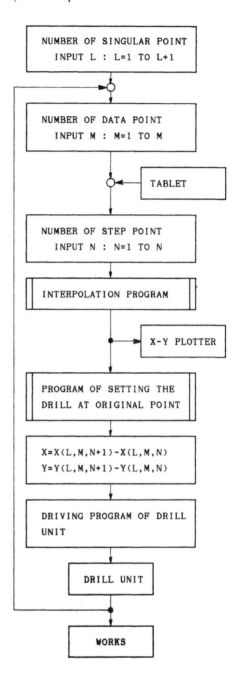

Fig. 5   Flow chart of the program

The notation N means the number of arbitrary step point between the L=1 and the
L=2 points.In the case of Fig.6, the values
of 50 to 100 are suitable for N.   Scale-
up and scale-down factors are included in
the program of the x-y plotter.   Then we
can see the amplified or reduced drawing
of the work. Therefor, we can easily make
the amplified or reduced work of the original size.

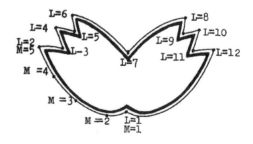

Fig. 6   Drawing of a work

PROCEDURE FOR OPERATION

First we place a paper,on which the shape
of a leaf is written on the tablet. Next,
we hold the contact-pen of the tablet and
set it at the original point, and mark the
point by it.   Subsquently, we mark the
selected other four points between the L=1
point and the L=2 point by the pen as shown
in Fig. 6. When we put the pen at the
final point of M, namely M equals 5, the
personal computer starts the Lagrange or
the Spline interpolation computation, and
the x-y ploter moves and draws the curve
between the point L=1 and the point L=2,
automatically. After we conform the curve
on the x-y plotter is as good as designed,
we put the return key on the keyboard of
the personal computer. Then the drill of
the drill unit begins to work moving along
the curve from the point L=1 to the point
L=2 of the work attached on the table of
the drill unit. In the same way, we mark
the selected points between L=2 point and
L=3 point by the contact-pen and so on.
Wooden material was used as experimental
works because the torque of the motors
used in the training drill unit was small.
In this training drill unit, D.C. motors
are used as servomotors, and both motors
of x and y axes move in series at any time,
the table of the drill unit moves slightly
in zigzags even the number of N is large.
In this manner, we can cut an arbitrary
plane shape model in only several or a few
tens of minutes in real time after starting
to draw the model.   An experimental result
is shown in Fig. 7.

CONCLUSIONS

It has been shown that arbitrary plane
shape works have been made in real time
at a job site using a small training
drill, a personal computer, a tablet and
a x-y plotter. The computer program was
written in BASIC except the part which
controls the movement of the drill unit,
where machine language was used. Although
the drill unit used in this study is a
small training one, this system will be
also applicable to other machine tools
such as milling machine tools or so.

ACKNOWLEDGMENTS

The auther wishes to acknowledge useful
advice given by the peoples in the
Pacific Industrial Co., Ltd. and the
Seiko Electronic Co., Ltd., and also
expresses his gratitude to Mr. H. Akikusa
for his assistance and cooperation in
this study.

REFERENCES

Servomechanism Laboratory, Massachusetts
     Institute of Technology. (1953) .
     A Numerically Controlld Milling Ma-
     chine. Final Report to the U.S. Air
     Force on construction and Initial
     Operation.

Hirose,T.,Horiki,Y.,Teramoto,T.,Takenaka,
     T.and Yamaguci,K.(1976). NC Machining
     of Turbine Rotor by Using EXAPT-2 Sys
     tem. in Japnese.MITSUBISHI JUKO GIHO,
     Vol.13, pp.967-972.

The Pacific Industrial Co.,Ltd. (1985).
     Training of the Mechanical Control by
     the Personal Computer PC-9801. in
     Japanese. Nikkan Kogyo Shinbun, Ltd.

Fig. 7 Photo of an experimental result

DYNAMIC DATA STRUCTURES FOR MANUFACTURING INFORMATION

R. R. Lamacraft

CSIRO Division of Manufacturing Technology, Woodville, Australia

Abstract: A memory-resident database management system for the dynamic construction, manipulation and display of directed acyclic graphs has been developed, to provide support for computer assisted design tools in facility planning and manufacturing management. The data structures permit access to information along any meaningful linked path. In the manufacturing context, these links provide the process-planning, resource allocation and time-ordering of events for decision support systems. Communication between an interactive design tool and the database is performed by message passing, while event queues provide synchronization between applications using the same database. This mechanism permits the automatic updating of windows of several design tools which are simultaneously active in a multi-window computing environment. Interactive software tools for designing manufacturing cells, balancing loads between cells and scheduling the operations of cells have been implemented using these concepts.

Keywords: Databases, dynamic structures, software tools, manufacturing cells, directed acyclic graphs, manufacturing information, decision support systems.

INTRODUCTION

Australian manufacturing industry is embracing the use of cellular manufacturing to increase its productivity, to improve the quality of its products and to make it price competitive in the international market place. The work described in this paper is part of the Australian Government's support for its manufacturing sector.

The facilities provided by the software tools described in this paper have made it possible to quickly examine numerous scenarios for cellularization of manufacturing plants. The general methodologies used by the project group have been described elsewhere (Wells *et al*, 1988; Wells 1989). Designs have been produced with predicted performance characteristics which have been achieved or bettered when implemented.

This paper examines a model for manufacturing information suitable for descision support systems. A general data structure is defined which can be extended and used recursively to represent all dynamic manufacturing information requirements. Criteria for database design are given and the use of the system with a suite of design tools is illustrated.

DATA CLASSES IN MANUFACTURING INFORMATION

When developing a model of the manufacturing operation it is necessary to identify fundamental data classes, the relationships between these classes, and the suite of operations that may be conducted on these classes and relationships.

The model used in this project identifies four fundamental data classes:

- *Resource* refers to machine, workcenter, tool, skilled staff, labour, transportation and other allocatable inputs
- *Step* refers to a logical point when resources are assigned to enable an operation to be performed
- *Task* refers to a sequence of one or more steps with a defined completion
- *Time* refers to the calendar of available time (eg number of shifts, length of shift, days available for work, etc)

Sometimes it is appropriate to collect together a group of items from a class and to label it as a compound-item in that same class. Sub-classes within a data class may also be dynamically created and destroyed as some items in a class achieve or lose a special status or position within that class as operations are performed on the class.

To complete the model six relationship classes also need to be defined:

- *Step-step* links which specify the order of performing steps in the completion of a task
- *Step-resource* links that collect together the

resources needed for a step

- *Resource-resource* links that specify tooling-machine and machine-operator constraints
- *Task-task* precedence links that represent the indented bill-of-materials for tasks that involve assemblies, associated tasks or prior completion of other tasks
- *Task-time* links that specify task completion date requirements
- *Resource-time* links that specify resource constraints for maintenance, critical scheduling paths, etc

DEFINING THE DIRECTED ACYCLIC GRAPH STRUCTURE

Any sequence of events within a manufacturing environment can be represented as a directed acyclic graph (DAG). A DAG (Figure 1) consists of *vertices*, A, B, C, and D, points at which events occur (eg. the step where two components are assembled to form one item) and *directed edges*, X, Y, and Z, paths from point to point (eg. the path a component travels as it is manufactured).

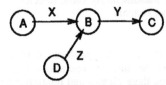

Figure 1: Simple Directed Acyclic Graph

It should be noted that common sub-components, use of common tooling, and multi-pod tooling, etc imply that these graphs may become a network rather than maintaining a simple hierarchical form (Figure 2).

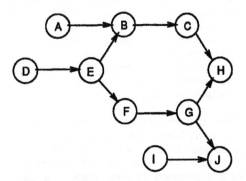

Figure 2: Directed Acyclic Graph Network

When representing manufacturing information by a DAG, it is necessary to be able to associate information with both vertices (eg. process setup and cycle times) and edges (eg. usage rates, transit times) of the graph.

The linked list structure (Figure 3) is a well-known method of implementing a list of elements which can be re-ordered, have elements inserted within it and deleted from it (Van Wyk, 1988).

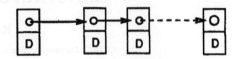

PARENT   LIST OF ELEMENTS

Figure 3: Singly-linked list

A linked list has a parent or head and a linked set of elements; data may be associated with the elements and/or with the parent. A doubly-linked list is one in which each element has pointers to both its pedecessor and successor, which means that it may be traversed efficiently in both directions from any point. The extra storage in the backward pointers may not be warranted in all applications. A pointer from the element to the parent may be used to provide rapid access to the parent from any element.

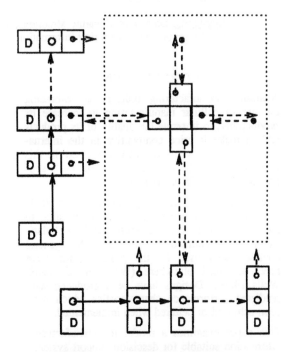

PARENT   LIST OF ELEMENTS

Figure 4: Two-dimensional linked list

A two-dimensional extension (Figure 4) of the linked-list structure where each element has two sets of links and two parents can be used to represent a rectangle of sparse elements each row and column of which has its own ordering, but where each element has a defined pair of parents. This principle may be extended to three or a higher number of dimensions to produce a class of structures denoted as hyper-lists (in the current implementation most cases are two dimensional with some cases involving time becoming three dimensional). The complexity of data linkages can be increased if each element of the structure may be the parent of a number of other list structures, and may also be a member of a number of linked-lists. The edge incidence matrix in the *step-step* links relationships is an example of the special case when the same parent can reference elements in both the row and column directions of the same sub-ordinate rectangular structure.

Each element in these data structures has a unique identification code, a type, and a status. Attached to each element is a linked-list of pointers to the first and last elements of the sub-ordinate linked-list for which this element is the common parent. Also each element has a linked-list of pointers to the data-blocks that contain the data of that element as managed by a particular application. The access, modification, and deletion of these data-blocks, and their supporting data structure elements, is controlled via the data base system kernel, according to the permissions set on the data-block by its creating application.

It is now possible to define a simplified form of structure to describe the DAG of the process-planning of a small set of parts (Figure 5) corresponding to the two tasks in Figure 1, T1 defined by steps A,B,C and T2 defined by step D, where A and B use resource R1, C and D use R2, and B also uses R3.

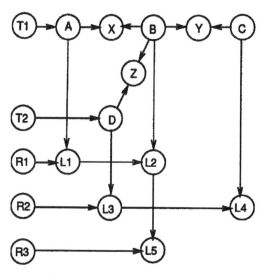

Figure 5: Example of combined data structures in manufacturing

In this schema, we commence with two uni-dimensional linked-lists, one for tasks, the other for resources. For each task in the task list there is a list of steps. The list of all steps (A, B, C, and D) and the resource list (R1, R2, and R3) become the parental lists for a rectangle of step-resource links (L1, L2, L3, L4 and L5). The list of all steps also becomes both the row and column parents to a rectangle of step-step links (X, Y, and Z) as illustrated in Table 1.

		Steps to			
Steps from	A	X			
	B		Y		
	C				
	D	Z			
Tasks	T1	A	B	C	
	T2				D
Resources	R1	L1	L2		
	R2			L4	L3
	R3		L5		

Table 1: Rectangles from Figure 5

DATA STRUCTURE OPERATIONS

The standard operations on linked-lists of inserting, deleting and reordering of elements are supported for these hyper-list structures. However, it is necessary to provide an additional set of primitive operations in order to support the DAG representation. The comparison of two linked chains of elements to identify the nature of their intersection, requires further elaboration. For some decisions it is necessary to be able to determine whether one is a super-graph of the other (considering only the vertex set, or including the edges). Hence this leads to the topological sorting under various constraints of a forest of directed acyclic graphs.

If the resources needed to satisfy a super-graph are grouped together within a manufacturing plant to form a cell, then all tasks whose graphs can be contained within that super-graph may be performed by that cell. When some of those graphs can be contained in several super-graphs those tasks may be arbitrarily allocated to those cells. Design tools are needed to resolve this allocation problem and optimize other criteria. For some operations, DAG's may need to be logically collapsed (eg. by ignoring some particular resources use, or by considering two or more resources to be equivalent, or by ignoring multiple visits to the same resource). Constraints on resources (eg. a particular resource cannot be moved in to a cell), or on the complexity of the graphs (eg. the cell formed is too large to be managed efficiently), may require that its defining DAG be cut in to several smaller ones. At each stage it is necessary that the topological sorting be re-evaluated to ensure that consistent information is available to the application tools.

CONCEPTS IN THE DYNAMIC SUPPORT OF DESIGN TOOLS

The database system is designed to dynamically support a suite of integrated design tools by considering the following factors:

- the database must be capable of supporting the information from real manufacturing environments, both in terms of problem size and complexity of constraints (capacity and capability)
- the database must be capable of self-maintenance to maintain internal consistency (propagation of consequential changes via internal procedure chains)
- the set of facilities within the database management support library should be an efficient, sufficient reduced set of primitives that spans the needs of the application tool (complete library)
- an application tool must not need to know the actual structure of the data storage (data-hiding and message-passing)
- the domain of access within the database of other application tools must be under the control the application tool that created the data (access control)
- the response time to a request for information, or the time to perform a global operation, must be short enough to permit effective use of interactive design tools (memory-resident versus disk-based)
- a common interface to the database for application tools to permit new application tools to be developed without interference with the operation of existing tools (modularity)
- the database must include a mechanism to invoke calls to application tools when identified events occur (event-notify queues)
- the whole environment must be able to be suspended, saved as disk files, then restored and resumed at a later time

The above concepts have been implemented as a memory-resident database management process in a virtual-memory computing environment (SUN / UNIX[†]). Each application runs as a separate process communicating to the database process through a bi-directional pipe. Each process makes use of the multi-window environment available on the SUN (Figure 6). Another similar problem was

---

[†]   SUN is a registered tradmark of Sun Microsystems, Inc., and UNIX is a registered trademark of AT&T Bell laboratories

addressed by Bregerson and Rochkind (1982).

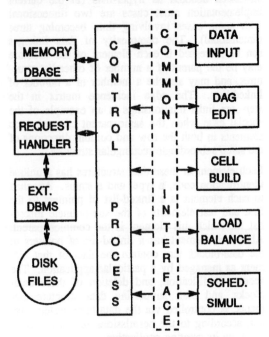

Figure 6: The dynamic database and its application tools

Each application tool is initiated from the control window of the database process and a pipeline between the database and the application processes is established (Rochkind 1985). Messages are exchanged along this pipeline. The application requests information from the existing data structures, and requests that new structures be created, modified or destroyed.

The request to the database process to create an entry in the notification queue associated with a specified data structure is critical for the cooperative use of the design tools. The existence of this entry is detected whenever the database management process modifies that data structure, causing an appropriate message to be sent to the creator of the entry, thus enabling it to invoke procedures to adjust its own state.

COMPUTER ASSISTED MANUFACTURING DESIGN TOOLS

The support of multiple design tools sharing the same database is illustrated by a system that:

- constructs the initial database from standard manufacturing information sources
- analyses the forest of graphs so formed
- builds a suite of manufacturing cells
- balances the loading on the processes between cells
- models these cells with a scheduler to

determine their performance characteristics

Database Construction and Edit

The modular design of the database interface permits the use of external software tools in UNIX to adapt data from manufacturer's own information systems into a form that can be input to the database construction phase.

As the design activity proceeds, the user returns to this phase to adjust the fundamental constraints on the database (eg. whether two resources are equivalent, whether the process-planning of a task is adjusted, etc).

Analysis of Directed Graphs

Topologically sorting and linking the data structures is computing intensive. Several data-storage models, comparison algorithms and updating strategies have been evaluated (the DAG comparison and manipulation tool was designed by R. R. Lamacraft). A typical problem size handled is 1500 tasks, utilizing 4200 steps, involving 150 resources and 900 task-task links, forming 210 distinct DAG, and where the largest DAG had 78 vertices, Profiling techniques enabled the first version execution of 50 minutes and 4.2Mbytes of virtual memory to be reduced to less than 4 minutes and 1.8Mbytes on a SUN 3/50. Further optimization has reduced this to about 90 seconds and 1.6Mbytes of virtual memory on a 8Mbyte SUN 3/60. Currently, the problems in extending the system to manipulate a manufacturer's production that involves 25,000 tasks and utilises about 400 resources are being investigated.

The support of the comparison of DAG's over the pipeline via messages has led to careful consideration to providing some higher level of the database manipulation facilities within its support library so as to minimize the message traffic. Initially, it was thought that progressive updating of the topological network as the database was being first constructed would be the most efficient method. However as the *task-task* relationship links and the *resource-resource* relationship links (which are often read very late in the standard manufacturing input), can considerably modify the relationships within the forest of DAG, many of the comparisons already performed become invalid. Maybe some further extension of the concepts to use petri nets would reduce this effect and make progressive updating more efficient.

Interactive Cell Design

With the topologically sorted DAG in the database, the cell design application tool (designed by D. H. Jarvis of this group) enables the designer to move groups of tasks around to form collections.

It computes loading on resources and identifies conflicts in the total resource inventory. Dependent on the strategy that the designer adopts for aggregation and division of collections, the final collections of resources are considered as candidate cells. Experience so far indicates that when different designers approach the same problem, some parts of the solution are independent of the designer's input, while other parts are distinctly different.

A version of the cell design tool was first built on top of a static database which had an incomplete knowledge of DAG manipulations. This proved inadequate for the following reasons. As the cell builder combined two DAGs, then various computations need to be done to maintain consistency in the database summaries. To do this the entire topological sorting of the DAG has to be scanned to determine whether the operation has reduced the available sites for performing the tasks, or created more alternative sites for other tasks. With a large number of tasks and complex combinations of resource usage it is not possible for a designer to predict the outcome without the dynamic re-compute and display facility.

Balancing Cell Loading

Assuming that the candidate cells have been defined, then there is usually a large number of less-complex tasks that may be assigned to one of a number of cells for execution. It is now necessary to distribute these tasks in the best way to statically balance the loading of these common resources, (an application tool to perform the automatic balancing of resource utilization was designed and implemented by A. T. Tharumarajah of the project group).

Scheduling of Cell Operations

Given the task load for a cell, the question becomes 'Can the cell be scheduled to perform all that is required of it in the available time, within the available number of operators, and maintaining adequate operator utilisation throughout the period?' (An application tool for cell scheduling was implemented by A. T. Tharumarajah, and is being continually extended to handle new scheduling heuristics and the more complex constraints that have arisen from the group's collaboration with Australian industry). Scheduling under contraints is computing intensive and needs rapid access to complex chains of information. Scheduling algorithms also need the ability to create and destroy local data and links within the database as the solution proceeds.

FURTHER EXTENSIONS

Facility Layout

Once a cell has been designed logically subject to the resource utilisation and scheduling constraints, then it is necesary to determine a physical layout for the cell. The dynamic database maintains the directed graphs for all the tasks assigned to each cell. By adding information about the physical space needed by each resource and the movement space needed for traffic between them, the physical space available, the position of services and the fixed elements, and the specification of a cost function for movement between resources, then the spatial layout of each cell may be optimized. (At the moment, this is being done by some third-party software, the generation of the input data for this software is time-consuming, and this software has a limited range of cost functions available.)

Process planning

The ability of a process planner to operate a design workstation that has access to the dynamic database of the manufacturing plant would permit the proposed process plans for new or revised products to be loaded on top of the current loads. The feasibility of the facility to accommodate the revision can be assessed, as well as as providing the ability to schedule the facility and to identify any new materials handling requirements. As an interactive design tool, it could lead to more balanced utilization of resources and could identify when a total re-evaluation of the current cell structure needs to be investigated.

Shop-floor monitoring

It would be ideal if information about the achievements and status of the real-world environment could be fed back to the database so that it could reflect the actual state of the manufacturing operation. This would enable the scheduling application tool to have an up-to-date basis on which to evaluate alternative "what-if" strategies proposed by the management. Integration of achievement reporting would enable statistics to be maintained on the setup/cycle/strip times used by the scheduler. Adaptive behaviour within the scheduler then could be enhanced.

Cooperative Scheduling

Where there are two or more manufacturing facilities whose tasks are related within the same database (eg. a number of manufacturing cells in a pressing shop, and the common blanking press line that supplies them), then the ability to schedule each cell separately, and then feed their combined requirements to the scheduler of the blanking line, would greatly assist in interactively tuning the total facility while balancing labour, maintenance, and output demands. Current experience indicates that this task has many interactions which would be well supported by the use of multi-window graphical displays to assist in the decision making.

SUMMARY

Experience with the dynamic database supporting an integrated suite of decision support tools operating in a multi-window workstation environment has shown that it greatly improves the productivity of the decision-makers, increases their understanding of the complex inter-relationships in the manufacturing plant and leads to a willingness to explore alternatives without disturbing the daily operations of the plant. Not all companies could afford this facility; but it is likely to be used by larger companies in-house, while its major use will most probably as a tool-set by manufacturing management consultants.

REFERENCES

Van Wyk, C. J. (1988) Data structures and C programs. Reading, Mass: Addison-Wesley
Bregerson, R. F, and Rochkind, M. J. (1982) Automated Repair Service Bureau: Software Tools and Components, *Bell System Technical Journal*, 61, no. 6 (July-August 1982), p. 1177-95.
Rochkind, M. K. (1985) Advanced UNIX programming. Englewood Cliffs: Prentice-Hall
Wells, A. J., Jarvis, D. H., Seabrook, T. D., and Lamacraft, R. R. (1988) Implementation of Cellular Manufacturing, *Fourth Int. Conf. Manuf. Engin.*, Brisbane, 11-13 May 1988, p. 190-193.
Wells, A. J. (1989) at INCOM89

ACKNOWLEDGEMENTS

Discussions with the members of the Integrated Manufacture Program of CSIRO (especially T. D. Seabrook, A. J. Wells, A. T. Tharumarajah, and D. H. Jarvis), and with various industrial collaborators has greatly assisted the effectiveness of the concepts to address real manufacturing issues.

DATABASE SUPPORT IN WORKCELL DESIGN

F. Prieto, F. Sastrón, A. Jiménez and E. A. Puente

Departamento de Automática, E.T.S.I.Industriales (DISAM), Madrid, Spain

Abstract. This paper deals with the information problems arising in manufacturing and design environments. These problems are caused by more complex data structures, more varied modes of user and tool interaction, and by the need to provide fast and flexible access means to the information.

The lack of tools to cope with these problems is one of the main reasons why integrated manufacturing is still a relatively far-fetched goal. Without trying to attain integration of the whole factory, information systems to support the communication and coordination of the activities used inside a department or engineering area, such as design, analysis, or software development, can be defined and implemented. One such system, used as a central database in a Flexible Assembly System design application, is presented in this paper. The system has been implemented and tested as part of ESPRIT Project 623, "Operational Control for Robot System Integration into CIM", demonstrating the feasibility of efficient information sharing and reliable data management.

Keywords. Database management systems; CIM; engineering databases; design information management.

INTRODUCTION

To achieve the goal of integration in manufacturing, an adequate information policy is needed. Three main requirements regarding overall factory operation avail the necessity of powerful information management systems:

1. Data must be shared among different factory functions, that present widely different information needs. An information interface is then necessary to perform this communication.
2. The overall state of the factory must be kept to allow activity coordination, resource assignment, and process monitoring.
3. The correction, validity and coherence of the information used in each factory function must be guaranteed, as well as the access means to this information.

The present lack of adequate information support has triggered research efforts to develop new Database Management Systems (DBMS) adapted to these information requirements: more complex, and more densely interrelated data need to be handled, different consistency management policies, new concurrency control mechanisms, and extended access capabilities must be provided. Systems offering these engineering-oriented capabilities are often termed Engineering DBMS (EDBMS).

In this paper a design database to support the activities in production engineering is presented. It has been developed as part of ESPRIT project no. 623, "Operational Control for Robot System Integration into CIM", where its task is to act as integrating kernel for the project's planning and programming utilities. The database provides overall information support, offers adequate information management capabilities, and allows the coordination of independently developed tools by means of standard exchange procedures.

In the following sections first the main concepts developed in the field of engineering databases are outlined, showing the main problems arising in their implementation; some related work on the subject is briefly discussed. Next, the way in which these concepts have been applied in the design of the manufacturing database is detailed, together with the additional constructs that have been implemented to facilitate operation and improve performance. Last the advantages of the designed system and its possible application to other engineering activities are briefly discussed.

ENGINEERING DATABASES

Among the main contributions in the research for EDBMS, the concepts of Complex Object -also called Aggregate or Molecular-, Version, and Long Transaction have been proposed as components of such systems. As a rather different approach, Persistence has been defined as one way to integrate databases with conventional programming languages. Another line of research involves extensions to the relational model to enable it to cope with more complex information. We now examine each concept in turn.

Object orientation is a means to encapsulate data as packages representing aspects of reality. These data appear outwardly as units that respond to predefined operations. This approach allows the relatively straightforward management of complex information and seems therefore well suited to support engineering applications. Coupled with generalization hierarchies and inheritance rules, object orientation can make up the basis of both data and knowledge management; it offers an adequate abstraction level and facilitates the maintenance of information consistency. It has been used as the basis of engineering databases in most of the recent research, including (Batory and Buchmann 1984; Lorie and Plouffe 1983; Stonebraker, Rubenstein and Guttman 1983; Kemper and Wallrath 1987; Ketabchi and Berzins 1988).

Design activities present two main features regarding information processing: first, the design process is stepwise; i.e., the final results are not obtained from the initial data in a single step, but following a predefined sequence of activities; second, the process is not straightforward, but iterative; this means that several alternatives have to be considered at each design phase to allow the process to reconsider previously discarded possibilities. The concept of version, as presented in (Neumann 1983; Katz and Lehman 1984; Dittrich and Lorie 1988) offers ways to maintain alternative designs, to handle incomplete information, and is useful to keep information about the general state of the project.

Classical DBMS enforce data consistency by means of the transaction mechanism. Since most engineering activities involve larger quantities of data and greater processing time than classical applications, a correspondingly more powerful tool must be devised to cope with this problem. The concept of extended transaction (Neumann and Hornung 1982; Lorie and Plouffe 1983; Ketabchi and Berzins 1987) has been introduced to avoid the excessive access restrictions imposed by the normally long processing time and to prevent the rollback mechanism from destroying large quantities of work if a deadlock or a system failure occur. Although actual implementations may differ, the idea is to separate the issues of recovery and consistency: using long transactions, the database does not reach a consistent state until the transaction completes, but exceptions no longer cause it to rollback to preserve the state previous to the transaction.

Persistence is a current research topic (Atkinson and Buneman 1987) whose aim is to allow variables in conventional programs to be permanent (i.e., existing beyond the scope of the program) and shareable (i.e., accessible from several programs). Clearly, the achievement of this goal would provide applications with information management mechanisms as powerful as programming languages are.

The relational model imposes the requirement that all elementary data be atomic; i.e., without structure. This is normally referred to as First Normal Form (1NF). The development of DBMS which accept data that do not comply with this requirement (so called Non-First Normal Form databases, or NF^2) is another research direction in the field of databases. Since relations need not be flat anymore, more complex data can be stored and, therefore, better information support is available. Research in this area includes the work of Dadam and co-workers (1986), Ozsoyoglu and Yuan (1987), and Roth and Korth (1987).

DATABASE IMPLEMENTATION

The work reported in this paper is the result of considering, on the one hand, those information requirements presented by engineering -and particularly, design activities, that are not met by conventional DBMSs. On the other hand, the actual needs of the applications to which the database was to give support and act as integrating element had to be considered: these include tools to aid the systems engineer in planning the operations, designing the layout, and programming the equipment of manufacturing workcells. Therefore, some of the work discussed here is of application only to this particular problem. Nonetheless, the authors believe that the methodology used in the database implementation is relevant in the general field of information systems for engineering.

The manufacturing system design process involves a number of steps in which the knowledge about the designed system is refined using auxiliary information -such as data about standards, technology, or available equipment- and human expertise (Fig. 1). Note that this functionality is shared by any design process. An information system to provide support to this kind of process must be able to:

1) Handle information at an abstraction level similar to the one used by the applications
2) Allow access to the needed information in a fast, reliable, and flexible way.
3) Provide the means to allow independent applications and/or users to share information, thus easing their integration (Fig. 2).

An information system with the above characteristics has been developed, and a version of it has been tested as part of a complete design system.

The database has been implemented using a commercial relational DBMS, enhanced by several software layers (Fig. 3) that are in charge of:

1) Accessing the database and performing conversion between relational and object level formats.
2) Managing the information consistency at the object level and checking user access.
3) Communicating with external applications and responding to data access requests.

The database has been configured so that access is only possible through the external interface, in this way ensuring that data consistency is maintained. To comply with the required information abstraction level, elementary data have been encapsulated as objects that are assigned unique, system-defined, identifiers. Objects are the information access unit; i.e., the minimum access operation involves a complete object.

Following Neumann (1983), the design information schema is configured as a directed acyclic graph (DAG) whose nodes represent different system aspects, called Design Levels. The arcs between nodes represent Design Actions; i.e., an arc linking design levels A and B represents the activity of refining the information about the system from one representation to another, more detailed, representation (for example, from the sequence of functions to be performed by a manufacturing system, lists of appropriate equipment can be selected: both Function Sequence and Proposed Device List are design levels; Equipment Pre-selection is a design action). See Fig. 4.

Data about a system at a certain development state is stored as a design object at the corresponding design level. Objects form the extension of the database. An object at a design level is related by derivation relationships to objects in all previous design levels. The design process interacts with the database by accessing existing objects and updating them or by using them to define new objects by means of design actions.

To overcome the transaction problem and to adapt the information system to the design activity, the version concept has been applied: versions are sets of objects representing portions of a system at some design level. Versions represent a complete manufacturing system at a design level. Thus, versions at the last design level hold (alternative descriptions of) the complete manufacturing system. As an example, consider Fig. 5: versions, and their associated objects, at four design levels have been represented. Versions V_{21} and V_{22} are design alternatives; i.e., each holds information about the whole manufacturing system. As can be seen, they share some information (object O_{21}). A designer willing to improve on the design at, say, design level #2 would either a) make modifications in an existing version by deleting, creating or updating design objects, or b) refine information in a version at design level #1, storing the results in a version -either existing or newly created- in level #2. These two procedures constitute the main ways of interaction between the design and information systems.

The version construct has been introduced for a number of reasons:

- Versions are used to hold project-related data (such as creation time, author, and development state). This is useful to separate technical and administrative information.

- The version hierarchy reflects the derivation relationships between system representations, allows the maintenance of alternative designs, and represents the development state of the overall project.

- Versions are the information consistency unit; i.e., after finishing work on a version, the database must be left on a consistent state (therefore, the period of time between the opening and closing of a version can be considered a long transaction, with any number of short transactions -object accesses- nested in it).

- Objects can be shared by several versions; thus, they can be accessed by several users at the same time without conflict.

CONCLUSIONS

The presented information system includes several major contributions that make it an improvement over present information systems and are general enough to be applied to other engineering areas. Among others, the following advantages can be pointed out:

- Information is structured as objects. This provides 1) an abstraction level similar to the one used by client applications, 2) efficient consistency management since operations on objects are predefined, 3) system-defined identifiers that avoid confusion and facilitate searches, and 4) an improvement in modularization that makes addition or modification of design tools easier.

- The information system is implemented as a version graph, each version holding system data at some design level, as well as project-oriented information. Versions are useful for 1) maintaining design alternatives and keeping track of relationships between design phases, 2) managing overall system consistency by preventing access conflicts not only on the versions being accessed, but on related ones as well, and enforcing orderly system updates, and 3) providing an extended transaction mechanism that does not demand excessive locking overhead, leaves a large amount of free resources, allows committing or aborting of all operations in the transaction and can be interrupted for any length of time to be resumed later.

- Long transactions, as needed by engineering applications, are provided by means of the version construct. This contributes to a more flexible operation.

- The interface to external applications is standardized. This 1) helps in the integration of different tools that can cooperate by making use of the same information, and 2) minimizes the possibility of human error by limiting the number of allowed operations.

As the main limitation of the described system, its lack of generality as it stands can be noted, since object definitions are adapted to a particular design environment. Nevertheless, the authors believe that the concepts developed for information management are easily extendable to other engineering areas. Work is currently under way to provide the system with an Object Definition Language to allow the user to specify the structure and behavior of the information he intends to use.

The work reported in this paper has been demonstrated jointly with a number of application programs, that used it both as an information centre and as a means to exchange information, as a preliminary result of the above mentioned ESPRIT project.

We have shown the necessity of powerful information management systems capable of acting as the integrating tool among independently developed applications that need to cooperate. The special information needs of such distributed systems have been pointed out, and some of the methods to fulfil these needs, together with a particular implementation, that presents a good compromise between development time, performance, and cost, have been shown. We believe the development of this kind of information systems will be a major contribution towards the CIM objective.

REFERENCES

Atkinson, M. P. and O. P. Buneman (1987). Types and persistence in database programming languages. *Computing Surveys of the ACM*. Vol 19(2). June 1987. pp. 105-190.

Batory, D. S. and A. P. Buchmann (1984). Molecular objects, abstract data types, and data models: a framework. *Proceedings 10th Intl. Conf. on Very Large Databases*. Singapore. August 1984. pp. 172-184.

Dadam, P., K. Küspert, F. Andersen, H. Blanken, R. Erbe, J. Günauer, V. Lum, P. Pistor and G. Walch (1986). A DBMS prototype to support extended NF^2 relations: an integrated view on flat tables and hierarchies. *Proceedings ACM SIGMOD Conference*. Washington. May 1986. pp. 356-367.

Dittrich, K. R. and R. A. Lorie (1988). Version support for engineering databases. *IEEE Transactions on Software Engineering*. Vol SE-14(4). April 1988. pp. 429-437.

Katz, R. H. and T. J. Lehman (1984). Database support for versions and alternatives of large design files. *IEEE Transactions on Software Engineering*. Vol SE-10(2). March 1984. pp. 191-200.

Kemper, A. and M. Wallrath (1987). An analysis of geometric modeling in database systems. *Computing Surveys of the ACM*. Vol 19(1). March 1987. pp. 47-91.

Ketabchi, M. A. and V. Berzins (1987). Modeling and managing CAD databases. *Computer*. Vol 20(2). February 1987. pp. 93-102.

Ketabchi, M. A. and V. Berzins (1988). Mathematical model of composite objects and its application for organizing engineering databases. *IEEE Transactions on Software Engineering*. Vol SE-14(1). January 1988. pp. 71-84.

Lorie, R. and W. Plouffe (1983). Complex objects and their use in design transactions. *Proceedings ACM/IEEE Database Week. Engineering design applications*. San José. May 1983. pp. 115-121.

Neumann, T. and C. Hornung (1982). Consistency and Transactions in CAD database. *Proceedings 8th Intl. Conf. on Very Large Databases*. Ciudad de México. September 1982. pp. 181-188.

Neumann, T. (1983). On representing the design information in a common database. *Proceedings ACM/IEEE*

Database Week. Engineering design applications. San José. May 1983. pp. 81-87.

Ozsoyoglu, Z. M. and L. Y. Yuan (1987). A new normal form for nested relations. *ACM Transactions on Database Systems.* Vol 12(1). March 1987. pp. 111-136.

Roth, M. A. y H. F. Korth (1987). The design of -1NF relational databases into nested normal form. *Pro-*

ceedings SIGMOD Annual Conference. San Francisco. May 1987. pp. 143-159.

Stonebraker, M., B. Rubenstein and A. Guttman (1983). Application of abstract data types and abstract indices to CAD databases. *Proceedings ACM/IEEE Database Week. Engineering design applications.* San José. May 1983. pp. 107-113.

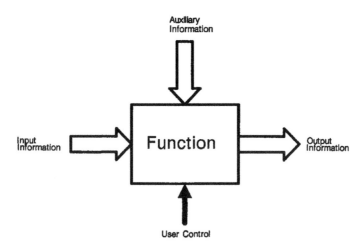

Fig. 1. Basic Design Function

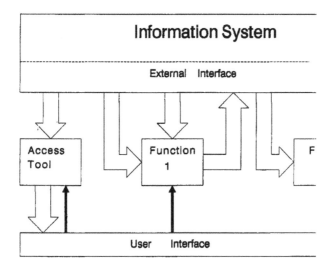

Fig. 2. Information System Functionality

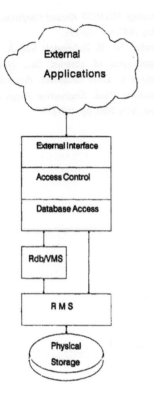

Fig. 3. Information System Implementation

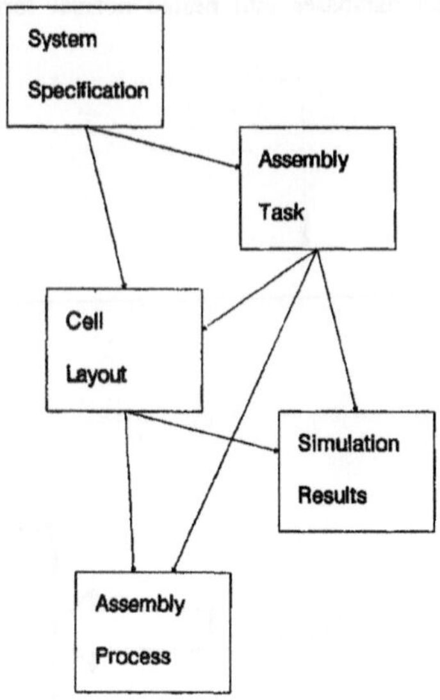

Fig. 4. Example of Design Hierarchy

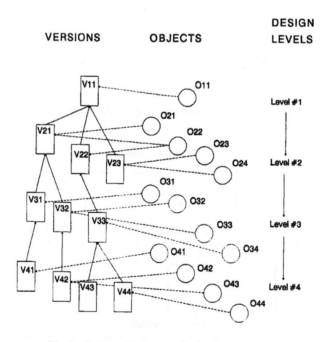

Fig. 5. Relationship between Design Levels,
Versions, and Design Objects

INTEGRATED MANUFACTURING USING CO-ORDINATE MEASURING MACHINES AND STATISTICAL PROCESS CONTROL SOFTWARE

W. Robertson

Department of Technical Sales, LK Tool Company Ltd., Derby, UK

Abstract. The adaptation of Statistical Process Control techniques on
co-ordinate measuring machines for on-line quality control of a group of CNC
machine tools is discussed (flexible manufacturing system).

An explanation of the system and computer software is provided. The specific
method of presenting measured results automatically to a database is given.
The system can be configured to suit any appropriate small-to-large batch
manufacturing process which utilises a mix of CNC machine tools and ancillary
equipment.

Although comprehensive failure and error detection routines are feasible,
automatic feed-back to a machining cell for on-line dimensional correction of
the workpiece is realistic but at this stage impractical.

The variety and complexity of the workpieces involved, together with the
response time required by the CNC machine tool system will dictate the level
of integration required for the co-ordinate measuring system and other
automation for workpiece handling etc.

Keywords. Statistical process control, co-ordinate measuring machines; quality
control; CNC; machine tools; computer software; manufacturing processes;
automation.

INTRODUCTION

The 3 dimensional co-ordinate measuring machine
(CMM) has long been accepted as a universal
solution to the measurement problems associated
with precision or complex workpieces produced by
the engineering manufacturing industries.

Until recently however it had been regarded as a
'tool' associated with the Metrology or standards
room under the sole jurisdiction of the Quality
Control department.

As such, it was to some extent isolated from
production and although accepted as providing an
extremely flexible, single reference inspection
facility, was not regarded as a very convenient
machine in everyday use. It is even true to say
that some production engineers viewed this type of
machine as a hinderance to their objectives of
economic/faster production. Many engineers
considered electronic touch-trigger probe
measurements on the actual production machines as
the answer to their inspection problems.

Whilst there is a clear advantage in less
double-handling of products by inspecting
workpieces in the machine tool, it is now widely
accepted that this practise should be limited to:-

- Measurement verification of critical dimensions
  only.
- Measurement of very large workpieces which impose
  handling difficulties.

The arguments against measurement within a machine
tool machine are that:-

1. Dimensional errors will result which could be
   attributal to thermal effects or workpiece
   contamination.
2. Production efficiency will be impaired as the
   machine is usually slow for inspection
   applications.
3. Heavy machine tools are designed for accuracy
   under 'cutting load' conditions.
4. There is no independent audit of the cutting
   machine.

INTEGRATION PRINCIPLES FOR A CMM

A gradual realisation of the limitations of
inspection by a machine tool places more emphasis
on solving how the universal characteristics and
flexibility associated with CMM's can be adapted to
provide instant response to inspection demands
imposed by faster CNC cutting technology which by
enabling shorter economic batch production
techniques to be adopted, limits the waiting time
available for approval of the 1st-off product.

The main requirements for the CMM are a faster,
simpler to operate machine and a machine which
could be resident close by or even integrated with
the actual production processes.

With this objective in mind, the approach during
the early 1980's was to not only pursue the policy
of sound design characteristics from a thermal
stability viewpoint, but to introduce new designs
of horizontal spindle measuring type robot machines
which have greater environmental tolerance. At the
same time to provide a control system and computer
system for the CMM which would enable true
integration into a production environment.

This concept has been developed and refined by LK through a 'learning curve' situation involving over 25 fully integrated installations in terms of automatic workpiece measurement which entails workpiece handling, error recovery, data retrieval and total management reporting.

There is no standard universal answer however to demands for this type of facility and although some standardisation is possible in the design of a complete system, much of the integration is customised because of the individual demands of the end user and the control systems of the associated machine tool and mechanical handling suppliers.

A typical flexible machining system (FMS) cell would highlight the main features to be considered to integrate a measuring machine which are:-

- Mechanical handling system (AGV, Robot, Rail/conveyor pallet system, etc.)
- Electronic interface (PLC controller etc.)
- Software requirements of the system (inspection requirements, message and programme handling etc.)

The details of individual flexible manufacturing systems are not illustrated or presented for discussion here as they form as separate topic.

STATISTICAL PROCESS CONTROL ASPECTS

Running parallel with the concept of integrated machines has been the search for ways of improving quality generally. In particular many smaller production units of the sub-contract type have faced increasing quality demands from the prime manufacturer who they supply. This theme of passing the responsibility for quality 'down the line' to the price part manufacturer was widely developed in Japan where it has proved to be of fundamental importance in improvements to quality and reliability and 'Just-in-Time' (JIT) techniques.

The main method of achieving this goal has been through the use of statistical process control (SPC).

Most companies are now compelled to instal some form of SPC to satisfy their customers demands and it does provide important documented evidence of part conformance.

Most CMM's are designed to enable various levels of SPC to be applied both off-line and on-line and the choice depends entirely on how the end-user wishes the system to integrate into the existing Quality System. Either way, the SPC facility can be run on stand alone machines or fully integrated machines.

Most engineers consider SPC's main application within the context of high volume production runs. While it is ideally suited to this, it can also be applied to short batch production runs when used in the machine capability sense (i.e. to verify the actual machines or processes).

To understand how SPC software is applied in CMM's, the method of how the CMM collects inspection data needs to be briefly explained.

The co-ordinate system relies on collecting inspected points relating to features on a workpiece through a touch sensing or non-contact device (probe) and this in turn is instigated by the CMM part program.

The actual method of constructing the part program can be done in a variety of ways, for instance:-

a. Direct part programming at the CMM computer terminal.
b. Through a simplified teach-learn user-friendly keyboard (step-by-step approach).
c. From a CAD system and via a post processor.

To utilise SPC effectively, other criteria needs including at the part program stage about the features to be considered for control.

SPC Control Of The System And Networking

The logical way to integrate SPC is to configure a universal database to which all data collected can be related. This includes information about the areas or departments concerned with manufacture of the product, as well as the information about the components, e.g.

- Department number or "workstation" number
- Part number, operation number, batch number and feature number

Subsets of inspection data, down to individual levels of responsibility can also be included if maximum control is to be required, e.g.

- Employee name and number and shift identification
- Fixture number and machine number
- Date of inspection

Each component will then be guaranteed having its own unique set of files containing the SPC data. The LK system is configured to save a number of historical sets of data before the first or oldest set is overwritten.

An in-house developed software command language is utilised for the construction of part programs and the SPC software package follows this general theme.

For instance, in order to add inspection data into the SPC database 'save feature' files are created.

These files are produced by using the following simple commands from the software language:-

- UF : Use feature
- SF : Save feature

Each inspected feature is then saved in the 'SF' file with its own unique lable attached.

The UF command is used at the end of the relevant portion of the part program to submit the data to the SPC database.

The feature itself is given a name from the library of features, e.g. HOLE or SLOT.

A 'Sample addition program' (SAP-MENU) is a menu-driven program that will prompt the operator to nominate all the information required by the SPC system as it is uniquely configured (for example, operator name may not be included).

The menu will appear on the VDU screen thus:-

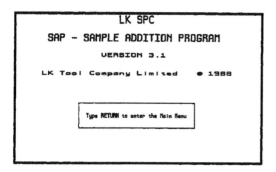

Fig. 1. Sample addition programme.

On keying 'return' the user is presented with a form which demands the additional information required by the database.

For instance, the series of menus will display lists of departments, nominated parts, operations etc. and by moving the cursor up or down on the screen the user selects the required field, then hits 'return' or can type in the appropriate number.

When sufficient selections have been made to uniquely define a component the system will display a new screen to enter key values by which to search the component database selected.

User configurable keys are defined on the terminal to assist at each stage in simplifying entry of data which is not covered by simple movements of the cursor for instance pressing F17 will allow the user to specify a range of a particular database to be search for. (It might be necessary to extract data which was entered between two dates and these upper/lower limits for the search will be displayed).

After working through the search menus, it is then possible to specify the report type.

The menu for this could appear as:-

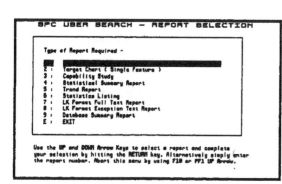

Fig. 2. Report selection.

Selection of the control chart option, will then produce a new screen containing the first 20 feature labels of the selected component, e.g.

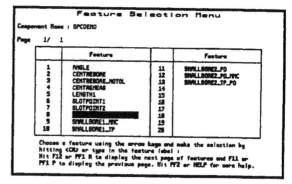

Fig. 3. Feature selection.

A further display will now present the dimensions available within the chosen feature along with any nominal and tolerance information. The operator must select which dimension is required for the control chart, e.g.

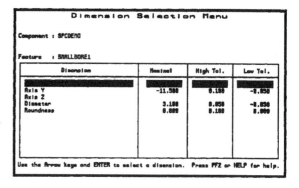

Fig. 4. Dimension selection menu.

A sub-group size needs then to be specified on the next screen (a default value is configurable). A number from 2 - 20 for instance might be the sub-group range but 2 could be the normal value for instance, e.g.

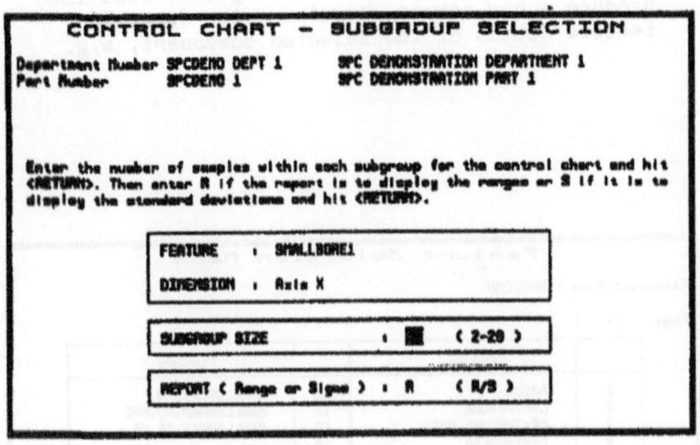

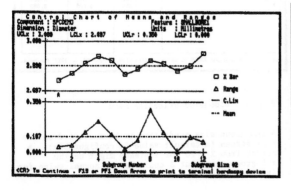

Fig. 5. Sub-group size selection.

This also prompts the user whether to plot an X-Bar vs Range chart or an X-Bar vs Sigma chart by simply entering S or R into the final screen option (Range chart being the normal default).

The system now searches the historical database to find all data that matches the search criteria and after displaying these on the screen will produce the appropriate chart (subject to no other selection modifications being made to the number of inspections displayed from the earliest inspection numbered 1 to the newest inspection with the highest number.

There is an 'error message check' included to ensure the operator has selected enough inspections to make up at least one sub-group of the data to construct a chart.

The system will then display the X-Bar and Range Chart (in this example) together with the upper and lower control limits for the X-Bar chart and the Mean of the Range.

Action codes for the control chart results can also be incorporated. Hard copies can also be printed.

In addition to the above example:

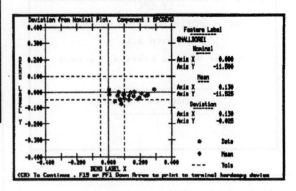

Fig. 7. Target chart.

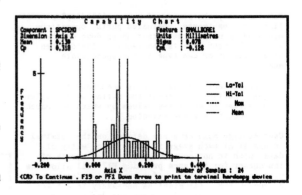

Fig. 8. Capability chart.

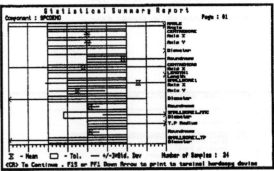

Fig. 9. Statistical summary report.

Fig. 6. Control chart.

The individual sub-group samples can be accessed at this stage together with the option of returning to the control chart.

Full text reports of the Statistical Summary can also be displayed or printed.

Trend reports allowing the user self-selection of how the control limits will be calculated for the report based on mean ± 3 Sigma or nomial ± 3 Sigma can be created.

The important point to remember is that on multi-task configured systems all this SPC activity can take place on a secondary VDU screen simultaneously while the inspection machine continues to work on other components.

This is only one very well developed SPC system for CMM applications and alternative off-line statistical analysis packages are of course widely used providing a similar degree of control chart capability but they are not as efficient in an automatic cell environment.

The reports generated automatically by this SPC system ensure sufficient data is available for corrective decisions and further investigations of the workpiece. Research is continuing into the possibility of automatic feed-back of trend data to the machine tools with the aim of zero-rejects. There are further difficulties however yet to be resolved with respect to the actual sources of errors before direct feed-back can be realistically installed in such systems.

INTEGRATION OF CAD SYSTEMS

As a final link in the chain, semi-direct part programming of the inspection machine should take place over a suitable network by using design data created on a CAD system and pre-processing to construct probe paths etc. before post processing at the measuring machine system into the measuring machine language. This obviously eliminates creation of part programmes at the inspection machine.

CONCLUSIONS

There are no barriers to fully integrated measuring systems in terms of speed and capability as a solution to total shop-floor control, given the correct range of workpieces. This usually only depends on having moderately complex prismatic or circular parts to manufacture.

The ultimate decision rests with the individual manufacturer to decide if the complexity of the workpieces, levels of production and demands of their own customers justifies the ultimate expenditure.

DYNAMIC PILOTING POLICY FOR FLEXIBLE ASSEMBLY LINES

O. K. Shin, J. P. Bourrieres and F. Lhote

Laboratoire d'Automatique, Institut de Productique, Besançon, France

ABSTRACT. *A new dynamic piloting policy to control in real time flexible assembly lines (FAL) is presented. The two objectives of this policy are to get rid of the combinatorial scheduling problem from the short time production planning and to take advantage of all of the available flexibilities : the flexibility of the FAL and the multiple orders of assembly operations in a given product. To achieve these goals, we propose in this paper a dynamic piloting approach where sequencing of operations as well as task assignments are driven by the actual state of the FAL. We begin by the product analysis then description of the FAL, and finally the piloting policy is presented. The policy is tested on a simulated FAL and the results are presented.*

KEYWORDS. Manufacturing; Flexible assembling; Real-time process control; Dynamic parts routing; Product analysis; Orders of assembly operations.

INTRODUCTION

The role of a conventional piloting system in a FMS consists mainly in the control of material flow by assigning tasks to machines. Usually this assignment of tasks as well as sequencing of processing operations is determined by short time production scheduling and thereafter these decisions are delivered to the piloting system.

In a job shop environment, this two-step 'scheduling-piloting' approach seems not to be suitable, because :

(1) The model of the production system rarely coincide with the real system, hence the accumulated gap between the model and the real system becomes larger as the time passes.

(2) The order of operations and parts routing in a work shop are usually pre-determined so that one can't take full advantage of neither multiplicity of possible orders of operations, flexibility of material handling system nor flexibility of manipulators.

(3) One can't face up to the pertubations of the system in real time.

(4) The establishment of the scheduling is very complicated, particularly in the case of assembly processes.

We propose here a new approach of dynamic real time piloting policy for a job shop type Flexible Assembly Line (FAL), where sequencing of operations as well as tasks assignments are driven by the actual state of the assembly line.

In this approach, the sequencing of tasks and tasks assignments are neither scheduled nor pre-determined ; on the other hand, all of these decisions are taken just when they are necessary.

The purpose of this piloting policy is to assemble, without detailed scheduling, a given quantity of products as rapid as possible by balancing (in real time) the amount of works allocated to each manipulator.

This can be achieved by means of the flexibility of the FAL (manipulation flexibily & routing flexibility) and the capability of the piloting system to adapt to the environment changes such as product quantity variations, development of new product types, equipment breakdowns, etc. (Browne,1984; Stecke,1985a)

Some of the following advantages of such approach were remarked by many authors concerning the design of a work shop or the flexibility of a production system (Matson,1982; Stecke,1985a,1985b; YAO, 1985) :

(1) One does not limit the available flexibility of the physical system.

(2) One can face up to pertubations in real time.

(3) One can avoid the efforts required for scheduling.

The piloting system presented in this paper is based on the following suppositions :

(1) The transfer system of the FAL provides the routing flexibilty, i.e., all the pallets on the system can be transfered to any destination in the FAL by at least more than one route.

(2) The manipulators are more or less flexible.

(3) The operations executed in the FAL are mainly assembly operations but not excluding drilling, marking, etc.

PRODUCT ANALYSIS

Once given a FAL and the product(s) to assemble, the first concern is to decompose the finished product into parts in order to study the product and then the order(s) of assembly operations.

The parts and the finished products are defined as the principal input and output material flow of the FAL respectively.

Moreover, we call constituent a set of parts assembled together which is on the way of evolution toward the finished product. This constituent is almost always holded up by a physical support. In this paper, we'll suppose that all of the constituents are supported by pallets and the association of a constituent and its pallet will be called an aggregate.

The aggregates are operated to reach the finished product by undergoing a series of operations according to certain order(s) of operations. We define two types of operations : the tasks and the manipulations.

A task is an abstract operation which contributes to evolve an aggregate toward the finished product by assembling a part (or a constituent) on the aggregate or by establishing a new character on the aggregate (for example, drilling, pleating, testing, etc...).

The task is independent of the manipulator by which it is realized and also independent of the state of the aggregate upon which this task is realized. The task is defined only by its contribution to an aggregate : for example, assembly task of the part A on a certain aggregate or drilling of the part B in an aggregate.

On the other hand, we call manipulation a concrete operation which has meaning only when it is defined with the manipulator and with the specific state of the aggregate upon which it is executed. Hence, the manipulation can be characterized by its execution time.

Following these definitions, a given task can be different manipulations according to the state of the aggregate and the manipulator which executes the task.

Meanwhile, we will keep the generic term 'operation' to designate all of the two terms defined above when the distinction is not necessary.

A ball pen and its parts which will be served as an example of product in this paper is depicted in Fig.1.

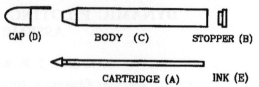

Fig.1. Decomposition of Ball pen into parts.

ORDERS OF ASSEMBLY OPERATIONS

In general, a given product can be assembled according to various orders of operations. A software to find all possible orders of operations for a given product is developed in our laboratory (Hoummady,1988).

Among these orders of operations, we define the sequential order of operations as the order in which all of the operations are composed of :

- the assembly operations of a part on an aggregate or

- the unary operations (drilling, pleating, palletization, depalletization etc.) on an aggregate.

For example, a sequential order of operations for the ball pen in Fig.1 would be : palletization of cartridge -> ink filling -> assembly of body -> assembly of cap -> assembly of stopper -> removal of the finished product from the pallet.

On the other hand, an order of operations is said to be non-sequential order if it contains at least one assembly operation of more than one constituents. In this case, these constituents are assembled independently, i.e., one can be built without chronological relation with the other(s).

For example, in the case of ball pen assembly, a non-sequential order would be : palletization of body -> assembly of stopper (= Constituent A) ; palletization of cartridge -> ink filling (= constituent B) ; assembly of constituents A & B -> assembly of cap -> removal of the finished product from the pallet.

To assemble a given product in a FAL, neither are we obliged to choose only one order of operations nor is it necessary to consider all of the possible orders of operations. By compromising these extremes, we define some orders of operations which are reasonnable and technically easy to realize. From now on, we will consider only the sequential orders of operations to simplify the assembly process.

In the sequential orders, the total number of assembly operations are fixed. (It can be shown that, for a given product, the number of assembly operations is invariant for all

sequential   orders   of   operations
(BOURRIERES,1989))

In fig.2, all the sequential orders of
operations to assemble the ball pen is
shown by means of Petri nets. Here, each
place represents a part or an aggregate
and each transition represents a task.

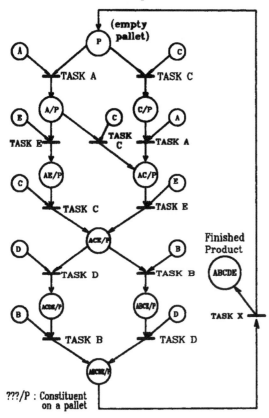

Fig.2. Representation of sequential
orders of tasks by Petri nets.

The **state of an aggregate** (or of a
constituent) is represented by all the
tasks that the aggregate has undergone.
For example, the state of the aggregate
ACE/P in Fig.2. is identified only by the
accumulated tasks A,C,E on the pallet P
independently of the order of these
tasks. This aggregate can evolve to
either ACDE/P or ABCE/P depending on the
decision.

DESCRIPTION OF THE F.A.L.

The state of a FAL can be represented by
many ways depending on the purpose or on
the point of view. For the purpose of
dynamic piloting of a FAL in real time,
we define three categories of
informations to model its state :
informations about physical resources of
the FAL, informations about the assembly
process & task assignment and finally the
informations about the aggregates in the
FAL.

The Physical Resources

Most of these informations are defined at
the moment of the FAL design and have
static characteristics. These are
furtherly classified into two types : the
characteristics of each manipulator and
the disposition of the manipulators on
the transfer network.

Characteristics of manipulators. Suppose
that there are N manipulators $(M_1,..M_N)$
in the FAL and that L manipulations
$(OP_1,..OP_L)$ can be performed by these
manipulators. Let :

$$E_M = \{M_1,..,M_i,..,M_N\} \quad \text{and}$$
$$E_P = \{OP_1,..,OP_J,..,OP_L\},$$

be the set of manipulators and the set of
feasible manipulations in the FAL
respectively.

Then each manipulator M_i is characterized
by :

- the subset of feasible
  manipulations,
  $\{OP_J,..,OP_k,..,OP_l\}$

- the execution time $(t_{iJ},..,t_{ik},..,t_{il})$
  of these operations

- the number of aggregates (or empty
  pallets) in the queue of M_i,
  $(q_{iJ},..,q_{ik},..,q_{il})$ for the purpose of
  manipulations $(OP_J,..,OP_k,..,OP_l)$ at a
  given instant.

- the work state WS_{iJ} (working/
  breakdown) of M_i for each operation
  OP_J $(WS_{iJ} = 1$ if M_i can realize OP_J
  ; $WS_{iJ} = 0$ otherwise)

Disposition of manipulators. An important
characteristic of the automated assembly
system is that the transfer times of
aggregates between work stations are
often much longer than manipulation
times, hence they must be taken into
account in the piloting policy.

We represent the disposition of
manipulators by a distance matrix D whose
elements d_{is} $(i,s = 1..N)$ are the
minimum distance (expressed in time unit)
from a manipulator M_i to an other M_s.

$$D = [d_{is}]$$

$$d_{is} > 0 ; \quad i, s = 1..N.$$

Here, d_{is} is infinite if no path exists
from M_i to M_s or if the path is
temporarily blocked.

Assembly Process & Task Assignment

Two kinds of informations are necessary :
the (sequential) orders of tasks, the
relation between tasks and manipulations.

Sequential orders of tasks. Here, our
concern is how to represent all of the
chosen sequential orders so as to :

- we don't pre-define an order of tasks
  neither for a given type of product
  nor for a given aggregate (or an empty
  pallet).

- when a task is completed, the next task to be performed is defined as a function of the state of the aggregate and of the current state of the FAL.

For any aggregate of a given product type, note that after the mth manipulation, whatever the sequential orders may be, (here, $1 \leq m < M$, M is the total number of manipulations of the product) :

- there will be a finite set of states of aggregates which have undergone m tasks.

- the next task ($(m+1)^{th}$ task) can be defined on the basis of only the current state of the constituent and is independent of the order of previous tasks.

This can be seen in Fig.2. and we can represent all the sequential orders of tasks as in TABLE 1. In this table, the next task of an aggregate is chosen among the realizable tasks which are classified, in their turn, according to the actual state of the given aggregate.

TABLE 1   Sequential orders of tasks for ball pen assembly.

Step	Actual state of constituent	Realizable tasks
1	ø (=ABCDEX)	A, C
2	A	E, C
	C	A
3	AC	E
	AE	C
4	ACE	B, D
5	ABCE	D
	ACDE	B
6	ABCDE	X

Relation between tasks and manipulations.
A given task does not always imply to only one manipulation. On the other hand, the manipulation(s), for a given task, are chosen as a function of the state of the aggregate. This relation can be represented by a table which assigns each task to manipulations according to the current state of the aggregate.

Information on the Aggregates

These informations represent the dynamic state of the FAL from the viewpoint of piloting. For each aggregate (or an empty pallet), we define the following attributes :

- the identification code.
- the type of product to which the aggregate belongs.
- the step (the number of tasks that the aggregate has undergone).

- the current state.
- the last manipulation that the aggregate has undergone.
- the next manipulation to be realized & its manipulator.

MODELIZATION OF PILOTING PROBLEM

The objective of the piloting is to produce the given quantity of products as quickly as possible. One simple and intuitive approach would be, although it may not guarantee the optimal solution, the attribution of each assembly task to a manipulator so that it can be executed as fast as possible.

Another approach would be to attribute each task to a manipulator in order that all the manipulators are loaded as equivalently as possible.

These two approaches are similar to each other. The only difference is that, in the first case, we take into consideration of the transfer time of aggregates whereas we ignore the effects of transfer in the second case. We formalize here the first approach and a heuristic algorithm is proposed in the next section.

Notation

$C_m(K)$: the amount of works in time unit which is attributed to the manipulator m at the instant K.

J : the set of realizable manipulations by the manipulator m.

$t_{(m,j)}$: the execution time of the manipulation j by manipulator m.

A : the set of preceding manipulators which the aggregates have left just before arriving at the manipulator m.

D : the set of next manipulators to where the aggregates can be destined from the manipulator m,

$V_{am}(K)$: the transfer command of an aggregate from manipulator a to the manipulator m.

$d_{(a,m)}$: the distance in time unit from manipulator a to m.

$u(K-\alpha)$: the unit step function with a delay α.

Model

Find the commands of task attribution V(K) at the end of each manipulation.

$$V(K) = \begin{vmatrix} v_{11}(K) & v_{12}(K) & \dots & v_{1N}(K) \\ v_{21}(K) & v_{22}(K) & \dots & v_{2N}(K) \\ \cdot & \cdot & \cdot & \cdot \\ \cdot & \cdot & \cdot & \cdot \\ v_{N1}(K) & v_{N2}(K) & \dots & v_{NN}(K) \end{vmatrix}$$

where

> K is the abbreviation of $K.\delta t$ and δt is a sampling interval. ($K = 1..T/\delta t$, T: a given piloting period)
>
> $V_{mn}(K) = 1$: if an aggregate out of the manipulator m is attributed to the manipulator n for the next manipulation.
>
> $V_{mn}(K) = 0$: otherwise.

Such as

$W(K) = \text{Max}(W_m(K))$ to be minimum, where

$$W_m(K) = \sum_{n=1}^{N} {}_{(n \neq m)} \mid C_m(K) - C_n(K) \mid, \quad (1)$$

and

$$C_m(K) = C_m(K-1) +$$

$$\sum_{(j \in J, a \in A)} \{ V_{am}(K-d_{(a,m)}).u(K-d_{(a,m)}).t_{(m,j)} \}$$

$$- \sum_{(j \in J, d \in D)} \{ V_{md}(K).t_{(m,j)} \}. \quad (2)$$

In Eq(1), $W_m(K)$ represents the sum of the differences of work amounts between the manipulator m and all the others.

The second term of Eq(2) is the the amount of work which is attributed to the manipulator m at the instant $(K\delta t - d_{(a,m)})$. The arrival of the work is delayed by the distance between maipulators a and m, that is, by $d_{(a,m)}$.

The last term of Eq(2) represents the aggregate which has been finished a manipulation at the manipulator m and leaves for the manipulator d at the moment K.

HEURISTIC ALGORITHM FOR DYNAMIC PILOTING

In the following heuristic algorithm, we will take into account principally the execution time of a manipulation by manipulator(s), the distance between manipulators and the amount of loaded works to the destined manipulator.

Algorithm of dynamic piloting

0. FOR each aggregate which has been finished a manipulation   DO

1. Update last manipulation, Success/Fail & manipulator ;

2. IF last manipulation failed, THEN alarm or evacuate it ;

3. Search for feasible tasks at the next step ;

4. FOR each feasible task DO

    41.   Search for manipulations corresponding to the task ;

42. FOR each manipulation found, search for manipulators capable of manipulation ;

43. FOR each manipulator, if it's not in breakdown DO

    431.   Get EXECUTION TIME of manipulation ;

    432.   Calculate amount of WORK in queue of manipulator ;

    433.   Get DISTANCE between last & next manipulators ;

44. FOR each found manipulator DO

    441.   SUM := α.(EXECUT TIME) + β.(WORK) + Γ.(DISTANCE) ; (α, β, Γ: weighting factors)

    442.   Keep manipulator whose SUM is minimum ;

5. Attribute the aggregate to manipulator found at step 442 ;

6. IF required quantity of product is not attained THEN GOTO step 0 ; ELSE Stop ;

CONCLUSION

The proposed algorithm was tested on a simulated FAL. In this simulation, the manipulators were not very flexible, i.e., most of the manipulator was capable of manipulate only one task (but capable of several manipulations).

On the other hand, the transfer system was simulated based on a real flexible transfer system. We have paid attention to the traffics of the aggregates by simulating the transfer system by a software developed in our laboratory (Lhote,1988).

The simulation results of the algorithm showed evidently that the local optimization of type 'the nearest neighborhood searching' is not efficient when the manipulators are not well aligned ; for example, in the case when the manipulators are disposed in the opposite direction of the most sequential orders of manipulations.

To overcome this inefficiency, we modified the proposed algorithm as to minimize the execution time of two succeeding manipulations of an aggregate.

The results of this algorithm with 'look-ahead' feature was very satisfactory, i.e., it was as good as an optimally cofigured non-flexible assembly line.

We have also applied these algorithms to observe the impacts of the pertubation on the FAL yields. In most cases, the FAL continued to assemble products with the varying yields whereas the non-flexible assembly line was not at all immune to these pertubations.

REFERENCES

BOURRIERES, J. P. (1989). Contribution à la modélisation des systèmes robotisés d'assemblage, Thèse d'état. to appear, Univ. de franche-comté Besançon, France.

BROWNE, J. et al. (1984, Apr.). Classification of flexible manufacturing systems. The FMS Magazine, 114-117.

HOUMMADY, A. (1988). Conception et développement d'un logiciel d'élaboration automatique des gammes d'assemblage, Thèse, Univ. de franche-comté Besançon, France.

LHOTE, F. et al. (1987). Modelization of transfer systems for flexible assembly lines, Proc. of the IASTED intl. symp., Paris, 398-402.

MATSON O. J. and WHITE, J. A. (1982). Operational research and material handling, European joural of operational research, 11, 309-318.

STECKE, K. E. and BROWNE, J. (1985a). Variations in flexible manufacturing systems according to the relevant types of automated material handling, Material flow, 2, 179 - 185.

STECKE, K. E. (1985b). Design, planning, scheduling and control problems of flexible manufacturing systems, Annals of operations research, 3, 3-12.

YAO D. D. (1985). Material and information flows in flexible manufacturing systems, Material flow, 2, 143-149.

MODELLING AND SIMULATION OF LARGE-SCALE MULTIPARAMETER COMPUTER AIDED DYNAMICAL CONTROL SYSTEMS

J. Jones, Jr.

*Air Force Institute of Technology, Department of Mathematics & Computer Science,
Ohio, USA*

Abstract. Modelling and simulation of large-scale
multidimensional multiparameter dynamical systems require the use
of large-scale computers to generate feed-back control laws,
especially when model uncertainties exist. Now robust control
theory is concerned with the problem of analyzing and synthesizing
control systems that provide and acceptable level of performance
where many model parameters or uncertainties may exist, since
mathematical models of physical systems are usually never exact
due to the presence of such parameters.

The need to be able to design robust feedback control laws is very
important in such systems. Usually a physical model will have
significant structural information about the interconnection of
components and subsytems but less information concerning their
integrated system performance. Hence, many variations of
parameters must be carried out on supercomputers in order to
determine the more significant and sensitive parameters which must
be adjusted very rapidly to accomplish a desired level of
performance.

Dynamical systems of the form:

$$\begin{cases} E(\theta)\dot{x}(t,\theta)=A(\theta)x(t,\theta)+B(\theta)u(t,\theta); \dot{x}(t,\theta)=\dfrac{dx(t,\theta)}{dt} \\ \\ y(t,\theta)=C(\theta)x(t,\theta)+D(\theta)u(t,\theta); t \geq 0 \end{cases}$$

are considered.

The elements of matrices $E(\theta)$, $A(\theta)$, $B(\theta)$, $C(\theta)$, $D(\theta)$ belong to
the ring of polynomials $R[\theta]$ where $\theta = (\theta_1,\theta_2,\theta_3,\ldots,\theta_q)$ is a
multiparameter and the coefficients of the polynomials belonging
to $C[\theta]$ have elements belonging to the field C of complex numbers,
or the elements of such multiparameter matrices may be the form
$f(\theta) = a(\theta)|b(\theta)$ where the polynomials $a(\theta)$, $b(\theta)$, ε $c[\theta]$, for
$b(\theta) \neq 0$. $E(\theta)$ may be a singular matrix for possible parameter
values of θ. The parameter θ may also be holomorphic function of
a single complex variable z for z belonging to simply-connected
bounded regions in the z-plane. The basic question of
stabilization, controllability, observability, etc., in the
presence of changes in subsystems as regards the overall dynamical
systems needs to be treated in response to changing parameters in
subsystems.

Fast numerical methods requiring parallel processing are necessary
to compute adjustments as time t changes. Transfer function
matrices, controllability matrices, observability matrices,
feedback control laws need to be recomputed as parameters change.
Such matrices may be multiparameter matrices and may allow for
improvement of control laws such as in cases where $E(\theta)$ may become
singular matrix and the dynamical systems require considerable
fast changes in feedback control laws.

Use is made of recent results of J. Jones, Jr. concerning
generalized inverses of such multiparameter matrices to aid in
computer aided changes to carry out modelling, simulation and
analysis of such dynamical systems.

1. INTRODUCTION

The main purpose of this work is to
establish conditions for the existence of
stationary solutions of nonlinear matrix
differential equations of the form

$$A(z)X(z) + X(z)B(z) + C(z) + X(z)D(z)X(z) \cdot$$

$$\begin{bmatrix} A & -UB \\ 0 & B \end{bmatrix} \begin{bmatrix} I & U \\ 0 & I \end{bmatrix} = \begin{bmatrix} I & -U \\ 0 & I \end{bmatrix}$$

$$\begin{bmatrix} A & 0 \\ 0 & B \end{bmatrix} \begin{bmatrix} I & U \\ 0 & I \end{bmatrix}$$

and the pairs of matrices above are similar. From the above it is seen that

$$\begin{bmatrix} I & U \\ 0 & I \end{bmatrix} \begin{bmatrix} A & C + XDX + \dfrac{dX}{dz} \\ 0 & B \end{bmatrix} \begin{bmatrix} I & -U \\ 0 & I \end{bmatrix} = $$

$$\begin{bmatrix} A & 0 \\ 0 & B \end{bmatrix}$$

Making use of the method J. Jones, Jr. [4] using elementary row and column operation and starting with the first matrix below to obtain the latter matrix:

z	$1-z^2$	1	$-2z$	1	0	0	0
1	$-z$	0	-1	0	1	0	0
0	0	0	0	0	0	1	0
0	0	0	0	0	0	0	1
1	0	0	0				
0	1	0	0				
0	0	1	0				
0	0	0	1				

elementary row (column)
→
operations

z	$1-z^2$	0	0	1	0	z	$-z^2-1$
1	$-z$	0	0	0	1	1	$-z$
0	0	0	0	0	0	1	0
0	0	0	0	0	0	0	1
1	0	$-z$	z^2+1				
0	1	-1	z				
0	0	1	0				
0	0	0	1				

Then by choosing the appropriate blocks of the latter matrix it follows that:

$$\begin{bmatrix} 1 & 0 & z & -z^2-1 \\ 0 & 1 & 1 & -z \\ 0 & 0 & 1 & 0 \\ 0 & 0 & 0 & 1 \end{bmatrix} \begin{bmatrix} z & 1-z^2 & 1 & -2z \\ 1 & -z & 0 & -1 \\ 0 & 0 & 0 & 0 \\ 0 & 0 & 0 & 0 \end{bmatrix}$$

$$\begin{bmatrix} 1 & 0 & -z & z^2+1 \\ 0 & 1 & -1 & z \\ 0 & 0 & 1 & 0 \\ 0 & 0 & 0 & 1 \end{bmatrix} = \begin{bmatrix} z & 1-z^2 & 1 & -2z \\ 1 & -z & 0 & -1 \\ 0 & 0 & 0 & 0 \\ 0 & 0 & 0 & 0 \end{bmatrix}$$

$$\begin{bmatrix} 1 & 0 & -z & z^2+1 \\ 0 & 1 & -1 & z \\ 0 & 0 & 1 & 0 \\ 0 & 0 & 0 & 1 \end{bmatrix} = \begin{bmatrix} z & 1-z^2 & 0 & 0 \\ 1 & -z & 0 & 0 \\ 0 & 0 & 0 & 0 \\ 0 & 0 & 0 & 0 \end{bmatrix}$$

5. SUMMARY

In this work, necessary conditions for the existence of a solution of the non-linear Riccati matrix dvifferential equation which occurs so frequently in the area of optimal control theory, stability theory, mathematical modelling of dynamical systems, tracking theory and model reduction, are considered. The approach used in this work also carries over to the case of matrix equations which includes the algebraic matrix equations whose coefficient matrices contain holomorphic parameters.

REFERENCES

[1] D.C. Babbitt and V.S. Varagarajan. Deformations of Nilpotent Matrices Over Rings and Reduction of Analytic Families of Meromorphic Differential Equations Memoirs of AMS, Vol. 55, No. 325 (1985), pp. 1-147.

[2] S. Friedland. Analytic Similarity of Matrices, Lectures in Applied Mathematics, AMS, Vol. 18 (1980), pp. 43-86.

[3] V.G. Leavitt. A Normal Form for Matrices Whose Elements are Holomorphic Functions, Duke Math. J., 15 (1985), pp. 463-472.

[4] J. Jones, Jr. (To appear).

$$\psi_1' \begin{bmatrix} T & U \\ V & W \end{bmatrix} = (V, W)$$

Then

$$Ker(\psi_1) = Ker(\psi_2) = \left\{ \begin{bmatrix} T & U \\ V & W \end{bmatrix} \middle| AT=TA; \ AU=UB \right\}$$

$$(3.6)$$

Also we have

$$Im(\psi_1) = L \qquad (3.7)$$

where Im = image. In order to show that

$$Im(\psi_1) = Im(\psi_2) = L$$

we see that $Im(\psi_1) = L$, for $BV = VA$, $BW = WB$ then

$$\begin{bmatrix} 0 & 0 \\ V & W \end{bmatrix} \varepsilon \ Ker(T_1); \ \psi_1 \begin{bmatrix} 0 & 0 \\ V & W \end{bmatrix} = (V, W)$$

and so

$$Im(\psi_2) \subseteq Im(\psi_1) \qquad (3.8)$$

Next to show set inclusion $\geq$ in (3.8) above we make use of the property of linear transformation T_1, T_2, ψ_1, ψ_2:

$$dim(Ker(\psi_1)) + dim(Im(\psi_1)) =$$
$$dim(Ker(T_1))$$

$$(3.9)$$

$$dim(Ker(\psi_2)) + dim(Im(\psi_2)) =$$
$$dim(Ker(T_2))$$

Since by (3.4) $dim(Ker(T_1)) = dim(Ker(T_2))$ and by (3.6) $Ker(\psi_1) = Ker(\psi_2)$ then $dim(Ker(\psi_1)) = dim(Ker(\psi_2))$ and by (3.9) $dim(Im(\psi_1)) = dim(Im(\psi_2))$. Also by (3.8) above $Im(\psi_2) = Im(\psi_1)$ and so $Im(\psi_1) = Im(\psi_2) = L$.

Now the matrix

$$\begin{bmatrix} I & 0 \\ 0 & -I \end{bmatrix} \varepsilon \ Ker(T_1), \ \psi_1 \begin{bmatrix} I & 0 \\ 0 & -I \end{bmatrix} = (0, -I)$$

and since $Im(\psi_1) = Im(\psi_2) = L$, there exists a matrix

$$\begin{bmatrix} T_0 & U_0 \\ V_0 & W_0 \end{bmatrix} \varepsilon \ Ker(T_2)$$

such that

$$\psi_2 \begin{bmatrix} T_0 & U_0 \\ V_0 & W_0 \end{bmatrix} = (0, \quad -I)$$

This implies that $V_0 = 0$, $W_0 = -I$ and thus there exists a matrix in $Ker(T_2)$ of

the form (3.5). Now the equation

$$AU_0 + \left(C + XDX + \frac{dX}{dz} \right)(-I) = U_0 B \quad (3.10)$$

is satisfied by U_0, X under the assumption that the pair of matrices in (3.2) are similar.

4. **EXAMPLE (THEOREM 3.1). A NECESSARY AND SUFFICIENT CONDITION THAT THE NONLINEAR MATRIX DIFFERENTIAL EQUATION**

$$AU - UB = C + XDX + \frac{dX}{dz}$$

has a solution pair $X \varepsilon C^{rxs}(z)$, $U(z) \varepsilon C^{rxs}(z)$ is that the following pair of matrices

$$\begin{bmatrix} A & C + XDX + \frac{dX}{dz} \\ 0 & B \end{bmatrix}, \ \begin{bmatrix} A & 0 \\ 0 & B \end{bmatrix}$$

are similar where

$$A(z) = \begin{bmatrix} z & 1-z^2 \\ 1 & -z \end{bmatrix}; \ B(z) = \begin{bmatrix} 0 & 0 \\ 0 & 0 \end{bmatrix};$$

$$C(z) = \begin{bmatrix} -1 & 0 \\ 0 & -1 \end{bmatrix}; \ D(z) = \begin{bmatrix} -1 & 0 \\ 0 & -1 \end{bmatrix}$$

are given matrices. A solution $X(z)$ is obtained by making use of associate matrix differential equations of the Riccati type nonlinear matrix differential equation above. Such a solution pair is given by

$$X(z) = \begin{bmatrix} z & -z^2-1 \\ 1 & -z \end{bmatrix} = U(z)$$

Also

$$\begin{bmatrix} z & 1-z^2 \\ 1 & -z \end{bmatrix} \begin{bmatrix} z & -z^2-1 \\ 1 & -z \end{bmatrix} - \begin{bmatrix} z & -z^2-1 \\ 1 & -z \end{bmatrix}$$

$$\begin{bmatrix} 0 & 0 \\ 0 & 0 \end{bmatrix} = \begin{bmatrix} -1 & 0 \\ 0 & -1 \end{bmatrix} + \begin{bmatrix} z & -z^2-1 \\ 1 & -z \end{bmatrix}$$

$$\begin{bmatrix} -1 & 0 \\ 0 & -1 \end{bmatrix} \begin{bmatrix} z & -z^2-1 \\ 1 & -z \end{bmatrix} + \begin{bmatrix} 1 & -2z \\ 0 & -1 \end{bmatrix}$$

and

$$\begin{bmatrix} 1 & -2z \\ 0 & -1 \end{bmatrix} = \begin{bmatrix} 1 & -2z \\ 0 & -1 \end{bmatrix}$$

Then if the nonlinear Riccati matrix differential equation above has a solution pair X, U

$$\begin{bmatrix} A & C+XDX+\frac{dX}{dz} \\ 0 & B \end{bmatrix} = \begin{bmatrix} A & AU-UB \\ 0 & B \end{bmatrix} =$$

Proof (Necessity). If (3.2) has a
solution X, U then matrix multiplication
shows that

$$\begin{bmatrix} I & -U \\ 0 & I \end{bmatrix} \begin{bmatrix} A & 0 \\ 0 & B \end{bmatrix} \begin{bmatrix} I & U \\ 0 & I \end{bmatrix} = \begin{bmatrix} A & -UB \\ 0 & B \end{bmatrix}$$

$$\begin{bmatrix} I & U \\ 0 & I \end{bmatrix} = \begin{bmatrix} A & AU-UB \\ 0 & B \end{bmatrix}$$

$$\begin{bmatrix} A & C + XDX + \dfrac{dX}{dt} \\ 0 & B \end{bmatrix}$$

However,

$$\begin{bmatrix} I & -X \\ 0 & I \end{bmatrix} = \begin{bmatrix} I & X \\ 0 & I \end{bmatrix}^{-1}$$

so

$$\begin{bmatrix} A & 0 \\ 0 & B \end{bmatrix} \quad \text{and} \quad \begin{bmatrix} A & C+XDX+\dfrac{dX}{dt} \\ 0 & B \end{bmatrix}$$

are similar with the similarity matrix

$$\begin{bmatrix} I & X \\ 0 & I \end{bmatrix} = S$$

Proof (Sufficiency). Let use suppose now
that the pair of matrices

$$\begin{bmatrix} A & 0 \\ 0 & B \end{bmatrix} \quad \text{and} \quad \begin{bmatrix} A & C + XDX + \dfrac{dX}{dt} \\ 0 & B \end{bmatrix}$$

are similar for some $X \epsilon C^{rxs}(z)$. Then

$$\begin{bmatrix} A & 0 \\ 0 & B \end{bmatrix} = S^{-1} \begin{bmatrix} A & C + XDX + \dfrac{dX}{dt} \\ 0 & B \end{bmatrix} S \quad (3.3)$$

for this non-singular matrix S.

Define two linear transformations T_1 and
T_2 on the set $C^{(r+s) \times (r+s)}(z)$ of complex
matrices of size $(r+s) \times (r+s)$, (which may
also be considered as an $(r+s)^2$
-dimensional vector space) as follows:

$$T_1(Z) = \begin{bmatrix} A & 0 \\ 0 & B \end{bmatrix} Z - Z \begin{bmatrix} A & 0 \\ 0 & B \end{bmatrix} ;$$

$$T_2(Z) = \begin{bmatrix} A & C + XDX + \dfrac{dX}{dt} \\ 0 & B \end{bmatrix} Z$$

$$- Z \begin{bmatrix} A & 0 \\ 0 & B \end{bmatrix}$$

where $Z \epsilon C^{(r+s) \times (r+s)}(z)$. Now using (3.3)
we have the following:

$$S^{-1} T_2(SZ) = S^{-1} \begin{bmatrix} A & C + XDX + \dfrac{dX}{dt} \\ 0 & B \end{bmatrix}$$

$$SZ - S^{-1} SZ \begin{bmatrix} A & 0 \\ 0 & B \end{bmatrix} = \begin{bmatrix} A & 0 \\ 0 & B \end{bmatrix}$$

$$Z - Z \begin{bmatrix} A & 0 \\ 0 & B \end{bmatrix} = T_1(z)$$

where $Z \epsilon C^{(r+s) \times (r+s)}(z)$. Now since S is
non-singular

$$Ker(T_2) = \left\{ SZ \mid Z \epsilon C^{(r+s) \times (r+s)}(z) \cap Ker(T_1) \right\}$$

$$(3.4)$$

where Ker = kernel. Then

$$\dim(Ker(T_1)) = \dim(Ker(T_1))$$

where dim = dimension.

The transformations T_1, T_2 will be used
in establishing the existence of a
solution of equation (3.2).

Now we have the following:

$$Ker(T_1) = \left\{ \begin{bmatrix} T & U \\ V & W \end{bmatrix} \middle| \begin{matrix} AT = TA; & AU = UB \\ BV = VA; & BW = WB \end{matrix} \right\}$$

$$Ker(T_2) = \left\{ \begin{bmatrix} T & U \\ V & W \end{bmatrix} \middle| \begin{matrix} AT + \left(C+XDX+\dfrac{dX}{dt}\right)V=TA; \\ BV=VA; \\ AU + \left(C + XDX + \dfrac{dX}{dt}\right)W = UB \\ BW = WB \end{matrix} \right\}$$

It is seen that it suffices to find a
matrix in $Ker(T_2)$ of the form

$$\begin{bmatrix} T & U \\ 0 & -I \end{bmatrix} \qquad (3.5)$$

and since the defining equation
$AU + \left(C + XDX + \dfrac{dX}{dt}\right)W = UB$, for $W = -I$
furnishes a solution of $AU - UB =$
$C + XDX + \dfrac{dX}{dt}$. Also for $W = 0$ the
$Ker(T_2)$ and $Ker(T_1)$ have two defining
equations in common.

In order to next establish the existence
of matrix U we introduce the following
linear space L:

$$L = \{(V, W) \mid V \epsilon C^{s \times r}(z), \ W \epsilon C^{s \times s}(z),$$

$$BV = VA, \ BW = WB\}$$

Then L is a linear space with
multiplication by any complex number α
and addition defined below:

$$\alpha(V, W) = (\alpha V, \alpha W), \ (V_1, V_2) + (W_2, V_2) =$$

$$(V_1 + V_2, \ W_1 + W_2)$$

We next define linear transformations ψ_1,
ψ_2 on $Ker(T_1)$ for $i = 1, 2$ as below:

are similar for $z \epsilon R$. Also, the determinant $|R(z) - \lambda I|$ is reducible to the product of a pair of polynomials $f_\alpha(z,\lambda)$, $g_\beta(z,\lambda)$ of degrees α, β in λ and whose coefficients are holomorphic functions of $z \epsilon R$. $f_\alpha(z,\lambda)$, $g_\beta(z,\lambda)$ are factors of the characteristic polynomial

$$\left[\left[\left(X(z)D(z) + A(z) + \left(\frac{dX}{dz}\right)X^{-1}(z)\right) - \lambda I\right]\right.,$$

$$\left.\left[\left(-D(z)X(z) - B(z)-X^{-1}\left(\frac{dX}{dz}\right)\right) - \lambda I\right]\right]$$

respectively, where $\alpha \leq n$, $\beta \leq n$. Also $f_\alpha(z,R)g_\beta(z,R) = 0$ where $f_\alpha(z,\lambda)g_\beta(z,\lambda)$ is not necessarily the minumum polynomials satisfied by $R(z)$ but is a divisor of $|R(z) - \lambda I|$ and such that:

$$f_\alpha\left[z, X(z)D(z) + \left(\frac{dX}{dz}\right)X^{-1}(z) + A(z)\right] = 0$$

$$g_\beta\left[z, -D(z)X(z) - B(z)-X^{-1}(z)\left(\frac{dX}{dz}\right)\right] = 0$$

Such polynomials $f_\alpha(z,\lambda)$ and $g_b(z,\lambda)$ of degree $\alpha \leq n$, and $\beta \leq n$ respectively, and such that $f_a(z) \cdot g_\beta(z)$ is a divisor of $|R(z) - \lambda I|$ and a multiple of the minimum polynomial satisfied by $R(z)$ will be called an admissible class of polynomials and are denoted by .

Theorem 2.5. If equation (1) has a non-singular holomorphic solution $X(z) \epsilon N_n(z)$ for $z \epsilon R$ then there exists a polynomial $f_\alpha(z,\lambda)$ in powers of λ of degree $\alpha \leq n$ with coefficients which are holomorphic functions of $z \epsilon R$ such that

$$[X(z) \quad I]f_\alpha(z,R) = [X(z) \quad I]$$

$$\begin{bmatrix} U(z) & M(z) \\ V(z) & N(z) \end{bmatrix} = [0 \quad 0]$$

where $N^{-1}(z)$ exists,

$$f_\alpha(z,R) = \begin{bmatrix} U(z) & M(z) \\ V(z) & N(z) \end{bmatrix}$$

and $U(z)$, $V(z)$, $M(z)$, $N(z)$ are polynomials in the matrices $A(z)$, $B(z)$, $C(z)$, $X^{-1}(z)$, $\frac{dX}{dz}$. If $N(z)$ is non-singular and also

$$\begin{bmatrix} U(z) & M(z) \\ V(z) & N(z) \end{bmatrix}\begin{bmatrix} I \\ -X(z) \end{bmatrix} = \begin{bmatrix} 0 \\ 0 \end{bmatrix} =$$

$$= g_\beta(z,R)\begin{bmatrix} I \\ -X(z) \end{bmatrix}$$

then the zeros of $f_\alpha(z,R)$ and $g_\beta(z,R)$ are similar for $z \epsilon R$ namely,

$$N(z)\left[X(z)D(z) + A(z) + \left(\frac{dX}{dz}\right)X^{-1}(z)\right] =$$

$$\left[-B(z)-D(z)X(z)-X^{-1}(z)\left(\frac{dX}{dz}\right)\right]N(z)$$

for $z \epsilon R$. The proof will appear elsewhere.

3. SOLUTIONS OF RICCATI MATRIX DIFFERENTIAL EQUATIONS AND OPTIMAL CONTROL THEORY AND TRANSPORT THEORY

Solutions of the nonlinear Riccati matrix differential equation

$$AX - XB = C + XDX + \frac{dX}{dt} \qquad (3.1)$$

are obtained which useful in the theory of robotics, optimal control, transport theory, and nonlinear optimization theory. The coefficient matrices A,B,C,D have elements which are holomorphic throughout a simply-connected, bounded region R in the finite z-plane except for a finite set of points. Necessary and sufficient conditions are also established along with an algorithm for obtaining such solutions. Use is made of associate equations to obtain solutions of Riccati matrix differential equations. Higher order nonlinear matrix theory arising in transport theory and nonlinear optimization theory will appear elsewhere.

Capital letters will denote matrices throughout this section. Let C be the field of complex numbers and $C^{mxn}(z)$ the vector space of complex mxn matrices. R will be a simply-connected, bounded region of the complex z-plane. The main purpose of this section is to consider the Riccati matrix differential equation

$$AX - XB = C + XDX + \frac{dX}{dt}$$

where A,B,C,D are given matrices and $X(t)$ is a matrix to be found. $A \epsilon C^{rxr}(z)$, $B \epsilon C^{sxs}(z)$, $C \epsilon C^{rxs}(z)$, $D \epsilon C^{sxr}(z)$, and $X \epsilon C^{rxs}(z)$. The coefficient matrices are holomorphic throughout R, a bounded simply-connected region of the z-plane.

The first result concerning the existence of solutions of (3.1) is as follows:

Theorem 3.1. A necessary and sufficient condition that the nonlinear matrix differential equation

$$AU - UB = C + XDX + \frac{dX}{dt} \qquad (3.2)$$

has a solution $X \epsilon C^{rxs}(z)$, $U \epsilon C^{rxs}(z)$ is that the pair of matrices

$$\begin{bmatrix} A & C + XDX + \frac{dX}{dz} \\ 0 & B \end{bmatrix}, \begin{bmatrix} A & 0 \\ 0 & B \end{bmatrix} \qquad (3.2)'$$

are similar, where $A \epsilon C^{rxr}(z)$, $B \epsilon C^{sxs}(z)$, $C \epsilon C^{rxs}(z)$, $D \epsilon C^{sxr}(z)$, are holomorphic except for a finite set of points for $z \epsilon R$, where R is a bounded simply-connected region in the z-plane.

where $H(z)$ is any holomorphic matrix for $z \varepsilon R$. Use of generalized inverses of holomorphic matices for $z \varepsilon R$ will be made in Section 2.

2. NECESSARY CONDITIONS FOR THE EXISTENCE OF A SOLUTION OF THE RICCATI MATRIX DIFFERENTIAL EQUATION (1)

Theorem 2.1. If the following matrix differential equation

$$A(z)X(z) - X(z)B(z) = C(z) + X(z)D(z)X(z) + \frac{dX(z)}{dz} \qquad (2)$$

for $z \varepsilon R$ and all matrices are holomorphic for $z \varepsilon R$ has a solution $X(z)$ holomorphic for $z \varepsilon R$, then the following pair of matrices

$$\begin{bmatrix} A(z) & C(z) + X(z)D(z)X(z) + \dfrac{dX(z)}{dz} \\ 0 & B(z) \end{bmatrix},$$

$$\begin{bmatrix} A(z) & 0 \\ 0 & B(z) \end{bmatrix} \qquad (3)$$

are similar for $z \varepsilon R$.

Proof. Let $X(z)$ be a holomorphic solution for $z \varepsilon R$ of equation (2), then

$$\begin{bmatrix} I & X(z) \\ 0 & I \end{bmatrix} \begin{bmatrix} A(z) & C(z) + X(z)D(z)X(z) + \dfrac{dX(z)}{dz} \\ 0 & B(z) \end{bmatrix}$$

$$\begin{bmatrix} I & -X(z) \\ 0 & I \end{bmatrix} = \begin{bmatrix} A(z) & 0 \\ 0 & B(z) \end{bmatrix}$$

for $z \varepsilon R$ and the matrices of (3) are similar for $z \varepsilon R$.

Theorem 2.2. Let equation (1) have a holomorphic non-singular matrix solution $X(z)$ for $z \varepsilon R$. Then the following pair of matrices

$$R(z) = \begin{bmatrix} -B(z) & D(z) + X(z)^{-1}\left(\dfrac{dX}{dz}\right)X^{-1}(z) \\ -C(z) & A(z) \end{bmatrix}$$

and

$$\begin{bmatrix} X(z)D(z) + A(z) + \left(\dfrac{dX}{dz}\right)X^{-1}(z) \\ D(z) + X^{-1}(z)\left(\dfrac{dX}{dz}\right)X^{-1}(z) \end{bmatrix}$$

$$\begin{matrix} 0 \\ -D(z)X(z) - B(z) - X^{-1}(z)\left(\dfrac{dX}{dz}\right) \end{matrix}$$

are similar for $z \varepsilon R$.

Proof. Let $X(z)$, $X^{-1}(z)\left(\dfrac{dX}{dz}\right)$ be holomorphic matrices and $X(z)$ be a solution of equation (1), then

$$\begin{bmatrix} X(z) & I \\ I & 0 \end{bmatrix} \begin{bmatrix} -B(z) & D(z) + X^{-1}(z)\left(\dfrac{dX}{dz}\right)X^{-1}(z) \\ -C(z) & A(z) \end{bmatrix}$$

$$\begin{bmatrix} 0 & I \\ I & -X(z) \end{bmatrix} = \begin{bmatrix} X(z)D(z) + A(z) + \left(\dfrac{dX}{dz}\right)(X^{-1}(z)) \\ D(z) + X^{-1}(z)\left(\dfrac{dX}{dz}\right)(X^{-1}(z)) \end{bmatrix}$$

$$\begin{matrix} 0 \\ -D(z)X(z) - B(z) - X^{-1}(z)\left(\dfrac{dX}{dz}\right) \end{matrix}$$

and the matrices above are similar.

Theorem 2.3. Let $X(z)$ be any holomorphic non-singular solution of (1) for $z \varepsilon R$, then $X(z)$ is also a solution of the following equations:

$$\begin{bmatrix} X(z) & I \end{bmatrix} \begin{bmatrix} -B(z) & D(z) + X^{-1}(z)\left(\dfrac{dX}{dz}\right)X^{-1}(z) \\ -C(z) & A(z) \end{bmatrix}$$

$$\begin{bmatrix} I \\ -X(z) \end{bmatrix} = 0$$

Proof. Let $X(z)$ be a non-singular holomorphic solutions of (1) for $z \varepsilon R$. Then,

$$\begin{bmatrix} X(z) & I \end{bmatrix} \begin{bmatrix} -B(z) & D(z) + X^{-1}\left(\dfrac{dX}{dz}\right)X^{-1}(z) \\ -C(z) & A(z) \end{bmatrix}$$

$$\begin{bmatrix} I \\ -X(z) \end{bmatrix} = -X(z)B(z) - C(z) - X(z)D(z)X(z) - \frac{dX}{dz} - A(z)X(z) = 0.$$

Theorem 2.4. Let $X(z)$ be any non-singular holomorphic solution of (1) for $z \varepsilon R$:

$$A(z)X(z) + X(z)B(z) + C(z) + X(z)D(z)X(z) + \frac{dX}{dz} = 0$$

then the following pair of matrices

$$\begin{bmatrix} X(z)D(z) + A(z) + \left(\dfrac{dX}{dz}\right)X^{-1}(z) \\ D(z) + X^{-1}\left(\dfrac{dX}{dz}\right)X^{-1}(z) \end{bmatrix}$$

$$\begin{matrix} 0 \\ -D(z)X(z) - B(z) - X^{-1}(z)\left(\dfrac{dX}{dz}\right) \end{matrix}$$

and

$$R(z) = \begin{bmatrix} -B(z) & D(z) + X^{-1}(z)\left(\dfrac{dX}{dz}\right)X^{-1}(z) \\ -C(z) & A(z) \end{bmatrix}$$

$$+ \frac{dX(z)}{dz} = 0 \qquad (1)$$

where $X(z_0) = X_0$ and the coefficient matrices $A(z)$, $B(z)$, $C(z)$, $D(z)$ belong to the matrix algebra $M_n(z)$ of n by n matrices of holomorphic functions of a single complex variable z for $z \varepsilon R$, a closed bounded simply-connected region in the complex z-plane.

The main purpose of this section is to establish necessary conditions for the existence of solutions of the following nonlinear matrix differential equation

$$A(z)X(z) + X(z)B(z) + C(z) + X(z)D(z)X(z)$$

$$+ \frac{dX(z)}{dz} = 0$$

where $X(z_0) = X_0$. The coefficient matrices $A(z)$, $B(z)$, $C(z)$, $D(z)$ are given matrices belonging to the matrix $M_n(z)$ of n by n matrices whose elements are holomorphic functions of a single complex variable z where $z \varepsilon R$, closed bounded simply-connected region in the z-plane, and $X(z) \varepsilon M_n(z)$ is to be determined.

For matrices $A(z) \varepsilon M_n(z)$ having elements belonging to the set of holomorphic functions of a single complex variable $z \varepsilon R$, a closed bounded simply-connected region of the z-plane and whose characteristic roots $\gamma_i(z)$ are also holomorphic functions of $z \varepsilon R$ there exists a similarity transformation $T(z)$ for which $|T(z)| \neq 0$ for $z \varepsilon R$ and such that $T^{-1}(z)A(z)T(z) = U(z)$. Thus, $A(z)$ may be reduced by a similarity transformation to the following normal form:

$$U(z) = \begin{bmatrix} \gamma_1(z) & \phi_{12}(z) & \cdots & \phi_{1n}(z) \\ & \cdot & \cdots & \cdot \\ & & & \cdot \\ & & & \cdot \\ & & & \gamma_m(z) \end{bmatrix}$$

where all of the elements below the main diagonal are zero, the elements $\lambda_i(z)$ and $\phi_{ij}(z)$ are holomorphic functions of z for $z \varepsilon R$ and each $\gamma_i(z)$ is repeated to its multiplicity h_i. The matrix $T(z)$ is called a unimodular matrix. For details a of constructive process of computing such a nonsingular matrix $T(z)$, see W.G. Leavitt [3]. The notion of similarity of pairs of such matrices has also been recently considered by D.G. Babitt and V.S. Varadarjan [1] and S. Friedland [2]. Use of the notion of similarity of pairs of matrices will play an important role in this paper.

Let $A(z)$ be m by n matrix whose elements $a_{ij}(z)$ are holomorphic functions of a complex variable $z \varepsilon R$ as above and let $A(z)$ have constant rank of q for $z \varepsilon R$. Then there exists square matrices $P(z)$ and $Q(z)$ which are holomorphic in R where the order $P(z) = \text{rank } P(z)$ in $R = m$, the order of $Q(z) = \text{rank } Q(z)$ in $R = n$, and

$P(z)A(z)Q(z)$ is the form

$$P(z)A(z)Q(z) = \begin{bmatrix} M(z) & K(z) \\ 0 & 0 \end{bmatrix}$$

where $M(z)$ is q by q matrix having constant rank q in R, $K(z)$ is a q by (n-q) matrix and all elements of $M(z)$, $K(z)$ are holomorphic for $z \varepsilon R$.

The matrices $P(z)$, $Q(z)$ may be obtained as products of elementary matrices representing elementary row and column operations, respectively, where the ordinary definitions of elementary operations are as follows:

(i)   Multiplication of a row (column) by arbitrary nonzero scalar constant),

(ii)   add to a row (column) another row (column) multiplied by an arbitrary non-identically zero holomorphic function of $z \varepsilon R$, and

(iii)   interchange of two rows (columns).

The n x m matrix

$$A_{1,2}(z) = Q(z) \begin{bmatrix} M^{-1}(z) & 0 \\ 0 & 0 \end{bmatrix} P(z)$$

is a (1,2) generalized inverse of $A(z)$, $z \varepsilon R$.

Let $A(z)$ be any m x n matrix for $z \varepsilon R$ and let $A_1(z)$ be any 1-generalized inverse of $A(z)$ for $z \varepsilon R$. Then every 1-generalized inverse of $A(z)$ for $z \varepsilon R$ is of the following form:

$$X(z) = A_1(z) + H(z) -$$

$$A_1(z)A(z)H(z)A(z)A_1(z)$$

where $H(z)$ is any arbitrary n x m matrix for $z \varepsilon R$.

Let $A(z)$ by any m x n holomorphic matrix for $z \varepsilon R$. If $A(z)$ has constant rank m for $z \varepsilon R$, then $A(z)A_1(z) = I_m$ for $z \varepsilon R$ for any 1-generalized inverse of $A(z)$, i.e., every 1-generalized inverse of $A(z)$ is a right-inverse of $A(z)$ for $z \varepsilon R$. If $A(z)$ has constant rank of n for $z \varepsilon R$ then $A_1(z)A(z) = I_m$ for $z \varepsilon R$, for any 1-generalized inverse of $A(z)$ is a left-inverse of $A(z)$ for $z \varepsilon R$.

Let $A(z)$ be any arbitrary holomorphic matrix for $z \varepsilon R$ and $F(z)$ any holomorphic p x q matrix for $z \varepsilon R$. Let A_1 and $F_1(z)$ be any 1-generalized inverse of $A(z)$ and $F(z)$, respectively, for $z \varepsilon R$. Then the matrix equation $A(z)X(z)F(z) = B(z)$ has a solution $X(z)$, an n x p holomorphic matrix for $z \varepsilon R$ if and only if $A(z)A_1(z)B(z)F_1(z)F(z) = B(z)$ for $z \varepsilon R$. This is a condition holds for $z \varepsilon R$, the general solution is given by

$$X(z) = A_1(z)B(z)F_1(z) + H(z) -$$

$$A_1(z)A(z)H(z)F(z)F_1(z)$$

THE SIMULATION LANGUAGE SIMIAN FOR DECISION SUPPORT SYSTEMS IN PRODUCTION PLANNING

E. Jordan and J. B. Evans

University of Hong Kong, Hong Kong

Abstract

The discrete-event simulation language SIMIAN is described along with its environmental supports. Some well-known problems when using simulation to create an effective decision-support system within an integrated production planning system are characterised together with detailed explanation of how these problems are overcome in the SIMIAN environment. An example is considered in some detail.

1. Introduction to SIMIAN

The SIMIAN working environment consists of a discrete-event simulation language, SIMIAN, an executive program to drive the SIMIAN simulation and an interactive environment to support simulation writing and running. The internal structure of the model is transformed from the well-established process-interaction (PI) description to one based on a Petri net (Reisig, 1985). This net (the DEVNET) becomes the basis for consistency-checking and activation of the model. Event-occurrence becomes transition firing and entities-in-state become tokens occupying a place. Entities joining together at a transition are termed an _engagement_, and these are of particular concern, allowing the parallelism inherent in the system to be dealt with explicitly. The DEVNET is based on the relation Petri net, with augmentation for simulation (Evans, 1988) as follows:

* starred transitions
which delay firing until the occurrence of a temporal event - in this way durational activities are expressed;

* coloured arcs
where the colour indicates which type of token (entity) can be carried;

* facts
which can inhibit or facilitate transition firing on the basis of whether a particular predicate holds;

* decision arcs
where a decision has to be made as to which of a set of alternatives to take.

The first transition in the example (see Fig. 1) is a starred transition, coloured arcs are shown by using different line patterns and a decision arc is shown in Fig. 2. The examples do not use "facts".

The translation of the PI description starts with the abstraction of a DEVNET from the SIMIAN text, although this may be facilitated if the interactive graphic mode of programme specification has been used (see below). The DEVNET structure is then used to direct the activation of the simulation, in analogy to the activation control of an algorithmic program through the run-time stack. More details of the DEVNET and the SIMIAN language will be given in Evans and Jordan (1989) and Evans (1990).

The PRIMATE algorithm is a seven-phase executive program for the engagement strategy using the net structure to control activation, detect unfairness, resolve conflicts and to enforce predicate-dependencies in the system. This is particularly significant for the pattern of engagements but these only make up one of the three separable domains of the simulation program, based on the second paradigm of discrete-event simulation (Evans, 1988):

$$\text{simulation} = \underline{\text{engagements}} + \text{allocation} \atop \text{data-probes}$$

Allocation comprises control flow which does not correspond to entity-flow, but to the logical choice between entities, especially in selection of entities prior to engagement. It is often allocation strategy which is the area for experimentation, the purpose of the simulation.

By keeping it separated, the structure of the model may be kept clear and the allocation changed as needed without disturbing the remaining program.

Data probes can be inserted at various transitions in the DEVNET to collect data from the token being transferred on transition firing. The location of the data probes is determined by the nature of the experiment as detailed through the environment. This domain of the simulation is able to be manipulated without affecting either of the other domains.

The SIMIAN environment consists of, as input facilities, an optional graphical network/program specification routine, the data-probe set-up facility and the experimental control. The output of the model can take a number of forms including one or more of the following:

    on-line real-time conflict highlighting,

    traditional summary and detailed data files,

    animation at various levels of complexity.

2. Problems in simulation of production systems

Coll, Brennan & Browne (1983)(CBB) describes five problems in making simulation an effective decision support system tool when integrated with an MRP system. These problems, which are widely known, are:

1. long lead time from design to implementation of the simulation model

2. validation of the model and verification of the program

3. definition of an objective function

4. interface to other production models (such as MRP, but including also management information systems and data base management systems)

5. man/machine interface

These problems are significantly overcome in the SIMIAN language, its experimental environment and through the changes in methodology that it allows. We consider each point in detail.

2.1 Lead-time

The lead-time will be drastically reduced in the SIMIAN working environment through two major factors

    (i) the graphical user-friendly model specification phase;

    (ii) a top-down, structured methodology.

(i) The graphical model specification entails using a WIMP (windows, icons, mouse, pull-down menus) interface to enter the augmented Petri-net specification of the model (DEVNET). Examples of the DEVNET are shown in Fig. 1 and 2 below. This model has a one-to-one correspondence with the engagement pattern abstracted from the SIMIAN text description of the model. Structural aspects of the model may be validated directly during the input phase but a significant factor is that the drawing program will be able to anticipate logical elements and relationships. That is, the model is being validated interactively as a logically coherent model during input. These features together with the ability to defer the experimental and data-probe specification (INIT and RUN sections of the SIMIAN program) and the independence of the allocation strategy will speed up the process of creation of the logical model with the physical model (SIMIAN code) being produced directly. Such features as queue-handling strategies, arrival and service distributions, etc., will be able to be default-selected or menu-selected with detailed user-specification as the final option.

(ii) The top-down structured methodology is made possible through the ability of any activity/process in the DEVNET to be "exploded" into a more detailed set of activities/processes. Thus a simple model may be quickly constructed and when this is validated a more detailed model may be developed without losing the work previously carried out. It was not necessary to use this methodology in the examples which follow as the problems are sufficiently simple.

2.2 Validation

CBB contrasts the validation of the model as a representation of the object system with the verification of the computer program as an expression of the model that has been developed. However both of these are addressed in the SIMIAN environment.

(i) Validation of the structure of the model. The structured top-down development methodology described above enables the end-user to be able to confirm visually the representation of the object system. The limited number of symbols in the DEVNET and their direct relationship to elements of the object system enables the involved non-computing personnel to be informed of the "designer's understanding of the system", this channel of communication facilitates correction of any errors in the model at a very early stage. In addition there are perceptual advantages in a two-dimensional representation compared with the one-dimension of a computer program. Structural relationships and parallel activities are the most prominent features of a two-dimensional representation but are generally obscured in a program.

(ii) Verification of the computer program as a representation (instantiation) of the model. Through the structured methodology, operation of the high-level model enables validation of the results at an aggregate level, these results need to build (or create) confidence in the minds of both the modeller and the end-user that the model reflects the reality of the object system. The potential to specify a detailed methodology exists and will be considered later.

2.3 Definition of the objective function

The choice of, or even necessity for, an objective function is subsidiary to the experimental methodology. The goal of the object system may or may not be known to the experimenter. The goal of the experiment may relate to the goal of the system in many different ways. In particular, the experimenter will have some characteristic of the object system that is of concern in the experiment. The goal of the experiment may be the elucidation of an objective function, it may be the enhancement of the experimenter's awareness of the contributing factors to a problem, it may be, in the case of a well-established and validated model, the direct comparison of alternative resource allocation strategies using an explicit objective function. As an experimental tool the SIMIAN environment may be used in any research methodology.

One goal of the SIMIAN environment is to enable a concentration of the experimentation on allocation to be realised. This is viewed as the "soft" part of the object system, capable of greater change than the net structure, which is deemed more "hard".

The independence of the data collection process from the structural model and from the allocation strategy in the SIMIAN environment enables the objective function to be even omitted in the earlier stages of the model development. As the user familiarises him/herself with the operation of the model then the object system characteristics of interest to the user may be parameterised and monitored.

The goal of the experiment can take many forms, from the need to quantitatively evaluate with precise statistics the consequences of a number of specific allocation strategies to a qualitative assessment, based on observing an animation, of a single design. The objective may even be the possibility of determining unresolved conflicts in allocation. A single numerical objective function is only appropriate in a restricted class of experiments.

2.4 Interface to other production models

When creating a simulation model to be used as a DSS within an existing environment it is critical that the simulation model be able to interface with the existing computer models such as MRP, MIS or whatever. For a general purpose package to exist it is necessary to have an import/export path that is capable of communicating with as wide a range of external models as possible.

The simplest path provided by the SIMIAN environment is that of a flat table of ASCII values such as in a microcomputer spreadsheet or word-processing application. This path allows almost any external system to communicate. The external system produces an ASCII table, this is read by the SIMIAN Import Data module which then sets up the INIT (initialisation) part of the SIMIAN program.

The simulation model is run, producing results into the SIMIAN data base which can be "saved" at any time. These results are then exported using the SIMIAN Export Data module in the same ASCII table format. This is not suitable for an online DSS which would require closer interaction but is quite satisfactory for strategic planning or major design projects.

The second path for interface provided by the SIMIAN environment is through an intermediate such as Oracle or SQL. The Oracle DBMS has been developed to include features so that it can intercept and recode enquiries and amendments to a variety of proprietary data bases such as IMS. Oracle also handles very effectively the distributed data base where the object system (or systems) being addressed by the model are in a variety of physical locations, using different computer hardware, operating systems and data base management systems. This path, when incorporated into a SIMIAN model performs as follows. SIMIAN model configuration data are passed to Oracle (on-line) as valid, formatted Oracle commands, the Oracle system returns the relevant data in its own (standard) format, this is then translated directly into SIMIAN INIT data structures as appropriate.

The third path consists of a fully-integrated operating environment where the SIMIAN model and the MIS, MRP, etc. models coexist. When, in the future, the SIMIAN model has been developed to include the second interface path, the necessity for proceeding to the third path will need to be re-evaluated in light of the then current state-of-the-art with respect to Open Systems Interface (OSI). This development of the third path may never be necessary as the expanding facilities provided by such software as Oracle could well, in the short-term, lead to an effective OSI for all future developed systems.

2.5 Man/machine interface

Model Specification

This is taken as a fundamental point in the design of the SIMIAN environment. The WIMP model construction facility and top-down development methodology are taken as being the most effective with currently available technology. Graphical output is provided in the form of animation of the model, in addition to the usual charts, tables and summary statistics. At its simplest level there is a default animation based on the DEVNET used as the input model specification. This shows circular tokens moving around the Petri net with simple statistics, such as queue size, being shown alongside appropriate model elements. This animation may be improved by editing the shape of the tokens and all the Petri net elements to form a more realistic representation of the object system. "Paintbrush" graphics techniques as well as the import of material from the existing situation (such as photographs, plans, drawings) can be used in this redrawing process. When animating it is essential for end-user model validation that the simulation clock be allowed to proceed proportionally to real time.

Experimental Control

There are a number of innovative features of the SIMIAN environment which add significantly to the range of experimentation that may be performed.

* interactive participation with the model - if the allocation strategy does not cover the current situation then the model run pauses so that the experimenter may input a decision. This allows training and what-if testing.

* the current situation may be "frozen" at any point, with saving of data values, so that the model may be reset, rerun or whatever. For example, saving the data in the above model at the point where the allocation strategy is undecided will allow the consequences of alternatives to be evaluated.

* limited look-ahead: in the above situation a general query by the experimenter, at the time that the allocation strategy fails to provide an answer, on the total set of current engagements over all places may allow the experimenter to determine the "best" allocation or strategy. Such a facility is the precursor of an Expert System approach to the scheduling/allocation problem as a set of look-ahead criteria are collected to form the basis of a set of rules.

* default to nil output. Simply running the model may highlight object system characteristics through observation of the animation or the occurrence of allocation exceptions. This could be the purpose of the experiment.

3. Machine Shop Example

This example is taken from Duggan and Browne (1988) and it is then extended. The original problem is discussed followed by more general cases.

This problem concerns a machine shop which has 3 machines M1, M2 and M3, and two operators, F1 and F2. Operator F1 can operate machines M1 and M2 while operator F2 can operate machines M1 and M3. Orders require two stages of processing, firstly they must be processed by machine M1 and then they may be processed by either M2 or M3. The DEVNET is shown in Fig. 1.

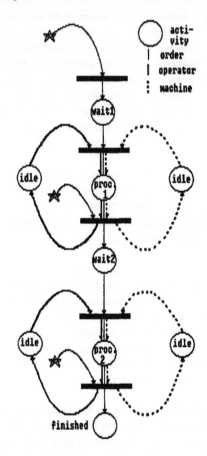

Figure 1 DEVNET for Machine Shop Problem (Duggan and Browne, 1988).

It is worth noting immediately certain features of this:

* simplicity
* clear sequential nature is highlighted
* flow of each entity is obvious
* allocation logic, initialisation and run-time data are excluded from a DEVNET

In SIMIAN this particular problem is programmed as follows:

```
SCHEMA
    01 operator;
        02 canuse SET OF machine;
    01 machine;
        02 canproc FROM (process1, process2);
    01 order

DYNA
order:
    (GEN order EVERY inter_arrival NEG;

    wait1: (ACQ operator | ACQ machine) WITH
        (operator.canuse = machine AND
        machine.canproc = process1) ARBIT;
    TIME process1;
    REL operator | REL machine;

    wait2: (ACQ operator | ACQ machine) WITH
        (operator.canuse = machine AND
        machine.canproc = process2) ARBIT;
    TIME process2;
    REL operator | REL machine;
    finished: TERM order)

INIT
    # operator := 2;
    # machine  := 3;
    operator[1].canuse:={machine[1],machine[2]};
    operator[2].canuse:={machine[1],machine[3]};
    machine[1].canproc := process1;
    machine[2].canproc := process2;
    machine[3].canproc := process2;
/* alternatively the first seven lines of the
INIT part can be described as follows, both of
these are acceptable SIMIAN:
    operator := {Fred, Harry};
    machine := {Bath, Lathe1, Lathe2};
    Fred.canuse:={Bath, Lathe1};
    Harry.canuse:={Bath, Lathe2};
    Bath.canproc:=process1;
    Lathe1.canproc:=process2;
    Lathe2.canproc:=process2;*/
    process1 := unif (9,11);
    process2 := unif (5,8);
    inter_arrival := expn (5)
RUN
    UNTIL 50 order IN finished;
    RESET;
    UNTIL 250 order IN finished;
    REPORT wait1; /* min,max,avg q., wait-time
    data */
    REPORT wait2;
    REPORT operator; /* counts, utilisation */
    REPORT machine
END
```

Generalisation 1

Allowing the production process to remain as two steps but extending to any number of operators and any number of machines with completely flexible sets of abilities and usages. That is the machines may be used for process 1 or process 2 or process 1 and process 2, and the operators can use any subset of the machines. The most significant observation is that there is at this stage no change to the DEVNET. In addition there is no change to the SIMIAN SCHEMA or DYNA blocks. The significant changes appear in the INIT block where we may have, for example:

```
operator := {Aristophanes, Brecht, ..., Zeus};
machine := {A101, A201, ..., Z909};
Aristophanes.canuse := {A304, B704, L209, Z404};
.

Zeus.canuse := {C202, E704, W202};
A101.canproc := {process1};
A201.canproc := {process1, process2};

.

Z909.canproc := {process2};
```

In addition, it is to be remarked that there are no changes to be made to the RUN block, as the resource class reporting (e.g. REPORT operator) will report all items of that class in the same way. This shows the way in which actual data from an external MIS needs to be formatted when it is imported into the SIMIAN model and the independence of the model structural specification from the initialisation and run-time data.

Generalisation 2

A very general class of production scheduling problems can be handled if we now introduce some complexity into the form of the order by allowing each order to have a "routeing" specifying the sequence of machines that must process it and allowing an unspecified number of process steps in this route. All orders do not necessarily use all machines. For the moment we will retain the time taken by any order on a particular machine as arising from a single distribution.

Thus the stream of orders will include items that need to be processed on a particular number of processes in a particular order. For example, an order of type 1 may proceed

process 1 -> process 7 -> process 4

while an item of type 2 may proceed:

process 4 -> process 1 -> process 3 -> process 5.

Note that there is no fixed sequence nor a fixed number of process steps. The DEVNET now becomes as shown in Fig. 2.

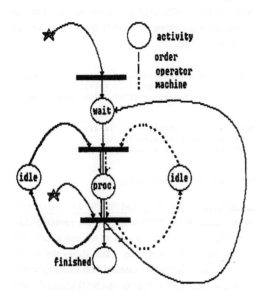

Figure 2 DEVNET for generalised machine shop

Notice, above all, that it is now simpler. The order will undergo a series of engagements until the next step in the routeing is null at which point it passes to the "finished" state. The two arcs along which orders may pass from the third transition are linked. The post-transition firing routine will determine where the token passes next. The SIMIAN now becomes:

```
SCHEMA
    01 order;
        02 item_type;
            03 route ORDERED SET OF process;
            03 pr;
    01 operator;
        02 canuse SET OF machine;
    01 machine;
        02 canproc SET OF process;
    01 process SET (P1, P2, ..., P999);
        02 duration

DYNA
order:
    (GEN order EVERY inter_arrival;
        item_type := SAMP dist_1
    NEG;

    FOR pr OVER item_type.route
    DO
        wait:(ACQ operator | ACQ machine ) WITH
            (operator.canuse = machine AND
             machine.canproc = pr) ARBIT;
        TIME duration;
        REL operator | REL machine
    OD
    finished: TERM order;)
```

```
INIT
    operator := (Alf, Beth, ... )
    machine := (A101, A102, ..., Z909)
    item_type := (widget, gizmo, thingummy)
    Alf.canuse := (B202, C303);
    Beth.canuse := (E404, L203);
    .
    .
    A101.canproc := (P3, P7, P201);
    A102.canproc := (P4, P12, P77);
    .
    widget.route:=<P4,P5,P6>; /* note:in order*/
    gizmo.route := <P1, P5, P2, P7>;
    thingummy.route := <P2, P4, P2, P91>;
    ...
    sampling distributions, RUN block, etc.
```

4. Conclusion

The SIMIAN environment describes a coherent, integrated, modelling environment where the focus on the experimenters' needs to be able to interact easily with the model leads to resolution (or bypassing) of some of the major current limitations of simulation software.

5. References

Coll A., Brennan L. and Browne J. (1985): "Digital Simulation Modelling of Production Systems", in Falster P. and Mazumder R.B.(eds.) Modelling Production Management Systems, North-Holland, Amsterdam, pp 175-195.

Duggan J., Browne J. (1988) "ESPNET: expert-system-based simulator of Petri Nets", IEE Proceedings Vol 135, Pt D, No. 4, July 1988.

Evans J.B. (1988): Structures of Discrete Event Simulation: An Introduction to the Engagement Strategy Ellis Horwood, Chichester, U.K.

Evans J.B. (1990) "A Primer on SIMIAN" to appear.

Evans J.B. and Jordan E. (1989) "An Interim Report on the Simulation Programming Environment SIMIAN" Proceedings of Beijing International Conference on Systems Simulation and Scientific Computing, Oct. 1989, Beijing P.R. China.

Reisig W. (1985): Petri Nets: An Introduction, Springer-Verlag, Berlin.

A KNOWLEDGE-BASED APPROACH TO PROCESS PLANNING: A CASE STUDY

J. L. Nealon* and P. Firth**

*Department of Computing & Mathematical Sciences, Oxford Polytechnic,
Oxford, UK
**School of Engineering, Oxford Polytechnic, Oxford, UK

Abstract. This case study examines the background, and
problems encountered building the EXCAST expert process
planning system for use in the casting room of a
collaborating manufacturing company. Knowledge-based
expert system prototyping and delivery using particular
computer software (the Prolog programming language), and
the facilities of both first and second generation expert
system shells are discussed. The paramount importance of
providing an effective interface between the computer and
user for a system sited in the work environment is
featured. This project was funded by Objex Ltd, UK and the
SERC/ACME directorate.

Keywords. Process planning, expert systems, expert system
shells, knowledge-based systems, computer software,
computer interfaces.

INTRODUCTION

There are many activities and
functions that must be
accomplished in the design and
manufacture of a product (Fig.
1). An idea for a product,
normally dictated by market
demand, is cultivated, analysed,
refined, improved and transformed
into a plan through the
engineering design process. The
plan is documented by drafting a
set of engineering drawings
showing how the product is to be
manufactured and providing a set
of specifications indicating how
the product should perform. The
first activity directly concerned
with the manufacture of the
product is the process plan which
specifies the sequence of
production operations required to
make the product. Scheduling
provides a plan that commits a
company to manufacture certain
quantities of the product by a
certain date.

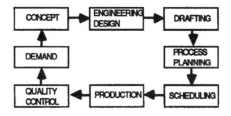

Fig. 1 Engineering Design and
Manufacturing Cycle.

Process planning is the critical
bridge between design and
manufacture (Chang and Wysk,
1985). It is concerned with the
sequence of individual
manufacturing operations needed
to produce a given part or
product. The resultant operating
sequence is documented on a
company route sheet. Process
planning has traditionally been a
task with a very high manual and
clerical content. Much variation
can be found in the level of
detail found in route sheets
between different companies and
between different industries. In
some small companies process

planning may be accomplished by releasing a sketch, drawing, or part print to the production shop with the instructions "make to drawing". More commonly a route sheet consists of a detailed list of steps describing each operation and each work centre. In a traditional manufacturing system a process plan is created by a manufacturing engineer who examines a part drawing and draws on his knowledge of similar plans, potential technologies and equipment to produce a short list of possible plans. Individual engineers each have their own opinions about what constitutes the best manufacturing procedures. Accordingly there are differences among the operating sequences developed by various planners. The previous experience of the process planner is critical to the success of the plan.

COMPUTER-AIDED PROCESS PLANNING

Early Computer-Aided Process Planning (CAPP) systems were primarily concerned with the storage, retrieval and editing of plans. Standard process plans could be recalled and edited to produce a plan for a new part reducing the clerical load on the manufacturing engineer. Retrieval process planning systems are also referred to as variant systems because they produce a new plan by varying an existing plan. The more sophisticated of these retrieval or variant CAPP systems use parts classification and coding as a foundation. Parts are grouped into families according to their geometric characteristics and issued with a code. In order to create process plans for new parts it is necessary to edit an existing plan of a similar component. Implementation of a variant-based system and the consequent standardization of a company's products can lead to a significant reduction in manufacturing costs (Hodgson and

Dinsdale, 1985). However, considerable time is required to create the necessary database and train staff before any benefits are realised. Despite the expense of building such systems it is still the most reliable and widely used type of CAPP system.

Recent developments in CAPP have focussed on eliminating the process planner from the entire planning function. The generative approach uses artificial intelligence techniques, particularly rule-based expert systems, to synthesize the optimum process plan based on an analysis of part geometry, material and other factors which influence manufacturing decisions (Tulkoff, 1981; Eversheim and Esch, 1983; Mill and Spraggett, 1984; Darbyshire and others, 1987).

On receiving a design model a generative process planning system can automatically create process plans from a manufacturing database without human intervention. In the ideal generative process planning package, any part design could be presented to the system for creation of the optimal plan. In practice, current generative type systems are not universal in application. They tend to fall short of a truly generative capability, and they are developed for a limited range of component geometries and a limited range of manufacturing processes, preventing their widespread adoption.

DEVELOPING EXCAST

At an early stage of knowledge acquisition it became apparent to the authors that the industrial collaborator's process planners did not always generate plans from first principles. Typically, their approach was to attempt to vary an existing plan. The process planners appeared to be drawing on their experience of similar products and repairing its associated plan accordingly. This is in accordance with a recent survey in which it was concluded that intelligent retrieval systems that can search for similar plans are highly desirable (Eversheim and Esch, 1985). It was this model of retrieval and repair of a similar product's plan that we employed. Such a mixed process planning system combines the approaches of both the variant and generative CAPP models. These systems may contain a decision logic to select sections of standard or previous process plans and combine them into a single process plan. Alternatively, they may contain a decision logic that selects several standard plan fragments, or have the ability to check some conditional requirements of a component and select a process.

The industrial collaborator manufactures products in acrylic plastic and their products appear as an embedment, such as a flower, oil drop or pencil, encapsulated in the centre of a clear plastic block. The processes involved in casting the acrylic monomer which encapsulates an embedment are particular to the company, and are the result of many years experience. Casting consists of pouring layers of a liquid plastic, at the correct temperature and time intervals into a simple mold and allowing the liquid to harden under pressure and temperature. The composition of the liquid plastic depends primarily on the characteristics of the component to be encapsulated and the size of the finished product.

The decision to build the casting room expert system was taken initially to capture, formalize and automate company's process planning expertise. Furthermore it was hoped that the expert system would:

a) improve the speed of decision making;
b) reduce the disruption when a key employee leaves;
c) reduce the training time for new employees; and
d) improve the quality and consistency of planning decisions.

The company required a system running on an inexpensive microcomputer that they could easily update to handle new products. At an early stage of the project it became clear that an isolated casting room expert system would not be entirely satisfactory. Many issues to the casting room were repeat or standard jobs. Fully describing a standard job to the expert system to obtain a casting plan would be wasteful of time and effort. Retrieval from a database of plans using either job number, job name, or a short descriptive key string provided a suitable mechanism for the creation of casting plans for a large proportion of the company's throughput.

EXCAST is a data driven expert system. That is, all geometric and other relevant features corresponding to the product's manufacture must be input before substantial inferencing takes place. Product information is prompted by a menu system with associated graphic explanation facilities. System users find it easier to relate geometric descriptions to pictures. It is possible to backtrack to previous menus to correct mistakes or modify the product description. However, several menu inputs are required in order to enter the fine detail required by EXCAST. This situation is satisfactory for a novice system user. In contrast, a more experienced user can speed up plan generation by volunteering information through a natural language interface. A simple natural language interpreter handles a range of product descriptions such as:

"A CAB incorporating a bronze coin embedded horizontally with a one colour external print"

A complete product description short circuits the menu system. However, the system will prompt the user if critical detail is absent from the volunteered product description. Product descriptions are created by sales staff for input to a micro-computer based quotation and database system. Feeding the product description field created by a quotation system to the natural language interpreter enables direct process plan generation on the receipt of a firm order.

The natural language interpreter consists of a grammar and a parser. A grammar of a language is a scheme for specifying the sentences allowed in the language, indicating the syntactic rules for combining words into well formed phrases and clauses. One particular grammar, known as a context-free grammar, has the advantage that all sentence structure derivations can be represented as a tree and practical parsing algorithms exist. Parsing assigns structures to sentences. Using transition nets, phrase structured grammars can be syntactically decomposed using a set of rewrite rules.

The development of the system passed through various stages. During the specification stage, adequate account must be given to the users' and clients' requirements. Initially, the company management and the potential users of the system within the company had to be educated as to the possible functionality of the proposed system. For example, managers were particularly interested in a system that could "learn". However, the current state of the art does not extend to providing this function, so the system specification had to include the best alternative. In the first instance, this was a facility to store and modify previous plans. Later, the system was re-implemented in an environment which would enable company staff to modify the system's knowledge-base themselves (see COMPUTER SOFTWARE below).

Knowledge acquisition is always a difficult stage in the development of an expert system, and this project was no exception. The company's expertise was held by a group of people, at different levels in the company and with different views of the process. Thus, the knowledge acquisition process had to take account of both the varying depths and applicability of the experts' knowledge but also some human psychological factors. Many human experts are very wary of the development of expert systems in their domain of expertise, and a knowledge engineer has to be fully aware and prepared for this attitude.

COMPUTER SOFTWARE

Expert system shells provide a basic framework for building an expert system; that is, a means of encoding the domain knowledge and an inferencing mechanism for making use of the encoded knowledge. First generation shells typically provided a knowledge representation language (KRL) in which to encode the domain knowledge. The intention was that these KRLs should be readily understood by domain experts so that they could develop their own knowledge bases. Early shells fell short of expectations. Their computer-user interfaces were poor, they were not comprehensible by a domain expert without assistance or training, and they provided no interfaces into other software. LISP and Prolog coded shells were particularly resource hungry. Moreover they only gave one knowledge representation paradigm; normally backward chaining rules. For the "retrieve and repair" model that we were attempting to simulate a richer knowledge representation language was required. Consequently, first generation expert system shells were found to be inappropriate as a delivery vehicle for a final product.

Prolog is programming language founded on a subset of classical logic, and is typically used in the definitions of relations. Prolog programs consist of a series of horn clauses[1]. The similarity of this modular form to the

IF situation THEN action rule

format gives a clear indication of its popularity with builders of rule-based expert systems. Prolog's pattern matching facility is an extremely useful device to replace selector and constructor functions for operating on structured data. An n-field data record may be represented by an n-ary Prolog structure. For example:

job(name,number,date,......).

Retrieval of a structure or record is achieved by matching a query with the corresponding field instantiations over the database. The following query finds a job which was issued on 12-5-87:

?job(Name,Number, '12-5-87',....)

Prolog's built in depth-first backtracking search strategy can be used to explore alternative branches of the search space to recall all jobs satisfying the field instantiations. Prolog is not an expert system in its own right, although it contains the basics of a backward-chaining expert system. It is possible to implement in Prolog the standard trace explanation found in most expert system shells.
Interfaces, such as mouse driven menus, are cumbersome to programme and the code is very

much dependent on the dialect of Prolog being used. Vendors of Prolog systems are making the programming of computer-user interfaces easier by implementing a multitude of non-standard predicates (functions). These facilities, together with the ease with which a relational database may be implemented, reveal Prolog as an ideal tool for coding an integrated database and expert system package.

Provision of user-friendly facilities to edit rules and extend the database is fairly easy on the Prolog based system. However, facilities to enhance the display forms in the light of experience proved to be resource hungry on a micro-computer. Moreover, tools facilitating the adaption of the natural language interpreter required an appreciation of parse trees and Prolog. It proved difficult to keep the base language hidden from the user. The trade off between expressiveness and intelligibility is never straightforward. The difficulty in providing updating facilities for the natural language interface and the collaborator's wish that all displays, databases, and knowledge bases be maintainable by intelligent non-computing staff led to a review of second generation of expert system shells with a view to transferring the expert system to such a shell.

Recent releases of shells provide many additional features compared to their predecessors. Such features include: frames, automatic menu generations, incorporation of graphics, and methods for performing both backward and forward chaining. However, the sophistication of these shells dictates the need for training courses or an appreciation of some aspects of knowledge engineering. Thus, many of the newer generation of shells enforce the division between domain specialist and knowledge engineer.

---

[1] Rules with single consequences.

In contrast, the expert system shell Crystal[1], one of the new generation of shells, was readily adopted by the company's casting room expert. Crystal is arguably the least sophisticated of the newer shells, but has a rule language syntax which is easy to understand, intelligible and case insensitive. Its knowledge representation language consists solely of rules. However, the shell's procedural nature was easily understood by the company staff, who were also impressed by its speed compared to the Prolog coded system. Furthermore, an interface into the spreadsheet package, Lotus-123[2], was available, freeing the company from the need for a computing specialist. The knowledge base has been completely restructured to accommodate the knowledge representations allowable in Crystal, and to render it easily understood and maintainable by the company. Work is ongoing to provide a natural language interpreter to interface this expert system into the company quotation system and to short circuit the menu system.

CONCLUSIONS

The new generation of expert system shells are a considerable improvement on their predecessors. However, the more sophisticated shells require an understanding of knowledge based systems and are aimed more at a knowledge engineer. Delivery of an expert system using a more sophisticated shell which has to be maintained solely by a client company employing little computing expertise may prove difficult. For the purposes of this project the Crystal shell appears to provide sufficient facilities to deliver an integrated expert system and database consistent with the company's wishes.

---

1   Crystal Expert System Shell, Intelligent Environments, UK.

2   Lotus 123 Spreadsheet, Lotus Development, UK.

Prolog appears to be an ideal language to implement an integrated database and expert system, although its micro computer performance is relatively slow due to its interpretive nature. A Prolog coded natural language interpreter potentially facilitates integration into existing company software, however a computing professional would almost certainly be needed if enhancements to the interpreter were required.

REFERENCES

Chang, T.C., and R.A. Wysk (1985) An Introduction to Automated Process Planning Systems, Prentice-Hall, U.S.A.

Darbyshire, I.L., A.J. Wright, and B.J. Davies (1987). Development of EXCAP: An Intelligent Knowledge-Based System for Generative Process Planning, Proc. 2nd UMIST/ACME Workshop on Advanced Research in CAM, Manchester.

Eversheim, W., and H. Esch (1983). Automatic Generation of Process Plans for Prismatic Parts, Annals of the CIRP, 32, 361-364.

Eversheim, W., and H. Esch (1985). Survey of Computer Aided Process Planning Systems, Annals of the CIRP, 34, 607-611.

Hodgson, T. and J. Dinsdale (1985). The Changing Manufacturing Environment in the USA, Annals of CIRP, 34, 615-620.

Mill, F. and S. Spraggett (1984). Artificial Intelligence for Production Planning, J. Computer-Aided Engineering, 1, 210-213.

Tulkoff, J. Lockheed's GENPLAN, Proc. 18th Annual Meeting and Technical Conf. of Num. Cntrl Soc., 417-421, Dallas.

AN ADAPTATION OF THE BLACKBOARD MODEL FOR THE MODELLING OF THE KNOW-HOW OF THE EXPERT IN AN AUTOMATION ENGINEERING WORKSTATION

J. P. Pouget, F. Le Gascoin, A. Saidi and J. P. Frachet

ISMCM, Laboratoire de Genie Automatique, Saint-Ouen, France

abstract

In realizing an Automation Engineering Workstation (PTA), a specific modelling concerns the know-how of the automation engineering expert.

This paper contains a second complementary model based on the Generic Macro Constituant (GMC) concept already presented [IMACS-B 87]. It consists in an adaptation of the blackboard model.

This adaptation contains a proposal of a blackboard oriented life-cycle, a definition of measures of complexity on the structure, and the use of the concepts of automatic classification and real-time for the realization respectively of the knowledge sources and the control structures.

keywords

Artificial intelligence, Automation engineering, Computer Aided Design, Expert systems, Knowledge engineering, Modelling

1/ General context

1.1/ The PTA (Automation Engineering Workstation) project

The global field aimed is that of the automated production equipment, unitary or semi-unitary. This restrictive field of studies can be define as that of Automation Engineering [FRA 87].

The goal of the PTA project , initiated by important industrial users of automation equipment such as RENAULT, MICHELIN and PSA, is to define the specifications of a complete CAD system for automated production equipment, in order to overcome the increasing complexity of the Automated Systems of Production (ASP) and to reduce the research costs (which represent on the average 10 per cent of the total cost of an ASP)

The assignment given to the Automation Engineering Laboratory of the ISMCM is to propose a modelling for the structuring of the data related to the ASP cycle life, from its conception to its final stage of development, and its maintenance [IMACS-P 88]. (fig. 1).

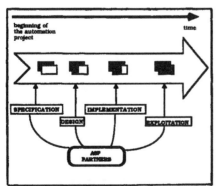

fig.1: *main steps of the ASP cycle life*

1.2/ General view on the model

The modelling of this data structure has been made using an object oriented approach on the reality of the ASP and has lead to the identification an the development of corresponding prototypes, through the object oriented language SMALLTALK, of four consistent and inter-related sub-universes: structure of the suppliers offer, structure of the automated production system, structure allowing the modelling of the automation expert and finally a technical documentation structure [GAM 87], [LEG 88]. (fig.2).

If we put the structure of the ASP, progressively instanciated following time on the horizontal axis, we can see the integration and the role of the three others sub-universes in the PTA chain :

- documentation basis
(technical documentation)

- components basis
(suppliers' offer)

- know-how basis (fig.3)

This first phase of the PTA project now has three extensions based upon the specification of the sub-universes:

- Techno-X project for the technical and the schematic documentation

- BASE-PTA (*) project for the specification of the conceptual model of the structure of the ASP data

- project for the modelling of the know-how of the automation expert, on which this paper is based

1.3/ Modelling of the know-how

Studies conducted with manufacturers have shown that 50 to 80 per cent of the production of an automation engineering department involves using already existing products either directly, or in some modified form. We propose to simulate this re-using process by:

565

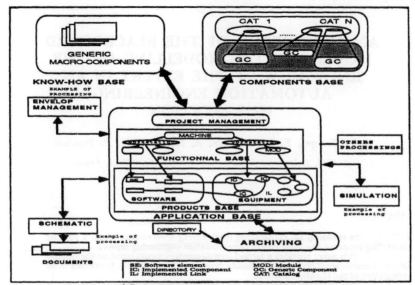

fig.2 : *data structuring in automation engineering*

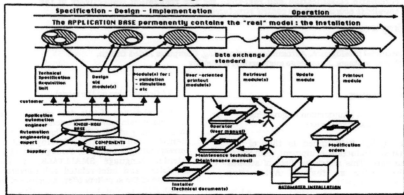

fig.3 : *the PTA: a CAD system for automation engineering*

- the Generic Macro Component (GMC) model [IMACS -B 87], [FRA 87], [POU 89], which can be considered as a generic technological solution (set of hardware and software) carrying out a given function

- the Blackboard model, conceptual support for an expert system helping with the technological choosing process, coming in between the specifications of the automation application and the GMC [LEG 88], [POU 89]. (fig.4).

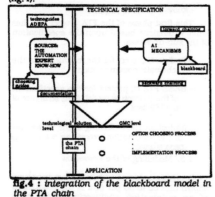

fig.4 : *integration of the blackboard model in the PTA chain*

(*) BASE-PTA: Application Base and Exchange Standart in the PTA

2/ Adaptation of the blackboard model

2.1/ Model presentation

The blackboard model can be classified as an expert system. According to the way it is presented in [PEN 86-1], [PEN 86-2] and [ERM 80], it is made up of three main elements:

-a data structure (the blackboard itself)

-a set of knowledge sources

-a control structure managing the operations on the data structure (fig.5)

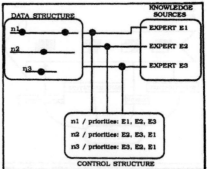

fig.5 : *blackboard model: the three main elements*

2.1.1/ The data structure

It deals with collecting the original data of the problem to be solved, as well as collecting the facts (intermediate or final solutions) produced by the various knowledge sources. The data structure is usually organized into several analysis levels of the problem to be solved.

2.1.2/ The knowledge sources

They present the distinctive characteristic of being able to interact only through the data structure. Each knowledge source can be considered as a 'Meta-rule', consisting of a conditional part and an action part. Thanks to this structure the action of the knowledge sources starts only when specific conditions are met.

2.1.3/ The control structure

It deals with managing the application of the knowledge sources on the data structure. It relies particularly on the the 'Meta-rule' structure of the knowledge source that will best make the reasoning process progress.

2.2/ Justification for using this model in automation engineering

The choice of the blackboard model seems justified mainly because of the inter-disciplinarity of the automation engineering field. As a matter of fact, the blackboard model is based on the concept of separating the knowledge of several experts working on a common data structure in order to solve a problem.

For example, the solving of a problem in the field of automation engineering may require the knowledge of an electrician, a mechanic and a computer scientist.

The ability to structure the blackboard according to the field of the automation engineering has also been a deciding factor in the choice of this model.

2.3/ Adaptation of the model

Starting with the knowledge gathered from various sources (catalogs, professionnal documents [TEC 83], [AP 87], interviews), the knowledge is organized in a way so that the rules are separated into different sets of knowledge sources, and a common structure exists for a blackboard suited for automation engineering [POU 89]. The name of this model is CEM (Coopération pour Experts Multiples - *Cooperation for Several Experts* -).

2.3.1/ Model suited for automation engineering

In the Automation Engineering Workstation suggested, the life-cycle orientation of the ASP has been stressed. For this reason, the suited blackboard model contains a level related to the life-cycle of the manufactured product and a level related to the life-cycle of the automated machine or the system that manufactures this product. Each of these levels is split into sub-levels. Particularly, the life-cycle of the machine or of the ASP contains a sub-level with all the elements related to outside entities (customers, users, maintenance, etc...) and another sub-level related to the evolution of the system through time from the specifications of the automation application to its exploitation and maintenance. (fig.6)

The blackboard is structured into levels, on which the *expressions* (name of the facts on the data structure) are grouped together into *reference elements*. (fig.7)

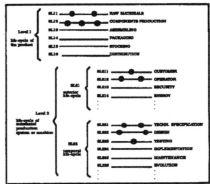

fig.6 : *a life-cycle oriented blackboard*

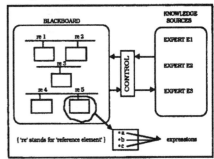

fig.7 : *data structuring in the blackboard model*

2.3.2/ Knowledge sources

The separated knowledge sources correspond to the experts identified during the knowledge extraction phase. Each rule is put in the corresponding knowledge source. For example, we show (fig.8) a structure made up of a blackboard and the knowledge sources stemming from an expert know-how expressed in the beginning of the design of a ASP dealing with the assembling of electric motors. This example [AP 87] is located near the specifications of the automation application.

2.3.3/ Methodology for knowledge gathering

To arrive at a structure such as the one in the former example, a specific methodology is compiled. It is made up of three phases of which we give their main elements:

PHASE 1
Gathering of raw information in example E selected and distribution of this knowledge to the various experts

phase 1.1 : Identification of the Experts: by type of expert

The blackboard model relies on separated knowledge sources, thus, each expert deals with a specific knowledge.Therefore, the first task consists of classifying by type the experts dealt in with example E.

phase 1.2 : Gathering of knowledge

Knowledge is gathered, shaped into rules, and distributed to the proper experts. Each rule is numbered in sequence, and the given number is indicated in the knowledge source in order to allow for checks and modifications.

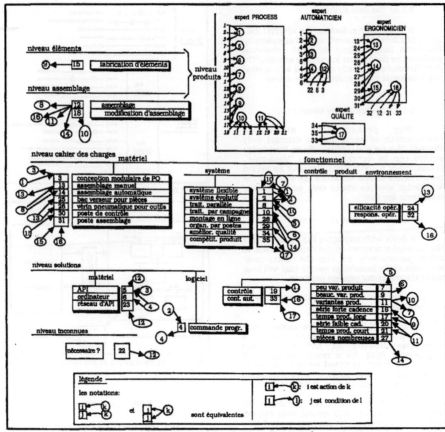

fig.8 : *an example of a life-cycle oriented blackboard*

PHASE 2

Organization of knowledge into tabular form

phase 2.1 : Installation of the rules related to the phase 1.1 with identification of the rules and the experts

phase 2.2 : Modification of the expressions (choice of better ones), elimination of redundancy, homogenization

phase 2.3 : Installation of links to signal identical expressions within a rule

PHASE 3

Elaboration of the blackboard

Installation of levels
Installation of eventual trees
Installation of reference elements with their identificators

PHASE 4

Parallelism of the following phases: construction of the inside of the blackboard, the knowledge sources and the links

phase 4.1 : construction of the inside of the blackboard element

phase 4.2 : construction of the inside of the knowledge sources elements

phase 4.3 : links between the blackboard and the knowledge sources

2.4/ Complements in the model

2.4.1/ Measurements

A measurement of complexity of the representation of knowledge is given with the model proposed. It consists of a measurement with respect to the knowledge sources and a similar measurement with respect to the blackboard element. We give the structure of the measurements defined on the blackboard element as an example.

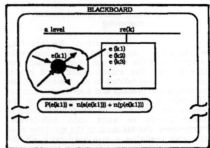

fig.9 : *informational weight of the blackboard element*

◊ The **weight** of an expression e_i is the sum of the number of source uses $n(s_i)$ and of the number of bin uses $n(p_i)$ of this expression by the knowledge sources:

$$P(e_i) = n(s_i) + n(p_i)$$

◊ The weight of a reference element re_k is the sum of the weights of the expressions that make it up:

$$P(re_k) = \sum_{e_j \in re_k} P(e_j)$$

◊ The weight of a level n_l is the sum of the weight of the reference elements that make it up:

$$P(n_l) = \sum_{re_k \in n_l} P(re_k)$$

◊ Finally, the weight of a complete blackboard BK_m is the sum of the weights of the levels that make it up:

$$P(BK_m) = \sum_{n_l \in BK_m} P(n_l)$$

$$= \sum_{n_e \in BK_m} \left(\sum_{re_k \in n_l} \left(\sum_{e_j \in e_k} (n(s_l) + n(p_l)) \right) \right)$$

The importance of the definition of such a measurement is that it gives the conditions required for mastering the complexity of the model. By offering smaller structures for the technological selection process based on the expert system, for the blackboard and for the knowledge sources, it is possible, through break-downs and refining of the structure, to obtain simpler expert systems.

For example (fig.10), it is possible to break the process down by setting a reasonable limit of complexity.

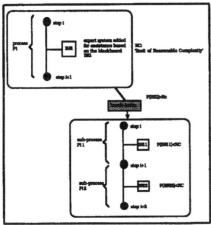

fig.10 : break-down of the assistance related to a reasonable limit of complexity

2.4.2/ Elaboration of internal knowledge sources through automatic classification

It is possible to internally modify the classification by type of the experts corresponding to separated knowledge sources in order to better match the organization to the structure selected for the blackboard.

To create this conjecture, a similarity index is defined on the rules, and an algorithm of the automatic classification field allows for a re-classification of the rules within new knowledge sources better suited to the structure of the blackboard. (fig.11)

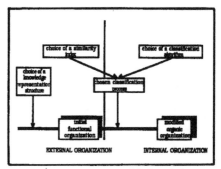

fig.11 : re-organization of the knowledge sources structure

2.4.3/ Control structure

The construction of the control structure scheduling the inferences of the knowledge sources on the common structure deals with real-time and managing process fields, with the use of priorities for the experts, the use of managing mechanisms for conflicts on the common ressources or the synchronizations (semaphores, for example).

A strong analogy can be observed between the notion of task in real-time and the notion of knowledge sources in the blackboard model. (fig.12)

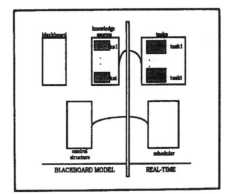

fig.12 : analogy between the blackboard model and the real-time field

conclusion

After the adaptation of the blackboard model in the automation engineering field, research has moved toward the design of prototypes via the object oriented concepts, of the elaborated concepts and their integration into all the projects aimed at the creation of an Automation Engineering Workstation.

references

1/blackboard

[ERM 80]
The HEARSAY II speech understanding project:
integrated knowledge to resolve uncertainty'
ACM Computing Survey, 1980
Erman L.D.F Hayes-Roth V.R Lesser & Raj Reddy

[PEN 86-1]
'Blackboard systems: the blackboard model of
problem solving and the evolution of blackboard
architecture'
The AI Magazine, Summer 1986
H. Penni Nii

[PEN 86-2]
'Blackboard systems: the blackboard model of
problem solving and the evolution of blackboard
architecture'
The AI Magazine, August 1986
H. Penni Nii

2/automation engineering
2.1/modelling

[BOR 88]
'Orientation objet et intelligence artificielle dans la
modélisation et le prototypage d'un univers de base en
Génie Automatique'
Colloque international de productique et robotique
Les apports de l'Intelligence Artificielle, 15-17 Mars
1988
Bordeaux (France)
J.P. Frachet, J.P. Pouget, A. Saïdi

[FRA 87]
'Une introduction au Génie automatique: faisabilité
d'une chaîne intégrée d'outils CAO pour la conception
et l'exploitation de machines automatiques
industrielles'
Thèse de Doctorat d'Etat, Université de Nancy I,
Juillet 1987
J.P. Frachet

[GAM 87]
'Contribution de l'approche orientée objet pour le
prototypage d'une application dans une CAO pour
automatismes industriels'
GAMI (Groupement pour l'Avancement de la
Mécanique Industrielle), Janvier 1987
3, rue F. Hainaut - 93407 St Ouen cédex (France)
J.P. Frachet, J.P. Pouget, A. Saïdi, C. Pailler

[IMACS-B 87]
'Modelling of the know-how of the automation expert
in the context of a CAD of automatic industrial
machines for discontinuous production'
International Symposium on AI expert systems and
languages in modelling and simulation
IMACS Barcelona, 1987
J.P. Frachet, J.P. Pouget, A. Saïdi

[IMACS-P 88]
'Object-oriented model for the specification of an
integrated set of tools for aiding choosing and
exploitation'
12th IMACS World Congress, July 18th-22nd, 1988,
Paris (France)
J.P. Frachet, J.P. Pouget, A. Saïdi

[LEG 88]
'Modélisation du savoir-faire de l'automaticien:
concepts d'Intelligence Artificielle appliqués au
processus de choix technologique dans la chaîne du
PTA'
Mémoire de recherche DEA de Production
Automatisée, Sept 1988
Université de Nancy I
F. Le Gascoin

[POU 89]
'Structuration d'un Poste de Travail de
l'Automaticien: application à l'intégration du
savoir-faire des experts'
Thèse de doctorat, 1989, Université de Nancy I
J.P. Pouget

2.2/ know-how origin

[AP 87]
'Les automatismes programmables'
CEPADUES Editions, 1987
D. Bouteille, N. Bouteille, S. Chantreuil, R. Collot, J.P.
Frachet, H. Le Gras, C. Merlaud, J. Selosse, A. Sfar

[TEC 83]
Les technoguides de l'ADEPA
ADEPA (Agence pour le Développement de la
Productique et de l'Automatique)
17, rue Périer
92120 Montrouge (France)

AN INTEGRATED CAD/CAM SYSTEM FOR A TEXTILE INDUSTRY WITH KNITTING MACHINERIES

A. Vaamonde*, R. Marin*, A. Ollero**

*Departamento de Ing. Eléctrica y de Computadores y Sistemas, E.T.S.I.Industriales,
Vigo, Spain
**Centro Superior de Tecnologías de la Información, Departamento de Ingeniería y
Ciencias de la Computación, Málaga, Spain

Abstract. A new integrated CAD/CAM system is presented, for use with knitting machines. Because of the absence of standards, designers must deal with many different systems to do the same work. The proposed CAD/CAM system makes posible design and manufacturing with a set of different knitting machines. It incorporates users experience, making easy the traditionally hard job of knitting designers and manufacturers.

Keywords. CAD, CAM, Data structures, Computer graphics.

INTRODUCTION

The textil machinery industry have reached an important development with a significative production increase in the last years (Akiyama, 1988).

Diversified consumer needs, development of high value-added products, and shortening of cyclical terms have created a tendency toward multi-items small lot production in the textil manufacturers.

Consequently, textile machinery incorporated into manufacturing process will be able to meet this trend by further adoption of computer systems which has already been incorporated in some machines (Ishida, 1988).

The role of integrated CAD/CAM systems is very important in this context (Kotchan, 1984). These systems can be very useful to overcome some common drawbacks when using the commercial available knitting machines.

Indeed, some recent computarized knitting machines have facilities for pattern design. However, they present several difficulties when design complicated tissue patterns including multi-colors Jacquard patterns, and shaping patterns. In these cases the facilities offered, at a reasonable cost, are very limited when considering the designers requirements. Moreover,

interfacing with friendly graphical design systems are not available.

Furthermore, the programmation systems of many machines with interesting quality and production characteristics, are relatively poor. The cost of computarized machinery with advanced design and programmation systems is usually too high for many small and medium size knitting manufacturers.

On the other hand, is very common that a manufacturer uses many different commercial knitting machines of different types (flat or circular machinery for instance). If we consider the absence of standars in this field, the designers and machine programmers must be familiarized with several systems, with quite different characteristics, for their jobs.

In this paper we present a new, relatively low cost, CAD/CAM system. This system has a man-machine interface specially conceived to promote the designers creative potencial and to minimize the time required to complete patterns and to prepare the machine programs. These programs are loading in different comercial available knitting machines including both flat and circular types. We describe the implementation of this system on a spanish textil industry.

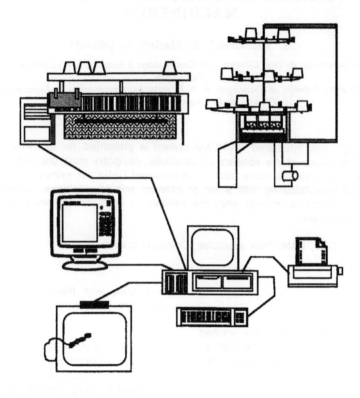

Figure 1. General scheme

SYSTEM ARCHITECTURE

Figure 1 shows the general scheme of the integrated CAD/CAM system.

The main characteristics are:

- Central Unit based on IBM AT with 2 Mbytes of RAM and a graphical card with additional graphical monitor giving a resolution of 1024x 1024, and 256 simultaneous colors selected from among 262,000.

- Tablet digitizer and color printer

- Serial ports for program loading in machines.

Suplementary IBM-PCs (or compatibles) can be included as systems terminals. These units can be used, concurrently, to complete or modify knitting machine programs. In Fig. 2 we show the system with a suplementary PC.

The system can include drivers for different comercial knitting machinery. In the actual implementation in a factory of the spanish firm Pili Carrera S. A. (Mos, Pontevedra) the system can generate programs for 8 flat knitting machines (Stoll CNCA-3B), and one circular knitting machine (Jumberca TLJ-5A).

The programs for the flat machines (see Fig. 3) are loaded automatically by serial ports, avoiding the conventional typing in the machine computer terminal. The programs for the circular machine must be loaded by using a floppy disk and no serial port is available. Thus, the CAD/CAM system generate a floppy disk to be read in the machine.

Moreover, by using the system, it is possible to centralize the machines operator control commands.Thus, operations such as starting, stopping, or machine parameter changes can be performed on line from the system.

Figure 2. Equipment

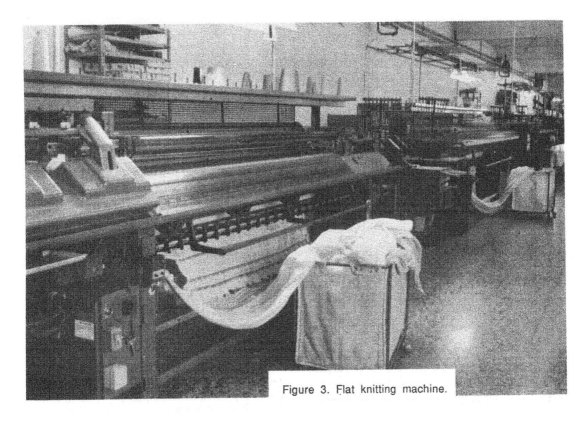

Figure 3. Flat knitting machine.

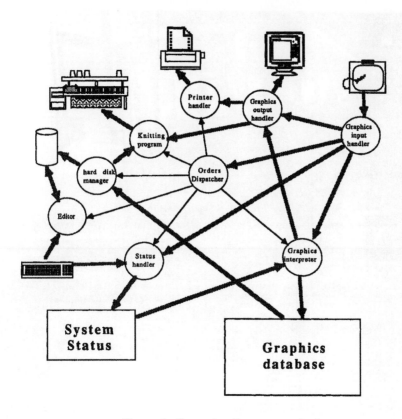

Figure 4. General software organization

SYSTEM SOFTWARE AND FUNCTIONS

The integrated CAD/CAM system software has been developped in Pascal. Near 20,000 lines of code in Pascal have been writen. Fig. 4 shows the general software organization. In this figure peripherics are represented by ikons. Squares and circles mean data structures and processes respectively. Thick lines represent data flow and thin ones control changes from the main program. So graphic input from the tablet, for example, is scanned by the orders dispatcher and then control is transferred to the graphics interpreter. In this way, the process get data from the input handler, and put data to the graphics data base, and to the graphics output handler too, taking into account the system status.

The man-machine graphical interface has been conceived and implemented according with the textile designers opinions. The commands are activated by using a table digitizer with cursor or light pen. Figure 5 shows the menu. This menu respond to ergonomical criteria and minimize the number of orders needed for graphical functions. The main idea is to put in the same only level all the orders, for designers never have to ask themselves where they are.

All interesting 2-D graphical functions to facilitate the designers job have been included. A valuable fuction, not included in the existing pattern design systems, is the realistic representation of the knitting characteristics. By using this function the designer can perceive the real look of knitting as will be produced by the machine (see Fig. 6).

Efficiency in graphical functions and graphical data base management have been obtained by using data structures which combine the properties of dynamic and static representations (Lings, 1986).

When the graphical design is completed, the user can select the type of knitting machine. Afterwards, the CAD/CAM system propose the most suitable knitting program considering the pattern design characteristics. In this way, most knitting machine programs can be generated automatically.

Moreover, a special purpose program editor has been developped and included in the system to modify the machine programs or to write new ones. A set of knitting routines and macros can be called.

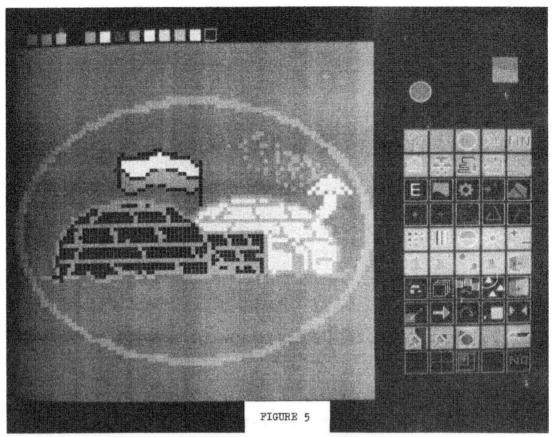

FIGURE 5

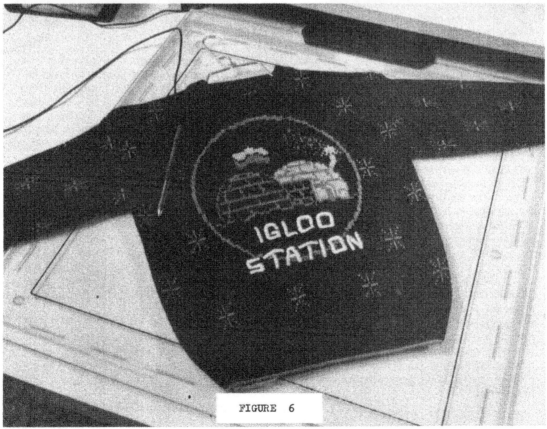

FIGURE 6

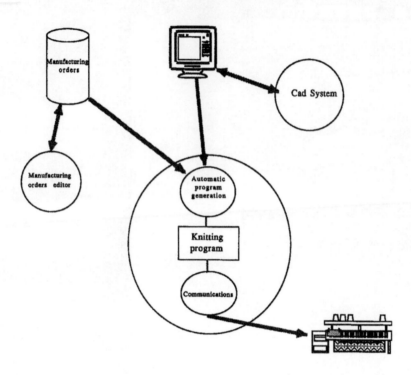

Figure 7. Knitting program generation.

It must be noticed that knitting programs have a certain number of parameters such as the number of lines for a pattern, the number of pattern repetitions, and so on. The system also propose values for these data from the accumulated knitting experience.

When the program with appropiated data is obtained, the user can select the most suitable particular machine for knitting. Then, the system load automatically the program into the knitting machine via serial line, or, alternatively, can write the program on a floppy disk with the appropiated format to be read for a machine. In figure 6, the result of knitting a jacquard with the pattern design in figure 5 is shown.

Figure 7 illustrate the functions of the integrated CAD/CAM system and the process followed for knitting on a flat machine. Knitting program is made by automatic program generator with data obtained from the graphics system. This program search in the hard disk the most suitable set of manufacturing orders, and propose this set to designer, who can change to another set.

A LANGUAGE FOR PROGRAMMING KNITTING MACHINES

A new language for programming knitting machines has been developped. It is based on previous languages for flat knitting machines. However, we include a set of new instructions to programe circular machines. Moreover, new sentences and syntactic modifications have been introduced to make easy the programmers job.

We have developed language processors for the implementation on the spanish textil factory. Particularly, a language processor has been defined to obtain a description which correspond to the codification used in the Jumberca circular machine programmation system. Moreover, we have developed and implemented the necesary functions to obtain floppy disks readables in this circular machine (TLJ-5E) from the above mentioned codification.

Consequently, the actual CAD/CAM system implementation take the following advantages:

- The programming process is much more simpler that when using the codification system a-vailable with the machines.

- The programmers only must know a knitting machine language.

- The programs are machine independent and can be adapted and loaded in any machine consider-ing criteria such us machine productivity, or quality.

CONCLUSIONS

In this paper we have presented the main cha-racteristics of a new integrated CAD/CAM sys-tem for knitting andits implementation on a spanish textil factory. The system is now used in this factory revealing a great number of ad-vantages when comparing with the precedent design and manufacturing process.

Thus,the system can save a great deal of time and simplify the work for pattern making and correction. Complex multi-color Jacquards can be designed graphically in a very simple way with a friendly interface specially conceived to acomplish the requirements of textil designers. From the results of the design, the knitting ma-chinery programs can be generated automatical-ly for flat and circular machines.

Moreover, it is possible to write programs by using a new machine independent language. A special purpose editor has been also developped for the system. Furthermore, the programs can be loaded automatically in 8 differents flat ma-chines, and by using floppy disks in the circular machine.

The full integration of all the functions a-bove mentioned has been obtained. In this way, the time required to complete Jacquard designs, programming patterns, and loading in machines have been drastically reduced. Time reductions from several days to some minutes have been obtained. Moreover, the number of high-valued complex Jacquard designs have been increased significatively.

The system can maintain the centralized operation control of the machines, and so, fur-ther labor reduction can be obtained.

The integration of data concerning design and manufacturing, lead us consider the produc-tion implications of a new design. Alternatively, it is possible to impose design restrictions considering conditions such as the avaibility of materials and knitting machines with the ne-cessary characteristics for knitting the de-signed patterns.

Advanced strategies of production control, including optimization of materials and time, and stock management can now be implement-ed.Thus, the sistem will simplify the implemen-tation of new advanced CIM strategies. Some of these strategies are actually been evaluated.

REFERENCES

Akiyama Y. (1988)."The present states and trend of textile machinery industry in Japanó. Digest of Japanese Industry and Technology, pp 10-16, No 239.

Ishida T. (1988). "Newly Developed Textile ma-chinery and its features".Digest of Japanese In-dustry and Technology,pp 3-9,No 239.

Kochan D.(editor), (1984). "Integration of CAD/ CAM". Proceeding of the IFIP Working Confer-ence on Integration of CAD/CAM. North Holland.

Lings B.(1986). "Information Structures". Chap-man and Hall, Computing, 1986.

AN INTEGRATED APPROACH TO SHOP-FLOOR INFORMATION SYSTEMS — A SUCCESSFUL IMPLEMENTATION IN A MODERN MANUFACTURING PROCESS

G. M. Geary

School of Engineering and Applied Science, Durham University, Durham, UK

Abstract. A joint project between Durham University and Philips Components to improve the yield of the television tube manufacturing process is discussed. The competitive manufacturing strategy of Philips can only be achieved with closely linked information systems covering the manufacturing process variables, the product parameters and equipment utilization and faults. A process management information system (PROMIS) was developed to obtain general information on the process and any areas that required further investigation were studied with a real time data logger/analysis system. An equipment utilization system (EQUIP) was devised to study manufacturing machinery downtime and help change maintenance attitudes from 'emergency only' to preventitive. A product information system is being set up to correlate product faults with manufacturing equipment problems identified by EQUIP. This will lead to a total product identification system being set up over the factory. The paper discusses the benefits of these systems and the problems of implementation and shows that the direct savings to date in the factory due to the project are about £120K/annum leading to savings of about £1.15M/ annum within the next four years. It has also led to a considerable improvement in the attitude of the workforce to the introduction of computer systems throughout the factory.

Keyword. CAM; production control; management systems; industrial control; industrial production systems; manufacturing processes; maintenance engineering; production control; quality control; manufacturing yield improvement.

INTRODUCTION

Philips Components are a division of the international Philips Group and concerned with the manufacture of colour tubes for both televisions and computer monitors. These are supplied to six other Philips factories worldwide, eleven UK manufacturers including Hitachi and Toshiba, and thirty European manufacturers.

The competitors in this market are mainly Japanese (eight manufacturers) led by Hitachi and Toshiba although more recently the Koreans have been making inroads with companies like Samsung. The strength of the competitors has been their efficient manufacturing, willingness to customise, close ties between manufacturing and development, and that they are also customers for their own product in that they also manufacture the television sets. Their weakness from Philips' point of view is that they are geographically a long distance from the European market and that the television set manufacturers (even Japanese) prefer to buy their tubes locally due to cost and reaction time to problems of quality.

The project between Philips and Durham University was set up in 1985 with the overall objective of improving the yield of the Flow Coat process in the Durham Factory (the only UK manufacturer of colour tubes). This is the area in which the three coloured phosphors of red, green and blue are flowed and fixed onto the glass screen. This area was recognised as the primary bottleneck in the manufacturing process in that its direct yield was in the region of 79% causing materials losses of £220K/year and lost opportunity of £2.6M/year with lost sales of £9.6M.

The project had to fit in with the manufacturing objectives of quality and cost effectiveness forced on the company by the competitors and also diversification and flexibility forced by the relatively short product cycle of about five years.

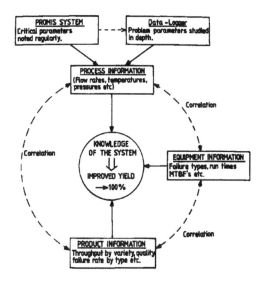

Fig. 1 The Integrated Shopfloor Information System

The manufacturing strategy necessary to achieve
this can only be carried out with closely linked
information systems covering the manufacturing
process variables, the product parameters, and
equipment utilization and faults. (See Fig. 1).
The ultimate objective is to integrate these
systems into a total shop-floor data collection
and analysis system but the lack of resources and
experience in this area made a step by step
approach much more sensible.

PROCESS INFORMATION SYSTEMS

The initial approach to understanding the process
more completely was to analyse the data from the
existing manual system. The manual system
required the operators to note process parameters
on a card and the information was then typed into
a computer and processed later. However this
information was found to be inadequate due to poor
control and accountability of the operators and
also the level of detail was limited.

A system was needed that would improve
considerably on the access and flexibility of the
manual system. To improve control and operator
accountability required fast data input, fast data
analysis and fast retrieval of results. It also
needed to monitor and note any corrective actions
that are taken. The Process Management
Information System (PROMIS) system was developed
for this purpose. A schematic diagram is shown in
Fig. 2.

The data is entered into a portable hand-held
computer once or twice a shift. This computer
then notifies the operator if any reading is out
of specification and informs him if any corrective
action is required. If corrective action is taken
this is also entered into the computer. Control
and monitoring of corrective actions are therefore
ensured. The time of entry is noted automatically
by the machine for display later.

The hand-held computer is down loaded through the
RS232 port to an IBM compatible personal computer
for data display and analysis. Figs. 3 and 4.

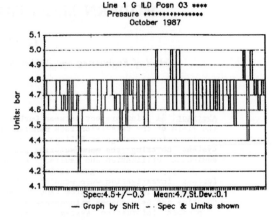

Fig. 3 PROMIS System - Water Pressure Readings
over One Month

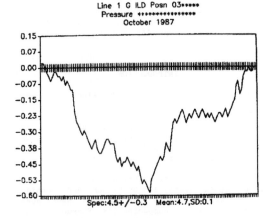

Fig. 4 PROMIS System - Cusum Analysis of Water
Pressure over One Month

Figure 3 is a set of readings of a single variable
water pressure (on Green in-line develop position
3) taken over October 1987, showing eight
deviation from specification together with mean
value of 4.7 bar and a standard deviation of 0.1.

Figure 4 shows a cusum analysis of the same
parameter indicating trends in the readings and
showing that some action was taken around the
middle of the month.

This system enables possible problem areas to be
located for more detailed continuous study using
the data-logging system that was also developed in
the first phase of the project.

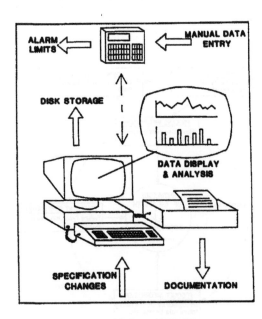

Fig. 2 Process Management Information System
(PROMIS)

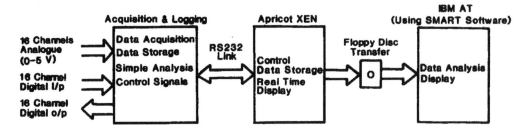

Fig. 5 Overall Real Time Data Logging and Analysis
         System

REAL TIME DATA LOGGING SYSTEM

The specification for this system was drawn up by the University together with the Production Department of Philips and it was then designed and constructed by Durham University and the Engineering Department of Philips with the software produced by a software consultant (Dr. T. Gleaves of Flexible Software, Whitby, North Yorkshire).

The system required 16 analogue channels and 16 digital channels of input operating at a minimum acquisition rate of 1000 readings/sec with 12 bit accuracy. Reading 'windows' of between 1 msec and 10 weeks were required with the start and stop times being shown as 'event', 'time of day' or 'delay time' driven. This gives the opportunity of starting the system on 'out of specification' readings as well as particular times of the day or week. The system permits graphical output in 'real time' with indicators for 'out of specification' readings and calculation of simple statistical analysis such as mean values standard deviation. More complex statistical information and hard copy are obtainable 'off line' on an IBM compatible PC (See Fig. 5).

Using the datalogger the process variations over active cycles can be studied in considerable detail. The variations of average and minimum values of a regulated water pressure over a 24 hour period is shown in Fig. 6 illustrating large deviations from specification. This information is then used to identify specific causes of problems and corrective action recommended.

The system has dramatically demonstrated its usefulness by highlighting parameters in the process that have been running out of specification for a considerable time. In one case a set of flowmeters were all reading 75% low, in another the pressure instrumentation had been located incorrectly, in another the increasing demands of the plant had outgrown the supply of the required flowrate demineralised water. The factory had to respond to the problems of acting on the recommendations of the University staff to improve their calibration procedures, relocate their instrumentation and boost their service supplies of water.

The yields in the flow-cost area of the factory have increased by about 2.8% since the project started with an overall saving in direct yield of £800K/annum. This point project contribution is estimated at about 15% giving a direct saving of £120K/annum.

The second phase of the project was aimed at the areas of equipment utilization and product information/identification.

PHASE 1
DATA INPUT-CARD SYSTEM

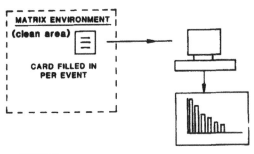

PHASE 2
DATA INPUT-HANDHELD COMPUTER

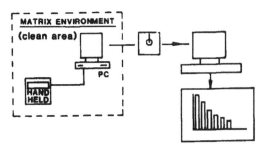

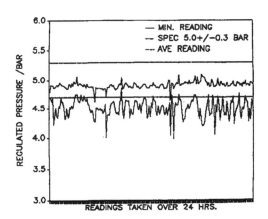

Fig. 6 Data Logger Output - Continuous 24 Hour
        Reading of Air Pressure

Fig. 7 Equipment Information System

EQUIPMENT INFORMATION SYSTEM

The improvement on utilization of manufacturing equipment is a key area of yield improvement and the EQUIP system has addressed this problem. It provides data on machine down-time and assists the Maintenance Department in planning its activities and also collects useful production control data. The introduction of IT in Maintenance has to be carefully phased to fit in with the manual system that has been in use for many years. The approach taken is shown in Fig. 7. Phase one input was onto card and then typed into a PC. Phase 2 provided hand-held computers for input with the data down-loaded to an IBM PC for immediate analysis. This is the system operating at present. Transfer of data to another machine for more complex analysis and display is by floppy disc. The reason for this second transfer via floppy disc is that the environment of the 'matrix area' for which the system was developed is a clean area with restricted access. The system is cheap to run and very flexible in terms of analysis of data and output. A typical output is shown in Fig. 8.

The traditional approach by Maintenance has been to 'react' to major one-off occurrences of products lost e.g. in Fig. 8, fault S88 caused 398

SUMMARY OF BREAKDOWNS   WEEK : 333
DEPT MATRIX
PRODUCTS LOST/FREQUENCY OF INCIDENT

M.CODE	MACHINE	SUN	MON	TUE	WED	THU	FRI	SAT	TOTAL
G11			48		48		48		144
			2		2		2		6
S22			3		0		0		3
			2		2		2		6
M33			39		39	39			117
			1		1	1			3
W55				0		50	50		100
				1		1	1		3
S46			19		4		19		42
			1		2		1		4
S88			199		199				398
			1		1				2
C21			5	4				4	13
			1	1				1	3
G22			13		18		18		49
			1		1		1		3
C22		20	20	1	1			28	70
		1	1	1	1			1	5

Fig. 8   Example of EQUIP System Output - Weekly Breakdown Summary

tubes lost in two occurrences, although a small and consistent fault such as C22 at 70 products lost in 5 occurrences may in the long term be much more of a problem. The first problem would probably produce a reaction whilst the second may not. This system allows the analysis of such problems and is therefore working towards the development of 'proactive' systems and true preventive maintenance.

The equipment cost of this system has been £14K with predicted savings of £500K/annum estimated by the factory.

SCREEN INFORMATION PROCESSING AND TRANSACTIONS SYSTEM (SCRIPT)

This IBM PC based system is a product information system that is being developed initially for use in one area of the factory. Due to its complexity and the interactions of the system with other areas it is intended only as a pilot study.

The data to be collected is:

good products by type;
reject products by type;
reject codes of rejected products;
process path of individual products;
operator data;

and this will provide information on:

product accounting;
throughput and yield analysis;
yield improvement;
process monitoring.

It will provide much closer on-line control of the process and ease fire-fighting. However the major problem is one of identification of the product which is also being tackled. The SCRIPT system will fit into the existing factory IT systems and is the starting point for the development of a complete product identification system in the factory.

The benefits of the system have been estimated as a saving of £500K/annum with a project equipment cost of £60K. It will also be a pilot scheme for future developments as the networking system etc will have to be researched and developed with all its implications for the future of the factory.

CONCLUSIONS

The paper has described the integrated approach taken by the Durham University/Philips Components joint project to solve some of the general problems of the factory. The commitment of everyone has assured its success but this has not been without a certain amount of pain and trial for all parties.

The PROMIS and data-logging system has been running for three years very satisfactorily largely due to the project personnel themselves maintaining and modifying the system as required by the Production Department. The hand-over to the factory is now complete but has required considerable organisational changes within the factory to handle IT based systems.

The direct cost savings over the past four years due to the project has been £120K/annum with £1M/annum of extra sales generated. The potential benefits of these systems are estimated at over £1.15M/annum which should be realised within the next four years.

The project has also successfully overcome many of the emotional barriers to the use of computers in the factory and is therefore acting as a catalyst to the introduction of advanced manufacturing methods into the factory. The attitude of the shop-floor operator has been changed by these projects from almost total opposition to total acceptance and "when can we have some more?" It is felt by Philips management and this has been a considerable achievement of the project in the factory as it has paved the way for other necessary IT developments. This is seen as an almost equal benefit in the long term as the present cost saving.

CIM CONCEPT FOR THE PRODUCTION OF WELDING TRANSFORMERS

P. Kopacek* and K. Fronius**

*University of Linz, Linz, Austria
**Fronius Welding Machine Corporation, Wels, Austria

Abstract. Modern production requires digital computers for various purposes. Especially for design, planning, control, and testing of products computers are absolutely necessary. These applications are summarized in the headline CIM. A lot of CIM packages are commercially available today but only some of them for the demands of small and medium sized enterprises. Therefore a CIM concept for a series of products of a small factory was developed. Manufacturing mainly consists of assembling operations. The automation of some of these is very difficult. As an example an assembling station equipped with two robots is described more detailled.

Keywords. Robots, assembling, control applications, supervisory control.

INTRODUCTION

Microelectronics are rapidly changing the face of manufacturing. Large and small companies alike actually adopt these new methods to improve the efficiency of their operations. Researchers develop advanced technologies suitable for application to manufacturing.

The automation of discrete processes is one of the classical application fields of digital computers today (mostly microcomputers). Especially in production or manufacturing automation NC (numerical controlled) or CNC (computer numerical controlled) controlled production machines have been in use for many years. Today DNC (direct numerical controlled) machines mainly equipped with microcomputers are available for production.

These machines might be the first step towards the so called "Factory of The Future" which will be characterized by total computer control. The elements of such a concept are

- CAD (Computer Aided Design)
- CAP (Computer Aided Production Planning)
- CAM (Computer Aided Manufacturing)
- CAQ/CAT (Computer Aided Quality Control, Computer Aided Testing)

For commercial purposes a PPS (Production Planning System) should be implemented, too. All these elements form a complete CIM (computer integrated manufacturing) concept. It allows a fully "computerized" production. Such systems fill the gap between inflexible high-production transfer lines and low production - but flexible - NC-machines. Computer integrated manufacturing systems are flexible enough to handle a variety of part designs as well as to have a sufficiently high production rate. Therefore they are suitable for small and medium sized factories.

A CIM system is very complex and therefore a well educated staff is necessary for an efficient operation. Unfortunately education for CIM includes a lot of knowledge from "classical" (Mechanical Engineering, Electrical Engineering, ...) as well as "modern" (Electronics, Computer Sciences, ...) engineering fields.

THE SPECIAL CASE - WELDING TRANSFORMERS

In a small sized factory - typical for Austria - producing mainly welding transformers the assembling of these devices is carried out by hand today. Parts of these transformers are manufactured by NC or CNC machine tools. 10 working stations and 3 test stations for assembling are necessary.

The idea was to develop a special CIM concept for the whole production of these devices taking into account the following boundary conditions:

from the technical point of view

- production of six different types of welding transformers with various dimensions in the next three years
- implementation of machine tools available today in the CIM concept
- network of low cost control computers (PCs or similar ones)

from the commercial side

- lower production time
- lower production costs
- increasing the quality of products
- increasing the quantity

Unfortunately from the technical point of view it is impossible to carry out all assembling operations automatically today. Some of the parts to be assembled are of a large size, some of them are "flexible" (cables...). Furthermore the necessary assembling operations include screwing, glueing, and soldering and some of the parts are not constructed for automatic assembling. Therefore only a part of the necessary assembling operations can be done automatically today - from the technical as well as from the commercial point of view.

Due to these facts a modular CIM concept is being developed now. The modularity offers the possibility of a step by step realization. In the first stage the automatic assembling of the primary part of the welding transformer was chosen. This will be described in the following.

ROBOT ASSEMBLING OF THE PRIMARY PART

The necessary assembling operations are described for the smallest of the six different types of the welding transformers, called "Pocket 140". The primary part consists of a heat sink whose dimensions are determined by the attached electronic parts as well as two printed circuits. Today assembling is carried out by hand at three assembling stations. The time consumption amounts approximately 3 to 5 minutes depending on the type of the welding transformer.

The necessary operations in the automatic assembling station are:

*   applying of a heat conductivity paste for the electronic parts onto distinct areas of the heat sink
*   attaching of two transistors, 4 diodes, 2 resistors and 1 thermocouple
*   screwing on of these electronic parts
*   attaching of the printed circuit 1 - the 8 pins of the diodes have to be inserted in the holes of the printed circuit 1
*   screwing of the printed circuit 1 onto the diodes
*   screwing of 3 power cables to the printed circuit 1
*   soldering of some cables and connections onto printed circuit 1
*   attaching and soldering of a capacitor onto print 1
*   attaching of a pressboard plate and the printed circuit 2
*   screwing on of these two parts

The primary part is now ready for further assembling in the welding transformer. These further assembling operations are too complicated for automation today.

The layout of the assembling device for these tasks is shown in figure 1. It consists of two robots, the storage units for the parts to be assembled, the screen printing machine, the necessary transportation devices, and the input-output station. First the head sink is fixed on a pallet. Together with the pallet it reaches the input storage unit. If the next part is needed for assembling the controller will give the command for feeding the screen printing machine. After having finished this process the head sink reaches the assembling station. In the way described earlier the part is assembled completely by the two robots. A lift - marked "L" in figure 1 - transports the part to the lower level of the transportation system and under the assembling station back to the output storage unit. The primary part of the welding transformer is now ready for further assembling while faulty assembled parts are removed by a locking out device.

The two robots used in the assembling stations have the following characteristics:

ROBOT 1:

*   Type RT 280 (IGM) with 6 rotational degrees of freedom - an Austrian product
*   diameter of the working space approximately 2.5 m
*   speed: 80 - 320 deg/sec
*   repeatability: +/- 0.2 mm

This robot is responsible mainly for tool handling operations.

ROBOT 2:

*   Type IRB 1000 with 5 rotational and one translatorial degree of freedom - a Swedish product
*   working space: ball segment; 0.95 x 0.95 x 0.35 m
*   speed: 86 - 480 deg/sec; 1 m/sec
*   repeatability: +/- 0.1 mm

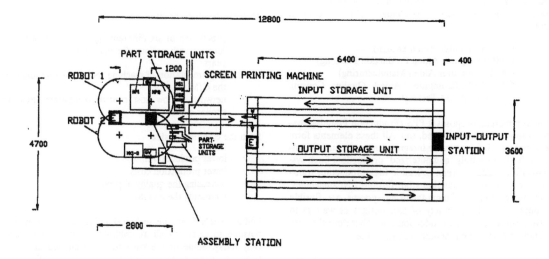

Figure 1: Layout of the assembling system

INPUTS

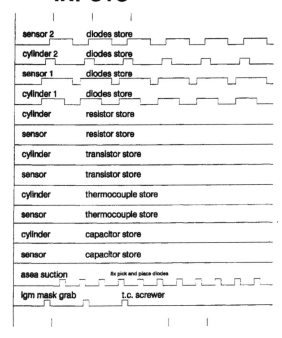

OUTPUTS

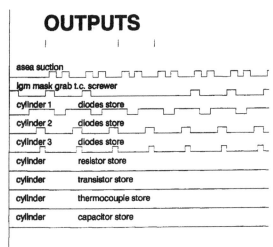

Figure 2: Diagram of the switching sequence (part)

This robot equipped with an additional degree of freedom is mainly responsible for part handling.

Both robots are equipped with various grippers changed automatically by a tool changing system.

CONTROL PROBLEMS

Figure 2 shows a part of the simplified control diagram.

The structure of the control system is hierarchically. On the highest level an IBM AT or 386 industrial computer is responsible for the coordination of

- the controllers of both robots
- the controller of the transportation system
- the controller for the part feeders
- the controller of the screen printing machine

as well as the optimization of the material flow. While the control of the transportation system and the screen printing machine can be implemented without any problems, the coordination of the movements of the two robots is more difficult.

Robot supervising and coordination are done on two levels:

Communication between robots over direct I/O-lines. These I/O-lines are responsible for time cycles in coordinated motion and for primary collision avoidance. Because of the high ability of the robot controls to support event handling depending on direct I/O-lines, this way of robot-robot communication is a very suitable one. Another reason why to choose this kind of communication is the very high safety of data transmission.

Communication robot - computer over standard interfaces (V24/RS232). Over these lines higher level collision avoidance algorithms can be supported. Depending on the actual task the host computer selects different programs for the robots and is responsible for the time management between robots and plant. As an additional feature it is possible to get statistical data about the robot's work and time balance over these lines.

One of the robots is programmable directly from the host computer while the other can be programmed in teach mode only. The robots are equipped with sensors necessary for the complex assembling tasks.

The controllers for the transportation system, the part feeders, and the screen printing machine are implemented as free programmable controllers. The whole plant is divided into autonomous cells from the viewpoint of control. Each autonomous cell has its own controller program. Control tasks are distributed on hard- and software to get optimal control results. Realization of this feature is possible because of the multiprocessor hierarchy of the used control unit. Signals are generated by various types of sensors like limit switches, approximity sensors (by inductivity or capacity), etc.

The free programmable controllers are connected to the host computer by serial lines (RS232/V24). These connections are used to coordinate plant and robots, to specify the task to be done and to transmit statistical data of the transportation system and the screen printing machine.

For the optimization of the material flow a special software package was developed. By simulation and variation of the assembling task critical process data could be found and optimized. For better recognition of bottlenecks the simulation can be run with graphical animation. This feature shows the working plant with all its material flows on the screen during simulation.

CONCLUSION

A modular realizable CIM concept for the assembling of the primary part of a welding transformer was created especially for a small-sized factory. Unfortunately only a few assembling operations can be realized today from the commercial as well as from the technical point of view. As a first step the assembling of the primary part of the welding transformer was chosen for realization. Parts to be assembled are electronic parts, "flexible" cables, printed circuits and a pressboard plate. Necessary operations are srewing, soldering and glueing.

Main parts of the assembling cell are two industrial robots for part and tool handling. As a consequence of the different sizes and geometrical forms of parts and tools both robots have to be equipped with flexible changing systems. Problems arise in developing a soldering device for the robots, construction of "intelligent" grippers for the handling of flexible parts and development of low cost sensors. The control of the whole cell is carried out by a hierarchical computer system. Inputs in the controllers consist of a great amount of signals from measurement devices. The main problem of control is the coordination of the movements of the two robots.

REFERENCES

Kopacek, P. (1987). Computer Aided Manufacturing And Construction. "Technik und Wirtschaft", IWI, Vienna, Vol.1., pp. 135-139.

Kopacek, P. , N. Girsule, R. Hittmair and R. Probst (1988). PCs in CIM Education. Proceedings of the "EMCSR'88", Vienna, Vol.2, pp .793-797.

CAM SYSTEMS WITH EXTREME
REQUIREMENTS OF QUALITY ASSURANCE —
TWO CASE STUDIES

T. Tempelmeier

Fachhochschule Rosenheim, Rosenheim, FRG

Abstract. Two case studies of advanced computer aided manufacturing systems are presented
briefly. In both cases the requirements of quality control & assurance are very high: a detailed
assembly protocol showing the origin of *every* assembled part is necessary as the manufactured
parts are safety-critical in their later environment. This documentation of origin can most easily
and in the safest way be obtained during the manufacturing process, or alternatively, it can be
defined in advance and serve as a very detailed work plan for the manufacturing process. The
possibilities and potential problems of both approaches for acquisition of these data for quality assurance (QA) are discussed.

Keywords. Computer aided manufacturing (CAM), quality control, documentation of origin,
assembly protocol, assembly automation, robotics, applications, Airbus fin shell manufacturing,
Airbag inflator manufacturing.

ASSEMBLY PROTOCOL FOR SAFETY-
CRITICAL PARTS AND DEVICES

When manufacturing safety-critical parts or devices
it is often required by the customer or the authorities to keep track of each assembled subpart in the
manufactured product. In this paper this is referred
to as *assembly protocol* or as *documentation of
origin*. There are various reasons for requiring such
an assembly protocol.

- As a provision of quality control it ensures the
  correct execution of the manufacturing procedures,
  especially with respect to manual operations.
- After manufacturing it enables users or authorities
  to check for correctness of the product - either in
  the course of preventive tests or in the case of
  malfunctions.
- When defective products are detected all assembly
  protocols can be inspected to identify those products which might for instance contain the same
  faulty subparts of a certain lot. By a recall and
  repair/replacement of all involved products a potential loss of life or material can be avoided.

This paper deals with the question how this vital
documentation of origin can be obtained in a safe,
consistent, and verifiable way in partly and fully automated manufacturing systems. Verification is especially important in fully automated systems as data

acquisition is done by the control system and the
related algorithms are thus "hidden" within the various computers. Quality assurance departments and
the authorities must nevertheless have the possibility
of verifying whether the acquisition scheme for the
assembly protocol data is implemented correctly.

The safety of the *design* of the product is not treated here because it can only be ensured in the design
and engineering phase, not during manufacturing.

The paper is based on two advanced CAM projects in
which the author was involved (Tempelmeier 1988b;
Tempelmeier and Gramatke, 1989). The first example
relates to a central manufacturing cell for the Airbus fin shells. Obviously, the safety of the aircraft
depends on correctly manufactured fin shells. The
second example describes the control system for the
assembly of the Airbag inflator, a device which is
used in automobiles as part of the Airbag active
seat restraint system. The severe requirements of
quality control - demanding an assembly protocol
- are also justified in this case as only an absolutely
safe and reliable product can be sold.

TWO CASE STUDIES OF ADVANCED
CAM-SYSTEMS

A Central Manufacturing Cell for the Airbus Fin Shells

The first example deals with the manufacturing of
the Airbus fin shells. These fin shells consist of
100% carbon fibre compound material. To achieve the
necessary strength the fin shells are structured into

(*) The basis for this contribution as described in
earlier works (Tempelmeier, 1988b; Tempelmeier
and Gramatke, 1989) was laid during the author's
employment at *MBB Automation Technology*,
P.O.B. 80 11 80, 8000 Munich 80, West Germany.

so-called modules, similar to a macroscopic honey-comb structure (see fig. 1).

An individual module is manufactured by laminating appropriate pieces of carbon fibre material around an aluminum core. Laminated cores are grouped into segments and placed on segment pallets. Combining the segment pallets finally yields the complete fin shell structure. These work steps are performed within the *fin shell laminating and setting up cell (L & S - cell)*, which is described here.

In the course of the further manufacturing process — outside the L & S - cell — the set up fin shell is tempered and the aluminum cores are removed and recycled to the L & S - cell.

All core transports within the L & S - cell are performed by a linear robot with a working space of 16·4·1.2 m, covering almost the entire cell. Segment pallets are moved by a special *segment pallet transportation and storage system* within the cell. Lamination of the module cores is semi-automatic: a worker places the pieces of carbon fibre material onto the core which is then compressed automatically. For that purpose four three-station rotary devices are available. The three stations are devised for

- transfer from/to the robot,
- manual folding,
- automatic compression.

The basic task of the L & S - cell is essentially the solution of an assignment problem: The individually cut and marked pieces of carbon fibre material assigned to a module core have to be laminated exactly onto that core. Each core must of course be placed only in its individual position of its assigned segment pallet. In view of about 1200 module cores (many of which look almost identical to the human eye), a large number of individual pieces of carbon fibre material, and more than 50 segment pallets, this is not a trivial task.

An assembly protocol showing the correct assignment

> carbon fibre material
> ♦♦
> module core
> ♦♦
> segment pallet

was filled up manually before automating the manufacturing process. In the automated manufacturing cell the correct assignment has to be guaranteed by the cell control system.

The Assembly System for the Airbag Inflator

The second example is the manufacturing system for the Airbag inflator (see fig. 2). Five assembly cells, fifteen robots, 12 minicomputers, and about 10 programmable controllers achieve the final assembly of inflators. In each cell some more parts are added to the subassembly of the preceding cell. Only in the welding cell, no parts are added but a transformation from the state "unwelded" to the the state "welded" is performed.. All parts are delivered to the appropriate cells on pallets or in safety containers

(for the pellets of the chemical propellent and the fuse). All subassemblies are transported between the cells on specific pallets. Apart from the insertion of the fuse all parts in all cells are handled automatically by the robots and other equipment. Figure 3 gives a schematic overview of this assembly system.

All components of the control system are connected via a fault-tolerant local area network, allowing for collection and storage of assembly protocol data (compare fig. 4).

In both projects the structuring principles for CAM systems according to Tempelmeier (1986) and for software systems, as described in Tempelmeier (1988a), have been followed.

ASSEMBLY PROTOCOL DATA IN THE TWO CASE STUDIES

Predefined Assembly Protocol for Manufacturing very complicated Parts with Batch Size 1

Case study I is very much influenced by the complexity of the fin shells. This results in a rather long assembly time for one fin shell (at least several hours). Thus orders to the L & S - cell are executed on a one-at-a-time basis (batch size 1). Furthermore, all module cores are individually marked for safe identification. In a similar way, every prefabricated, i.e individually cut, piece of carbon fibre material is marked during cutting by a number (barcode). If necessary, a documentation of origin can easily be associated with each of the pieces.

All parts being put together in the L & S - cell during the execution of one order can thus definitely and individually be identified. This makes it feasible to work with a predefined, order-specific bill of material. For every order this bill of material is sent to the L & S - cell computer and the whole manufacturing process is driven by and continuously checked against that bill of material. As for the module cores, only "allowed" transport commands are issued to the robot thus guaranteeing correct placement of cores on the segment pallets. It can also easily be tracked which core is in the manual work station for laminating. Before laminating a piece of carbon fibre material the worker has to feed the barcode of the piece into the cell computer, which checks for the correctness of the operation.

The cell computer thus ensures the correct and complete composition of all parts of the fin shell according to the predefined, order-specific bill of material. Automation of this assembly cell thus does not solely result in a rationalization effect, but also fulfills the requirements of quality assurance in a more trustworthy form — even though manual operations are involved.

Organization in fully automated Systems for Mass Production Parts

In case study II the simple scheme from above cannot be applied:

- Due to mass production it is no longer *feasible* to work with predefined, orderspecific bills of material for every manufactured item.
- It would not be *economic* to mark each subpart individually. Instead, subparts can only be identified via the identification of their delivery box. Clearly, all parts within one delivery box must be of the same origin as they cannot be distinguished from each other in that scheme.
- In some cases it is *impossible* to mark the subparts individually. This holds for instance for the pellets of the chemical propellent.
- Fortunately, it is not *necessary* to follow the simple scheme from case study 1 as a more complicated flow of material can still be tracked exactly, due to complete automation through advanced robot technology.

These considerations naturally lead to the following scheme for acquiring the assembly protocol data (compare fig. 5).
- Each flange is marked individually by a bar code as a leading part for the manufacturing process.
- Subparts to be assembled are identified via the identification of their delivery box.
- A record for the documentation of origin is set up for every marked flange.
- In the course of the manufacturing process this record is successively filled with the information of origin of the assembled subparts, yielding one complete record for each completed inflator.

Unfortunately, this method does not work without further arrangements. It would only work with a simple flow of material as shown in figure 6a. In practice the flow of material may be as complicated as shown in figure 6b. Internal buffers and the flow of rejected parts render the tracking of all subparts a nontrivial task.

As an example, a model for the flow of materials in the cleaning & commissioning cell is depicted in figure 7. It may seem that this model is detailed enough to allow for implementing the tracking of all parts in this cell by the cell computer. But still some vital information is missing:

- *Missing information for synchronisation between cell computer and the flow of material.*
  It has not yet been defined what flow of material occurs during the time interval from the start of a robot by the cell computer until the end of the robot operation, i.e. what happens during one robot cycle.
- *Missing information concerning functionality*
  Will the robot automatically take a new part after rejecting a defective one until it finds a good part? What has to be done in case of errors? What is the flow of material during start-up and shutdown of the cell like?
- *Missing information for synchronisation between parallel branches in the flow of material.*
  Incomplete sets of parts must not be deposited. For instance, in the case of a rejected casing in FIFO buffer position 2, no filters must be deposited.

This missing information can easily be included by using Petri nets instead of drawings as in figure 7. Petri nets are useful in three aspects:
- They allow for an exact specification of the flow of material in a cell as it is necessary for the cell computer to trace all subparts.
- They can be used for a highly modular, flexible implementation of the tracing algorithms in the cell computer.
- They can serve as a documentation aid for the programs in the robot controllers. Even difficult tasks as shown in the presented example, where the operations of two cooperating robots had to be specified, can be described unambiguously.

The exact Petri net specification is the common basis for guaranteeing consistency between the model of the flow of material in the cell computer and the real flow as implemented in the robot programs. To reach this consistency a careful implementation of the computer and robot programs according to the Petri net specification is necessary.

Tempelmeier (1988b) reports some more details on this.

INTEGRATING OTHER TEST RESULTS INTO THE ASSEMBLY PROTOCOL

In both case studies results of various additional quality checks are obtained. These checks may have been performed in advance or during the described assembly processes. Some examples from the two case studies are:
- advance a hundred per cent testing of subparts,
- advance sampling and testing of the used material, e.g. of the propellant and the carbon fibre material,
- shape control of subparts by an image system during assembly,
- acquisition of process parameters and results, e.g. during and after welding,
- final visual inspection.

Negative test results cause the related parts to be rejected (compare e.g. figure 7). All test results can naturally be integrated into the assembly protocol by adding some more record fields for the results or for references to other test reports.

CONCLUSION

One effect of automation — besides rationalization — is an improvement in the reliability of the manufactured product. Especially with safety-critical products the aspects of quality control and assurance almost come up to the significance of rationalization when automating a manufacturing process. However, as shown in this paper, careful design of the automated manufacturing process and especially of the control system is necessary to achieve this goal.

REFERENCES

Tempelmeier, T. (1986). Microprocessors in Factory Automation - A Case Study of an Automated Guided Vehicle System and its Integration into a Hierarchical Control Structure. In: Proceedings EUROMICRO '86, Venice, September 15-18. *Microprocessing and Microprogramming*, **18**, 647-656.

Tempelmeier, T. (1988a). The Creative Step in Designing Real-Time Software. In A. Crespo, J. de la Puente (Eds.), *Real-Time Programming 1988*, 15th IFAC/IFIP Workshop on Real-Time Programming. Valencia, Spain, May 25-27. Pergamon Press, Oxford.

Tempelmeier, T. (1988b). Das Steuerungskonzept für die Fertigung des Airbag-Gasgenerators. - Einsatz von Montagerobotern unter extremen Anforderungen der Qualitätssicherung. (The Control Concept for the Assembly of the Airbag Inflator. - Using Assembly Robots under extreme Requirements of Quality Control. (in German)). *Robotersysteme* **4**, 129 - 138.

Tempelmeier, T., and H.-P. Gramatke, (1989). Eine Roboterzelle für die Fertigung der Airbus-Seitenleitwerksschalen. (A Robot Cell for Manufacturing the Airbus Fin Shells (in German)). *Robotersysteme* **5**. (To appear in issue 4).

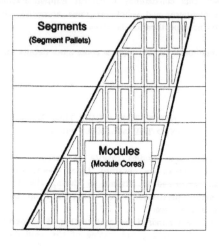

Fig. 1: Airbus Fin Shell Structure

Fig. 2: Airbag Inflator

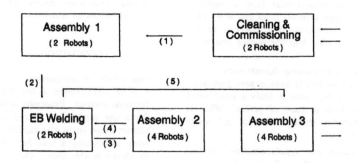

Fig. 3: Schematic Layout for the Assembly
System of the Airbag Inflator

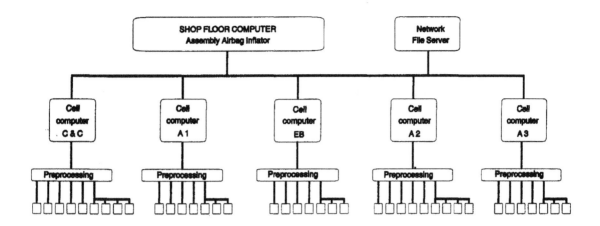

Fig. 4: Communication Structure in the Control System for Airbag Manufacturing

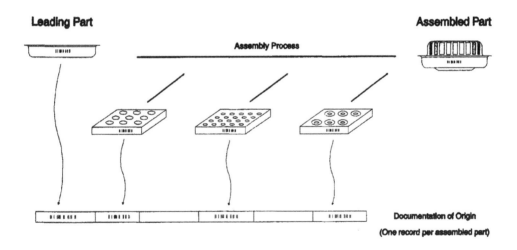

Fig. 5: Acquiring Assembly Protocol Data during the Manufacturing Process

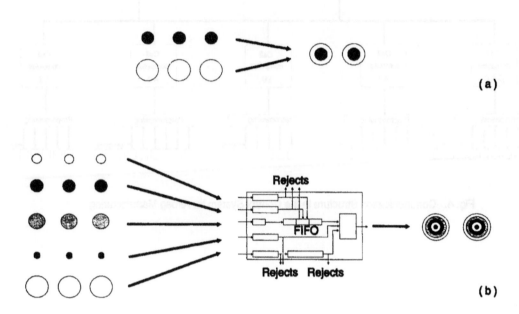

Fig. 6: Simple (a) and Complex (b) Flow of Material

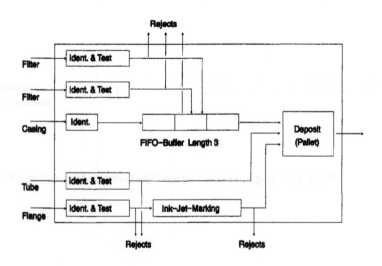

Fig. 7: Flow of Material in the Cleaning & Commissioning Cell

Copyright © IFAC Information Control Problems in
Manufacturing Technology, Madrid, Spain 1989

APPLICATION OF A NOVEL FORCE-TORQUE SENSOR IN ADVANCED ROBOT ASSEMBLY AND MACHINING TASKS

J. Wahrburg

Institute of Control Engineering, University of Siegen, Siegen, FRG

Abstract. Important fields in production processes are only to a very small extent automated by use of industrial robots, because most of today's robots are not yet equipped with appropriate sensors. The reasons for this are above all non-satisfying performance characteristics of available sensors and unsolved questions concerning the interface between sensors and robot control systems. Using the example of a novel force-torque sensor the present paper discusses approaches to the solution of both problem areas. They are based on a well suited architecture to improve the integration of sensors and industrial robots. The most important features of the force-torque sensor are pointed out as well as the coupling of the sensor and robot control systems is described. First experimental results illustrate the presented concept.

Keywords. Robots; sensors; automation; control applications; machining.

INTRODUCTION

The outfit of industrial robots with appropriate sensors is a prerequisite for the flexible solution of more complex manufacturing tasks which exceed the demands of simple applications in industrial mass production as for example in case of spot welding tasks. However, only a very small part of all robots that are in use in industry today is equipped with sensors, although various papers concerning this field have already been published. The main reasons for this fact reside in two areas: At first the performance characteristics of available sensors partly do not yet fulfill the demands that are made on them, and secondly, there are no satisfactory solutions to many problems concerning the interface between sensors and robot control systems as well in physical as in logical layers.

This applies particularly to force-torque sensors, which still are rarely found in industrial applications, but can significantly contribute to the solution of unsolved problems in the areas of assembly and machining. Mounted as a wrist sensor, 6-component force-torque sensors facilitate the threedimensional measurement of interaction forces and torques between robot and workpieces. By this way position tolerances and deviations can be compensated in combination with an appropriate control strategy, which is not possible when using only pure positional control of the robot.

This paper presents some essential features of a novel force-torque sensor and discusses approaches to improve its integration into robotic systems by investigating a suitable system architecture.

FEATURES OF THE 6-COMPONENT FORCE-TORQUE SENSOR

Our investigations are carried out on the basis of a novel force-torque sensor which is in development at the swiss firm Kistler. The measuring

principle is based on the detection of elastic deformations by strain gages. A novel mechanical design of the sensor, as illustrated in Fig. 1, achieves high sensivity and excellent decoupling of the load components (Cavalloni, 1988, 1989). The mechanical precision, including cross coupling errors, is better than 2% without any external compensation.

The excellent decoupling leads to a fairly simple form of the decoupling matrix of the sensor that transforms the measuring signals of the eight strain gage bridges into the six force-torque components. Due to zero elements and pairwise equal elements no time consuming digital matrix calculation has to be carried out, as in case of other force-torque sensor designs, but all forces and torques can be determined directly by analog amplification of the bridge voltages. This facilitates the development of the accompanying sensor electronics which we designed for full integration into the sensor body. Apart from analog amplification elements a built-in microprocessor is included. It provides a digital interface and allows the implementation of additional features such as offset adjustment, self calibration and compensation of unwanted payloads. A photo of the sensor without case is given in Fig. 2. Due to the close connection between transducer and amplification electronics the effects of noise, voltage drops etc. are minimized.

The integrated electronic frees the user from dealing with specific sensor problems and turns the sensor into a flexible, programmable measuring device. It features a high signal processing rate: All force-torque components can be determined and digitally transmitted via a serial RS 485 - interface within 0.5 msec. This assures small measuring delays and allows the use of the sensor in applications with high dynamics.

ARCHITECTURE FOR THE INTEGRATION OF SENSORS AND ROBOT CONTROLS

Problem Formulation

The coupling of sensors and robot controls demands the analysis of the information flow from detecting the sensor signals up to the execution of the robot movements. If the sensor information is used to perform a fine tuning of preprogrammed robot motions, as in case of a force-torque sensor, the robot control must on-line, that is during the movement of the robot, synthetize the motion commands given in its internal program store and the associated sensor signals. This results in a closed sensor control loop, the analysis of which is a prerequisite for the integration of sensors and robot controls (Wahrburg, 1988).

A simplified block diagram of a conventional sensor feedback system is given in Fig. 3. It turns out that the dynamical characteristics of the system is dominated by a large deadtime which is due to the computational burden of the arithmetic operations that have to be performed by the robot control. As clarification two main information processing stages, represented by their associated time delays, are depicted in Fig. 3, whereas the samplers are introduced to symbolize the periodicity of the computations. Computed results are passed from one stage to the other at fixed sampling intervals Ts that for industrial robots range from 20 to 50 msec.

As sensor controlled motions usually are not very fast, the dynamical behaviour of the robot itself is not as critical as in high speed applications. The joint movements approximately may be regarded as decoupled having a first or second order dynamics.

The resulting time delay of the robot system reaches a magnitude of up to 100 msec without taking into account sensor signal processing and transmission that further increase the deadtime. Therefore the sensor loop gain is restricted to small values to maintain system stability and the attainable motion speed of the robot is rather slow. Various proposals for the control of sensor based robots lead to different control loops in comparison to the basic structure given in Fig. 3 (Salisbury, 1980; Craig, 1981; Mason, 1983; Hirzinger, 1986; Inigo, 1987). However, they are often closely adapted to the specific problems being discussed, or based on sophisticated control strategies that impose difficulties to a verification with standard industrial robots. Due to these reasons robot-sensor integration is still characterized by a big gap between scientific research and industrial application.

Information Processing in Closed Sensor Loops

In order to increase the bandwidth of closed loop sensor systems we first define a uniform system architecture which tries to avoid these disadvantages. It serves as a basic platform providing important guidelines for the following design of appropriate control algorithms. Our investigations issue from two main objectives:

- The system architecture should be as generally valid as possible to be adaptable to different sensors and applications without the necessity of fundamental changes.

- The interfaces to the robot control are designed in such a way that standard industrial controls can be used with no or only minor modifications. This avoids the complete redesign of existing systems which may be desirable but is very uneconomic.

The main features of the architecture are illustrated in Fig. 4 and can be summarized as follows:

1. By introducing additional underlying signal paths the bandwidth of the system can be increased in all applications where certain stages of information processing in the robot control can be bypassed. The additional signal inputs are fed into the robot control at those points where temporary data are generated also within the control, and must be supplied at the corresponding sampling intervals.

2. In order to impose no further arithmetic burden on the robot control due to sensor related tasks it is presupposed that these tasks are carried out by external processing units. This leads to the use of intelligent sensors which are equipped with their own dedicated microcomputers. Every sensor must perform a preprocessing of the original transducer signals, adapting all corresponding data formats to the internal data representation and a coordinate system of the robot control.

3. A separate sensor control unit is introduced as a common interface between sensors and robot control. Keeping the uniform system architecture unchanged it is the only part of the system where the specific adaptations to different applications are carried out. Its main tasks are:
   - Providing a common physical interface to the robot control, that is, analog and digital input/output - ports must be implemented.
   - Temporal synchronization of information processing in the sensor electronics and the robot control.
   - Selection of the active signal processing stages and the corresponding control algorithms.
   - Data fusion in multiple sensor systems.

ADAPTATION AND APPLICATION OF THE FORCE-TORQUE SENSOR IN ROBOTIC SYSTEMS

Interface between Sensor and Robot Control

The adaptation of the force-torque sensor to a robot control in accordance with the presented system architecture requires the design of an appropriate sensor control unit, because the versatility of this approach cannot be achieved by a direct connection between sensor and robot control. All computations of the sensor signal processing tasks are performed by the microcomputer of the sensor control unit, which is not subjected to the spatial limitations that hold for the microcomputer built into sensor, and thus may be designed to be more powerful. Futhermore the signal outputs of the sensor control unit are adaptable to digital as well as to analog inputs of a robot control without causing any wiring problems, as sensor control unit and robot control are located close to each other. Fig. 5 illustrates the connections between the different components, showing a high speed serial data link between sensor and sensor control unit which offers high noise immunity and is easily to wire.

The firmware of the sensor system offers two important features that greatly simplify its use.

1. The desired characteristics of the force-torque sensor are fully programmable via its bidirectional robot control interface. All

sensor commands are embedded within the robot motion program, that is, the sensor appears as an integral part of the robot. It does not demand any external initialization or adjustments, e.g., in course of the start-up phase. The sensor characteristics may be altered dynamically during the motion of the robot.

2. Several functions for on-line signal processing are implemented. At first they include the possibility to select those force-torque components that are actually needed to perform a given task. The smaller the number of selected components, the faster they can be updated. The sampling frequency ranges from 2 kHz for processing all components to more than 10 kHz when monitoring only a single one. A threshold may be programmed for every component, the crossing of which causes a binary output signal. At present we are adding additional features, including the performance of filter operations as well as a coordinate transformation which maps the measuring system of the sensor into a robot related coordinate system.

The versatility of this design allows the application of the force-torque sensor in a wide range of applications, including the determination of the absolut robot position referenced to the outer world of the robot, handling, assembly and machining tasks, and monitoring tasks due to safety considerations.

Experimental Results

First experimental results have been obtained by carrying out two projects in cooperation with industrial and scientific partners. One task is given by a deburring problem. An industrial robot picks up metal parts which are sprayed by a rubber tread. The parts are conducted along a stationary grinding wheel in order to remove the rubber burrs, see Fig. 6. Due to the abrasion of the grinding wheel the starting point of the robot trajectory must be continously adapted to the decreasing wheel diameter.

A solution to this problem is obtained by use of the force-torque sensor. The robot approaches the grinding wheel until the sensor detects that a given contact force between workpiece and grinding wheel has been reached. Then the corresponding position is taken as the actual starting point of the trajectory. The monitoring of the threshold is performed by the sensor system which supplies a binary stop motion signal to the robot control. As the robot control is not involved in in sensor signal processing, a very fast response time of the system results which only depends on the dynamics of the joint movements. The robot stops within about 15 msec after the crossing of the force-threshold has been sensed and can approach the grinding wheel with a velocity of more than 200 mm/sec.

Another application deals with a commision task (Krächter 1988). An industrial robot picks up objects (boxes, sacks) from incoming pallets and stacks them onto an outgoing pallet. If the robot works without sensors a troublefree operation demands very narrow positioning tolerances of the incoming objects which in practice cannot be fulfilled with sufficient reliability. Assuming uniform filling of the boxes the force-torque sensor, mounted as shown in Fig. 7, allows to detect inadmissible deviations. After picking up the box the sensor will generate output signals that represent the magnitude and direction of the wrong gripper position with a resolution of about

1 cm. This is a sufficient accuracy for this task and enables the robot to modify its preprogrammed trajectory accordingly. In addition, the sensor may be used to monitor forces and torques when stacking the objects onto the outgoing pallet.

CONCLUSION

In this work we explain a strategy for the integration of sensors, especially force-torque sensors, into robotic systems. As sophisticated control algorithms in this area hardly found their way to industrial applications so far, at first a basic system architecture is introduced to handle the information flow in sensor-based robotic systems. It is adapted to the use of commercially available robots and supplies guidelines to the design and the application of intelligent sensors. This particularly includes the definition of data interfaces and the distribution of computing tasks.

The realization of the concept is demonstrated by the design of the signal processing system of a novel force-torque sensor. A built-in microcomputer and a dedicated sensor control unit turn the sensor into an adaptable and powerful measuring device. Based on the successful results of first industrial applications we are now working on the development of enlarged control algorithms as well as on the extension of the concept to multi-sensor systems.

REFERENCES

Cavalloni, C., L. Schmieder, and J. Wahrburg(1988). A novel static six-component force-torque sensor with integrated fast electronics. Proc. Sensor '88, Nürnberg/W.Germany, May 3-5. (in German)

Cavalloni, C.(1989). A novel 6-component force-torque sensor for robotics. Proc. Int. Conf. on Advanced Mechatronics, Tokyo, May 21-24.

Craig, J.J., and M.H. Raibert(1981). Hybrid position/force control of manipulators. Trans. ASME, J. Dynamic Syst., Meas., Contr., vol. 102, pp. 126-133.

Hirzinger, G.(1986). Robot systems completely based on sensory feedback. IEEE Trans. Ind. Electr., vol. IE-33, pp. 105- 109.

Inigo, R.M., and R.M. Kossey(1987). Closed-loop control of a manipulator using a wrist force sensor. IEEE Trans. Ind. Electr, vol. IE-34, pp. 371-378.

Krächter, R.D.(1988). Description of a sensor-based commission task using a force-torque sensor. Packung und Transport, pp. 42-44. (in German)

Mason, M.T.(1983). Compliance and force control for computer controlled manipulators. IEEE Trans. Syst., Man, Cybern., vol. SMC-13, pp. 298- 316.

Salisbury, J.K.(1980). Active stiffness control of a manipulator in cartesian coordinates. Proc. 19th IEEE Conf. Decision and Control, Albuquerque Nov. 1980.

Wahrburg, J.(1988). Integrating multiple sensors and industrial robots: System architecture and control aspects. Proc. 3rd IEEE Symposium on Intelligent Control, Arlington, Aug. 1988.

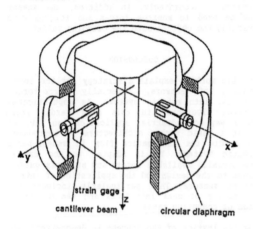

Fig. 1. Mechanical design of a novel force-torque
sensor (Cavalloni, 1989).

Fig. 2. Prototyp of the Kistler force-torque
sensor with integrated electronics.

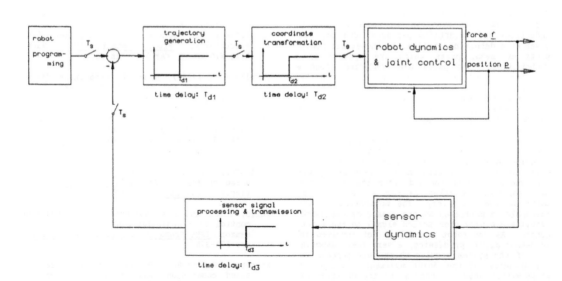

Fig. 3. Block diagram of a conventional sensor feedback system.

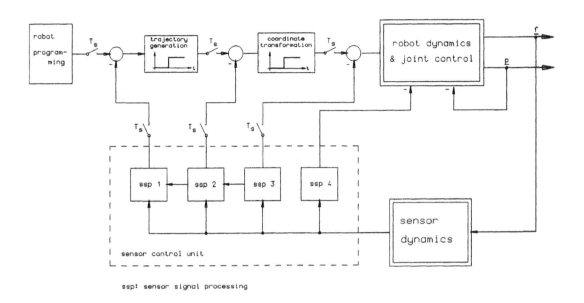

Fig. 4. Flexible system architecture to integrate industrial robots and sensors.

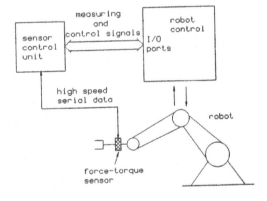

Fig. 5. Sensor control unit as
interface between force-torque
sensor and robot control.

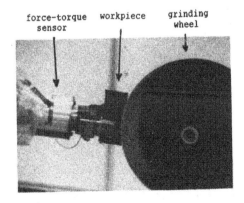

Fig. 6. Application of the force-torque sensor in
a deburring task.

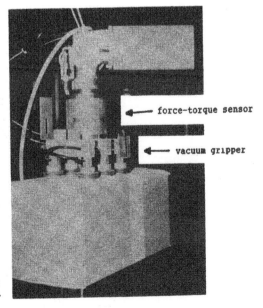

Fig. 7. Force-torque sensor mounted
on a commission task robot.

Fig. 4. Flexible system architecture to integrate robot(s), robots and sensors

A CAD BASED VISION SYSTEM FOR
IDENTIFYING INDUSTRIAL WORKPIECES

A. Sanfeliu and M. Añaños

Instituto de Cibernética, Barcelona, Spain

Abstract :   In this work we present a vision system for identifying 3D industrial workpieces, which uses models designed in a CAD system. The system only uses information extracted from the geometric model, and does not use any specific perspective projection. Moreover, the system is designed to identify models in any orientation and position, and which can be partially occluded by other objects. At present, the system is prepared for models with plane surfaces and which have explicit or implicit parallel straight segments (Sanfeliu, 1987).

Keywords :   Image processing, pattern recognition, artificial intelligence, CAD

1 Introduction

The identification of objects in industrial tasks is one of the important goals for the manipulation of workpieces by robots. The industrial objects are three-dimensional (3D), and this fact produces enormous problems for the identification of workpieces when they can come in any position and orientation, or partially occluded by other objects. Most of the existing industrial vision systems assume that the objects are isolated and that the objects can be described by few known perpectives. When the objects appear partially occluded, the last assumption is almost never true, and the objects must be identified without knowing the position and orientation in advance. This implies that, from the models, is necessary to have many more perspective projections, or to work directly with the 3D model without any commitment about the perpectives. Moreover, the occlusion creates an additional problem, because of the main features for identifying an object must be independent of rotation, position, scale, perspective projection, and very robust against noise or shadow problems (typical in images with overlapped objects). In last years, there have been several studies on the topic for example (Ayache and Faugeras, 1986; Bolles and Cain, 1982; Brooks, 1981; Grimson and Lozano-Perez, 1985; Oshima and Shirai, 1983; Sanfeliu, 1984).

In this work we describe a system which can overcome the aforementioned problems for workpieces which have plane surfaces, and which have parallel straight segments. The system works with models designed in a CAD system and from where the main geometric features are extracted. Moreover, the system does the matching from the geometric model to the 2D image without using any specific perpective projection.

Although the system is very ambiguous and it can identify a large number of objects, the restriction of the system is that the objects must have parallel straight segments which can be extracted. Other features have been considered, for example, segments with constant curvature, elemental surfaces, ... , but, for this work, we have simplified the problem in order to discover the process related problems.

The complete system consists of the following modules:

1. A CAD system to design the 3D reference models.

2. A structure transformer module which converts the CAD format into several structures used in the analysis process.

3. An initial hypothesis module, which creates the best partial matchings between the objects extracted from the 2D image and each one of the reference models.

4. A position and orientation module which infers from the 2D coordinates, the 3D coordinates (depth and translation) and the three angles of rotation.

5. A verification module which compares the candidate models with the objects of the image. The result of this module is the identification of the image objects.

In this article, we will describe the three first modules, since the last two have been described previously (Sanfeliu, 1987; Sanfeliu, 1988a; Sanfeliu, 1988b; Añaños y Sanfeliu, 1988). The principle of the method has been described in (Sanfeliu, 1987) and in (Sanfeliu, 1988b).

2 The CAD modeling and the structure transfomer module

We have used the CAD modeling PADL2 developed by the University of Rochester to design the reference models. The PADL2 allows us to generate the models by CGS modeling, and then, has the possibility to transform the CGS model into a boundary representation. The CGS modeling uses some primitive objects defined in the system, and other constructed by the user, for the design of the models. The reason to adopt the PADL2 came from the fact that the source code can be easily obtained and that the PADL2 low cost.

In our system, we generate the models in PADL2, transform them into boundary representation and create the structures needed for the analysis processes. Fig.1 shows the two basic representations obtained from the PADL2 models. The module which transforms the boundary representation into the analysis structures is the structure transformer module. The two derived structures obtained from PADL2 model are described below.

---

*This work has been partially supported by a grant from the Fundación Areces.

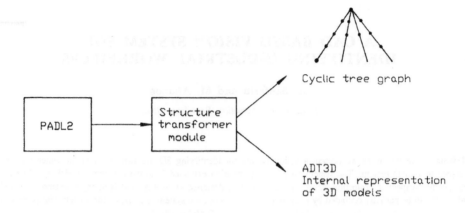

Fig. 1 *From PADL2 to internal representation of 3D models.*

The first structure contains only the information of the parallel straight segments (PSS) grouped in form of a cyclic tree graph (Sanfeliu, 1987). The cyclic tree graph has in every branch all the straight segments which are parallel, ordered following a convention (Fig.2). Each node is a straight segment and has associated attributes which consist of geometric and numeric information. The tree is cyclic because the order of the nodes may change although the relative precedence is always mantained. The same happens with the branches, but in this case, is the precedence of the branches which is mantained.

The second structure consists of the complete information about the relations between the vertices, arcs and surfaces. Moreover, it incorporates information about the orientation of the surfaces and additional information for the analysis processes. Fig.3 shows the structure adopted for the representation. The representation has been built as an abstract data type, from where we can do all the required operations. From this representation, named ADT3D, we derive other data structures for the processes of initial hypothesis, and verification. Using ADT3D we can obtain the information for any of the analysis procedures.

3   Initial hypothesis module

This module is one of the most importants of the system, since we have to try to identify the objects of a 2D image from a series of 3D reference models, without using any specific perspective projection. This fact implies that the hypothesis generation is done without information of depth, either the position and orientation, which increases the number of ambiguities which can be produced. Moreover, the objects in the image can be partially occluded and part of the main features for the identification may be lost.

The initial hypothesis generation is based on the following property (Sanfeliu, 1987): the ratio between the lengths of two parallel straight segments is approximately independent of translation, rotation, scale and perspective projection, if the distance of the object to the camera is much greater than the maximun distance between two points of the object. Using this property we have built the cyclic tree graph, from where we make the generation of the initial hypothesis. The process of hypothesis generation is the following one:

1. We extract from the image the straight segments using a new algorithm developed at Instituto de Cibernetica (Añaños and Sanfeliu, 1988), which is very robust againts noise, and from where we obtain very good results of this length and angle (we try before other methods based on Hough transform but the results were too sensitive to external conditions like light, noise, ... -).

2. We build the cyclic tree graph of the image, computing the ratio between consecutive parallel straight segments.

3. For every model, we match every branch of the model with every branch of the image, computing the cost of identification. If the cost is higher than a threshold, the matching is rejected. The computation of this matching is done as follows:

   - From the lists of straight segments of the branch of either the model and the image, we generate a list of lists which asociates for every segment of the model list, all possible segments of the image, and for every model list all possible segments of the image. The only restriction imposed in this process comes from the assumption that the object is far away from the camera. This assumption allows us prune the generation of the tree by means of the condition:

     $$l_k^{I_i} \leq l_j^{M_p}(1 + \epsilon_l) \qquad (1)$$

     If this condition is accomplished, $l_k^{I_i}$ is accepted, otherwise is rejected. ($l_k^{I_i}$ is the length of the straight segment $r_k^{I_i}$ of the image - segment k of the branch i of the image I - and $l_j^{M_p}$ is the length of a segment of model M_p). The error ϵ_l is incorporated to overcome the problems due to noise.

   - Once all the straight segments of the image has been associated with the segments of the model, we construct a data structure for the process of computing the costs of branch identification. The computation of matching a sequence of the image and a sequence of one model must take into account the following issues (Fig.4):

     - Since, in the image, straight segments which does not belong to the model can appear, the

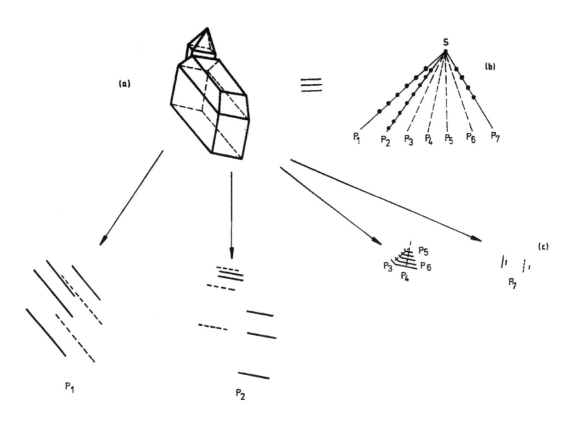

Fig. 2 (a) A 3D model, (b) cyclic tree graph, (c) sets of parallel straight segments of the model.

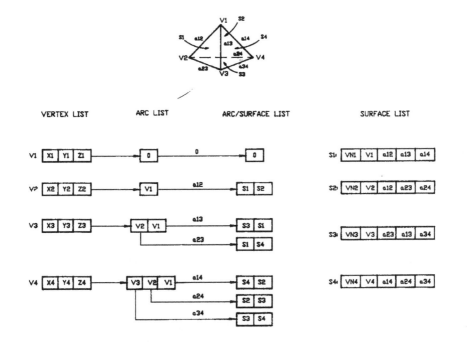

Fig. 3 Internal representation of a 3D model.

cost of deleting this straight segment must be included in the computation cost.

- Since some straight segments of the model could not appear in the image, the cost of inserting it must be included in the computation cost.

- In the computation of matching , the precedence between straight segments is always mantained, that is, the sequence of straight segments a-b-c-d never can generate a-c-b-d.

• The cost of matching two sequence of parallel lines, $d(B_i^I, B_j^{M_p})$ where B_i^I is the branch i of the image and $B_j^{M_p}$ is the branch j of the model M_p, is computed for every combination of node (each straight segment) matchings, where the deletion and insertion of nodes is included. The cost of matching is computed as follows (see Fig.4):

$$d(B_i^I, B_j^{M_p}) = \min_{conf.} \frac{1}{N} \sum_{j=1}^{N} C_{ij} \qquad (2)$$

where C_{ij} is computed as follows:

(a) If one node, that is a straight segment, of the image is missing, then $C_{ij} = 1$. This operation is denominated insertion of a node.

(b) If there is one node in the image which has not correspondence with a node of the model, then $C_{ij} = 1.01$. This operation is denominated delation of a node. In this case, the cost is greater than 1, since we have realized that is better to punish the cost of deletion that the cost of insertion.

(c) Otherwise

$$C_{ij} = 2* \mid 1 - \frac{C_j^M}{C_i^I} \mid \qquad (3)$$

where

$$C_j^M = \frac{l_j^M}{l_{j+1}^M} \quad \text{and} \quad C_i^I = \frac{l_i^I}{l_{i+1}^I} \qquad (4)$$

The structure of the generation of the initial hypothesis is, at present, completely open and there is not any heuristic taken into account, and is prepared for including new restrictions for pruning the tree of possible solutions. These restrictions are now being considered to decrease the computation time.

4   Position and orientation module

This module infers the position and orientation in three-dimensional coordinates, from the two-dimensional coordinates of the image object. This calculation is done with the candidate solutions obtained in the process of generation of the initial hypothesis. The computation of the position and orientation is guided by the candidate reference model. The calculation is explained in detail in (Sanfeliu, 1987) and in (Añaños y Sanfeliu, 1988) and a brief resume is shown below.

In the computation of the depth we require at least 4 points which must not be coplanar (we use 6 points). With three points and the focus point of the camera (we simplify the camera model to a point-hole camera), we obtain the following equation:

$$as^8 + bs^6 + cs^4 + ds^2 + e = 0$$

From the solution of this equation (Sanfeliu, 1987), we obtain eight solutions. Four solutions are negative and can be rejected. From the other four we keep the solutions which has the real part large and the imaginari part zero or very small (due to the uncertainties of the location of the points). In order to select the best solutions, we recompute again the depth with three new points. By matching the solutions we obtain one or two solutions through computing the minimum mean cuadratic error of the solutions.

The results of the depth are not very precise, we allow depth solutions which are in between ±5%, but good enough for identifying the object. With the results, we compute the orientation and position of the model and we keep the best results for the verification process. The results that we get are quite good and they do not depend too strongly on the computed depth. However, we realize that the selection of the candidate points for computing the depth must be done carefully. We have used several strategies, and one of the best ones is to select the first two points (of three) close together and the other one, as far as possible.

5   Verification module

In this module, the verification process is performed. From the results of the position and orientation module, we obtain the perspective projection which will be used for computing the cost of identifying the image object by a model. The computation of the verification process is based on a similarity measure between graphs explained in (Sanfeliu, 1984; Sanfeliu, 1987). The object identified is which has the minimun distance value between the image object and every candidate model obtained in the generation of the initial hypothesis.

For the present work, we have only used straight segments in the verification process. Every straight segment of the model is compared with the segments of the image, and only the segments of the object image which are very close to the model, in orientation and position, are admitted. For these segments, the cost of identification is computed and included in the value of the similarity distance. In the verification, there are several parameters taken into account: the length of the segment, the relative orientation between both segments and the relative position of both segments. Moreover, also it is included the case that a segment of the model be matched to several in the image or viceversa. This verification process allows to make flexible matchings overcoming small problems in:

• Position and orientation of the model in the object.

• Errors in the extraction of the straight segments.

• Errors in the straight segments due when an object overlaps another one.

• Colinearity of several straight segments after computing the perspective projection.

A brief description of the verification process is explained below. The verification consists of two steps: (1) the association of image segments to model segments, and (2) the computation of the cost to identify the object by each one of the candidate models. The association between a straight segment, $\vec{r_i}$, from the image and a straight segment, $\vec{r_m}$, consists of verifying that the orientations of both segments, θ_i, θ_m, are into a predefined interval, and that the geometric centers are also into a predefined interval.

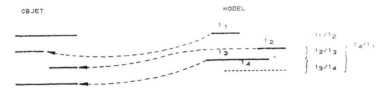

Fig. 4 *Initial hypothesis process matching.*

Fig. 5 *3D model of an electrical transformer.*

The computation of the identification cost is done by calculating the cost of recognizing each one of the visible surfaces. This is done as follows:

$$C_T = \frac{1}{N_{ST}} \sum_{j=1}^{N_{Ts}} C_{Sj}$$

where N_{ST} is the number of visible surfaces of a model and C_{Sj} is the associated cost to the surface j. The computation of each surface is:

$$C_{Sj} = \frac{1}{N_{Tr_j}} \sum_{j=1}^{N_{Tr_j}} C_{r_i}$$

where N_{Tr_j} is the total number of segments which belong to a surface j and C_{r_i} is the cost of verifying every one of the segments. This last cost is computed as follows:

$$C_{Sj} = \frac{l_{mj} - \sum_{i=1}^{N_i} l_{ijp}(1 - \frac{(\theta_{mj}-\theta_{ij})^2}{\epsilon_\theta^2})(1 - \frac{d_{ijm}^2}{\epsilon_d^2})}{l_{mj}}$$

where N_i is the number of image segments associated to the model segment j, θ_{mj} and θ_{ij} are the respective angles of associated segments, d_{ijm} is the minimum distance between the model segment and the others, l_{ijp} is the length of the image segments, l_{mj} is the length of the model segment, ϵ_θ is the maximum diference between angles and ϵ_d is the maximun allowed diference between segment distances. When $C_{Sj} < 0$ we take $C_{Sj} = 0$.

A sequence of the recognition process can be seen in Fig.6-9. Fig.5 shows the model used for identification. Fig.6 shows the original image. The extraction of straight segments can be seen in Fig.7. Fig.8 shows the best position and orientation of the object over the identified image object, and Fig.9 shows the straight segments considered for the computation of the similarity distance.

6 Conclusions

In this work we present a system to identify objects in a two-dimensional image from a three-dimensional reference model, without using in advance any specific perspective projection. The only restriction imposed is that the object must be far

Fig. 6 *Original image.*

Fig. 7 *Extraction of straight segments from the image.*

away from the camera. The system is based on the identification of three parallel lines of the image object, to generate the initial hypothesis. The results obtained show that the system can identify objects although they are partially occluded, and that the recognition looks quite robust. This last issue is now being studied carefully in order to verify the robustness of every process of the system. The research is now conducted to relax the restriction of three parallel lines to only two, without increasing the computation time. On the other side, we are now considering other features to prune the tree generated in the initial hypothesis process.

7 References

Añaños M. y Sanfeliu A.; 1988 Posicionamiento y verificación de objetos 3D parcialmente ocultos representados por fronteras conocidos 4 puntos de su proyección. *III Simposium Nacional de Reconocimiento de Formas y Análisis de Imágenes, Oviedo 27-30, Sept..*

Ayache N. and Faugeras O.D.; 1986 HYPER: a new approach for the recognition and position of two dimensional objects. *IEEE Trans. on Pattern Analysis and Machine Intelligence, Vol PAMI-8, NO. 1, Jan..*

Bolles R.C. and Cain R.A.; 1982 Recognizing and locating partially visible workpieces. *Proc. IEEE Comput. Soc. Conf. Pattern Recognition and Image Processing, Las Vegas, NV, Jun., pp.498-503.*

Brooks R.; 1981 Model-based three dimensional interpretations of two dimensional images. *Proc. 7th INt. Joint Conf. Artificial Intell., Vancouver, B.C., Canada, Aug., pp.619-624.*

Grimson W.E.L. and Lozano-Perez T.; 1985 Recognition abd location of overlapping parts from sparse data in two and three dimensions. *IEEE Comput. Soc. Int. Conf. Robotics, St. Louis, MO, Mar..*

Oshima M. and Shirai Y.; 1983 Object recognition using three dimensional information. *IEEE Trans. on Pattern Analysis and Machine Intelligence, VOL. PAMI-5, NO.4, Jul., pp.353-361.*

Sanfeliu A.; 1984 A distance measure based on tree-graph grammars: a way of recognizing hidden and deformed 3D complex objects. *7th Int. Conf. on Pattern Recognition, Montreal, Canada Jul. 30- Aug. 2, 1984; also in Instituto de Cibernetica Technical Report IC-DT-1984.01.*

Sanfeliu A.; 1987 Parallel straight segments in the recognition of 3D objects. *Instituto de Cibernética, Technical document IC-DT-1987.04.*

Sanfeliu A.; 1988a Matching on complex structures: the cyclic tree representation (invited paper). *IAPR Syntactic and Structural Pattern Recognition Workshop, Pont-a-Mousson, France, Sep. 12-14.*

Sanfeliu A.; 1988b Un enfoque general para identificar objetos representados tridimensionalmente en imágenes bidimensionales. *III Simposium Nacional de Reconocimiento de Formas y Análisis de Imágenes, Oviedo 27-30, Sept..*

Turney J.L., Mudge T.N. and Volz R.A.; 1985 Recognizing partially occluded parts. *IEEE Trans. PAMI-7, NO.4, Jul., pp.410-421.*

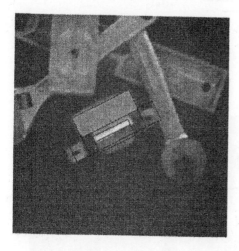

Fig. 8 *The identified 3D model once the orientation and translation has been computed.*

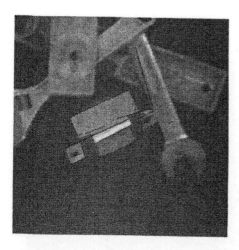

Fig. 9 *Associated image straight segments for computing the cost of identification of the model.*

ASSEMBLY PERFORMANCE OF A ROBOTIC
VIBRATORY WRIST

K. Won Jeong* and H. Suck Cho**

*Mechatronics Department, Research Institute of Industrial Science & Technology,
Kyungbuk, Korea
**Department of Production Engineering, Korea Advanced Institute of Science &
Technology, Seoul, Korea

Abstract. In precision assembly, even small misalignment between two mating parts can cause large
reaction forces and thus make the assembly task be impossible. This paper considers a vibratory
assembly method which can compensate such misalignment for successful insertion task. For this
purpose a PWM controller–based pneumatic vibratory wrist is used to investigate assembly
performance such as search time, assembly force and correctable error range. Since they are critically
dependent upon various system parameters, they were experimentally investigated for a chamferless
peg–in–hole assembly task. Experimental results show that the proposed method yields good assembly
performance; compensation of large initial error, fast searching and small reaction forces.

Keywords. Assembly robot; vibratory wrist; pneumatics; PWM; search time.

INTRODUCTION

For the precision assembly process, many studies have
been conducted and various assembly devices, as a result,
have been developed utilizing industrial robot system
(Cho, Warnecke and Gweon, 1987). However, the
application fields of most of the methods were restricted
because of the geometrical constraints. A typical method is
the one using RCC (Whitney, 1982) device as a passive
method. The device accomplishes assembly task very fast,
but it can be only applied to parts having a chamfer which
guides a male part. On the other hand, active assembly
methods also have been developed, which have sensory
systems for the purpose of measuring the initial lateral
error and controlling the active device. These methods
perform precision assembly task even with large initial
positioning error, but they require expensive devices and
take rather long assembly time. To overcome these
problems, vibratory assembly methods have been
developed (Hoffman, 1985; Savishchenko, 1965; Unimation,
1976; Warnecke, 1988). This method is very attractive in
that it does not use any sensors or search algorithm to
detect the hole. At the instance when the alignment is
established while the peg is vibrating, the insertion force
pushes the male part into the hole. This method can be
applied for assembly of non–standard components or
mating chamferless parts. Most of the previous works of
the vibratory assembly devices have some limitations such
as a fixed vibratory trajectory or lack of adaptability to
changing assembly environments. Therefore, a vibratory
assembly wrist which has adaptability and
programmability has been developed (Jeong and Cho,
1989) based upon a new concept. The vibration of the peg
is induced by a PWM–based pneumatic vibration control
system. Therefore, the desired vibration magnitude and
vibration trajectory can be obtained by changing carrier
frequency ratios. Furthermore, the center position of the
vibration is controlled through a vibration control system.

A series of insertion experiments were conducted to
examine the performance of the vibratory assembly
method through a precise peg–in–hole insertion process.
The compensable initial lateral error, search time and
reaction forces were experimentally obtained for the
various carrier frequency ratios.

VIBRATORY ASSEMBLY WRIST

The schematic diagram of the vibratory wrist used in this
paper is shown in Fig.1, which is composed of three parts;
(1) z–axis compliance, (2) x–y vibrator, and (3) gripper
which can hold a peg. Since the detailed operation
principle and characteristics were described in reference
(Jeong and Cho, 1989), they are briefly described in this
paper. To prevent from any damages in emergency state, a
z–axis compliant device and a limit switch were equipped.
The x–y vibrator, which forced the peg to move in x–y
horizontal plane, was composed of two axes perpendicular
to each other. They had same construction. Each axis was
operated by a pneumatic system which was controlled by a
PI controller based upon PWM method. The horizontal

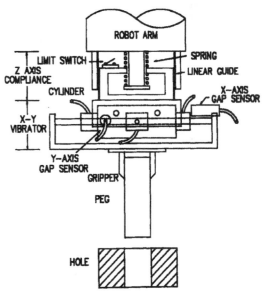

Fig.1 Schematic diagram of the pneumatic vibratory wrist

position of the gripper flange was measured by a gap sensor and fed back to the PI controller, whose output became the input signal to the pulse width modulator. The modulator, then, compared the control input with the carrier voltage and produced the solenoid valve driving signals. When the solenoid valve in the left hand side was energized, the other solenoid valve in the right hand side was deenergized. In this situation the pneumatic force raised in the cylinders pushed the gripper and the peg to the positive direction. On the contrary, the situation was reversed when the solenoid valve in the right hand side was energized. Therefore, the vibration magnitude was dependent upon the carrier frequency. When the carrier frequency was high relative to the natural frequency of the system, the oscillatory motion of the peg disappeared and the position of the peg was accurately controlled by the PI controller. But if the carrier frequency was low, the peg was vibrated with larger amplitude of the same frequency as the carrier frequency. In this case the center position of the oscillating peg was also regulated at the reference position. Because the wrist had perpendicular two axes, it generated search area on the horizontal plane according to the carrier frequency ratios of both axes. Therefore, the initial lateral error within search area could be compensated. The performances of this method were examined through a series of experiments described in the next section.

VIBRATORY INSERTION EXPERIMENTS

The assembly system can be assessed by several factors. For high productivity, insertion time should be as short as possible regardless of the magnitude of initial lateral error. In addition, the reaction forces should be reduced as small as possible to prevent from any damages. Therefore, the performance of the vibratory assembly method were examined experimentally from several view points; the search area within which the peg can be successfully inserted, search time required to find the hole and reaction forces generated during search and insertion period.

The performances of the vibratory assembly method were studied on the experimental rig shown in Fig.2. The major components of the rig were as follows; a vibratory wrist described in the previous section, a microcomputer with data acquisition system (A/DC and D/AC), function generators which generates carrier waves. A 6 axis force sensor (Barry Wright Co.) was installed in order to measure the reaction forces during assembly. The supplementary pneumatic components to operate the vibratory wrist were also included. The hole was positioned to have prescribed initial lateral error using two, x and y directions, linear magnescales, while the insertion depth was detected by a linear position

transducer(LVDT). The center of the wrist vibration (x_d) was specified through the D/A converter of the microcomputer. And the actual position of the peg (x) was acquired through the A/D converter while it was fed back to the vibratory wrist controller. In the experiments, the pneumatic supply pressure was 5.585Kgf/cm^2 (abs.) and the effective orifice areas of the flow control valves for x and y axes were fixed at 5.375×10^{-4}cm^2 and 5.682×10^{-4}cm^2, respectively. The assembly task was inserting a cylindrical peg into a hole, all of which were made of steel. Either of them has no chamfer. The diameter of the peg and the hole were d_p=19.99mm and d_h=20.0mm, respectively. Therefore, the clearance was only 0.005mm and the corresponding clearance ratio ($C_r=(d_h-d_p)/d_h$) was

$$C_r=5\times10^{-4}.$$

The vibrating peg was inserted into the hole by z-directional movement of the hole with a constant insertion speed of 6.72mm/sec. For all experiments, the peg was set to have no tilting angular error. The search time (t_s) was obtained by investigating the vibration signal while the reaction forces (F_x and F_z) from search stage to early insertion stage were measured by the force sensor.

EXPERIMENTAL RESULTS AND DISCUSSIONS

Vibration and Reaction force characteristics

The position and force responses are shown in the Fig.3, in which the peg had a lateral error of (e_x, e_y) = (0.4mm, 0.4mm) and no tilting angular error. The figure shows (a) position response of x axis, (b) insertion depth, (c) reaction force in x axis and (d) insertion force along z axis. The results of y axis are not shown here because it showed similar responses. In the figure, the positive insertion depth indicates that the tip of the peg is not in contact with upper surface of the part, while the negative insertion depth indicates that the tip of the peg is being inserted into the hole. As shown in the figure, the responses can be divided into three distinct stages; approach stage, search stage, and insertion stage. The approach stage is the time interval during which the peg vibrates freely in the air. The reaction forces are not generated in this interval. In the search stage, the vibrating peg is searching the hole in contact with upper surface of the part having hole. The insertion depth in this case is zero. It is noted that the position response in the stage is slightly changed due to the friction between the interface of the peg and hole as compared with that observed in the approach stage. It is notable that the F_z in insertion direction is caused by

impact and increased as the search time elapses because the peg is being pressed in the insertion direction due to the z axis compliance spring while the vibrating peg searches the hole. At the end of the stage the peg is instantaneously inserted into the hole, and thereafter, the peg goes on being inserted while vibrating. This is the insertion stage. The position of the peg is shown to be shifted toward the center position of the hole. The center of vibration, however, shifts towards the zero position i.e. the initial peg position, as time elapses. This is because the PI controller causes the center of vibration to move towards the original peg position. Due to this phenomenon, the mean value of the lateral force F_x was also shifted. In this stage the reaction forces in the lateral direction (F_x) shows oscillation whose frequency component is the same as the carrier frequency of the vibrator. On the contrary to the lateral force, the reaction force in insertion direction F_z shows small magnitude because it is caused only by friction between the peg and the hole.

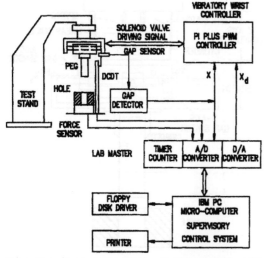

Fig.2 Experimental setup for the vibratory assembly wrist

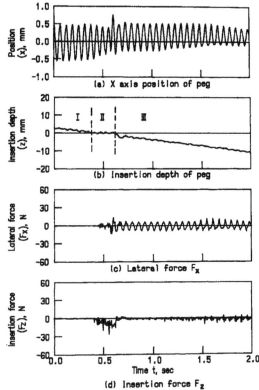

(a) X axis position of peg

(b) Insertion depth of peg

(c) Lateral force F_x

(d) Insertion force F_z

I:Approach stage II:Search stage III:Insertion stage

Fig.3   Position and force responses during insertion
$(e_x, e_y)=(0.4,0.4)$, $(\theta_x, \theta_y)=(0.,0.)$,
$f_x/f_y=16\text{Hz}/16.5\text{Hz}$

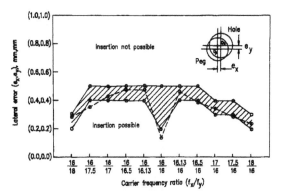

Fig.4   Variation of insertion error range
with carrier frequency ratios

to be gradually decreased, because the vibrating speed gets faster. Fast search means that assembly task can be accomplished within short period. Another observation is that initial lateral error does not greatly influence the search time. This is attributable to the fact that the peg vibrates in a random manner to search the hole center.

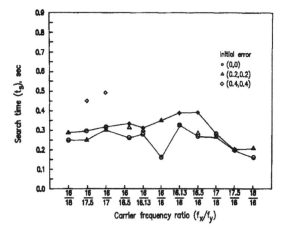

Fig.5   Variation of search time
with carrier frequency ratios

The insertion error range

The insertion error ranges within which the peg insertion is always possible, are shown for the various carrier frequency ratios in the Fig.4. The solid line indicates that the peg having the initial lateral error less than that of the line can be successfully inserted. The dotted line denotes the computed results from the vibration magnitude at the given carrier frequency ratios. The peg can be inserted in the region under the dashed band and can not be inserted in the region greater than the band. Therefore, the band is an uncertain range. The experimental and computed results are relatively in good agreements. When both axes have same carrier frequencies, the correctable error range is small, compared with those of other frequency ratios. The insertion range depends very much upon the hole position because the vibration path is very limited, as discussed in reference (Jeong and Cho, 1989). On the other hand, when the carrier frequencies are different from each other, the possible insertion ranges are expected to be larger than the former case since the vibration trajectory covers larger area so that the search areas become larger. However, the correctable error ranges are not substantially increased, because, although the search motion covers larger area, the vibration magnitude gets small with the combination of higher carrier frequencies.

The search time

Fig.5 shows the search time for the various carrier frequency ratios. The experimental results show that the search time ranges from about 0.2sec. to 0.35sec. As the frequency of one axis is increased, the search time is seen

The impact force

Maximum value of the peak–to–peak impact force in x direction F_x generated during search stage and insertion stage, is shown in Fig.6. As shown in the figure, the impact forces are ranged from about 10N to 35N. As the carrier frequency increases, the maximum impact force is decreased slightly. This is because the pneumatic force in the cylinder becomes small with increase of the carrier frequency. As in the case of the search time, the impact force is slightly increased when the both axes have same carrier frequencies. This is because the peg moves only along specified path. Although the results for y axis has not been shown here, they showed similar trend.

Fig.7 shows the maximum value of peak–to–peak impact force F_z generated during search stage and insertion stage.

In the figure, the insertion forces are ranged almost from 20N to 30N irrespective of the carrier frequencies.

CONCLUSIONS

A vibratory insertion method were experimentally examined for a chamferless peg–in–hole assembly task

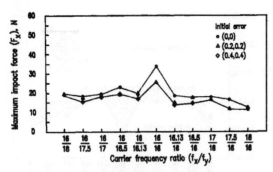

Fig.6   Variation of lateral impact force
          with carrier frequency ratios

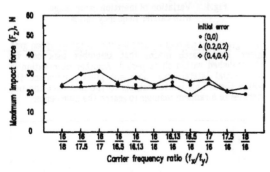

Fig.7   Variation of insertion impact force
          with carrier frequency ratios

using a vibratory wrist. From these experimental results the following major conclusions can be drawn:

(1) The vibratory assembly stage can be divided into three stages, i.e. *approach stage*, *search stage*, and *insertion stage*.

(2) In search stage, the lateral force is small but oscillating due to impact from the vibrating peg. On the contrary, the reaction force in insertion direction is dominant and closely related to the search time. As search time increases, the reaction force F_z is increased due to the spring force pressing towards insertion direction. The magnitude is found to be 27N on average.

(3) The increase in the carrier frequency decrease the search time. Hole searching is accomplished very fast within about 0.27sec.

(4) In insertion stage, there is oscillating lateral force whose major frequency is the same as the carrier frequency of the corresponding axis. The average value of force is found to be 18N.

(5) If the carrier frequencies of both axes are identical to each other, the assembly performance is deteriorated; small search area, long search time, large reaction force, since the vibration trajectory, in this case, forms a very narrow elliptic path.

REFERENCES

Cho, H.S., Warnecke, H.J. and Gweon, D.G. (1987). Robotic assembly: a synthesizing overview. *Robotica*, 5, 153–165.

Whitney, D.E. (1982). Quasi–static assembly of compliantly supported rigid parts. *J. Dynam. Systems, Measur. Control.*, 104, 65–77.

Savishchenko, V.M. and Bespalov, V.G. (1965). The orientation of components for automatic assembly, *Russian Engineering Journal*, 45, 50–52.

Unimation Inc. (1976). Programmed manipulator apparatus for assembly parts. *UK patent* spec.1437003.

Hoffman, B.D., Pollack, S.H. and Weissman, B. (1985) Vibratory Insertion Process: A new approach to nonstandard component insertion. *Robot 8*, 8–1~8–10.

Warnecke, H.J., Frankenhauser, B., Gweon, D.G. and Cho, H.S. (1988) Fitting of crimp contacts to connectors using industrial robots supported by vibrating tools. *Robotica*, 6, 123–129.

Jeong, K.W. and Cho, H.S. (1989). Development of a pneumatic vibratory wrist for robotic assembly, *Robotica*, 7, 9–16.

SENSOR-BASED ROBOTS IN COMPUTER-AIDED MANUFACTURING

K. Fuchs

RWTH Aachen, Process Control in Welding, Aachen, FRG

Abstract. The structure of a flexible sensor aided robot system is described. The
system makes use of CAD and CAM data to reduce the programming time for different
manufacturing tasks. The system permits the integration of serveral robots and sensors
working at the same workpiece within one manufacturing cell.

Keywords. Robot; sensor system; CAD; CAM; coordination; adaptive control.

INTRODUCTION

The increase of productivity, production speed,
flexibility, quality and reliability makes high
demands on the next generation of our factories.
In order to realize the factory of the future
with computer-aided information systems and computer-controlled manufacturing procedures, conventional production structures and technologies
are not sufficient anymore. New production
structures in the field of manufacturing technology are required which at the same time, also
account for the problem of rising multitude of
variants, small lot sizes and innovation times
that are becoming shorter.

The following technological-economical objectives
are relevant for automation in nearly all fields
of industrial manufacturing:

- possibility to construct complex systems
- application of modern control methods
- minimization of cost/profit relation
- short manufacturing starting times
  with new products
- detection and monitoring of the processes
- assurance and increase of product quality
- improvement of the condition of the plant and
  characteristics
- increase of productivity

The most important demand on the new generation
of the so-called intelligent, sensor-aided robot
welding system as <u>integral</u> component of computer
integrated manufacturing will be to fulfill the
requirements made by future manufacturing tasks.
The availability of flexible, modular manufacturing devices allowing the coupling to different
process components is therefore an absolutely
necessary premise for flexible automation.

By developing sensor-aided robot welding systems,
an approach to the automation of welding processes as well as to the integration of the
systems into computer-integrated manufacturing
shall be outlined in this paper.

CONCEPTION OF A MODULAR, SENSOR-AIDED WELDING ROBOT

If man is to be taken out of particular stages of
the production process, the typical sensorical,
intelligent and motorial performances must be
copied as close as possible with the technical
means available.

In order to keep the systems complexity as small
as possible, it is attempted to have the information about process and surrounding (sensorical
performance) acquired by sensors. The evaluation
and interpretation of the information taken in as
well as the data provided by CAD/CAM-systems
("intelligent" performance) is generally undertaken by one or more coupled process computers.
The transformation of the processing results into
action (motorial performance) is carried out by
handling systems (robots).

The wide range of applications of arc welding
technology requires suitable system configurations
allowing for the automation in industrial scale
manufacturing as well as in small and medium lot
production. In this context, the advantage of
industrial robotic welding doesn't lie in the
degree of automation itself, but in the obtainable flexibility. This is of special interest
particularly with small lot sizes and great multitude of types. Still, the great breakthrough
in the treatment of small series has been denied
to welding robots. This is basically due to technical and economical difficulties arising with
the automation of arc welding. On the one hand
this is in covering deviations of the groove from
the pre-defined path and, on the other hand, the
time-consuming programming of welding path course
is considered problematic. A sensor-guided robot
welding system contributes to the solution of
these problems. It has to fulfill the following
basic functions in order to assure the quality of
seam: - reliable detection of the joint or
       joint end, respectively
     - tracking the varying trajectory
     - adaption of the welding parameters to
       the varying geometry along the joint
     - detection of the joint end

First of all, a system like this requires suitable
sensor systems providing information about the
location of the joint and its geometries. Appropriate controlling functions in the robot control
unit as well as in the welding periphery are
indispensable, too.

Sensor systems are used to detect the process
state and its surrounding. By applying suitable
controlling techniques, modern automated manufacturing lines must additionally take informaion
about the geometry of the object to be worked at
into account, in order to reduce extensive production preparatory processes. These information
are available in the existing CAD-devices and
provide the possibility of controlling the
fabrication.

During CAD-construction information about the workpiece, workpiece geometry and demands on the seam are put in or generated by the system. For the automated fabrication of the designed product, these stored data cannot only be used by the individual processing facilities, but also for the controlling of the manufacturing process.
In this context, the task of the planning component (CAP) is fabrication planning and disposition. A component closely cooperating with this planning system is the computer-aided manufacturing (CAM), conducting construction and planning data procession.

The workpiece model data provided by a CAD-system do not correspond to the geometry of the real component due to workpiece tolerances and isuf ficient workpiece preparation. With the sensor guided welding of toleranced workpieces with industrial robots CAD-data should therefore be taken into account as prior information for the process, on the one hand. On the other hand, measure, shape and cut deviations should be corrected during the process by the application of sensor systems.

Industrial applications in fields where in-process correction of the programmed robot movement is performed in dependence on sensor information are not widely spread for the time being. Systems having the additional possibility of carrying out an automated adaption of the welding parameters to the joint geometry are frequently to be found only in research laboratories. Welding robots taking model data from CAD-systems as supporting values into account are seldom to be found accordingly.

The reasons for this are manifold. The available senor systems must be adjusted to the conditions of industrial applications such as price, reliability and precision. Robot control systems have only confined possibilities to process the correction data provided by a sensor system or to integrate them into the handling task, respectively. Common welding systems are to be extended to "intelligent" welding components by the employment of micro- processors. CAD-systems must be equipped with interfaces allowing for the required exchange of information with process level partly even under realtime conditions.

SYSTEM STRUCTURE, TASKS AND COUPLING (INTERFACING)

Within the framework of this paper, a conception has been designed to realize a system for sensor-aided robotic welding under consideration of CAD/CAP/CAM-information which account for the requirements outlined above. Fig. 1 illustrates the structure of a sensor-aided robot welding system, consisting of the individual components of sensor system, robot, welding system as well as the coordinator connected to a superordinated CAD/CAM-system.

The coordinator represents the central unit coordinating all the components involved in the process. It furthermore takes over the communication with superordinated systems for the exchange of construction data from the CAD-system, planning data from the CAP-system as well as production data from the CAM-system. Thus the coordinator comprises all the hardware and software means necessary to reconcile the sensor system, robot, welding system and CAD/CAD-system to an operating total system. These are the main duties of the coordinator:

- user guidance
- programming of robot, sensor system, welding system
- data procession
- transformation as regards different coordination systems
- distribution of information
- synchronization
- data storage on the communication of the subsystems
- documentation
- archivization
- system monitoring (supervision)

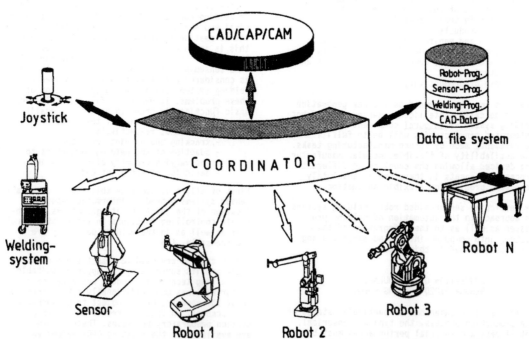

Fig. 1. Structure of a flexible manufacturing component

The tasks of a sensor system basically consist of calculating the distance between measuring sensor and workpiece as well as the deviation of the sensor from the mid of the joint. In context with these information the sensor system supplies details about measuring incertainties of the values resulting from the number of error measurements related to the number of all the measurements within a given time interval. Apart from this, the sensor system additionally supplies messages such as "workpiece in sight","joint in sight". These messages are used for the detection of the joint start and end. Error messages from the sensor systems are available also at the sensor interface for statistical documentary and process supervising purposes.

In general, the task of an industrial robot is the tool guidance at a given velocity and orientation along a preprogrammed path. In this context, six motion axes are necessary. In order to adapt the motion of the tool to the real course of the contour, it is particularly necessary to have a communication with the coordinator via suitable interfaces. This requires a particulary powerful interface within the robot as during the process the external path corrections must be taken into account on path planning in the robot control system, on the one hand and, on the other hand, also on-line information about the actual velocity and the position of the tool must be available on request at the interface.

The flexible welding periphery basically consists of the two components, controllable welding source and a micro computer for the determination of the welding parameters, housekeeping of welding programmes as well as providing knowledge of the parts to be welded. Before starting the process, the welding computer determines the required starting values for the process out of its own information base depending on the workpiece chosen by the user and the shape of the groove. In case the welding task specified by the user is unknown to the system, there is a possibility to convey the necessary information during a learning phase. The data obtained this way are stored in the information system of the welding computer and thus kept available for a similar welding task at some other time.

During the welding process, the welding computer permanently determines the parameters current, voltage and velocity in dependence on the joint geometry and the type of joint. With regard to the effective values of welding current and voltage, a control algorithm implemented into the welding system is responsible for the set-values being sticked to. This assures a reproducable quality of the seam. Furthermore, the welding system takes over the supervision of the actual welding process with special reference to shielding gas and wire. The welding processor also supervises given tolerance limits and generates appropriate messages for the user in the case of error or causes the interruption of the process if the predefined error boundaries are exceeded, for reasons of security.

Before the beginning of the process, the coordinator has the task of generating robot, sensor and welding programmes or parameters, respectively, under consideration of the CAD, CAP and CAM-data and the user inputs and transferring them to the respective subcomponents. In general the coordinator is responsible for data procession and the distribution of information for the subcomponents presented in figure 1 as well as also for the internal file system. All the system data necessary for the process, such as number and type of robots, type of the sensor used and interface specification are stored within this file system.

During the process the coordinator takes over the procession of the joint position data determined by the advancing sensor in dependence on contour and velocity. It also carries out the transformations of these positions into robot-specific coordinates and the transmission of these data as pathvalues to the robot. During the entire process the coordinator suvervises and records significant states in the process and triggers appropriate actions if necessary.

Suitable interfaces are required for the communication between coordinator and the information processing subsystems. These interfaces are producer-specifically designed. For the time being, there is only a confined industrial interface standard according to which the individual components can communicate with each other. Standardization of the interface, however, facilitates the coupling of the individual components and it is a premise for the flexible and economical application of such a total system.

The structure presented in Fig. 1 illustrates the logical separation among the subsystems. This logical separation has to be strictly distinguished from the physical division. So, for example, the hardware components of a welding robot are often integrated into a control box and present themselves to the user as one individual unit. The logical separation is only to stress the technical interrelationship between the subsystems for the exchange of information.

CONCLUSION

The modularity of the total system allows individual subsystems such as, for example, the sensor system to be replaced by a different system with similar performance characteristics and interface convention. Moreover, the chosen system structure and the capacity of the process computer permits the integration and coordination of several robots working at the same component within one manufacturing cell.

REFERENCES

Drews, P. - Fuchs, K. (1989). Welding Automation. Proceedings "2nd International Conference on Trends Welding Research". Gatlinburg, Tennessee, USA.

Drews, P. - Fuchs, K. (1989). Coordination of CAD/CAM information in robotic applications. Proceedings Vol 2 No 1. Butterworth & Co., London. Pp. 35 - 37.

Fuchs, K. (1987). Flexible, sensorgesteuerte Roboterschweißsysteme, Dissertation, TH Aachen.

COMPUTER AIDED PLANNING OF ASSEMBLY SYSTEMS

G. Seliger, V. Gleue, H. J. Heinemeier and S. Kruger

Department of Assembly Technology, Production Technology Centre, Berlin, FRG

Abstract. The integrated computer aided planning system is developed for assembly systems with special reference to automated material handling. Based on the specification of planning tasks, assembly functions and equipment, modules for assembly sequence planning, layout design and system behaviour simulation have been realized. The open architecture is characterized by an infrastructure of data base and network, integrating instruments as CAD and expert systems for consistent information processing. The paper describes system development and application in industrial assembly.

Keywords. Assembly planning; CAD; configuration; computer simulation; database; expert systems; network.

INTRODUCTION

Increasingly it is necessary to assemble in optional sequence a variety of products in small lot sizes to meet the market requirements. Flexible assembly systems with a maximum of standardized components are used to realize a long service life of equipment in spite of shorter product life cycles. The extension of assembly system performance effects a large number of parts, tools and joining materials to be placed at disposal. A high degree of automation causes more data exchange between the system components.

The increased complexity of kinematics, material flow and communication raises the necessary planning efforts. Companies are forced to faster execute planning tasks to put new products on the market as early as possible. Computer aided planning tools offer the chances to plan more complex systems with a higher level of detail in shorter periods of time.

At the Production Technology Centre Berlin several task-specific software tools have been developed based on an integrated architecture of systematic assembly planning. All tools independently access a common, consistent data base. A coherent information flow through the

entire system assures the automatic transfer of planning results without any deficiency of information.

POTENTIAL OF INFORMATION TECHNOLOGY

The progress in information technology opens new application areas in assembly planning. Basic instruments as

- CAD-systems,
- knowledge-based systems,
- data base systems and
- communication systems

offer the potential for assisting planning engineers in creating more solutions in a higher level of detail.

Assembly tasks are indicated by a variety of operations which can be realized with various equipment in different arrangements. Movements in assembly cells must be optimized regarding the dependencies of kinematic configurations and product-specific components e. g. grippers and fixtures. This asks too much of the planners three-dimensional imagination. Thus, CAD-systems are essential instruments for the geometric modelling of product and equipment. Tridimensional graphical representation and motion simulation improve the transparency of assembly processes and respective evaluations.

Based on his specific experience the planner tends to solve planning tasks by modifying known solutions. This prejudicial behaviour may exclude suitable approaches and lead to less than optimal solutions. Creativity and inspiration can be supported by expert systems holding a large knowledge base and committing the planner to leave his established procedures. By taking into account all relevant criteria a larger solution space can be achieved and leads to approximating an optimal result.

Database systems offer the opportunity of acquiring knowledge and information from external departments in realtime. Moreover, it is possible to store large data sets according to specific criteria or planning steps. Recorded solutions may be used as a basis for solving future planning tasks.

The efficiency of planning systems depends on the availability of information. Network communication systems ensure data interchange between different departments of a factory.

ASSEMBLY PLANNING SYSTEM

Structure

Figure 1 represents the structure of the assembly planning system which has been developed at the Production Technology Centre in Berlin. The system is to support the user during the entire planning process. The programs are implemented on VAX-computers and IBM Personal Computers which are connected via an Ethernet. At present, the planning system consists of seven tools:

- assembly sequence planning,
- solution generation,
- evaluation,
- detailing,
- assembly process planning,
- simulation and
- communication.

An information system has been implemented by means of a relational data base, which ensures an efficient interchange of data between the different tools.

Assembly Sequence Planning

Based on the product structure the alternatives of

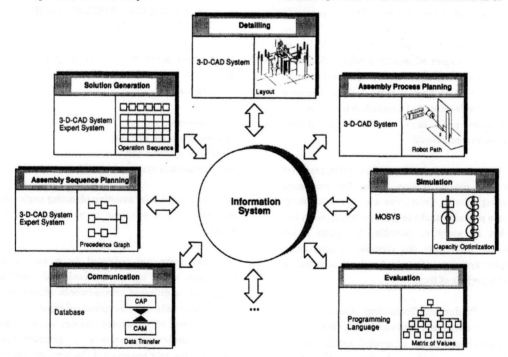

Fig. 1   Structure of Realized Planning
         System

the sequence for the execution of the assembly operations are elaborated. Documents resulting from this planning tool are

- an operation plan and
- a precedence graph.

The developed tool is based on the programming language PROLOG. Figure 2 depicts the general procedure of the assembly sequence planning.

Starting from a list of predecessor-successor entities resulted from an analysis of mating parts the system generates the precedence graph. Then, the tool supports the planner in breaking down the assembly task into closed sub-tasks which serve as basis for the definition of assembly cells in the subsequent step. After the determination of basic parts, their orientation for each station and the estimation of assembly times the operation plan can be generated by the system. Finally, the assembly operations are divided into handling and joining elements as the interface to the solution generation tool.

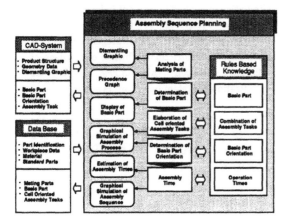

Fig. 2   Procedure of Assembly Sequence
         Planning

Solution Generation

This tool aims at the elaboration of principle solutions. A principle solution is the concept of an assembly system already determined by essential characteristics like system structure, assembly operations to be executed and standard devices.

The quality of the planning result mainly depends on the execution of the conceptual planning phase. In this stage of the planning process the main costs of the assembly system are determined. Therefore, a large number of different principle

solutions should be outlined to approximate an optimum result.

A computer-aided tool for the design of principle solutions has to support the user's creativity and to inspire the planner to new solution concepts. Figure 3 represents the single steps of the developed tool. It is based on the morphological method. Devices are assigned to the single operations of the assembly task described by the functional handling symbols according to the guideline 2860 of the Verein Deutscher Ingenieure (VDI). The system generates a morphological matrix listing the devices in columns under the assembly operations. An expert system supports the planner in identifying suitable devices and analyzes the compatibility of the components. A final rough evaluation according to technical criteria results a rank of principle solutions which are to be detailed in the following planning steps.

Detailing

Main task of the detailing module is the concretion of the principle solution elaborated in the previous planning step. From a library of tridimensional graphic models of components available on the market the user selects equipment for the execution of the assembly operations. The system automatically arranges the components to a first layout suggestion. Figures 4 and 5 show assembly cells for industrial applications generated by the detailing planning tool. Afterwards, the planner can modify the layout interactively on a graphic workstation. A complete geometrical description of the layout alternatives is stored in the database.

Evaluation

To value the elaborated system alternatives a complex system of user defined objectives has to be regarded. Not only monetary factors but also quantifyable criterea like security, organization, ergonomics and flexibility determine the worth of a solution. Therefore, in addition to traditional investment calculations a benefit-analysis is realized as a planning tool. The system guides the planner to define complex objective structures and provide each objective with a special value. The system evaluates the elaborated alternatives based on the created objective system and gives as result a list of the ranked solutions.

G. Seliger *et al.*

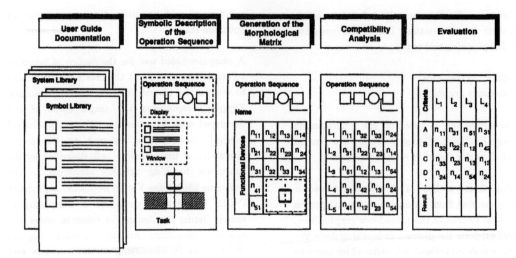

Fig. 3 Process of Solution Generation

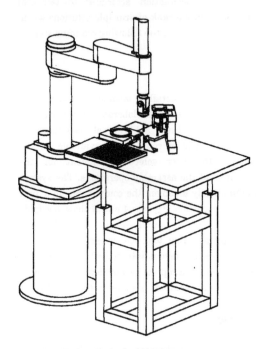

Fig. 4 Layout of an Assembly Cell

Assembly Process Planning

This planning tool enables the planner to check arrangement and functionality of the components by visualizing the entire assembly process. The user defines in detail movements in the system (Fig. 6). Significant trajectories can be stored in the database by a teach-in similar method. The mating processes are displayed on screen. The kinematic requirements of the robot as well as the gripper design can be examined. If necessary, the layout of the cell components can be modified by the user. Additional information like robot speed or accuracy can be inserted. The resulting process description is stored in the database in pseudocode and can be used for subsequent offline-programming tasks.

Simulation

During the planning process of assembly systems simulation is applied to find out independencies in the time behaviour of the components to investigate different control strategies and to look upon capacity restrictions. Thus, a comparison of different solution alternatives and the optimization of system behaviour is achieved.

The simulation is done using a model of the planned system (Fig. 7). After the completion of a simulation experiment the following data are available:

- data per product type
  * number of processed parts,
  * average turn-around time,
  * average processing, testing and transport time.
- data per building block
  * number of utilized capacity units,
  * utilization,
  * processing time,
  * machine down time,
  * set-up time,
  * queuing length.

OPEN ARCHITECTURE

In the integrated planning system formalized description tools will be used to develop a common

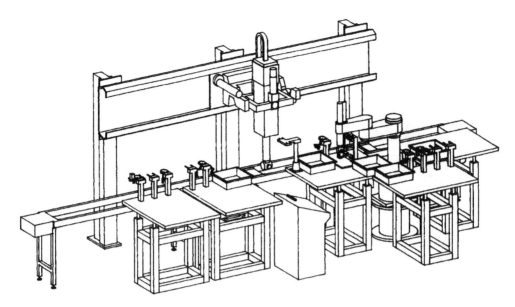

Fig. 5   Layout of a Two-Robot Assembly Cell

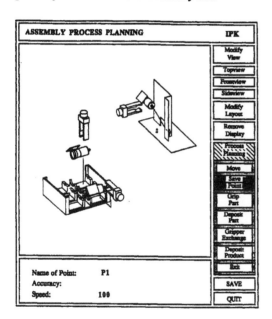

Fig. 6   User Interface for Assembly
            Process Planning

access to stored experience and ensure that he has regarded all relevant design aspects. The architecture is illustrated in Fig. 8. The computer aided tools can be classified into the following three groups:

- tools for formalized description,
- tools for analysis and
- tools for synthesis.

The realised planning system is characterized by an open architecture. Thus, it can be enlarged with task-specific tools and implemented on user-specific environment. Precondition is the availability of the information in different departments of a factory. The realisation of integrated information flow depends on the development of standardized interfaces to transfer data without deficiency of information.

integrated model of an assembly system which will contain all details relevant for design. Various analysis methods can be applied to this model to study its static and dynamic performance and cost behaviour. The planner will be supported in his modelling and analysis efforts by knowledge based systems. Furthermore, a module for knowledge based design will open up access to expert knowledge for defining the configuration and corresponding control strategies. Such knowledge based support will give the planner systematic

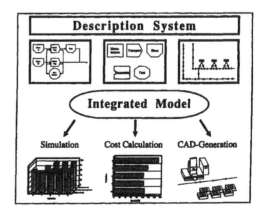

Fig. 7   Structure of Planning Tool
            for Simulation

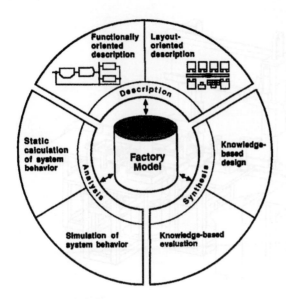

Fig. 8   Factory Model

The presented planning system is used for research projects on industrial applications. Based on the gained experience the system will be continously improved and augmented.

REFERENCES

Bullinger, H.J. (1986). Systematische Montageplanung. München, Wien.

Deutschländer, A., and F. Severin (1986). Rechnerunterstützte Layout-Planung für Industrieroboteranwendungen. ZwF 81, H. 10, S. 515-522.

Seliger, G. (1988). CIM - was ist das? - Grundkonzept. DIN - Mitteilungen 67. Nr. 6, S. 325-330.

Seliger, G., B. Wieneke, and M. Rabe (1987). Integrated Modelling of Manufacturing Systems for Intelligent Purposes. Proc. ASME Symposium on Integrated Manufacturing. West Lafayette, Indiana.

Seliger, G., I. Furgac, and A. Deutschländer (1987). Flexibles Montagesystem. ZwF 82, H. 3, S. 133-136.

Spur, G., F.-L. Krause, and G. Seliger (1987). Software Structure for Factory Integration. Proc. IFIP Working Conference on Software for Factory Automation, Tokyo.

INTEGRATION OF A MACHINE VISION SYSTEM IN A FLEXIBLE WORKSHOP FITTED OUT WITH CAD/CAM TOOLS

Y. Lucas, T. Redarce and M. Betemps

Laboratoire d'Automatique Industrielle, Villeurbanne, France

Abstract . CAD/CAM tools make up an essential component of computer integrated factory. Up to now, they were used for tasks such as simulation and path programming of numeric command machine-tools and sometimes industrial robots. The CAD-VISION connecting described here enables, to program parts learning on the workstation, to download pieces features in the vision system for production line inspection, to simulate recognition process on a number of parts stored in the computer and to update vision files after modifications in the CAD system database.

Keywords . CAD/CAM ,off-line programming, industrial vision, pattern recognition, simulation.

INTRODUCTION

Why talking about Computer Integrated Manufacturing today

It is because the governing idea is integration (Scheer, 1988). To ensure that productivity, competitiveness, flexibility, profitability, which are actual preoccupations become a reality, all advanced technology equipments must operate and communicate together. Production automation, seen from the computer science angle, is the information exchange between CAD, production management and machines like robots, numeric command machine-tools, automata, vision and peripherical robotic systems (Batchelor, 1987). This data must circulate within all the levels of the decision hierarchy : from sensors, drives and man-machine dialog, to programmable automata and numeric commands integrated in the machines (machine-tools, robots, conveyors). In a higher level, micro-computers supervise the production line on the workshop, under control of mini-computers. At the highest level, the factory central computer coordinates all tasks. To make this dream of complete integration become true, it will be necessary to improve its lingpin : data bases, to centralize information, and local networks to make it circulate (Gardarin ,1988 ;Lepage ,1988).

Using databases for robotic systems

Different specialized products have appeared in the middle of eighties allowing the potential of CAD systems to be profitable for production tools like robots. In that case, CAD can act on three distinct levels (Rembold ,1986; Henderson ,1987 ;Kak ,1987) :

- the modeling : the geometrical and kinematic models are defined for each element of robotic work cell ;

- the simulation : primordial to optimizing a layout. Obstacle avoidance, cycle-time analysis, selection of a robot in a library are performed (Coiffet ,1982) ;

- the programming (Dombre ,1983) : to make

the set CAD-Robot operational, the CAD system must generate data assimilated by the robot controller.

Therefore, we are faced to compatibility problems. If the robot is usually programmed by teaching, it will need the points along the path. On the contrary, if it is programmed by textual language, path will be provided by literal instructions. In both cases, a translator will manage data conversion which is downloaded by the local network.

Using databases for vision systems

Use of databases for vision systems remains at the state of research . (Gruver,1984 ; Henderson,1985 ; Crosnier,1987). The aim being to recognize three dimensional parts by comparison of the geometric model contained in the CAD database with the one obtained from camera images of three dimensional sensors.

However, these models are different

- at a functional level : the CAD model favours interpretation and set operation while vision model performs camera raw data analysis ;

- at a physical level : CAD systems help to define new shapes while vision systems analyse real parts.(Dhome,1984; Kasvan,1986 ; Nurre,1986; Horaud,1988).

Our ambition is to retrieve CAD data :

- to integrate the vision system in the robotic work cell;
- to simulate recognition on a collection of parts ;
- to carry out parts learning by off-line programming.

In the second paragraph we describe hard and soft architecture that we have chosen . Then, in the third paragraph ,we deal with a typical development of an application that concerns parts recognition by a vision system. Finally, we examine the CAD-VISION connecting contribution in a CIM approach.

CAD-VISION CONNECTING

MATERIAL CONNECTING

The system we have worked on (Fig. 1) includes :

- a workstation (DEC VAXSTATION II GPX) on

which resides a. CAD/CAM software (MCDONNELL DOUGLAS UNIGRAPHICS II) and all the modules developped for the vision system programming. Several workstations are linked (by ETHERNET network under DECNET control) together and to a host computer (DEC VAX 8250) ;

- an industrial vision system (ALLEN-BRADLEY SERVOVISION VISIOMAT). This machine is dedicated to general vision purpose and is equipped with console and video-cameras. It allows inspection (quantitative or qualitative), parts recognition and sorting, and robot guiding. Specialized processors, programmed functions,and vision oriented language help to build applications supporting high industrial rates. The vision system communicates by serial, parallel or switched links ;

- a numeric command machine-tool (GRAFFENSTADEN) for machining of parts later inspected on the production line ;

- a local network (ETHERNET and soon FACTOR) to transmit learning programs to the vision system and machining programs to the machine-tool.

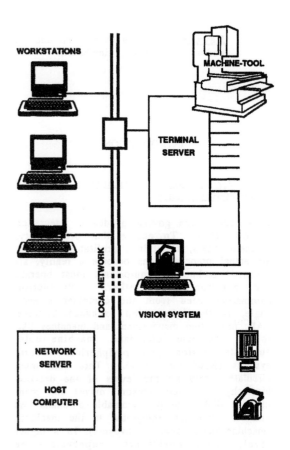

Fig. 1 Material connecting.

What to note ?

The link between the CAD/CAM system and the machine-tool is in fact traditional. But, what is new in workshop is the connection between a vision system and this CAD/CAM system. This link, installed with the local network, allows to download data assigned to the vision system on the production line from the computer department where the CAD/CAM system is located . This vision system can at the same time use the network for coordination with robots, automata, sensors and drives.

SOFTWARE INTERFACING

Software available on the CAD/CAM system and the vision system (Fig. 2) are interactive enough to cooperate. The CAD module involves several modeling units : surfacing, finite elements, solid modeling, schematics, plastic moulds design, printed circuits shaping, etc... The CAM module is composed of machining, shape cutting, and workshop data managing units. Optional modules allow to manage production. All information related to part models are centralized in a single database (Mac Donnel Douglas ,1988).

The vision software involves vision oriented language (LV), (as (LM) is a manipulation oriented language). Specialized modules can be run interactively or by program : pattern recognition, zone control, dimension control, mathematical morphology, digital display control, and luminosity control (Allen Bradley ,1987).

Interfacing problems arise at different levels :

Models agreement . CAD/CAM systems have extensive graphics display capabilities but there are large differences between image synthesis and analysis. Then, it is necessary to establish a link between the geometric model of parts (realistic reproduction of colour is only a coating) contained in the CAD system database and the model obtained out of camera image processing using classical learning procedure. Part model is composed of elementary entities which may be segments, circle arcs, conics, B-splines curves, Bézier surfaces, etc... To reproduce scene observation conditions given by the camera, it is necesary to project different part boundaries on the image plane and to eliminate the non visible boundaries, whether hidden or external. Remaining boundaries (consisting of graphic entities) are selected using a mouse.These ones and those extracted out of a grey levels camera image (consisting of pixels) are processed by the computer in the same way. If necessary, the boundaries can be converted into points to obtain certain geometric parameters, not calculated by the CAD system analysis module.

Frame grabing adjustement .During classical parts learning using the camera, the grabing parameters are practically set (digitizer gain and offset, binary thresholds). These experimental values depending on the lighting device and the inspected part appearance must be downloaded on the workstation at the time of object files creation. The vision system will read it over to configure the camera automatically before parts recognition process.

Files compatibility . Numeric command machine-tools or robots off-line programming produce incompatible data with those assimilated by the machine. Using a postprocessor allows to execute these file conversions respecting coding format adapted to each machine. In the same way, learning, collection, or calibrating files will be read over properly by the vision system only after particular coding imposed by the machine.

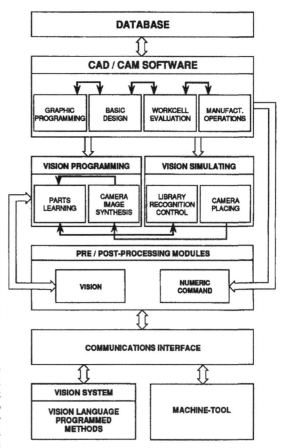

Fig. 2 Software interfacing.

Data conformity . Using a single database for the whole CAD/CAM system offers many advantages : non redundant data, compatibility with different software modules, and easy up-

.dating. When a new part is first manufactured or is modified, the joint learning files must be available. The library consisting of the objects known by the vision system must also be regularly up-dated (creation, modification, deletion).

APPLICATION

By following the successive stages of a part recognition application development, we can appreciate the interest of this CAD-VISION connecting, in particular in a CIM approach. The steps are the following:

Robotic work cell conception

The graphics capabilities of the CAD system allows display, and even animation of the moving elements of the robotic work cell. This tool is useful to determine camera position and orientation in the scene, lens type to mount, distance between worksurface and the camera optical center, and cycle time which governs image processing period. At this level camera characteristics and calibration files may be established .

Definition of the parts to be inspected

New parts are introduced in the CAD database. Stable state faces are defined for each part . Then, using camera information, the projection of three dimensional part visible boundaries can be calculated to obtain the two dimensional image captured by the camera. Other ways to define a camera model are investigated in our laboratory .

Parts learning off-line programming

The vision system we consider is able to recognize objects in an image by working on edges extracted from this image. Objects are defined like closed edges sets topologically linked together (i.e inclusions, distances) and owning several geometrical attributes (perimeter surface, minor and major ray, inertia). Classic learning consists of presenting parts under the camera in different positions and orientations on the worksurface plane. Here, the geometric model of parts stored in the CAD system database is directly retrieved. So, object files are created after interactive learning using the workstation .

Recognition simulation on a parts library

We mean by library a set of parts likely to be at the same moment on the production line controled by the vision system. At this level we check whether recognition algorithm succeeds in discriminating the different objects of the library. If partial or total confusions are detected among some parts, geometric tolerance values must be adjusted or recognition options configuration revised .

Data coding and transfer

Coding is performed to make the files compatible with those of the vision system. Then, objects and library files are downloaded on the vision system.A transmission utility is used to dispatch the files to the local network .Concurrently, machining programs postprocessing and transmission to the numeric command machine-tool are achieved to start parts manufacturing. To ensure data permanence, the database updating must induce the updating of the parts libraries intended to the vision system .

Parts recognition

The vision system initiates the recognition of learned parts, whether in interactive manner or by running a particular application program.

CONCLUSION

The explosion of the industrial vision market is today confirmed by turnover increase, diversification of products available and industries concerned with these techniques .In the field, the vision system has proved its economical profitabilities. It has won its place in the flexible workshop (Bonetto, 1985) (Laurgeau, 1985). So, CAD-VISION connecting represents an additional link to complete computer integrated manufacturing.

The vision system integration inside the factory CAD/CAM system offers a lot of advantages that are advisable to mention :

Using the vision system

Parts manual learning under vision systems cameras is suppressed. It is a supplementary step to producing complete automation. The learning quality is improved due to the fact that object file contains real parts geometrical parameters. Classical learning which works on edges extracted from a binary image is inevitably rough due to acquisition and data processing degradation. Productivity is enhanced because it is no longer necessary to immobilise the vision system and therefore the production line for learning tasks.

Using the CAD/CAM system

The single database for the whole CAD/CAM system allows at every moment to obtain the last release of designed parts. Hence manufacturing delays can be reduced because the learning file intended to the vision system is available even before the part has been machined. Vision information reliability is strengthened by recognition simulation allowing to detect possible confusions between library objects. In addition, robotic workcell display ensures optimal placing of the camera in the scene .

Using the local network

It allows to keep a minimum number of data items in the vision system memory for high speed execution of the vision program. On the other hand, regular updating preserves conformity of parts stored in the vision system with their latest modifications,insuring further flexibility.Object libraries and associated vision programs are daily downloaded according to factory orders promoted by the production management unit.

An implementation of this has been carried out: a family of parametered parts has been designed using a graphical programming language. Learning files have been created using two dimensional views of part edges in stable state situation. Likewise, machining programs have been established. The working library has been constituted from these objects, after a stage of simulation to validate it. Afterwards, corresponding information has been dowloaded to the vision system and machine-tool. When manufacturing was over, the vision system was able to accomplish the recognition of machined parts.

REFERENCES

Allen Bradley Servovision (1987). Manuel de l'Expert.V3.1.

Batchelor, B.G. , D.A .Hilland and D.C Hogdson (1987). Automated visual inspection..Ifs publications. United Kindom.

Bonetto, R. (1985) .Les ateliers flexibles de production. Hermès. France.

Coiffet, P. (1982) Les robots. Interaction avec l'environnement. Hermès. France.

Crosnier, A. (1987,a) Intégration d'une fonction de perception dans un système de conception assistée par ordinateur pour la robotique.Thèse Académie de Montpellier. France.

Crosnier,A. (1987,b).Simulation of cameras and proximity sensors for computer aided design and the off-line robot programming. Proceedings of the international workshop on industrial applications of machine vision and machine intelligence. Tokyo.

Dhome, M. (1984).Contours et reconnaissance de formes pour l'analyse de formes en robotique. Thèse Academie de Clermont II. France.

Dombre, E. and P. Borrel. (1983).Off-line robot programming. Robotics systems course. Ibm europe institute West Germany.

Gardarin, G. (1982). Bases de données, Eyrolles.

Gruver, W.A., C.T. Thompson, S.D. Chawla and L.A.Schmitt. (1984) .Off-line vision system programming by computer-aided design Robot 8 conference proceedings.

Gruver,W.A, C.T. Thompson,S.D. Chawla and L.A. Schmitt.(1985). CAD off-line programming for robot vision .Robotics.vol 1. Netherlands.

Henderson, T.C. (1985) Intrinsic characteristics as the interface between Cad and machine vision systems.Pattern recognition letters vol.3.

Henderson, T.C., E.Weitz, E. Hansen, R.Grupen, C.C Ho, and B.Bhanu(1987).Cad-based robotics. IEEE transactions on pattern analysis and machine intelligence.vol 1.

Horaud, R. (1988) .Modèles géométriques et images numériques, p.132-144. Séminaire, université Lyon I.

Kak, A.C., A.J.Vayda, R.L.Cromwell, W.Y.Kim and Chen C.H. (1987). Knowledge-based robotics.IEEE transactions on pattern nalysis and machine intelligence.vol.1.

Kasvan.T., R.Oka, M.Dhome and M.Rioux .(1986) object detection and recognition from 3-D images.Microcomputer Applications.vol 5, n° 2,p.55-8.

Laurgeau, C. and M. Parent (1985). Les machines de vision en productique.ETA. Agence de l'informatique. Afri.

Lepage,F. (1988). Les réseaux locaux industriels. Hermès. Mac Donnel Douglas (1988).Unigraphics 2 documentation .v5.0.

Nurre,J.H and E.L. Hall.(1986).Computer integrated engineering system using vision.Proceedings of the IEEE 1986 national Aerospace and Electronics conference. Dayton,USA.

Rembold, U. and R. Dilmann (1986) Computer-aided design and manufacturing. methods and tools Springer-Verlag.

Scheer, A.-W. (1988). Computer integrated manufacturing. Computer steered industry Springer-Verlag.

NEUTRAL INTERFACES THAT WORK:
APPLICATION FOR ROBOT WELDING

U. Kroszynski, B. Palstrom and E. Trostmann

Control Engineering Institute, Technical University of Denmark, Lyngby, Denmark

Abstract. The integration of sub-systems, leading from geometrical design in a CAD system to a weld planning module and a robot simulation module, is realized by employing the neutral geometry description developed in the CAD²I project.

The offline robot program resulting from the simulation system is transferred for execution by a welding robot via the neutral IRDATA job description file.

A real time remote monitoring system, driven by the same signals that are sent from the robot controller to the robot, is used as a supervisory tool.

The actual set-up was successfully demonstrated at the Fifth ESPRIT Technical Week in Brussels in November 1988 and serves to illustrate how the data transfer via neutral interfaces integrates five otherwise independent modules for a specific application.

Keywords. CAD; Industrial robots; Welding; Computer interfaces; Standards; Data transmission; Modeling; Simulation; Monitoring; Local area networks.

INTRODUCTION

One of the most advanced industrial applications of robots concerns the welding of parts using offline programming techniques. Unlike spot-welding tasks or pick-and-place tasks for assembly, which normally allow for comparatively large tolerances, this type of application requires particular accuracy.

Besides more powerful path control capabilities, based on real-time sensory information, the demand for a very precise trajectory definition implies that the robot programming must be performed with the help of computerized tools that fulfill the functions of geometry modeling, task planning, path generation, simulation, and program verification.

Since the same functional components for design, planning and manufacturing are also present in all other types of production equipment that work independently or in coordination with the robots, they can be regarded as true sub-systems in a Computer Integrated Manufacturing (CIM) environment.

Although dedicated interfaces may prove more efficient in any particular CIM implementation, the multiplicity of alternative computerized tools, employed for design, planning and manufacturing of mechanical parts, strongly speaks in favor of employing standard, neutral interfaces between the major functional components in the production process.

A conceptual model is shown in Fig. 1, where the design function is performed in a generic CAD system, the manufacturing planning in a generic CAM system, and the actual production in a generic machine tool, robot, etc.

This article describes an implementation of the conceptual model for the particular application of robot welding.

A first, trial implementation of a demonstration facility, featuring a geometry modeller, a Weld Planning Module (WPM), a robot simulation system and a welding robot, was formerly described in Palstrøm (1988), and served to illustrate the feasibility of the approach.

In the following sections we describe a more advanced and streamlined configuration of the facility, where the design module was substituted by a commercial CAD system, the WPM by a more versatile one, and a new module for real-time monitoring of the welding task was added.

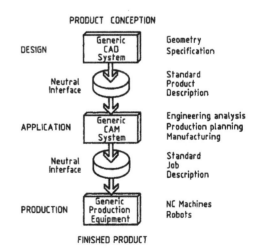

Fig. 1: A conceptual model for a CIM environment employing neutral interfaces.

The relative easyness with which individual components could be replaced and a new one incorporated, proves that the approach is not only feasible, but also general and advantageous.

LAYOUT OF THE DEMONSTRATION FACILITY

The detailed architecture of the CIM environment for the offline programming of a welding robot is sketched in Fig. 2. The small dark rectangles represent pre- and post-processor programs.

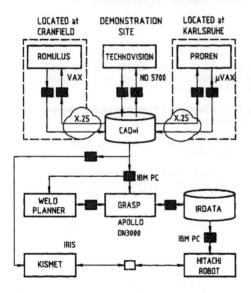

Fig. 2.   Demonstration facility architecture

The different functional components can be identified as follows:

The design component

The geometry of the object to be welded is generated as a Boundary representation (B-rep) solid model in the TECHNOVISION CAD system. This commercial system substitutes the Constructive Solid Geometry (CSG) based solid modeller GDS (Kroszynski, 1983) of the implementation described in Palstrøm (1988). The model is translated to the neutral representation. This neutral interface was developed in the framework of Project 322, CAD Interfaces (CAD*I) of the European Strategic Program for Research in Information Technology (ESPRIT). It proved to be reliable for describing CSG solid models, as shown in the trial implementation, and also for B-rep solid models, as shown in detail in a separate section, later on.

The CAD*I interface allows other CAD/CAM systems to recover the model and activate other applications, like for instance engineering analysis (Kroszynski, 1986) in other locations, as shown in Fig. 2. The final geometry, after eventual design modifications, is recovered in neutral format.

Manufacturing Planning

A WPM accepts the neutral geometry file as input. A welding specialist determines, via graphic interaction, the technological data needed for welding. This information concerns the optimal orientation of the object when taking account of gravity, the sequence and direction in which the different curve

segments have to be welded, optimal torch orientations along these paths for continuous arc welding, when to turn the torch on and off, with what speed to advance, and the voltages and currents associated with each path.

At this stage, the welding specialist does not necessarily have to know where the geometry was generated on the one side, or which robot is going to perform the welding on the other side.

Weld planning modules were coded at the Control Engineering Institute (IFS) of the Technical University of Denmark for CSG based solid geometry, as outlined in Hansen (1985), and for neutral B-rep descriptions. The latter is more flexible and features a more advanced user interface, where the welding paths can be selected by graphical interaction. This is possible, as the paths are offsets of the intersection curve segments between the model surfaces, which are already part of the B-rep description.

Robot simulation

The original geometry, augmented with technological data from the WPM is automatically translated, in that program, to the GRASP input language. GRASP (Bonney, 1984) is a commercial robot simulation system featuring a Polyhedral solid modeler of rather limited capabilities. Originally designed for simulating assembly tasks, it does not provide comfortable tools for simulating continuous arc (MIG/MAG) welding along curved paths. This is the main reason for having the WPM. The system is to be used primarily by robot specialists. A library of robot models is created beforehand. Kinematic models of new robots can be created by the operator and added to the library.

After processing the information from the WPM, the robot and tool models are selected from the library. Several alternatives of object position and clamp configurations are tested by graphical simulation (Fig. 3) of the robot movements, checking for collisions with the object and environment, and optimizing the performance for time of execution.

The output from GRASP is a GRASP Data (GRDATA) file describing the robot job. This is converted to the Industrial Robot Data (IRDATA) format, representing the other neutral interface in the conceptual model of Fig. 1. Details on this interface are given in Palstrøm (1988).

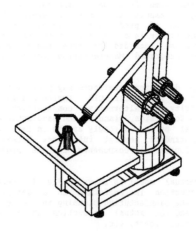

Fig. 3.   Simulation of robot program in GRASP

Job Execution

IRDATA is a proposal for a standard of the german engineering society VDI. An increasing number of case studies, where IRDATA is used for robots of different makes, gives some evidence that it can be considered a "neutral interface that works". A post-processor for the HITACHI PW10 robot was coded at IFS for a subset of the IRDATA specification.

This program performs an exact inverse kinematic transformation and produces a file in HITACHI codes which marks the conclusion of the offline programming task.

The file is finally downloaded to the robot controller for execution. This part of the layout was described in greater detail in Palstrøm (1988).

The welding itself was tried at IFS in various occasions, for different objects. For the actual demonstration, the MIGATRONIC welding equipment was not activated. Instead, a small red diode lamp, attached to the tip of the welding gun used as the robot tool, was turned on and off, imitating the torch.

Real-time surveillance

The nominal trajectories and orientation of the welding tool mounted on the robot are subject to real-time, sensor based corrective control for deviations. In the demonstration facility, however, these are not included.

On the other hand, a real-time surveillance system was incorporated, as shown in Fig. 2.

The KISMET system was developed at Kernforschungszentrum Karlsruhe (KfK) as a tool for remote monitoring of robot in hostile environments. It features a high resolution Silicon Graphics IRIS workstation with advanced graphical capabilities. It is currently used for operating an articulated boom inside the vacuum vessel of the Joint European Torus (JET) (Kühnapfel, 1987).

The advantages of employing this system over traditional video camera monitoring stems from its capability to visualize any portion of the scene from an arbitrary position. It also allows zooming in and out, making surfaces transparent, etc.

The kinematics and the geometric shapes of the robot, object and environment are modelled interactively in the system. This information can also be defined as a CSG oriented CAD*I neutral description.

The signals sent by the robot controller to the robot, are also sent, via an RS232 serial connection to the IRIS workstation where a very impressive real-time animation of the scene, with shading of surfaces and realistic colour imaging effects, can be visualized.

HARDWARE ARCHITECTURE

The hardware configuration for the demonstration facility is shown in Fig. 4.

The components are:

- A Norsk Data ND5700 supermini and TECHNOVISION workstation supporting the software for TECHNOVISION and the corresponding CAD*I pre- and post-processors.

- An Apollo DN3000 supporting GRASP, its CAD*I post-processor and its IRDATA pre-processor.

- A Silicon Graphics IRIS workstation for performing the real-time monitoring of the robot movements.

- An IBM-PS80 supporting the WPM with its CAD*I post-processor, the IRDATA post-processor to the HITACHI (Hipp) and the local and modem connections.

- A HITACHI PW10 welding robot with its control unit.

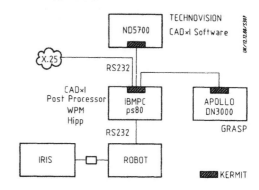

Fig. 4.   Hardware architecture

The communication between most of the hardware components was realized via RS232 connections (V24) using the public domain KERMIT protocol for file transfer. The connection with remote design systems was realized by accessing the public wide area net (WAN X.25) via a modem.

The demonstration facility was presented in the CAD*I exhibition booth of the Fifth ESPRIT Technical Week, held in Brussels in November 1988, raising great interest. Some pictures of the facility are shown in Fig. 5.

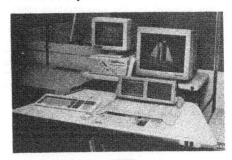

Fig. 5.   The demonstration facility

A CAD GEOMETRY INTERFACE
THAT WORKS

The most important feature to be propagated from the design phase to all applications in a CIM environment is undoubtedly the product geometry. Although in any particular application system a module is provided for geometry definition, the most advanced design capabilities are to be found in general purpose, commercial CAD systems.

Attempts to provide neutral interfaces for geometry data transfer have resulted in various standards. The best known is IGES, primarily employed for planar and spatial wire-frame model descriptions. Surface models are covered by the VDAFS standard, employed by the german car manufacturers and the SET standard, in the french aerospace industry.

Solid modeling is increasingly being used in the design of mechanical parts. Specifications for the transfer of solid geometry within the above standards did not prove general or flexible enough to be used on an industrial level.

One of the most remarkable achievements of the CAD*I project is the specification of a neutral description for geometry Schlechtendahl (1988), and the coding of pre- and post-processors for a series of commercial CAD systems.

In the framework of its participation in this project, the IFS has contributed to the specification in the area of solid models, and centralized an extensive program of solid model transfer tests between different CAD systems, as described in Trostmann (1988).

Of particular relevance were the inter-system transfer tests between the B-rep oriented systems ROMULUS, PROREN and TECHNOVISION, located at the Cranfield Institute of Technology (CIT) in England, at KfK in Germany, and at the IFS in Denmark, respectively. The accuracy of the recovered models and the reliability and stability of the transfer were demonstrated at the Third International CAD*I Workshop, held in Copenhagen in October 1988.

The object selected for the demonstration consists of a cylindrical tube inserted across a square plate and reinforced with four triangular ribs. The object was modelled in TECHNOVISION (Fig. 6) and integral properties (surface area, volume, center of mass) were recorded. The model was pre-processed, and the resulting neutral file was sent to both ROMULUS and PROREN, employing the public X.25 network sketched in Fig. 7.

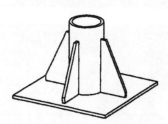

Fig. 6.   Test object model

A letter was attached to the neutral file in the CAD*I envelope, requesting the operations to be performed on the model at the receiving sites.

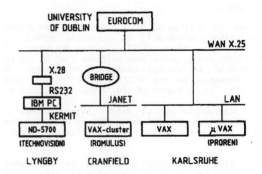

Fig. 7.   Network connections

The first request was to calculate the integral properties on the model recovered after post-processing in their own systems. The results are given in TABLE 1, where S is the total surface area in mm^2, V the volume in mm^3, and Yc the ordinate of the center of mass in mm.

TABLE 1   Integral properties of the test model

	TECHNOVISION	ROMULUS	PROREN
S	3.57116 E 05	357116.211883	–
V	1.38755 E 06	1.38755 E 06	1.3875496 E 06
Xc	0.	0.	0.
Yc	0.	0.000043	0.6 E –05
Zc	0.	0.	0.

The next request was to modify the model by inserting two 20 mm diameter cylindrical holes at given locations in the square plate, record the new integral properties, pre-process the modified model and send the resulting neutral file back to TECHNOVISION. The models recovered after post-processing the neutral files from both systems were identical as shown in Fig. 8 and TABLE 2.

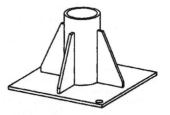

Fig. 8.   Modified model recovered in TECHNOVISION

TABLE 2   Integral properties of modified model

	TECHNOVISION	ROMULUS	PROREN
S	3.56865 E 05	356864.884471	–
V	1.38252 E 06	1.38252 E 06	1.38252 E 06
Xc	0.	0.	0.
Yc	0.15294	0.15292	0.15294
Zc	0.	0.	0.

Finally, the original model was requested to be cut in two and the parts separated. The results from ROMULUS and PROREN were sent back and recovered in TECHNOVISION as shown in Fig. 9.

A glueing operation on the cut parts in both the recovered models resulted in objects indistinguishable from the original one, both visually and with respect to integral properties. Differences could only be seen for the coordinates of the center of mass and were of the order of 10^{-5} mm.

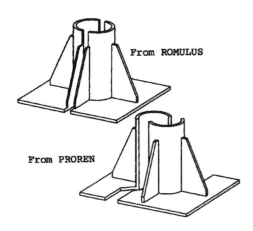

From ROMULUS

From PROREN

Fig. 9. Model cut in two pieces, recovered in TECHNOVISION

The above tests show that the CAD*I neutral description, with proper processor programs, can be considered as reliable enough for use in the transfer of geometry from CAD to applications.

The CAD*I effort was credited by the International Standardization Organization (ISO) by the inclusion of its neutral geometry description in the forthcoming STandard for Exchange of Product model data (STEP) specification (Wilson, 1987), with almost no modification.

CONCLUSIONS

A demonstration facility for robot welding was presented as an example of the use of standard, neutral interfaces in CIM implementations.

Of particular relevance are the implications of employing a neutral interface for CAD. The ability to access geometry information from an individual design sub-system, or more generally, from any CAD system that complies with the standard description, gives obvious advantages over other strategies.

Similarly, production machinery driven with standard job descriptions in neutral formats imply increased compatibility, greater flexibility and, in the end, reduced costs.

A prerequisite for the success of this type of implementations is the easy access to data transmission tools, and the availability of powerful neutral product information descriptions that result in stable standards, recognized by a broad base of vendors and users, and with a life span of, say, a decade. We have called these "neutral interfaces that work" and demonstrated how they can be implemented.

ACKNOWLEDGEMENTS

The demonstration facility is the result of a team effort. IFS students and staff contributed in its realization. The authors are particularly grateful to D. Welner for his magnificent WPM, to T. Sørensen, our GRASP and HITACHI specialist, who also contributed the GRASP pre-processor, and to A. Sølby.

The ND5700 and TECHNOVISION installation in Brussels were provided by courtesy of Norsk Data. K. Hansen of ND Denmark, D. Andersen and D. Moerkens of ND Holland and G. Niephaus of ND Germany are the ones who made this possible.

In Germany, our CAD*I project partners I. Bey and E.G. Schlechtendahl merit our deepest gratitude, as does W. Weick for operating PROREN, and Mrs. U. Frey for making all the administrative arrangements. We are particularly in debt with U. Kühnapfel, who provided and operated the KISMET system.

Finally, special thanks to S. Hailstone who operated ROMULUS at Cranfield.

REFERENCES

Bonney, M.C., and others (1984). The Simulation of Industrial Robot Systems. OMEGA, Int. J. of Mgmt. Sci., 12, 3, 273-281.

Hansen, P.B., and L. Frøslev-Nielsen (1985). A Geometrical Module for Offline Programming of Robots. Report Nr. S85.52. Control Eng. Inst., Technical Univ. of Denmark, (in Danish).

Kroszynski, U.I., and A. Sorgen (1983). Tailoring Intelligent and Versatile CAD Systems for Different Needs. Proc. Int. Conf. on Eng. Design, ICED'83. Copenhagen, 649-654.

Kroszynski, U.I., E. Trostmann, and B. Palstrøm (1986). Standard Interfaces for CAD Data Exchange. Preprints 2nd. Int. Conf. on Computer Applications in Production and Engineering, CAPE'86. Copenhagen, 537-549.

Kühnapfel, U., and colleagues (1984). Graphics Support for JET Boom Control. Proc. Remote Systems and Robotics in Hostile Environments. Int. Pub. Meet., American Nuclear Society. Pasco, Washington, 28-34.

Palstrøm, B., U.I. Kroszynski, and E. Trostmann (1988). CAD Data Transfer to Robot Programming and Control. Proc. 4th. CIM Europe Conf., Madrid, 139-158.

Schlechtendahl, E.G. (Ed.) (1988). Specification of a CAD*I Neutral File for CAD Geometry, Version 3.3. ESPRIT Res. Rep., Vol. 1, Springer Verlag, Berlin.

Trostmann, E., and co-workers (1988). CAD Data Exchange Via Neutral Interface. Proc. Enterprise Networking Event, ENE 88, SME Technical Paper MS88-376. Baltimore, Maryland, 2.95-2.108.

Wilson, P.R., and Ph.R. Kennicott (1987) STEP/PDES Testing Draft: St. Louis Edition. ISO TC184/SC4/WG1. Document Nr. N165.

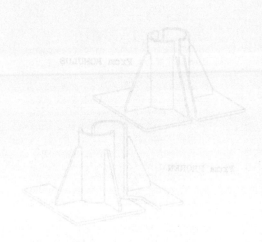

Fig. 9. Model cut in two pieces, recovered in
TECHNOVISION

ISOMATIC PLUS — LOW COST DISTRIBUTED CONTROL SYSTEM

V. Lazarov, G. Nachev and D. Lakov

Institute of Engineering Cybernetics & Robotics, Bulgarian Academy of Sciences, Sofia, Bulgaria

<u>Abstract</u>. In this article the implementation of a low cost DCS based on the Intel Bitbus specification, named ISOMATIC PLUS is discussed. This system uses programmable controllers based on INTEL and MOTOROLA microprocessors. A case with redundant master is considered and some results for system performance are obtained. The design of described system is one approach to connect existing control equipment with newly developed one using low cost manufacturing network. Analyses of system requirements and hierarchical structure of manufacturing control system shows that Bitbus is appropriate. It is very important system to be able to connect different types of control devices. In Isomatic Plus some of services defined in MMS are implemented, and by using bridge or simply changing physical and link communication layers can easily move to another type of network. Performance tests show that future work have to be directed towards decreasing switching times.

<u>Keywords</u>. Distributed control system; manufacturing networks; master-slave; redundant master; programmable controllers.

INTRODUCTION

In the last years programmable controllers have been largely used in manufacturing as a cheap and reliable device for machine and process control. The first step towards CIM is connecting of distinct control devices in a whole distributed control system (DCS). Such a system can provide significant advantages in optimal load distribution between control units, in achieving higher reliability and effectivenes. Very important element of this system is the communication network, allowing data transfer between control units. Recent developments of highly integrated hardware with acceptable prices and international standards for communication contributes in spreading this new technology.

In this article we discuss the implementing of a low cost DCS based on the Intel Bitbus specification, named ISOMATIC PLUS. This system uses programmable controllers based on INTEL and MOTOROLA microprocessors. A case with redundant master is considered and some results for system performance are obtained.

SYSTEM REQUIREMENTS

The system requirements can be divided into two main classes - technical and economical. The technical ones are timely processing of data, media access control mechanism and system reliability. The economical requirements are primarily the cost the controllers, interconnection costs, cost of end-user system implementation and maintenance. In most cases the final decision is a trade-off between technical and economic requirements.

For real-time control it is necessary data to be processed within a predefined time limit. If this requirement is not met a failure can occurre. For instance using of 'old' data may disturb the process to be controlled. The time limits are different for different control levels. For robot control the critical time is 1 - 10ms, for robot cell 10 -50ms, and for a production line critical time may be in order of 200ms. Our applications are mainly the latter two types of production units. Therefore considered system consists mainly of inteligent sensors and actuators and programmable controllers.

For manufacturing control and monitoring the media access control mechanism (MAC) is an important issue. A suitable access technique should guarantee reliable operation and bounded real-time access for all nodes in the system.

There are various methods for media access control but for real-time applications deterministic methods are preffered. One of these methods is token passing, defined in IEEE 802.4 standard and used in MAP. Most researchers consider this method suitable for real-time applications, because of its determinism. On the other hand if a single failure occurs the token will not reach the desired node within a predefined time interval. Hence it is necessary to develop special additional software to handle cases of token loss or duplication. The complexity of such protocols depends on the faults to be tolerated. The more fault-tolerant the protocol is, the more complex and un-dependable it becomes because of dif-ficulty to verify it design and implemen-tation. Moreover token losses and dupli-cates can impact access delay sig-nificantly. Finally token passing needs equal complexity in every node and in-creases significantly the price of the system. On the other hand current im-plementations of MAP protocols are ineffi-cient for our system because their address fields are too long (16 or 48 bits) and CRC field - 32bits, which decrease infor-mation troughput. Similarily specified LLC services are not satisfactory, for example they do not allow the transfer of informa-tion in both directions and do not provide cyclic services. The ISA (1989) proposal for Fieldbus overcomes the most of out-lined drawbacks, but it is still under devolopment.

Another possible method is centralized MAC. The most popular methods are using bus arbiter and polling with master/slave relationship. The polling is more flexible than token passing, because access to com-mon bus is granted by the master in accor-dance with the application program. The protocol is easy to implement and the sys-tem is cost effective. A drawback of the method is necessity of communication through the master when two station want to exchange messages and the overall sys-tem dependence on the master reliability. The method with bus arbiter is proposed by French Fieldbus group (FIP, 1989), but it is not still implemented.

We choose master/slave relationship due to the following reasons:

(i) The control structure of almost any industrial object cannot be totally distributed but will preferably have some kind of hierarchy. Therefore master/slave relationship is natural,

(ii) As the bounded response time is the major critical issue deterministic media access control is prefferable,

(iii) Distributed control system is easy to implement and maintain, and has a low cost.

Another important issue is system reliability. It depends on the reliability of various system components including the communication link. Since the only way for communication between the nodes is message passing an important requirement is achieving fault-free and timely delivery of messages.

The problem can be resolved by using a redundant master and redundant media which will naturally increase system cost.

SYSTEM IMPLEMENTATION

The structure of ISOMATIC PLUS system is shown on fig.1.

The network fully supports the Bitbus specification (Intel, 1987) and has some additional features: communication media is optically isolated and some user defined commands are used for performing application tasks in remote controllers in addition to the RAC functions defined by Intel. On the application level we choose some of the services defined in MMS (ISO/DIS 9506, 1988). These are:

Connection/Context Managment
- Initiate, Conclude, Abort, Cancel, Reject

VMD Suport
- Status, Unsolicited-Status, Identify

Domain Maanagement
- Initiate-Download-Sequence,
Download segment, Terminate-Download-Sequence, Upload-Segment
Initiate-Upload-Sequence
Terminate-Upload-Sequence
Request-Domain-Download,
Request-Domain-Upload,
Get-Domain-Attributes

Remote Program Execution
- Create-Program-Invocation,
Delete-Program-Invocation,
Start, Stop, Resume, Reset,
Get-Program-Invocation-Attributes

Remote Variable Access
-Read, Write, Phys-Read, Phys-Write,
Define-Variable-List, Delete-Variable-List
Get-Variable-Access-Atributes,

Event Management
- Alter-Event-Condition-Monitoring
Event-Notification,
Acknowledge-Event-Notification,
Report-Event-Condition-Status

We suppose these services cover most of the applications in manufacturing and process control. Programs or program seg-ments can be downloaded or uploaded to/from remote controllers. Tasks can be started or halted and the whole memory and I/O space of the remote controllers is ac-cessible from the master or other control-lers. Devices can be identified and vari-ables can be defined and properly used. On the LLC level services as cyclic request with/without data are under development.

Isomatic Plus gives an opportunity to con-nect two different types of controllers - based on Motorola and Intel microproces-

sors. Isomatic 1001UC programmable controller, based on MC6800 microprocessor is largely used in manufacturing in Bulgaria and USSR. It consists of matherboard and of the following interface boards:
- K115 - 16 digital inputs, 24V/20mA
- K120 - 8 differential analog inputs
- K210 - 16 digital outputs, 24V/100mA
- K220 - 4 analog outputs +/- 10V/20mA
- K240 - 8 relay outputs 220V/1A
- K540 - 4 current loops up to 9.6Kbaud
- K550 - Bitbus adapter 62.5/375Kbaud
- K815 - 2 counter inputs up to 100KHz

Different configurations can be arranged depending on the application. Implementation of the Bitbus adapter is cheap, because we use Motorola's communication controller MC6854 to connect existing controllers to the network. Additional logic is used for automatic address recognition. Thus the processor is interrupted only if a message addressed to it is received. All the RAC functions are executed in the interrupt and immediate response is returned to the master, thus minimizing the response time. The user defined commands are transferred to the operating system in a way similar to those in iRMX-51. We use two diferent types of operating systems. The former is based on FORTH core, and the latter is real-time operating system specially developed for Isomatic controllers. Isomatic IN44 programmable controller is based on I8344 microcontroller and has a subset of the interface capabilities of Isomatic 1001UC. In Isomatic Plus system IBM XT/AT is used as a network master. The communication adapter for personal computers has two versions. In the former connection to the host microprocessor is implemented by means of two byte FIFO, and in the latter by means of dual port RAM. The operation of the system can be programmed using C, Forth or Chof languages from the master station. Remote controllers can be programmed, also using programming unit which can be connected to them using paralel port of controller. This allows users to implement some of the tasks on the separate controllers and some of them to download from the master during system operation. In some applications where higher reliability is required redundant master can be used. There exist two major cases of failures:
(j) slave node failure
(jj) master node failure
The first case takes place when the master node stop receiving messages from a slave. In this situation the master has to inform and change working conditions for these nodes which are functionally connected with the failed one. In the second case a recovery algorithm should take place to switch from basic master to the redundant one (Nachev, 1989).
The system has been tested assuming the following conditions:

- Poisson distribution of requests in each of the controllers with equal arrival rate 1 or different l_i,
- The time it takes the master to switch from one slave to another and the service time for each request are exponentially distributed with the mean r and b respectively. These assumption are very close to the real working conditions e.g. in an assembly line. Delay time for the request in each slave is calculated using results in (Takagi, 1985) and (Lazarov, 1989). Delay time of a request is defined as the time spent by it waiting for service in a queue and the transmition time. On fig. 2 delay times in ms for five stations with service times corresponding to the three communication speeds defined in Bitbus - 62.5Kb, 375Kb, 2.4Mb and r = 10ms as a function of system load are plotted.
On fig.3 delay times are plotted as a function of number of slaves for two switching intervals - 10ms, and 2ms. Bitbus specification offers switchin times of maximum 5 bit times or 2us in synchronouse mode, but the longer switching time is caused by the host microprocessor, which manages all message transfers. It takes about 3.5 ms to transfer one message of maximal lenght trough byte FIFO. Service times are assumed to be 0.5ms (375Kb) and interarrival times of requests equal to 50ms and 200ms.
It follows from the figures that in the case of interarival times equal to 50ms the maximum number of controllers which can be connected in a system is equal to 4 and in the case of 200ms interarival time maximum 17 controllers can be used although the network can support much more controllers. Very important conclusion is that increasing the communication speed over some limit for example in this case 375Kb will not affect the system performance significantly due to the long switching time.

CONCLUSION

The design of described system is one approach to connect existing control equipment with newly developed one using low cost manufacturing network. Analyses of system requirements and hierarchical structure of manufacturing control system shows that Bitbus is appropriate. It is very important system to be able to connect different types of control devices. We think that using of standardized descriptions of objects to be controlled will enable user to write application programs independently of particular controller. In Isomatic Plus we implemented some of services defined in MMS and by using bridge or simply changing physical and link communication layers can easily

move to another type of network. Performance tests show that future work have to
be directed towards decreasing switching
times. In applications where additional
reliability is required extra communication cable can be added.

REFERENCES

FIP, 1988. Data link layer. ISA-SP50-
1988-159.

Intel (1987). Distributed control data
book.

Lazarov V. T., and Lakov D. V. (1987).
Configurable distributed control
sistem for robotic cell. Proc. of
AEDTP, Plovdiv (in bulgarian).

Lazarov V. T. (1989). Analysis of LAN with
cyclic service. Automatic and computer sciences, 2, 12-18. (in
bulgarian).

ISO/DIS 9506, (1988). Manufacturing message specification - service definition
and protocol specification, Draft International standard Part 1 and Part 2.

Nachev G. N., V. Lazarov, and D. Lakov
(1989). Fault-tolerant real-time distributed control system. Proc. of
Fault-tolerant systems and diagnostics, Praha. (to be published).

Takagi H. (1985). Mean message waiting
times in symmetric multiqueue systems
with cyclic service. Performance
evaluation, Vol.5, 271-277.

Unified centralized and decentralized
Fieldbus. (1989). Proposal SP50-1989/U1
to ISA SP50&IEC/SC6/WG6.

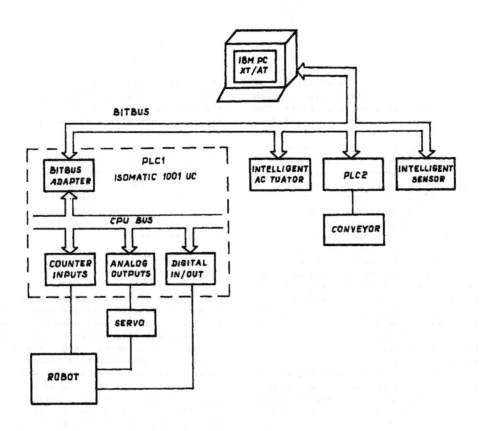

Fig. 1

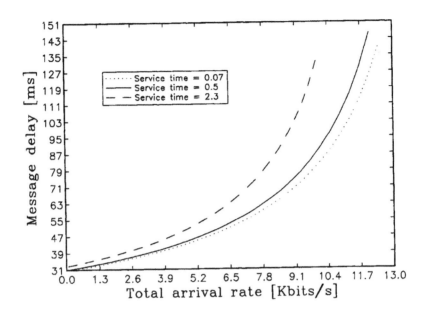

Fig. 2

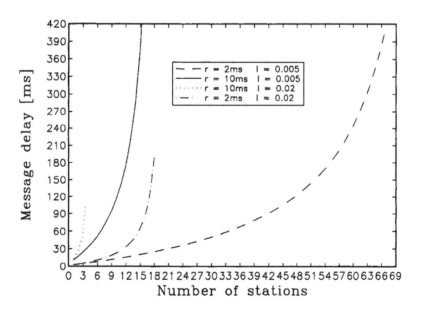

Fig. 3

Fig. 2

Fig. 3

Copyright © IFAC Information Control Problems in
Manufacturing Technology, Madrid, Spain 1989

SEMANTIC NETWORK PROCESSING SYSTEM USED FOR SIMULATION, FAST PROTOTYPING AND CONFIGURATION OF LARGE SOFTWARE SYSTEMS

D. M. Sotirovski

Computer Engineering Department, Institute "Mihajlo Pupin", Beograd, Yugoslavia

Abstract. Since the time DENDRAL, MYCIN and SAINT have surpassed the performance of most human experts in the corresponding areas, the inceasing role that AI techniques, and in particular expert systems, will play in a variety of domains has been recognised. These techiques can capture the knowledge about the structure and behaviour of complex systems such as military C3 systems and space exploration systems. The required knowledge may comprise definitions of hundreds of objects and even thousands of specialised rules describing their relationships. This paper addresses the way that SeNet, an existing semantic network processing system implemented in Ada®, is used to: (1) model the behaviour of a large hardware/software system, (2) support fast prototyping and (3) provide configuration data for the on-line software.

Keywords. Artificial intelligence; Cognitive systems; Computer simulation; Configuration;

INTRODUCTION

In certain respects, the software part of a hardware/software system may be considered as a mapping from the end-user view of the system to the functionality of the existing hardware. This mapping is so complex that it has to be decomposed into a large number of small, manageable pieces of software, called applications for brevity. From a very abstract point of view, the applications can be grouped into three classes (a somewhat modified model from Harmon and Brandenburg, 1981):

- DEVICE applications, which encapsulate the hardware units (e.g. radar, functional keyboard) into suitable abstract objects and hide the details of the implementation from other components of the system,
- STORAGE applications, which store and distribute interface (data carrier) objects that can be exchanged between the applications, and
- CONTROLLER applications, which organize the Device and Storage applications in order to supply the required overall system behaviour.

Applications from Device and Storage classes are almost always hard-coded (existing products are often re-used). The problem which always faces designers and implementors is how to organize the existing applications (or those currently under development) in order to fulfil the end-user view of the system. This problem is even more important when a family of systems is being produced, which differ mainly in the way the same underlying functionality is organized. Finally, the specification of the overall system behaviour is often unstable for a long time after implementation had started. It is not suprising that each of the family members often finish with a number of "system-tailored", low-level solutions. This situation seems to be common within the current engineering practices.

It seems that these problems can be avoided, or at least relieved, with the usage of available a priori knowledge of each individual application (Smith, 1984). The required knowledge comprises:

- the attributes and the behaviour of the objects implemented with Device and Storage applications and
- rules which describe the organisation scheme that Controller applications should perform in order to provide the required overall behaviour of the system.

This article contains a short, informal description of the syntax and the capacities of a semantic network processing system

---

called SeNet. In particular, this paper describes (on a "toy" example) the way SeNet has been used to simulate the behaviour of a large hardware/software system, support fast prototyping and to supply the on-line software with configuration data.

TOY EXAMPLE

The toy example used in this paper (see Fig. 1) is a system consisting of a TV camera, TV monitor, functional keyboard and a computer which controls these devices. The functional keyboard provides the following function keys for controlling the system: Power_On, Power_Off, Increment_Zoom and Decrement_Zoom. The Power_On and Power_Off keys are used to connect and disconnect the TV Monitor and TV Camera to the power supply. The Increment_Zoom and Decrement_Zoom keys are used to set the zoom factor of the TV Camera. It is assumed that:

- the video signal from the TV Camera is connected to the TV Monitor,
- the TV Camera and the Functional Keyboard are encapsulated into Device type of applications,
- a single Controller application organizes the system to provide the required behaviour,
- no Storage applications are needed, and
- shared memory provides the Controller with insight of the current state of the Devices.

SENET

SeNet is an existing semantic network processing system implemented in Ada which can be used interactively by a human, or as a collection of functions encapsulated in a single generic Ada package. SeNet is a frame-based system (Fikes and Kehler, 1985) with some features of a rule-based system (Hayes-Roth, 1985; Hayes-Roth, Waterrman and Lenat, 1983), since it incorporates rules and provides truth maintenance of the rules. Details of the SeNet representation of rules and truth maintenance capability are outside the scope of this article (Sotirovski, 1989a). In certain respects, SeNet can be considered as a system for building directed graphs with labelled nodes and links, and locating nodes and links in such graphs according to graph patterns. Therefore, SeNet can be considered to be a member of the same class of devices as for instance SNePS (Shapiro, 1979). However, since nodes represent objects from the real world being modelled and links stand for their relationships, SeNet is not intended for processing labelled graphs, but is assigned to the purpose of "cognitive" modelling.

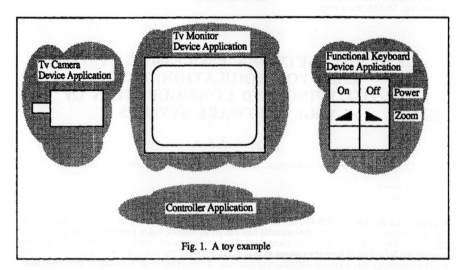

Fig. 1. A toy example

Syntax

SeNet can accept and process requests expressed in a simple but powerful syntax. The basis of the syntax is atom, which is the text equivalent of a graph pattern. An atom is a sequence of node and link names, the latter surrounded with symbols ":" and ">" associated with the link's direction. Node and link names in an atom are either constants or variables. To differentiate between them, variables begin with "*". The following atoms are easily associated with the structure of the semantic network in Fig. 2:

        Key <Isa: Power_On;
        Tv_Camera :Zoom_Factor> *Z :Isa> Zoom_Value;

Atoms, optionally preceeded with "Not", are combined into expressions using "And" ,"Or", "(" and ")". Expressions further enhance the ability to describe graph patterns. The following expressions have an obvious interpretation in the scope of the semantic network in Fig. 1:

        Tv_Camera :Power> On And
        Tv_Camera :Zoom_Factor> *Z;

        Zoom_Value :Min> *Min_Value And
        Zoom_Value :Max> *Max_Value;

Request, the final SeNet syntax element, is composed of an expression called "guard" optionally succeeded by either "=>" or "<=>" and another expression called "action". Request can be preceded with a name surounded by "/" characters. For example:

        Tv_Camera :Power> On <=>
        Key <Isa: Power_Off :Functionality> Available;

        Tv_Camera :Power> Off <=>
        Key <Isa: Power_On :Functionality> Available;

When used interactively, each SeNet request is followed with a "tellback" which expresses (in the same syntax) the results of the evaluation of the request, including the effects of the truth maintenance of the rules. In the examples, tellbacks are printed italic.

Processing Properties

Following a request, if the guard is proven true, the semantic network is updated to match the patterns of the action part. Any variables that may be present in the request are treated as if in the scope of a universal quantifier. Hence, the semantic network is searched for all constants that match identical constants or unbounded variables in the guard pattern.

Requests of the form " True => ..." are used to build the desired semantic network, while requests with no action part make provision for examining the network structure. Part of the semantic network in Fig. 2 can be created with the following sequence of requests:

        True => Power_On :Isa> Key And
                 Power_Off :Isa> Key;

        True => Increment_Zoom :Isa> Key And
                 Decrement_Zoom :Isa> Key;

        True => Available :Isa> Key_State And
                 Unavailable : Isa > Key_State;

        True => On :Isa> Tv_Power And
                 Off :Isa> Tv_Power;

        True => 1X :Isa> Zoom_Value And
                 3X :Isa> Zoom_Value And
                 9X :Isa> Zoom_Value;

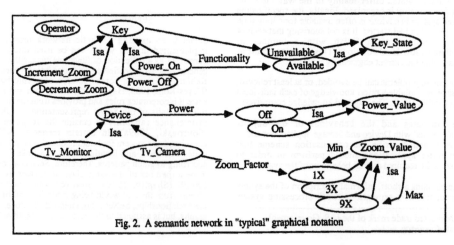

Fig. 2. A semantic network in "typical" graphical notation

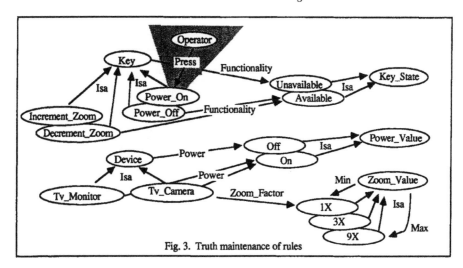

Fig. 3. Truth maintenance of rules

True => Zoom_Value :Min> 1X And
        Zoom_Value :Max> 9X;

True => Tv_Camera :Isa> Device And
        Tv_Monitor :Isa> Device;

True => Tv_Camera :Zoom_Factor> 1X :Isa> Zoom_Value;

The first time that a relationship is introduced into the semantic network, the domain (set of starting nodes) and codomain (set of ending nodes) of that relationship are implicitly defined. SeNet performs strong type-checking and will refuse any attempt to install a relationship with an inappropriate domain or codomain. The request:

True => Device :Power> Off :Isa> Power_Value;

will set the domain of the "Power" relation to all ancestors of the class "Device" and the codomain to all ancestors of the class "Power_Value". Examples of erroneous requests are:

True => Increment_Zoom :Power> On;
*** Semantic error: Increment_Zoom is not in the domain
    (Device) of relation Power.

True => Tv_Monitor : Power > Going_Up;
*** Semantic error: Value Going_Up not in codomain
    (Power_Value) of relation Power.

In common with all frame-based systems, SeNet supports inheritance associated with the membership relation ":Isa>". The SeNet approach to the many-sided semantics of the ":Isa>" link (Brachman, 1983 and 1985) is outside the scope of this paper. The SeNet data-retreival capability is illustrated with the following examples:

        Tv_Monitor :Power> *X;
(True) Tv_Monitor :Power> Off

        Tv_Camera :Zoom_Factor> *Z :Min> *M;
(True) Tv_Camera :Zoom_Factor> 1X :Min> 9X

Requests with non-trivial guards and actions (called rules for short) capture the domain dependent knowledge of the modelled world. For the semantic network in Fig. 2, the following are sound rules:

/Rule_1/ Tv_Camera :Power> On
          <=>
          Key <Isa: Power_Off :Functionality> Available;

/Rule_2/ Tv_Camera :Power> Off
          <=>
          Key <Isa: Power_On :Functionality> Available;

/Rule_3/ Operator:Press> Power_On :Functionality> Available
          =>
          Device <Isa: *Unit :Power> On;

The SeNet truth maintenance mechanism is activated whenever the network structure is changed. When a new fact is entered into the semantic network, SeNet triggers all the rules whose validity may be endangered and re-evaluates them. Note that evaluation of the triggered rules can further modify the network and trigger other rules. The execution is suspended when the list of triggered rules is empty. Conflicting rules will, therefore, make SeNet run forever.

For example, the request:

        True => Operator : Press > Power_On;

will result in:

(True) Operator :Press> Power_On :Functionality>Available =>
(New) Device <Isa: Tv_Camera :Power> On;

(True) Operator :Press> Power_On :Functionality>Available =>
(New) Device <Isa: Tv_Monitor :Power> On;

(True) Tv_Camera :Power> On =>
(New) Key <Isa: Power_Off :Functionality> Available;

(True) Not Tv_Camera :Power> Off =>
(New) Not Key <Isa: Power_On :Functionality> Available;

and the resulting network is displayed in Fig. 3.

A "Not" operator can be applied to atoms both in the guard and in the action part of a rule. The proof procedure of a negated atom in a guard is conducted under the "closed world assumption". SeNet performs non-monotonic inference implied by negated atoms in the action part of the rules, i.e. what is denied with a negated atom is withdrawn from the semantic network.

SYMBOLIC LEVEL SIMULATION

SeNet processing capacity and, particularly, the truth maintenance mechanism which propagates the consequences of a new fact (or withdrawal of an existing fact) is the foundation of the symbolic level simulation of large systems. Since SeNet is exceptionally efficient in dealing with a large number of highly specialized rules (Sotirovski, 1989a and 1989b), it has been used to model and support experiments with a system similar to the toy example, but with several hundred keys. The SeNet model captures the structure and the behavioural characteristics of the objects supported by the Device, Storage and Controller classes of applications. Experiments with the model are used to prove that the overall functionality of the system being developed will be fulfilled. Another important benefit is that the need to express the knowledge of the system in a formalised manner provides the designers with more insight of the problem space. It also raises a number of questions early which would otherwise appear at a later, and therefore less favorable, stage.

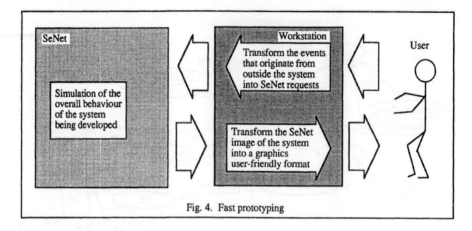

Fig. 4. Fast prototyping

FAST POTOTYPING

Fast prototyping is another use for a SeNet model of a system to be developed. The principal scheme of an environment that may be used for fast prototyping is given in Fig. 4. Motives for the proposed architecture are to use the SeNet symbolic level simulation as an overall behavioural model of the system, and the sophisticated graphical (and other) capability of contemporary workstations to transform the SeNet image of the system into a user-friendly form. Prototyping of the toy example would probably make a graphical presentation for each of the devices. The operator would use a mouse to "press a key" and the graphical presentation of the devices would change according to the action he has initiated.

CONFIGURATION DATA FOR THE ON-LINE SOFTWARE

The data stored in SeNet, i.e. the structure of the objects and the rules describing the behaviour of the system can be used as a unique source of configuration data to be given to the on-line software components. Device applications are seldom significant consumers of configuration data. On the contrary, an important part of the knowledge stored in the SeNet model describes the functionality of the Device applications. Storage applications are sometimes configured with a description of the data with which they deal. Since Storage applications are not interested in the data themselves, they usually need knowledge of the number and underlying base types of the storage objects in order to provide appropriate storage capacity. Controller applications are the most thorough, and therefore the most important, consumers of the configuration data. In principle, Controllers need answers to the following questions:

- What am I good for ?
- How and when will my own fuctionality be employed ?
- How and when can the functionality of the other Controllers, Devices and Storages be employed ?

Since SeNet models all objects implemented in the Device and Storage class and is capable of simulating the behaviour of the system, it is certain that the SeNet model contains all data any Controller may need. SeNet represents the rules in the same fashion as the objects to which they are applied. Therefore, it is possible to retrieve not only the structure of the objects but also the rules describing the desired system behaviour. The mere fact that a single source is used to supply the configuration data to all Controllers provides each and all cooperating Controllers with the same image of the software environment.

Controllers which use large amounts of configuration data may endanger the performance of a real-time system. However, these same Controllers make early releases of the on-line software less sensitive to unstable system requirements. Performance of the critical Controllers can, afterwards, be enhanced with hard-coded solutions. The knowledge stored in the SeNet model (the same that has been used to configure the data-driven Controllers and has proven to meet the end-user requirements) can be used as an excellent guideline for implementing the hard-coded solutions.

CONCLUSION

SeNet, a semantic network processing system implemented in Ada, has been used successfully to model and simulate the behaviour of a large software system which controls dozens of devices and utilizes hundreds of forms, functional keys and graphics objects. The symbolic level simulation capability of SeNet supports experiments which prove that the system being developed conforms to the end-user requirements. The model serves also as a unique source of configuration data for the on-line software.

The author is denied permission to publicate informations about concrete applications, underlying hardware and performance of the system which would make the paper complete.

ACKNOWLEDGEMENTS

The author thanks Vladimir Kukic, Nebojsa Vuksanovic and Slobodan Jovanovic for their valuable comments on early drafts of this paper.

REFERENCES

Brachman, R.J. (1983). What IS-A is and isn't: an analysis of taxonomic links in semantic networks, *IEEE Computer, 16* (No. 10), 30-36.

Brachman, R.J. (1985). "I lied about the trees" or defaults and definitions in knowledge representation, *The AI Magazine*, Fall 1985, pp.80-93.

Fikes, R. and Kehler, T. (1985). The role of frame-based representation in reasoning, *Communications of the ACM*, *28*, 904-920.

Harmon, S.Y. and Brandenburg, R.L. (1981). Concepts for description and evaluation of military C3 systems, *Proc. of the CDC Conference*, San Diego, December 1981.

Hayes-Roth, F. (1985). Rule-based systems, *Communications of the ACM, 28*, 921-932.

Hayes-Roth, F., Waterman, D.A. and Lenat, D.B (Eds.) (1983). Building Expert Systems, *Addison-Wesley*

Shapiro, S.C. (1979). The SNePS semantic network processing system, in N.Findler (Ed.), *Associative Networks - The Representation and Use of Knowledge by Computers*, Academic Press, New York, pp. 179 - 203.

Smith, R.G., Lafue, G.M.E, Shoen, E. and Vestal, S.C. (1984). Declarative task description as a user-interface structuring mechanism, *IEEE Computer, 17* (No. 9), 29-38.

Sotirovski, D.M (1989a). Representation of rules and truth maintenance in semantic networks, *in preparation*.

Sotirovski, D.M (1989b). Implementation of a semantic network processing system in ADA, *Proc. of the Expert Systems, Theory and Applications*, IASTED International Conference, Zurich, Switzerland, June 26-28, 1989.

A NEW CONTROL STRATEGY BASED ON THE CONCEPT OF NON INTEGER DERIVATION: APPLICATION IN ROBOT CONTROL

A. Oustaloup, P. Melchior and A. El Yagoubi

Equipe Systèmes et Commande d'Ordre Non Entier, Université de Bordeaux I, Talence, France

ABSTRACT

This paper deals with the robustness as far as damping is concerned, and more particularly the robustness as for control damping versus the parameters of the plant.

After defining robustness in time domain, it presents the non integer approach of the CRONE control, a french abbreviation of "Commande Robuste d'Ordre Non Entier", namely "Non Integer Order Robust Control". This approach uses the mathematical principle which insures the robustness of stability degree in nature, namely non integer derivation.

An open loop frequency template is deduced from the non integer order differential equation which describes the relaxation of the ebb and flow on a porous dyke, this phenomenon being robust as for stability degree since the damping factor is independent of the motion water mass. This template illustrates robustness in frequency domain.

The general expressions of the damping factor and the resonance factor are proved. These one indeed translate the robustness of the control.

The last part of the paper deals with the synthesis of a CRONE regulator insuring a precorrection of the plant, in the general case of an indifferent frequency placement of the template. The study plant is achieved by an inclining polar table which constitutes an elementary manipulator with three degrees of freedom. This one is chosen because of its structure which insures large inertia variations, strong dynamic couplings and a great number of non-linearities.

KEYWORDS

Control ; damping robustness ; non integer derivation ; non-stationarity of the plant ; non-linearity of the plant ; robot control.

I-INTRODUCTION

For many years, it has been common to speak of robustness. But this concept is very wide, even in a same domain such as the automatic control one. In fact, robustness is a notion which always translates the same idea, namely insensitivity.

In automatic control, it is frequent to consider the robustness as far as stability is concerned.

In the non integer approach, the considered robustness is much stricter, that is to say the robustness as for stability degree. More precisely, the robustness which is at stake translates the insensitivity of the damping factor or the stability degree of the control to the plant parameters ; at least, in so far as they remain within given ranges.

Although time domain makes it possible to illustrate the definition of robustness, particularly from the transient of the step response, it is not a priviliged domain for specifying robustness, not in terms of response performances in closed loop, but in terms of control performances in open loop.

It is true that frequency domain is a domain in which robustness can be illustrated by a characteristic transfer of the control in open loop.

I. HOROWITZ (1) is the one among others who has tried to synthesize a robust control through the frequency approach. From tolerances on the control damping, that is to say time specifications which he translates in frequency specifications on the closed loop gain, he develops a synthesis method of the Black locus of the open loop frequency response of the control.

The approach we propose is more based on the concept of non integer derivation, in so far as it uses the mathematical principle which constitutes the origin of the robustness of stability degree in nature, that is to say non integer derivation. It is true that the relaxation of the ebb and flow on a porous dyke, described by a non integer order differential equation, is characterized by a damping which is independent of the motion water mass.

From such a differential equation, it is possible to determine an "open loop frequency template" (or more simply "template") which illustrates robustness in frequency domain, in this case, a vertical straight line segment lying between the abscissae $-\pi/2$ and $-\pi$ in the Black plane.

By associating an asymptotic frequency behaviour to this template, the consideration of a non integer derivation order is quite obvious. Indeed, the template defines an open loop asymptotic behaviour of non integer order n' between 1 and 2 ; it is true that such a behaviour is between the asymptotic behaviours of orders 1 an 2 which are defined respectively by the angular boundaries $-\pi/2$ and $-\pi$.

This paper presents a new synthesis technique of the template in the case of an indifferent frequency placement of this one. The corresponding process consists in synthesizing a CRONE regulator which insures both a precorrection of the plant and the correction in open loop.

Such a synthesis is carried out in the case of a polar table whose configuration insures large inertia variations, strong dynamic couplings and a great number of non-linearities.

II-REPRESENTATION OF ROBUSTNESS IN TIME DOMAIN

In time domain, the principle of robustness is translated by a step response which presents the same overshoots independently of the parameters of the plant ; only the natural frequency changes ; so, the transient keeps its form with only a time scale changing (figure 1).

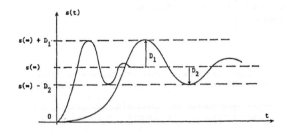

Fig.1. Illustration of robustness in time domain

III-FROM THE ROBUSTNESS OF STABILITY DEGREE IN NATURE TO A NEW ROBUST CONTROL STRATEGY : THE NON INTEGER APPROACH OF THE CRONE CONTROL

Our approach, whose aim is the conception and the application of a new robust control strategy, uses the observation of a natural phenomenon, the one of the ebb and flow on a porous dyke. Already in the 17rd century, the constructors of dykes had noted the damping properties of the very disturbed dykes and particularly that forming air pockets which can be compressed by the advance of water. Otherwise, an attentive observation of the ebb and flow phenomenon consecutive to the damping of water on fluvial or coastal dykes, shows that in the case of very damping (or absorbing) dykes through a porous volumic structure and a rough* surfacic structure :

- the natural frequency of the relaxation is different whether the dyke is fluvial or coastal ;

- the damping of the relaxation seems to be independent of the dyke, whether it is fluvial or coastal.

Given that the fluvial and coastal tests can be distinguished by very different motion water masses, the observation seems to show that the relaxation is characterized by a natural frequency which depends on the motion water mass and by a damping which is independent of it. Although it should be paradoxal when one knows the properties of a pendular relaxation, this result is as well remarkable as fundamental in so far as its reveals the insensitiveness of the damping factor to a parameter of the process, in this case the motion water mass ; in automatic language, this translates the phenomenon

---

* The consideration of a rough surfacic structure (or very disturbed in the sense of B.MANDELBROT), permits to minimize the reflections on the dyke faces and so, to free oneself from stationary wave phenomena which stems from them ; that is to say that the observation turns, not on water motions consecutive to reflections, but on the motion of the water which rushes into the dykes through their faces.

robustness as for stability degree.

After trying to determine the mathematical origin of this type of natural robustness, it appears that it resides in non integer derivation. Indeed, by taking into account the fractality of porosity and the corresponding recursivity, we show (25) that the process is described by a differential equation of non integer order n' between 1 and 2, namely :

$$\tau^{n'}(\frac{d}{dt})^{n'} P(t) + P(t) = 0 , \qquad (1)$$

P(t) designating the dynamic pressure at the water-dyke interface.

The corresponding characteristic equation is of the form :

$$(\tau s)^{n'} + 1 = 0 . \qquad (2)$$

Finally, the purpose is to obtain the same thing in automatic control, that is to say a control which should be characterized by such a characteristic equation. Indeed, it seems interesting to use a so fundamental result for synthesizing a robust control strategy : it is the approach (said non integer) that the CRONE control uses.

IV-REPRESENTATION OF ROBUSTNESS IN FREQUENCY DOMAIN : OPEN LOOP FREQUENCY TEMPLATE

IV-1. Transfer in closed loop

As "synthesis transmittance in closed loop", one considers a transfer function of the form :

$$F(s) = \frac{1}{1 + (\tau s)^{n'}} , \qquad (3)$$

whose characteristic equation is indeed that given by relation (2).

IV-2. Transfer in open loop

Let us designate by E(s) and S(s) the Laplace transforms of the input and output of the control. Relation (3) permits then to write :

$$\frac{S(s)}{E(s)} = \frac{1}{1 + (\tau s)^{n'}} , \qquad (4)$$

from where one draws :

$$S(s) = \frac{1}{(\tau s)^{n'}} \left[E(s) - S(s) \right] , \qquad (5)$$

a symbolic equation which is translated by the functional diagram proposed in figure 2 and in which the transmittance of the direct chain determines the open loop transfer function, namely :

$$\beta(s) = (\frac{1}{\tau s})^{n'} \qquad (6)$$

this one can be considered as a "synthesis transmittance in open loop" of a robust control.

The corresponding open loop frequency response, namely :

$$\beta(j\omega) = (\frac{1}{j\tau\omega})^{n'} \qquad (7)$$

admits, as Black locus, a vertical straight line of abscissa between -90° and -180°C (figure 3).

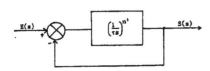

Fig.2. Functional diagram of a non integer
order robust control

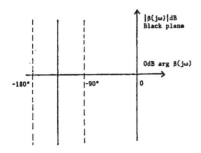

Fig.3. Frequency representation of $\beta(j\omega)$

IV-3. Open loop frequency template

Given that the dynamic behaviour in closed loop is essentially linked to the behaviour in open loop close to the unit gain frequency ω_u, a vertical straight line segment is sufficient to insure the robustness of damping. This segment, called "open loop frequency template" (or more simply "template"), illustrates robustness in frequency domain (figure 4) ; the longer the segment, the greater the robustness.

If the parameters of the plant vary, the segment AB slides vertically on itself. This insures a constant phase margin (independent of the plant parametric state) and, consequently, the invariance of the corresponding damping factor in time domain.

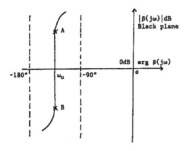

Fig.4. Illustration of robustness in frequency domain : AB is the segment to be synthesized

V. NATURAL FREQUENCY AND DAMPING FACTOR

The poles of the transmittance $F(s)$ of the control satisfy the equation :

$$(\tau s)^{n'} + 1 = 0 , \tag{8}$$

namely $(\tau s)^{n'} = -1 = e^{j(\pi + 2k\pi)}$, $\tag{9}$

from where one draws :

$$s_k = \tau^{-1} e^{j\frac{1+2k}{n'}\pi} . \tag{10}$$

The non integer character of n' involves the multiformity of equation (8). Indeed, if the operational variable s is written under the form $s = |s| e^{j\theta}$ with $\theta = \theta_0 + 2k\pi$, it is possible to write :

$$s^{n'} = |s|^{n'} e^{jn'\theta_0} e^{j2n'k\pi} . \tag{11}$$

This form dictates the consideration of two distinct cases : if n' is integer, $e^{j2n'k\pi} = 1 \; \forall \; k$, which expresses that $s^{n'}$ has only one meaning, so translating the uniformity of equation (8) ; in the case when n' is non integer, $e^{j2n'k\pi}$ depends on k, which expresses that $s^{n'}$ has several meanings, so translating the multiformity of equation (8). In order to make this equation uniform, it is necessary to avoid that the argument of s should describe a complete turn, which is possible by cutting the plane s, generally along R^-. Such a cut imposes the determination $]-\pi, +\pi[$ for the argument of s. This allows to write :

$$- \pi < \frac{1+2k}{n'}\pi < +\pi$$

namely : $- \dfrac{n'+1}{2} < k < \dfrac{n'-1}{2}$; $\tag{12}$

this translates that the problem consists in trying to find the values of k which satisfy this double inequality.

There exists exclusively two poles corresponding respectively to $k = 0$ and $k = -1$, namely :

$$s_0 = \tau^{-1} e^{j(\pi/n')} \quad \text{and} \quad s_{-1} = \tau^{-1} e^{-j(\pi/n')} ; \tag{13}$$

there are complex, conjugate and form a centre angle 2θ with $\theta = \pi - \pi/n'$; this one is constant and is fixed by the non integer order n'.

The natural frequency and the damping factor are directly deduced from the pole through their modulus and the half-centre angle θ that they form :

$$\omega_p = \tau^{-1} \sin\theta = \tau^{-1} \sin(\pi - \frac{\pi}{n'}) = \tau^{-1} \sin\frac{\pi}{n'}$$

and $\tag{14}$

$$\zeta(n') = \cos\theta = \cos(\pi - \frac{\pi}{n'}) = -\cos\frac{\pi}{n'} . \tag{15}$$

Relation (15) clearly reveals the robustness as for damping in so far as the damping is essentially linked with the non integer order n' of the control. In other hand, the natural frequency given by (14) is a function of τ.

VI-RESONANCE FREQUENCY AND RESONANCE FACTOR

To the transmittance given by relation (3) corresponds the closed loop frequency response :

$$F(j\omega) = \frac{1}{1 + (j\frac{\omega}{\omega_u})^{n'}} , \tag{16}$$

in which the transitional frequency in closed loop, ω_u, is nothing but the unit gain frequency in open loop.

By replacing j by $e^{j\pi/2}$, relation (16) becomes :

$$F(j\omega) = \frac{1}{1 + (\frac{\omega}{\omega_u})^{n'} \cos n'\frac{\pi}{2} + j(\frac{\omega}{\omega_u})^{n'} \sin n'\frac{\pi}{2}} , \tag{17}$$

or, putting $a = \omega/\omega_u$:

$$F(j\omega) = \frac{1}{1 + a^{n'} \cos n' \frac{\pi}{2} + j a^{n'} \sin n' \frac{\pi}{2}} , \quad (18)$$

whose modulus is given by :

$$|F(j\omega)| = \frac{1}{(1 + 2 a^{n'} \cos n' \frac{\pi}{2} + a^{2n'})^{1/2}} \quad (19)$$

$|F(j\omega)|$ reaches a maximum when the quantity under the radical is minimum, which corresponds to :

$$\frac{d}{da} \left[1 + 2 a^{n'} \cos n' \frac{\pi}{2} + a^{2n'} \right] = 0 , \quad (20)$$

namely $a^{n'} + \cos n' \frac{\pi}{2} = 0$. $\quad (21)$

The resonance frequency so satisfies the relation

$$(\frac{\omega_r}{\omega_u})^{n'} + \cos n' \frac{\pi}{2} = 0 , \quad (22)$$

from which one draws :

$$\omega_r = (- \cos n' \frac{\pi}{2})^{1/n'} \omega_u , \quad (23)$$

a result which translates the existence of a resonance when $\cos n' \pi/2 < 0$, namely for $1 < n' < 3$, then for the CRONE control since $1 < n' < 2$.

The corresponding resonance factor is expressed by the relation :

$$Q(n') = \frac{|F(j\omega_r)|}{|F(jo)|} = (1 - \cos^2 n' \frac{\pi}{2})^{-1/2}$$

$$= \frac{1}{\sin n' \frac{\pi}{2}} , \quad (24)$$

whose from makes it possible to introduce the notion of "robust resonance", given that Q is exclusively linked to the non integer order n' of the control, then, as ζ, independent of the transitional frequency ω_u.

VII-FREQUENCY PLACEMENT OF THE TEMPLATE

VII-1. Asymptotic frequency behaviour of the plant

An "asymptotic frequency behaviour" (or more simply "asymptotic behaviour") is characterized by a locked phase which is defined by a phase independent of frequency ; it corresponds to a range of frequencies in which the phase diagram is comparable to the corresponding asymptotic diagram. A plant presents one or several asymptotic behaviours. An order n asymptotic behaviour corresponds to a locked phase at $-n\pi/2$ (figure 5).

VII-2. Asymptotic frequency placement of the template

In such a placement, the template belongs to a frequency range which corresponds to an asymptotic behaviour of the plant (figure 5).

Given that such an asymptotic behaviour corresponds often to high frequencies, the template is also situated towards the high frequencies, which imposes to fixing a relatively high value of the frequency ω_u, then a fast dynamic in closed loop. So, in

practice, given the existence of noise and disturbances, an asymptotic frequency placement can lead to excessive promptings of the power devices.

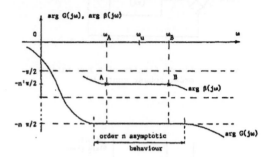

Fig. 5. Illustration of an asymptotic frequency placement of the template in the Bode plane : $G(j\omega)$ designates the frequency response of a plant which presents an order n asymptotic behaviour.

To pass from the argument of $G(j\omega)$ to the argument of $\beta(j\omega)$ for $\omega_A < \omega < \omega_B$, the observation of figure 5 shows that the regulator placed in cascade with the plant (figure 6), must provide an advance of phase equal to $m'\pi/2$, with $m' = n-n'$, in the frequency range corresponding to the template, namely $[\omega_A, \omega_B]$.

Such a phase advance can be obtained with an order m' frequency response of the form :

$$C_{m'}(j\omega) = C_0 \left[\frac{1 + j \frac{\omega}{\omega_b}}{1 + j \frac{\omega}{\omega_h}} \right]^{m'} , \quad (25)$$

in which the transitional frequencies ω_b and ω_h satisfy the conditions :

$$\omega_b << \omega_A \quad \text{and} \quad \omega_h >> \omega_B . \quad (26)$$

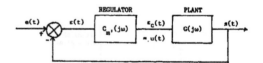

Fig. 6. Localization of the regulator $C_{m'}(j\omega)$ in the control diagram

Otherwise, one knowns (13) that the ideal frequency response of non integer m' (25) can be approximated by the real frequency response of finite integer order N :

$$C_N(j\omega) = C_0 \prod_{i=1}^{N} \frac{1 + j\omega/\omega'_i}{1 + j\omega/\omega_i} , \quad (27)$$

in which :

$$\frac{\omega_i}{\omega'_i} = \alpha \quad \text{and} \quad \frac{\omega'_{i+1}}{\omega_i} = \eta , \quad (28)$$

where α and η are superior to tunity and called "recursive factors". Moreover, m' is given by :

$$m' = \frac{1}{1 + \log \eta / \log \alpha} . \quad (29)$$

In practice, the ratio between two consecutive zeros or poles must make a compromise between two conflicting requirements :

- it should be close enough to unity to limit the undulations associated to the order m' asymptotic behaviour, due to the frequency connections of the various steps and crenels of the asymptotic diagrams

- it must be sufficiently greater than unity to cover a wide range of frequencies with a reasonable number of zeros and poles.

This compromise is achieved by taking a ratio between 5 and 10 and a number of zeros or poles also between 5 and 10. Satisfactory results are obtained with a ratio of 6 and 8 zeros and 8 poles ; in robotics, or more precisely in manipulator control, 4 zeros and 4 poles are often sufficient.

VII-3. Indifferent frequency placement of the template

In the general case of an indifferent frequency placement of the template, this one belongs to a frequency range which corresponds to an indifferent behaviour of the plant. In fact, such a placement can be asymptotic, non asymptotic or both asymptotic and non asymptotic.

The interest of this type of placement comes from the fact that it makes it possible to position the template around a given frequency ω_u, fixed by the conceptor, generally inferior to that imposed by an asymptotic frequency placement.

Although the placement should be indifferent, it is possible to synthesize the template from an asymptotic behaviour if this one is, not that of the plant $G(j\omega)$, but that of a new plant $G'(j\omega)$ obtained through a precorrection of the plant. In this case, the problem consists in embedding the plant into a new plant in conformity with the asymptotic frequency specifications (figure 7).

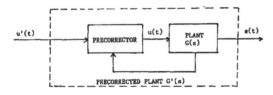

Fig. 7. Embedding of the plant into a new plant (or precorrected plant) whose frequency placement of the asymptotic behaviour obeys the conceptor's requirements.

VIII-APPLICATION IN MANIPULATOR CONTROL

VIII-1. Study plant : inclining polar table

The study plant is achieved by an inclining polar table which constitutes an elementary manipulator with three degrees of freedom q_1, q_2 and q_3 (figure 8). It is achieved from an electrical motor m_1 driving in rotation a ruler (m_1, l_1) on which a guided lead m_2 is moved in translation by a motor M_2 ; the whole is drived in rotation by a motor M_3.

Given that the phenomena are sufficiently interesting without taking into account the gravitational forces, this paper exclusively deals with the case of the horizontality of the table. In this configuration, the equations describing the dynamic behaviour of the table are the following ones :

$$(I_1 + m_2 q_2^2)\ \ddot{q}_1 + \left[f_1 + \frac{\alpha_1 \beta_1}{R_1} + 2m_2\ q_2\ \dot{q}_2 \right]\dot{q}_1$$
$$= \frac{\alpha_1}{R_1}\ u_1 \qquad (30)$$

and

$$\frac{I_2}{R^2}\ \ddot{q}_2 + \left[f_2 + \frac{\alpha_2 \beta_2}{R^2 R_2} \right]\dot{q}_2 - m_2\ \dot{q}_1^2\ q_2 = \frac{\alpha_2}{R\ R_2}\ u_2\ , \qquad (31)$$

in which the different magnitudes are defined as follows :

- I_1 : inertia in relation to q_1 of the ruler, the motor M_1 rotor and the motor M_2
- I_2 : inertia of the motor M_2 and the lead in relation to q_2
- f_1 and f_2 : viscous friction coefficients associated to q_1 and q_2
- α_1 and α_2 : proportionality coefficients between motor torque and armature current
- β_1 and β_2 : proportionality coefficients between f.c.e.m. and rotor angular speed
- R_1 and R_2 : armature resistances
- u_1 and u_2 : armature voltages (controls)
- R : radius of driving pulleys of the lead

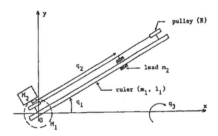

Fig. 8. Configuration of the table

VIII-2. Setting of the problem and adopted control scheme

- Setting of the problem : the control of such a table is very difficult, particularly if one wants to have mastery of the damping of each degree of freedom q_i, independently of the temporal configuration of the table. Nevertheless, the aim of the non integer approach is to insure this mastery.

- Adopted control scheme : the corresponding control strategy consists in elaborating the control u_i of the freedom degree q_i from the error signal $\varepsilon_i = q_{c_i} - q_i$ through a non integer order robust regulator, q_{c_i} designating the input reference of rank i.

VIII-3. Interpretation of the non linear model of the table

The non linear model of the table given by relations (30) and (31) may be interpreted as a linear non stationary model whose coefficients can vary as fast as the variables, namely :

$$a_1(t)\ \ddot{q}_1(t) + b_1(t)\ \dot{q}_1(t) = d_1\ u_1(t) \qquad (32)$$

and

$$a_2\ \ddot{q}_2(t) + b_2\ \dot{q}_2(t) + c_2(t)\ q_2(t) = d_2\ u_2(t), \qquad (33)$$

the corresponding coefficients being functions of time.

VIII-4. Frequency approach in the case of a weak non-stationarity of the table : local (or instantaneous) frequency representation

In the case of a weak non-stationarity, the coefficients of the dynamic model of the plant vary relatively slowly in comparison with the variables :

+ Their variations are not sufficiently slow for continuing to apply the results obtained in stationary state, that is to say for justifying the approximation of the near stationary states.

+ Their variations are sufficiently slow in order that the response to a sinusoïdal input keeps a near sinusoïdal form, or more precisely that the response to a sinusoïdal stationary oscillation should be a sinusoïdal non-stationary oscillation (amplitude and phase angle depending on time).

In the particular case of the table, a weak non-stationarity corresponds to a slow evolution of q_2 in comparison with that of q_1 when the dynamic behaviour in relation to q_1 is at stake, and vice versa when the dynamic behaviour in relation to q_2 is at stake.

After defining a generalized frequency response $G(j\omega,t)$ which insures an instantaneous frequency representation (29), it appears that non-stationarity does not modify asymptotic behaviours. For instance, independently of non-stationarity, the generalized frequency response relative to q_1 is always characterized by an order 1 asymptotic behaviour for $\omega \ll \omega_o(t)$ and an order 2 asymptotic behaviour for $\omega \gg \omega_o(t)$, where $\omega_o(t)$ is given by $\omega_o(t) = b_1(t)/a_1(t)$ (figure 9).

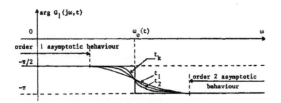

Fig. 9. Non-stationarity does not modify asymptotic behaviours.

VIII-5. Adopted control strategy : CRONE regulator insuring a precorrection of the plant

In a previous paper, we have shown how to synthesize two CRONE regulators in the case of an asymptotic frequency placement of the template, which leads to fixing a relatively high value of the frequency ω_u, then a fast dynamic in closed loop.

In this paper we propose to show how it is possible to synthesize the control in the general case of an indifferent frequency placement of the template, although we want to synthesize the template from an asymptotic behaviour.

In this case, the problem consists in synthesizing a new asymptotic behaviour of the plant situated towards intermediate frequencies, particularly around the transitional frequency $\omega_o(t)$ that the phase asymptotic diagram defines (figure 9).

To present the method which permits to reach this aim, it suffices to consider one degree of freedom, in this case the most important, namely q_1.

The CRONE regulator insures both the precorrection of the plant and the correction in open loop.

The precorrector permits to synthesize an order 2 asymptotic behaviour of the precorrected plant around ω_u, through a decrease of $\omega_o(t)$, which corresponds to a translation of the phase asymptotic diagram towards the low frequencies (figure 10).

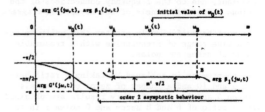

Fig. 10. Synthesis process of the template.

The corrector $C_{m'}(j\omega)$ or really $C_N(j\omega)$ makes it possible to insure the phase placement of the template from the order 2 asymptotic behaviour of $G'_1(j\omega,t)$, by providing a constant phase advance equal to $m'\pi/2$ in the frequency interval (ω_A, ω_B). The form of its frequency response is given for the ideal corrector of order m' :

$$C_{m'}(j\omega) = C_o \left[\frac{1 + j\omega/\omega_b}{1 + j\omega/\omega_h} \right]^{m'} \text{ with } m' = 2 - n' \quad (34)$$

and for the corresponding real corrector of order N:

$$C_N(j\omega) = C_o \prod_{i=1}^{N} \frac{1 + j\omega/\omega'_i}{1 + j\omega/\omega_i} \quad , \quad (35)$$

with $\omega_i/\omega'_i = \alpha$ and $\omega'_{i+1}/\omega_i = \eta$ (36)

and then $m = \dfrac{1}{1 + \log\eta/\log\alpha}$ (37)

VIII-6. Synthesis of the precorrector

From the expression of $\omega_o(t)$, a decrease of $\omega_o(t)$ can be obtained through an increase of $a_1(t)$. So, the precorrection of the plant must insure such an increase. To that effect, consider again the differential equation of the plant, namely :

$$a_1(t) \ddot{q}_1(t) + b_1(t) \dot{q}_1(t) = d_1 u_1(t), \quad (38)$$

then impose that the precorrected plant should be described by a differential equation of the form :

$$a'_1(t) \ddot{q}_1(t) + b_1(t) \dot{q}_1(t) = d_1 u'_1(t) \quad (39)$$

with $a'_1(t) = a_1(t) + d_1 \tau_d$,   (40)

$a'_1(t)$ being a new coefficient of the angular acceleration, superior to $a_1(t)$, in which τ_d designates a differentiation time constant.

Combining relations (39) and (40) determines the equation :

$$a_1(t) \ddot{q}_1(t) + b_1(t) \dot{q}_1(t) = d_1 [u'_1(t) - \tau_d \ddot{q}_1(t)]$$
$$= d_1 [u'_1(t) - \tau_d \, d/dt \, \dot{q}_1(t)] \quad (41)$$

from which one draws, by identifying with (38) :

$$u_1(t) = u'_1(t) - \tau_d \, d/dt \, \dot{q}_1(t) \quad , \quad (42)$$

which is the equation of the control of the plant. The precorrection is obtained through a freeback precorrector which insures an acceleration correction of the plant.

IX-SIMULATION OF THE CONTROL

We have simulated the dynamic behaviour of the control through a numerical processing of the differential equations of the table and the regulator. The used numerical integration method is that of Runge-Kutta (order 4). The simulation has been executed from the following data.

A - First kind of data : features of the motorized table
- ruler : $l_1 = 0.5$ m
- lead : $m_2 = 0.5$ kg
- driving pulleys : $R = 0.5$ cm
- motor M_1 : CEM-AXEM-F9M4
  $\alpha_1 = 5.92$ N cm/A ; $\beta_1 = 6.2$ V/1000 tr/mn
  $R_1 = 1.1$ Ω; rotor inertia : 350 g cm^2 ;
  motor mass : 2.3 kg
- motor M_2 : CEM-AXEM-UGP MEE 09B12
  $\alpha_2 = 3.1$ N cm/A ; $\beta_2 = 3.3$ V/1000 tr/mn
  $R_2 = 1.1$ Ω; rotor inertia : 340 g cm^2 ;
  motor mass : 0.6 kg
- $I_1 = 2.5 \ 10^{-3}$ kg m^2 ; $I_2 = 5.10^{-5}$ kg m^2.

B - Second kind of data : features of the CRONE regulator

The corrector $C_N(j\omega)$ corresponding to an asymptotic frequency placement of the template is defined by the frequency response :

$$C_2(j\omega) = 6433 \frac{(1 + j\omega/38.7) \ (1+ j\omega/477)}{(1 + j\omega/190.7) \ (1 + j\omega/2350.6)} \quad (43)$$

For an indifferent frequency placement, the corrector $C_N(j\omega)$ is defined by :

$$C_2(j\omega) = 14 \frac{(1 + j\omega/0.0612)(1 + j\omega/1.213)}{(1 + j\omega/0.404) \ (1 + j\omega/8)} \quad (44)$$

C - Performances

The recorded step responses corresponding to the freedom degree q_1 are shown in figure 11.

This figure shows the influence of q_2 on the dynamics of q_1. In each case, the input reference of the control of q_1 is submitted to a step whose amplitude is equal to 90° . q_2 is successively defined as follows : (a) $q_2 = 0$; (b) $q_2 = 25$ cm ; (c) $q_2 = 50$ cm ; (d) q_2 increases from 0 to 50 cm : (e) q_2 decreases from 50 cm to 0. Although the inertia in relation to q_1 varies by a factor of 50, the first overshoot of q_1 remains practically constant ; only the natural frequency changes in the case of an asymptotic frequency placement of the template (figure 11.a) ; in other hand, it remains practically constant for an indifferent frequency placement (figure 11.b).

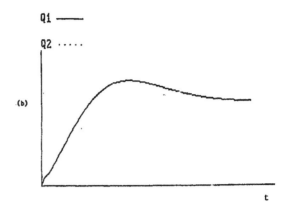

Fig. 11. Step responses of the freedom degree q_1 : (a) without precorrection ; (b) with precorrection

X-CONCLUSION

In non integer order robust control, the CRONE regulator can be adapted to parameter ranges of the plant, and this for a given adjustable of the regulator parameters. So, when the plant parameters vary within these ranges, a real time adaptation of the regulator parameters is not necessary, whereas it is with the self-adaptative regulator.

The proposed regulator is different from the self-adaptative one. Its interest derives from the fact that it does not require an on line identification of the plant. It is true that if the plant parameters vary within ranges compatible with the regulator robustness, an off line identification of the regulator parameters.

The idea of such a control strategy comes from the observation of a natural phenomenon, that of the ebb and flow on a porous dyke. This phenomenon indeed presents a stability degree which is independant of the motion water mass, so translating the robustness of the relaxation as far as damping is concerned. The non integer order differential equation which describes it, makes it possible to introduce an "open loop frequency template" in automatic control which illustrates simply robustness in frequency domain.

In this paper, the template is synthesized through a CRONE regulator which insures a precorrection of the plant, this one permiting to fix the open loop unit gain frequency to a given value, chosen by the conceptor.

Moreover, such a precorrection makes the natural frequency robust versus the parameters of the plant, which is an interesting property in tracking, given that it suffices to filter the reference input with a constant parameter filter. The constance of the natural frequency is due to that of the unit gain frequency of the precorrected plant, which insures that of the open loop unit gain frequency since the cascade corrector possesses constant parameters.

The considered plant is a polar table. This one is chosen because of its structure which insures large inertia variations, strong dynamic couplings and a great number of non-linearities.

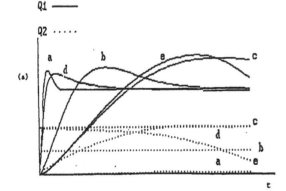

The performances obtained through a numerical simulation of the control are remarkable since the first overshoot of the step response remains practically constant independently of the parameters of the plant.

REFERENCES

(1) Horowitz I.M., Sidi M. (1972). Synthesis of feedback systems with large plant ignorance for prescribed time domain tolerances. Int. J. Control, vol. 16, n°2, 287-309.

(2) Horowitz I.M. (1975). A synthesis theory for linear time varying feedback systems with plant uncertainty. IEEE T-AC, 20, 454-464.

(3) Horowitz I.M. (1979). Quantitative synthesis of uncertain multiple input-output feedback. Int. J. Control, 30, 81.

(4) Horowitz I.M. and Liao Y.K. (1984). A limitation of non-minimum phase feedback systems. Int. J. Control, 40, 1003-1013.

(5) Oustaloup A. (1975). Etude et réalisation d'un système d'asservissement d'ordre 3/2 de la fréquence d'un laser à colorant continu. Thèse de Docteur-Ingénieur, Université de Bordeaux I.

(6) Oustaloup A. (1979). Etude comparative de systèmes asservis d'ordre 1/2, 1, 3/2 et 2. L'Onde Electrique, vol. 59, n°2, pp. 41-47.

(7) Oustaloup A. (1980). Contribution du bruit de l'électronique sur la grandeur de sortie de systèmes asservis d'ordre 1/2, 1, 3/2 et 2. L'Onde Electrique, vol. 60, n°4, pp. 40-44.

(8) Oustaloup A. (1980). Active programmable fractional order filter ; performances of non integer order systems. European Conf. on Circuits Theory and Design, Warsow (Poland), September 2-5.

(9) Oustaloup A. (1981). Fractional order sinusoïdal oscillators : optimization and their use in highly linear FM modulation. IEEE Transactions on Circuits and Systems, vol. cas 28, n°10, pp. 1007-1009.

(10) Oustaloup A. (1981). Linear feedback control systems of fractional order between 1 and 2. IEEE Int. Symp. on Circuits and Systems, Chicago (Illinois), April 27-29.

(11) Oustaloup A. (1981). Systèmes asservis linéaires d'ordre fractionnaire. Thèse de Doctorat ès Sciences, Université de Bordeaux I.

(12) Oustaloup A., Pistre J.D. and Mora A. (1982). Fractional order systems with localized and distributed parameters. IASTED Conference, Davos (Switzerland), March 2-5.

(13) Oustaloup A. (1983). Systèmes asservis linéaires d'ordre fractionnaire : théorie et pratique. Edit. Masson, Paris.

(14) Oustaloup A. and Levron F. (1986). Robustesse par l'approche temporelle. Symposium IMACS, Lille, 3-6 juin.

(15) Oustaloup A. and Mouss L.H. (1986). Synthèse en frequence d'un régulateur robuste : une approche originale utilisant le concept de fractale. Colloque "RAI/IPAR'86", Toulouse, June 18-20.

(16) Oustaloup A. and Bergeon B. (1987). Frequency space synthesis of a robust dynamic command. IFAC'87, 10th World Congress on Automatic Control, Munich (Fed. Rep. of Germany) July 27-31.

(17) Landau I.D. and Tomizuka M. (1981). Theory and practice of adaptive control systems, Ohm, Tokyo.

(18) Landau I.D. (19). Model reference adaptive controllers and stochastic self-tuning regulators : an unified approach. Trans. ASME J. of Dyn.Syst.Meas. and Cont., vol. 103, n°4, pp.404-416.

(19) Astrom K.J. (1983). Theory and applications of adaptive control. A survey automatica, vol.19, n°5, pp.471-486.

(20) Goodwin and Sin K.S. (1984). Adaptive filtering prediction and control. Englewood Cliffs, Prentice Hall.

(21) Mandelbrot B. (1975). Les objets fractals, forme, hasard et dimension. Edit. Flammarion, Paris.
    - Fractal, form, chance and dimension. Freeman, San Francisco (1977).
    - The fractal geometry of nature. Freeman, San Francisco (1983).

(22) Le Mehaute A. and Crepy G. (1983). Introduction to transfer and motion in fractal media : the geometry of kinetics. Solid State Ionics 9 & 10, North-Holland Publishing Company.

(23) Oustaloup A. et Bergeon B. (1986). Support d'étude au passage de géométries récursive et fractale à un ordre de dérivation non entier: Effet de la rugosité de surface sur la réponse en fréquences d'un système capacitif. Commun. presented at the Colloque "Des fractales en math. et physique", Marseille, 13-18 janvier.

(24) Roques Carmes C. et Wehbi D. (1986). Dimension fractale de graphes. "Réunion des théoriciens des circuits". Arcachon (France) 2-3 octobre.

(25) Oustaloup A. (1988). From fractality to non integer derivation : a fundamental idea for a new process control strategy. 8th Int. Conf. "Analysis and optimization of systems". INRIA, Antibes (France), June 8-10.

(26) Oustaloup A. (1987). Frequency synthesis of a non integer order linear robust regulator : an original approach using fractal concept. Paper presented and published in the Proc. "1987 Workshop on Robust and Adaptive Control", Oaxaca (Mexico), December 2-4.

(27) Oustaloup A. (1988). Conception d'un pilote automatique d'aéroglisseur. Epreuve d'informatique -automatique de l'Agrégation de Mécanique, 14 avril. Sujet remis au Minist. Educ. Nat. 7 décembre 1987.

(28) Oustaloup A. (1988). From fractality to non integer derivation through recursivity, a property common to these two concepts. Survey, session "Fractality and non integer derivation", 12th IMACS World Congress on Scientific Computation, Paris, July 18-22.

(29) Oustaloup A. (1988). From the robustness of stability degree in nature to the control of highly non linear manipulators. Colloque Int. CNRS "Automatique Non Linéaire", Nantes (France), 13-17 juin.

POWER OPTIMIZATION OF MULTI-FLUID
TRANSPORTATION SYSTEMS

E. F. Camacho, M. A. Ridao and J. A. Ternero

Departamento de Automática, E.T.S.I.Industriales, Sevilla, Spain

Abstract. This paper presents an algorithm for optimizing the energy operating
costs in multi-fluid transportation systems. The algorithm works in two steps. The
first one consists of the computation of a function that measures the estimated
minimum cost to the goal node. This computation involves the use of Bellman's
optimality principle and some heuristic rules in order to avoid the combinatorial explosion. During the second step the optimum trajectory is obtained with the help of
the function mentioned above and using an accurate simulation of the transportation system. The algorithm is applied to a model of an oil pipeline system.

Keywords. Optimization; multifluid transportation systems; dynamic programming.

INTRODUCTION

Multi-fluid transportation systems consist of a
pipeline network, a set of valves, pumps and
storage tanks. The products are transported
through the pipeline network from the input nodes
to the terminal nodes where there are normally
stored in tanks. The product is then extracted
from the storage tanks by the consumers in a
continous or semi continous way. The fluid has
to be transported, taking into account the system's capabilities, in order to satisfy the consumers' needs, that is, making sure that there
is always enough products in the storage tanks,
that the tanks at each terminal can receive the
product delivered at any given moment and that
the product can be delivered throughout the system as required.

The energy optimization problem consists in minimizing the power bill while satisfying the restrictions mentioned above. The interest of such
an optimization is evident as power costs are increasing in respect to the total operating costs of
fluid transportation systems as shown in (Basavaraj, 1984).

In order to obtain a realistic solution to the optimization problem, the following should be taken
into account:

- Topology and topography of the network.

- Non-linear characteristics of pumps and pipelines.

- Variable demands of consumers.

- Time changing prices of electrical energy.

- Hydraulic equations throughout the systems.

The problem has been dealt with in the literature, especially for mono-fluid transportation
systems, as is the case of water distribution systems (Coulbeck, 1978), (Millaoka et al. 1984).
However, even in this case, the optimization problem leads to a dynamic and non-linear problem
of high dimension as stated in (Kemplous et al.
1986). In this last reference an optimization algorithm that works at two levels has been proposed. At the upper level the dynamic problem is solved and some data passed to the lower
level where the static problem is solved giving
the number of pumps turnned-on in the pumping station.

The qualitative and quantitative complexity of the problem is substantially increased by taking into account multi-fluid capabilities. This problem has been dealt with in literature, in (Burham, 1983) the potencial savings of an adequate operation are shown.

This article presents an algorithm for power cost optimization in multi-fluid transportation systems. The algorithm is basically a dynamic programming algorithm and works in two steps. In the first step a function that measures the optimum expected cost to the goal node is computed by the use of some simplifying assumptions. In the second step, the optimum trajectory is determined by taking into account the previously mentioned function and using a more accurate simulation of the transportation system. The algorithm has been applied to different oil pipeline models.

The paper is organized as follows: Section 2 describes the way in which multi-fluid transportation systems are operated and how the optimization problem is formulated. Section 3 is devoted to the optimization algorithm while an application of the algorithm to an oil pipeline is presented in section 4. Finally the conclusions are presented in section 5.

PROBLEM DESCRIPTION

In multi-fluid transportation systems, the different fluids are injected into the system at input nodes and then transported to the destination nodes through the pipeline network. At any given time different products (separated by the corresponding interfaces) can be transported. The products are routed through the network by opening or closing the appropiate set of valves.

The mechanical energy needed for the transportation is supplied by a set of electrically powered pumps. The transportation velocity, for a particular network and a given pipeline state, depends on the number and the flow-head characteristics of the pumps working at each of the pumping stations.

The operation of a multi-fluid transportation system consists in determining the order and the way in which the different fluids are going to be transported to satisfy the demand. This task can be divided into two different problems. The first one involves the determination of the approximate transportation needs. The solution of this problem will consist in

- a batch sequence that will minimize the number of interfaces and will cover the consumption needs at each destination node.

- an approximate pumping schedule.

The batch sequence consists of a list of batches each of which is defined by the product identification, approximate departure time, the volume destined to each output node and approximate arrival time to each terminal node.

The second problem involves determining how to set the different pumps and valves at each time interval in such a way that the batch sequence given above is carried out in a given period of time. This can be accomplished in different ways, that is, using different pump combinations and valve positions (and therefore flows at all points in the network) in the different time intervals. This will turn into different batch arrival times, and different power operating costs as the price of electricity changes throughout the day and also friction losses and pump performances depend on the flow.

The time intervals mentioned above are defined by certain events. These are produced when significant changes occur within the system. The main events are:

- Changes in electrical tariffs.
- Batch arrival to nodes.
- Tank levels reaching limits.
- Shut-down of the system.
- Periodic events.

The periodic events are introduced to guarantee that the time interval between two consecutive events is small. In this case, the pipeline state does not change substancially from one event to the next, and the pipeline state can be considered constant between two consecutive states. Computations are greatly simplified this way.

The power cost optimization problem consists in finding out which way of operating the system (that is, how to set the pumps and valves at each time interval) is the cheapest and fulfils the needs mentioned above.

In the following it will be consider that the system is operated in the "stripping" mode. That is, each output node is working with an input flow proportional to the amount of product that the terminal has to receive from the current batch. This method of operating is widely considered as the best way of using the transportation capabilities of the pipelines.

Notice that if the stripping mode is used, the amount of product stripped at each terminal depends on the batch passing in that moment. Therefore, for a given pipeline state the flow at each terminal, and at each bifurcation node, can be computed knowing the flow at the entry node. The pipeline state can be computed from the volume injected into the entry node and the batch sequence.

When operating in the stripping mode the future evolution of the system can be computed if the following items are known.

- amount of product which has been pumped (v)

- the time used (t)

- the setting of the pumps

- the flow at the entry node.

The state of the system can, therefore, be resumed by the pair (v,t). The control variables are the pump setting and the flow at the entry point. By looking at typical characteristics of pump power consumption - flow, it is clear that the power consumption of the pumps grows slowly with the flow and depends mainly on the pump being connected or not. Therefore, from the power optimization point of view, the best results will be obtained when the maximum flow is chosen.

Taking into account these considerations, the control variables will be a vector of dimension equal to the number of pumps in the system containing a logical value which will indicate if the corresponding pump is active. The number of combinations of the previously mentioned vector can be reduced for pumping stations with identical pumps. In this last case, knowing the number of pumps connected (in series or parallel) is enough.

The working space will be the state plane v-t and the optimum control problem will consist in determining an optimum trajectory in the plane v-t from point (0,0) to point (target-volume,target-time).

The feasible trajectories are constrained by two curves as indicated in figure 1. The upper one is the curve resulting when pumping the batch sequence using the maximum capabilities of the system. Notice the horizontal pieces of the curve corresponding to intervals where the pumps have

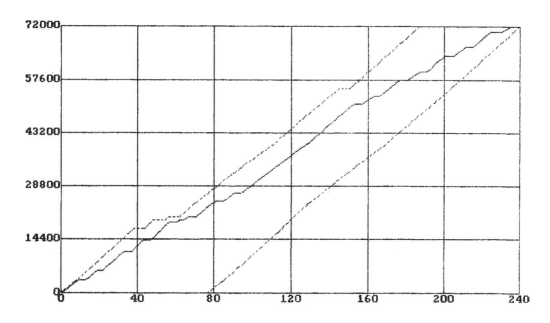

Fig. 1. Feasible Optimization Space State.

to stop either to avoid tank overflow or due to maintenance shut-down of the system. The lower curve is computed taking into account the minimum pumped amount at each interval in order to satisfy the needs and taking into account the system pumping capabilities.

OPTIMIZATION ALGORITHM

The optimization algorithm uses the cost function:

$$f(x_{v,t}) = g(x_{v,t}) + h(x_{v,t}) \qquad (1)$$

where $x_{v,t}$ is the state vector and consists of a pair (volume,time), $g(x_{v,t})$ is a function representing the minimum cost from the origin to node $x_{v,t}$ and $h(x_{v,t})$ is a function that estimates the minimum cost from $x_{v,t}$ to a goal node. That is, $f(x_{v,t})$ is an estimate of the cost of the minimal cost path constrained to go through node $x_{v,t}$.

The optimization algorithm works in two steps. In the first one the function $h(x_{v,t})$ is computed on a grid by using Bellman's Optimality Principle and in the second one the optimum path is found by using this function and an accurate simulation of the transportation system.

In the first step of the algorithm, function $h(x_{v,t})$ is made equal to zero for all $x_{v,t}$ with volume equal or greater than the target volume. The next step for estimating the function $h(x_{v,t})$ is to compute all states x_{v_1,t_1} from where $x_{v,t}$ can be reached. Notice that if the volume difference $\triangle v = v - v_1$ is small, the pipe state will be practically the same. Therefore, for a given pump combination the flow will be very similar and can be considered constant.

The time t_1 for each pump combination can be determined by the intersection of the line L1, see figure 2, that goes through state $x_{v,t}$ with a slope equal to the flow for the corresponding pumping combination and the horizontal line L2 of volume equal to v_1. If the line L1 crosses a shut down period the computation of time t_1 should be made as indicated in figure 2.

In order to avoid the combinatorial explosion, a procedure for eliminating states is used. At first, the procedure eliminates the trivial states, and then a set of heuristic rules are used to decide

which states are "the best" for the next step of the algorithm.

A state $x_{v,t}$ is considered to be trivial if there is another state x_{v,t_i} with
$t_i \geq t$ and
$h(x_{v,t}) \geq h(x_{v,t_i})$.

Notice that state $x_{v,t}$ could be reached from x_{v,t_i} by switching off all the pumps between t_i and t (with cost zero) and the function $h(x_{v,t})$will therefore be equal to $h(x_{v,t_i})$.

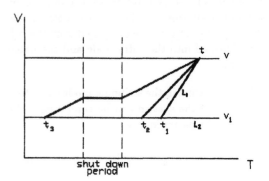

Fig. 2. State Transitions.

Once all trivial state have been eliminated, a second elimination procedure takes place in order to obtain a feasible amount of states for the line L2. The segment of L2 including the limit curves (figure 1) is divided into a number of regions. At each region the point with "the best" characteristics is chosen. This guarantiees that the function $h(x_{v,t})$ is computed over a grid covering the feasible space.

Once the function $h(x_{v,t})$ has been computed over the grid, the second step of the algorithm can be carried out. It consists of computing, at each interval and for all possible pumping combinations, the accumulated cost to the next event $g(x_{v,t})$ plus the estimated cost to the goal node $h(x_{v,t})$. The function $h(x_{v,t})$ is calculated by a linear interpolation in the grid computed in the first part of the algorithm.

The accumulated cost is computed by using an accurate simulation of the pipeline network. It involves the computation of pressure drops, flow, power consumption etc. and it takes into account the non-linear characteristics of pumps and pipelines as well as different electricity tariffs throughout the day.

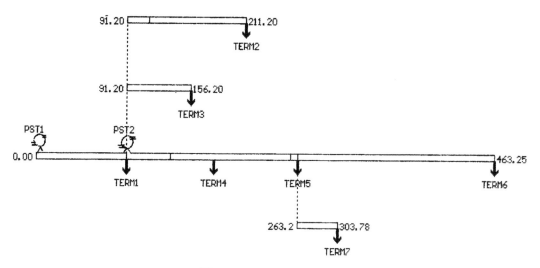

Fig. 3. Pipeline Model.

APPLICATION TO OIL PIPELINES

The optimization algorithm has been included in a program developed for CAMPSA. The program can be applied to any pipeline transportation system with only one entry node. The algorithm has been tested with the model of an oil pipeline shown in figure 3.

The pipeline has seven terminals and two pump stations. The first one has three booster pumps operating in parallel and four main pumps operating in series. The second one has three main pumps operating in series. The three booster pumps have identical characteristics, and so do

the main pumps at the first pumping station. The three main pumps at the second pumping station are also identical. The pipeline has a main pipe and three branches. Two of them leave the main pipe at Km 91 and the other one at Km 263.

Five types of products, two types of gasoline, two types of diesel oil and kerosene, were considered.

The algorithm has been tested with different cases. Table I shows one of the batch sequences used in the tests. Figure 1 corresponds to the solution in a case covering the batch schedule during ten days. Notice that the upper curve has a few horizontal pieces corresponding to points where the

TABLE 1 Batch Sequence.

NAME	VOLUMEN (M³)	TERM1 (%)	TERM2 (%)	TERM3 (%)	TERM4 (%)	TERM5 (%)	TERM6 (%)	TERM7 (%)
P1/018/00C	10000.00	0.00	20.00	35.00	5.00	0.00	0.00	40.00
P1/013/090	7000.00	10.00	40.00	0.00	0.00	0.00	50.00	0.00
P1/019/00C	9000.00	0.00	30.00	20.00	0.00	0.00	40.00	10.00
P1/026/00A	6000.00	0.00	0.00	0.00	0.00	66.67	0.00	33.33
P1/014/090	15000.00	0.00	0.00	30.00	0.00	0.00	40.00	30.00
P1/004/096	8000.00	0.00	0.00	0.00	25.00	0.00	0.00	75.00

TABLE 2 Optimal Solutions for Different Cases.

INITIAL DATE	TARGET TIME (DAYS)	TARGET VOLUME (M³)	NON-TARIFFS CHANGE DAYS	COST (PTS.)
0:00 1/8/89	7	55000	2	1.235.318
0:00 1/8/89	10	55000	2	1.066.495
0:00 1/8/89	15	55000	5	940.510
0:00 11/8/89	7	55000	3	1.111.791

tanks would overflow if the pumping continued at maximum flow.

The different slopes in the curves correspond to different flows due to different pumping conditions as fluids in the pipeline and flows in the output terminals change with state.

The lower curve corresponds, as mentioned before, to the minimum volume that has to be pumped at each time interval in order to satisfy the consumers' needs. Notice that the pipeline is kept switch off and that the maximun flow is used in order to pump the target volume in the prescribed time.

The thicker line corresponds to the optimum trajectory. It can be seen that the solution involves shutting down all pumps during peak hours, and using high flow combinations during valley periods.

Good economical results have been achieved using the algorithm in the management of the oil pipeline. Table II shows the different solutions obtained when using different target times. As the target time increases the products can be transported during off-peak hours and the cost therefore decreases.

The energy cost has been reduced in approximately 20 per cent from an initial solution.

CONCLUSIONS

An algorithm for the power cost optimization in multi-fluid transportation systems has been presented. The algorithm uses a function that measures the estimated minimum cost to the goal node and an accurate simulation of the transportation system. The algorithm has been tested with success in an oil pipeline model.

ACKNOWLEDGEMENT

The authors would like to acknowledge the finantial support of CAMPSA for this work and also to thank S. Heras, J.M Rodriguez and R. LLorente of the Departamento de Oleoductos y Operaciones of CAMPSA for their valuable comments. Further thanks is due to F. J. Vidal, C. del Valle, J. Gómez and R. Millán for their collaboration.

REFERENCES

Basavaraj. B.H. (1984) Cut pipeline power cost by demand control. *Oil & Gas Journal*, Jan, pp. 88-90.

Burham C.G. (1983) Rate Schedule Application - Off Peak Operation of Pump Station. *Proc. of the Pipeline Engineering Symposium*, pp. 11-16, Houston, Texas.

Coulbeck B. and M. Sterling. (1978) Optimizated Control of Water Distribution Systems. *Proc. IEEE*, Vol. 122, N. 2.

Klempous R, J. Kotwosky, J. Nikodem and J. Ulasiewicz. (1986), Water Distribution Systems. *Components, Instruments and Techniques for Low Cost Automation & Applications*, pp. 541-551, Valencia (Spain).

Miyaoka S., M. Funabashi. (1984) Optimal Control of Water Distribution Systems by Network Flow Theory". *IEEE Transaction on Automatic Control*, Vol. AC-29, N. 4, pp. 303-311.

VISUAL FEEDBACK APPLIED TO AN AUTONOMOUS VEHICLE

D. Maravall, M. Mazo, V. Palencia, M. Pérez and C. Torres

Grupo de Control Avanzado y Visión Artificial, Facultad de Informática, Madrid, Spain

Abstract:

The symbiosis between control engineering and computer vision is the topic of this paper, in which the foundations of visual feedback are presented and discussed as well as the description of a mobile robot with artificial vision, developed by the authors.

Keywords:

Mobile robots; computer vision; adaptive control.

Introduction. Foundations of Visual Feedback

The mobile robot described in this paper is based on the concept of visual feedback control. Two very important disciplines meet here: control engineering and artificial vision.

Visual feedback differs from other types of control in that the signals are of "visible" nature; that is, if the output of the process or the system to control is obtained through image processing, then we can talk about visual feedback control.

Although Norbert Wiener, pioneer of Cybernetics, used in his work "Cybernetics, or Control and Communication in the Animal and the Machine" an example of visual feedback to present and develop the principles of negative feedback, most control system designed and built since that date (1.948) do not use visual feedback. In fact, visual feedback control is a branch of control engineering which is beginning to develop. For this reason there are only a few systems implemented and a very limited number of papers published.

Focusing on the prototype of mobile robot described in this communication the control system with visual feedback that we have implemented is represented in figure 1.

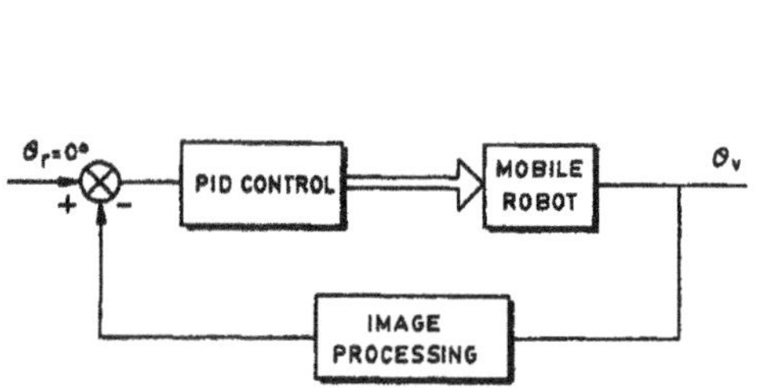

In this figure the physical variable to control, θ_v, is the deviation angle of the robot longitudinal axis which coincides with the camera in relation to the tracking line. Obviously, the reference signal is $\theta_r = 0^\circ$, as the objetive of the robot is to follow the line.

The control physical signal, u, is the rotation angle applied to the front wheels which determine the robot direction. There exist other possible control signals: the speed applied to the rear wheels and the distance between the middle points of the tracking line and the mobile robot longitudinal axis.

It would take us long to explain the techical aspects of the image processing techniques involved in this application, such as: image capture speed (1/30 second); procedures to eliminate noise (ground characteristics, tracking line quality...) computation of the line slope (carried out by digital image thresholding, from the histogram); total computation time (this aspect does not offer any difficulty as it is lesser than the necessary to keep a typical industrial mobile robot velocity, which is usually about 1 m/s), etc.

In this paper, we are going to focus mainly on three problems of general interest in visual feedback and particularly in this application: a) A hierarchy of visual and non-visual controls; b) Identification in real time of the robot dynamics and c) Elimination of the delay time effects of the robot actuators.

Hierarchy of visual and non - visual controls.

The main idea is that the robot, as it works in an structured and a priori known universe, can make use of a non-visual level which we will refer to as the "superior" control, based on the universe model. If the universe is very complex, advanced techniques of representation will have to be used and therefore the "inferior" or visual control level will go into action with a high frequency; on the contrary, if the universe is simple, as it happens for the tracking lines system in a factory, it will suffice with simple geometrical representations and a moderate interaction between the two control levels: visual and non-visual.

The knowledge of the universe of the robot's actions may be prior in an strict sense (that is the case of a factory where the system of tracks or routes is known) or can be learnt by the robot. In any case, we start here from the hypothesis of the previous knowledge of the robot universe of action.

Figure 1.- Block diagram of visual feedback used in the protopype described in the paper.

Identification of the robot dynamics.

If in the control loop of figure 1 the robot transference function θ_v (s)/U(s) was known, mathematical and optimal control techniques could be applied.

Unfortunately, the dynamics of our prototype is unknown. For this reason, an identification module in real time, Maravall (1987), and Mazo (1988), has beeen implemented using Taylor's series development of the variable to be controlled, θ_v which for a generic instant of sampling is:

$$\theta_v [(K + 1) T_s] = \theta_v (k T_s) + T_s \ddot{\theta}_v (K T_s) + \frac{1}{2} T_s^2 .$$

$$\ddot{\theta}_v (K T_s) + ... + \frac{1}{n!} T_s^n \theta_v^{(n)} (K T_s) + ... \qquad (1)$$

That determines the real dymamics of the robot. It is clear that the problem lies in the fact that the derivatives of the variable to control are not completely known.

By approximating:

$$\dot{\theta}_{v} = \frac{\theta_v [K T_s] - \theta_v [(k - 1) T_s]}{T_s}$$

$$\ddot{\theta}_{v} = \frac{\dot{\theta}_v [K Ts] - \dot{\theta}_v [(k - 1) T_s]}{T_s} \qquad (2)$$

. . . . .

then the θ_v dynamics without making any explicit reference to the sampling time would be given by:

$$\theta_v (K + 1) \sim a_1 \theta_v (k) + a_2 \theta_v (k - 1) + ... + a_{N+1} \theta_v (k - 1)$$

The a_1 , a_2 .... a_{N+1} coefficients which determine the mobile robot dynamics can be calculated by means of well-known estimation techniques.

Predictive control.

In the prototype considered in this paper, developed by the GCAVA (in Spanish, Group of Advanced Control and Artificial Vision) an additional problem appears originated by the fact that the time to compute the control algorithm, the time of processing the images and the time of obtaining the deviation angle θ_v is notably inferior to the operating time of the front wheel servos (u control signal).

This peculiarity leads to the problem of delay systems control . An advisable solution and very popular in the control of industrial processes, succesfully tested in our prototype, is the so-called Smith's predictor. It is based on the previous knowledge of the system transference function and particularly of its delay time.

2. DESCRIPTION OF A MOBILE ROBOT PROTOTYPE DEVELOPED BY THE GROUP OF ADVANCE CONTROL AND ARTIFICIAL VISION OF THE POLYTHECNIC UNIVERSITY OF MADRID

The Group of Advanced Control and Artificial Vision (GCAVA) of the Polythecnic University of Madrid has developed a prototype of a mobile robot guided by computer vision.

At present, this prototype is capable of following all kinds of trajectories with speeds of 1m. per second. The tracking lines in this case are defined by a white cello-tape.

Our system configuration consists mainly of:

- Mobile robot structure.
- Black and white video camera.
- PC/AT computer, with image digitizer.
- Electronic board to control the servomotors of the robot and to communicate with the image processor.

This prototype (fig.3) is a 44x25 cm platform on four wheels placed in two parallel axes (rear and front). The driving wheels are the rear ones which have a differential gear and the front wheels act as directrices. The velocity control, taken on the rear axis or driving axis, and the rotation control, taken on the front wheels, are implemented by two servomechanisms. The maximum rotation angle of the front wheels is 45º with a 1.04º resolution and the velocity range between both directions is from 0 to 4m/sc. with a 0.125m/sc. resolution.

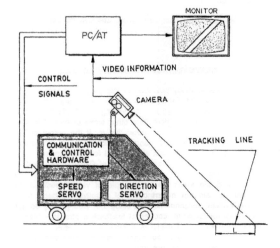

Fig 2. Configuration of the system now operating

Fig 3. Experimental Prototype used by GCAVA

The prototype includes an electronic equipment on board designed by the GCAVA which permits to control the direction, velocity and communication of the servomotors with the PC/AT computer.

In the future, this electronic equipment on board will also be in charge of the other possible sensors (velocity, acceleration, proximity, etc).

The information relative to the trajectory to follow (fig. 4) is obtained by the on-board video-camera which is lined up with the mobile robot longitudinal axis. This information consists mainly of:

- Type of trajectory on which the robot is moving.
- Distance x, between the middle point of the tracking line and the robot longitudinal axis.
- Angle θ formed by the robot longitudinal axis and the tangent to the tracking line.
- Crossings, etc, etc.

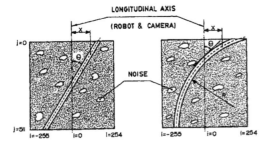

Fig. 4. Control variables

The distance and the angle (fig.4) can be obtained, inside the field of vision, by means of points more or less distant from the robot. The horizontal lines to explore on the digitized image plane can be voluntarily chosen from Y=0, the most distant point from the robot, to Y=521, which is the nearest point. This permits to control the robot with information obtained from different distances to the mobile. The importance of this peculiarity lies on the fact that a human being when driving a mobile gets information from points more or less distant and this depends basically on the kind of tracking and the velocity at which he or she is moving.

The control variables (x and θ) can be obtained either on the digitized image plane or on the ground after a previous camera calibration, and the corresponding transformation of coordinates, Cerceda del Rio (1988).

An important point to be considered in this type of tracking is the line width. In Mazo (1988) the optimum value of this width was obtained in order to minimize the errors derived from the limited resolution of the camera and its geometrical layout.

Although the purpose of controlling the line tracking could be to maintain x=0 from a specific distance to the robot (this has already been successfully tested), the use of x as the only control variable causes that the movements are carried out making little "S's". Even though this problem can be satisfactorily solved by introducing a dead band in the control algorithm, the most intresting solution consists in introducing the angle as an additional control variable, resulting in a "double channel" control algorithm.

The prototype now in use, after analizing a picture frame verifies on which type of tracking is moving (straight or curve), and it also verifies whether there are or not crossings Mazo (1988)

In the case of turns, the algorithm updates the velocity to a suitable value according to the curvature radio and uses as variables the x obtained from the nearest point (inside the vision field) to the mobile (this permits the mobile to move over the tracking line) and the θ angle obtained from the most distant point.

In the case of straight lines, the variables to control will be the x, obtained from the most distant point (inside the vision zone, again) together with θ.

The control algorithm used in our prototype is the PID (consult the paragraph of this paper on visual feedback). Its parameters are updated in real time, based on the identification algorithm of the mobile dymamics (self-tuning adaptive control).

The capability to recuperate the tracking line, in case of possible losts, consists in making motions based on the last orders sent to the robot and also on the position of the line tracking instants before it got lost.

Conclusions

We have very briefly presented the main characteristics of a mobile robot with visual feedback and we have focused on the control aspects of the mobile.

The first point to considerer is the robot's dynamical model, which we have obtained through the Taylor's serie development of the mobile's controlled variable (the rotation angle applied to the direction wheels). This approach permits to tune the real accuracy of the model as much as desired, by taking sufficient terms in the Taylor's development. Once modelled the controlled variable, an appropiate control strategy must be implemented provided that the visual feedback information is available to the control algorithm (the paper also describes the image processing techniques used in our mobile robot).

In this sense, the necessity to use adaptive control is clear: the robot's transference function (although it has been modeled) is not only unknown a priori, but it is also liable to changes because of the different loads it can carry, of the changing velocities it moves along and of the different resistances to the motion it may meet.

The control strategy that we have successfully implemented for the mobile's guidance consists of identifying in real-time the robot's dynamics and tuning the control law (PID) consequently. For a more detailed discussion of this control strategy the reader is referred to our monographies, Maravall (1987) and Mazo (1988).

Another different approach to this self-tuning regulation (STR) is the so-called model reference adaptive control (MRAC), Astrom and Wittenmark (1989), which we are testing in our prototype as well, although its performance with the self-tuning control has been very satisfactory, as it is able to follow with standard industrial speeds (about 1 m/s) trajectory lines of any shape embedded in highly noisy backgrounds.

References

Astrom, K.J. and B. Wittenmark (1989).
  Adaptive Control. Addison-Wesley, Reading, Mass.

Cerceda del Rio, R. (1988).
  "Digital Image Processing Applied to the Guidance of an Autonomous Vehicle in an Industrial Enviroment". (In Spanish) Master's Dissertation Polytechnical University of Madrid.

Maravall, D. (1987)
"Identification and Control of a Vehicle Guided by Computer Vision". (In Spanish). GCAVA, Madrid.

Mazo, M. (1988)
"Contributions to the Control and Guidance of an Autonomous Vehicle with visual Feedback". (In Spanish). Ph D's Dissertation. Polytechnical University of Madrid.

A PRODUCTION SYSTEM FOR AGVS CONTROL

L. Moreno, M. A. Salichs, R. Aracil and P. Campoy

Departamento Automática, E.T.S.I.Industriales (DISAM), Madrid, Spain

Abstract

This work present an AGV Expert System (AES) which has been developed to control an automated guided vehicle system. It has been implemented in OPS5, a production system language. The AES structure is shown in two main aspects: the element classes used to model an AGV system and the organization of the AES knowledge base. Finally, rules used to deal with the load assignment problem in systems whose vehicles can transport more than one load each time have been considered.

Keywords

Expert systems; Heuristic programming; Industrial production systems; Vehicles; Transportation control

1. INTRODUCTION

Automated guided vehicle systems present a combined routing and scheduling problem. These type of problems are extremely complex to solve optimally and heuristic methods have been developed to reduce the processing time to feasible levels.

Currently a high number of installations use dispatching rule methods to control an automated guided vehicle system.

The work presented uses an expert system based in OPS5 (production system) to control an AGV system. AES (AGV Expert System) is designed to control, detect, diagnose and respond if it is possible to situations in the system domain.

The AES was designed to deal with real time AGV system operations. It receives messages from the AGV system and analyzes the state of the target system. Based on that analysis AES send commands to optimize the system operations.

The knowledge base has the expert Knowledge encoded in a declarative format, production rules. Since production rules are highly modulars, it is easier to modify than would be in a procedural method of knowledge encoding.

2. PROBLEM OVERVIEW

An AGV System Controller responds to request for transport loads from various stations to their destinations by assigning them to vehicles available in the system. The assignments are made in order to reach a goal. Different goals can be desirable according with the application of the system, for instance:

- Minimize the number of vehicles required in the system.
- Minimize the transportation costs.
- Maximize the utilization of the workstations.
- Minimize the job execution time.
- Minimize the work in process.

The solution must meet different constraints to be feasible. The constraints are related to each problem and can generally included in the next groups:

- Time windows to serve a task.
- Number of vehicles in the system.
- Type of vehicles, that can be heterogeneous.
- Vehicle capacity constraints, that can include volume

and weight.
- Operations, involve pickups and drop-offs.
- Load size.
- Buffers length at workstations.
- Location of demands.

It is complex to model all the characteristics presents in these type of systems. Artificial Intelligence techniques permit the easy modelization of the system and the development of scheduling heuristic, that can be continually improved.

3. SYSTEM MODELLING

The AES has been implemented in OPS5, a production system language developed at Carnegie Mellon University. OPS5 has several advantages for the AGV System modelling and control:

1.- It is flexible and modifiable.

2.- Its use of production rules as a knowledge based representation seems suited to AGV Systems domain.

3.- Its data-driven mechanism of inference is appropriate to be used the in real time operation of these systems.

An expert system written in OPS5 consists of three parts:

1. Working memory, which contains the data elements. Each one is an attribute-value element.

2. Production memory, where the expert knowledge about the domain is encoded in the form of if-then production rules.

3. Inference engine, which control the process of firing a production rule by matching, selecting and executing the rules.

The AGV System model structure is defined as different element class (in each class definition an explicit list of attributes is given) and each working memory element is a member of a class.

The main element classes used in the modelling are refereed to:

1. Messages

The controller communications with the overall system is message based. Different types of messages are defined in the model. An example of a message from AES to a vehicle :

```
(literalize msg-ct.c
    agv-id
    station
    priority
    action
    load-type
    load-id
    path
        )
```

2. Vehicles

The vehicle characteristics and status at each moment are considered in the vehicle element class. Different types of vehicles can be defined in the system. The vehicle element class is shown below:

```
(literalize vehicle
    veh-id
    veh-type
    element-id
    element-type
    stopping-point
    enter-time
    veh-status
    load-id
    battery-time
    maint.-time
    remaining-capacity
    veh-length
        )
```

3. Loads

The load element class defines the characteristics and the status of the transport demands generated in the system. This element class is illustrated below:

```
(literalize load
    load-id
    generation-time
    load-type
    origin
    destination
    priority
```

size
)

4. Network

The installation network is defined using two types of element class: links and nodes. The link element class is defined as:

```
(literalize link
      link-id
      length
      origin
      destination
      bidirectional
      current-direction
      AGVS-number
      AGVS-order
      )
```

and the node element class is:

```
(literalize node
      node-id
      node-type
      link-number
      veh-number
      strategy
      )
```

5. Workstations

Another element class is used to define a workstation of the system. This element class has the following structure:

```
(literalize workstation
      workst-id
      workst-type
      load-times
      unload-times
      machining-times
      node-position
      load-types
      load-input-number
      load-input-order
      load-output-number
      load-output-order
      max-input
      max-output
      mainten-time
      )
```

4. CONTROL KNOWLEDGE ORGANIZATION

The AGV System control strategies are defined like if-then production rules in the production memory. Each production rule has two parts: LHS (left hand side) and RHS (right hand side). The LHS specifies the conditions which must be satisfied before the actions specified in the RHS can be executed.

The expert system receives messages from the local controllers and the vehicles existing in the system, analyzes the situation and takes actions by sending message to the vehicle or the local controllers in the system. Five main groups of OPS-rules have been implemented in the expert system:

1. Communication rules

This group contains the messages generation rules and the communication control rules. The basic communication control rules have been developed in other language, because VAX OPS5 do not have communications routines, these are:

WAIT, SEND, RECEIVE and SET.

2. Model creation and updating rules

This group of rules create the AGV system model at the starting time and it is continuously updating the status of the different elements contained in the model. These rules create and eliminate loads in the model, update position and status of the vehicles, stations and network.

3. Assignment rules

The assignment rules depend on the goals and the constraints imposed to the system elements. The rules implemented had taken into account the possibility of multiload accommodation in the vehicles. The accommodation has been restricted to a maximum of two load at a given time.

4. Routing rules

The routing group of rules, decide the path to be followed by each vehicle. In a first approach precalculated shortest paths are used, assuming the system has been designed such that the conflicts do not affect strongly the predeterminated travel time.

5. Event management rules

The last group of rules take actions when special situations happen in the system, for instance battery requirements in vehicle, maintenance times, or in case of unexpected events like obstacle in a point of the network.

5. ASSIGNMENT RULES

There are a high number of goals that can be implemented in this group of rules according to the requirements of the system. In this work we have focus the system to manage systems whose vehicles can be loaded with more than one load. The load accommodation possibilities have been restricted two a maximum of two loads per vehicle (if the total weight is less or equal to the vehicle capacity) in

order to keep the search and decision time into operative limits.

An important point to consider is that the system must take decisions on line, and the decision time grows with the number of vehicles, loads and accommodation possibilities.

Assuming this limitation, a short term scheduling rules have been developed to assign loads to vehicles. These set of rules are shown in figure 1. The vehicle assignment status before and after each rule accomodate a load to a vehicle plan is shown in it. For instance, rule r1 evaluates the feasibility to assign a new load, from A to C, to a vehicle that is currently at point P and it's going toward point A to take a load A —> B, that had been assigned to it. It supposes that the vehicle goes to point A takes both loads A —> C and A —> B and goes to C to deposit a load and continues to B to deposit the other one.

The rules used to assign loads to vehicles are shown below:

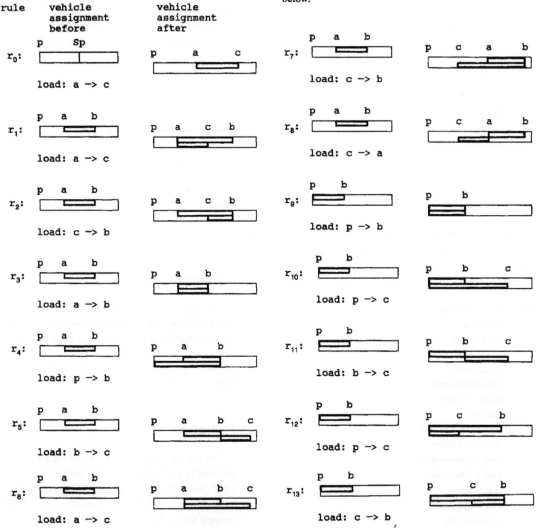

Figure 1. Assignment rules.

In order to select between the different possibilities to assign the load, the rule that lead to the shortest travel distance for the new planified load is selected.

The assigned loads wait until a vehicle finishes the transport of one or more load, and then the waiting load are intended. The waiting load are ordered by arrival time. If all the vehicles have two loads assigned no rules are fired. In case a load can not fire any rule and not all vehicles have two loads planified, the next waiting load is considered.

6. CONCLUSIONS

The AES presents many different advantages for the AGVS control. The choice of an Artificial Intelligence technique has enabled a quick modelization of an AGV system and a flexible and easy way to modify control rules.

The selection of a production system to control an AGV system has suited to this industrial domain.

There are some OPS5 limitations that have been solved using external routines developed in Pascal.

The AES has been checked in different installation models, and today is being used to develop control rules for a real installation.

The assignment rules used to accomodate two loads per vehicle have been tested with different models. Even thouth the performance depend stronly of the network topology, it shows positive results in most cases.

7. REFERENCES

1. Automated Guided Vehicles
   Müller, T.
   IFS and Springer-Verlag 1983

2. A continuous real-time expert system for computer operations
   Ennis, R.L. et AL
   IBM J. Res Develop. Vol 30 No.1 Jan.1986

3. Programming Expert Systems in OPS5
   Brownston L.
   Addison-Wesley 1985

4. A practical guide to Designing Expert Systems
   Weiss, S.M y Kulikowski C.A
   Rowman and Allanheld 1984

5. Rule-based Expert System
   Buchanan, B. G. y Shortliffe E. H.
   Addison-Wesley 1984

6. Knowledge-based Expert Systems in Industry
   Kriz, J.
   Ellis Horwood 1987

7. Introduction to Artificial Intelligence
   Charniak E. and Mc Dermott D.
   Addison-Wesley 1985

8. Artificial Intelligence
   Winston, P. H.
   Addison-Wesley 1984

9. The uses of Artificial Intelligence techniques in AGV systems.
   Wing, M. A. and Rezvski, G.
   Proc. of the 5th I. C. on A.G.V. Systems

10. Artificial Intelligence
    Rich, E.
    Mc Graw-Hill 1983

11. AGVS at Work
    Hammond, G.
    IFS and Springer-Verlag 1986

12. Expert Systems
    Harmon, P. and King, D.
    John Wiley and sons 1985

Copyright © IFAC Information Control Problems in
Manufacturing Technology, Madrid, Spain 1989

OPTIMAL DESIGN OF BUFFER STORAGE IN AN ASSEMBLY LINE

W. Hyun Kwon, H. S. Park and B. J. Chung

Department of Control & Instrumentation Engineering, Seoul National University, Seoul, Korea

Abstract. In this paper, we consider an optimization problem via the perturbation analysis technique for the discrete parameter such as the buffer storage. The creation and the elimination of Full Out(FO) and Null Input(NI) with respect to the buffer storage is presented in the tandem queue. Perturbation propagation rules are presented, through which the exact perturbed path can be obtained. Also the perturbation analysis algorithm for the on_line usage is suggested. The optimal buffer storage in the assembly line is obtained by using the new performance measure consisted of the buffer storage cost and the throughput. The proposed perturbation analysis technique is validated by the brute force simulations.

Keywords. Discrete event system; sample path; simulation; perturbation analysis; buffer storage; assembly line; optimization.

INTRODUCTION

Many systems such as FMS(Flexible Manufacturing System) and communication networks are activated by events, which are considered as the arrival or the service completion of customers. The dynamic systems activated by these events are called Discrete Event Dynamic Systems(DEDS). In the analysis on these systems, the difficulty is due to the complex interactions between these events with respect to time. Two general methods using the analytic and the simulation model exist for the analysis of DEDS. The analytic model has minimal efforts and costs but too many unrealistic assumptions, while the simulation model has less assumptions but is too costly. Ho and colleagues (1979,1983a) suggest the perturbation analysis (PA) technique, which is considered to be the combination of the first and the second methods. In this model, the observed data in a sample path can be used to estimate the sensitivity of the performance measure with respect to parameters in the DEDS. Thus this method is less costly and has less assumptions. Many researchers(Ho,1983a, 1983b; Cao,1986; Cao,1987; Suri,1987) have successfully applied the PA method to the sensitivity analysis of some performance measures such as the system throughput and the waiting time.

Costs of buffer storages, servers, and in-process inventory are important parameters in the design of assembly lines. To find the best buffer storage configuration leads to an optimization problem. To solve this problem, it is essential to quantify the relations between performance measures and assembly line design parameters such as service rates of servers, arrival rates of jobs, and the buffer storage capacity.

This paper presents another type of performance measures consisted of the throughput and the buffer storage cost and the on_line perturbation analysis technique with respect to the discrete parameter such as the buffer storage. The creation and the elimination of Full Out(FO) and Null Input(NI) with respect to the buffer storage is presented in the tandem queue. The suggested propagation rules are represented by mathmatical expressions, through which the exact perturbed path can be obtained. Also a design method of the optimal buffer storage in the assembly line is presented. Assembly line systems with finite buffers have been studied by the approximation method(Ancelin,1987; Lim,1985; Mak,1986; Wijngaard,1979). Ho(1979) studied the off_line PA technique w.r.t. the buffer storage in the production line and Cassandras(1987) studied the on_line PA technique w.r.t. the buffer storage in the system with the single server(G/G/1/T). In this paper, the on_line PA technique for the buffer

storage will be discussed in the assembly line with mutiple servers. In order to validate the proposed algorithm, a numerical example is given.

BASIC MODELS

For simplicity, we consider the event as the service completion of customers. Let's recall that the ith server S_i is always in one of three possible states : 1) busy(BY) ; 2) blocked, also called full output(FO) ; 3) idle, also termed no input(NI). The sample path $E_i(w)$ consists of the event sequences, the element of which is the sample service time of each server(i: server i, w : sample realization). The sample tableau consists of $E_i(w)$'s. This tableau, an example of which is shown in Fig.1, represents the result of the sample path of DEDS.

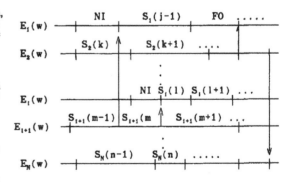

Fig.1. Event sequence tableau of general queueing network

If a perturbation occurs at time t in $E_i(w)$, this perturbation may propagate to $E_j(w)$ of another server S_j. When two event sequences are interactive in intervals that FO or NI occurs, perturbations propagate to another event sequence through these events. Perturbations may also cancel in event sequences. A PA method is shown in Fig. 2.

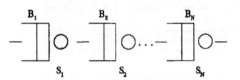

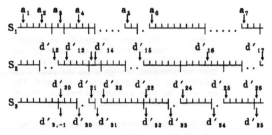

Fig.2. Method of PA

Our model of the assembly line, which is a specific case of the general queueing networks, is shown in Fig.3.

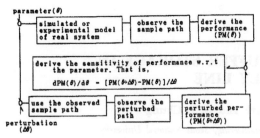

Fig. 3.   Queueing network of the assembly line
          S_i : the ith server(or server i)
          B_i : length of buffer storage of server i

We have the following facts for the tandem queue of the assembly line.

Fact 1(FO case) : Assume that S_i services the jth customer and S_{i+1} services the $(j-B_{i+1})$th customer, where B_i is the length of the buffer in the sever S_i. Then, if S_i completes the service of the jth customer before S_{i+1} completes the service of the $(j-B_{i+1})$th customer, the jth customer in S_i is blocked until S_{i+1} completes the service of the $(j-B_{i+1})$th customer.

Fact 2(NI case) : If S_{i+1} completes the service of the $(j-1)$th customer before S_i completes the service of the jth customer, the state of S_{i+1} is NI and the service completion time of jth customer of S_i is the time that NI state expires.

PA FOR THE BUFFER STORAGE

First we will consider the activity of the tandem queue with 3 servers, in which $B_1 = B_2 = B_3 = 1$ as a model of Fig.3.

The nominal sample path of the above system is shown in Fig.4. When B_2 is perturbed by $\Delta B_2 = 1$, the perturbed path is shown in Fig.5. Also, when B_3 is perturbed by $\Delta B_3 = 1$, the perturbed path is shown in Fig.6. Assume that the buffer B_i is perturbed by $\Delta B_i (>0)$. We define that d_{ik} is the time that the kth customer departs the ith server in the nominal sample path and d'_{jk} is the time that the kth customer departs the jth server in the perturbed path. And $d'_{ij,v}$ is the service completion time or the virtual departure time and may be different from d'_{ij} since FO may exist. Let's observe the dynamic phenomenon of the example through Fig.4 - Fig.6. In Fig.4 - Fig.6 the a_i is the ith arrival time from the external environment and $'$. represent NI or FO state and $\dashv$ denotes the service completion of customers in servers(i.e. $d_{ij,v}$). We observe the following facts.

Creation of NI
Consider the relation between a_5, d_{14} in Fig.4 and a_5, d'_{14} in Fig.5. Since $d_{14} > a_5$ in Fig.4, NI isn't created . But as $d'_{14} < a_5$ in the perturbed path due to $\Delta B_2 = 1$, NI is created and its time duration is $(a_5 - d'_{14})$.

Elimination of NI
If we observe the relation between d_{13}, d_{22} in Fig.4 and d'_{13}, d'_{22} in Fig.5, we can know the following: if $d_{13} > d_{22}$

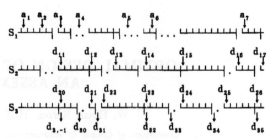

Fig.4. The nominal sample path

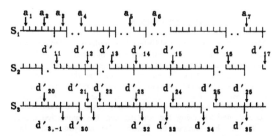

Fig.5. The perturbed path ($\Delta B2=1$)

Fig.6. The perturbed path ($\Delta B3=1$)

in the nominal path, NI of the time duration $(d_{13} - d_{22})$ is created. But, as $d'_{13} < d'_{22}$ in the perturbed path, NI isn't created.

Creation of FO
In the relation between d_{23}, d_{32} in Fig.4 and d'_{23}, d'_{32} in Fig.5, we know the following :because $d_{23,v} < d_{32}$, FO of the time duration $(d_{32} - d_{23,v})$ is created. By the effect of $\Delta B2 = 1$, $d'_{23,v}$ is prior to $d_{23,v}$. Because $d'_{23,v} < d'_{32}$, FO is created and its time duration is $(d'_{32} - d'_{23,v})$ time, which is longer than FO time duration created in the nominal path.

Elimination of FO
If we observe the relation between d_{23}, d_{32} in Fig.4 and d'_{23}, d'_{32} in Fig.6. We can know the following: if $d_{23,v} < d_{32}$, FO of the time duration $(d_{32} - d_{23,v})$ is created. But by the effect of $\Delta B_3 = 1$, FO isn't created in the perturbed path.

From this example we can extend the above results to the general tandem queue, which are stated in the following theorems.

Theorem 1 : (Creation of FO)   In the case that the jth customer completed in S_i is about to enter in S_{i+1}, if $d_{ij,v} < d_{i+1,j-B(i+1)}$, FO is created and its time duration is $(d_{i+1,j-B(i+1)} - d_{ij,v})$ where $B(i+1)$ denotes B_{i+1} and B_i is the maximum buffer length of the server i.
Proof: This follows from Fact 1.

Because of interaction of adjacent servers in the tandem queue, FO is created though FO doesn't occur between two servers, or the FO's time duration is changed in case that more than 3 servers are interactive.

Corollary : (Creation of FO due to another FO and Variation of its time duration) Consider the interaction of three servers which is S_i, S_{i+1}, and S_{i+2}. Assume that
$$d_{i+1,j-B(i+1),v} < d_{i+2,j-B(i+1)-B(i+2)}.$$

case 1: if $d_{ij,v} > d_{i+1,j-B(i+1),v}$, FO is created in S_i.
case 2: if $d_{ij,v} < d_{i+1,j-B(i+1),v}$, FO's time duration is changed in S_i. For case 1 and case 2, FO's time duration is $(d_{i+1,j-B(i+1)-B(i+2)} - d_{ij,v})$.

Proof :

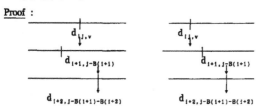

Fig.7. Creation of FO   Fig.8. Variation of FO's
                                      time duration

Consider the case of Fig.7. By Fact 2, the interaction between S_i and S_{i+1} creates NI in the S_i. By Fact 1, the interaction between the server $(i+1)$ and the server $(i+2)$ creates FO in the server $(i+1)$ and $d_{i+1,j-B(i+1)} = d_{i+2,j-B(i+1)-B(i+2)}$. Thus by the interaction of the server i and the server $i+1$, FO is created in the server i and its time duration is $(d_{i+2,j-B(i+1)-B(i+2)} - d_{ij,v})$ and $d_{ij} = d_{i+2,j-B(i+1)-B(i+2)}$. Similarly, it is proved in the case of Fig.8.

Theorem 2 : (Creation of NI) If $d_{ij} > d_{i+1,j-1}$, NI is created in the server $(i+1)$ and its time duration is $(d_{ij} - d_{i+1,j-1})$.
Proof : This follows from Fact 2.

For the perturbation propagation, we define the terminology as follows :

NIT_i : amount of time that NI occurs before the ith server services a customer in the nominal sample path.

FOT_i : amount of time that FO occurs after the ith server services a customer in the nominal sample path.

$PNIT_i$: amount of time that NI occurs before the ith server services a customer in the perturbed sample path.

$PFOT_i$: amount of time that FO occurs after the ith server services a customer in the perturbed sample path.

$\Delta_i(t)$: amount of perturbations accumulated till time t in the ith server(or Δ_i).

$\delta(t)$: amount of perturbation generated at t

Then we obtain the following relations for the propagation of the perturbation.

$$\Delta(d_{ik}) = \Delta(d_{i,k-1}) - \delta(d_{i,k}) \quad , \Delta(0)=0 \quad (1)$$

$$\delta(d_{i,k}) = (PNIT_i + PFOT_i) - (NIT_i + FOT_i) \quad (2)$$

The departure time of the jth customer of server i in the perturbed path is

$$d'_{ij} = d_{ij} + \Delta(d_{ij}) \quad (3)$$

In the next section the PA technique w.r.t. the buffer storage will be presented by using eq.(1) - eq.(3).

OPTIMAL BUFFER STORAGE VIA PA

A performance measure is usually the throughput, defined as follows :

$$T_p = K/d_K \quad (4)$$

where K is the number of customers and d_K is the departure time of Kth customer. We know that costs of

buffer storages, servers, and in_process inventory are also important parameters in the optimal design of the assembly line. The throughput depends on service rates and the number of servers and buffer storages. As the buffer size increase, both the throughput(T_p) and the cost of the buffer storage increase. By combining costs of the buffer storage with the throughput in this paper, it is possible to obtain an optimal buffer size to maximize a new performance measure(P), which is defined by

$$P = -\alpha B + \beta T_p \quad (5)$$

where α is the cost per one buffer, B is the amount of the buffer storage, and β is the cost per the throughput. The perturbation of the throughput, ΔT_p, resulting from $\Delta B_i > 0$ is easily obtained at the end of the observation interval from the following eq. :

$$\Delta T_p = N/(d_N + \Delta d_N) - N/d_N \quad (6)$$

The perturbation(ΔP) in the presented performance measure w.r.t. ΔB is obtained from

$$\Delta P = -\alpha \Delta B + \beta \Delta T_p \quad (7)$$

where ΔT_p is given in (5). The sensitivity of the throughput w.r.t the buffer storage is

$$\Delta T_p / \Delta B = \{N/(d_N + \Delta d_N) - N/d_N\}/\Delta B$$

and the sensitivity of the presented performance measure is

$$\Delta P/\Delta B = -\alpha + \beta \Delta T_p / \Delta B$$

The optimal buffer storage by using ΔP w.r.t. each buffer storage is obtained which is one to maximize the ΔP.

We introduce the on_line perturbation analysis technique w.r.t. the buffer storage by using theorems and eq.(1) - eq.(3) in the previous section. In this PA algorithm the departure time of the current customer is easily obtained from the recursive form consisted of the amount of accumulated perturbations and the departure time of the previous customer.

Perturbation Algorithm For the buffer change ΔB_m (= T) : Assume that in the nominal sample path all events are known and the total number of servers is N and $B_i = 1$ (i=1,2,...N).

1) $\Delta_i = 0$; $i = 1,2,.....,N$
2) Calculate the amount of accumulated perturbations and the perturbed departure time.

$$\Delta_i = \Delta_i - (NIT_i + FOT_i) + PNIT_i$$
$$d'_{ij} = d_{ij} + \Delta_i$$

3) Calculate the FO time in the perturbed path

For $k = N,....,m+1,m-1,....,2$

if $d'_{k-1,j-k+2} < d'_{k,j-k+1}$

$\quad PFO_{k-1} = d'_{k,j-k+1} - d'_{k-1,j-k+2}$

else $\quad PFO_{k-1} = 0$

$\Delta_{k-1} = \Delta_{k-1} + PFO_{k-1}$

$d'_{k-1,j-k+2} = d'_{k-1,j-k+2} + PFO_{k-1}$

For $k = m$

if $d'_{m-1,j-m+2} < d'_{m,j-m+1-T}$

$\quad PFO_{m-1} = d'_{m,j-m+1-T} - d'_{m-1,j-m+2}$

else $\quad PFO_{m-1} = 0$

$\Delta_{m-1} = \Delta_{m-1} + PFO_{m-1}$

$d'_{m-1,j-m+2} = d'_{m-1,j-m+2} + PFO_{m-1}$

4) Evaluate the NI time in the perturbed path

For k = 0,1,2,....,N-1 (in k=0, $d'_{0,1}$ is the lth
        arrival from the external environment.)

if $d'_{k,J-k+1} > d'_{k+1,J-k}$

$$PNI_{k+1} = d'_{k,J-k+1} - d'_{k+1,J-k}$$

else $PNI_{k+1} = 0$

5) if j = M customers, evaluate the performance
              measures(eq.(6) and eq.(7)).
    else   repeat the step 2) - step 4).

We obtain the optimal buffer storage via the results from
this PA algorithm which can be used for the on_line
optimaization problem.

The PA technique for the sensitivity of the system
performance function J w.r.t. continuous parameters is to
assume

$$\frac{\partial E[L(\theta,\xi)]}{\partial \theta} = E[\frac{\partial L(\theta,\xi)}{\partial \theta}] \quad (8)$$

which holds under some conditions, one of which has been
introduced by Cao(1985). It is noted that unlike the
continuous parameter case the sensitivity estimator w.r.t. the
discrete parameter can always be given by

$$\frac{\Delta E[L(\theta,\xi)]}{\Delta \theta} = E[\frac{\Delta L(\theta,\xi)}{\Delta \theta}] \quad (9)$$

The properity of eq. (9) can be utilized if the sensitivity
estimator is necessary for the buffer storage.

SIMULATION

We consider an example of the tandem queue with three
servers, in which $B_1=B_2=B_3 = 1$. Assume that the
distribution functions of the interarrival time and the service
time of each server are exponential. The average time of the
interarrival is 4 unit times and average service times of each
server are all 4 unit times. Also assume that $\alpha=2$ and
$\beta=1000$ in eq. (5).

When B_2 and B_3 are perturbed, the results are shown in
Table 1 and Table 2. From these tables, the optimal buffer
storage is obtained. That is, $\Delta B_2=3$ is the optimal buffer
storage. Table 3 shows the results of the proposed algorithm
and the brute force simulation results. Comparison of these
two results shows that the proposed algorithm is sufficiently
valid in the analysis on performance measures w.r.t. discrete
parameters.

CONCLUSION

In this paper, we have attempted to provide extensions of
PA for the buffer storage. The suggested perturbation analysis
algorithm is easily implemented and used to the on_line
optimization problem. The creation and the elimination of Full
Out(FO) and Null Input(NI) with respect to the buffer
storage is presented in the tandem queue. The exact
perturbed path can be obtained through the suggested PA
method. For the optimal design for the buffer storage size
in the tandem queue, the new performance measure, which
consists of buffer storages and the throughput, is suggested.
The above suggested algorithm is used for getting the total
amount of perturbations accumulated in the observation period
w.r.t. the buffer storage. The perturbed amount of the
presented performance measure w.r.t. the buffer storage are
obtained, from which the optimal buffer size is obtained. The
above suggested algorithm is validated by the brute force
simulation. The method suggested in this paper may also be
applied to the other performance measures w.r.t. other discrete
parameters. And it can be extended to the analysis on the

TABLE 1 Sensitivity of Throughput and
        Productivity(B_2 is Perturbed;
        Number of Customers : 5000)

ΔB_2	ΔT_p	$\Delta T_p/\Delta B_2$	ΔP	$\Delta P/\Delta B_2$
1	0.0127	0.01270	10.7	10.70
2	0.0189	0.00945	14.9	7.45
3	0.0218	0.00727	15.8	5.30
4	0.0234	0.00585	15.4	3.85
5	0.0241	0.00482	14.1	2.82
6	0.0247	0.00412	12.7	2.12
7	0.0250	0.00357	11.0	1.57
8	0.0251	0.00314	9.1	1.14
9	0.0251	0.00279	7.1	0.79
10	0.0251	0.00229	5.1	0.51
11	0.0252	0.00210	3.2	0.29
12	0.0252	0.00210	1.2	0.10

TABLE 2 Sensitivity of Throughput and
        Productivity(B_3 is Perturbed;
        Number of Customers : 5000)

ΔB_3	ΔT_p	$\Delta T_p/\Delta B_3$	ΔP	$\Delta P/\Delta B_3$
1	0.0120	0.01200	10.0	10.00
2	0.0177	0.00885	13.7	6.85
3	0.0208	0.00693	14.8	4.93
4	0.0225	0.00563	14.5	3.63
5	0.0236	0.00472	13.6	2.72
6	0.0240	0.00400	12.0	2.00
7	0.0244	0.00349	10.4	1.49
8	0.0246	0.00308	8.6	1.08
9	0.0247	0.00274	6.7	0.74
10	0.0247	0.00247	4.7	0.47
11	0.0247	0.00225	2.7	0.25
12	0.0248	0.00207	0.8	0.07

TABLE 3 Comparison of Proposed Algorithm
        and the Brute Force Simulation
        (Number of Customers(N) : 5000)

	Proposed algorithm Δd_N	Brute force simulation Δd_N
$\Delta B_2=1$	-2889.12	-2874.29
$\Delta B_3=1$	-2799.67	-2825.42

general queueing network with respect to the discrete
parameter.

REFERENCES

Ancelin, B.,and A. Semery. (1987). Calcul de la
    productivite d'une ligne integree de
    fabrication : CALIF, un logiciel industriel base
    sur une nouvelle heuristique. *AUTO. PROD.
    INFO. INDUS.*,21,209-238.

Cao,X. (1985). Convergence of Parameter Sensitivity
    Estimates in a Stochastic Experiment.
    IEEE Trans. AC.,AC-30,846-855.

Cao,X.,and Y.Dallery. (1986). Sensitivity Analysis of
    Closed Queueing Networks : An Operation
    Approaches. *Proc. of ACC.*,2034-2039.

Cao,X.,and Y.C.Ho. (1987). Estimating the Sojourn
    Time Sensitivity in Queueing Network using
    Perturbation Analysis. *J. Optimiz. Theory Appl.*
    ,53,353-375.

Cassandras,C.G. (1987). On-Line Optimization for a
    Flow Control Strategy. *IEEE Trans. AC.*,AC-32,
    1014-1017.

Ho,Y.C.,M.A.Eyler,and T.T. Chien (1979). A gradient
    technique for general buffer storage design
    in a production line. *Int. J. Prod. Res.*,17,
    557-580.

Ho,Y.C. (1983a). A New Approach to Analysis of
    Discrete Event Dynamic Systems. *Automatica,*
    19,149-167.
Ho,Y.C.,and M.A.Eyler. (1983b). A New Approach to
    determine Parameter Sensitivities on   Transfer
    Line. *Mgt. Sci.,*29,700-714.
Lim,J.-T. and S.M. Meekov. (1985). Asymptotic
    Analysis of A Simple Model of An Assembly
    Line. *Proc. of 24th Conf. Decision and Control,*
    2012-2015.
Mak,K.L. (1986). The Allocation of Interstage Buffer
    Storage Capacity in Production Lines.
    *Comput.& Indust. Engng.,*10,163-169.
Suri,R. (1987). Infinitesimal Perturbation Analysis
    of Discrete Event Dynamic System - A General
    Theory. *J. of ACM.,*34,686-717.
Wijngaard,J. (1979). The Effect of Interstage Buffer
    Storage on the Output of Two Unrelieable
    Production Units in Series, with Different
    Production Rates. *AIIE TRANSACTIONS,*11,42-47.

COMBINED SCHEDULING AND ROUTING IN DISCRETE MANUFACTURING SYSTEMS

M. Aicardi, A. Di Febbraro and R. Minciardi

Department of Communications, Computer and System Sciences, University of Genoa, Genova, Italy

Abstract. The paper considers the problems of real-time scheduling and routing in discrete manufacturing systems which are modelled as queueing networks with deterministic service times. The determination and the application of closed-loop scheduling/routing strategies based on real-time information referring to the whole queueing network are far beyond practical feasibility even for quite simple networks. For this reason, we present possible approaches for determining local scheduling/routing strategies, i.e. strategies that each node (machine) in the network implements in a decentralized fashion, and which are based on local on-line information only. Then, the need for the coordination of the local strategies is taken into account by considering a mechanism for the exchange of the information condensed in quantities called expected downstream delays.

Keywords. Manufacturing processes; queueing networks; discrete-event systems; routing and scheduling.

INTRODUCTION

In the field of optimization and control of discrete manufacturing systems, the problems of routing and scheduling of workpieces are solved, as a rule, by means of "static" strategies, i.e., not depending on real-time information. This is mainly due to the considerable complexity characterizing such problems even in presence of relatively simple queueing networks. This fact by itself prevents the possibility of real-time updating of scheduling and routing policies. Secondly, the necessity of dealing with a large amount of information relevant to the whole state of the network makes optimal closed-loop scheduling/routing strategies practically infeasible, even whenever their determination could appear conceivable.

Nevertheless, the necessity of making use of real-time information is generally recognized, at least in the fact that real-time strategies are sought in order to face transient problems and phenomena. Thus, the most common practical solution in discrete manufacturing systems is that of setting up a static routing/scheduling strategy, with the possibility of corrections based on real-time information, whenever significant transient phenomena are recognized.

As a matter of fact, real-time control approaches for queueing networks have already been attempted and reported in the literature. However, the most common case dealt with refers to real-time control of stochastic queues characterized by particular topologies (see, for instance: Wu et al., 1988; Hajek, 1984; Ephremides

et al., 1980). Recently, Beutler and Teneketzis (1987) considered a control problem for stochastic queueing networks (modelling, in this case, telecommunication networks) based on imperfect information about the network state. Finally, Perkins and Kumar (1989) considered a class of scheduling policies implementable in real-time in a purely distributed way at the various machines, and proved the stability of such policies, with reference to flexible manufacturing assembly/disassembly systems.

In this paper, we are interested in developing feasible approaches for the control of queueing networks with general topology, and characterized by the assumption of deterministic services times. This assumption can be considered as quite realistic if we are interested in discrete manufacturing processes. Moreover, we are interested in routing/scheduling strategies which are based on individual decisions relevant to any single workpiece. In order words, we do not look for randomized strategies based, e.g., on routing frequencies. Thus, we are considering the development of strategies such that for each workpiece the decision is taken on the basis of the present state of the network, to some extent. Since we intend to come to feasible strategies, we are led to consider strategies constrained to be based on a subset of the overall real-time information regarding the network state. Then, the most reasonable choice for this subset is that corresponding to the state of the machines which are "directly" (i.e., in the close future) influenced by the decision taken with reference to the considered workpiece. We shall come back later to this point. For the moment, it is sufficient to say that this leads to the development of "local" routing/scheduling strategies which are based on an arbitrarily constrained information set, relevant to the machine where the considered workpiece is presently, and to the immediate downstream machines.

Even if the routing/scheduling strategies which are used are of local type, nevertheless some degree of coordination among them is needed in order to improve the significance of the overall solution. Then, we present the possibility of coordinating these strategies by means of some information exchange among the nodes (machines) in the queueing network. This information is relevant to expected downstream delays which can be used as additional arguments of the local routing strategies. Thus, the local decisions are dependent not only on local real-time information but also on aggregated and estimated (but still real-time) information about the state of the overall queueing network.

The paper is organized in the following way. In the next section, the characteristics of the considered local problem are reported in detail. Then, in the third section, feasible strategies for this problem are presented and discussed. Finally, in the fourth section, the coordination of the strategies on the basis of the exchange of real-time aggregate information is discussed.

THE PHYSICAL MODEL AND THE LOCAL DECISION PROBLEM

In this section, let us consider the characteristics of the decision problem to be solved at every node (machine) in the network, with reference to any workpiece. In the description of this problem, we will follow the model presented by Aicardi et al. (1988). First of all, we point out that we consider a discrete manufacturing system affected by some basic simplifications. Namely, we do not consider assembly/disassembly processes, that is, we consider any workpiece as having its own individuality from the beginning to the end of its manufacturing process. Moreover, the transportation system is assumed to be of ideal type, i.e., with infinite capacity and no delay. Finally, buffers are considered as unlimited. In addition, we suppose that every machine is monoserver and that service times are deterministic.

Thus, we are dealing with (deterministic) queueing networks where P part types are circulating, each one of them characterized by a fixed sequence of operations $o(p,1),\ldots, o(p,k_p)$, $p=1,\ldots,P$, $o(s,1) \in O$. For the sake of simplicity, we assume that no operation is repeated in a given sequence; if this is not the case, clearly, the problem may be overcome by defining additional operations only formally different from the basic one. In the network there are N machines $(M_1,\ldots, M_N)$, each one with its own queue $q(i)$, $i=1,\ldots,N$. $T(p,o,m)$ indicates the service time which characterizes the execution of operation o on workpieces of the p-th class, on machine M_m.

Having so depicted the model we are referring to, let us specify the decision problems we are considering. Actually, the only characteristics that are still to be specified are the routing and scheduling policies. As already pointed out, in this paper we are interested in developing purely real-time scheduling and routing policies. Here we use the term scheduling in order to designate merely the local scheduling, i.e., the choice of the sequence of the services for the workpieces queued, at a certain instant, at a certain machine. This is not a single-machine scheduling problem, since, as we shall see later, the above choice is made taking into account the services and the state of the machines which are downstream with respect to the considered one.

On the other hand, we use the term routing in order to designate the decision which has to be taken, at the completion of every service at every machine, about the destination of the workpiece just processed. Of course, this decision must be in accordance with the specifications in the operation sequence of the workpiece to be routed.

Clearly, the routing and scheduling decisions should be taken in accordance with a common objective, being parts of an integrated strategy for the real-time control of the queueing network. Unfortunately, as one can easily understand, any possible significant choice of the optimization objective (e.g., system throughputs, average workpieces delay, overall completion time of a given workload) readily leads to untractable combinatorial optimization problems, whose solution is not conceivable in a real-time context.

On the basis of the previous observation, we are led to introduce some basic simplifications in our control problems. The fundamental one is that of considering a decentralized decision making structure, in which each node of the network makes use of real-time information consisting only of a local subset of the system state. Nevertheless, we do not consider decisional problems based on the information relevant to a single machine. On the contrary, we consider a local information set related to the state of the interested node (machine), and to the states of its immediate successors (in a sense that we will explain later on).

More specifically, we will refer to the basic situation represented in Fig. 1, where the decision that has to be taken is relevant to the routing of a generic workpiece w_x, of the p-th class, which is presently at machine M_i, to one of the machines in the set $\{M^1,\ldots, M^{\sigma(p)}\} \triangleq S(i,p)$, which are supposed all compatible with the operation $s(o(w_x),p)$ which follows, in the sequence of operations relevant to the p-th class, operation $o(w_x)$, which is executed on M_i. The above set of machines $S(i,p)$ is that of the successors of M_i, with respect to the p-th class. Let us further define $S(i) \triangleq \bigcup_p S(i,p) \triangleq \{M^1,\ldots, M^{\sigma}\}$, that is, the set of machines successors of M_i. No specific problem arises whenever $M_i \in S(i)$, so this case will not be excluded in our treatment. Finally, if there is any machine, say M_1, such that $M_1 \in S(i)$, but $M_1 \notin S(i,p)$ for some p, then we simply put $T(p, s(o(w_x),p),1) = \infty$.

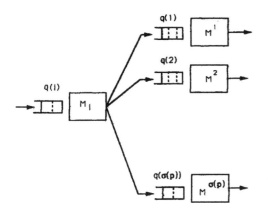

Fig. 1. The local decision problem

The basic characteristics of the decisional situation under concern are the following:

i)   the routing decision for workpiece w_x is taken on the basis of the states of the machines $M_i, M^1, \ldots, M^\sigma$; here the state of a machine designates the contents of the queue and the remaining processing time for the service under execution, in case of a currently active machine;

ii)   the routing decision for workpiece w_x is taken as a part of the solution of the global routing problem for the contents of queue q(i) (relevant to machine M_i);

iii) in the solution of the above routing problem, the service discipline at machines $M^4, \ldots, M^\sigma$ is supposed to be simply FIFO; a consequence of this assumption is that the solution of the routing problem is based on the supposition that workpiece w_x will be served by the machine M_k to which it is finally routed, after all workpieces already present in q(k);

iv)   we deliberately neglect the possibility of future incoming workpieces entering q(i), q(1),..., q(σ); so the evolution of the subsystem under concern is completely deterministic and foreseeable;

v)   even if the routing decision regarding a workpiece w_x is obtained by solving the whole routing problem relevant to all customers in q(i), only the decision regarding w_x (i.e., the first one in temporal order) is actually implemented; the following routing decision will be taken by repeating the whole procedure on the basis of updated information.

Having so defined the environment in which the decision has to be taken, we can consider two different problems:

a) in the first problem the decision concerns the routing of workpiece w_x after the completion of its service on M , supposing that the service discipline in M_i is FIFO, too;

b) in the second problem the decision regards the choice of the workpiece which has to be picked up in the queue q(i), and immediately served by M_i, as well as the determination of its destination after the service completion; in other words, we no longer assume that the service discipline in M_i is FIFO.

Clearly, the choice between the two above problems depends on their significance in the real context under examination. The choice of the second problem seems more appropriate whenever buffer reordering at the different machines is not a complicate operation from the physical point of view. Of course, if the second problem is actually considered, the assumption that the downstream (successor) machines are characterized by FIFO disciplines introduces some degree of suboptimality in the solution of the local problem. Nevertheless, we shall retain such hypothesis and consider problem b) in the rest of the paper. In any case, if one is interested in the first problem, then a feasible approach for its solution can be easily derived from the solution that we propose for the second problem (see also Aicardi et al., 1988).

THE LOCAL SCHEDULING ROUTING STRATEGY

The choice to consider problem b) introduced in the previous section is not yet completely defined, unless we specify the objective of our problem. A reasonable choice for the objective to take into account in the routing decision is the following:

minimize the total completion time of the services corresponding to workpieces presently in q(i), q(1),..., q(σ), i.e., in the whole set of machines considered.

The above defined problem can be viewed as a complication of the problem of scheduling independent tasks on nonidentical processors, which has been considered, for instance, by Ibarra and Kim (1977), who have provided some heuristic solutions. Actually, the complication mainly consists in the fact that the tasks (workpieces) in M_i are not yet available for service in $M^4, \ldots, M^\sigma$. In addition, the sequence of services in M_i is not yet established, so that it is not possible to consider the scheduling of tasks in q(i) as a problem characterized by fixed release dates.

A heuristic algorithm in order to find a scheduling/routing strategy implementable at machine M_i is provided and discussed by Aicardi et al. (1988). Clearly, it does not provide an optimal solution to the problem of minimizing the overall completion time of the considered tasks,

even if its development has been carried
out with this objective in mind. The
algorithm essentially relies on the
decomposition of the overall
scheduling/routing problem into two
distinct phases, namely:

i) a first phase where the tasks in $q(i)$
   are assigned to $M^1,\ldots, M^\sigma$, as if they
   were immediately available;

ii) a second phase where the same tasks
    are scheduled for their service on M_i,
    on the basis of the results of the
    previous assignment procedure.

Algorithm for the local scheduling/routing strategy

Define as J the set of workpieces queued
in $q(i)$. Let $lq(l)$ be the time needed by
machine $M_l \in S(i)$ to empty its queue, at
the time instant in which the decision has
to be taken. Moreover, let w_r be the r-th
workpiece in J, of class $p(r)$. The
relevant processing times for services
in downstream machines are thus $T(p(r),
s(o(w_r), p(r)),l)$, $\forall M_l \in S(i)$.

Step 1. For every w_r in J, set up the
vector $\tau(w_r)$, in which the indexes of the
machines belonging to $S(i)$ are ordered
according to nondecreasing $T(p(r),
s(o(w_r),p(r)),l)$. Moreover, order the
workpieces in J in a list $\mathcal{L}$ according to
non increasing minimum processing times,
i.e., non increasing $T(p(r), s(o(w_r),
p(r)), \tau_1(w_r))$, where $\tau_1(w_r)$ is the
first element of vector $\tau(w_r)$.

Step 2. For the first workpiece w_z in
$\mathcal{L}$, define

$$\overline{T}(w_z) \triangleq \min_{M_l \in S(i)} T(p(z), s(o(w_z), p(z)), l) \quad (1)$$

Then assign workpiece w_z to machine
M_h, where $h \triangleq \tau_j(w_z)$, being j the
smallest index such that

$$\left| T(p(z), s(o(w_z), p(z)), h) - \overline{T}(w_z) \right| < \Delta \quad (2)$$

where Δ is a fixed threshold. Update
$\mathcal{L} \leftarrow \mathcal{L} - \{w_z\}$. Define the due date of
workpiece w_z as

$$dd(w_z) \triangleq lq_h$$

Step 3. Set

$$lq_h \leftarrow lq_h + T(p(z), s(o(w_z), p(z)), h)$$

Step 4. If J is empty go further;
otherwise go to step 2.

Step 5. Apply a single-machine scheduling
procedure which minimizes the maximum
tardiness, i.e., schedule the tasks
corresponding to workpieces in $q(i)$
according to non decreasing due-dates (see
Bellmann et al., 1982).

Step 6. Start the execution of the first
scheduled task; at the end of this
execution, the interested workpiece is
routed to the machine $\in S(i)$ to which it
has been assigned to at step 2.

COORDINATION OF LOCAL STRATEGIES BY MEANS OF AGGREGATE INFORMATION

Although the use of the algorithm
presented in the previous section entirely
relies on heuristic justifications, it can
be considered as a reasonable
scheduling/routing rule whenever the
objective of minimizing the total
completion time of the services of the
tasks under consideration is meaningful.
Nevertheless, it is just this objective
that needs some further remarks. Actually,
a correct modification of this objective,
when referring to a machine inserted in a
network with a general topology is that of
considering the minimization of the
overall completion time regarding the
whole processing sequences of the
workpieces taken into consideration. This
may be accomplished by making use of the
downstream expected delays with respect to
the machines $\in S(i)$.

More specifically, let us define as
$D(p,o,m)$ the downstream expected delay,
i.e. the expected remaining time before
the overall completion of the sequence of
operations relevant to workpieces of class
p, provided that operation o has been
executed by machine M_m. Then, an obvious
modification in the above algorithm, in
order to take into account the information
provided by these expected delays, is that
of considering in Step 2 of the above
algorithm, instead of the execution times
$T(p(r),s(o(w_r),p(r)),l),\forall M_l \in S(i),\forall w_r \in J$,
the modified execution times given by

$$T'(p(r),s(o(w_r),p(r)),l) \triangleq$$

$$T(p(r),s(o(w_r),p(r)),l) +$$

$$+ D(p(r),s(o(w_r),p(r)),l)$$

$$\forall M_l \in S(i), \quad \forall w_r \in J$$

and then to apply the same algorithm as
before, where, of course, Step 3 still
takes into account the unmodified
execution times.

The use of the previously introduced
downstream expected delays is clearly
conditioned to a practical way to compute
them in real-time. We discard the trivial
procedure consisting in the collection of
the delays experienced by the most
recently routed workpieces and in the
subsequent averaging on these delays (this
procedure would involve a considerable
amount of information flow in the network,
and would not be based on properly real-
time information). Instead, we choose to
consider a distributed way to generate and
to propagate this information in the
network.

Actually, let us suppose, with reference
to machine M_i, the expected downstream
delays

$$D(p(r), s(o(w_r),p(r)), l)$$

$$\forall M_l \in S(i), \quad \forall w_r \in J$$

to be known. Each of these quantities
represents a measure of the overall delay
which is encountered after machine M_l
(with respect to the considered class and
operation). Then, we can consider the
problem of updating the values of the

expected downstream delays

$$D(p(r), o(w_r), i) \quad \forall w_r \in J$$

i.e., the delays which are encountered <u>after</u> <u>machine</u> M_i. These delays can be generated as follows:

$$D(p(r), o(w_r), i) = \widetilde{D}(p(r), o(w_r), i) +$$

$$+ \sum_{M_1 \in S(i)} [\lambda(p(r), o(w_r), i, 1) \cdot$$

$$\cdot D(p(r), s(o(w_r), p(r)), 1)] \qquad (3)$$

where

$\lambda(p(r), o(w_r), i, 1) \triangleq$ the fraction of the set of workpieces present in $q(i)$, of class $p(r)$, on which operation $o(w_r)$ has to be executed by M_i, which are assigned to machine M_1 for the next operation;

$\widetilde{D}(p(r), o(w_r), i) \triangleq$ estimated delay which is experienced by workpieces present in $q(i)$, of class $p(r)$, on which operation $o(w_r)$ has to be executed by M_i, between the end of the operation on M_i and the end of the subsequent operation.

Clearly the quantities $\lambda(p(r), o(w_r), i, 1)$ can be easily determined on the basis of the results of the application of the local scheduling/routing strategy described in the previous section. Remember that, in fact, this strategy requires that every workpiece in $q(i)$ is formally assigned to one of the downstream machines.

The same can also be said for the determination of the quantities $\widetilde{D}(p(r), o(w_r), i)$, since these estimated delays can be readily obtained by simply averaging on the individual delays which will affect each workpiece of class $p(r)$ on which operation $o(w_r)$ has to be performed by M_i. These individual delays, in turn, can be immediately estimated as a by-product of the application of the considered local strategy.

By means of the previous considerations, we have demonstrated that every $D(p(r), o(w_r), i)$ can be determined as in (3). Of course, this determination has to be accomplished only for each different pair (p, o) corresponding to workpieces presently in $q(i)$. Thus, the estimated downstream delays can be easily propagated along the network (from downstream to upstream, beginning with zero delay for output nodes), on the basis of the topology of the network and of simple elaboration of on-line information, which, in turn, depends on the application of the local strategy.

Clearly, the propagation and the updating of the information relevant to expected downstream delays have to be performed in a distributed asynchronous way, since these operations are based on the results of the applications of the local scheduling/routing strategies, which become active only at specified discrete occurrences.

CONCLUDING REMARKS

In this paper we have presented a possible approach in order to develop feasible scheduling/routing policies based on pure real-time information, in a queueing network with deterministic processing times. Due to the last assumption, the approach seems well-suited for application to discrete manufacturing environments. The feasibility of the proposed policies comes out from the adopted decentralized decision making scheme. More specifically, the scheduling/routing decisions are taken at each machine on the basis of a strategy which makes use of: i) "detailed" local information, relevant to the states of the machine plus the immediate downstream machines; ii) "aggregate" information about the rest of the network state, consisting in expected downstream delays.

Both types of information are real-time; the proposed strategies derive from a heuristic solution of the purely local scheduling/routing problem. The use of aggregate information does not perturb the structure of the local strategy, while preserving its meaning. The feasibility of the overall approach is due to the following facts:

a) each node (machine) applies its strategy on the basis of a limited information set, and by means of a computationally simple procedure;

b) the aggregate information is exchanged and updated at the different nodes with a slight computational effort; in addition, the information exchange mechanism does not imply considerable communication requirements in the network.

The use of aggregate information makes the strategies at the various nodes <u>not purely local</u>, since the entire state of the network is somehow taken into account. That introduces some degree of coordination among the decision makers (nodes) in the network, which otherwise would act in a purely decentralized way.

Of course, the approach proposed in this paper is just the first step towards the development and the analysis of feasible and yet significant scheduling/routing policies in deterministic queueing networks. The main idea of introducing strategies depending on a twofold real-time information needs further investigation, from both theoretical and experimental points of view. Further research will be first aimed at analyzing the degree of suboptimality of the proposed solution (especially as far as it regards the "solution" of the local problem).

In addition, conditions assuring the overall stability of the network governed by the considered policies should be determined.

REFERENCES

Aicardi, M., Di Febbraro, A., and R. Minciardi (1988). On Closed-loop Routing Policies in Discrete Manufacturing Systems Modelled as Queueing Networks. Proc. of the CAA on Computers and Factory Automation, Turin, Italy, 391-400.

Bellmann, R., Esogbue, A.O., and I. Nabeshima (1982). Mathematical Aspects of Scheduling and Applications. Pergamon Press, Oxford.

Beutler, F.J., and D. Teneketzis (1987). Threshold Properties of Optimal Policies in Queueing Networks with Imperfect Information. Proc. 26th IEEE Conf. on Decision and Control, Los Angeles, California, 1508-1513.

Ephremides, A., Varaiya, P., and J. Walrand (1980). A Simple Dynamic Routing Problem. IEEE Trans. on Automat. Control, 25, 698-693.

Hajek, B. (1984). Optimal Control of Two Interacting Service Stations. IEEE Trans. on Automat. Control, 29, 491-499.

Ibarra, O.H., and C.E. Kim (1972). Heuristic Algorithms for Scheduling Independent Tasks on Nonidentical Processors. Journal of the A.C.M., 24, 2, 280-289.

Perkins, J.R., and P.R. Kumar (1989). Stable, Distributed, Real-Time Scheduling of Flexible Manufacturing / Assembly/Disassembly Systems. IEEE Trans. on Automat. Control, 34, 139-148.

Wu, Z.J., Luh, P.B., Chang, S.C., and D.A. Castanon (1988). Optimal Control of a Queueing System with Two Interacting Service Stations and Three Classes of Impatient Tasks. IEEE Trans. on Automat. Control, 33, 42-49.

PROPAGATION OF SPATIAL UNCERTAINTIES
BETWEEN ASSEMBLY PRIMITIVES

J. H. M. van der Drift and C. J. M. Heemskerk

*Laboratory for Manufacturing Systems, University of Technology, Delft,
The Netherlands*

Abstract

Planning a product assembly requires a geometric model of the assembly cell, the parts and the product to
be assembled, that can also represent spatial uncertainties. To guarantee successful execution of an assembly
plan, one needs to be able to reason about uncertain relations. In this paper, we discuss an algorithm to
compute the net spatial uncertainty between any two features in the assembly cell in six dimensions. The
implemented application of the program is automatic Sensor Action Planning.

Keywords

Assembling, Simulation, Error Compensation, Estimation Theory, Flexible Manufacturing, Industrial Robots,
Models, Multidimensional Systems, Predictor-corrector Methods, Sensors

1 Introduction

In recent years, there has been a large effort towards automatic programming systems for assembly. In assembly, geometric reasoning is essential. Path planning, Grasp planning and Sensor planning all require extensive geometric
reasoning capabilities. A problem is that the environment,
the dimensions of the parts and the distances between objects are never known precisely, but only within uncertainties. This paper discusses a way to model and reason about
spatial uncertainties to support the planners mentioned.

The first paragraph describes a way to model spatial uncertainties. Then a way to determine the net uncertainty
in any relation in the assembly cell is discussed. The judg-
ing of uncertainties and an application in Sensor Action
Planning are described. Finally, a simple test is presented.

2 Modeling the Assembly Cell

For automatic assembly planning, the environment and the
products need to be modeled. A convenient tool to describe
relative positions and orientations is the frame concept.
Each device and each important feature is assigned a reference frame, and relations are described by homogeneous
transformation matrices. With simple matrix arithmetic,
it is possible to determine the relative position and orientation of any frame with respect to any other frame in the
cell.

With the frame concept however, spatial uncertainties
cannot be modeled. Therefore, *approximate transformations (ATs)* are introduced. An AT consists of an *Estimated Mean (EM)* relation of one reference frame to another and an *uncertainty*. Like in the frame concept, the

EM is described by a 4x4 homogeneous transformation
matrix. Two different ways to model uncertainties can be
denoted, namely stochastic (Smith and Cheeseman 1986)
and worst case (Taylor 1976, Brooks 1982). The stochastic
method represents uncertainties with a covariance matrix
C, the worst case approach describes them with a 6x1 uncertainty vector $\vec{u}$. This vector puts hard bounds on the
coordinates. Both representations can be used in the algorithm described in this paper, but the stochastic method
cannot model systematic errors. This is a problem, since
systematic errors do occur in a flexible assembly cell. E.g.
it is known that most robots have a moderate absolute
positioning uncertainty, but a good repeatability (Reijers
and de Haas 1986). This means that the systematic error in
the position of the robot is relatively large compared to the
stochastic error. For this reason, worst case uncertainties
will be used henceforth.

The assembly cell will be modeled as a network of *ATs*
(see fig. 1). Now it becomes important to reason about
specific relations, e.g. the net relation between the Gripper
and $Part_A$ in fig. 1. This is a difficult problem, because
this relation is not known directly, but only via other paths
of *ATs* in the network. Every AT on such a path has
uncertainties with six degrees of freedom, symbolized by
rectangles in the figure.

3 Network Reduction

In this paragraph, an algorithm to compute net relations
in a network of *ATs* is presented. The algorithm does not
determine a critical path, but combines all the *ATs* into
one net AT between the *start frame* and the *finish frame*,
e.g. between the reference frame of a peg that is to be

inserted into a hole, and the frame of the hole. In this network reduction algorithm, five basic steps are used:

1. Compounding
2. Merging
3. Elimination of Deadends
4. Reduction of Loops
5. Inversion

These steps will be treated now.

3.1 Compounding

A compound step is depicted in fig. 2. A compound action can be executed when a frame that is not the start or finish frame, takes part in exactly two relations. Compounding is used to combine two adjacent relations into one net relation. In this step, the frame mentioned is eliminated.

Mathematics of Compounding

A compound action consists of two steps, i.e. computing the net EM transformation and computing the net uncertainty vector. Let

$$AT_i \quad = \quad (EM_i, \vec{u}_i),$$
where
$$AT_1, AT_2 \quad = \quad \text{Two adjacent } ATs,$$
$$AT_3 \quad = \quad \text{The net } AT.$$
Then
$$EM_3 \quad = \quad EM_1 \bullet EM_2$$
and
$$\vec{u}_3 \quad = \quad |J| \bullet (\vec{u}_1 \, \vec{u}_2)^T,$$

where J is the Jacobian of the transformation, evaluated at the mean values of the variables. For a three-dimensional world, the Jacobian is derived in the Appendix.

3.2 Merging

A merge step is depicted in fig. 3. A merge action can be executed when there are two parallel transformations between two frames. The two parallel ATs are combined into one net AT. In contradiction to a compound step, no frame is eliminated here.

Mathematics of Merging

Like compounding, merging consists of two steps: The net EM transformation and the net uncertainty vector have to be computed. Because of the worst case approach, merging is very simple mathematically. Let

$$AT_i \quad = \quad (EM_i, \vec{u}_i),$$
where
$$AT_1, AT_2 \quad = \quad \text{Two parallel } ATs,$$
$$AT_3 \quad = \quad \text{The net } AT,$$
Also, let
$$U_{ij} \quad = \quad j^{th} \text{ entry of } \vec{u}_i.$$
Then:
$$EM_3 \quad = \quad (EM_1 + EM_2)/2$$
$$U_{3j} \quad = \quad MIN(U_{1j}, U_{2j}) \quad j \in [0, 5]$$

3.3 Elimination of Deadends

A deadend in a network of ATs is a frame that is not the start or finish frame, and that takes part in only one relation. An example of a network with a deadend is depicted

in fig. 4. The net relation between the start frame and the finish frame is not influenced by a deadend. Therefore, a deadend can be eliminated without any computations.

3.4 Reduction of Loops

Most networks can be reduced with the three basic operations described in the previous paragraphs. Sometimes however, a situation as depicted in fig. 5 occurs. No compounding or merging can be done, and no deadends can be eliminated. This situation is called a *loop*.

Because of the worst case approach, the loop problem can be dealt with easily. The first step is to translate the irreducable situation into an equivalent, but reducable situation, as depicted in fig. 5. This partial network can now be reduced with merging and compounding.

3.5 Inversion

A problem that will occur frequently is that transformations do not point into the right direction. This can be solved by inverting the AT. No reduction of the network is accomplished, but the invert operation is required to continue the reduction.

Mathematics of Inversion

Let:

$$AT_i \quad = \quad (EM_i, \vec{u}_i),$$
where
$$AT_1 \quad = \quad \text{Original } AT.$$
$$AT_2 \quad = \quad \text{Inverted } AT.$$
Then
$$EM_2 \quad = \quad EM_1^{-1}$$
and one would expect that
$$\vec{u}_2 \quad = \quad |R| * \vec{u}_1,$$

where R is the Jacobian of the inversion, estimated at the mean values of the variables. It appears however that inverting an AT twice with these formulas does not return the original AT. This is caused by the fact that the formulas presented treat positional and angular uncertainties *in one* AT as if they are dependent, which they are not. Therefore, the correct inversion formulas are very simple:

$$EM_2 \quad = \quad EM_1^{-1}$$
$$\vec{u}_2 \quad = \quad \vec{u}_1$$

4 Judging the Known Critical Relation

In the previous paragraphs, an algorithm to compute the net uncertainty of any relation in a network of ATs has been presented. Once the net uncertainty is known, it must be decided whether it is acceptable or not. One possibility is to provide the algorithm with the uncertainty constraints, e.g. in a data base or in the input command:

$$MOUNT(PEG, Part_1(Hole), \Delta_{max}(X, Y, Z, \Theta_x, \Theta_y, \Theta_z));$$

The vector Δ_{max} contains the maximum net uncertainties in the AT between the bottom of the peg and the top of the hole for each coordinate for succesful assembly. Now one only has to test the computed net uncertainties against the entries of Δ_{max}. Hardly any process knowledge is required, and the product model can be very simple.

When the maximum tolerable uncertainties are not known, checking the net relation is much harder. The program needs to recognize what the major constraints for succesful insertion are (collision of the peg with the surroundings of the hole, jamming, wedging, etc.), and it has to test against them. Extensive process knowledge and an accurate geometric model are required. Advantages are that the input can be much simpler, and that the results will be better: In the first algorithm, the net uncertainties for each degree of freedom are treated independently, which is not correct. E.g. a rotational uncertainty about the z-axis influences the maximum allowable translational uncertainties along the x- and y-axes. The constraints put in Δ_{max} will therefore have to be more strict than is actually required.

For simplicity, our system uses the first algorithm. Separate Grasp and Assembly planners will have to be developed first. In a later stage the various planners can be integrated, and the algorithm presented in this paper can then be used as a framework for testing uncertainties.

5   An Application: Sensor Action Planning

The algorithm described so far can be used as a basis for grasp planning, fine motion planning, sensor planning, etc. The application that we have implemented is sensor action planning. This algorithm determines the sensor actions that have to be performed to guarantee successful execution of an assembly plan. The first step in this algorithm is to determine which relation is critical, e.g. the relation between a peg and a hole just before an insertion. The network reduction algorithm is used to compute the net uncertainty in this relation, and then it is checked whether this relation is known sufficiently accurate. If it is not, a sensor action will be added to guarantee successful insertion.

A sensor action can be modeled merely by adding to the network an AT between the sensor frame and the frame of the object to be measured (see fig. 6). The EM can be computed with the reduction algorithm, and since the characteristics of the sensor will be known, the uncertainty vector will be known. With the network reduction algorithm, the improved net uncertainty can now be computed. This cycle can be repeated until the critical relation is known accurately enough, or until no more sensors are available. In the latter case, the assembly plan cannot be executed.

6   Implementation and Tests

To test the idea behind the algorithm, a prototype Sensor Action Planner operating in two dimensions (X, Y, Θ) was built. One very simple test will be presented here. The example is two-dimensional and without rotational uncertainties, to enable the reader to check for correctness. Note that EM transformations do not influence uncertainties when there are no rotations. Also note that compounding without rotations comes down to mere vector addition of the uncertainties, since the Jacobian for compounding in this case is (in 2D):

$$J = \begin{pmatrix} 1 & 0 & 0 & | & 1 & 0 & 0 \\ 0 & 1 & 0 & | & 0 & 1 & 0 \\ 0 & 0 & 1 & | & 0 & 0 & 1 \end{pmatrix}$$

and therefore

$$\begin{aligned} \vec{u}_3 &= J * [\vec{u}_1\ \vec{u}_2]^T \\ &= \vec{u}_1 + \vec{u}_2. \end{aligned}$$

Consider the situation depicted in fig. 6. The uncertainties in the ATs are:

Robot	− Gripper	$(0.1, 0.1, 0.0)^T$
Robot	− Feeder	$(0.4, 0.7, 0.0)^T$
Table	− Robot	$(0.3, 0.4, 0.0)^T$
Feeder	− Camera	$(0.0, 0.0, 0.0)^T$

Feeder feeds parts with uncertainty $(0.3, 0.3, 0.0)^T$
The camera senses parts with uncertainty $(0.1, 0.05, 0.0)^T$

The assembly plan to be tested is:

FEED($Part_A$, Feeder)
GRASP($Part_A$) $\Delta_{max}(0.7, 0.9, 0.0)^T$

One can verify that the net uncertainty between the gripper and $Part_A$ without sensing is $[0.8, 1.1, 0.0]^T$, by summing the uncertainties along the path Gripper-Robot-Feeder-$Part_A$ (no merge actions are required in this example). When the camera senses $Part_A$, an AT between the camera and $Part_A$ is added to the network. The net uncertainty then becomes $[0.6, 0.85, 0.0]^T$, which is sufficiently accurate. Therefore, the output of the Sensor Action Planner is:

FEED($Part_A$, Feeder)
SENSE($Part_A$, camera)
GRASP($Part_A$)

After thorough testing, a version operating in three dimensions $(X, Y, Z, \Theta_x, \Theta_y, \Theta_z)$ was built and tested successfully. Tests in three dimensions with rotational uncertainties are much harder to verify by hand. Therefore, no examples are presented here.

7   Conclusions

A way to model and reason about spatial uncertainties in a flexible assembly cell has been presented. An algorithm to determine the net uncertainties between two arbitrary features in a cell model with six degrees of freedom has been developed, implemented and tested succesfully. This algorithm can be used as a framework for automatic sensor action planning.

There are still many issues for future work:

- The current algorithm has been implemented as a stand-alone algorithm, that adds sensor actions to a fixed assembly plan. It only has limited geometric reasoning capabilities. Integration with other high level planners, like grasp and assembly planners, can greatly improve its power.

- The algorithm adds virtual sensor actions, that return ATs between sensor frames and object frames, to the assembly plan. High level drivers between real world sensors and these virtual sensors have to be implemented.

- The cell and product model still consist of low level data (reference frames with transformations). A better user interface is desirable.

A Derivation of the Jacobian for Compounding

Let H_1 be a homogeneous transformation matrix:

$$\begin{pmatrix} \cos\Theta_y\cos\Theta_z & \sin\Theta_x\sin\Theta_y\cos\Theta_z - \cos\Theta_x\sin\Theta_z & \cos\Theta_x\sin\Theta_y\cos\Theta_z + \sin\Theta_x\sin\Theta_z & X \\ \cos\Theta_y\sin\Theta_z & \sin\Theta_x\sin\Theta_y\sin\Theta_z + \cos\Theta_x\cos\Theta_z & \cos\Theta_x\sin\Theta_y\sin\Theta_z - \sin\Theta_x\cos\Theta_z & Y \\ -\sin\Theta_y & \sin\Theta_x\cos\Theta_y & \cos\Theta_x\cos\Theta_y & Z \\ 0 & 0 & 0 & 1 \end{pmatrix}$$

The rotation angles can be determined unambiguously from this matrix. First, the coordinate system is moved to obtain a zero fourth column. The matrix can now be written as:

$$H_2 = \begin{pmatrix} n_x & o_x & a_x & 0 \\ n_y & o_y & a_y & 0 \\ n_z & o_z & a_z & 0 \\ 0 & 0 & 0 & 1 \end{pmatrix}$$

This matrix can also be written as H_1, with $X = Y = Z = 0$. For H_1 and H_2 to be equal, all the corresponding terms have to be equal, which gives nine non-trivial simultaneous equations. The problem of finding the explicit rotation angles seems easy to be solved by equating the 1,3 terms of both matrices, but this method introduces inverse sine and cosine functions, and requires division by cosine. This causes ambiguity and numerical instability. The solution to this problem is to use the arctan2 function (Snyder 1985). One can check that:

$$\Theta_x = \arctan 2(a_x\sin\Theta_z - a_y\cos\Theta_z, o_y\cos\Theta_z - o_x\sin\Theta_z)$$
$$\Theta_y = \arctan 2(-n_z, n_x\cos\Theta_z + n_y\sin\Theta_z)$$
$$\Theta_z = \arctan 2(n_y, n_x)$$

When compounding, there are three matrices involved (see Section 3.1). The rotation angles of EM_1 and EM_2 first have to be computed explicitly. The rotation angles of EM_3 can be expressed explicitly in the rotation angles of EM_1 and EM_2, since

$$EM_3 = EM_1 \bullet EM_2$$

Now one can say that

$$X_3 = f(X_1, Y_1, \ldots \Theta_{z2})$$
$$\vdots \qquad \vdots$$
$$\Theta_{z3} = k(X_1, Y_1, \ldots \Theta_{z2})$$

Therefore, all the terms of the (6x12) Jacobian can be filled in:

$$J = \begin{pmatrix} \frac{\partial f}{\partial X_1} & \frac{\partial f}{\partial Y_1} & \cdots & \cdots & \frac{\partial f}{\partial \Theta_{z2}} \\ \vdots & \vdots & & & \vdots \\ \frac{\partial k}{\partial X_1} & \frac{\partial k}{\partial Y_1} & \cdots & \cdots & \frac{\partial k}{\partial \Theta_{z2}} \end{pmatrix}$$

References

[1] Boneschanscher, N., van der Drift, J.H.M., Buckley, S.J. and Taylor, R.H. 1987. "Subassembly Stability". *AAAI 1988, St.Paul, Minnesota, also IBM Research Report RC 13569(60682)3/4/88*

[2] Brooks, R.A. 1982. "Symbolic Error Analysis and Robot Planning". *Int. Journal of Robotics Research*, Vol. 1.,No. 4.

[3] Heemskerk, C.J.M. 1987. "Programming an Intelligent Assembly Cell." *First European Symposium on Assembly Automation*, march 1987, Veldhoven, The Netherlands.

[4] Reijers, Prof.ir. L.N. and de Haas, ir. H.J.L.M. 1986. "Flexibele Produktie Automatisering deel 3: Industriele Robots". *Technische Uitgeverij De Vey Mestdag BV*, the Netherlands. Dutch.

[5] Smith, R. and Cheeseman, P. 1986. "On the Representation and Estimation of Spatial Uncertainty". *International Journal of Robotics Research 5(4)*, pp. 56-68.

[6] Snyder, W.E. 1985. "Industrial Robots: Interfacing and Control". *Prentice Hall, Industrial Robots Series, W.E. Snyder, Series Editor.*

[7] Taylor, R.H. 1976. "A synthesis of Manipulator Control Programs from Task-Level Specifications". *Ph.D. dissertation, Stanford University, also AIM-282, Stanford Artificial Intelligence Laboratory.*

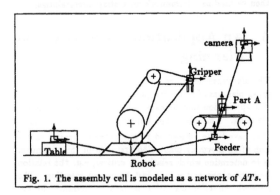

Fig. 1. The assembly cell is modeled as a network of *ATs*.

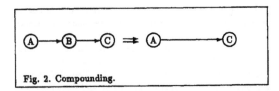

Fig. 2. Compounding.

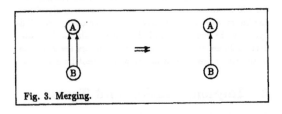

Fig. 3. Merging.

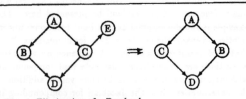

Fig. 4. Elimination of a Deadend.
A is the start node, *D* is the finish node.

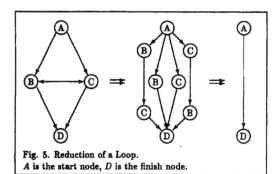

Fig. 5. Reduction of a Loop.
A is the start node, D is the finish node.

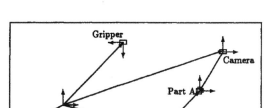

Fig. 6. Network of AT_s

Fig. 5. Reduction of a Loop.
A is the start node, D is the finish node.

Fig. 6. Network of AT.

AUTHOR INDEX

KEYWORD INDEX

SYMPOSIA VOLUMES

ADALI & TUNALI: Microcomputer Application in Process Control

AKASHI: Control Science and Technology for the Progress of Society, 7 Volumes

ALBERTOS & DE LA PUENTE: Components, Instruments and Techniques for Low Cost Automation and Applications

ALONSO-CONCHEIRO: Real Time Digital Control Applications

AMOUROUX & EL JAI: Control of Distributed Parameter Systems (1989)

ATHERTON: Multivariable Technological Systems

BABARY & LE LETTY: Control of Distributed Parameter Systems (1982)

BALCHEN: Automation and Data Processing in Aquaculture

BANKS & PRITCHARD: Control of Distributed Parameter Systems (1977)

BAOSHENG HU: Analysis, Design and Evaluation of Man–Machine Systems (1989)

BARKER & YOUNG: Identification and System Parameter Estimation (1985)

BASANEZ, FERRATE & SARIDIS: Robot Control "SYROCO '85"

BASAR & PAU: Dynamic Modelling and Control of National Economies (1983)

BAYLIS: Safety of Computer Control Systems (1983)

BEKEY & SARIDIS: Identification and System Parameter Estimation (1982)

BINDER & PERRET: Components and Instruments for Distributed Computer Control Systems

CALVAER: Power Systems, Modelling and Control Applications

Van CAUWENBERGHE: Instrumentation and Automation in the Paper, Rubber, Plastics and Polymerisation Industries (1980) (1983)

CHEN HAN-FU: Identification and System Parameter Estimation (1988)

CHEN ZHEN-YU: Computer Aided Design in Control Systems (1988)

CHRETIEN: Automatic Control in Space (1985)

CHRISTODULAKIS: Dynamic Modelling and Control of National Economies (1989)

COBELLI & MARIANI: Modelling and Control in Biomedical Systems

CUENOD: Computer Aided Design of Control Systems†

DA CUNHA: Planning and Operation of Electric Energy Systems

DE CARLI: Low Cost Automation

De GIORGIO & ROVEDA: Criteria for Selecting Appropriate Technologies under Different Cultural, Technical and Social Conditions

DUBUISSON: Information and Systems

EHRENBERGER: Safety of Computer Control Systems (SAFECOMP '88)

ELLIS: Control Problems and Devices in Manufacturing Technology (1980)

FERRATE & PUENTE: Software for Computer Control (1982)

FLEISSNER: Systems Approach to Appropriate Technology Transfer

FLORIAN & HAASE: Software for Computer Control (1986)

GEERING & MANSOUR: Large Scale Systems: Theory and Applications (1986)

GENSER, ETSCHMAIER, HASEGAWA & STROBEL: Control in Transportation Systems (1986)

GERTLER & KEVICZKY: A Bridge Between Control Science and Technology, 6 Volumes

GHONAIMY: Systems Approach for Development (1977)

HAIMES & KINDLER: Water and Related Land Resource Systems

HARDT: Information Control Problems in Manufacturing Technology (1982)

HERBST: Automatic Control in Power Generation Distribution and Protection

HRUZ & CICEL: Automatic Measurement and Control in Woodworking Industry — Lignoautomatica '86

HUSSON: Advanced Information Processing in Automatic Control

ISERMANN: Automatic Control, 10 Volumes

ISERMANN: Identification and System Parameter Estimation (1979)

ISERMANN & KALTENECKER: Digital Computer Applications to Process Control

ISIDORI: Nonlinear Control Systems Design

JANSSEN, PAU & STRASZAK: Dynamic Modelling and Control of National Economies (1980)

JELLALI: Systems Analysis Applied to Management of Water Resources

JOHANNSEN & RIJNSDORP: Analysis, Design, and Evaluation of Man–Machine Systems

JOHNSON: Adaptive Systems in Control and Signal Processing

JOHNSON: Modelling and Control of Biotechnological Processes

KAYA & WILLIAMS: Instrumentation and Automation in the Paper, Rubber, Plastics and Polymerization Industries (1986)

KLAMT & LAUBER: Control in Transportation Systems (1984)

KOPACEK et al.: Skill Based Automated Production

KOPACEK, TROCH & DESOYER: Theory of Robots

KOPPEL: Automation in Mining, Mineral and Metal Processing (1989)

KUMMEL: Adaptive Control of Chemical Processes (ADCHEM '88)

LARSEN & HANSEN: Computer Aided Design in Control and Engineering Systems

LEININGER: Computer Aided Design of Multivariable Technological Systems

LEONHARD: Control in Power Electronics and Electrical Drives (1977)

LESKIEWICZ & ZAREMBA: Pneumatic and Hydraulic Components and Instruments in Automatic Control†

LINKENS & ATHERTON: Trends in Control and Measurement Education

MACLEOD & HEHER: Software for Computer Control (SOCOCO '88)

MAHALANABIS: Theory and Application of Digital Control

MANCINI, JOHANNSEN & MARTENSSON: Analysis, Design and Evaluation of Man–Machine Systems (1985)

MARTOS, PAU, ZIERMANN: Dynamic Modelling and Control of National Economies (1986)

McGREAVY: Dynamics and Control of Chemical Reactors and Distillation Columns

MLADENOV: Distributed Intelligence Systems: Methods and Applications

MUNDAY: Automatic Control in Space (1979)

NAJIM & ABDEL-FATTAH: System Approach for Development (1980)

NIEMI: A Link Between Science and Applications of Automatic Control, 4 Volumes

NISHIKAWA & KAYA: Energy Systems, Management and Economics

NISHIMURA: Automatic Control in Aerospace

NORRIE & TURNER: Automation for Mineral Resource Development

NOVAK: Software for Computer Control (1979)

O'SHEA & POLIS: Automation in Mining, Mineral and Metal Processing (1980)

OSHIMA: Information Control Problems in Manufacturing Technology (1977)

PAUL: Digital Computer Applications to Process Control (1985)

PERRIN: Control, Computers, Communications in Transportation

PONOMARYOV: Artificial Intelligence

PUENTE & NEMES: Information Control Problems in Manufacturing Technology (1989)

RAMAMOORTY: Automation and Instrumentation for Power Plants

RANTA: Analysis, Design and Evaluation of Man–Machine Systems (1988)

RAUCH: Applications of Nonlinear Programming to Optimization and Control†

RAUCH: Control of Distributed Parameter Systems (1986)

REINISCH & THOMA: Large Scale Systems: Theory and Applications (1989)

REMBOLD: Robot Control (SYROCO '88)

RIJNSDORP: Case Studies in Automation Related to Humanization of Work

RIJNSDORP et al.: Dynamics and Control of Chemical Reactors (DYCORD '89)

RIJNSDORP, PLOMP & MÖLLER: Training for Tomorrow— Educational Aspects of Computerized Automation

ROOS: Economics and Artificial Intelligence

SANCHEZ: Fuzzy Information, Knowledge Representation and Decision Analysis

SAWARAGI & AKASHI: Environmental Systems Planning, Design and Control

SINHA & TELKSNYS: Stochastic Control

SMEDEMA: Real Time Programming (1977)†

STRASZAK: Large Scale Systems: Theory and Applications (1983)

SUBRAMANYAM: Computer Applications in Large Scale Power Systems

TAL': Information Control Problems in Manufacturing Technology (1986)

TITLI & SINGH: Large Scale Systems: Theory and Applications (1980)

TROCH, KOPACEK & BREITENECKER: Simulation of Control Systems

UHI AHN: Power Systems and Power Plant Control (1989)

VALADARES TAVARES & EVARISTO DA SILVA: Systems Analysis Applied to Water and Related Land Resources

van WOERKOM: Automatic Control in Space (1982)

WANG PINGYANG: Power Systems and Power Plant Control

WESTERLUND: Automation in Mining, Mineral and Metal Processing (1983)

YANG JIACHI: Control Science and Technology for Development

YOSHITANI: Automation in Mining, Mineral and Metal Processing (1986)

ZWICKY: Control in Power Electronics and Electrical Drives (1983)

WORKSHOP VOLUMES

ASTROM & WITTENMARK: Adaptive Systems in Control and Signal Processing

BOULLART et al.: Industrial Process Control Systems

BRODNER: Skill Based Automated Manufacturing

BULL: Real Time Programming (1983)

BULL & WILLIAMS: Real Time Programming (1985)

CAMPBELL: Control Aspects of Prosthetics and Orthotics

CHESTNUT: Contributions of Technology to International Conflict Resolution (SWIIS)

CHESTNUT et al.: International Conflict Resolution using Systems Engineering (SWIIS)

CHESTNUT, GENSER, KOPACEK & WIERZBICKI: Supplemental Ways for Improving International Stability

CICHOCKI & STRASZAK: Systems Analysis Applications to Complex Programs

CRESPO & DE LA PUENTE: Real Time Programming (1988)

CRONHJORT: Real Time Programming (1978)

DI PILLO: Control Applications of Nonlinear Programming and Optimization

ELZER: Experience with the Management of Software Projects

GELLIE & TAVAST: Distributed Computer Control Systems (1982)

GENSER et al.: Safety of Computer Control Systems (SAFECOMP '89)

GOODWIN: Robust Adaptive Control

HAASE: Real Time Programming (1980)

HALME: Modelling and Control of Biotechnical Processes

HARRISON: Distributed Computer Control Systems (1979)

HASEGAWA: Real Time Programming (1981)†

HASEGAWA & INOUE: Urban, Regional and National Planning—Environmental Aspects

JANSEN & BOULLART: Reliability of Instrumentation Systems for Safeguarding and Control

KOTOB: Automatic Control in Petroleum, Petrochemical and Desalination Industries

LANDAU, TOMIZUKA & AUSLANDER: Adaptive Systems in Control and Signal Processing

LAUBER: Safety of Computer Control Systems (1979)

LOTOTSKY: Evaluation of Adaptive Control Strategies in Industrial Applications

MAFFEZZONI: Modelling and Control of Electric Power Plants (1984)

MARTIN: Design of Work in Automated Manufacturing Systems

McAVOY: Model Based Process Control

MEYER: Real Time Programming (1989)

MILLER: Distributed Computer Control Systems (1981)

MILOVANOVIC & ELZER: Experience with the Management of Software Projects (1988)

MOWLE: Experience with the Management of Software Projects

NARITA & MOTUS: Distributed Computer Control Systems (1989)

OLLUS: Digital Image Processing in Industrial Applications—Vision Control

QUIRK: Safety of Computer Control Systems (1985) (1986)

RAUCH: Control Applications of Nonlinear Programming

REMBOLD: Information Control Problems in Manufacturing Technology (1979)

RODD: Artificial Intelligence in Real Time Control (1989)

RODD: Distributed Computer Control Systems (1983)

RODD: Distributed Databases in Real Time Control

RODD & LALIVE D'EPINAY: Distributed Computer Control Systems (1988)

RODD & MULLER: Distributed Computer Control Systems (1986)

RODD & SUSKI: Artificial Intelligence in Real Time Control

SIGUERDIDJANE & BERNHARD: Control Applications of Nonlinear Programming and Optimization

SINGH & TITLI: Control and Management of Integrated Industrial Complexes

SKELTON & OWENS: Model Error Concepts and Compensation

SOMMER: Applied Measurements in Mineral and Metallurgical Processing

SUSKI: Distributed Computer Control Systems (1985)

SZLANKO: Real Time Programming (1986)

TAKAMATSU & O'SHIMA: Production Control in Process Industry

UNBEHAUEN: Adaptive Control of Chemical Processes

VILLA & MURARI: Decisional Structures in Automated Manufacturing

†Out of stock—microfiche copies available. Details of prices sent on request from the IFAC Publisher.

IFAC Related Titles

BROADBENT & MASUBUCHI: Multilingual Glossary of Automatic Control Technology

EYKHOFF: Trends and Progress in System Identification

NALECZ: Control Aspects of Biomedical Engineering

Printed and bound by CPI Group (UK) Ltd, Croydon, CR0 4YY

03/10/2024

01040320-0016